Ma

10	11 IB	12 IIB	13 IIIA	14 IVA	15 VA	16 VIA	17 VIIA	
								2 **He** 4.002602
			5 **B** 10.811	6 **C** 12.0107	7 **N** 14.0067	8 **O** 15.9994	9 **F** 18.9984032	10 **Ne** 20.1797
			13 **Al** 26.9815386	14 **Si** 28.0855	15 **P** 30.973762	16 **S** 32.065	17 **Cl** 35.453	18 **Ar** 39.948
28 **Ni** 58.6934	29 **Cu** 63.546	30 **Zn** 65.409	31 **Ga** 69.723	32 **Ge** 72.64	33 **As** 74.92160	34 **Se** 78.96	35 **Br** 79.904	36 **Kr** 83.798
46 **Pd** 106.42	47 **Ag** 107.8682	48 **Cd** 112.411	49 **In** 114.818	50 **Sn** 118.710	51 **Sb** 121.760	52 **Te** 127.60	53 **I** 126.90447	54 **Xe** 131.293
78 **Pt** 195.084	79 **Au** 196.966569	80 **Hg** 200.59	81 **Tl** 204.3833	82 **Pb** 207.2	83 **Bi** 208.98040	84 **Po** (209)	85 **At** (210)	86 **Rn** (222)
110 **Ds** (281)	111 **Rg** (280)	112 **Cn** (285)	113 **Uut** (284)	114 **Uuq** (289)	115 **Uup** (288)	116 **Uuh** (293)	117 **Uus** (294)	118 **Uuo** (294)

Inner Transition Metals

62 **Sm** 150.36	63 **Eu** 151.964	64 **Gd** 157.25	65 **Tb** 158.92535	66 **Dy** 162.500	67 **Ho** 164.93032	68 **Er** 167.259	69 **Tm** 168.93421	70 **Yb** 173.54
94 **Pu** (244)	95 **Am** (243)	96 **Cm** (247)	97 **Bk** (247)	98 **Cf** (251)	99 **Es** (252)	100 **Fm** (257)	101 **Md** (258)	102 **No** (259)

A value in parentheses is the mass number of the isotope of the longest half-life.

Permanent names are not yet assigned for elements 112 to 118 (element 117 has not yet been synthesized). These elements are assigned temporary names based on their atomic numbers. See www.webelements.com for more information.

STUDENT SOLUTIONS MANUAL

About the Authors

DARRELL D. EBBING

Darrell Ebbing became interested in chemistry at a young age when he tried his hand at doing magic tricks for friends, such as turning water to wine and having (nitrated) handkerchiefs disappear in a poof. Soon, however, his interests turned to the chemistry behind the magic and he even set up a home laboratory. After briefly becoming interested in botany in high school (having gathered several hundred plant specimens), his interest in chemistry was especially piqued when he managed to isolate white crystals of caffeine from tea. From that point, he knew he would go on to major in chemistry. During college, he helped pay his expenses by working at the USDA lab in Peoria, Illinois, as an assistant to a carbohydrate chemist, where he worked on derivatives of starch. As a graduate student at Indiana University, his interests gravitated to the theoretical—to understand the basis of chemistry—and he pursued a PhD in physical chemistry in the area of quantum chemistry.

Professor Ebbing began his professional career at Wayne State University where he taught courses at the undergraduate and graduate level and was for several years the Head of the Physical Chemistry Division. He soon became especially involved in teaching general chemistry, taking the position of Coordinator of General Chemistry. In his teaching, he used his knowledge of "chemical magic" to do frequent lecture demonstrations. He has written a book for introductory chemistry as well as this one for general chemistry (where you will see many of those lecture demonstrations). Although retired from active teaching, he retains a keen interest in frontier topics of science and in the history and philosophy of physical science, interests he hopes to turn into another book.

Having grown up in farm country, surrounded by fields and woods, Professor Ebbing has always maintained a strong interest in the great outdoors. He enjoys seeing nature up close through hiking and birding. His interests also include concerts and theater, as well as world travel.

STEVEN D. GAMMON

Steve Gammon started on his path to becoming a chemist and science educator in high school where he was captivated by a great instructor. After receiving a PhD in inorganic chemistry and chemical education from the University of Illinois-Urbana, he worked for two years at the University of Wisconsin-Madison, serving as the General Chemistry Laboratory Coordinator and becoming immersed in the field of chemical education. Professor Gammon then went on to join the faculty at the University of Idaho as the Coordinator of General Chemistry. In this role, he taught thousands of students, published instructional software, directed federally funded projects involving K-12 teachers, and began his work on *General Chemistry* (then going into its sixth edition). During his 11 years at the University of Idaho he was honored with both university and national (Carnegie Foundation) teaching awards.

In his current appointment at Western Washington University in Bellingham, Washington, Professor Gammon's focus has expanded from chemical education into science education, where he has the opportunity to work with a wide variety of students including undergraduate science majors, preservice K-12 teachers, and practicing K-12 teachers. Some of his current work has been funded by grants from the National Science Foundation. Throughout all of these changes, Professor Gammon has maintained his keen interest in the learning and teaching of introductory chemistry; his science education experiences have only made him better at understanding and addressing the needs of general chemistry students. Professor Gammon's motivation in working with all groups is to create materials and methods that inspire his students to be excited about learning chemistry and science.

When Professor Gammon isn't thinking about the teaching and learning of chemistry, he enjoys doing a variety of activities with his family, including outdoor pursuits such as hiking, biking, camping, gold panning, and fishing. Scattered throughout the text you might find some examples of where his passion for these activities is used to make connections between chemistry and everyday living.

1

Chemistry and Measurement

One of the forms of SiO_2 in nature is the quartz crystal. Optical fibers that employ light for data transmission use ultrapure SiO_2 that is produced synthetically.

Contents and Concepts

An Introduction to Chemistry

We start by defining the science called chemistry and introducing some fundamental concepts.

Physical Measurements

Making and recording measurements of the properties and chemical behavior of matter is the foundation of chemistry.

OWL Sign in to OWL at **www.cengage.com/owl** to view tutorials and simulations, develop problem-solving skills, and complete online homework assigned by your professor.

go Chemistry Download mini lecture videos for key concept review and exam prep from OWL or purchase them from **www.cengagebrain.com**.

In 1964 Barnett Rosenberg and his coworkers at Michigan State University were studying the effects of electricity on bacterial growth. They inserted platinum wire electrodes into a live bacterial culture and allowed an electric current to pass. After 1 to 2 hours, they noted that cell division in the bacteria stopped. The researchers were very surprised by this result, but even more surprised by the explanation. They were able to show that cell division was inhibited by a substance containing platinum, produced from the platinum electrodes by the electric current. A substance such as this one, the researchers thought, might be useful as an anticancer drug, because cancer involves runaway cell division. Later research confirmed this view, and the platinum-containing substances, *cisplatin, carboplatin*, and *oxaliplatin* are all current anticancer drugs. (Figure 1.1).

This story illustrates three significant reasons to study chemistry. First, chemistry has important practical applications. The development of lifesaving drugs is one, and a complete list would touch upon most areas of modern technology.

Second, chemistry is an intellectual enterprise, a way of explaining our material world. When Rosenberg and his coworkers saw that cell division in the bacteria had ceased, they systematically looked for the chemical substance that caused it to cease. They sought a chemical explanation for the occurrence.

Finally, chemistry figures prominently in other fields. Rosenberg's experiment began as a problem in biology; through the application of chemistry, it led to an advance in medicine. Whatever your career plans, you will find that your knowledge of chemistry is a useful intellectual tool for making important decisions.

An Introduction to Chemistry

All of the objects around you—this book, your pen or pencil, and the things of nature such as rocks, water, and plant and animal substances—constitute the *matter* of the universe. Each of the particular kinds of matter, such as a certain kind of paper or plastic or metal, is referred to as a *material*. We can define chemistry as the science of the composition and structure of materials and of the changes that materials undergo.

One chemist may hope that by understanding certain materials, he or she will be able to find a cure for a disease or a solution for an environmental ill. Another chemist may simply want to understand a phenomenon. Because chemistry deals with all materials, it is a subject of enormous breadth. It would be difficult to exaggerate the influence of chemistry on modern science and technology or on our ideas about our planet and the universe. In the section that follows, we will take a brief glimpse at modern chemistry and see some of the ways it has influenced technology, science, and modern thought.

Figure 1.1 ▶

Barnett Rosenberg and Cisplatin

1

Photo of the discoverer of the anti-cancer activity of Cisplatin.

2

Photo of Cisplatin crystals as seen with a microscope.

1.1 Modern Chemistry: A Brief Glimpse

For thousands of years, human beings have fashioned natural materials into useful products. Modern chemistry certainly has its roots in this endeavor. After the discovery of fire, people began to notice changes in certain rocks and minerals exposed to high temperatures. From these observations came the development of ceramics, glass, and metals, which today are among our most useful materials. Dyes and medicines were other early products obtained from natural substances. For example, the ancient Phoenicians extracted a bright purple dye, known as Tyrian purple, from a species of sea snail. One ounce of Tyrian purple required over 200,000 snails. Because of its brilliant hue and scarcity, the dye became the choice of royalty.

Although chemistry has its roots in early technology, chemistry as a field of study based on scientific principles came into being only in the latter part of the eighteenth century. Chemists began to look at the precise quantities of substances they used in their experiments. From this work came the central principle of modern chemistry: the materials around us are composed of exceedingly small particles called *atoms,* and the precise arrangement of these atoms into *molecules* or more complicated structures accounts for the many different characteristics of materials. Once chemists understood this central principle, they could begin to fashion molecules to order. They could *synthesize* molecules; that is, they could build large molecules from small ones. Tyrian purple, for example, was eventually synthesized from the simpler molecule aniline; see Figure 1.2. Chemists could also correlate molecular structure with the characteristics of materials and so begin to fashion materials with special characteristics.

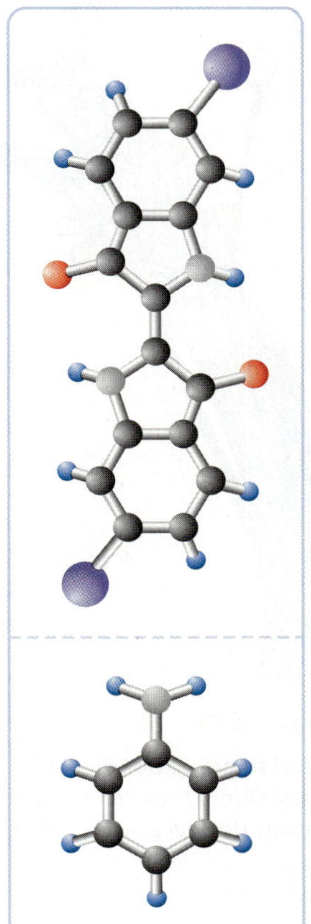

Figure 1.2 ▲

Molecular models of Tyrian purple and aniline Tyrian purple (*top*) is a dye that was obtained by the early Phoenicians from a species of sea snail. The dye was eventually synthesized from aniline (*bottom*). In molecular models the balls represent atoms; each element is represented by a particular color. The lines between the balls indicate that there is a connection holding the atoms together.

The liquid-crystal displays (LCDs) that are used on everything from watches and cell phones to computer monitors and televisions are an example of an application that depends on the special characteristics of materials (Figure 1.3). The liquid crystals used in these displays are a form of matter intermediate in characteristics between those of liquids and those of solid crystals—hence the name. Many of these liquid crystals are composed of rodlike molecules that tend to align themselves something like the wood matches in a matchbox. The liquid crystals are held in alignment in layers by plates that have microscopic grooves. The molecules are attached to small electrodes or transistors. When the molecules are subjected to an electric charge from the transistor or electrode, they change alignment to point in a new direction. When they change direction, they change how light passes through their layer. When the liquid-crystal layer is combined with a light source and color filters, incremental changes of alignment of the molecules throughout the display allow for images that have high contrast and millions of colors. ▶ Figure 1.4 shows a model of one of the molecules that

Liquid crystals and liquid-crystal displays are described in the essay at the end of Section 11.7.

Figure 1.3 ▶

An iPad© that uses a liquid-crystal display These liquid-crystal displays are used in a variety of electronic devices.

© PSL Images/Alamy

Figure 1.4 ▶

Model of a molecule that forms a liquid crystal Note that the molecule has a rodlike shape. In this model each atom in the structure is represented by a colored shape. Atoms that are touching are connected to each other.

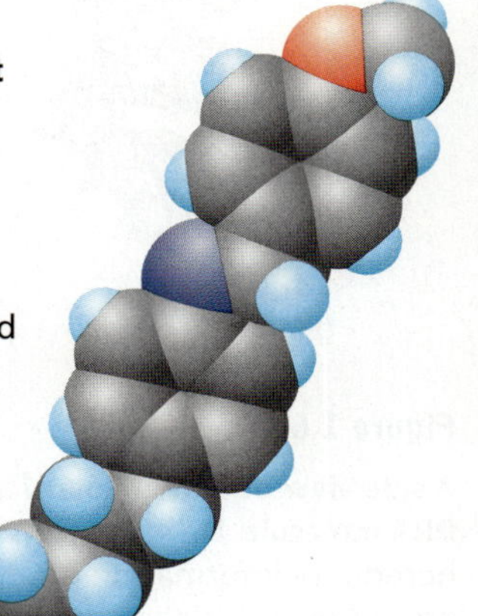

© Paul Orr/Shutterstock.com

Figure 1.5 ▲

Optical fibers A bundle of optical fibers that can be used to transmit data via pulses of light.

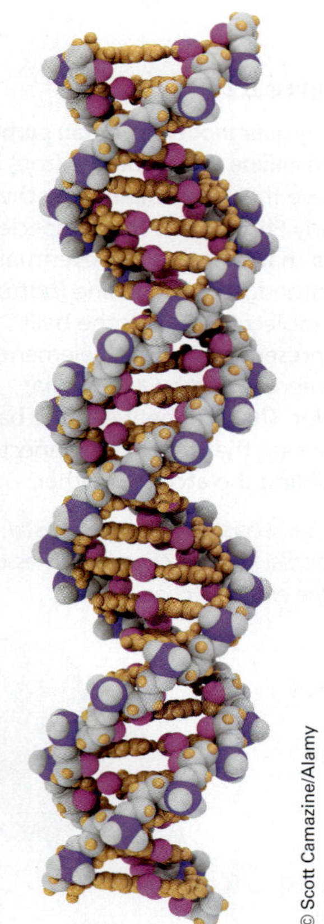
© Scott Camazine/Alamy

Figure 1.6 ▲

A side view of a fragment of a DNA molecule DNA contains the hereditary information of an organism that is passed on from one generation to the next.

forms a liquid crystal; note the rodlike shape of the molecule. Chemists have designed many similar molecules for liquid-crystal applications.

Chemists continue to develop new materials and to discover new properties of old ones. Electronics and communications, for example, have been completely transformed by technological advances in materials. Optical-fiber cables have replaced long-distance telephone cables made of copper wire. Optical fibers are fine threads of extremely pure glass. Because of their purity, these fibers can transmit laser light pulses for miles compared with only a few inches in ordinary glass. Not only is optical-fiber cable cheaper and less bulky than copper cable carrying the same information, but through the use of different colors of light, optical-fiber cable can carry voice, data, and video information at the same time (Figure 1.5). At the ends of an optical-fiber cable, devices using other new materials convert the light pulses to electrical signals and back, while computer chips constructed from still other materials process the signals.

Chemistry has also affected the way we think of the world around us. For example, biochemists and molecular biologists—scientists who study the molecular basis of living organisms—have made a remarkable finding: all forms of life appear to share many of the same molecules and molecular processes. Consider the information of inheritance, the genetic information that is passed on from one generation of organism to the next. Individual organisms, whether bacteria or human beings, store this information in a particular kind of molecule called deoxyribonucleic acid, or DNA (Figure 1.6).

DNA consists of two intertwined molecular chains; each chain consists of links of four different types of molecular pieces, or bases. Just as you record information on a page by stringing together characters (letters, numbers, spaces, and so on), an organism stores the information for reproducing itself in the order of these bases in its DNA. In a multicellular organism, such as a human being, every cell contains the same DNA.

One of our first projects will be to look at this central concept of chemistry, the atomic theory of matter. We will do that in the next chapter, but first we must lay the groundwork for this discussion. We will need some basic vocabulary to talk about science and to describe materials; then we will need to discuss measurement and units, because measurement is critical for quantitative work.

1.2 Experiment and Explanation

Experiment and explanation are the heart of chemical research. A chemist makes observations under circumstances in which variables, such as temperature and amounts of substances, can be controlled. An **experiment** is *an observation of natural phenomena carried out in a controlled manner so that the results can be duplicated and rational conclusions made.* In the chapter opening it was mentioned that Rosenberg studied the effects of electricity on bacterial growth. Temperature and amounts of nutrients in a given volume of bacterial medium are important variables in such experiments. Unless these variables are controlled, the work cannot be duplicated, nor can any reasonable conclusion be drawn.

After a series of experiments, perhaps a researcher sees some relationship or regularity in the results. For instance, Rosenberg noted that in each experiment in which an electric current was passed through a bacterial culture by means of platinum wire electrodes, the bacteria ceased dividing. If the regularity or relationship is fundamental and we can state it simply, we call it a law. A **law** is *a concise statement or mathematical equation about a fundamental relationship or regularity of nature.* An example is the law of conservation of mass, which says that the mass, or quantity of matter, remains constant during any chemical change.

At some point in a research project, a scientist tries to make sense of the results by devising an explanation. Explanations help us organize knowledge and predict future events. A **hypothesis** is *a tentative explanation of some regularity of nature.*

A CHEMIST Looks at...

Daily Life

The Birth of the Post-it Note®

Have you ever used a Post-it and wondered where the idea for those little sticky notes came from? You have a chemist to thank for their invention. The story of the Post-it Note illustrates how the creativity and insights of a scientist can result in a product that is as common in the office as the stapler or pen.

In the early 1970s, Art Fry, a 3M scientist, was standing in the choir at his church trying to keep track of all the little bits of paper that marked the music selections for the service. During the service, a number of the markers fell out of the music, making him lose his place. While standing in front of the congregation, he realized that he needed a bookmark that would stick to the book, wouldn't hurt the book, and could be easily detached. To make his plan work, he required an adhesive that would not *permanently* stick things together. Finding the appropriate adhesive was not as simple as it may seem, because most adhesives at that time were created to stick things together permanently.

Still thinking about his problem the next day, Fry consulted a colleague, Spencer Silver, who was studying adhesives at the 3M research labs. That study consisted of conducting a series of tests on a range of adhesives to determine the strength of the bond they formed. One of the adhesives that Silver created for the study was an adhesive that always remained sticky. Fry recognized that this adhesive was just what he needed for his bookmark. His first bookmark, invented the day after the initial idea, consisted of a strip of Silver's tacky adhesive applied to the edge of a piece of paper.

Part of Fry's job description at 3M was to spend time working on creative ideas such as his bookmark. As a result, he continued to experiment with the bookmark to improve its properties of sticking, detaching, and not hurting the surface to which it was attached. One day, while doing some paperwork, he wrote a question on one of his experimental strips of paper and sent it to his boss stuck to the top of a file folder. His boss then answered the question on the note and returned it attached to some other documents. During a later discussion over coffee, Fry and his boss realized that they had invented a new way for people to communicate: the Post-it Note was born. Today the Post-it Note is one of the top-selling office products in the United States.

■ See Problems 1.143 and 1.144.

Having seen that bacteria ceased to divide when an electric current from platinum wire electrodes passed through the culture, Rosenberg was eventually able to propose the hypothesis that certain platinum compounds were responsible. If a hypothesis is to be useful, it should suggest new experiments that become tests of the hypothesis. Rosenberg could test his hypothesis by looking for the platinum compound and testing for its ability to inhibit cell division.

If a hypothesis successfully passes many tests, it becomes known as a theory. A **theory** is *a tested explanation of basic natural phenomena.* An example is the molecular theory of gases—the theory that all gases are composed of very small particles called molecules. This theory has withstood many tests and has been fruitful in suggesting many experiments. Note that we cannot prove a theory absolutely. It is always possible that further experiments will show the theory to be limited or that someone will develop a better theory. For example, the physics of the motion of objects devised by Isaac Newton withstood experimental tests for more than two centuries, until physicists discovered that the equations do not hold for objects moving near the speed of light. Later physicists showed that very small objects also do not follow Newton's equations. Both discoveries resulted in revolutionary developments in physics. The first led to the theory of relativity, the second to quantum mechanics, which has had an immense impact on chemistry.

The two aspects of science, experiment and explanation, are closely related. A scientist performs experiments and observes some regularity; someone explains this regularity and proposes more experiments; and so on. From his experiments, Rosenberg explained that certain platinum compounds inhibit cell division. This explanation led him to do new experiments on the anticancer activity of these compounds.

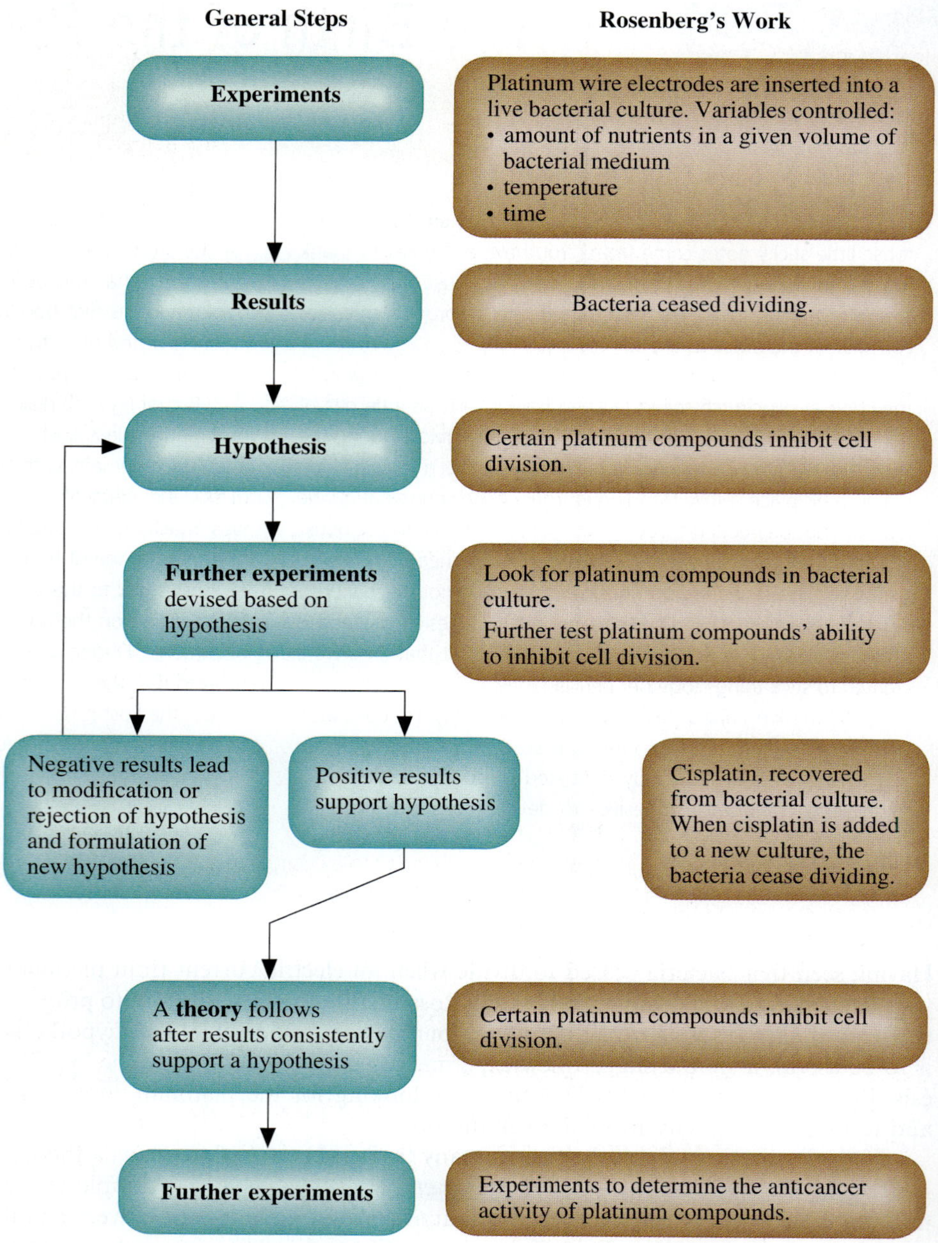

Figure 1.7 ▶

A representation of the scientific method This flow diagram shows the general steps in the scientific method. At the right, Rosenberg's work on the development of an anticancer drug illustrates the steps.

General Steps	Rosenberg's Work
Experiments	Platinum wire electrodes are inserted into a live bacterial culture. Variables controlled: • amount of nutrients in a given volume of bacterial medium • temperature • time
Results	Bacteria ceased dividing.
Hypothesis	Certain platinum compounds inhibit cell division.
Further experiments devised based on hypothesis	Look for platinum compounds in bacterial culture. Further test platinum compounds' ability to inhibit cell division.
Negative results lead to modification or rejection of hypothesis and formulation of new hypothesis	
Positive results support hypothesis	Cisplatin, recovered from bacterial culture. When cisplatin is added to a new culture, the bacteria cease dividing.
A **theory** follows after results consistently support a hypothesis	Certain platinum compounds inhibit cell division.
Further experiments	Experiments to determine the anticancer activity of platinum compounds.

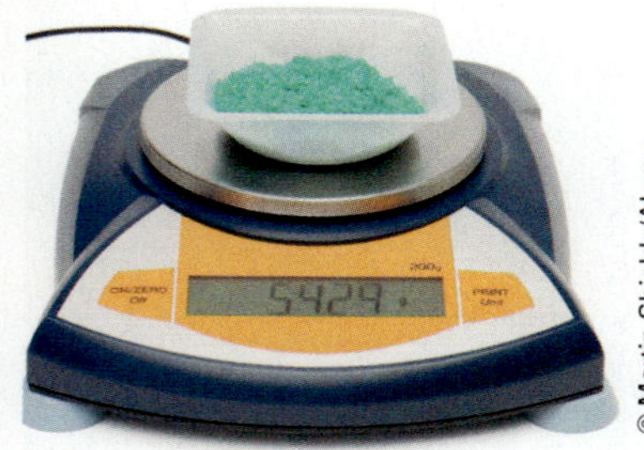

Figure 1.8 ▲

Laboratory balance A modern single-pan balance. The mass of the material on the pan appears on the digital readout.

The *general* process of advancing scientific knowledge through observation; the framing of laws, hypotheses, or theories; and the conducting of more experiments is called the *scientific method* (Figure 1.7). It is not a method for carrying out a *specific* research program, because the design of experiments and the explanation of results draw on the creativity and individuality of a researcher.

1.3 Law of Conservation of Mass

Modern chemistry emerged in the eighteenth century, when chemists began to use the balance systematically as a tool in research. Balances measure **mass,** which is *the quantity of matter in a material* (Figure 1.8). **Matter** is the general term for the material things around us; we can define it as *whatever occupies space and can be perceived by our senses.*

Figure 1.9 ▶

Heating mercury metal in air Mercury metal reacts with oxygen to yield mercury(II) oxide. The color of the oxide varies from red to yellow, depending on the particle size.

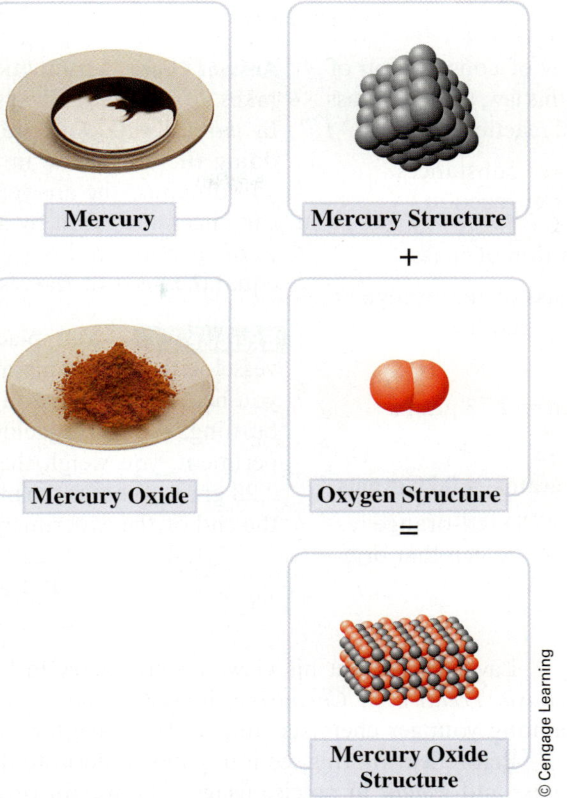

Figure 1.10 ▲

Heated mercury(II) oxide When you heat mercury(II) oxide, it decomposes to mercury and oxygen gas.

Antoine Lavoisier (1743–1794), a French chemist, was one of the first to insist on the use of the balance in chemical research. By weighing substances before and after chemical change, he demonstrated the **law of conservation of mass,** which states that *the total mass remains constant during a chemical change (chemical reaction).* ▶

Chemical reactions may involve a gain or loss of heat and other forms of energy. According to Einstein, mass and energy are equivalent. Thus, when energy is lost as heat, mass is also lost, but such small changes in mass in chemical reactions (billionths of a gram) are too small to detect.

In a series of experiments, Lavoisier applied the law of conservation of mass to clarify the phenomenon of burning, or combustion. He showed that when a material burns, a component of air (which he called oxygen) combines chemically with the material. For example, when the liquid metal mercury is heated in air, it burns or combines with oxygen to give a red-orange substance, whose modern name is mercury(II) oxide. We can represent the chemical change as follows:

$$\text{Mercury} + \text{oxygen} \longrightarrow \text{mercury(II) oxide}$$

The arrow means "is changed to." See Figure 1.9.

By strongly heating the red-orange substance, Lavoisier was able to decompose it to yield the original mercury and oxygen gas (Figure 1.10). The following example illustrates how the law of conservation of mass can be used to study this reaction.

OWL Interactive Example 1.1 Using the Law of Conservation of Mass

Gaining Mastery Toolbox

Critical Concept 1.1

The law of conservation of mass applies to chemical reactions. Whenever a chemical reaction occurs, the mass of the substances before the reaction (reactants) is identical to the mass of the newly formed substances after the reaction (products).

Solution Essentials:

- Definition of mass
- Definition of matter
- Law of conservation of mass

You heat 2.53 grams of metallic mercury in air, which produces 2.73 grams of a red-orange residue. Assume that the chemical change is the reaction of the metal with oxygen in air.

$$\text{Mercury} + \text{oxygen} \longrightarrow \text{red-orange residue}$$

What is the mass of oxygen that reacts? When you strongly heat the red-orange residue, it decomposes to give back the mercury and release the oxygen, which you collect. What is the mass of oxygen you collect?

(continued)

(continued)

Problem Strategy You apply the law of conservation of mass to the reaction. According to this law, the total mass remains constant during a chemical reaction; that is,

$$\text{Mass of substances before reaction} = \text{mass of substances after reaction}$$

Solution From the law of conservation of mass,

$$\text{Mass of mercury} + \text{mass of oxygen} = \text{mass of red-orange residue}$$

Substituting, you obtain

$$2.53 \text{ grams} + \text{mass of oxygen} = 2.73 \text{ grams}$$

or

$$\text{Mass of oxygen} = (2.73 - 2.53) \text{ grams} = \mathbf{0.20 \text{ grams}}$$

The mass of oxygen collected when the red-orange residue decomposes equals the mass of oxygen that originally reacted **(0.20 grams).**

Answer Check Arithmetic errors account for many mistakes. You should always check your arithmetic, either by carefully redoing the calculation or, if possible, by doing the arithmetic in a slightly different way. Here, you obtained the answer by subtracting numbers. You can check the result by addition: the sum of the masses of mercury and oxygen, 2.53 + 0.20 grams, should equal the mass of the residue, 2.73 grams.

Exercise 1.1 You place 1.85 grams of wood in a vessel with 9.45 grams of air and seal the vessel. Then you heat the vessel strongly so that the wood burns. In burning, the wood yields ash and gases. After the experiment, you weigh the ash and find that its mass is 0.28 gram. What is the mass of the gases in the vessel at the end of the experiment?

■ See Problems 1.37, 1.38, 1.39, and 1.40.

Lavoisier set out his views on chemistry in his *Traité Élémentaire de Chimie* (*Basic Treatise on Chemistry*) in 1789. The book was very influential, especially among younger chemists, and set the stage for modern chemistry.

Before leaving this section, you should note the distinction between the terms *mass* and *weight* in precise usage. The weight of an object is the force of gravity exerted on it. The weight is proportional to the mass of the object divided by the square of the distance between the center of mass of the object and that of the earth. ◀ Because the earth is slightly flattened at the poles, an object weighs more at the North Pole, where it is closer to the center of the earth, than at the equator. The mass of an object is the same wherever it is measured.

The force of gravity F between objects whose masses are m_1 and m_2 is Gm_1m_2/r^2, where G is the gravitational constant and r is the distance between the centers of mass of the two objects.

1.4 Matter: Physical State and Chemical Constitution

We describe iron as a silvery-colored metal that melts at 1535°C (2795°F). Once we have collected enough descriptive information about many different kinds of matter, patterns emerge that suggest ways of classifying it. There are two principal ways of classifying matter: by its physical state as a solid, liquid, or gas, and by its chemical constitution as an element, compound, or mixture.

Solids, Liquids, and Gases

Commonly, a given kind of matter exists in different physical forms under different conditions. Water, for example, exists as ice (solid water), as liquid water, and as steam (gaseous water) (Figure 1.11). The main identifying characteristic of solids is their rigidity: they tend to maintain their shapes when subjected to outside forces. Liquids and gases, however, are *fluids;* that is, they flow easily and change their shapes in response to slight outside forces.

What distinguishes a gas from a liquid is the characteristic of *compressibility* (and its opposite, *expansibility*). A gas is easily compressible, whereas a liquid is not. You can put more and more air into a tire, which increases only slightly in volume. In fact, a given quantity of gas can fill a container of almost any size. A small quantity would expand to fill the container; a larger quantity could be compressed to fill the same space. By contrast, if you were to try to force more liquid water into a closed glass bottle that was already full of water, it would burst.

These two characteristics, rigidity (or fluidity) and compressibility (or expansibility), can be used to frame definitions of the three common states of matter:

solid *the form of matter characterized by rigidity;* a solid is relatively incompressible and has fixed shape and volume.

liquid *the form of matter that is a relatively incompressible fluid;* a liquid has a fixed volume but no fixed shape.

gas *the form of matter that is an easily compressible fluid;* a given quantity of gas will fit into a container of almost any size and shape.

The term *vapor* is often used to refer to the gaseous state of any kind of matter that normally exists as a liquid or a solid.

These three forms of matter—solid, liquid, gas—comprise the common **states of matter.**

Elements, Compounds, and Mixtures

To understand how matter is classified by its chemical constitution, we must first distinguish between physical and chemical changes and between physical and chemical properties. A **physical change** is *a change in the form of matter but not in its chemical identity.* Changes of physical state are examples of physical changes. The process of dissolving one material in another is a further example of a physical change. For instance, you can dissolve sodium chloride (table salt) in water. The result is a clear liquid, like pure water, though many of its other characteristics are different from those of pure water. The water and sodium chloride in this liquid retain their chemical identities and can be separated by some method that depends on physical changes.

Distillation is one way to separate the sodium chloride and water components of this liquid. You place the liquid in a flask to which a device called a *condenser* is attached (see Figure 1.12). The liquid in the flask is heated to bring it to a boil. (Boiling entails the formation of bubbles of the vapor in the body of the liquid.) Water vapor forms and passes from the flask into the cooled condenser, where the vapor changes back to liquid water. The liquid water is collected in another flask, called a *receiver.* The original flask now contains the solid sodium chloride. Thus, by means of physical changes (the change of liquid water to vapor and back to liquid), you have separated the sodium chloride and water that you had earlier mixed together.

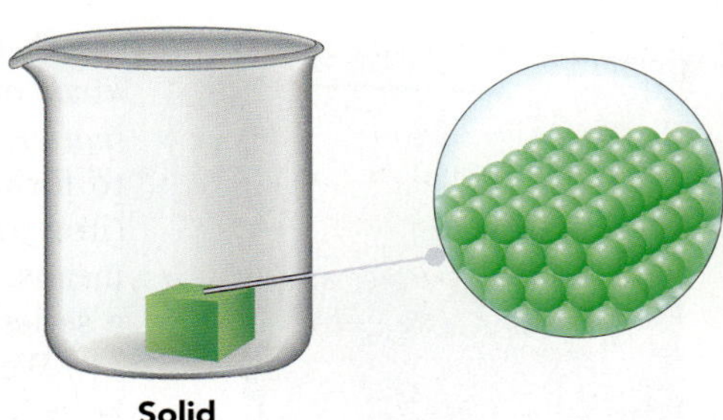

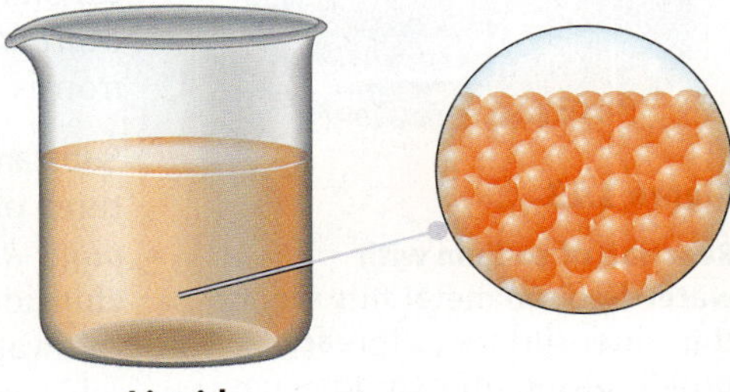

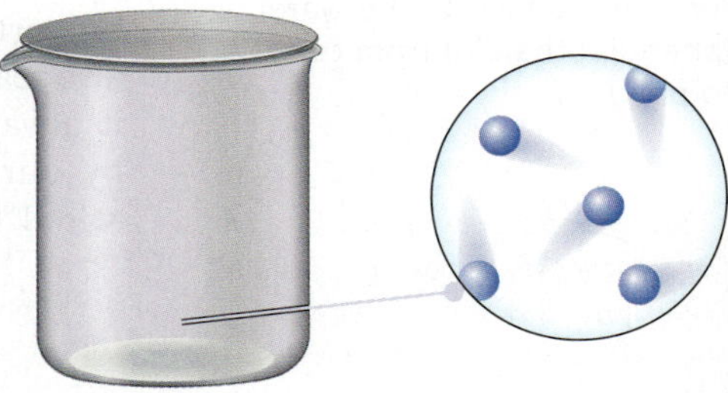

Figure 1.11 ▲

Molecular representations of solid, liquid, and gas The top beaker contains a solid with a molecular view of the solid; the molecular view depicts the closely packed, immobile atoms that make up the solid structure. The middle beaker contains a liquid with a molecular view of the liquid; the molecular view depicts atoms that are close together but moving freely. The bottom beaker contains a gas with a molecular view of the gas; the molecular view depicts atoms that are far apart and moving freely.

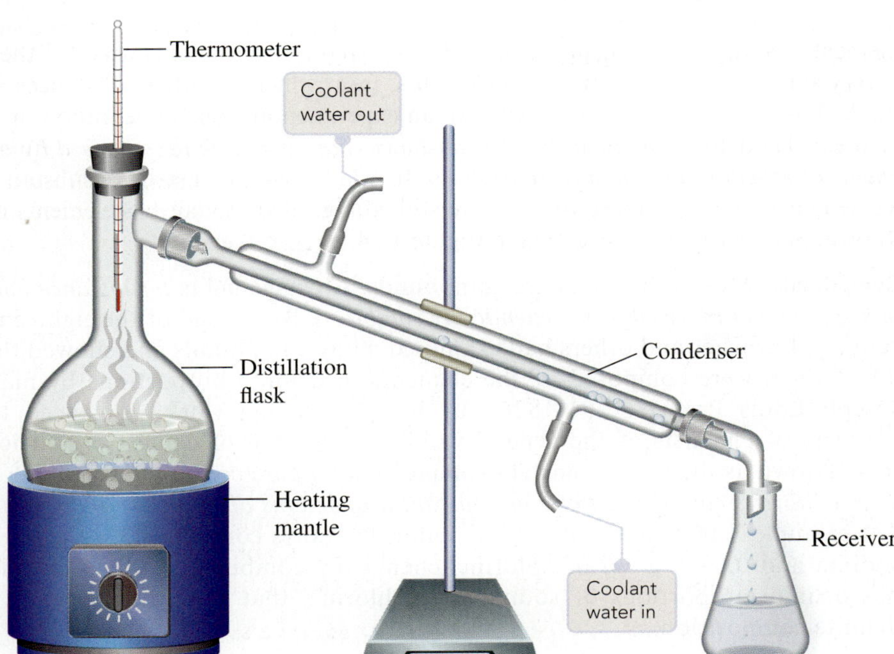

Figure 1.12 ◀

Separation by distillation You can separate an easily vaporized liquid from another substance by distillation.

© Cengage Learning

Figure 1.13 ▲

Reaction of sodium with water Sodium metal flits around the water surface as it reacts briskly, giving off hydrogen gas. The other product is sodium hydroxide, which changes a substance added to the water (phenolphthalein) from colorless to pink.

A **chemical change,** or **chemical reaction,** is *a change in which one or more kinds of matter are transformed into a new kind of matter or several new kinds of matter.* The rusting of iron, during which iron combines with oxygen in the air to form a new material called rust, is a chemical change. The original materials (iron and oxygen) combine chemically and cannot be separated by any physical means. To recover the iron and oxygen from rust requires a chemical change or a series of chemical changes.

We characterize or identify a material by its various properties, which may be either physical or chemical. A **physical property** is *a characteristic that can be observed for a material without changing its chemical identity.* Examples are physical state (solid, liquid, or gas), melting point, and color. A **chemical property** is *a characteristic of a material involving its chemical change.* A chemical property of iron is its ability to react with oxygen to produce rust.

Substances The various materials we see around us are either substances or mixtures of substances. A **substance** is *a kind of matter that cannot be separated into other kinds of matter by any physical process.* Earlier you saw that when sodium chloride is dissolved in water, it is possible to separate the sodium chloride from the water by the physical process of distillation. However, sodium chloride is itself a substance and cannot be separated by physical processes into new materials. Similarly, pure water is a substance.

No matter what its source, a substance always has the same characteristic properties. Sodium is a solid metal having a melting point of 98°C. The metal also reacts vigorously with water (Figure 1.13). No matter how sodium is prepared, it always has these properties. Similarly, whether sodium chloride is obtained by burning sodium in chlorine or from seawater, it is a white solid melting at 801°C.

Exercise 1.2 Potassium is a soft, silvery-colored metal that melts at 64°C. It reacts vigorously with water, with oxygen, and with chlorine. Identify all of the physical properties given in this description. Identify all of the chemical properties given.

■ See Problems 1.47, 1.48, 1.49, and 1.50.

Elements Millions of substances have been characterized by chemists. Of these, a very small number are known as elements, from which all other substances are made. Lavoisier was the first to establish an experimentally useful definition of an element. He defined an **element** as *a substance that cannot be decomposed by any chemical reaction into simpler substances.* In 1789 Lavoisier listed 33 substances as elements, of which more than 20 are still so regarded. Today 118 elements are known. Some elements are shown in Figure 1.14. ◀

In Chapter 2, we will redefine an element in terms of atoms.

Compounds Most substances are compounds. A **compound** is *a substance composed of two or more elements chemically combined.* By the end of the eighteenth century, Lavoisier and others had examined many compounds and showed that all of them were composed of the elements in definite proportions by mass. Joseph Louis Proust (1754–1826), by his painstaking work, convinced the majority of chemists of the general validity of the **law of definite proportions** (also known as the **law of constant composition**): *a pure compound, whatever its source, always contains definite or constant proportions of the elements by mass.* For example, 1.0000 gram of sodium chloride always contains 0.3934 gram of sodium and 0.6066 gram of chlorine, chemically combined. Sodium chloride has definite proportions of sodium and chlorine; that is, it has constant or definite composition. ◀

It is now known that some compounds do not follow the law of definite proportions. These nonstoichiometric compounds, as they are called, are described briefly in Chapter 11.

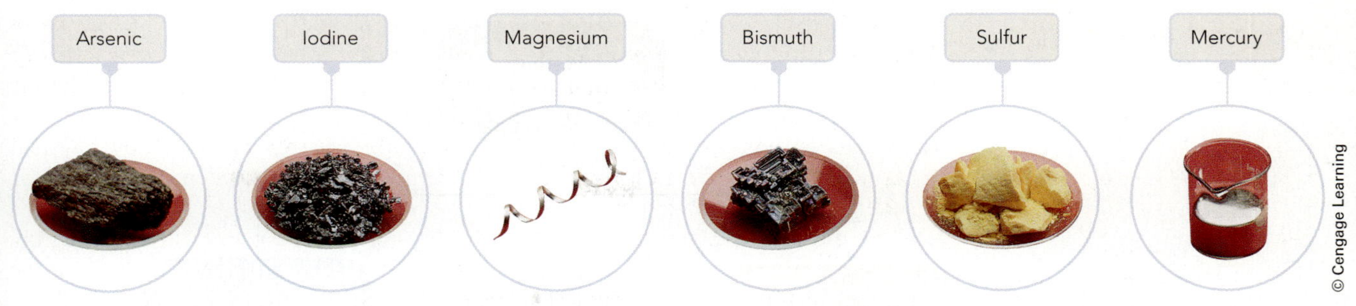

Figure 1.14 ▲
Some elements

Mixtures Most of the materials around us are mixtures. A **mixture** is *a material that can be separated by physical means into two or more substances.* Unlike a pure compound, a mixture has variable composition. When you dissolve sodium chloride in water, you obtain a mixture; its composition depends on the relative amount of sodium chloride dissolved. You can separate the mixture by the physical process of distillation. ▶

Chromatography, another example of a physical method used to separate mixtures, is described in the essay at the end of this section.

Mixtures are classified into two types. A **heterogeneous mixture** is *a mixture that consists of physically distinct parts, each with different properties.* Figure 1.15 shows a heterogeneous mixture of potassium dichromate and iron filings. Another example is salt and sugar that have been stirred together. If you were to look closely, you would see the separate crystals of sugar and salt. A **homogeneous mixture** (also known as a **solution**) is *a mixture that is uniform in its properties throughout given samples.* When sodium chloride is dissolved in water, you obtain a homogeneous mixture, or solution. Air is a gaseous solution, principally of two elementary substances, nitrogen and oxygen, which are physically mixed but not chemically combined.

A **phase** is *one of several different homogeneous materials present in the portion of matter under study.* A heterogeneous mixture of salt and sugar is said to be composed of two different phases: one of the phases is salt; the other is sugar. Similarly, ice cubes in water are said to be composed of two phases: one phase is ice; the other is liquid water. Ice floating in a solution of sodium chloride in water also consists of two phases, ice and the liquid solution. Note that a phase

Figure 1.15 ▼
A heterogeneous mixture

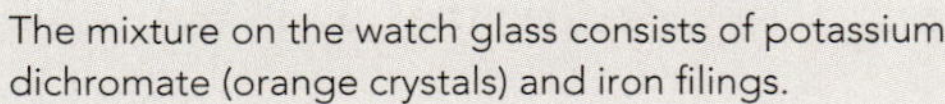
The mixture on the watch glass consists of potassium dichromate (orange crystals) and iron filings.

A magnet separates the iron filings from the mixture.

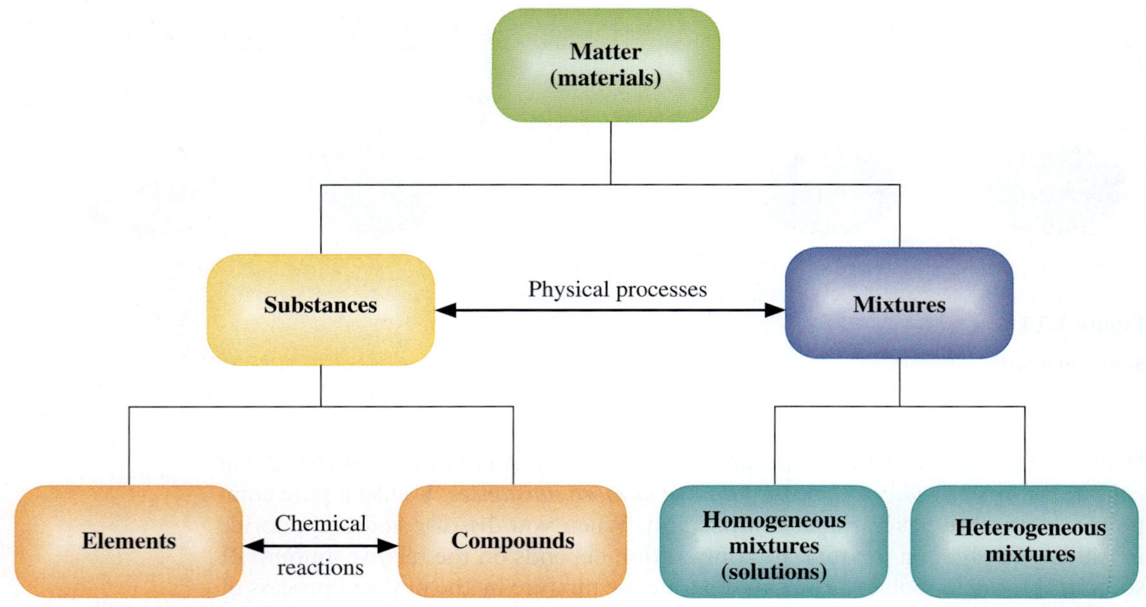

Figure 1.16

Relationships among elements, compounds, and mixtures Mixtures can be separated by physical processes into substances, and substances can be combined physically into mixtures. Compounds can be separated by chemical reactions into their elements, and elements can be combined chemically to form compounds.

may be either a pure substance in a particular state or a solution in a particular state (solid, liquid, or gaseous). Also, the portion of matter under consideration may consist of several phases of the same substance or several phases of different substances.

Figure 1.16 summarizes the relationships among elements, compounds, and mixtures. Materials are either substances or mixtures. Substances can be mixed by physical processes, and other physical processes can be used to separate the mixtures into substances. Substances are either elements or compounds. Elements may react chemically to yield compounds, and compounds may be decomposed by chemical reactions into elements.

CONCEPT CHECK 1.1

Matter can be represented as being composed of individual units. For example, the smallest individual unit of matter can be represented as a single circle, •, and chemical combinations of these units of matter as connected circles, ••, with each element represented by a different color. Using this model, place the appropriate label—element, compound, or mixture—below each representation.

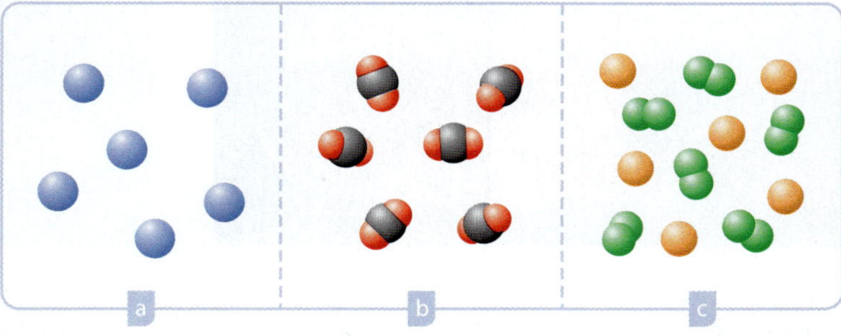

Instrumental Methods

Separation of Mixtures by Chromatography

Chromatography is a group of similar separation techniques. Each depends on how fast a substance moves, in a stream of gas or liquid, past a stationary phase to which the substance may be slightly attracted. An example is provided by a simple experiment in *paper chromatography* (see Figure 1.17). In this experiment, a line of ink is drawn near one edge of a sheet of paper, and the paper is placed upright with this edge in a solution of methanol and water. As the solution creeps up the paper, the ink moves upward, separating into a series of different-colored bands that correspond to the different dyes in the ink. All the dyes are attracted to the wet paper fibers, but with different strengths of attraction. As the solution moves upward, the dyes less strongly attracted to the paper fibers move more rapidly.

The Russian botanist Mikhail Tswett was the first to understand the basis of chromatography and to apply it systematically as a method of separation. In 1906 Tswett separated pigments in plant leaves by *column chromatography*. He first dissolved the pigments from the leaves in petroleum ether, a liquid similar to gasoline. After packing a glass tube or column with powdered chalk, he poured the solution of pigments into the top of the column (see Figure 1.18). When he washed the column by pouring in more petroleum ether, it began to show distinct yellow and green bands. These bands, each containing a pure pigment, became well separated as they moved down the column, so that the pure pigments could be obtained. The name *chromatography* originates from this early separation of colored substances (the stem *chromato-* means "color"), although the technique is not limited to colored substances.

Gas chromatography (GC) is a more recent separation method. Here the moving stream is a gaseous mixture of vaporized substances plus a gas such as helium, which is called the *carrier*. The stationary material is either a solid or a liquid adhering to a solid, packed in a column. As the gas passes through the column, substances in the mixture are attracted differently to the stationary column packing and thus are separated. Gas chromatography is a rapid, small-scale method of separating mixtures. It is also important in the analysis of mixtures because the time it takes for a substance at a given temperature to travel through the column to a detector (called the *retention time*) is fixed. You can therefore use retention times to help identify substances. Figure 1.19 shows a gas chromatograph and a portion of a computer plot (*chromatogram*). Each peak on the chromatogram corresponds to a specific substance. The peaks were automatically recorded by the instrument as the different substances in the mixture passed the detector. Chemists have analyzed complicated mixtures by gas chromatography. Analysis of chocolate, for example, shows that it contains over 800 flavor compounds.

Figure 1.17 ▶

An illustration of paper chromatography A line of ink has been drawn along the lower edge of a sheet of paper. The dyes in the ink separate as a solution of methanol and water creeps up the paper.

(continued)

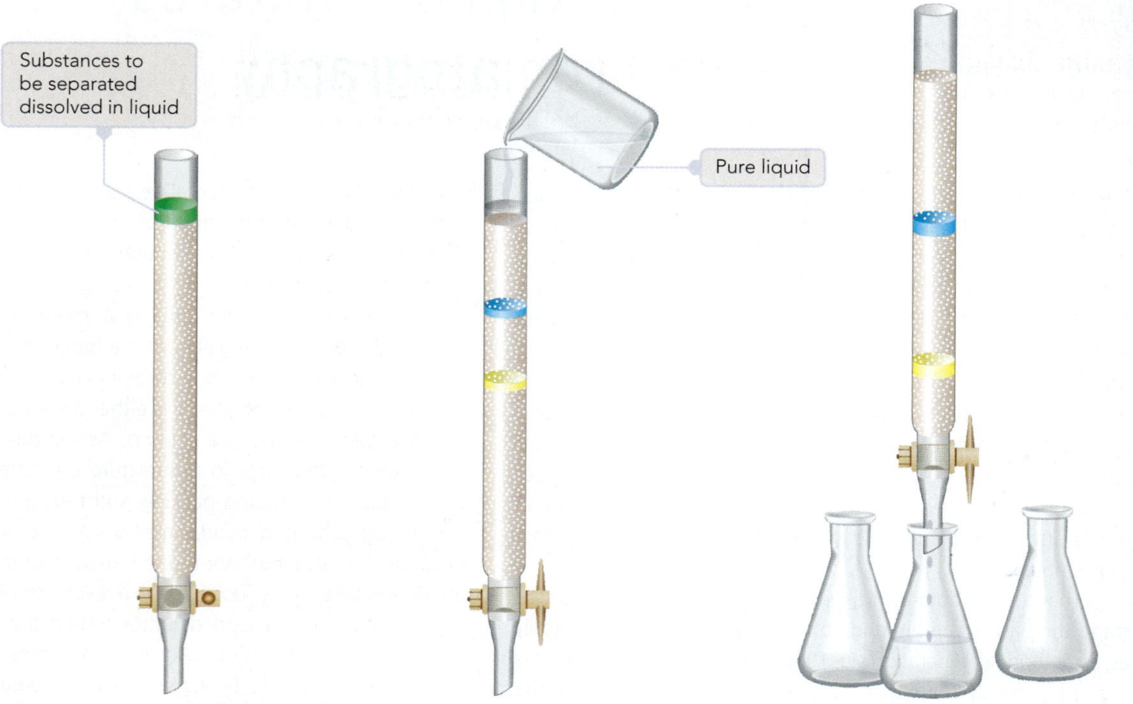

Substances to be separated dissolved in liquid

Pure liquid

a A solution containing substances to be separated is poured into the top of a column, which contains powdered chalk.

b Pure liquid is added to the column, and the substances begin to separate into bands.

c The substances separate further on the column. Each substance is collected in a separate flask as it comes off the column.

Figure 1.18 ▲

Column chromatography

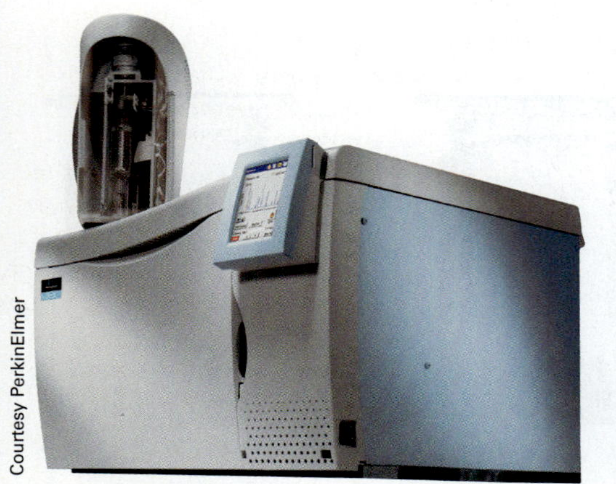

Courtesy PerkinElmer

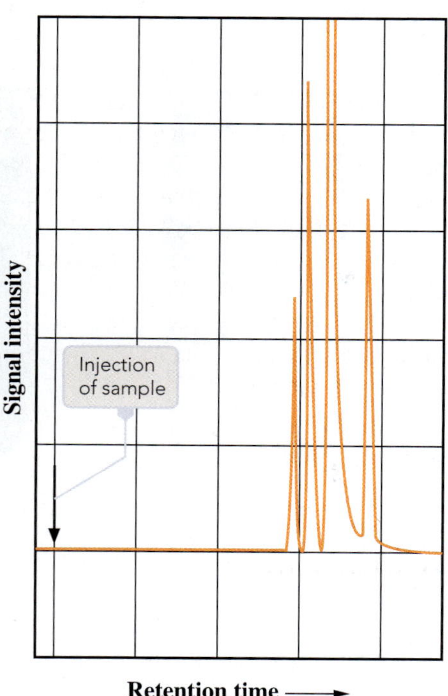

Figure 1.19 ◀

Gas chromatography *Left:* A modern gas chromatograph. *Right:* This is a chromatogram of a hexane mixture, showing its separation into four different substances. Such hexane mixtures occur in gasoline; hexane is also used as a solvent to extract the oil from certain vegetable seeds.

■ See Problems 1.145 and 1.146.

Physical Measurements

Chemists characterize and identify substances by their particular properties. To determine many of these properties requires physical measurements. Cisplatin, the substance featured in the chapter opening, is a yellow substance whose solubility in water (the amount that dissolves in a given quantity of water) is 0.252 gram in 100 grams of water. You can use the solubility, as well as other physical properties, to identify cisplatin. In a modern chemical laboratory, measurements often are complex, but many experiments begin with simple measurements of mass, volume, time, and so forth. In the next few sections, we will look at the measurement process. We first consider the concept of *precision* of a measurement.

1.5 Measurement and Significant Figures

Measurement is the comparison of a physical quantity to be measured with a **unit** of measurement—that is, with *a fixed standard of measurement.* On a centimeter scale, the centimeter unit is the standard of comparison. ▶ A steel rod that measures 9.12 times the centimeter unit has a length of 9.12 centimeters. To record the measurement, you must be careful to give both the *measured number* (9.12) and the *unit* (centimeters).

2.54 centimeters = 1 inch

If you repeat a particular measurement, you usually do not obtain precisely the same result, because each measurement is subject to experimental error. The measured values vary slightly from one another. Suppose you perform a series of identical measurements of a quantity. The term **precision** refers to *the closeness of the set of values obtained from similar measurements of an identical quantity.* **Accuracy** is a related term; it refers to *the closeness of a single measurement to its true value.* To illustrate the idea of precision, consider a simple measuring device, the centimeter ruler. In Figure 1.20, a steel rod has been placed near a ruler subdivided into tenths of a centimeter. You can see that the rod measures just over 9.1 cm (cm = centimeter). With care, it is possible to estimate by eye to hundredths of a centimeter. Here you might give the measurement as 9.12 cm. Suppose you measure the length of this rod twice more. You find the values to be 9.11 cm and 9.13 cm. Thus, you record the length of the rod as being somewhere between 9.11 cm and 9.13 cm. The spread of values indicates the precision with which a measurement can be made by this centimeter ruler. ▶

Measurements that are of high precision are usually accurate. It is possible, however, to have a systematic error in a measurement. Suppose that, in calibrating a ruler, the first centimeter is made too small by 0.1 cm. Then, although the measurements of length on this ruler are still precise to 0.01 cm, they are accurate to only 0.1 cm.

To indicate the precision of a measured number (or result of calculations on measured numbers), we often use the concept of significant figures. **Significant figures** are *those digits in a measured number (or in the result of a calculation with measured numbers) that include all certain digits plus a final digit having some uncertainty.* When you measured the rod, you obtained the values 9.12 cm, 9.11 cm, and 9.13 cm. You could report the result as the average, 9.12 cm. The first two digits (9.1) are certain; the next digit (2) is estimated, so it has some uncertainty. It would be incorrect to write 9.120 cm for the length of the rod. This would say that the last digit (0) has some uncertainty but that the other digits (9.12) are certain, which is not true.

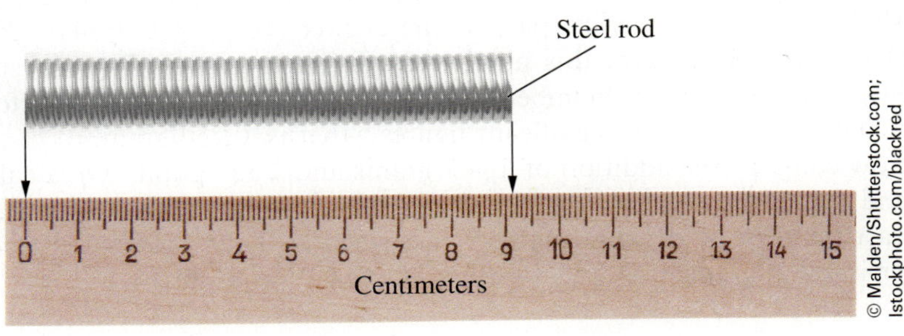

Figure 1.20 ◀

Precision of measurement with a centimeter ruler The length of the rod is just over 9.1 cm. On successive measurements, we estimate the length by eye at 9.12, 9.11, and 9.13 cm. We record the length as between 9.11 cm and 9.13 cm.

Number of Significant Figures

The term **number of significant figures** refers to *the number of digits reported for the value of a measured or calculated quantity, indicating the precision of the value.* Thus, there are three significant figures in 9.12 cm, whereas 9.123 cm has four. To count the number of significant figures in a given measured quantity, you observe the following rules:

1. All digits are significant except zeros at the beginning of the number and possibly terminal zeros (one or more zeros at the end of a number). Thus, 9.12 cm, 0.912 cm, and 0.00912 cm all contain three significant figures.
2. Terminal zeros ending at the right of the decimal point are significant. Each of the following has three significant figures: 9.00 cm, 9.10 cm, 90.0 cm.
3. Terminal zeros in a number without an explicit decimal point may or may not be significant. If someone gives a measurement as 900 cm, you do not know whether one, two, or three significant figures are intended. If the person writes 900. cm (note the decimal point), the zeros are significant. More generally, you can remove any uncertainty in such cases by expressing the measurement in scientific notation.

Scientific notation is *the representation of a number in the form $A \times 10^n$, where A is a number with a single nonzero digit to the left of the decimal point and n is an integer, or whole number.* In scientific notation, the measurement 900 cm precise to two significant figures is written 9.0×10^2 cm. If precise to three significant figures, it is written 9.00×10^2 cm. Scientific notation is also convenient for expressing very large or very small quantities. It is much easier (and simplifies calculations) to write the speed of light as 3.00×10^8 (rather than as 300,000,000) meters per second. ◀

See Appendix A for a review of scientific notation.

© Cengage Learning

Figure 1.21 ▲

Significant figures and calculators Not all of the figures that appear on a calculator display are significant. In performing the calculation 100.0 × 0.0634 ÷ 25.31, the calculator display shows 0.250493875. We would report the answer as 0.250, however, because the factor 0.0634 has the least number of significant figures (three).

Significant Figures in Calculations

Measurements are often used in calculations. How do you determine the correct number of significant figures to report for the answer to a calculation? The following are two rules that we use:

1. **Multiplication and division.** When multiplying or dividing measured quantities, give as many significant figures in the answer as there are in the measurement with *the least number of significant figures.*
2. **Addition and subtraction.** When adding or subtracting measured quantities, give the same number of decimal places in the answer as there are in the measurement with *the least number of decimal places.*

Let us see how you apply these rules. Suppose you have a substance believed to be cisplatin, and, in an effort to establish its identity, you measure its solubility (the amount that dissolves in a given quantity of water). You find that 0.0634 gram of the substance dissolves in 25.31 grams of water. The amount dissolving in 100.0 grams is

$$100.0 \text{ grams of water} \times \frac{0.0634 \text{ gram cisplatin}}{25.31 \text{ grams of water}}$$

Performing the arithmetic on a pocket calculator (Figure 1.21), you get 0.250493875 for the numerical part of the answer (100.0 × 0.0634 ÷ 25.31). It would be incorrect to give this number as the final answer. The measurement 0.0634 gram has the least number of significant figures (three). Therefore, you report the answer to three significant figures—that is, 0.250 gram.

Now consider the addition of 184.2 grams and 2.324 grams. On a calculator, you find that 184.2 + 2.324 = 186.524. But because the quantity 184.2 grams has the least number of decimal places—one, whereas 2.324 grams has three—the answer is 186.5 grams.

Exact Numbers

So far we have discussed only numbers that involve uncertainties. However, you will also encounter exact numbers. An **exact number** is *a number that arises when you count items or sometimes when you define a unit.* For example, when you say there are 9 coins in a bottle, you mean exactly 9, not 8.9 or 9.1. Also, when you say there are 12 inches to a foot, you mean exactly 12. Similarly, the inch is defined to be exactly 2.54 centimeters. The conventions of significant figures do not apply to exact numbers. Thus, the 2.54 in the expression "1 inch equals 2.54 centimeters" should not be interpreted as a measured number with three significant figures. In effect, the 2.54 has an infinite number of significant figures, but of course it would be impossible to write out an infinite number of digits. You should make a mental note of any numbers in a calculation that are exact, because they have no effect on the number of significant figures in a calculation. The number of significant figures in a calculation result depends only on the numbers of significant figures in quantities having uncertainties. For example, suppose you want the total mass of 9 coins when each coin has a mass of 3.0 grams. The calculation is

$$3.0 \text{ grams} \times 9 = 27 \text{ grams}$$

You report the answer to two significant figures because 3.0 grams has two significant figures. The number 9 is exact and does not determine the number of significant figures.

Rounding

In reporting the solubility of your substance in 100.0 grams of water as 0.250 gram, you *rounded* the number you read off the calculator (0.2504938). **Rounding** is *the procedure of dropping nonsignificant digits in a calculation result and adjusting the last digit reported.* The general procedure is as follows: Look at the leftmost digit to be dropped. Then . . .

1. If this digit is 5 or greater, add 1 to the last digit to be retained and drop all digits farther to the right. Thus, rounding 1.2151 to three significant figures gives 1.22.
2. If this digit is less than 5, simply drop it and all digits farther to the right. Rounding 1.2143 to three significant figures gives 1.21.

In doing a calculation of two or more steps, it is desirable to retain nonsignificant digits for intermediate answers. This ensures that accumulated small errors from rounding do not appear in the final result. If you use a calculator, you can simply enter numbers one after the other, performing each arithmetic operation and rounding only the final answer. ▶ To keep track of the correct number of significant figures, you may want to record intermediate answers with a line under the last significant figure, as shown in the solution to part d of the following example.

The answers for Exercises and Problems will follow this practice. For most Example Problems where intermediate answers are reported, all answers, intermediate and final, will be rounded.

Example 1.2 Using Significant Figures in Calculations

Gaining Mastery Toolbox

Critical Concept 1.2

The final result of any calculation must be reported to the correct number of significant figures. When performing numerical operations (addition, subtraction, multiplication, and division), rules must be applied to determine the correct reporting of the final answer. Rounding should *only* be applied to the final result.

Solution Essentials:

- Determination of the number of significant figures in a measured value
- Tracking the number of significant figures of each measured and calculated value throughout the calculation process
- Reporting a calculated value (often the "answer") to the correct number of significant figures by applying the appropriate rules for the operations (multiplication/division and addition/subtraction) and the rule for rounding

Perform the following calculations and round the answers to the correct number of significant figures (units of measurement have been omitted).

a $\dfrac{2.568 \times 5.8}{4.186}$ **b** $5.41 - 0.398$

c $3.38 - 3.01$ **d** $4.18 - 58.16 \times (3.38 - 3.01)$

Problem Strategy Parts a, b, and c are relatively straightforward; use the rule for significant figures that applies in each case. In a multistep calculation as in d, proceed step by step, applying the relevant rule to each step.

(continued)

(continued)

First, do the operations within parentheses. By convention, you perform multiplications and divisions before additions and subtractions. Do not round intermediate answers, but do keep track of the least significant digit—say, by underlining it.

Solution

a The factor 5.8 has the fewest significant figures; therefore, the answer should be reported to two significant figures. Round the answer to **3.6.** **b** The number with the least number of decimal places is 5.41. Therefore, round the answer to two decimal places, to **5.01.** **c** The answer is **0.37.** Note how you have lost one significant figure in the subtraction. **d** You first do the subtraction within parentheses, underlining the least significant digits.

$$4.1\underline{8} - 58.1\underline{6} \times (3.3\underline{8} - 3.0\underline{1}) = 4.1\underline{8} - 58.1\underline{6} \times 0.3\underline{7}$$

Following convention, you do the multiplication before the subtraction.

$$4.1\underline{8} - 58.1\underline{6} \times 0.3\underline{7} = 4.1\underline{8} - 2\underline{1}.5192 = -1\underline{7}.3392$$

The final answer is **−17.**

Answer Check When performing any calculation, you should always report answers to the correct number of significant figures. If a set of calculations involves *only* multiplication and division, it is not necessary to track the significant figures through intermediate calculations; you just need to use the measurement with the least number of significant figures as the guide. However, when addition or subtraction is involved, make sure that you have noted the correct number of significant figures in each calculated step in order to ensure that your answer is correctly reported.

Exercise 1.3 Give answers to the following arithmetic setups. Round to the correct number of significant figures.

a $\dfrac{5.61 \times 7.891}{9.1}$ **b** $8.91 - 6.435$

c $6.81 - 6.730$ **d** $38.91 \times (6.81 - 6.730)$

■ See Problems 1.61 and 1.62.

CONCEPT CHECK 1.2

a When you report your weight to someone, how many significant figures do you typically use?

b What is your weight with two significant figures?

c Indicate your weight and the number of significant figures you would obtain if you weighed yourself on a truck scale that can measure in 50 kg or 100 lb increments.

1.6 SI Units

The first measurements were probably based on the human body (the length of the foot, for example). In time, fixed standards developed, but these varied from place to place. Each country or government (and often each trade) adopted its own units. As science became more quantitative in the seventeenth and eighteenth centuries, scientists found that the lack of standard units was a problem. ◀ They began to seek a simple, international system of measurement. In 1791 a study committee of the French Academy of Sciences devised such a system. Called the *metric system,* it became the official system of measurement for France and was soon used by scientists throughout the world. Most nations have since adopted the metric system or, at least, have set a schedule for changing to it.

In the system of units that Lavoisier used in the eighteenth century, there were 9216 grains to the pound (the livre). English chemists of the same period used a system in which there were 7000 grains to the pound—unless they were trained as apothecaries, in which case there were 5760 grains to the pound!

SI Base Units and SI Prefixes

In 1960 the General Conference of Weights and Measures adopted the **International System of units** (or **SI,** after the French *le Système International d'Unités*), which is *a particular choice of metric units.* This system has seven **SI base units,** *the SI units from which all others can be derived.* Table 1.1 lists these base units and the symbols used to represent them. In this chapter, we will discuss four base quantities: length, mass, time, and temperature. ◀

The amount of substance is discussed in Chapter 3, and the electric current (ampere) is introduced in Chapter 19. Luminous intensity will not be used in this book.

One advantage of any metric system is that it is a decimal system. In SI, a larger or smaller unit for a physical quantity is indicated by an **SI prefix,** which is *a prefix used in the International System to indicate a power of 10.* For example,

Table 1.1 SI Base Units

Quantity	Unit	Symbol
Length	meter	m
Mass	kilogram	kg
Time	second	s
Temperature	kelvin	K
Amount of substance	mole	mol
Electric current	ampere	A
Luminous intensity	candela	cd

Table 1.2 Selected SI Prefixes

Prefix	Multiple	Symbol
mega	10^6	M
kilo	10^3	k
deci	10^{-1}	d
centi	10^{-2}	c
milli	10^{-3}	m
micro	10^{-6}	μ*
nano	10^{-9}	n
pico	10^{-12}	p

*Greek letter mu, pronounced "mew."

the base unit of length in SI is the meter (somewhat longer than a yard), and 10^{-2} meter is called a *centi*meter. Thus 2.54 centimeters equals 2.54×10^{-2} meters. The SI prefixes used in this book are presented in Table 1.2.

Length, Mass, and Time

The **meter (m)** is *the SI base unit of length.* ▶ By combining it with one of the SI prefixes, you can get a unit of appropriate size for any length measurement. For the very small lengths used in chemistry, the nanometer (nm; 1 nanometer = 10^{-9} m) or the picometer (pm; 1 picometer = 10^{-12} m) is an acceptable SI unit. *A non-SI unit of length* traditionally used by chemists is the **angstrom (Å),** which equals 10^{-10} m. (An oxygen atom, one of the minute particles of which the substance oxygen is composed, has a diameter of about 1.3 Å. If you could place oxygen atoms adjacent to one another, you could line up over 75 million of them in 1 cm.)

The meter was originally defined in terms of a standard platinum–iridium bar kept at Sèvres, France. In 1983, the meter was defined as the distance traveled by light in a vacuum in 1/299,792,458 seconds.

The **kilogram (kg)** is *the SI base unit of mass,* equal to about 2.2 pounds. ▶ This is an unusual base unit in that it contains a prefix. In forming other SI mass units, prefixes are added to the word *gram* (g) to give units such as the *milli*gram (mg; 1 mg = 10^{-3} g).

The present standard of mass is the platinum–iridium kilogram mass kept at the International Bureau of Weights and Measures in Sèvres, France.

The **second (s)** is *the SI base unit of time* (Figure 1.22). Combining this unit with prefixes such as *milli-, micro-, nano-,* and *pico-,* you create units appropriate for measuring very rapid events. The time required for the fastest chemical processes is about a picosecond, which is the current limit of timekeeping technology. When you measure times much longer than a few hundred seconds, you revert to *minutes* and *hours,* an obvious exception to the prefix–base format of the International System.

Exercise 1.4 Express the following quantities using an SI prefix and a base unit. For instance, 1.6×10^{-6} m = 1.6 μm. A quantity such as 0.000168 g could be written 0.168 mg or 168 μg.

a. 1.84×10^{-9} m
b. 5.67×10^{-12} s
c. 7.85×10^{-3} g
d. 9.7×10^{3} m
e. 0.000732 s
f. 0.000000000154 m

■ See Problems 1.65 and 1.66.

Temperature

Temperature is difficult to define precisely, but we all have an intuitive idea of what we mean by it. It is a measure of "hotness." A hot object placed next to a cold one becomes cooler, while the cold object becomes hotter. Heat energy passes from a hot object to a cold one, and the quantity of heat passed between the objects

Figure 1.22 ▲

NIST F-1 cesium fountain clock First put into service in 1999 at the National Institute of Standards and Technology (NIST) in Boulder, Colorado, this clock represents a new generation of super-accurate atomic clocks. The clock is so precise that it loses only one second every 20 million years.

depends on the difference in temperature between the two. Therefore, temperature and heat are different, but related, concepts.

A thermometer is a device for measuring temperature. The common type consists of a glass capillary containing a column of liquid whose length varies with temperature. A scale alongside the capillary gives a measure of the temperature. The **Celsius scale** (formerly the centigrade scale) is *the temperature scale in general scientific use.* On this scale, the freezing point of water is 0°C and the boiling point of water at normal barometric pressure is 100°C. However, *the SI base unit of temperature* is the **kelvin (K),** a unit on an *absolute temperature* scale. (See the first margin note on the next page.) On any absolute scale, the lowest temperature that can be attained theoretically is zero. The Celsius and the Kelvin scales have equal-size units (that is, a change of 1°C is equivalent to a change of 1 K), where 0°C is a temperature equivalent to 273.15 K. Thus, it is easy to convert from one scale to the other, using the formula

$$T_K = \left(t_C \times \frac{1\text{ K}}{1°\text{C}}\right) + 273.15\text{ K}$$

where T_K is the temperature in kelvins and t_C is the temperature in degrees Celsius. A temperature of 20°C (about room temperature) equals 293 K.

The Fahrenheit scale is at present the common temperature scale in the United States. Figure 1.23 compares Kelvin, Celsius, and Fahrenheit scales. As the figure shows, 0°C is the same as 32°F (both exact), and 100°C corresponds to 212°F

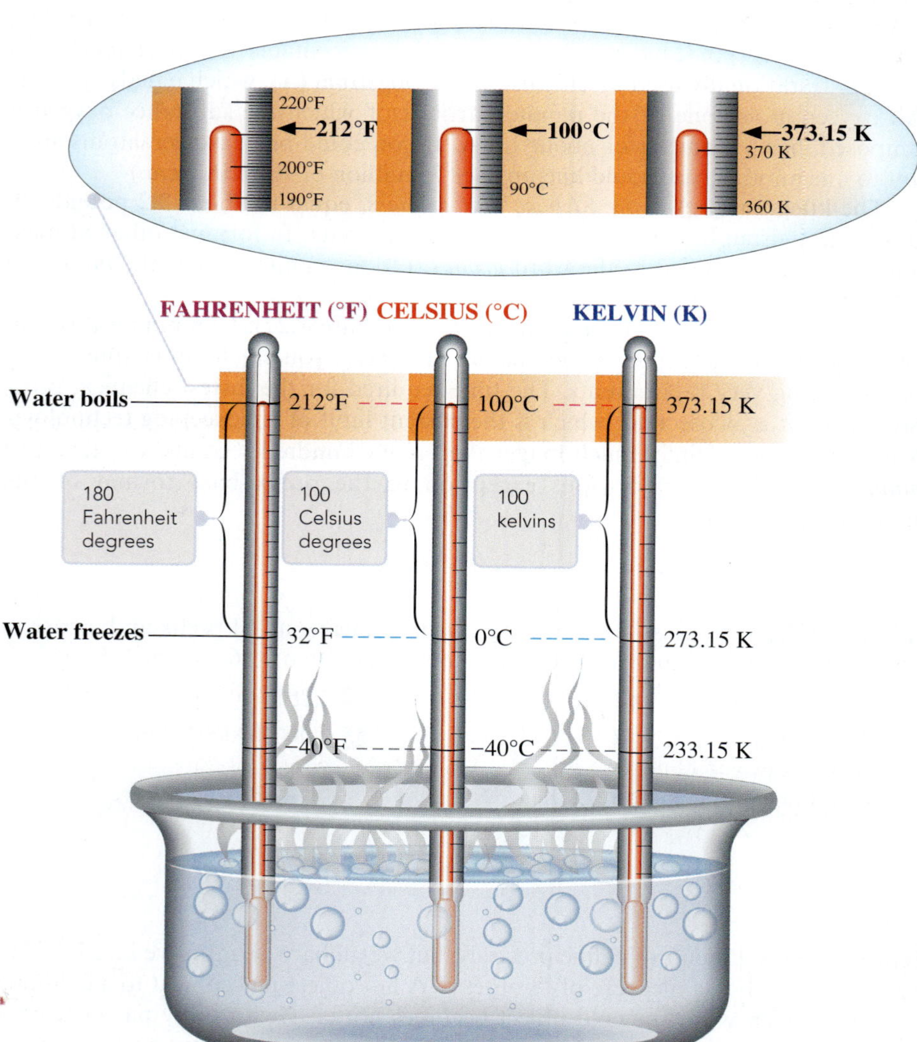

Figure 1.23 ▶

Comparison of temperature scales Room temperature is about 68°F, 20°C, and 293 K. Water freezes at 32°F, 0°C, and 273.15 K. Water boils under normal pressure at 212°F, 100°C, and 373.15 K.

(both exact). Therefore, there are 212 − 32 = 180 Fahrenheit degrees in the range of 100 Celsius degrees. That is, there are exactly 9 Fahrenheit degrees for every 5 Celsius degrees. Knowing this, and knowing that 0°C equals 32°F, we can derive a formula to convert degrees Celsius to degrees Fahrenheit. ▶

$$t_F = \left(t_C \times \frac{9°F}{5°C}\right) + 32°F$$

where t_F is the temperature in Fahrenheit and t_C is the temperature in Celsius. By rearranging this, we can obtain a formula for converting degrees Fahrenheit to degrees Celsius:

$$t_C = \frac{5°C}{9°F} \times (t_F - 32°F)$$

Note that the degree sign (°) is not used with the Kelvin scale, and the unit of temperature is simply the kelvin (not capitalized).

These two temperature conversion formulas are often written K = °C + 273.15 and °F = (1.8 × °C) + 32. Although these yield the correct number, they do not take into account the units.

Example 1.3 Converting from One Temperature Scale to Another

Gaining Mastery Toolbox

Critical Concept 1.3
There are several commonly used temperature scales: Celsius, Fahrenheit, and Kelvin. The Celsius temperature scale is used most frequently in the laboratory. Using mathematical relationships, temperature values can be converted between the Celsius, Kelvin, and Fahrenheit scales.

Solution Essentials:
- Relationship (conversion formula) between Kelvin and Celsius temperatures
- Relationship (conversion formula) between Celsius and Fahrenheit temperatures

The hottest place on record in North America is Death Valley in California. It reached a temperature of 134°F in 1913. What is this temperature reading in degrees Celsius? in kelvins?

Problem Strategy This calculation involves conversion of a temperature in degrees Fahrenheit (°F) to degrees Celsius (°C) and kelvins (K). This is a case where using formulas is the most reliable method for arriving at the answer.

Solution Substituting, we find that

$$t_C = \frac{5°C}{9°F} \times (t_F - 32°F) = \frac{5°C}{9°F} \times (134°F - 32°F) = \mathbf{56.7°C}$$

In kelvins,

$$T_K = \left(t_C \times \frac{1\,K}{1°C}\right) + 273.15\,K = \left(56.7°C \times \frac{1\,K}{1°C}\right) + 273.15\,K = \mathbf{329.9\,K}$$

Answer Check An easy way to check temperature conversions is to use the appropriate equation and convert your answer back to the original temperature to see whether it matches.

Exercise 1.5 **a** A person with a fever has a temperature of 102.5°F. What is this temperature in degrees Celsius? **b** A cooling mixture of dry ice and isopropyl alcohol has a temperature of –78°C. What is this temperature in kelvins?

■ See Problems 1.69, 1.70, 1.71, and 1.72.

CONCEPT CHECK 1.3

a Estimate and express the length of your leg in an appropriate metric unit.
b What would be a reasonable height, in meters, of a three-story building?
c How would you be feeling if your body temperature was 39°C?
d Would you be comfortable sitting in a room at 23°C in a short-sleeved shirt?

1.7 Derived Units

Once base units have been defined for a system of measurement, you can derive other units from them. You do this by using the base units in equations that define other physical quantities. For example, *area* is defined as length times length. Therefore,

$$\text{SI unit of area} = (\text{SI unit of length}) \times (\text{SI unit of length})$$

Table 1.3 Derived Units

Quantity	Definition of Quantity	SI Unit
Area	Length squared	m^2
Volume	Length cubed	m^3
Density	Mass per unit volume	kg/m^3
Speed	Distance traveled per unit time	m/s
Acceleration	Speed changed per unit time	m/s^2
Force	Mass times acceleration of object	$kg \cdot m/s^2$ (= newton, N)
Pressure	Force per unit area	$kg/(m \cdot s^2)$ (= pascal, Pa)
Energy	Force times distance traveled	$kg \cdot m^2/s^2$ (= joule, J)

From this, you see that the SI unit of area is meter × meter, or m^2. Similarly, *speed* is defined as the rate of change of distance with time; that is, speed = distance/time. Consequently,

$$\text{SI unit of speed} = \frac{\text{SI unit of distance}}{\text{SI unit of time}}$$

The SI unit of speed is meters per second (that is, meters divided by seconds). The unit is symbolized m/s or $m \cdot s^{-1}$. The unit of speed is an example of an **SI derived unit,** which is *a unit derived by combining SI base units.* Table 1.3 defines a number of derived units. Volume and density are discussed in this section; pressure and energy are discussed later (in Sections 5.1 and 6.1, respectively).

Volume

Volume is defined as length cubed and has the SI unit of cubic meter (m^3). This is too large a unit for normal laboratory work, so we use either cubic decimeters (dm^3) or cubic centimeters (cm^3, also written cc). Traditionally, chemists have used the **liter (L),** which is *a unit of volume equal to a cubic decimeter* (approximately one quart). In fact, most laboratory glassware (Figure 1.24) is calibrated in liters or milliliters (1000 mL = 1 L). Because 1 dm equals 10 cm, a cubic decimeter, or one liter, equals $(10\ cm)^3 = 1000\ cm^3$. Therefore, a milliliter equals a cubic centimeter. In summary

$$1\ L = 1\ dm^3 \qquad \text{and} \qquad 1\ mL = 1\ cm^3$$

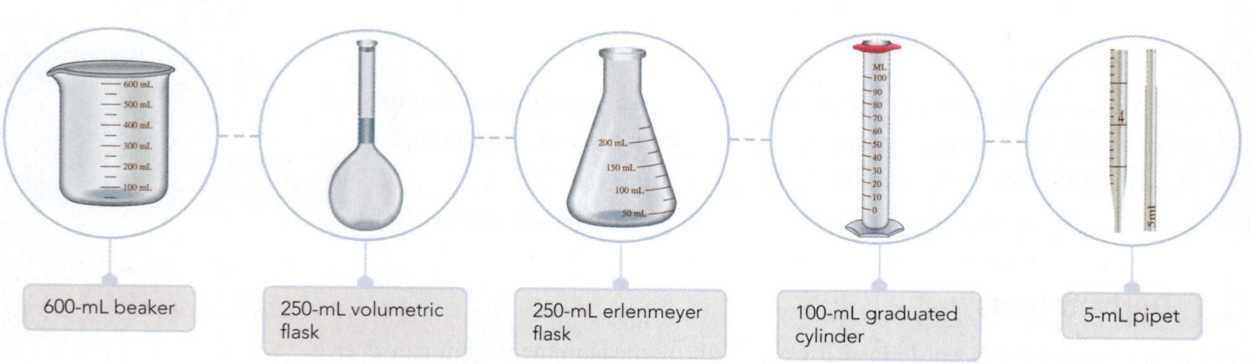

Figure 1.24 ▲
Some laboratory glassware

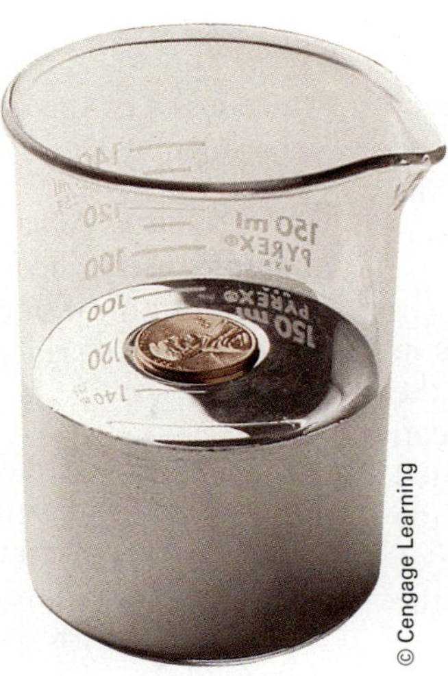

Figure 1.25 ▶

The relative densities of copper and mercury Copper floats on mercury because the density of copper (8.95 g/cm^3) is less than the density of mercury (13.5 g/cm^3). (This is a copper penny; recently minted pennies are copper-clad zinc.)

Figure 1.26 ▲

Relative densities of some liquids Shown are three liquids (dyed so that they will show up clearly): *ortho*-xylene, water, and 1,1,1-trichloroethane. Water (blue layer) is less dense than 1,1,1-trichloroethane (colorless) and floats on it. *Ortho*-xylene (top, golden layer) is less dense than water and floats on it.

Density

The **density** of an object is its *mass per unit volume*. You can express this as

$$d = \frac{m}{V}$$

where d is the density, m is the mass, and V is the volume. Suppose an object has a mass of 15.0 g and a volume of 10.0 cm^3. Substituting, you find that

$$d = \frac{15.0\ \text{g}}{10.0\ \text{cm}^3} = 1.50\ \text{g/cm}^3$$

The density of the object is 1.50 g/cm^3 (or 1.50 g·cm^{-3}).

Density is an important physical property of a material. Water, for example, has a density of 1.000 g/cm^3 at 4°C and a density of 0.998 g/cm^3 at 20°C. Lead has a density of 11.3 g/cm^3 at 20°C. (Figures 1.25 and 1.26 dramatically show some relative densities.) ▶ Oxygen gas has a density of 1.33 × 10^{-3} g/cm^3 at normal pressure and 20°C. (Like other gases under normal conditions, oxygen has a density that is about 1000 times smaller than those of liquids and solids.) Because the density is characteristic of a substance, it can be helpful in identifying it. Example 1.4 illustrates this point. Density can also be useful in determining whether a substance is pure. Consider a gold bar whose purity is questioned. The metals likely to be mixed with gold, such as silver or copper, have lower densities than gold. Therefore, an adulterated (impure) gold bar can be expected to be far less dense than pure gold.

The density of solid materials on earth ranges from about 1 g/cm^3 to 22.5 g/cm^3 (osmium metal). In the interior of certain stars, the density of matter is truly staggering. Black neutron stars—stars composed of neutrons, or atomic cores compressed by gravity to a superdense state—have densities of about 10^{15} g/cm^3.

Example  1.4 Calculating the Density of a Substance

Gaining Mastery Toolbox

Critical Concept 1.4

Density is the relationship of the mass to volume (m/V) of a substance. Unlike mass, density is a characteristic (physical property) that can be used to identify and distinguish pure substances from each other.

Solution Essentials:

- Density of a substance is calculated by dividing the mass of substance (m) by the volume (V) that the mass of substance occupies
- Rules for significant figures and rounding
- The metric system

A colorless liquid, used as a solvent (a liquid that dissolves other substances), is believed to be one of the following:

Substance	*Density (in g/mL)*
n-butyl alcohol	0.810
ethylene glycol	1.114
isopropyl alcohol	0.785
toluene	0.866

(continued)

(*continued*)

To identify the substance, a chemist determined its density. By pouring a sample of the liquid into a graduated cylinder, she found that the volume was 35.1 mL. She also found that the sample weighed 30.5 g. What was the density of the liquid? What was the substance?

Problem Strategy The solution to this problem lies in finding the density of the unknown substance. Once the density of the unknown substance is known, you can compare it to the list of known substances presented in the problem and look for a match. Density is the relationship of the mass of a substance per volume of that substance. Expressed as an equation, density is the mass divided by the volume: $d = m/V$.

Solution You substitute 30.5 g for the mass and 35.1 mL for the volume into the equation.

$$d = \frac{m}{V} = \frac{30.5 \text{ g}}{35.1 \text{ mL}} = \mathbf{0.869 \text{ g/mL}}$$

The density of the liquid equals that of **toluene** (within experimental error).

Answer Check Always be sure to report the density in the units used when performing the calculation. Density is not always reported in units of g/mL or g/cm^3, for example; gases are often reported with the units of g/L.

Exercise 1.6 A piece of metal wire has a volume of 20.2 cm^3 and a mass of 159 g. What is the density of the metal? We know that the metal is manganese, iron, or nickel, and these have densities of 7.21 g/cm^3, 7.87 g/cm^3, and 8.90 g/cm^3, respectively. From which metal is the wire made?

■ See Problems 1.73, 1.74, 1.75, and 1.76.

In addition to characterizing a substance, the density provides a useful relationship between mass and volume. For example, suppose an experiment calls for a certain mass of liquid. Rather than weigh the liquid on a balance, you might instead measure out the corresponding volume. Example 1.5 illustrates this idea.

OWL Interactive Example 1.5 Using the Density to Relate Mass and Volume

Gaining Mastery Toolbox

Critical Concept 1.5

Density is a mathematical equation expressed as $d = m/V$. Because this relationship is an equation with three variables, it can be manipulated to solve for one variable when two of the variables are known.

Solution Essentials:

- Identify the unknown quantity in the $d = m/V$ equation.
- Rearrange the equation to solve for the unknown quantity
- Substitute known values into the rearranged equation to solve for unknown quantity
- The metric system
- Rules for significant figures and rounding

An experiment requires 43.7 g of isopropyl alcohol. Instead of measuring out the sample on a balance, a chemist dispenses the liquid into a graduated cylinder. The density of isopropyl alcohol is 0.785 g/mL. What volume of isopropyl alcohol should he use?

Problem Strategy The formula that defines density provides a relationship between the density of a substance and the mass and volume of the substance: $d = m/V$. Because this relationship is an equation, it can be used to solve for one variable when the other two are known. Examining this problem reveals that the mass (m) and density (d) of the substance are known, so the equation can be used to solve for the volume (V) of the liquid.

Solution You rearrange the formula defining the density to obtain the volume.

$$V = \frac{m}{d}$$

Then you substitute into this formula:

$$V = \frac{43.7 \text{ g}}{0.785 \text{ g/mL}} = \mathbf{55.7 \text{ mL}}$$

Answer Check Note that the density of the alcohol is close to 1 g/mL, which is common for many substances. In cases like this, you can perform a very quick check of your answer, because the numerical value for the volume of the liquid should not be much different from the numerical value of the mass of the liquid.

Exercise 1.7 Ethanol (grain alcohol) has a density of 0.789 g/cm^3. What volume of ethanol must be poured into a graduated cylinder to equal 30.3 g?

■ See Problems 1.77, 1.78, 1.79, and 1.80.

CONCEPT CHECK 1.4

You are working in the office of a precious metals buyer. A miner brings you a nugget of metal that he claims is gold. You suspect that the metal is a form of "fool's gold," called marcasite, that is composed of iron and sulfur. In the back of your office, you have a chunk of pure gold. What simple experiments could you perform to determine whether the miner's nugget is gold?

1.8 Units and Dimensional Analysis (Factor-Label Method)

In performing numerical calculations with physical quantities, it is good practice to enter each quantity as a number with its associated unit. Both the numbers and the units are then carried through the indicated algebraic operations. The advantages of this are twofold:

1. The units for the answer will come out of the calculations.
2. If you make an error in arranging factors in the calculation (for example, if you use the wrong formula), this will become apparent because the final units will be nonsense.

Dimensional analysis (or the **factor-label method**) is *the method of calculation in which one carries along the units for quantities.* As an illustration, suppose you want to find the volume V of a cube, given s, the length of a side of the cube. Because $V = s^3$, if $s = 5.00$ cm, you find that $V = (5.00\text{ cm})^3 = 5.00^3\text{ cm}^3$. There is no guesswork about the unit of volume here; it is cubic centimeters (cm^3).

Suppose, however, that you wish to express the volume in liters (L), a metric unit that equals 10^3 cubic centimeters (approximately one quart). You can write this equality as

$$1\text{ L} = 10^3\text{ cm}^3$$

If you divide both sides of the equality by the right-hand quantity, you get

$$\frac{1\text{ L}}{10^3\text{ cm}^3} = \frac{\cancel{10^3\text{ cm}^3}}{\cancel{10^3\text{ cm}^3}} = 1$$

Observe that you treat units in the same way as algebraic quantities. Note too that the right-hand side now equals 1 and no units are associated with it. Because it is always possible to multiply any quantity by 1 without changing that quantity, you can multiply the previous expression for the volume by the factor $1\text{ L}/10^3\text{ cm}^3$ without changing the actual volume. You are changing only the way you express this volume:

$$V = 5.00^3\,\cancel{\text{cm}^3} \times \underbrace{\frac{1\text{ L}}{10^3\,\cancel{\text{cm}^3}}}_{\substack{\text{converts}\\ \text{cm}^3\text{ to L}}} = 125 \times 10^{-3}\text{ L} = 0.125\text{ L}$$

In order to solve problems using dimensional analysis, you may find it helpful to diagram a solution pathway using units as a guide. For example, in the problem above:

In this diagram the ⇒ represents the ratio $1\text{ L}/10^3\text{ cm}^3$, which we use to convert from units of cm^3 to L.

The ratio $1\text{ L}/10^3\text{ cm}^3$ is called a **conversion factor** because it is *a factor equal to 1 that converts a quantity expressed in one unit to a quantity expressed in another unit.* ▶ Note that the numbers in this conversion factor are *exact,* because 1 L equals exactly 1000 cm^3. Such exact conversion factors do not affect the number of significant figures in an arithmetic result. In the previous calculation, the

It takes more room to explain conversion factors than it does to use them. With practice, you will be able to write the final conversion step without the intermediate algebraic manipulations outlined here.

quantity 5.00 cm (the measured length of the side) does determine or limit the number of significant figures.

The next two examples illustrate how the conversion-factor method may be used to convert one metric unit to another.

Example 1.6 Converting Units: Metric Unit to Metric Unit

Gaining Mastery Toolbox

Critical Concept 1.6
Dimensional analysis can be used to convert metric units. Unit conversion factors are the key to applying dimensional analysis to solve problems.

Solution Essentials:
- Metric conversion factors
- The metric system
- Rules for significant figures and rounding

Nitrogen gas is the major component of air. A sample of nitrogen gas in a glass bulb weighs 243 mg. What is this mass in SI base units of mass (kilograms)?

Problem Strategy This problem requires converting a mass in milligrams to kilograms. Finding the correct relationship for performing such a conversion directly in one step might be more difficult than doing the conversion in multiple steps. In this case, conversion factors for converting milligrams to grams and from grams to kilograms are relatively easy to derive, so it makes sense to do this conversion in two steps. First convert milligrams to grams; then convert grams to kilograms. To convert from milligrams to grams, note that the prefix *milli-* means 10^{-3}. To convert from grams to kilograms, note that the prefix *kilo-* means 10^3.

Diagramming the solution clearly shows how two conversion factors (2 ⇒) are used to solve the problem.

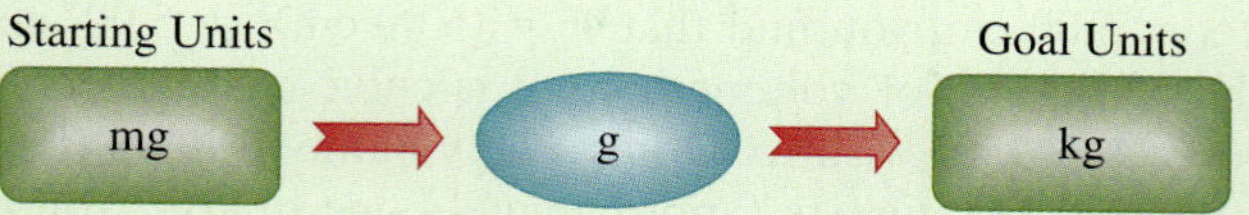

Nitrogen molecular model

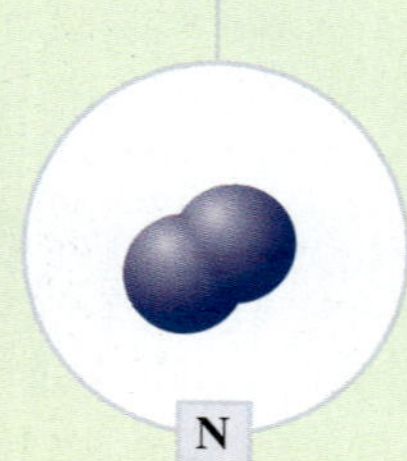

Solution Since 1 mg = 10^{-3} g, you can write

$$243\ \cancel{\text{mg}} \times \frac{10^{-3}\ \text{g}}{1\ \cancel{\text{mg}}} = 2.43 \times 10^{-1}\ \text{g}$$

Then, because the prefix *kilo-* means 10^3, you write

$$1\ \text{kg} = 10^3\ \text{g}$$

and

$$2.43 \times 10^{-1}\ \cancel{\text{g}} \times \frac{1\ \text{kg}}{10^3\ \cancel{\text{g}}} = \mathbf{2.43 \times 10^{-4}\ kg}$$

Note, however, that you can combine the two conversion steps as follows:

$$243\ \cancel{\text{mg}} \times \underbrace{\frac{10^{-3}\ \cancel{\text{g}}}{1\ \cancel{\text{mg}}}}_{\substack{\text{converts}\\ \text{mg to g}}} \times \underbrace{\frac{1\ \text{kg}}{10^3\ \cancel{\text{g}}}}_{\substack{\text{converts}\\ \text{g to kg}}} = 2.43 \times 10^{-4}\ \text{kg}$$

Answer Check Often, mistakes lead to answers that are easily detected by giving them a *reasonableness check*. For example, if while solving this problem you mistakenly write the conversion factors in inverted order (although this should not happen if you use units in your calculation), the final answer will be very large. Or perhaps you inadvertently copy the exponent from your calculator with the wrong sign, giving the answer as 2.43×10^4 kg. A large answer is not reasonable, given that you are converting from small mass units (mg) to relatively large mass units (kg). Both of these errors can be caught simply by taking the time to make sure that your answer is reasonable. After completing any calculations, be sure to check for reasonableness in your answers.

Oxygen molecular model

121 pm

O

Exercise 1.8 The oxygen molecule (the smallest particle of oxygen gas) consists of two oxygen atoms a distance of 121 pm apart. How many millimeters is this distance?

■ See Problems 1.81, 1.82, 1.83, and 1.84.

Example 1.7 Converting Units: Metric Volume to Metric Volume

Gaining Mastery Toolbox

Critical Concept 1.7
Dimensional analysis can be used to convert between volumes. Volume measurements typically involve cubed conversion factors.

Solution Essentials:
- Derive cubic (volume) conversion factors from metric conversion factors
- Metric conversion factors
- The metric system
- Rules for significant figures and rounding

The world's oceans contain approximately 1.35×10^9 km^3 of water. What is this volume in liters?

Problem Strategy This problem requires several conversions. A relationship that will prove helpful here is the definition of the liter: 1 dm^3 = 1 L. This provides a clue that if we can convert cubic kilometers (km^3) to cubic decimeters (dm^3), we can use this relationship to convert to liters. When solving problems that involve cubic units such as this one, it is often helpful to think of the conversion factors that you would use if the units were not cubed. In this case, convert from kilometers to meters and then from meters to decimeters. Next, think about how these conversion factors would look if they were cubed. For example, 10^3 m = 1 km, so $(10^3 \text{ m})^3 = (1 \text{ km})^3$ is the cubic relationship. Likewise, 10^{-1} m = 1 dm, so $(10^{-1} \text{ m})^3 = (1 \text{ dm})^3$ is the cubic relationship. Now these two cubic conversion factors can be combined into three steps to solve the problem.

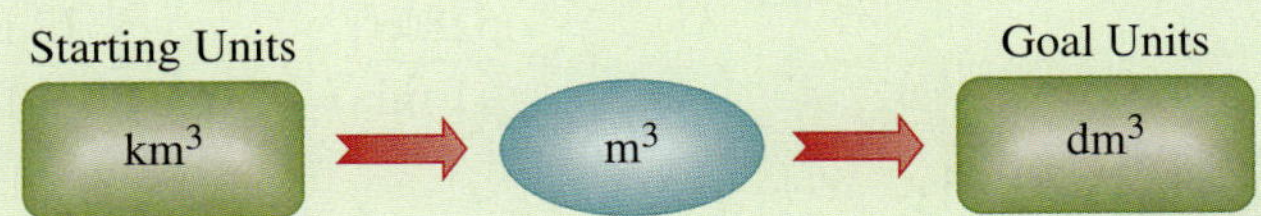

Solution

$$1.35 \times 10^9 \text{ km}^3 \times \underbrace{\left(\frac{10^3 \text{ m}}{1 \text{ km}}\right)^3}_{\text{converts km}^3 \text{ to m}^3} \times \underbrace{\left(\frac{1 \text{ dm}}{10^{-1} \text{ m}}\right)^3}_{\text{converts m}^3 \text{ to dm}^3} = 1.35 \times 10^{21} \text{ dm}^3$$

Because a cubic decimeter is equal to a liter, the volume of the oceans is

$$\mathbf{1.35 \times 10^{21} \text{ L}}$$

Answer Check One of the most common errors made with this type of problem is a failure to cube both the numerator and the denominator in the conversion factors. For example, $(10^3 \text{ m}/1 \text{ km})^3 = 10^9 \text{ m}^3/1 \text{ km}^3$.

Exercise 1.9 A large crystal is constructed by stacking small, identical pieces of crystal, much as you construct a brick wall by stacking bricks. A unit cell is the smallest such piece from which a crystal can be made. The unit cell of a crystal of gold metal has a volume of 67.6 Å^3. What is this volume in cubic decimeters?

■ See Problems 1.85 and 1.86.

The conversion-factor method can be used to convert any unit to another unit, provided a conversion equation exists between the two units. Relationships between certain U.S. units and metric units are given in Table 1.4. You can use these to convert between U.S. and metric units. Suppose you wish to convert 0.547 lb to grams. From Table 1.4, note that 1 lb = 0.4536 kg, or 1 lb = 453.6 g, so the conversion factor from pounds to grams is 453.6 g/1 lb. Therefore,

$$0.547 \text{ lb} \times \frac{453.6 \text{ g}}{1 \text{ lb}} = 248 \text{ g}$$

The next example illustrates a conversion requiring several steps.

Table 1.4 Relationships of Some U.S. and Metric Units

Length	Mass	Volume
1 in. = 2.54 cm (exact)	1 lb = 0.4536 kg	1 qt = 0.9464 L
1 yd = 0.9144 m (exact)	1 lb = 16 oz (exact)	4 qt = 1 gal (exact)
1 mi = 1.609 km	1 oz = 28.35 g	
1 mi = 5280 ft (exact)		

Example 1.8 Converting Units: Any Unit to Another Unit

Gaining Mastery Toolbox

Critical Concept 1.8
Dimensional analysis can be used to perform any type of unit conversion. As long as conversion factors between units exist, dimensional analysis can be applied to solve the problem.

Solution Essentials:
- U.S. to metric conversion factors
- Metric conversion factors
- The metric system
- Rules for significant figures and rounding

© Spencer Grant/Photo Edit

How many centimeters are there in 6.51 miles?

Problem Strategy This problem involves converting from U.S. to metric units: specifically, from miles to centimeters. One path to a solution is using three conversions: miles to feet, feet to inches, and inches to centimeters. Table 1.4 has the information you need to develop the conversion factors for U.S. to metric units.

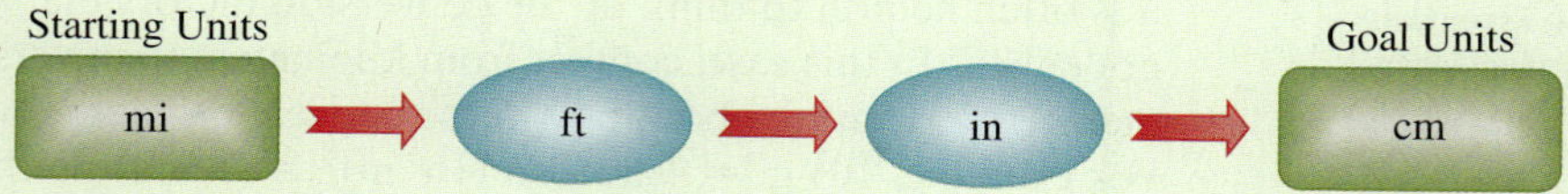

Solution From the definitions, you obtain the following conversion factors:

$$1 = \frac{5280 \text{ ft}}{1 \text{ mi}} \qquad 1 = \frac{12 \text{ in.}}{1 \text{ ft}} \qquad 1 = \frac{2.54 \text{ cm}}{1 \text{ in.}}$$

Then,

$$6.51 \cancel{\text{mi}} \times \underbrace{\frac{5280 \cancel{\text{ft}}}{1 \cancel{\text{mi}}}}_{\text{converts mi to ft}} \times \underbrace{\frac{12 \cancel{\text{in.}}}{1 \cancel{\text{ft}}}}_{\text{converts fi to in.}} \times \underbrace{\frac{2.54 \text{ cm}}{1 \cancel{\text{in.}}}}_{\text{converts in. to cm}} = \mathbf{1.05 \times 10^6 \text{ cm}}$$

All of the conversion factors used in this example are exact, so the number of significant figures in the result is determined by the number of significant figures in 6.51 mi.

Answer Check Because miles are a relatively large unit of length and centimeters are a relatively small unit of length, one would expect that converting from miles to centimeters would yield a very large number. This is the case here, so the answer is reasonable.

Exercise 1.10 Using the definitions 1 in. = 2.54 cm and 1 yd = 36 in. (both exact), obtain the conversion factor for yards to meters. How many meters are there in 3.54 yd?

■ See Problems 1.87, 1.88, 1.89, and 1.90.

A Checklist for Review

Summary of Facts and Concepts

Chemistry is an experimental science in that the facts of chemistry are obtained by *experiment.* These facts are systematized and explained by *theory,* and theory suggests more experiments. The *scientific method* involves this interplay, in which the body of accepted knowledge grows as it is tested by experiment.

Chemistry emerged as a *quantitative science* with the work of the eighteenth-century French chemist Antoine Lavoisier. He made use of the idea that the mass, or quantity of matter, remains constant during a chemical reaction *(law of conservation of mass).*

Matter may be classified by its physical state as a *gas, liquid,* or *solid.* Matter may also be classified by its chemical constitution as an *element, compound,* or *mixture.* Materials are either substances or mixtures of substances. Substances are either elements or compounds, which are composed of two or more elements. Mixtures can be separated into substances by physical processes, but compounds can be separated into elements only by chemical reactions.

A quantitative science requires the making of measurements. Any measurement has limited *precision,* which you convey by writing the measured number to a certain number of *significant figures.* There are many different systems of measurement, but scientists generally use the metric system. The International System (SI) uses a particular selection of metric units. It employs seven *base units* combined with *prefixes* to obtain units of various size. Units for other quantities are derived from these. To obtain a *derived unit* in SI for a quantity such as the volume or density, you merely substitute base units into a defining equation for the quantity.

Dimensional analysis is a technique of calculating with physical quantities in which units are included and treated in the same way as numbers. You can thus obtain the *conversion factor* needed to express a quantity in new units.

Learning Objectives	Important Terms
1.1 Modern Chemistry: A Brief Glimpse	
■ Provide examples of the contributions of chemistry to humanity.	
1.2 Experiment and Explanation	
■ Describe how chemistry is an experimental science. ■ Understand how the scientific method is an approach to performing science.	**experiment** **law** **hypothesis** **theory**
1.3 Law of Conservation of Mass	
■ Explain the law of conservation of mass. ■ Apply the law of the conservation of mass. Example 1.1	**mass** **matter** **law of conservation of mass**
1.4 Matter: Physical State and Chemical Constitution	
■ Compare and contrast the three common states of matter: solid, liquid, and gas. ■ Describe the classifications of matter: elements, compounds, and mixtures (heterogeneous and homogeneous). ■ Understand the difference between chemical changes (chemical reactions) and physical changes. ■ Distinguish between chemical properties and physical properties.	**solid** **liquid** **gas** **states of matter** **physical change** **chemical change (chemical reaction)** **physical property** **chemical property** **substance** **element** **compound** **law of definite proportions (law of constant composition)** **mixture** **heterogeneous mixture** **homogeneous mixture (solution)** **phase**
1.5 Measurement and Significant Figures	
■ Define and use the terms *precision* and *accuracy* when describing measured quantities. ■ Learn the rules for determining significant figures in reported measurements. ■ Know how to represent numbers using scientific notation. ■ Apply the rules of significant figures to reporting calculated values. ■ Be able to recognize exact numbers. ■ Know when and how to apply the rules for rounding. ■ Use significant figures in calculations. Example 1.2	**unit** **precision** **accuracy** **significant figures** **number of significant figures** **scientific notation** **exact number** **rounding**

1.6 SI Units	
■ Become familiar with the SI (metric) system of units, including the SI prefixes. ■ Convert from one temperature scale to another. Example 1.3	**International System of units (SI)** **SI base units** **SI prefix** **meter (m)** **angstrom (Å)** **kilogram (kg)** **second (s)** **Celsius scale** **kelvin (K)**
1.7 Derived Units	
■ Define and provide examples of derived units. ■ Calculate the density of a substance. Example 1.4 ■ Use density to relate mass and volume. Example 1.5	**SI derived unit** **liter (L)** **density**
1.8 Units and Dimensional Analysis (Factor-Label Method)	
■ Apply dimensional analysis to solving numerical problems. ■ Convert from one metric unit to another metric unit. Example 1.6 ■ Convert from one metric volume to another metric volume. Example 1.7 ■ Convert from any unit to another unit. Example 1.8	**dimensional analysis (factor-label method)** **conversion factor**

Key Equations

$$T_K = \left(t_C \times \frac{1\text{ K}}{1°\text{C}}\right) + 273.15\text{ K}$$

$$t_C = \frac{5°\text{C}}{9°\text{F}} \times (t_F - 32°\text{F})$$

$$d = m/V$$

Questions and Problems

OWL Interactive versions of these problems may be assigned in OWL.

Self-Assessment and Review Questions

Key: These questions test your understanding of the ideas you worked with in the chapter. These problems vary in difficulty and often can be used for the basis of discussion.

1.1 Discuss some ways in which chemistry has changed technology. Give one or more examples of how chemistry has affected another science.
1.2 Define the terms *experiment* and *theory*. How are theory and experiment related? What is a hypothesis?
1.3 Illustrate the steps in the scientific method using Rosenberg's discovery of the anticancer activity of cisplatin.
1.4 Define the terms *matter* and *mass*. What is the difference between mass and weight?
1.5 State the law of conservation of mass. Describe how you might demonstrate this law.
1.6 A chemical reaction is often accompanied by definite changes in appearance. For example, heat and light may be emitted, and there may be a change of color of the substances. Figure 1.9 shows the reactions of the metal mercury with oxygen in air. Describe the changes that occur.
1.7 Characterize gases, liquids, and solids in terms of compressibility and fluidity.
1.8 Choose a substance and give several of its physical properties and several of its chemical properties.
1.9 Give examples of an element, a compound, a heterogeneous mixture, and a homogeneous mixture.
1.10 What phases or states of matter are present in a glass of bubbling carbonated beverage that contains ice cubes?
1.11 What distinguishes an element from a compound? Can a compound also be an element?

1.12 What is meant by the precision of a measurement? How is it indicated?

1.13 Two rules are used to decide how to round the result of a calculation to the correct number of significant figures. Use a calculation to illustrate each rule. Explain how you obtained the number of significant figures in the answers.

1.14 Distinguish between a measured number and an exact number. Give examples of each.

1.15 How does the International System (SI) obtain units of different size from a given unit? How does the International System obtain units for all possible physical quantities from only seven base units?

1.16 What is an absolute temperature scale? How are degrees Celsius related to kelvins?

1.17 Define *density*. Describe some uses of density.

1.18 Why should units be carried along with numbers in a calculation?

1.19 When the quantity 12.9 g is added to 2×10^{-02} g, how many significant figures should be reported in the answer?

a one b two c three d four e five

1.20 You perform an experiment in the lab and determine that there are 36.3 inches in a meter. Using this experimental value, how many millimeters are there in 1.34 feet?

a 4.43×10^2 mm
b 4.05×10^2 mm
c 44.3 mm
d 4.43×10^5 mm
e 4.05×10^8 mm

1.21 A 75.0-g sample of a pure liquid, liquid A, with a density of 3.00 g/mL is mixed with a 50.0-mL sample of a pure liquid, liquid B, with a density of 2.00 g/mL. What is the total volume of the mixture? (Assume there is no reaction upon the mixing of A and B and volumes are additive.)

a 275 mL
b 175 mL
c 125 mL
d 100. mL
e 75 mL

1.22 Which of the following represents the *smallest* mass?

a 23 cg
b 2.3×10^3 μg
c 0.23 mg
d 0.23 g
e 2.3×10^{-2} kg

Concept Explorations

Key: Concept explorations are comprehensive problems that provide a framework that will enable you to explore and learn many of the critical concepts and ideas in each chapter. If you master the concepts associated with these explorations, you will have a better understanding of many important chemistry ideas and will be more successful in solving all types of chemistry problems. These problems are well suited for group work and for use as in-class activities.

1.23 Physical and Chemical Changes

Say you are presented with two beakers, beaker A and beaker B, each containing a white, powdery compound.

a From your initial observations, you suspect that the two beakers contain the same compound. Describe, in general terms, some experiments in a laboratory that you could do to help prove or disprove that the beakers contain the same compound.

b Would it be easier to prove that the compounds are the same or to prove that they are different? Explain your reasoning.

c Which of the experiments that you listed above are the most convincing in determining whether the compounds are the same? Justify your answer.

d A friend states that the best experiment for determining whether the compounds are the same is to see if they both dissolve in water. He proceeds to take 10.0 g of each compound and places them in separate beakers, each containing 100 mL of water. Both compounds completely dissolve. He then states, “Since the same amount of both substances dissolved in the same volume of water, they must both have the same chemical composition.” Is he justified in making this claim? Why or why not?

1.24 Significant Figures

Part 1:

a Consider three masses that you wish to add together: 3 g, 1.4 g, and 3.3 g. These numbers represent measured values. Add the numbers together and report your answer to the correct number of significant figures.

b Now perform the addition in a stepwise fashion in the following manner. Add 3 g and 1.4 g, reporting this sum to the correct number of significant figures. Next, take the number from the first step and add it to 3.3 g, reporting this sum to the correct number of significant figures.

c Compare your answers from performing the addition in the two distinct ways presented in parts a and b. Does one of the answers represent a “better” way of reporting the results of the addition? If your answer is yes, explain why your choice is better.

d A student performs the calculation $(5.0 \times 5.143 \text{ g}) + 2.80$ g and, being mindful of significant figures, reports an answer of 29 g. Is this the correct answer? If not, what might this student have done incorrectly?

e Another student performs the calculation $(5 \times 5.143 \text{ g}) + 2.80$ and reports an answer of 29 g. Is this the correct answer? If not, what might this student have done incorrectly?

f Yet another student performs the calculation $(5.00 \times 5.143 \text{ g}) + 2.80$ and reports an answer of 28.5 g. Is this the correct answer? If not, what did this student probably do incorrectly?

g Referring to the calculations above, outline a procedure or rule(s) that will always enable you to report answers using the correct number of significant figures.

Part 2:

a. A student wants to determine the volume of 27.2 g of a substance. He looks up the density of the material in a reference book, where it is reported to be 2.4451 g/cm³. He performs the calculation in the following manner:

$$27.2 \text{ g} \times 1.0 \text{ cm}^3/2.4 \text{ g} = 11.3 \text{ cm}^3$$

Is the calculated answer correct? If not, explain why it is not correct.

b. Another student performs the calculation in the following manner:

$$27.2 \text{ g} \times 1.00 \text{ cm}^3/2.45 \text{ g} = 11.1 \text{ cm}^3$$

Is this a "better" answer than that of the first student? Is this the "best" answer, or could it be "improved"? Explain.

c. Say that you have ten ball bearings, each having a mass of 1.234 g and a density of 3.1569 g/cm³. Calculate the volume of these ten ball bearings. In performing the calculation, present your work as unit conversions, and report your answer to the correct number of significant figures.

d. Explain how the answer that you calculated in part c is the "best" answer to the problem?

Conceptual Problems

Key: These problems are designed to check your understanding of the concepts associated with some of the main topics presented in each chapter. A strong conceptual understanding of chemistry is the foundation for both applying chemical knowledge and solving chemical problems. These problems vary in level of difficulty and often can be used as a basis for group discussion.

1.25

a. Sodium metal is partially melted. What are the two phases present?

b. A sample of sand is composed of granules of quartz (silicon dioxide) and seashells (calcium carbonate). The sand is mixed with water. What phases are present?

1.26 A material is believed to be a compound. Suppose you have several samples of this material obtained from various places around the world. Comment on what you would expect to find upon observing the melting point and color for each sample. What would you expect to find upon determining the elemental composition for each sample?

1.27 You need a thermometer that is accurate to ±5°C to conduct some experiments in the temperature range of 0°C to 100°C. You find one in your lab drawer that has lost its markings.

a. What experiments could you do to make sure your thermometer is suitable for your experiments?

b. Assuming that the thermometer works, what procedure could you follow to put a scale on your thermometer that has the desired accuracy?

1.28 Imagine that you get the chance to shoot five arrows at each of the targets depicted below. On each of the targets, indicate the pattern that the five arrows make when

a. You have poor accuracy and good precision.

b. You have poor accuracy and poor precision.

c. You have good accuracy and good precision.

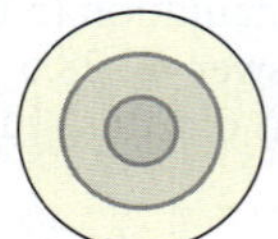 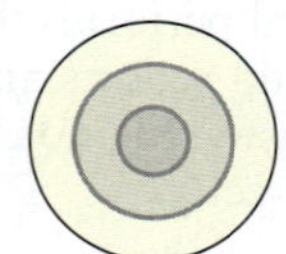 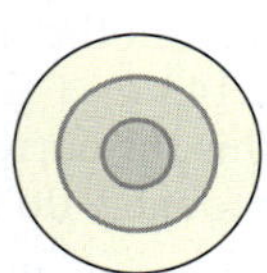

1.29 Say you live in a climate where the temperature ranges from −100°F to 20°F and you want to define a new temperature scale, YS (YS is the "Your Scale" temperature scale), which defines this range as 0.0°YS to 100.0°YS.

a. Come up with an equation that would allow you to convert between °F and °YS.

b. Using your equation, what would be the temperature in °F if it were 66°YS?

1.30 You are presented with a piece of metal in a jar. It is your job to determine what the metal is. What are some physical properties that you could measure in order to determine the type of metal? You suspect that the metal might be sodium; what are some chemical properties that you could investigate? (See Section 1.4 for some ideas.)

1.31 You have two identical boxes with interior dimensions of 8.0 cm × 8.0 cm × 8.0 cm. You completely fill one of the boxes with wooden spheres that are 1.6 cm in diameter. The other box gets filled with wooden cubes that are 1.6 cm on each edge. After putting the lid on the filled boxes, you then measure the density of each. Which one is more dense?

1.32 Consider the following compounds and their densities.

Substance	*Density (g/mL)*	*Substance*	*Density (g/mL)*
Isopropyl alcohol	0.785	Toluene	0.866
n-Butyl alcohol	0.810	Ethylene glycol	1.114

You create a column of the liquids in a glass cylinder with the most dense material on the bottom layer and the least dense on the top. You do not allow the liquids to mix.

a. First you drop a plastic bead that has a density of 0.24 g/cm³ into the column. What do you expect to observe?

b. Next you drop a different plastic bead that has a volume of 0.043 mL and a mass of 3.92×10^{-2} g into the column. What would you expect to observe in this case?

c. You drop another bead into the column and observe that it makes it all the way to the bottom of the column. What can you conclude about the density of this bead?

1.33

a. Which of the following items have a mass of about 1 g?
a grain of sand
a paper clip
a nickel
a 5.0-gallon bucket of water
a brick
a car

b. What is the approximate mass (using SI mass units) of each of the items in part a?

1.34 What is the length of the nail reported to the correct number of significant figures?

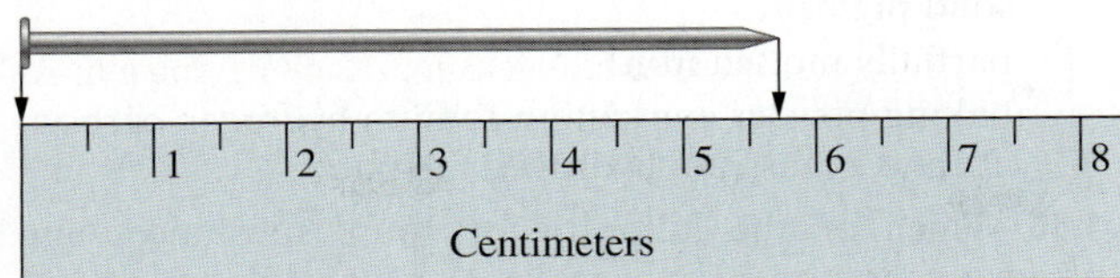

1.35 For these questions, be sure to apply the rules for significant figures.

a You are conducting an experiment where you need the volume of a box; you take the length, height, and width measurements and then multiply the values together to find the volume. You report the volume of the box as 0.310 m^3. If two of your measurements were 0.7120 m and 0.52145 m, what was the other measurement?

b If you were to add the two measurements from the first part of the problem to a third length measurement with the reported result of 1.509 m, what was the value of the third measurement?

1.36 You are teaching a class of second graders some chemistry, and they are confused about an object's mass versus its density. Keeping in mind that most second graders don't understand fractions, how would you explain these two ideas and illustrate how they differ? What examples would you use as a part of your explanation?

Practice Problems

Key: These problems are for practice in applying problem-solving skills. They are divided by topic, and some are keyed to exercises (see the ends of the exercises). The problems are arranged in matching pairs; the odd-numbered problem of each pair is listed first, and its answer is given in the back of the book.

Conservation of Mass

1.37 A 15.9-g sample of sodium carbonate is added to a solution of acetic acid weighing 20.0 g. The two substances react, releasing carbon dioxide gas to the atmosphere. After reaction, the contents of the reaction vessel weighs 29.3 g. What is the mass of carbon dioxide given off during the reaction?

1.38 Some iron wire weighing 5.6 g is placed in a beaker and covered with 15.0 g of dilute hydrochloric acid. The acid reacts with the metal and gives off hydrogen gas, which escapes into the surrounding air. After reaction, the contents of the beaker weighs 20.4 g. What is the mass of hydrogen gas produced by the reaction?

1.39 Zinc metal reacts with yellow crystals of sulfur in a fiery reaction to produce a white powder of zinc sulfide. A chemist determines that 65.4 g of zinc reacts with 32.1 g of sulfur. How many grams of zinc sulfide could be produced from 20.0 g of zinc metal?

1.40 Aluminum metal reacts with bromine, a red-brown liquid with a noxious odor. The reaction is vigorous and produces aluminum bromide, a white crystalline substance. A sample of 27.0 g of aluminum yields 266.7 g of aluminum bromide. How many grams of bromine react with 18.1 g of aluminum?

Solids, Liquids, and Gases

1.41 Give the normal state (solid, liquid, or gas) of each of the following.

a sodium hydrogen carbonate (baking soda)
b isopropyl alcohol (rubbing alcohol)
c carbon monoxide
d lead

1.42 Give the normal state (solid, liquid, or gas) of each of the following.

a potassium hydrogen tartrate (cream of tartar)
b tungsten
c carbon (graphite)
d mercury

Chemical and Physical Changes; Properties of Substances

1.43 Which of the following are physical changes and which are chemical changes?

a melting of sodium chloride
b pulverizing of rock salt
c burning of sulfur
d dissolving of salt in water

1.44 For each of the following, decide whether a physical or a chemical change is involved.

a dissolving of sugar in water
b rusting of iron
c burning of wood
d evaporation of alcohol

1.45 A sample of mercury(II) oxide was heated to produce mercury metal and oxygen gas. Then the liquid mercury was cooled to −40°C, where it solidified. A glowing wood splint was thrust into the oxygen, and the splint burst into flame. Identify each physical change and each chemical change.

1.46 Solid iodine, contaminated with salt, was heated until the iodine vaporized. The violet vapor of iodine was then cooled to yield the pure solid. Solid iodine and zinc metal powder were mixed and ignited to give a white powder. Identify each physical change and each chemical change.

1.47 The following are properties of substances. Decide whether each is a physical property or a chemical property.

a Chlorine gas liquefies at −35°C under normal pressure.
b Hydrogen burns in chlorine gas.
c Bromine melts at −7.2°C.
d Lithium is a soft, silvery-colored metal.
e Iron rusts in an atmosphere of moist air.

1.48 Decide whether each of the following is a physical property or a chemical property of the substance.

a Salt substitute, potassium chloride, dissolves in water.
b Seashells, calcium carbonate, fizz when immersed in vinegar.
c The gas hydrogen sulfide smells like rotten eggs.
d Fine steel wool (Fe) can be burned in air.
e Pure water freezes at 0°C.

1.49 Iodine is a solid having somewhat lustrous, blue-black crystals. The crystals vaporize readily to a violet-colored gas. Iodine combines with many metals. For example, aluminum combines with iodine to give aluminum iodide. Identify the physical and the chemical properties of iodine that are cited.

1.50 Mercury(II) oxide is an orange-red solid with a density of 11.1 g/cm^3. It decomposes when heated to give mercury and oxygen. The compound is insoluble in water (does not dissolve in water). Identify the physical and the chemical properties of mercury(II) oxide that are cited.

Elements, Compounds, and Mixtures

1.51 Consider the following separations of materials. State whether a physical process or a chemical reaction is involved in each separation.

a Sodium chloride is obtained from seawater by evaporation of the water.
b Mercury is obtained by heating the substance mercury(II) oxide; oxygen is also obtained.
c Pure water is obtained from ocean water by evaporating the water, then condensing it.
d Iron is produced from an iron ore that contains the substance iron(III) oxide.
e Gold is obtained from river sand by panning (allowing the heavy metal to settle in flowing water).

1.52 All of the following processes involve a separation of either a mixture into substances or a compound into elements. For each, decide whether a physical process or a chemical reaction is required.

a Sodium metal is obtained from the substance sodium chloride.
b Iron filings are separated from sand by using a magnet.
c Sugar crystals are separated from a sugar syrup by evaporation of water.
d Fine crystals of silver chloride are separated from a suspension of the crystals in water.
e Copper is produced when zinc metal is placed in a solution of copper(II) sulfate, a compound.

1.53 Label each of the following as a substance, a heterogeneous mixture, or a solution.

a seawater b sulfur
c fluorine d beach sand

1.54 Indicate whether each of the following materials is a substance, a heterogeneous mixture, or a solution.

a milk b bromine
c gasoline d aluminum

1.55 Which of the following are pure substances and which are mixtures? For each, list all of the different phases present.

a bromine liquid and its vapor
b paint, containing a liquid solution and a dispersed solid pigment
c partially molten iron
d baking powder containing sodium hydrogen carbonate and potassium hydrogen tartrate

1.56 Which of the following are pure substances and which are mixtures? For each, list all of the different phases present.

a a sugar solution with sugar crystals at the bottom
b ink containing a liquid solution with fine particles of carbon
c a sand containing quartz (silicon dioxide) and calcite (calcium carbonate)
d liquid water and steam at 100°C

Significant Figures

1.57 How many significant figures are there in each of the following measurements?

a 57.00 g b 0.0503 kg
c 6.310 J d 0.80090 m
e 5.06×10^{-7} cm f 2.010 s

1.58 How many significant figures are there in each of the following measurements?

a 4.0100 mg b 0.05930 g
c 0.035 mm d 3.100 s
e 8.91×10^1 L f 9.100×10^4 cm

1.59 The circumference of the earth at the equator is 40,000 km. This value is precise to two significant figures. Write this in scientific notation to express correctly the number of significant figures.

1.60 The astronomical unit equals the mean distance between the earth and the sun. This distance is 150,000,000 km, which is precise to three significant figures. Express this in scientific notation to the correct number of significant figures.

1.61 Assuming all numbers are measured quantities, do the indicated arithmetic and give the answer to the correct number of significant figures.

a $\dfrac{8.71 \times 0.0301}{0.031}$
b $0.71 + 89.3$
c $934 \times 0.00435 + 107$
d $(847.89 - 847.73) \times 14673$

1.62 Assuming all numbers are measured quantities, do the indicated arithmetic and give the answer to the correct number of significant figures.

a $\dfrac{08.71 \times 0.57}{5.871}$
b $8.937 - 8.930$
c $8.937 + 8.930$
d $0.00015 \times 54.6 + 1.002$

1.63 One sphere has a radius of 5.15 cm; another has a radius of 5.00 cm. What is the difference in volume (in cubic centimeters) between the two spheres? Give the answer to the correct number of significant figures. The volume of a sphere is $(4/3)\pi r^3$, where $\pi = 3.1416$ and r is the radius.

1.64 A solid circular cylinder of iron with a radius of 1.500 cm has a ruler etched along its length. What is the volume of iron contained between the marks labeled 3.20 cm and 3.50 cm? The volume of a circular cylinder is $\pi r^2 l$, where $\pi = 3.1416$, r is the radius, and l is the length.

SI Units

1.65 Write the following measurements, without scientific notation, using the appropriate SI prefix.

a 5.89×10^{-12} s
b 0.2010 m
b 2.560×10^{-9} g
d 6.05×10^{3} m

1.66 Write the following measurements, without scientific notation, using the appropriate SI prefix.

a 4.851×10^{-6} g
b 3.16×10^{-2} m
b 2.591×10^{-9} s
d 8.93×10^{-12} g

1.67 Using scientific notation, convert:

a 6.15 ps to s
b 3.781 μm to m
c 1.546 Å to m
d 9.7 mg to g

1.68 Using scientific notation, convert:

a 6.20 km to m
b 1.98 ns to s
b 2.54 cm to m
d 5.23 μg to g

Temperature Conversion

1.69 Convert:

a 68°F to degrees Celsius
b −23°F to degrees Celsius
c 26°C to degrees Fahrenheit
d −81°C to degrees Fahrenheit

1.70 Convert:

a 51°F to degrees Celsius
b −11°F to degrees Celsius
c −41°C to degrees Fahrenheit
d 22°C to degrees Fahrenheit

1.71 Salt and ice are stirred together to give a mixture to freeze ice cream. The temperature of the mixture is −20.0°C. What is this temperature in degrees Fahrenheit?

1.72 The reaction of oxygen and hydrogen is used to propel rockets. Liquid oxygen has a boiling point of –222.7°C. What is this temperature in degrees Fahrenheit?

Density

1.73 A certain sample of the mineral galena (lead sulfide) weighs 12.4 g and has a volume of 1.64 cm^3. What is the density of galena?

1.74 A flask contains a 30.0 mL sample of acetone (nail polish remover) that weighs 23.6 grams. What is the density of the acetone?

1.75 A liquid with a volume of 8.5 mL has a mass of 6.71 g. The liquid is either octane, ethanol, or benzene, the densities of which are 0.702 g/cm^3, 0.789 g/cm^3, and 0.879 g/cm^3, respectively. What is the identity of the liquid?

1.76 A mineral sample has a mass of 5.94 g and a volume of 0.73 cm^3. The mineral is either sphalerite (density = 4.0 g/cm^3), cassiterite (density = 6.99 g/cm^3), or cinnabar (density = 8.10 g/cm^3). Which is it?

1.77 Platinum has a density of 21.4 g/cm^3. What is the mass of 5.9 cm^3 of this metal?

1.78 What is the mass of a 43.8-mL sample of gasoline, which has a density of 0.70 g/cm^3?

1.79 Ethanol has a density of 0.789 g/cm^3. What volume must be poured into a graduated cylinder to give 19.8 g of alcohol?

1.80 Bromine is a red-brown liquid with a density of 3.10 g/mL. A sample of bromine weighing 88.5 g occupies what volume?

Unit Conversions

1.81 Sodium hydrogen carbonate, known commercially as baking soda, reacts with acidic materials such as vinegar to release carbon dioxide gas. An experiment calls for 0.480 kg of sodium hydrogen carbonate. Express this mass in milligrams.

1.82 The acidic constituent in vinegar is acetic acid. A 10.0-mL sample of a certain vinegar contains 611 mg of acetic acid. What is this mass of acetic acid expressed in micrograms?

1.83 The different colors of light have different wavelengths. The human eye is most sensitive to light whose wavelength is 555 nm (greenish-yellow). What is this wavelength in centimeters?

1.84 Water consists of molecules (groups of atoms). A water molecule has two hydrogen atoms, each connected to an oxygen atom. The distance between any one hydrogen atom and the oxygen atom is 0.96 Å. What is this distance in millimeters?

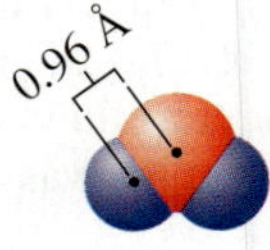

1.85 The total amount of fresh water on earth is estimated to be 3.73×10^8 km^3. What is this volume in cubic meters? in liters?

1.86 A submicroscopic particle suspended in a solution has a volume of 1.3 μm^3. What is this volume in liters?

1.87 How many grams are there in 3.58 short tons? Note that 1 g = 0.03527 oz (ounces avoirdupois), 1 lb (pound) = 16 oz, and 1 short ton = 2000 lb. (These relations are exact.)

1.88 The calorie, the Btu (British thermal unit), and the joule are units of energy; 1 calorie = 4.184 joules (exact), and 1 Btu = 252.0 calories. Convert 3.15 Btu to joules.

1.89 The first measurement of sea depth was made in 1840 in the central South Atlantic, where a plummet was lowered 2425 fathoms. What is this depth in meters? Note that 1 fathom = 6 ft, 1 ft = 12 in., and 1 in. = 2.54×10^{-2} m. (These relations are exact.)

1.90 The estimated amount of recoverable oil from the field at Prudhoe Bay in Alaska is 1.3×10^{10} barrels. What is this amount of oil in cubic meters? One barrel = 42 gal (exact), 1 gal = 4 qt (exact), and 1 qt = 9.46×10^{-4} m^3.

1.91 A fish tank is 20.0 in. long, 20.0 in. deep, and 10.0 in. high. What is the maximum volume of water, in liters, that the fish tank can hold?

1.92 The population density of worms in a particular field is 33 worms per cubic meter of soil. How many worms would there be in the top meter of soil in a field that has dimensions of 1.00 km by 2.0 km?

General Problems

Key: These problems provide more practice but are not divided by topic or keyed to exercises. Each section ends with essay questions, each of which is color coded to refer to the *A Chemist Looks at Daily Life* (orange) or *Instrumental Methods* (brown) chapter essay on which it is based. Odd-numbered problems and the even-numbered problems that follow are similar; answers to all odd-numbered problems except the essay questions are given in the back of the book.

1.93 Sodium metal reacts vigorously with water. A piece of sodium weighing 19.70 g was added to a beaker containing 126.22 g of water. During reaction, hydrogen gas was produced and bubbled from the solution. The solution, containing sodium hydroxide, weighed 145.06 g. How many grams of hydrogen gas were produced?

1.94 An antacid tablet weighing 0.853 g contained calcium carbonate as the active ingredient, in addition to an inert binder. When an acid solution weighing 56.519 g was added to the tablet, carbon dioxide gas was released, producing a fizz. The resulting solution weighed 57.152 g. How many grams of carbon dioxide were produced?

1.95 When a mixture of aluminum powder and iron(III) oxide is ignited, it produces molten iron and aluminum oxide. In an experiment, 5.40 g of aluminum was mixed with 18.50 g of iron(III) oxide. At the end of the reaction, the mixture contained 11.17 g of iron, 10.20 g of aluminum oxide, and an undetermined amount of unreacted iron(III) oxide. No aluminum was left. What is the mass of the iron(III) oxide?

1.96 When chlorine gas is bubbled into a solution of sodium bromide, the sodium bromide reacts to give bromine, a red-brown liquid, and sodium chloride (ordinary table salt). A solution was made by dissolving 20.6 g of sodium bromide in 100.0 g of water. After passing chlorine through the solution, investigators analyzed the mixture. It contained 16.0 g of bromine and 11.7 g of sodium chloride. How many grams of chlorine reacted?

1.97 A beaker weighed 53.10 g. To the beaker was added 5.348 g of iron pellets and 56.1 g of hydrochloric acid. What was the total mass of the beaker and the mixture (before reaction)? Express the answer to the correct number of significant figures.

1.98 A graduated cylinder weighed 66.5 g. To the cylinder was added 58.2 g of water and 5.279 g of sodium chloride. What was the total mass of the cylinder and the solution? Express the answer to the correct number of significant figures.

1.99 Describe each of the following as a physical or chemical property of each listed chemical substance.

a. baking soda reacts with vinegar
b. ice melts at 0°C
c. graphite is a soft, black solid
d. hydrogen burns in air

1.100 Describe each of the following as a physical or chemical property of each listed chemical substance.

a. chlorine is a green gas
b. zinc reacts with acids
c. magnesium has a density of 1.74 g/cm^3
d. iron can rust

1.101 Analyses of several samples of a material containing only iron and oxygen gave the following results. Could this material be a compound?

	Mass of Sample	*Mass of Iron*	*Mass of Oxygen*
Sample A	1.518 g	1.094 g	0.424 g
Sample B	2.056 g	1.449 g	0.607 g
Sample C	1.873 g	1.335 g	0.538 g

1.102 A red-orange solid contains only mercury and oxygen. Analyses of three different samples gave the following results. Are the data consistent with the hypothesis that the material is a compound?

	Mass of Sample	*Mass of Mercury*	*Mass of Oxygen*
Sample A	1.0410 g	0.9641g	0.0769 g
Sample B	1.5434 g	1.4293 g	0.1141 g
Sample C	1.2183 g	1.1283 g	0.0900 g

1.103 A cubic box measures 39.3 cm on an edge. What is the volume of the box in cubic centimeters? Express the answer to the correct number of significant figures.

1.104 A cylinder with circular cross section has a radius of 2.56 cm and a height of 56.32 cm. What is the volume of the cylinder? Express the answer to the correct number of significant figures.

1.105 An aquarium has a rectangular cross section that is 47.8 in. by 12.5 in.; it is 19.5 in. high. How many U.S. gallons does the aquarium contain? One U.S. gallon equals exactly 231 in^3.

1.106 A spherical tank has a radius of 175.0 in. Calculate the volume of the tank in cubic inches; then convert this to Imperial gallons. The volume of a sphere is $(4/3)\pi r^3$, where r is the radius. One Imperial gallon equals 277.4 in^3.

1.107 Obtain the difference in volume between two spheres, one of radius 5.61 cm, the other of radius 5.85 cm. The volume V of a sphere is $(4/3)\ \pi r^3$, where r is the radius. Give the result to the correct number of significant figures.

1.108 What is the difference in surface area between two circles, one of radius 7.98 cm, the other of radius 8.50 cm? The surface area of a circle of radius r is πr^2. Obtain the result to the correct number of significant figures.

1.109 Perform the following arithmetic setups and express the answers to the correct number of significant figures.

a $\dfrac{56.1 - 51.1}{6.58}$ b $\dfrac{56.1 + 51.1}{6.58}$

c $(9.1 + 8.6) \times 26.91$ d $0.0065 \times 3.21 + 0.0911$

1.110 Assuming all of the numbers are measured quantities, perform the following arithmetic setups and report the answers to the correct number of significant figures.

a $\dfrac{9.345 - 9.005}{9.811}$ b $\dfrac{9.345 + 9.005}{9.811}$

b $(7.50 + 7.53) \times 3.71$ d $0.71 \times 0.36 + 17.36$

1.111 For each of the following, write the measurement in terms of an appropriate prefix and base unit.

a The mass of calcium per milliliter in a sample of blood serum is 0.0912 g.

b The radius of an oxygen atom is about 0.000000000066 m.

c A particular red blood cell measures 0.0000071 m.

d The wavelength of a certain ultraviolet radiation is 0.000000056 m.

1.112 For each of the following, write the measurement in terms of an appropriate prefix and base unit.

a The mass of magnesium per milliliter in a sample of blood serum is 0.0186 g.

b The radius of a carbon atom is about 0.000000000077 m.

c The hemoglobin molecule, a component of red blood cells, is 0.0000000065 m in diameter.

d The wavelength of a certain infrared radiation is 0.00000085 m.

1.113 Write each of the following in terms of the SI base unit (that is, express the prefix as the power of 10).

a 1.07 ps b 5.8 μm

b 319 nm d 15.3 ms

1.114 Write each of the following in terms of the SI base unit (that is, express the prefix as the power of 10).

a 6.6 mK b 275 pm

b 22.1 ms d 45 μm

1.115 Tungsten metal, which is used in lightbulb filaments, has the highest melting point of any metal (3410°C). What is this melting point in degrees Fahrenheit?

1.116 Titanium metal is used in aerospace alloys to add strength and corrosion resistance. Titanium melts at 1677°C. What is this temperature in degrees Fahrenheit?

1.117 Calcium carbonate, a white powder used in toothpastes, antacids, and other preparations, decomposes when heated to about 825°C. What is this temperature in degrees Fahrenheit?

1.118 Sodium hydrogen carbonate (baking soda) starts to decompose to sodium carbonate (soda ash) at about 50°C. What is this temperature in degrees Fahrenheit?

1.119 Gallium metal can be melted by the heat of one's hand. Its melting point is 29.8°C. What is this temperature in kelvins? in degrees Fahrenheit?

1.120 Mercury metal is liquid at normal temperatures but freezes at −38.9°C. What is this temperature in kelvins? in degrees Fahrenheit?

1.121 Zinc metal can be purified by distillation (transforming the liquid metal to vapor, then condensing the vapor back to liquid). The metal boils at normal atmospheric pressure at 1666°F. What is this temperature in degrees Celsius? in kelvins?

1.122 Iodine is a bluish-black solid. It forms a violet-colored vapor when heated. The solid melts at 236°F. What is this temperature in degrees Celsius? in kelvins?

1.123 An aluminum alloy used in the construction of aircraft wings has a density of 2.70 g/cm^3. Express this density in SI units (kg/m^3).

1.124 Vanadium metal is added to steel to impart strength. The density of vanadium is 5.96 g/cm^3. Express this in SI units (kg/m^3).

1.125 The density of quartz mineral was determined by adding a weighed piece to a graduated cylinder containing 51.2 mL water. After the quartz was submerged, the water level was 65.7 mL. The quartz piece weighed 38.4 g. What was the density of the quartz?

1.126 Hematite (iron ore) weighing 70.7 g was placed in a flask whose volume was 53.2 mL. The flask with hematite was then carefully filled with water and weighed. The hematite and water weighed 109.3 g. The density of the water was 0.997 g/cm^3. What was the density of the hematite?

1.127 Some bottles of colorless liquids were being labeled when the technicians accidentally mixed them up and lost track of their contents. A 15.0-mL sample withdrawn from one bottle weighed 22.3 g. The technicians knew that the liquid was either acetone, benzene, chloroform, or carbon tetrachloride (which have densities of 0.792 g/cm^3, 0.899 g/cm^3, 1.489 g/cm^3, and 1.595 g/cm^3, respectively). What was the identity of the liquid?

1.128 Providing no reaction occurs, a solid will float on any liquid that is more dense than it is. The volume of a piece of calcite weighing 35.6 g is 12.9 cm^3. On which of the following liquids will the calcite float: carbon tetrachloride (density = 1.60 g/cm^3), methylene bromide (density = 2.50 g/cm^3), tetrabromoethane (density = 2.96 g/cm^3), or methylene iodide (density = 3.33 g/cm^3)?

1.129 Platinum metal is used in jewelry; it is also used in automobile catalytic converters. What is the mass of a cube of platinum 4.40 cm on an edge? The density of platinum is 21.4 g/cm^3.

1.130 Ultrapure silicon is used to make solid-state devices, such as computer chips. What is the mass of a circular cylinder of silicon that is 12.40 cm long and has a radius of 4.00 cm? The density of silicon is 2.33 g/cm^3.

1.131 Vinegar contains acetic acid (about 5% by mass). Pure acetic acid has a strong vinegar smell but is corrosive to the skin. What volume of pure acetic acid has a mass of 35.00 g? The density of acetic acid is 1.053 g/mL.

1.132 Ethyl acetate has a characteristic fruity odor and is used as a solvent in paint lacquers and perfumes. An experiment requires 0.070 kg of ethyl acetate. What volume is this (in liters)? The density of ethyl acetate is 0.902 g/mL.

1.133 Convert:
- a 8.45 kg to micrograms
- b 318 μs to milliseconds
- c 93 km to nanometers
- d 37.1 mm to centimeters

1.134 Convert:
- a 127 Å to micrometers
- b 21.0 kg to milligrams
- c 1.09 cm to millimeters
- d 4.6 ns to microseconds

1.135 Convert:
- a 5.91 kg of chrome yellow to milligrams
- b 753 mg of vitamin A to micrograms
- c 90.1 MHz (megahertz), the wavelength of an FM signal, to kilohertz
- d 498 mJ (the joule, J, is a unit of energy) to kilojoules

1.136 Convert:
- a 7.19 μg of cyanocobalamin (vitamin B_{12}) to milligrams
- b 104 pm, the radius of a sulfur atom, to angstroms
- c 0.010 mm, the diameter of a typical blood capillary, to centimeters
- d 0.0605 kPa (the pascal, Pa, is a unit of pressure) to centipascals

1.137 The largest of the Great Lakes is Lake Superior, which has a volume of 12,230 km^3. What is this volume in liters?

1.138 The average flow of the Niagara River is 3.50 km^3 per week. What is this volume in liters?

1.139 A room measures 10.0 ft × 11.0 ft and is 9.0 ft high. What is its volume in liters?

1.140 A cylindrical settling tank is 5.0 ft deep and has a radius of 15.0 ft. What is the volume of the tank in liters?

1.141 The masses of diamonds and gems are measured in carats. A carat is defined as 200 mg. If a jeweler has 275 carats of diamonds, how many grams does she have?

1.142 One year of world production of gold was 49.6 × 10^6 troy ounces. One troy ounce equals 31.10 g. What was the world production of gold in metric tons (10^6 g) for that year?

■ **1.143** What are some characteristics of the adhesive used for Post-it Notes?

■ **1.144** All good experiments start with a scientific question. What was the scientific question that Art Fry was trying to answer when he embarked on finding the adhesive for the Post-it Note?

■ **1.145** What do the various chromatographic separation techniques have in common?

■ **1.146** Describe how gas chromatography works.

Strategy Problems

Key: As noted earlier, all of the practice and general problems are matched pairs. This section is a selection of problems that are not in matched-pair format. These challenging problems require that you employ many of the concepts and strategies that were developed in the chapter. In some cases, you will have to integrate several concepts and operational skills in order to solve the problem successfully.

1.147 When the quantity 5 × 10^{-2} mg is subtracted from 4.7 mg, how many significant figures should be reported in the answer?

1.148 A 33.0-g sample of an unknown liquid at 20.0°C is heated to 120°C. During this heating, the density of the liquid changes from 0.854 g/cm^3 to 0.797 g/cm^3. What volume would this sample occupy at 120°C?

1.149 A 175-g sample of a pure liquid, liquid A, with a density of 3.00 g/mL is mixed with a 50.0-mL sample of a pure liquid, liquid B, with a density of 2.00 g/mL. What is the total volume of the mixture? (Assume there is no reaction upon the mixing of A and B, and volumes are additive.)

1.150 On a long trip you travel 832 miles in 21 hours. During this trip, you use 31 gallons of gasoline.
- a Calculate your average speed in miles per hour.
- b Calculate your average speed in kilometers per hour.
- c Calculate your gas mileage in kilometers per liter.

1.151 The density of lead at 20°C is 11.3 g/cm^3. Rank the volumes of the following quantities of lead from least to greatest: 50.0 mL, 0.100 qt, 50.0 g, and 0.0250 lb.

1.152 A 2010 oil spill in the Gulf of Mexico was estimated to dump as much as 7.6 × 10^8 L of crude oil into the water. Calculate the volume of this spill in cubic feet.

1.153 Catalytic converters are used in automobiles to convert harmful gaseous substances in the exhaust to more benign substances, for example, carbon monoxide to carbon dioxide. Beads used in these converters have extremely large surface areas. For instance, a single 3.0-mm porous bead can have a surface area of 1.0 × 10^6 cm^2. If a car has two catalytic converters, each with 5.0 × 10^3 beads, calculate the surface area of the two catalytic converters in km^2.

1.154 The density of liquid water at 80°C is 972 kg/m^3 and at 20°C is 998 kg/m^3. If you have 200.0 mL of water at 20°C, what volume (mL) will the water occupy at 80°C? Which will contain more water molecules, 1.0 L of water at 80°C or 1.0 L of water at 20°C?

1.155 A 55.0-cm^3 sample of ocean water has a density of 1025 kg/m^3. Assuming all of the solids stay in solution, what will be the density of this water sample in units of g/cm^3 after 5.0 mL of water has evaporated?

1.156 At 20°C liquid gasoline gas has a density of 0.75 g/cm^3. If a 5.5-mL sample of gasoline is placed into a sealed 3.00-L container at 50°C and allowed to completely evaporate, what will be the density of the gasoline vapor in the container?

1.157 The figures below represent a gas trapped in containers. The orange balls represent individual gas atoms. Container A on the left has a volume that is one-half the volume of container B on the right.

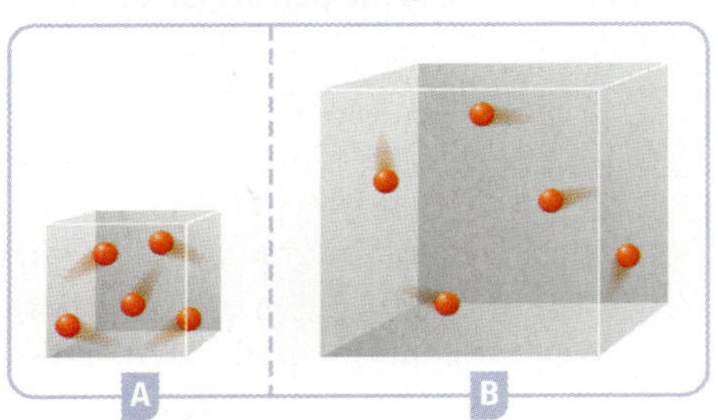

a How does the mass of gas in one container compare with the mass of gas in the other?

b Which container has the greater density of gas?

c If the volume of container A were increased to the same volume as container B, how would the density of the gas in container A change?

1.158 An ice cube measures 3.50 cm on each edge and weighs 39.45 g.

a Calculate the density of ice.

b Calculate the mass of 400.4 mL of water in an ice cube.

1.159 The total length of all the DNA molecules contained in a human body is 1 × 10^{10} miles. The population of the United States is about 300 million. What is the total length of all the DNA of the U.S. population in lightyears? (A lightyear is the distance that light travels in a year and is 9.46 × 10^{15} m.)

1.160 Prospectors are considering searching for gold on a plot of land that contains 1.31 g of gold per bucket of soil. If the volume of the bucket is 4.67 L, how many grams of gold are there likely to be in 2.38 × 10^3 cubic feet of soil?

1.161 A solution is prepared by dissolving table salt, sodium chloride, in water at room temperature.

a Assuming there is no significant change in the volume of water during the preparation of the solution, how would the density of the solution compare to that of pure water?

b If you were to boil the solution for several minutes and then allow it to cool to room temperature, how would the density of the solution compare to the density in part a?

c If you took the solution prepared in part a and added more water, how would this affect the density of the solution?

1.162 Water and saline (salt) solution have in common that they are both homogeneous. How do these materials differ? Be specific and use chemical terms to describe the two systems.

Cumulative-Skills Problems

Key: The problems under this heading combine skills introduced in the previous chapter with those given in the current one.

1.163 When 10.0 g of marble chips (calcium carbonate) is treated with 50.0 mL of hydrochloric acid (density 1.096 g/mL), the marble dissolves, giving a solution and releasing carbon dioxide gas. The solution weighs 60.4 g. How many liters of carbon dioxide gas are released? The density of the gas is 1.798 g/L.

1.164 Zinc ore (zinc sulfide) is treated with sulfuric acid, leaving a solution with some undissolved bits of material and releasing hydrogen sulfide gas. If 10.8 g of zinc ore is treated with 50.0 mL of sulfuric acid (density 1.153 g/mL), 65.1 g of solution and undissolved material remains. In addition, hydrogen sulfide (density 1.393 g/L) is evolved. What is the volume (in liters) of this gas?

1.165 A steel sphere has a radius of 1.58 in. If this steel has a density of 7.88 g/cm^3, what is the mass of the sphere in grams?

1.166 A weather balloon filled with helium has a diameter of 3.50 ft. What is the mass in grams of the helium in the balloon at 21°C and normal pressure? The density of helium under these conditions is 0.166 g/L.

1.167 The land area of Greenland is 840,000 mi^2, with only 132,000 mi^2 free of perpetual ice. The average thickness of this ice is 5000 ft. Estimate the mass of the ice (assume two significant figures). The density of ice is 0.917 g/cm^3.

1.168 Antarctica, almost completely covered in ice, has an area of 5,500,000 mi^2 with an average height of 7500 ft. Without the ice, the height would be only 1500 ft. Estimate the mass of this ice (two significant figures). The density of ice is 0.917 g/cm^3.

1.169 A sample of an ethanol–water solution has a volume of 54.2 cm^3 and a mass of 49.6 g. What is the percentage of ethanol (by mass) in the solution? (Assume that there is no change in volume when the pure compounds are mixed.) The density of ethanol is 0.789 g/cm^3 and that of water is 0.998 g/cm^3. Alcoholic beverages are rated in *proof,* which is a measure of the relative amount of ethanol in the beverage. Pure ethanol is exactly 200 proof; a solution that is 50% ethanol by volume is exactly 100 proof. What is the proof of the given ethanol–water solution?

1.170 You have a piece of gold jewelry weighing 9.35 g. Its volume is 0.654 cm^3. Assume that the metal is an alloy (mixture) of gold and silver, which have densities of 19.3 g/cm^3 and 10.5 g/cm^3, respectively. Also assume that there is no change in volume when the pure metals are mixed. Calculate the percentage of gold (by mass) in the alloy. The relative amount of gold in an alloy is measured in *karats.* Pure gold is 24 karats; an alloy of 50% gold is 12 karats. State the proportion of gold in the jewelry in karats.

1.171 A sample of vermilion-colored mineral was weighed in air, then weighed again while suspended in water. An object is buoyed up by the mass of the fluid displaced by the object. In air, the mineral weighed 18.49 g; in water, it weighed 16.21 g. The densities of air and water are 1.205 g/L and 0.9982 g/cm^3, respectively. What is the density of the mineral?

1.172 A sample of a bright blue mineral was weighed in air, then weighed again while suspended in water. An object is buoyed up by the mass of the fluid displaced by the object. In air, the mineral weighed 7.35 g; in water, it weighed 5.40 g. The densities of air and water are 1.205 g/L and 0.9982 g/cm^3, respectively. What is the density of the mineral?

1.173 A student gently drops an object weighing 15.8 g into an open vessel that is full of ethanol, so that a volume of ethanol spills out equal to the volume of the object. The experimenter now finds that the vessel and its contents weigh 10.5 g more than the vessel full of ethanol only. The density of ethanol is 0.789 g/cm^3. What is the density of the object?

1.174 An experimenter places a piece of a solid metal weighing 255 g into a graduated cylinder, which she then fills with liquid mercury. After weighing the cylinder and its contents, she removes the solid metal and fills the cylinder with mercury. She now finds that the cylinder and its contents weigh 101 g less than before. The density of mercury is 13.6 g/cm^3. What is the density of the solid metal?

2 Atoms, Molecules, and Ions

The probe of a scanning tunneling microscope (as sharp as a single atom!) was used to arrange iron atoms in an oval shape on a surface layer of copper atoms and collect data to produce this image.

Contents and Concepts

Atomic Theory and Atomic Structure

The key concept in chemistry is that all matter is composed of very small particles called atoms. We look at atomic theory, discuss atomic structure, and finally describe the periodic table, which organizes the elements.

Chemical Substances: Formulas and Names

We explore how atoms combine in various ways to yield the millions of known substances.

Chemical Reactions: Equations

Sodium is a soft, silvery metal. This metal cannot be handled with bare fingers, because it reacts with any moisture on the skin, causing a burn. Chlorine is a poisonous, greenish yellow gas with a choking odor. When molten sodium is placed into an atmosphere of chlorine, a dramatic reaction occurs: the sodium bursts into flame and a white, crystalline powder forms (see Figure 2.1). What is particularly interesting about this reaction is that this chemical combination of two toxic substances produces sodium chloride, the substance that we commonly call table salt.

Sodium metal and chlorine gas are particular forms of matter. In burning, they undergo a chemical change—a chemical reaction—in which these forms of matter change to a form of matter with different chemical and physical properties. How do we explain the differences in properties of different forms of matter? And how do we explain chemical reactions such as the burning of sodium metal in chlorine gas? This chapter and the next take an introductory look at these basic questions in chemistry. In later chapters we will develop the concepts introduced here.

Atomic Theory and Atomic Structure

As noted in Chapter 1, all matter is composed of atoms. In this first part of Chapter 2, we explore the atomic theory of matter and the structure of atoms. We also examine the *periodic table,* which is an organizational chart designed to highlight similarities among the elements.

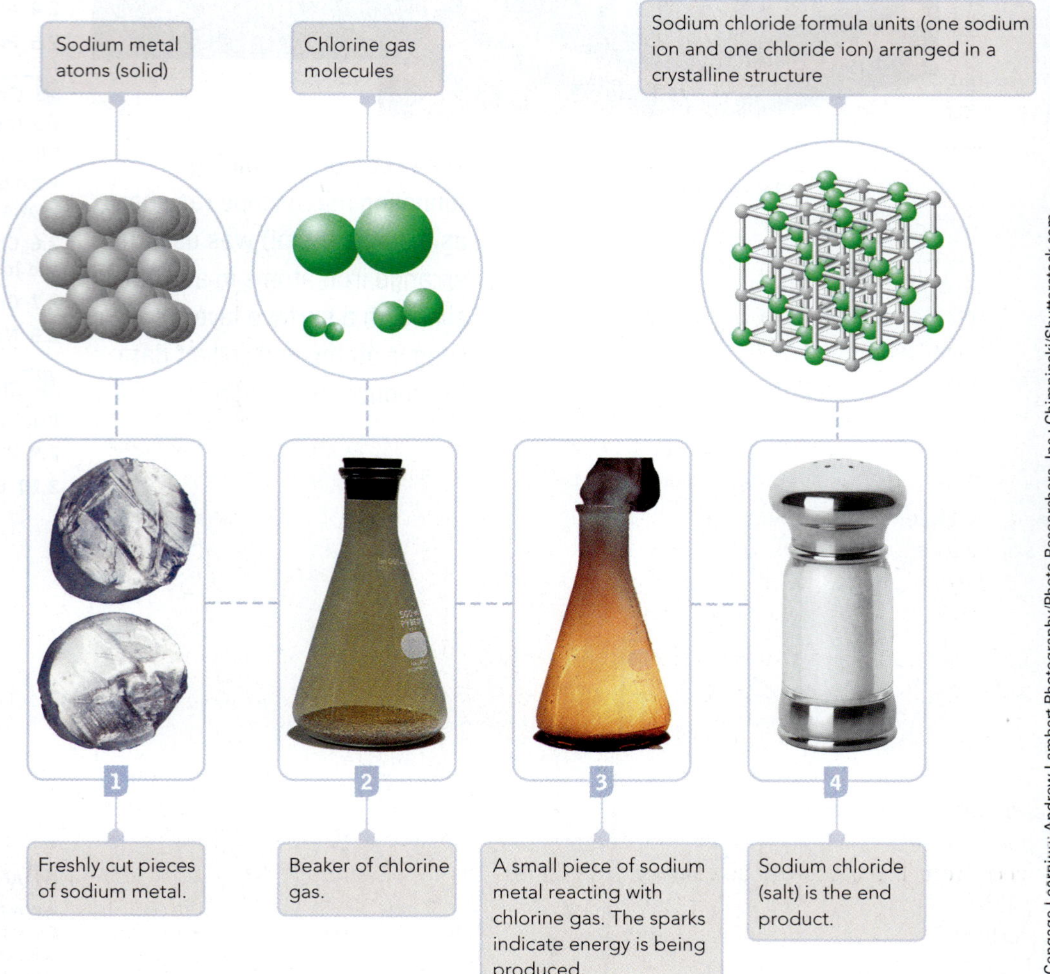

Figure 2.1 ▲

Reaction of sodium and chlorine *Left to Right:* Sodium metal, chlorine gas, a small piece of sodium metal reacting with chlorine, the sodium chloride product of the reaction. The sparks in the flask of the reaction indicate that energy is produced.

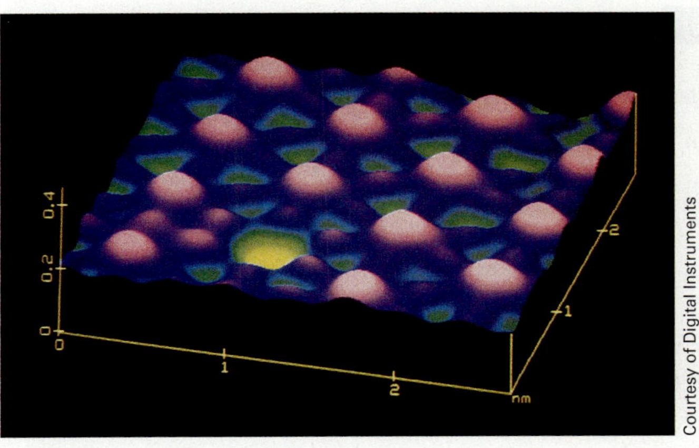

Courtesy of Digital Instruments

Figure 2.2 ◀

Image of iodine atoms on a platinum metal surface This computer-screen image was obtained by a scanning tunneling microscope (discussed in Chapter 7); the color was added to the image by computer. Iodine atoms are the large peaks with pink tops. Note the "vacancy" in the array of iodine atoms. A scale shows the size of the atoms in nanometers.

2.1 Atomic Theory of Matter

As we noted in Chapter 1, Lavoisier laid the experimental foundation of modern chemistry. But the British chemist John Dalton (1766–1844) provided the basic theory: all matter—whether element, compound, or mixture—is composed of small particles called atoms. The postulates, or basic statements, of Dalton's theory are presented in this section. Note that the terms *element, compound,* and *chemical reaction,* which were defined in Chapter 1 in terms of matter as we normally see it, are redefined here by the postulates of Dalton's theory in terms of atoms.

Postulates of Dalton's Atomic Theory

The main points of Dalton's **atomic theory,** *an explanation of the structure of matter in terms of different combinations of very small particles,* are given by the following postulates:

1. All matter is composed of indivisible atoms. An **atom** is *an extremely small particle of matter that retains its identity during chemical reactions.* (See Figure 2.2.)
2. An **element** is *a type of matter composed of only one kind of atom,* each atom of a given kind having the same properties. Mass is one such property. Thus, the atoms of a given element have a characteristic mass. (We will need to revise this definition of element in Section 2.3 so that it is stated in modern terms.) ▶

As you will see in Section 2.4, it is the average mass of an atom that is characteristic of each element on earth.

3. A **compound** is *a type of matter composed of atoms of two or more elements chemically combined in fixed proportions.* The relative numbers of any two kinds of atoms in a compound occur in simple ratios. Water, for example, a compound of the elements hydrogen and oxygen, consists of hydrogen and oxygen atoms in the ratio of 2 to 1.
4. A **chemical reaction** consists of *the rearrangement of the atoms present in the reacting substances to give new chemical combinations present in the substances formed by the reaction.* Atoms are not created, destroyed, or broken into smaller particles by any chemical reaction.

Today we know that atoms are not truly indivisible; they are themselves made up of particles. Nevertheless, Dalton's postulates are essentially correct.

Atomic Symbols and Models

It is convenient to use symbols for the atoms of the different elements. An **atomic symbol** is *a one- or two-letter notation used to represent an atom corresponding to a particular element.* ▶ Typically, the atomic symbol consists of the first letter, capitalized, from the name of the element, sometimes with an additional letter from the name in lowercase. For example, chlorine has the symbol Cl. Other symbols are derived from a name in another language (usually Latin). Sodium is given the symbol Na from its Latin name, *natrium.* Symbols of selected elements are listed in Table 2.1.

The atomic names (temporary) given to newly discovered elements are derived from the atomic number (discussed in Section 2.3) using the numerical roots: nil *(0),* un *(1),* bi *(2),* tri *(3),* quad *(4),* pent *(5),* hex *(6),* sept *(7),* oct *(8), and* enn *(9) with the ending* -ium. *For example, the element with atomic number 116 is ununhexium; its atomic symbol is Uuh.*

Early in the development of his atomic theory, Dalton built spheres to represent atoms and used combinations of these spheres to represent compounds. Chemists continue to use this idea of representing atoms by three-dimensional models, but because we now know so much more about atoms and molecules than Dalton did, these models have become more refined. Section 2.6 describes these models in more detail.

Table 2.1 Some Common Elements

Name of Element	Atomic Symbol	Physical Appearance of Element*
Aluminum	Al	Silvery-white metal
Barium	Ba	Silvery-white metal
Bromine	Br	Reddish-brown liquid
Calcium	Ca	Silvery-white metal
Carbon	C	
Graphite		Soft, black solid
Diamond		Hard, colorless crystal
Chlorine	Cl	Greenish-yellow gas
Chromium	Cr	Silvery-white metal
Cobalt	Co	Silvery-white metal
Copper	Cu (from *cuprum*)	Reddish metal
Fluorine	F	Pale yellow gas
Gold	Au (from *aurum*)	Pale yellow metal
Helium	He	Colorless gas
Hydrogen	H	Colorless gas
Iodine	I	Bluish-black solid
Iron	Fe (from *ferrum*)	Silvery-white metal
Lead	Pb (from *plumbum*)	Bluish-white metal
Magnesium	Mg	Silvery-white metal
Manganese	Mn	Gray-white metal
Mercury	Hg (from *hydrargyrum*)	Silvery-white liquid metal
Neon	Ne	Colorless gas
Nickel	Ni	Silvery-white metal
Nitrogen	N	Colorless gas
Oxygen	O	Colorless gas
Phosphorus (white)	P	Yellowish-white, waxy solid
Potassium	K (from *kalium*)	Soft, silvery-white metal
Silicon	Si	Gray, lustrous solid
Silver	Ag (from *argentum*)	Silvery-white metal
Sodium	Na (from *natrium*)	Soft, silvery-white metal
Sulfur	S	Yellow solid
Tin	Sn (from *stannum*)	Silvery-white metal
Zinc	Zn	Bluish-white metal

*Common form of the element under normal conditions.

Deductions from Dalton's Atomic Theory

Note how atomic theory explains the difference between an element and a compound. Atomic theory also explains two laws we considered earlier. One of these is the law of conservation of mass, which states that the total mass remains constant during a chemical reaction. By postulate 2, every atom has a definite mass. Because a chemical reaction only rearranges the chemical combinations of atoms (postulate 4), the mass must remain constant. The other law explained by atomic theory is the law of definite proportions (constant composition). Postulate 3 defines a compound as a type of matter containing the atoms of two or more elements in definite proportions. Because the atoms have definite mass, compounds must have the elements in definite proportions by mass.

A good theory should not only explain known facts and laws but also predict new ones. The **law of multiple proportions,** deduced by Dalton from his atomic theory, states that *when two elements form more than one compound, the masses of one element in these compounds for a fixed mass of the other element are in ratios of small whole numbers.* To illustrate this law, consider the following. If we take a fixed mass of carbon, 1.000 gram, and react it with oxygen, we end up with two compounds: one that contains 1.3321 grams of oxygen for each 1.000 gram of carbon, and one that contains 2.6642 grams of oxygen per 1.000 gram of carbon. Note that the ratio of the amounts of oxygen in the two compounds is 2 to 1 (2.6642 ÷ 1.3321), which indicates that there must be twice as much oxygen in the second compound. Applying atomic theory, if we assume that the compound that has 1.3321 grams of oxygen to 1.000 gram of carbon is CO (the combination of one carbon atom and one oxygen atom), then the compound that contains twice as much oxygen per 1.000 gram of carbon must be CO_2 (the combination of one carbon atom and two oxygen atoms). The deduction of the law of multiple proportions from atomic theory was important in convincing chemists of the validity of the theory.

CONCEPT CHECK 2.1

Like Dalton, chemists continue to model atoms using spheres. Modern models are usually drawn with a computer and use different colors to represent atoms of different elements. Which of the models below represents CO_2?

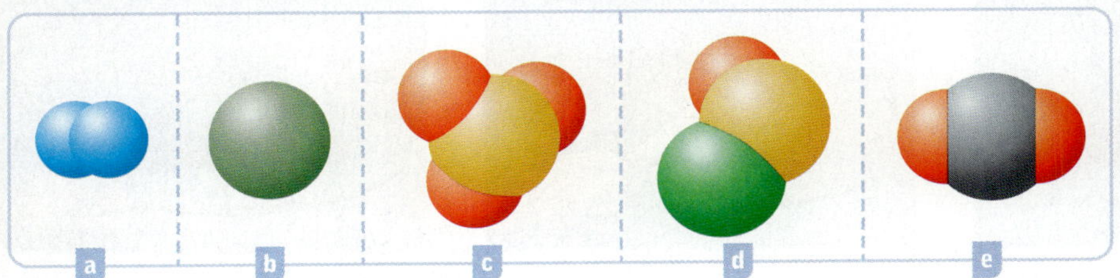

2.2 The Structure of the Atom

Although Dalton had postulated that atoms were indivisible particles, experiments conducted around the beginning of the last century showed that atoms themselves consist of particles. These experiments showed that an atom consists of two kinds of particles: a **nucleus,** *the atom's central core, which is positively charged and contains most of the atom's mass,* and one or more electrons. An **electron** is *a very light, negatively charged particle that exists in the region around the atom's positively charged nucleus.*

Discovery of the Electron

In 1897 the British physicist J. J. Thomson (Figure 2.3) conducted a series of experiments that showed that atoms were not indivisible particles. Figure 2.4 shows an experimental apparatus similar to the one used by Thomson. In this apparatus, two electrodes from a high-voltage source are sealed into a glass tube from which the air has been evacuated. The negative electrode is called the *cathode;* the positive one, the *anode.* When the high-voltage current is turned on, the glass tube emits a greenish light. Experiments showed that this greenish light is caused by the interaction of the glass with *cathode rays,* which are rays that originate from the cathode. ▶

After the cathode rays leave the negative electrode, they move toward the anode, where some rays pass through a hole to form a beam (Figure 2.4). This beam bends away from the negatively charged plate and toward the positively charged plate. (Cathode rays are not directly visible, but do cause certain materials such as zinc sulfide to glow so you can observe them.) Figure 2.5 shows a similar experiment, in which cathode rays are seen to bend when a magnet is brought toward them. Thomson showed that the characteristics of cathode rays are independent of the material making up the cathode. From such evidence, he concluded that a cathode ray consists of a beam of negatively charged particles (or electrons) and that electrons are constituents of all matter.

From his experiments, Thomson could also calculate the ratio of the electron's mass, m_e, to its electric charge, e. He could not obtain either the mass or the charge

Figure 2.3 ▲

Joseph John Thomson (1856–1940) J. J. Thomson's scientific ability was recognized early with his appointment as professor of physics in the Cavendish Laboratory at Cambridge University when he was not quite 28 years old. Soon after this appointment, Thomson began research on the discharge of electricity through gases. This work culminated in 1897 with the discovery of the electron. Thomson was awarded the Nobel Prize in physics in 1906.

A television tube uses the deflection of cathode rays by magnetic fields. A beam of cathode rays is directed toward a coated screen on the front of the tube, where by varying the magnetism generated by electromagnetic coils, the beam traces a luminescent image. Television tubes are becoming much less common with the advent of flat screen displays. Flat screen displays use a much different technology.

Figure 2.4 ▶

Formation of cathode rays Cathode rays leave the cathode, or negative electrode, and are accelerated toward the anode, or positive electrode. Some of the rays pass through the hole in the anode to form a beam, which then is bent by the electric plates in the tube.

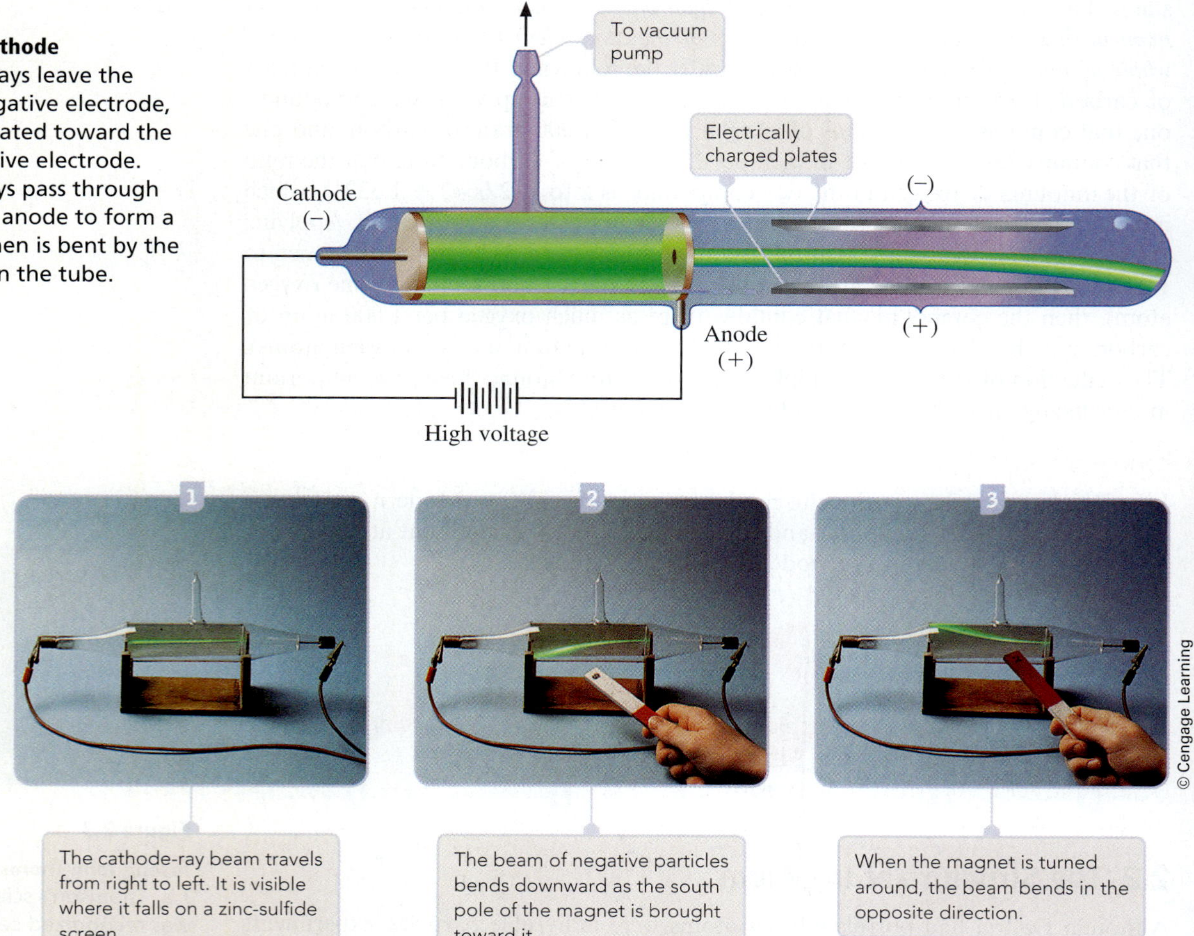

Figure 2.5 ▲

Bending of cathode rays by a magnet

separately, however. In 1909 the U.S. physicist Robert Millikan performed a series of ingenious experiments in which he obtained the charge on the electron by observing how a charged drop of oil falls in the presence and in the absence of an electric field (Figure 2.6). From this type of experiment, the charge on the electron is found to be 1.602×10^{-19} coulombs (the *coulomb,* abbreviated C, is a unit of electric charge). If you use this charge with the most recent value of the mass-to-charge ratio of the electron, you obtain an electron mass of 9.109×10^{-31} kg, which is more than 1800 times smaller than the mass of the lightest atom (hydrogen). This shows quite clearly that the electron is indeed a subatomic particle.

The Nuclear Model of the Atom

Ernest Rutherford (1871–1937), a British physicist, put forth the idea of the *nuclear model* of the atom in 1911, based on experiments done in his laboratory by Hans Geiger and Ernest Marsden. These scientists observed the effect of bombarding thin gold foil (and other metal foils) with alpha radiation from radioactive substances such as uranium (Figure 2.7). Rutherford had already shown that alpha rays consist of positively charged particles. Geiger and Marsden found that most of the alpha particles passed through a metal foil as though nothing were there, but a few (about 1 in 8000) were scattered at large angles and sometimes almost backward.

According to Rutherford's model, most of the mass of the atom (99.95% or more) is concentrated in a positively charged center, or nucleus, around which the negatively charged electrons move. Although most of the mass of an atom is in its nucleus, the nucleus occupies only a very small portion of the space of the atom. Nuclei have diameters of about 10^{-15} m (10^{-3} pm), whereas atomic diameters are about 10^{-10} m (100 pm), a hundred thousand times larger. If you were to use a golf ball to represent the nucleus, the atom would be about 3 miles in diameter!

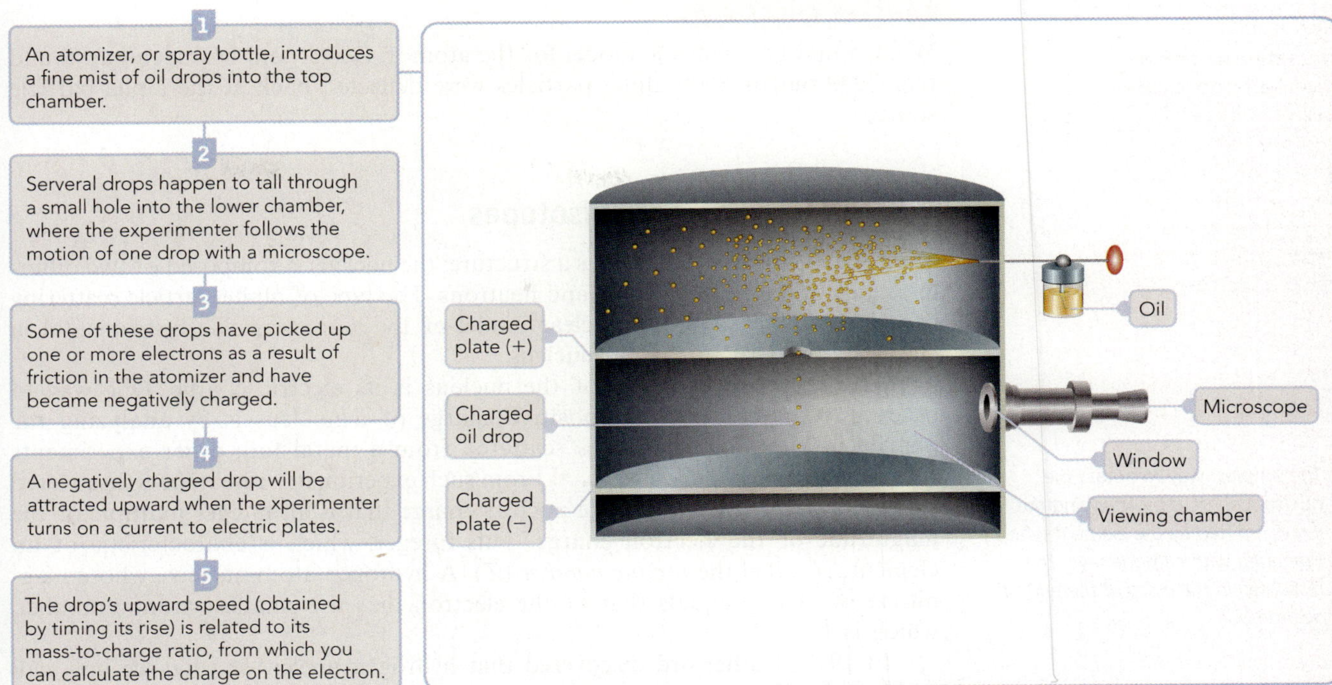

Figure 2.6 ▲

Millikan's oil-drop experiment

The nuclear model easily explains the results of bombarding gold and other metal foils with alpha particles. Alpha particles are much lighter than gold atoms. (Alpha particles are helium nuclei.) Most of the alpha particles pass through the metal atoms of the foil, undeflected by the lightweight electrons. When an alpha particle does happen to hit a metal-atom nucleus, however, it is scattered at a wide angle because it is deflected by the massive, positively charged nucleus (Figure 2.8).

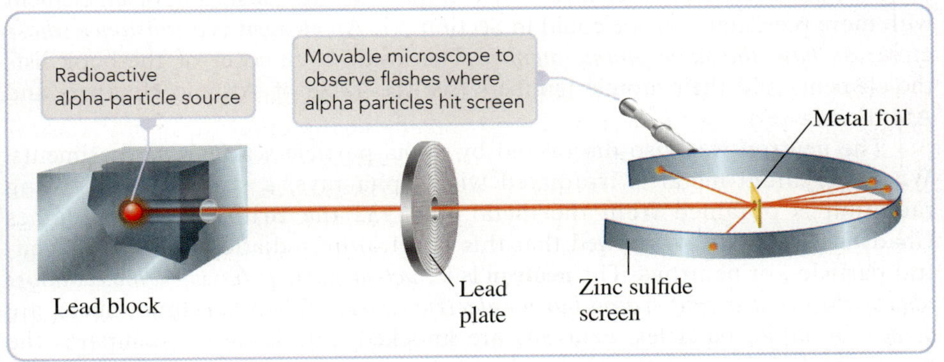

Figure 2.7 ◄

Alpha-particle scattering from metal foils Alpha radiation is produced by a radioactive source and formed into a beam by a lead plate with a hole in it. (Lead absorbs the radiation.) Scattered alpha particles are made visible by a zinc sulfide screen, which emits tiny flashes where particles strike it. A movable microscope is used for viewing the flashes.

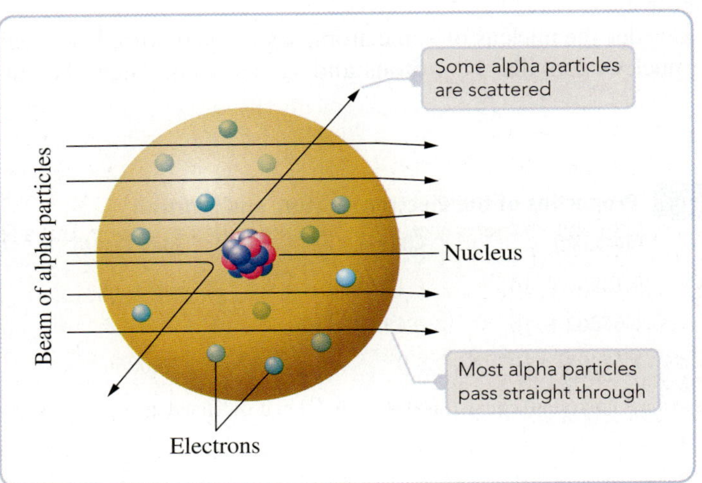

Figure 2.8 ◄

Representation of the scattering of alpha particles by a gold foil Most of the alpha particles pass through the foil barely deflected. A few, however, collide with gold nuclei and are deflected at large angles. (The relative sizes of nuclei are smaller than can be drawn here.)

CONCEPT CHECK 2.2

What would be a feasible model for the atom if Geiger and Marsden had found that 7999 out of 8000 alpha particles were deflected back at the alpha-particle source?

2.3 Nuclear Structure; Isotopes

The nucleus of an atom also has a structure; the nucleus is composed of two different kinds of particles, protons and neutrons. The type of alpha-particle scattering experiment that led to the nuclear model of the atom was also instrumental in clarifying this structure of the nucleus.

An important property of the nucleus is its electric charge. One way to determine the value of the positive charge of a nucleus is by analyzing the distribution of alpha particles scattered from a metal foil; other experiments also provide the nuclear charge. ◀ From such experiments, researchers discovered that each element has a unique nuclear charge that is an integer multiple of the magnitude of the electron charge. This integer, which is characteristic of an element, is called the *atomic number* (Z). A hydrogen atom nucleus, whose magnitude of charge equals that of the electron, has the smallest atomic number, which is 1.

The simplest way to obtain the nuclear charge is by analyzing the x rays emitted by the element when irradiated with cathode rays. This is discussed in the essay at the end of Section 8.2.

In 1919, Rutherford discovered that hydrogen nuclei, or what we now call protons, form when alpha particles strike some of the lighter elements, such as nitrogen. A **proton** is *a nuclear particle having a positive charge equal in magnitude to that of the electron and a mass more than 1800 times that of the electron.* When an alpha particle collides with a nitrogen atom, a proton may be knocked out of the nitrogen nucleus. The protons in a nucleus give the nucleus its positive charge.

The **atomic number** (Z) is therefore *the number of protons in the nucleus of an atom.* Because you can determine the atomic number experimentally, you can determine unambiguously whether or not a sample is a pure element or if perhaps you have discovered a new element. We can now state the definition of an element with more precision than we could in Section 2.1. An **element** is *a substance whose atoms all have the same atomic number.* The inside back cover of the book lists the elements and their atomic numbers (see the Table of Atomic Numbers and Atomic Masses).

The neutron was also discovered by alpha-particle scattering experiments. When beryllium metal is irradiated with alpha rays, a strongly penetrating radiation is obtained from the metal. In 1932 the British physicist James Chadwick (1891–1974) showed that this penetrating radiation consists of neutral particles, or neutrons. The **neutron** is *a nuclear particle having a mass almost identical to that of the proton but no electric charge.* When beryllium nuclei are struck by alpha particles, neutrons are knocked out. Table 2.2 compares the masses and charges of the electron and the two nuclear particles, the proton and the neutron.

Now consider the nucleus of some atom, say of the naturally occurring sodium atom. The nucleus contains 11 protons and 12 neutrons. Thus, the charge on the

Table 2.2 Properties of the Electron, Proton, and Neutron

Particle	Mass (kg)	Charge (C)	Mass (amu)*	Charge (e)
Electron	9.10938×10^{-31}	-1.60218×10^{-19}	0.00055	-1
Proton	1.67262×10^{-27}	$+1.60218 \times 10^{-19}$	1.00728	$+1$
Neutron	1.67493×10^{-27}	0	1.00866	0

*The atomic mass unit (amu) equals 1.66054×10^{-27} kg; it is defined in Section 2.4.

sodium nucleus is $+11e$, which we usually write as simply $+11$, meaning $+11$ units of electron charge e. Similarly, the nucleus of a naturally occurring aluminum atom contains 13 protons and 14 neutrons; the charge on the nucleus is $+13$.

We characterize a nucleus by its atomic number (Z) and its mass number (A). The **mass number** (A) is *the total number of protons and neutrons in a nucleus.* The nucleus of the naturally occurring sodium atom has an atomic number of 11 and a mass number of 23 (11 + 12).

A **nuclide** is *an atom characterized by an atomic number and mass number.* The shorthand notation for any nuclide consists of the symbol of the element with the atomic number written as a subscript on the left and the mass number as a superscript on the left. You write the *nuclide symbol* for the naturally occurring sodium nuclide as follows:

$$\begin{array}{r} \text{Mass number} \longrightarrow \\ \text{Atomic number} \longrightarrow \end{array} {}^{23}_{11}\text{Na}$$

An atom is normally electrically neutral, so it has as many electrons about its nucleus as the nucleus has protons; that is, the number of electrons in a neutral atom equals its atomic number. A sodium atom has a nucleus of charge $+11$, and around this nucleus are 11 electrons (with a charge of -11, giving the atom a charge of 0).

All nuclei of the atoms of a particular element have the same atomic number, but the nuclei may have different mass numbers. **Isotopes** are *atoms whose nuclei have the same atomic number but different mass numbers; that is, the nuclei have the same number of protons but different numbers of neutrons.* Sodium has only one naturally occurring isotope, which we denote by the same symbol we used for the nuclide (${}^{23}_{11}\text{Na}$); we also call the isotope sodium-23. Naturally occurring oxygen is a mixture of isotopes; it contains 99.759% oxygen-16, 0.037% oxygen-17, and 0.204% oxygen-18. ▶

Figure 2.9 may help you to visualize the relationship among the different subatomic particles in the isotopes of an element. The size of the nucleus as represented in the figure is very much exaggerated in order to show the numbers of protons and neutrons. In fact, the space taken up by the nucleus is minuscule compared with the region occupied by electrons. Electrons, although extremely light particles, move throughout relatively large diffuse regions, or "shells," about the nucleus of an atom.

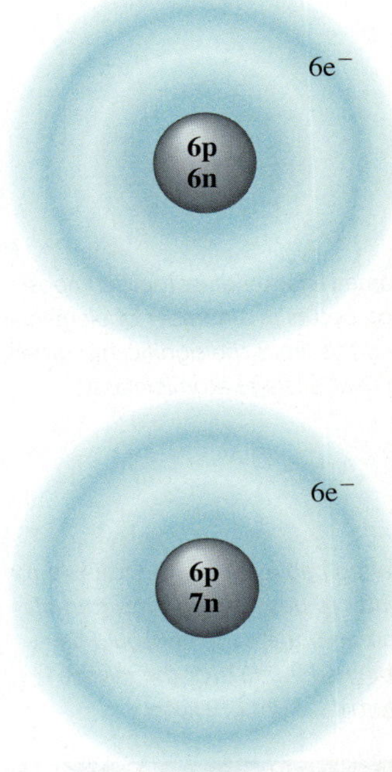

Figure 2.9 ▲

A representation of two isotopes of carbon The drawing shows the basic particles making up the carbon-12 and carbon-13 isotopes. (The relative sizes of the nuclei are much exaggerated here.)

Isotopes were first suspected in about 1912 when chemically identical elements with different atomic masses were found in radioactive materials. The most convincing evidence, however, came from the mass spectrometer, discussed in Section 2.4.

Example 2.1 **Writing Nuclide Symbols**

Gaining Mastery Toolbox

Critical Concept 2.1
The atomic number defines an element. The atomic number is the number of protons in the nucleus of an atom.

Solution Essentials:
- Nuclear particles: protons, neutrons, and electrons
- Atomic number
- Mass number

What is the nuclide symbol for a nucleus that contains 38 protons and 50 neutrons?

Problem Strategy To solve this problem, we need to keep in mind that the number of protons in the nucleus (the atomic number) is what uniquely identifies an element and that the mass number is the sum of the protons and neutrons.

Solution From the Table of Atomic Numbers and Atomic Masses on the inside back cover of this book, you will note that the element with atomic number 38 is strontium, symbol Sr. The mass number is 38 + 50 = 88. The symbol is ${}^{88}_{38}\textbf{Sr}$.

Answer Check When writing elemental symbols, as you have done here, make sure that you always capitalize *only* the first letter of the symbol.

Exercise 2.1 A nucleus consists of 17 protons and 18 neutrons. What is its nuclide symbol?

■ See Problems 2.47 and 2.48.

2.4 Atomic Masses

A central feature of Dalton's atomic theory was the idea that an atom of an element has a characteristic mass. Now we know that a naturally occurring element may be a mixture of isotopes, each isotope having its own characteristic mass. However, the percentages of the different isotopes in most naturally occurring elements have remained essentially constant over time and in most cases are independent of the origin of the element. Thus, what Dalton actually calculated were *average* atomic masses (actually average *relative* masses, as we will discuss). Since you normally work with naturally occurring mixtures of elements (whether in pure form or as compounds), it is indeed these average atomic masses that you need in ordinary chemical work. ◀

Some variation of isotopic composition occurs in a number of elements, and this limits the significant figures in their average atomic masses.

Relative Atomic Masses

Dalton could not weigh individual atoms. What he could do was find the average mass of one atom relative to the average mass of another. We will refer to these relative atomic masses as *atomic masses.* To see how Dalton obtained his atomic masses, imagine that you burn hydrogen gas in oxygen. The product of this reaction is water, a compound of hydrogen and oxygen. By experiment, you find that 1.0000 gram of hydrogen reacts with 7.9367 g of oxygen to form water. To obtain the atomic mass of oxygen (relative to hydrogen), you need to know the relative numbers of hydrogen atoms and oxygen atoms in water. As you might imagine, Dalton had difficulty with this. Today we know that water contains two atoms of hydrogen for every atom of oxygen. So the atomic mass of oxygen is $2 \times 7.9367 = 15.873$ times that of the mass of the average hydrogen atom. ◀

Dalton had assumed the formula for water to be HO. Using this formula and accurate data, he would have obtained 7.9367 for the relative atomic weight of oxygen.

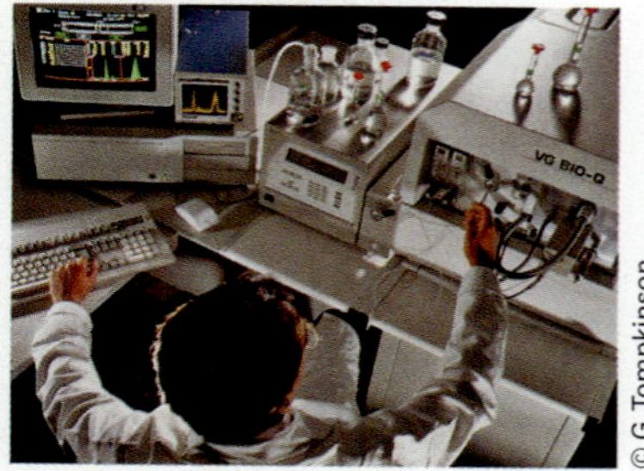

© G. Tompkinson

Figure 2.10 ▲

A mass spectrometer This instrument measures the masses of atoms and molecules.

Atomic Mass Units

Dalton's hydrogen-based atomic mass scale was eventually replaced by a scale based on oxygen and then, in 1961, by the present carbon-12 mass scale. This scale depends on measurements of atomic mass by an instrument called a *mass spectrometer* (Figure 2.10), which we will describe briefly later in this section. You make accurate measurements of mass on this instrument by comparing the mass of an atom to the mass of a particular atom chosen as a standard. On the present atomic mass scale, the carbon-12 isotope is chosen as the standard and is arbitrarily assigned a mass of exactly 12 atomic mass units. One **atomic mass unit (amu)** is, therefore, *a mass unit equal to exactly one-twelfth the mass of a carbon-12 atom.*

On this modern scale, the **atomic mass** of an element is *the average atomic mass for the naturally occurring element, expressed in atomic mass units.* A complete table of atomic masses appears on the inside back cover of this book. (Atomic mass is sometimes referred to as atomic weight. Although the use of the term *weight* in *atomic weight* is not strictly correct, it is sanctioned by convention to mean "average atomic mass.")

Mass Spectrometry and Atomic Masses

Earlier we noted that you use a mass spectrometer to obtain accurate masses of atoms. The earliest mass spectrometers measured the mass-to-charge ratios of positively charged atoms using the same ideas that Thomson used to study electrons. Modern instruments may use other techniques, but all measure the mass-to-charge ratios of positively charged atoms (and molecules as well). Figure 2.11 shows a simplified sketch of a mass spectrometer running a neon sample (see the essay on page 98 for a discussion of modern mass spectrometry).

Mass spectrometers produce a *mass spectrum,* which shows the relative numbers of atoms for various masses. If a sample of neon, which has the three isotopes depicted in Figure 2.12, is introduced into a mass spectrometer, it produces the mass spectrum shown in Figure 2.13. This mass spectrum gives us the information we need to calculate the atomic mass of neon: all of the isotopic masses (neon-20, neon-21, and neon-22) and the relative number, or fractional abundance, of each isotope. The **fractional abundance** of an isotope is *the fraction*

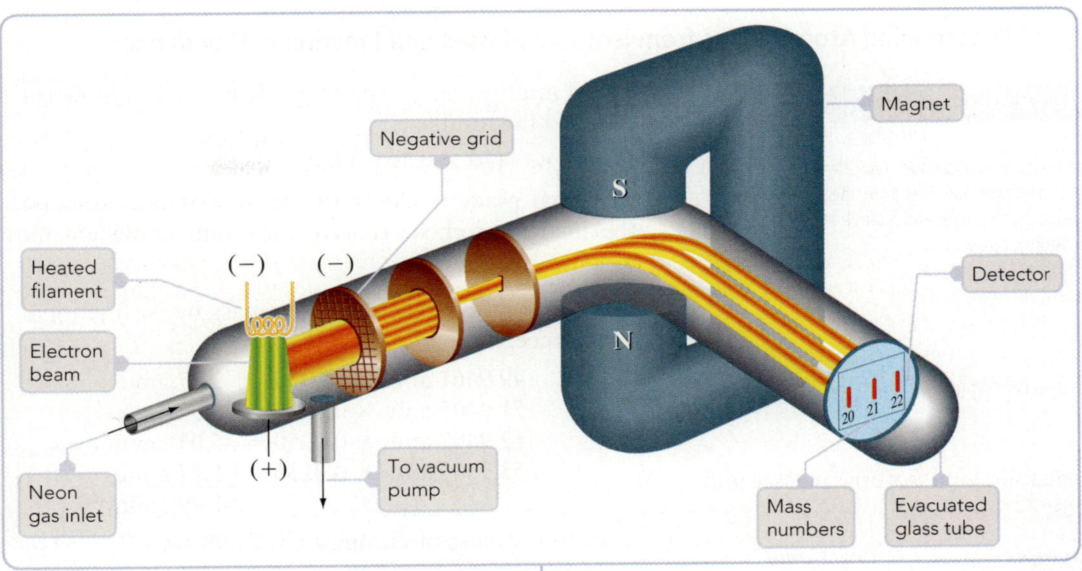

Figure 2.11 ▲

Diagram of a simple mass spectrometer, showing the separation of neon isotopes

Figure 2.12 ▼

Representations of the three naturally occurring isotopes of Ne *Top to bottom:* Ne-20, Ne-21, and Ne-22. Ne-20 is the most abundant isotope. Representations are not drawn to scale.

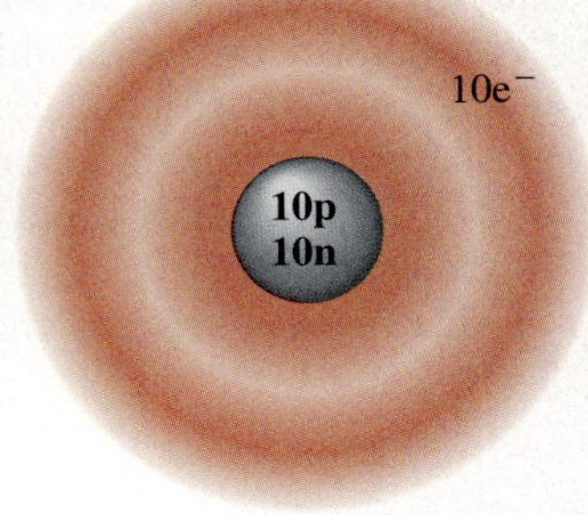

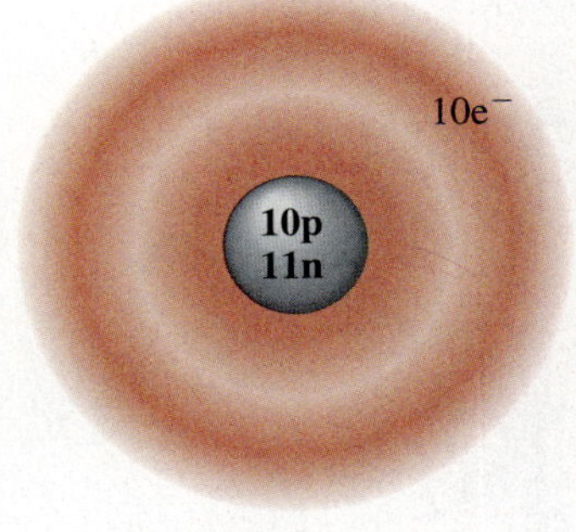

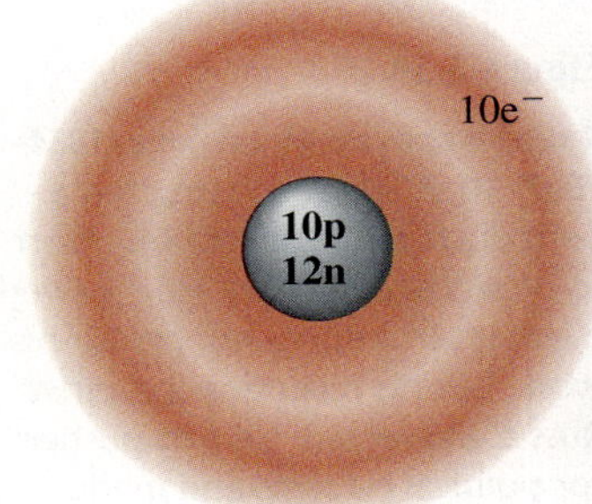

of the total number of atoms that is composed of a particular isotope. The fractional abundances of the neon isotopes in naturally occurring neon are neon-20, 0.9051; neon-21, 0.0027; and neon-22, 0.0922.

You calculate the atomic mass of an element by multiplying each isotopic mass by its fractional abundance and summing the values. If you do that for neon using the data given here, you will obtain 20.179 amu. The next example illustrates the calculation in full for chromium.

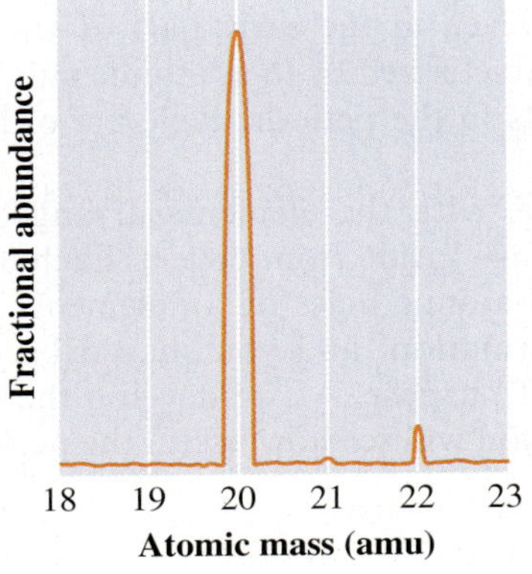

Figure 2.13 ◄

The mass spectrum of neon Neon is separated into its isotopes Ne-20, Ne-21, and Ne-22. The height at each mass peak is proportional to the fraction of that isotope in the element.

Example 2.2 Determining Atomic Mass from Isotopic Masses and Fractional Abundances

Gaining Mastery Toolbox

Critical Concept 2.2
The atomic masses of the elements are calculated. Typically each element has more than one naturally occurring isotope. Therefore, the mass of each element must be calculated based on the mass and relative abundance of each isotope; it is the average atomic mass.

Solution Essentials:
- Isotope
- Atomic mass unit
- Fractional abundance
- Rules for significant figures and rounding

Chromium, Cr, has the following isotopic masses and fractional abundances:

Mass Number	*Isotopic Mass (amu)*	*Fractional Abundance*
50	49.9461	0.0435
52	51.9405	0.8379
53	52.9407	0.0950
54	53.9389	0.0236

What is the atomic mass of chromium?

Problem Strategy The type of average used to calculate the atomic mass of an element is similar to the "average" an instructor might use to obtain a student's final grade in a class. Suppose the student has a total exam grade of 76 and a total laboratory grade of 84. The instructor decides to give a weight of 70% to the exams and 30% to the laboratory. How would the instructor calculate the final grade? He or she would multiply each type of grade by its weight factor and add the results:

$$(76 \times 0.70) + (84 \times 0.30) = 78$$

The final grade is closer to the exam grade, because the instructor chose to give the exam grade greater weight.

Solution Multiply each isotopic mass by its fractional abundance, then sum:

$$\begin{aligned}
49.9461 \text{ amu} \times 0.0435 &= 2.17 \text{ amu} \\
51.9405 \text{ amu} \times 0.8379 &= 43.52 \text{ amu} \\
52.9407 \text{ amu} \times 0.0950 &= 5.03 \text{ amu} \\
53.9389 \text{ amu} \times 0.0236 &= \underline{1.27 \text{ amu}} \\
& 51.99 \text{ amu}
\end{aligned}$$

The atomic mass of chromium is **51.99 amu.**

Answer Check The average mass (atomic mass) should be near the mass of the isotope with greatest abundance: in this case, 51.9405 amu with fractional abundance of 0.8379. This provides a quick check on your answer to this type of problem; any "answer" that is far from this will be incorrect.

Exercise 2.2 Chlorine consists of the following isotopes:

Isotope	*Isotopic Mass (amu)*	*Fractional Abundance*
Chlorine-35	34.96885	0.75771
Chlorine-37	36.96590	0.24229

What is the atomic mass of chlorine?

See Problems 2.51, 2.52, 2.53, and 2.54.

© Edgar Fahs Smith Collection, University of Pennsylvania

Figure 2.14

Dmitri Ivanovich Mendeleev (1834–1907) Mendeleev constructed a periodic table as part of his effort to organize chemistry. He received many international honors for his work, but his reception in czarist Russia was mixed. He had pushed for political reforms and made many enemies as a result.

2.5 Periodic Table of the Elements

In 1869 the Russian chemist Dmitri Mendeleev (1834–1907) (Figure 2.14) and the German chemist J. Lothar Meyer (1830–1895), working independently, made similar discoveries. They found that when they arranged the elements in order of atomic mass, they could place them in horizontal rows, one row under the other, so that the elements in each vertical column have similar properties. *A tabular arrangement of elements in rows and columns, highlighting the regular repetition of properties of the elements,* is called a **periodic table.**

Eventually, more accurate determinations of atomic masses revealed discrepancies in this ordering of the elements. However, in the early part of the last century, it was shown that the elements are characterized by their atomic numbers, rather than atomic masses. When the elements in the periodic table are ordered by atomic number, such discrepancies vanish.

A modern version of the periodic table, with the elements arranged by atomic number, is shown in Figure 2.15 (see also inside front cover). Each entry lists the atomic number, atomic symbol, and atomic mass of an element. This is a convenient way of tabulating such information, and you should become familiar with using the periodic table for that purpose. As we develop the subject matter of chemistry throughout the text, you will see how useful the periodic table is.

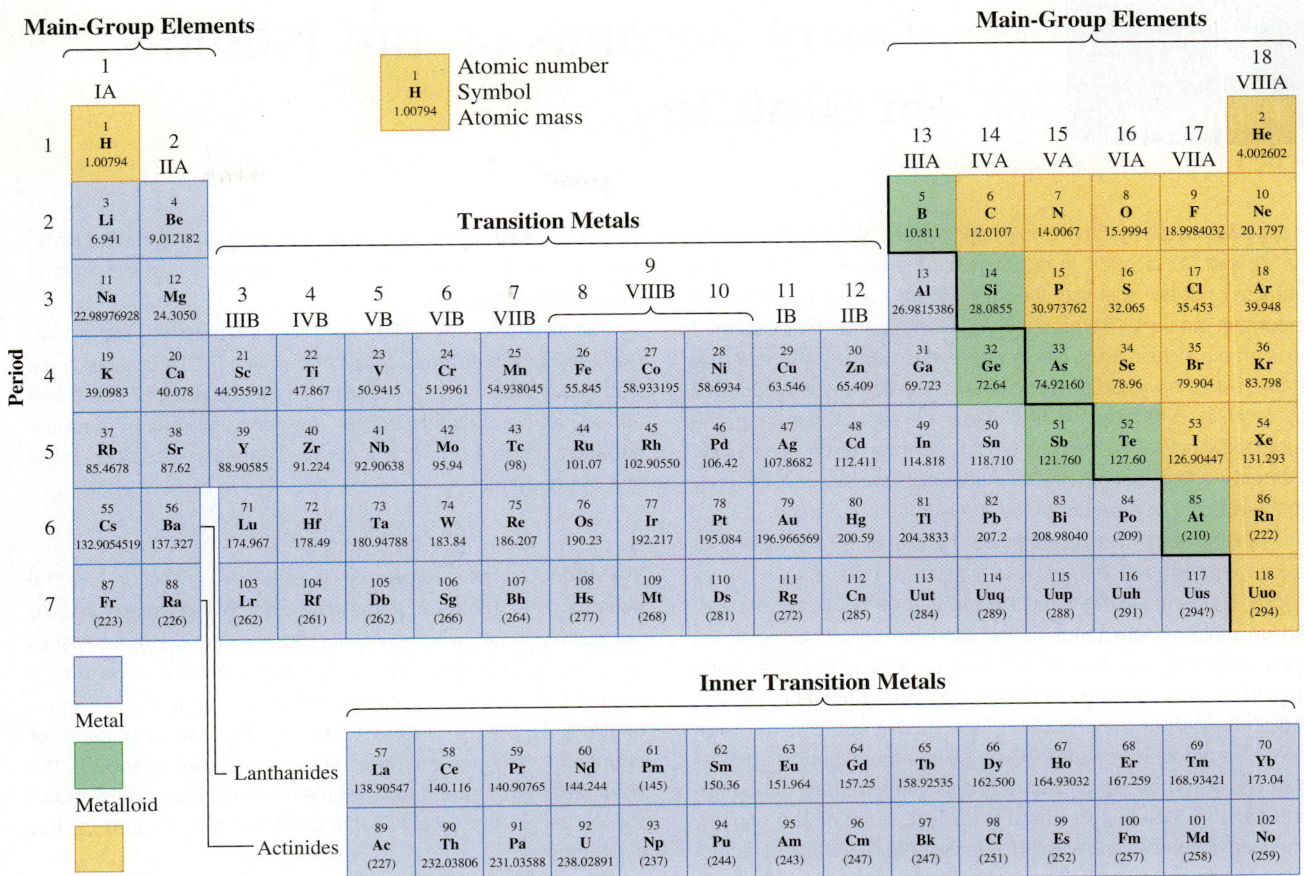

Period	1 IA	2 IIA	3 IIIB	4 IVB	5 VB	6 VIB	7 VIIB	8	9 VIIIB	10	11 IB	12 IIB	13 IIIA	14 IVA	15 VA	16 VIA	17 VIIA	18 VIIIA
1	1 H 1.00794																	2 He 4.002602
2	3 Li 6.941	4 Be 9.012182											5 B 10.811	6 C 12.0107	7 N 14.0067	8 O 15.9994	9 F 18.9984032	10 Ne 20.1797
3	11 Na 22.98976928	12 Mg 24.3050											13 Al 26.9815386	14 Si 28.0855	15 P 30.973762	16 S 32.065	17 Cl 35.453	18 Ar 39.948
4	19 K 39.0983	20 Ca 40.078	21 Sc 44.955912	22 Ti 47.867	23 V 50.9415	24 Cr 51.9961	25 Mn 54.938045	26 Fe 55.845	27 Co 58.933195	28 Ni 58.6934	29 Cu 63.546	30 Zn 65.409	31 Ga 69.723	32 Ge 72.64	33 As 74.92160	34 Se 78.96	35 Br 79.904	36 Kr 83.798
5	37 Rb 85.4678	38 Sr 87.62	39 Y 88.90585	40 Zr 91.224	41 Nb 92.90638	42 Mo 95.94	43 Tc (98)	44 Ru 101.07	45 Rh 102.90550	46 Pd 106.42	47 Ag 107.8682	48 Cd 112.411	49 In 114.818	50 Sn 118.710	51 Sb 121.760	52 Te 127.60	53 I 126.90447	54 Xe 131.293
6	55 Cs 132.9054519	56 Ba 137.327	71 Lu 174.967	72 Hf 178.49	73 Ta 180.94788	74 W 183.84	75 Re 186.207	76 Os 190.23	77 Ir 192.217	78 Pt 195.084	79 Au 196.966569	80 Hg 200.59	81 Tl 204.3833	82 Pb 207.2	83 Bi 208.98040	84 Po (209)	85 At (210)	86 Rn (222)
7	87 Fr (223)	88 Ra (226)	103 Lr (262)	104 Rf (261)	105 Db (262)	106 Sg (266)	107 Bh (264)	108 Hs (277)	109 Mt (268)	110 Ds (281)	111 Rg (272)	112 Cn (285)	113 Uut (284)	114 Uuq (289)	115 Uup (288)	116 Uuh (291)	117 Uus (294?)	118 Uuo (294)

Lanthanides	57 La 138.90547	58 Ce 140.116	59 Pr 140.90765	60 Nd 144.244	61 Pm (145)	62 Sm 150.36	63 Eu 151.964	64 Gd 157.25	65 Tb 158.92535	66 Dy 162.500	67 Ho 164.93032	68 Er 167.259	69 Tm 168.93421	70 Yb 173.04
Actinides	89 Ac (227)	90 Th 232.03806	91 Pa 231.03588	92 U 238.02891	93 Np (237)	94 Pu (244)	95 Am (243)	96 Cm (247)	97 Bk (247)	98 Cf (251)	99 Es (252)	100 Fm (257)	101 Md (258)	102 No (259)

Figure 2.15 ▲

A modern form of the periodic table This table is also given on the inside front cover of the book.

Periods and Groups

The basic structure of the periodic table is its division into rows and columns, or periods and groups. A **period** consists of *the elements in any one horizontal row of the periodic table.* A **group** consists of *the elements in any one column of the periodic table.*

The first period of elements consists of only hydrogen (H) and helium (He). The second period has 8 elements, beginning with lithium (Li) and ending with neon (Ne). There is then another period of 8 elements, and this is followed by a period having 18 elements, beginning with potassium (K) and ending with krypton (Kr). The fifth period also has 18 elements. The sixth period actually consists of 32 elements, but in order for the row to fit on a page, part of it appears at the bottom of the table. Otherwise the table would have to be expanded, with the additional elements placed after barium (Ba, atomic number 56). The seventh period, though not complete, also has some of its elements placed as a row at the bottom of the table.

The groups are usually numbered. The numbering frequently seen in North America labels the groups with Roman numerals and A's and B's. In Europe a similar convention has been used, but some columns have the A's and B's interchanged. To eliminate this confusion, the International Union of Pure and Applied Chemistry (IUPAC) suggested a convention in which the columns are numbered 1 to 18. Figure 2.15 shows the traditional North American and the IUPAC conventions. When we refer to an element by its periodic group, we will use the traditional North American convention. The A groups are called *main-group* (or *representative*) *elements;* the B groups are called *transition elements.* The two rows of elements at the bottom of the table are called *inner transition elements* (the first row is referred to as the *lanthanides;* the second row, as the *actinides*).

A **CHEMIST** Looks at...

Frontiers

Thirty Seconds on the Island of Stability

It appears that scientists have now been on the "Island of Stability"—for at least 30 seconds. In recent years, the island of stability has attracted nuclear scientists the way Mount Everest has attracted mountain climbers. Like mountain climbers, scientists have wanted to surmount their own tallest peak—discovery of the relatively stable, superheavy nuclide at the peak of stability that is surrounded by foothills composed of less stable nuclides. Discovering the superheavy nuclide at the peak would also mean discovering a new element—a task that has had its allure since Lavoisier gave currency to the concept of elements in the late eighteenth century.

Theoretical physicists predicted the existence of the island of stability, centered at element 114 with mass number 298, in the 1960s. The term *stability* refers here to that of the atomic nucleus. An unstable nucleus tends to fall apart by radioactive decay—a piece spontaneously flies off the nucleus, leaving a different one behind. Approximately 275 nuclides are completely stable, or nonradioactive. All of these nuclides have atomic numbers (or numbers of protons) no greater than 83 (the atomic number for the element bismuth). Beyond bismuth, all elements are radioactive and become increasingly unstable. In fact, none of the original, primeval elements past uranium (element 92)—the *transuranium* elements—exists any longer; they have long since vanished by radioactive decay. Scientists have made transuranium elements in the laboratory, however.

In February 1996, for example, scientists at Darmstadt, Germany, produced a few atoms of element 112 by bombarding a lead target with zinc atoms. Each atom of element 112 lasted only about 240 microseconds. We can represent the result by a nuclear equation, which is similar to a chemical equation (described at the end of the chapter).

$$^{70}_{30}\text{Zn} + ^{208}_{82}\text{Pb} \longrightarrow ^{277}_{112}\text{Uub} + ^{1}_{0}\text{n}$$

We introduced the nuclide symbols used here in Section 2.3. The equation is a summary of the following. A high-speed zinc-70 nucleus (as an ion, which is an electrically charged atom) smashes into a lead-208 target where, as the arrow denotes, it yields new products, a nucleus of element 112 (mass number 277) and a neutron, denoted $^{1}_{0}$n. Element 112 has the provisional name of ununbium until several other laboratories can reproduce the work, after which the discoverer may name it. Figure 2.16 shows the ion accelerator (UNILAC) that produced the zinc ions for this experiment.

As noted earlier, the elements in any one group have similar properties. For example, the elements in Group IA, often known as the *alkali metals,* are soft metals that react easily with water. (Hydrogen, a gas, is an exception and might better be put in a group by itself.) Sodium is an alkali metal. So is potassium. The Group VIIA elements, known as *halogens,* are also reactive elements. Chlorine is a halogen. We have already noted its vigorous reaction with sodium. Bromine, which is a red-brown liquid, is another halogen. It too reacts vigorously with sodium.

Metals, Nonmetals, and Metalloids

The elements of the periodic table in Figure 2.15 are divided by a heavy "staircase" line into metals on the left and nonmetals on the right. A **metal** is *a substance that has a characteristic luster, or shine, and is generally a good conductor of heat and electricity.* Except for mercury, the metallic elements are solids at room temperature (about 20°C). They are more or less *malleable* (can be hammered into sheets) and *ductile* (can be drawn into wire).

A **nonmetal** is *an element that does not exhibit the characteristics of a metal.* Most of the nonmetals are gases (for example, chlorine and oxygen) or solids (for example, phosphorus and sulfur). The solid nonmetals are usually hard, brittle substances. Bromine is the only liquid nonmetal.

Most of the elements bordering the staircase line in the periodic table (Figure 2.15) are metalloids, or semimetals. A **metalloid,** or **semimetal,** is *an element having both metallic and nonmetallic properties.* These elements, such as silicon (Si) and germanium (Ge), are usually good *semiconductors*—elements that, when pure, are poor conductors of electricity at room temperature but become moderately good conductors at higher temperatures. ◀

When these pure semiconductor elements have small amounts of certain other elements added to them (a process called doping), they become very good conductors of electricity. Semiconductors are the critical materials in solid-state electronic devices.

© A. Zschau/GSI

Figure 2.16 ▲

Within the UNILAC ion accelerator The accelerator consists of a series of electrodes arranged one after the other in a copper-plated steel container. Each electrode accelerates the ion. Eventually, the ion gains sufficient speed, and therefore energy, to cause its nucleus to fuse on collision with a target atom nucleus giving a new nucleus.

Scientists in Dubna, north of Moscow in Russia, and their American collaborators from Lawrence Livermore Laboratory in California produced one atom of element 114 (provisional name ununquadium) in December 1998, placing scientists on the shore of the island of stability. In this experiment, scientists bombarded a plutonium-244 target with calcium-48 atoms.

$$^{48}_{20}\text{Ca} + ^{244}_{94}\text{Pu} \longrightarrow ^{289}_{114}\text{Uuq} + 3^{1}_{0}\text{n}$$

The atom of element 114 lasted about 30 seconds. While a 30-second lifetime may seem brief, it is many times longer than that seen in the previously discovered element 112. In 2001 this same group of scientists reported the discovery of element 116 (ununhexium).

As a rather bizarre footnote to this story, the "discovery" of element 118 was reported in 1999 by scientists at Lawrence Berkeley National Laboratory in California. Two years later, these scientists reported that they were unable to reproduce their results. The laboratory fired one of the scientists, who they alleged had faked the data. Fortunately, fraud is rather unusual in chemistry; when it does occur, it is discovered, as in this case, when the work cannot be reproduced. Notably, element 118 was rediscovered in 2006 by the same Russian and American team who discovered elements 114 and 116, bringing scientists one step closer to the island of stability.

■ See Problems 2.131 and 2.132.

Exercise 2.3 By referring to the periodic table (Figure 2.15 or inside front cover), identify the group and period to which each of the following elements belongs. Then decide whether the element is a metal, nonmetal, or metalloid.

a Se b Cs c Fe d Cu e Br

■ See Problems 2.57 and 2.58.

CONCEPT CHECK 2.3

Consider the elements He, Ne, and Ar. Can you come up with a reason why they are in the same group in the periodic table?

Chemical Substances: Formulas and Names

Atomic theory has developed steadily since Dalton's time and has become the cornerstone of chemistry. It results in an enormous simplification: all of the millions of compounds we know today are composed of the atoms of just a few elements. Now we look more closely at how we describe the composition and structure of chemical substances in terms of atoms.

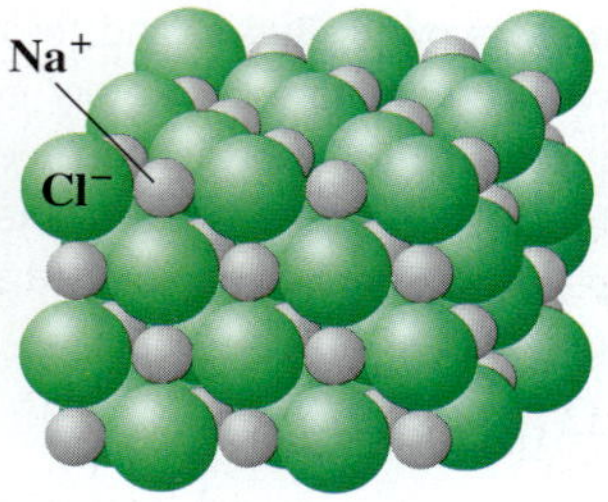

Na$^+$

Cl$^-$

Figure 2.17 ▲

Sodium chloride model This model of a sodium chloride crystal illustrates the 1 : 1 packing of the Na^+ and Cl^- ions.

2.6 Chemical Formulas; Molecular and Ionic Substances

The **chemical formula** of a substance is *a notation that uses atomic symbols with numerical subscripts to convey the relative proportions of atoms of the different elements in the substance.* Consider the formula of aluminum oxide, Al_2O_3. This means that the compound is composed of aluminum atoms and oxygen atoms in the ratio 2 : 3. Consider the formula for sodium chloride, NaCl. When no subscript is written for a symbol, it is assumed to be 1. Therefore, the formula NaCl means that the compound is composed of sodium atoms and chlorine atoms in the ratio 1 : 1 (Figure 2.17).

Additional information may be conveyed by different kinds of chemical formulas. To understand this, we need to look briefly at two main types of substances: molecular and ionic.

Molecular Substances

A **molecule** is *a definite group of atoms that are chemically bonded together—that is, tightly connected by attractive forces.* The nature of these strong forces is discussed in Chapters 9 and 10. A *molecular substance* is a substance that is composed of molecules, all of which are alike. The molecules in such a substance are so small that even extremely minute samples contain tremendous numbers of them. One billionth (10^{-9}) of a drop of water, for example, contains about 2 trillion (2×10^{12}) water molecules. ◄

Another way to understand the large numbers of molecules involved in relatively small quantities of matter is to consider 1 g of water (about one-fifth teaspoon). It contains 3.3×10^{22} water molecules. If you had a penny for every molecule in this quantity of water, the height of your stack of pennies would be about 300 million times the distance from the earth to the sun.

A **molecular formula** *gives the exact number of different atoms of an element in a molecule.* The hydrogen peroxide molecule contains two hydrogen atoms and two oxygen atoms chemically bonded. Therefore, its molecular formula is H_2O_2. Other simple molecular substances are water, H_2O; ammonia, NH_3; carbon dioxide, CO_2; and ethanol (ethyl alcohol), C_2H_6O. In Chapter 3, you will see how we determine such formulas.

The atoms in a molecule are not simply piled together randomly. Rather, the atoms are chemically bonded in a definite way. A *structural formula* is a chemical formula that shows how the atoms are bonded to one another in a molecule. For example, it is known that each of the hydrogen atoms in the water molecule is bonded to the oxygen atom. Thus, the structural formula of water is H—O—H. A line joining two atomic symbols in such a formula represents the chemical bond connecting the atoms. Figure 2.18 shows some structural formulas. Structural formulas are sometimes condensed in writing. For example, the structural formula of ethanol may be written CH_3CH_2OH or C_2H_5OH, depending on the detail you want to convey.

The atoms in a molecule are not only connected in definite ways but exhibit definite spatial arrangements as well. Chemists often construct molecular models as an aid in visualizing the shapes and sizes of molecules. Figure 2.18 shows molecular models for several compounds. While the ball-and-stick type of model shows the bonds and bond angles clearly, the space-filling type gives a more realistic feeling of the space occupied by the atoms. Chemists also use computer-generated models of molecules, which can be produced in a variety of forms in addition to those shown here.

Some elements are molecular substances and are represented by molecular formulas. Chlorine, for example, is a molecular substance and has the formula Cl_2, each molecule being composed of two chlorine atoms bonded together. Sulfur consists of molecules composed of eight atoms; its molecular formula is S_8. Helium and neon are composed of isolated atoms; their formulas are He and Ne, respectively. Other elements, such as carbon (in the form of graphite or diamond), do not have a simple molecular structure but consist of a very large, indefinite number of atoms bonded together. These elements are represented simply by their atomic symbols. (An important exception is a form of carbon called buckminsterfullerene, which was discovered in 1985 and has the molecular formula C_{60}.) Models of some elementary substances are shown in Figure 2.19.

	WATER	AMMONIA	ETHANOL
Molecular formula	H_2O	NH_3	C_2H_6O
Structural formula	H–O–H	H–N–H | H	H H | | H–C–C–O–H | | H H
Molecular model (ball-and-stick type)			
Molecular model (space-filling type)			
Electrostatic potential map			

Figure 2.18 ◀

Examples of molecular and structural formulas, molecular models, and electrostatic potential maps Three common molecules—water, ammonia, and ethanol—are shown. The electrostatic potential map representation at the bottom of the figure illustrates the distribution of electrons in the molecule using a color spectrum. Colors range from red (relatively high electron density) all the way to blue (low electron density).

An important class of molecular substances is the polymers. **Polymers** are *very large molecules that are made up of a number of smaller molecules repeatedly linked together.* **Monomers** are *the small molecules that are linked together to form the polymer.* A good analogy for the formation of polymers from monomers is that of making a chain of paper clips. Say you have boxes of red, blue, and yellow paper clips. Each paper clip represents a monomer. You can make a large chain of paper clips, representing the polymer, with a variety of patterns and repeating units. One chain might involve a repeating pattern of two red, three yellow, and one blue. Another chain might consist of only blue paper clips. Just like the paper clips, the creation of a particular polymer in a laboratory involves controlling both the monomers that are being linked together and the linkage pattern.

Polymers are both natural and synthetic. Hard plastic soda bottles are made from chemically linking two different monomers in an alternating pattern. Wool and silk are natural polymers of amino acids linked by peptide bonds. Nylon® for fabrics, Kevlar® for bulletproof vests, and Nomex® for flame-retardant clothing all contain a CONH linkage. Plastics and rubbers are also polymers that are made from carbon- and hydrogen-containing monomers. Even the Teflon® coating on cookware is a polymer that is the result of linking CF_2CF_2 monomers (Figure 2.20). From these examples, it is obvious that polymers are very important molecular materials that we use every day in a wide variety of applications. For more on the chemistry of polymers, refer to Chapter 24.

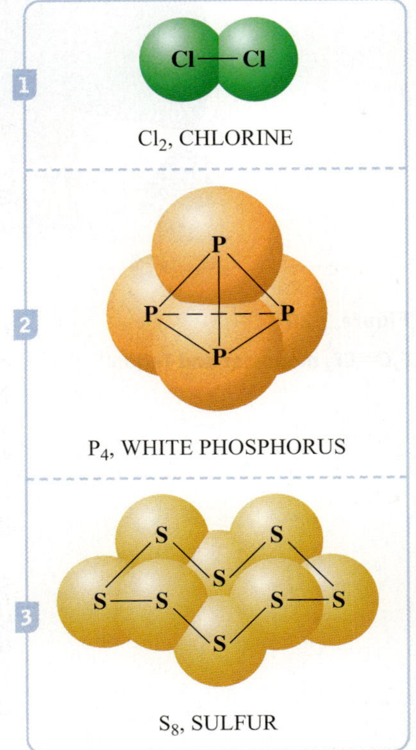

Figure 2.19 ▲

Molecular models of some elementary substances

Ionic Substances

Although many substances are molecular, others are composed of ions (pronounced "eye'-ons"). An **ion** is *an electrically charged particle obtained from an atom or chemically bonded group of atoms by adding or removing electrons.* Sodium chloride is a substance made up of ions.

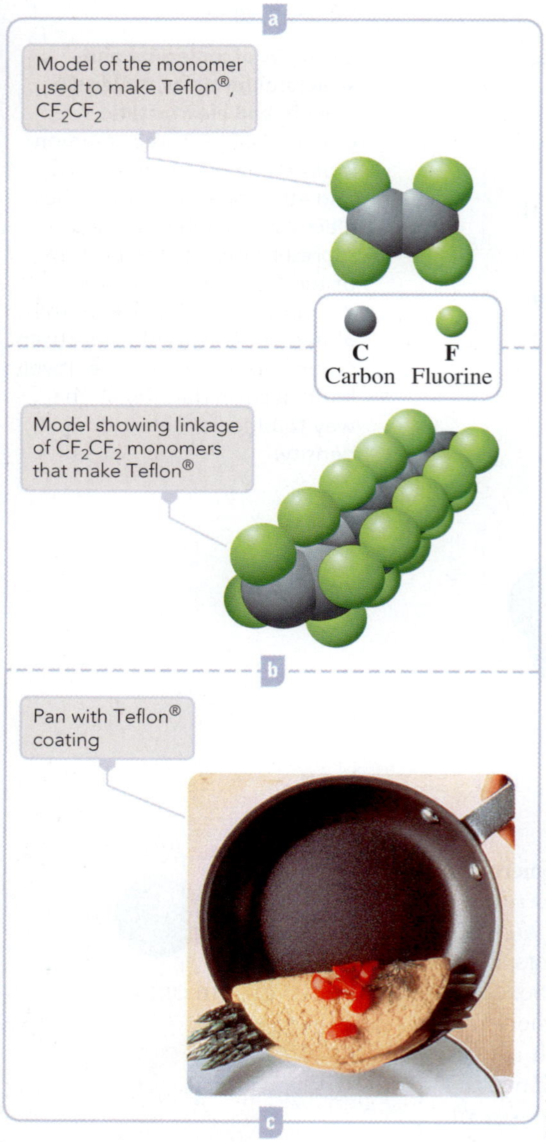

a

Model of the monomer used to make Teflon®, CF_2CF_2

C Carbon F Fluorine

Model showing linkage of CF_2CF_2 monomers that make Teflon®

b

Pan with Teflon® coating

c

Teflon® is a registered trademark of DuPont

Figure 2.20 ▲

$F_2C{=}CF_2$ monomer and Teflon®

Although isolated atoms are normally electrically neutral and therefore contain equal numbers of positive and negative charges, during the formation of certain compounds atoms can become ions. Metal atoms tend to lose electrons, whereas nonmetals tend to gain electrons. When a metal atom such as sodium and a nonmetal atom such as chlorine approach one another, an electron can transfer from the metal atom to the nonmetal atom to produce ions.

An atom that picks up an extra electron becomes *a negatively charged ion,* called an **anion** (pronounced "an'-ion"). An atom that loses an electron becomes *a positively charged ion,* called a **cation** ("cat'-ion"). A sodium atom, for example, can lose an electron to form a sodium cation (denoted Na^+). A chlorine atom can gain an electron to form a chloride anion (denoted Cl^-). A calcium atom can lose two electrons to form a calcium cation, denoted Ca^{2+}. Note that the positive-two charge on the ion is indicated by a superscript 2+.

Some ions consist of two or more atoms chemically bonded but having an excess or deficiency of electrons so that the unit has an electric charge. An example is the sulfate ion, SO_4^{2-}. The superscript 2− indicates an excess of two electrons on the group of atoms.

An **ionic compound** is *a compound composed of cations and anions.* Sodium chloride consists of equal numbers of sodium ions, Na^+, and chloride ions, Cl^-. The strong attraction between positive and negative charges holds the ions together in a regular arrangement in space. For example, in sodium chloride, each Na^+ ion is surrounded by six Cl^- ions, and each Cl^- ion is surrounded by six Na^+ ions. The result is a *crystal,* which is a kind of solid having a regular three-dimensional arrangement of atoms, molecules, or (as in the case of sodium chloride) ions. Figure 2.21 shows sodium chloride crystals and two types of models used to depict the arrangement of the ions in the crystal. The number of ions in an individual sodium chloride crystal determines the size of the crystal.

The formula of an ionic compound is written by giving the smallest possible integer number of different ions in the substance, except that the charges on the ions are omitted so that the formulas merely indicate the atoms involved. For example, sodium chloride contains equal numbers of Na^+ and Cl^- ions. The formula is written NaCl (not Na^+Cl^-). Iron(III) sulfate is a compound consisting of iron(III) ions, Fe^{3+}, and sulfate ions, SO_4^{2-}, in the ratio 2 : 3. The formula is written $Fe_2(SO_4)_3$, in which parentheses enclose the formula of an ion composed of more than one atom (again, omitting ion charges); parentheses are used only when there are two or more such ions.

Although ionic substances do not contain molecules, we can speak of the smallest unit of such a substance. The **formula unit** of a substance is *the group of atoms or ions explicitly symbolized in the formula.* For example, the formula unit of water, H_2O, is the H_2O molecule. The formula unit of iron(III) sulfate, $Fe_2(SO_4)_3$, consists of two Fe^{3+} ions and three SO_4^{2-} ions. The formula unit is the smallest neutral unit of such substances.

All substances, including ionic compounds, are electrically neutral. You can use this fact to obtain the formula of an ionic compound, given the formulas of the ions. This is illustrated in the following example.

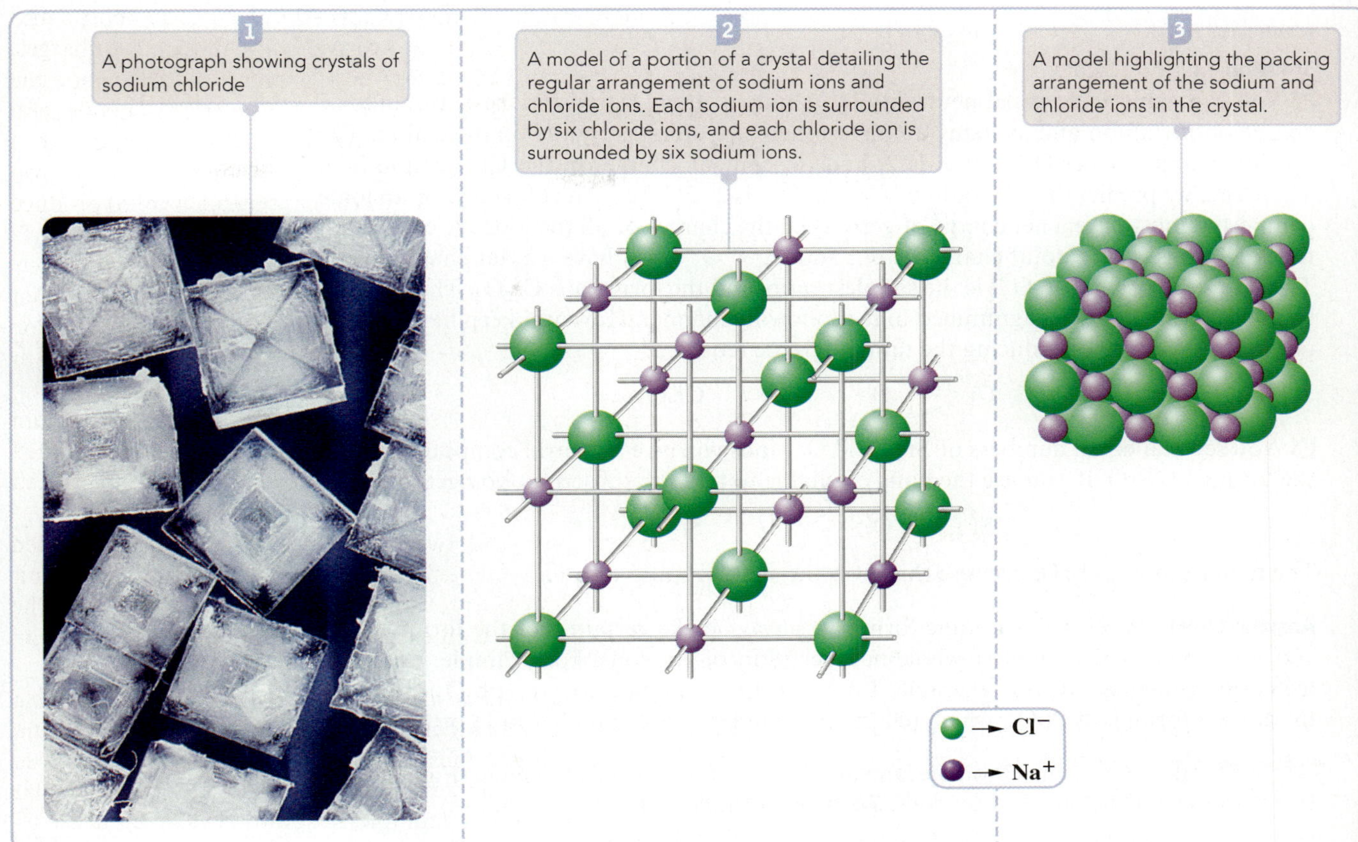

Figure 2.21 ▲
The sodium chloride crystal

Example 2.3 Writing an Ionic Formula, Given the Ions

Gaining Mastery Toolbox

Critical Concept 2.3
Ionic compounds are formed by the combination of positive and negatively charged particles. When the charged particles (ions) combine, they combine in a ratio that yields a compound with an overall charge of zero. Ionic compounds are often combinations of metal and nonmetal atoms.

Solution Essentials:
- Compound
- Ion, anion, and cation
- Ionic compound
- Formula unit

a Chromium(III) oxide is used as a green paint pigment (Figure 2.22). It is a compound composed of Cr^{3+} and O^{2-} ions. What is the formula of chromium(III) oxide?

b Strontium oxide is a compound composed of Sr^{2+} and O^{2-} ions. Write the formula of this compound.

© Cengage Learning

Figure 2.22 ▲
Chromium(III) oxide The compound is used as a green paint pigment.

Problem Strategy Because a compound is neutral, the sum of positive and negative charges equals zero. Consider the ionic compound $CaCl_2$, which consists of one Ca^{2+} ion and two Cl^- ions. The sum of the charges is

$$\underbrace{1 \times (+2)}_{\text{Positive charge}} + \underbrace{2 \times (-1)}_{\text{Negative charge}} = 0$$

Note that the number of calcium ions in $CaCl_2$ equals the magnitude of charge on the chloride ion (1), whereas the number of chloride ions in $CaCl_2$ equals the magnitude of charge on the calcium ion (2). In general, you use the magnitude of charge on one ion to obtain the subscript for the other ion. You may need to simplify the formula you obtain this way so that it expresses the simplest ratio of ions.

(continued)

(continued)

Solution

a You can achieve electrical neutrality by taking as many cations as there are units of charge on the anion and as many anions as there are units of charge on the cation. The unit of charge on the O^{2-} anion is 2, and the unit of charge on the Cr^{3+} cation is 3. Therefore, we predict that the ratio of Cr^{3+} to O^{2-} is 2 : 3. The correct formula of an ionic compound must have a net charge of zero when the charges of all the ions are combined; two Cr^{3+} ions have a total charge of 6^+, and three O^{2-} ions have a total charge of 6^-. The 2 : 3 ratio of Cr^{3+} to O^{2-} is the simplest ratio, and the formula is $\mathbf{Cr_2O_3}$. The solution to this problem can be diagrammed in the following manner. However, keep in mind that one additional step of reducing the ratio might be required.

$$Cr^{③+} \quad O^{②-} \quad \text{or} \quad Cr_2O_3$$

b You see that equal numbers of Sr^{2+} and O^{2-} ions will give a neutral compound. Thus, the formula is SrO. If you use the units of charge to find the subscripts, you get

$$Sr^{②+} \quad O^{②-} \quad \text{or} \quad Sr_2O_2$$

The final formula is **SrO,** because this gives the simplest ratio of ions.

Answer Check When writing ionic formulas, always make certain that the formula that you write reflects the smallest whole-number ratio of the ions. For example, using the technique presented in this example, Pb^{4+} and O^{2-} combine to give Pb_2O_4; however, the correct formula is PbO_2, which reflects the simplest whole-number ratio of ions.

Exercise 2.4 Potassium chromate is an important compound of chromium (Figure 2.23). It is composed of K^+ and CrO_4^{2-} ions. Write the formula of the compound.

See Problems 2.75 and 2.76.

Figure 2.23 ▲

Potassium chromate Many compounds of chromium have bright colors, which is the origin of the name of the element. It comes from the Greek word *chroma,* meaning "color."

CONCEPT CHECK 2.4

Classify each of the following as either an ionic or molecular compound.

CH_4 Br_2 $CaCl_2$ KNO_3 CH_4O LiF

Which is the best statement regarding molecular compounds?

- **a** Molecular compounds always contain at least two elements. For example, H_2O is a molecular compound because it contains two elements: hydrogen and oxygen.
- **b** Molecular compounds usually contain carbon.
- **c** Molecular compounds consist of cations and anions.
- **d** A molecular compound is a combination of nonmetal atoms.
- **e** Molecular compounds are gases at room temperature.

Figure 2.24 ▲

Molecular model of urea (CH_4N_2O) Urea was the first organic molecule deliberately synthesized by a chemist from non-organic compounds.

2.7 Organic Compounds

An important class of *molecular substances that contain carbon combined with other elements, such as hydrogen, oxygen, and nitrogen,* is **organic compounds.** Organic chemistry is the area of chemistry that is concerned with these compounds (Chapter 23 is devoted to this topic). Historically, organic compounds were restricted to those that could be produced only from living entities and were thought to contain a "vital force" based on their natural origin. The concept of the vital force was disproved in 1828 when a German chemist, Friedrick Wöhler, synthesized urea (a molecular compound in human urine, CH_4N_2O, Figure 2.24) from the molecular compounds ammonia (NH_3) and cyanic acid (HNCO). His work clearly demonstrated that a given compound is exactly the same whether it comes from a living entity or is synthesized.

Organic compounds make up the majority of all known compounds. Since 1957, more than 13 million (60%) of the recorded substances in an international materials registry have been listed as organic. You encounter organic compounds in both living and nonliving materials every day. The proteins, amino acids, enzymes, and DNA that make up your body are all either individual organic molecules or contain organic molecules. Table sugar, peanut oil, antibiotic medicines, and methanol (windshield washer) are all examples of organic molecules as well. Organic chemistry and the compounds produced by the reactions of organic molecules are probably responsible for the majority of the materials that currently surround you as you read this page of text.

The simplest organic compounds are hydrocarbons. **Hydrocarbons** are *those compounds containing only hydrogen and carbon.* Common examples include methane (CH_4), ethane (C_2H_6), propane (C_3H_8), acetylene (C_2H_2), and benzene (C_6H_6). Hydrocarbons are used extensively as sources of energy for heating our homes, for powering internal combustion engines, and for generating electricity. They also are the starting materials for most plastics. Much of the mobility and comfort of our current civilization is built on the low cost and availability of hydrocarbons.

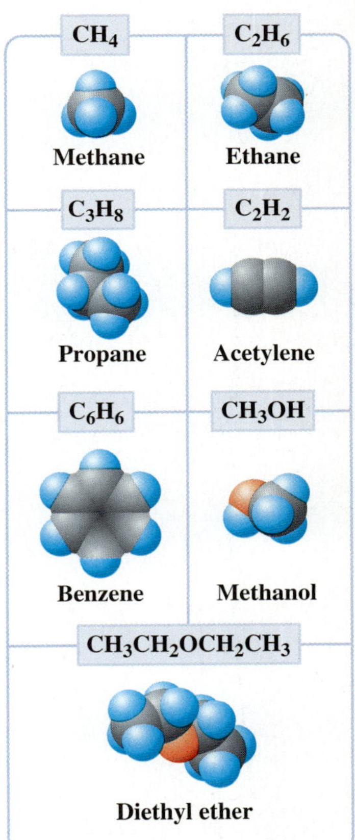

The chemistry of organic molecules is often determined by groups of atoms in the molecule that have characteristic chemical properties. A **functional group** is *a reactive portion of a molecule that undergoes predictable reactions.* When you use the term *alcohol* when referring to a molecular compound, you actually are indicating a molecule that contains an —OH functional group. Methyl alcohol has the chemical formula CH_3OH. The term *ether* indicates that an organic molecule contains an oxygen atom between two carbon atoms, as in diethyl ether ($CH_3CH_2OCH_2CH_3$). Table 2.3 contains a few examples of organic functional groups along with example compounds.

Chapter 23 contains a much more extensive treatment of organic chemistry. Due to its importance and its relation to living organisms, a substantial part of your future chemistry experience will likely be in the field of organic chemistry.

2.8 Naming Simple Compounds

Before the structural basis of chemical substances became established, compounds were named after people, places, or particular characteristics. Examples are Glauber's salt (sodium sulfate, discovered by J. R. Glauber), sal ammoniac (ammonium chloride, named after the ancient Egyptian deity Ammon from the temple near which the substance was made), and washing soda (sodium carbonate, used for softening wash water). Today several million compounds are known and thousands of new ones are discovered every year. Without a system for naming compounds, coping with this multitude of substances would be a hopeless task. **Chemical nomenclature** is *the systematic naming of chemical compounds.*

Table 2.3 Examples of Organic Functional Groups

Functional Group	Name of Functional Group	Example Molecule	Common Use
—OH	Alcohol	Methyl alcohol (CH_3OH)	Windshield washer
—O—	Ether	Dimethyl ether (CH_3OCH_3)	Solvent
—COOH	Carboxylic acid	Acetic acid (CH_3COOH)	Acid in vinegar

CONCEPT CHECK 2.5

Identify the following compounds as being a hydrocarbon, an alcohol, an ether, or a carboxylic acid.

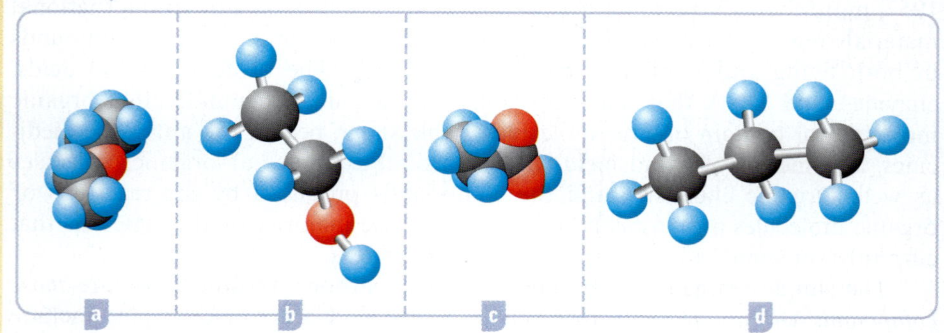

If a compound is not classified as organic, as discussed in Section 2.7, it must be inorganic. **Inorganic compounds** are *composed of elements other than carbon.* A few notable exceptions to this classification scheme include carbon monoxide, carbon dioxide, carbonates, and the cyanides; all contain carbon and yet are generally considered to be inorganic.

In this section, we discuss the nomenclature of some simple inorganic compounds. We first look at the naming of ionic compounds. Then, we look at the naming of some simple molecular compounds, including binary molecular compounds (molecular compounds of two elements) and acids. Finally, we look at hydrates of ionic compounds. These substances contain water molecules in loose association with ionic compounds.

Ionic Compounds

Ionic compounds, as we saw in the previous section, are substances composed of ions. Most ionic compounds contain metal and nonmetal atoms—for example, NaCl. (The ammonium salts, such as NH_4Cl, are a prominent exception.) You name an ionic compound by giving the name of the cation followed by the name of the anion. For example,

$$\underbrace{\text{potassium}}_{\text{cation name}}\ \underbrace{\text{sulfate}}_{\text{anion name}}$$

Before you can name ionic compounds, you need to be able to write and name ions.

The simplest ions are monatomic. A **monatomic ion** is *an ion formed from a single atom.* Table 2.4 lists common monatomic ions of the main-group elements. You may want to look at the table while you read first the rules for predicting the charges on such ions and then the rules for naming the monatomic ions.

Table 2.4 Common Monatomic Ions of the Main-Group Elements*

	IA	IIA	IIIA	IVA	VA	VIA	VIIA
Period 1							H^-
Period 2	Li^+	Be^{2+}	B	C	N^{3-}	O^{2-}	F^-
Period 3	Na^+	Mg^{2+}	Al^{3+}	Si	P	S^{2-}	Cl^-
Period 4	K^+	Ca^{2+}	Ga^{3+}	Ge	As	Se^{2-}	Br^-
Period 5	Rb^+	Sr^{2+}	In^{3+}	Sn^{2+}	Sb	Te^{2-}	I^-
Period 6	Cs^+	Ba^{2+}	Tl^+, Tl^{3+}	Pb^{2+}	Bi^{3+}		

*Elements shown in color do not normally form compounds having monatomic ions.

Rules for Predicting the Charges on Monatomic Ions

1. Most of the main-group metallic elements have one monatomic cation with a charge equal to the group number in the periodic table (the Roman numeral). Example: aluminum, in Group IIIA, has a monatomic ion Al^{3+}.
2. Some metallic elements of high atomic number are exceptions to the previous rule; they have more than one cation. These elements have common cations with a charge equal to the group number minus 2, in addition to having a cation with a charge equal to the group number. Example: The common ion of lead is Pb^{2+}. (The group number is 4; the charge is 4 − 2.) In addition to compounds containing Pb^{2+}, some lead compounds contain Pb^{4+}.
3. Most transition elements form more than one monatomic cation, each with a different charge. Most of these elements have one ion with a charge of +2. Example: Iron has common cations Fe^{2+} and Fe^{3+}. Copper has common cations Cu^{+} and Cu^{2+}.
4. The charge on a monatomic anion for a nonmetallic main-group element equals the group number minus 8. Example: Oxygen has the monatomic anion O^{2-}. (The group number is 6; the charge is 6 − 8.)

Rules for Naming Monatomic Ions

1. Monatomic cations are named after the element if there is only one such ion. Example: Al^{3+} is called aluminum ion; Na^{+} is called sodium ion.
2. If there is more than one monatomic cation of an element, Rule 1 is not sufficient. The *Stock system* of nomenclature names the cations after the element, as in Rule 1, but follows this by a Roman numeral in parentheses denoting the charge on the ion. Example: Fe^{2+} is called iron(II) ion and Fe^{3+} is called iron(III) ion. ▶

   In an older system of nomenclature, such ions are named by adding the suffixes *-ous* and *-ic* to a stem name of the element (which may be from the Latin) to indicate the ions of lower and higher charge, respectively. Example: Fe^{2+} is called ferrous ion; Fe^{3+}, ferric ion. Cu^{+} is called cuprous ion; Cu^{2+}, cupric ion.

   Table 2.5 lists some common cations of the transition elements. Most of these elements have more than one ion, so require the Stock nomenclature system or the older suffix system. A few, such as zinc, have only a single ion that is normally encountered, and you usually name them by just the metal name. You would not be wrong, however, if, for example, you named Zn^{2+} as zinc(II) ion.
3. The names of the monatomic anions are obtained from a stem name of the element followed by the suffix *-ide*. Example: Br^{-} is called bromide ion, from the stem name *brom-* for bromine and the suffix *-ide*.

The Roman numeral actually denotes the oxidation state, or oxidation number, of the atom in the compound. For a monatomic ion, the oxidation state equals the charge. Otherwise, the oxidation state is a hypothetical charge assigned in accordance with certain rules; see Section 4.5.

A **polyatomic ion** is *an ion consisting of two or more atoms chemically bonded together and carrying a net electric charge.* Table 2.6 lists some common polyatomic ions. The first two are cations (Hg_2^{2+} and NH_4^{+}); the rest are anions. There are

Table 2.5 Common Cations of the Transition Elements

Ion	Ion Name	Ion	Ion Name	Ion	Ion Name
Cr^{3+}	Chromium(III) or chromic	Co^{2+}	Cobalt(II) or cobaltous	Zn^{2+}	Zinc
Mn^{2+}	Manganese(II) or manganous	Ni^{2+}	Nickel(II) or nickel	Ag^{+}	Silver
Fe^{2+}	Iron(II) or ferrous	Cu^{+}	Copper(I) or cuprous	Cd^{2+}	Cadmium
Fe^{3+}	Iron(III) or ferric	Cu^{2+}	Copper(II) or cupric	Hg^{2+}	Mercury(II) or mercuric

Table 2.6 Some Common Polyatomic Ions

Name	Formula	Name	Formula
Mercury(I) or mercurous	Hg_2^{2+}	Permanganate	MnO_4^-
Ammonium	NH_4^+	Nitrite	NO_2^-
Cyanide	CN^-	Nitrate	NO_3^-
Carbonate	CO_3^{2-}	Hydroxide	OH^-
Hydrogen carbonate (or bicarbonate)	HCO_3^-	Peroxide	O_2^{2-}
Acetate	$C_2H_3O_2^-$	Phosphate	PO_4^{3-}
Oxalate	$C_2O_4^{2-}$	Monohydrogen phosphate	HPO_4^{2-}
Hypochlorite	ClO^-	Dihydrogen phosphate	$H_2PO_4^-$
Chlorite	ClO_2^-	Sulfite	SO_3^{2-}
Chlorate	ClO_3^-	Sulfate	SO_4^{2-}
Perchlorate	ClO_4^-	Hydrogen sulfite (or bisulfite)	HSO_3^-
Chromate	CrO_4^{2-}	Hydrogen sulfate (or bisulfate)	HSO_4^-
Dichromate	$Cr_2O_7^{2-}$	Thiosulfate	$S_2O_3^{2-}$

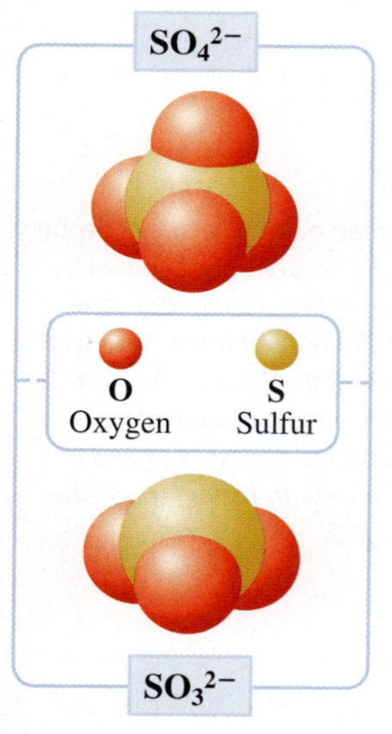

Figure 2.25 ▲

Sulfate and sulfite oxoanions Molecular models of the sulfate *(top)* and sulfite *(bottom)* oxoanions.

no simple rules for writing the formulas of such ions, although you will find it helpful to note a few points.

Most of the ions in Table 2.6 are *oxoanions* (also called *oxyanions*), which consist of oxygen with another element (called the characteristic or central element). Sulfur, for example, forms the oxoanions sulfate ion, SO_4^{2-}, and sulfite ion, SO_3^{2-} (Figure 2.25). The oxoanions in the table are grouped by the characteristic element. For instance, the sulfur ions occur as a group at the end of the table.

Note that the names of the oxoanions have a stem name from the characteristic element, plus a suffix *-ate* or *-ite.* These suffixes denote the relative number of oxygen atoms in the oxoanions of a given characteristic element. The name of the oxoanion with the greater number of oxygen atoms has the suffix *-ate;* the name of the oxoanion with the lesser number of oxygen atoms has the suffix *-ite.* The sulfite ion, SO_3^{2-}, and sulfate ion, SO_4^{2-}, are examples. Another example are the two oxoanions of nitrogen listed in Table 2.6:

NO_2^- nitr<u>ite</u> ion
NO_3^- nitr<u>ate</u> ion

Unfortunately, the suffixes do not tell you the actual number of oxygen atoms in the oxoanion, only the relative number. However, if you were given the formulas of the two anions, you could name them.

The two suffixes *-ite* and *-ate* are not enough when there are more than two oxoanions of a given characteristic element. The oxoanions of chlorine are an example. Table 2.6 lists four oxoanions of chlorine: ClO^-, ClO_2^-, ClO_3^-, and ClO_4^-. In such cases, two prefixes, *hypo-* and *per-,* are used in addition to the two suffixes. The two oxoanions with the least number of oxygen atoms (ClO^- and ClO_2^-) are named using the suffix *-ite,* and the prefix *hypo-* is added to the one of these two ions with the fewer oxygen atoms:

ClO^- <u>hypo</u>chlor<u>ite</u> ion
ClO_2^- chlor<u>ite</u> ion

The two oxoanions with the greatest number of oxygen atoms (ClO_3^- and ClO_4^-) are named using the suffix *-ate,* and the prefix *per-* is added to the one of these two ions with the greater number of oxygen atoms:

ClO_3^- chlor<u>ate</u> ion
ClO_4^- <u>per</u>chlor<u>ate</u> ion

Some of the polyatomic ions in Table 2.6 are oxoanions bonded to one or more hydrogen ions (H^+). They are sometimes referred to as *acid anions,* because acids are substances that provide H^+ ions. As an example, monohydrogen phosphate ion, HPO_4^{2-}, is essentially a phosphate ion (PO_4^{3-}) to which a hydrogen ion (H^+) has bonded. The prefix *mono-*, from the Greek, means "one." Similarly, *di-* is a prefix from the Greek meaning "two," so dihydrogen phosphate ion is a phosphate ion to which two hydrogen ions have bonded. In an older terminology, ions such as hydrogen carbonate and hydrogen sulfate were called bicarbonate and bisulfate, respectively.

The last anion in the table is thiosulfate ion, $S_2O_3^{2-}$. The prefix *thio-* means that an oxygen atom in the root ion name (sulfate, SO_4^{2-}) has been replaced by a sulfur atom.

There are only a few common polyatomic cations. Those listed in Table 2.6 are the mercury(I) ion (also called mercurous ion) and the ammonium ion. Mercury(I) ion is one of the few common metal ions that is not monatomic; its formula is Hg_2^{2+}. The ion charge indicated in parentheses is the charge per metal atom. The ammonium ion, NH_4^+, is one of the few common cations composed of only nonmetal atoms.

Now that we have discussed the naming of ions, let us look at the naming of ionic compounds. Example 2.4 illustrates how you name an ionic compound given its formula.

OWL Interactive Example 2.4 Naming an Ionic Compound from Its Formula

Gaining Mastery Toolbox

Critical Concept 2.4

There are rules that can be applied to naming ionic compounds. In order to be successful in naming ionic compounds, you have to know both the names and charges of the cation and anion that form the compound. Ionic compounds are often metal-nonmetal combinations of atoms.

Solution Essentials:

- Monatomic ions and charges
- Polyatomic ions and charges
- Cation names (monatomic and polyatomic)
- Anion names (monatomic and polyatomic)
- The naming rules of three possible combinations of ions: monatomic ions, monatomic and polyatomic ions, and polyatomic ions.
- Metal and nonmetal
- Main group and transition metal

Name the following: a Mg_3N_2, b $CrSO_4$.

Problem Strategy First, you need to identify the type of compound that you are dealing with. Because each of these compounds contains both metal and nonmetal atoms, we expect them to be ionic, so we need to apply the rules for naming ionic compounds. When naming ionic compounds, start by writing the formulas of the cations and anions in the compound, then name the ions. Look first for any monatomic cations and anions whose charges are predictable. Use the rules given in the text to name the ions. For those metals having more than one cation, you will need to deduce the charge on the metal ion (say the ion of Cr in $CrSO_4$) from the charge on the anion. You will need to memorize or have available the formulas and names of the polyatomic anions to be able to write them from the formula of the compound.

Solution

a Magnesium, a Group IIA metal, is expected to form only a 2+ ion (Mg^{2+}, the magnesium ion). Nitrogen (Group VA) is expected to form an anion of charge equal to the group number minus 8 (N^{3-}, the nitride ion). You can check that these ions would give the formula Mg_3N_2. The name of the compound is **magnesium nitride** (the name of the cation followed by the name of the anion).

b Chromium is a transition element and, like most such elements, has more than one monatomic ion. You can find the charge on the Cr ion if you know the formula of the anion. From Table 2.6, you see that the SO_4 in $CrSO_4$ refers to the anion SO_4^{2-} (the sulfate ion). Therefore, the Cr cation must be Cr^{2+} to give electrical neutrality. The name of Cr^{2+} is chromium(II) ion, so the name of the compound is **chromium(II) sulfate.**

Answer Check Whenever you have to name a compound, you can check your answer to see if the name will lead you back to the correct formula.

Exercise 2.5 Write the names of the following compounds: a CaO, b $PbCrO_4$.

■ See Problems 2.77 and 2.78.

Example 2.5 Writing the Formula from the Name of an Ionic Compound

Gaining Mastery Toolbox

Critical Concept 2.5
The name of the ionic compound provides the names of the ions that form the compound. In the formula the cation name will always be listed first, followed by the anion name. For metals that can have more than one charge, Roman numerals following the cation name indicate the charge.

Solution Essentials:
- Monatomic ions and charges
- Polyatomic ions and charges
- Cation names (monatomic and polyatomic)
- Anion names (monatomic and polyatomic)
- Ionic compound
- Example 2.3
- Metal and nonmetal
- Main group elements and transition metal

Write formulas for the following compounds: a iron(II) phosphate, b titanium(IV) oxide.

Problem Strategy As in all nomenclature problems, first determine the type of compound you are working with. The compounds presented in this problem contain both metal and nonmetal atoms, so they are ionic. Use the name of the compound, obtain the name of the ions, and then write the formulas of the ions. Finally, use the method of Example 2.3 to obtain the formula of the compound from the formulas of the ions.

Solution

a Iron(II) phosphate contains the iron(II) ion, Fe^{2+}, and the phosphate ion, PO_4^{3-}. (You will need to have memorized the formula of the phosphate ion, or have access to Table 2.6.) Once you know the formulas of the ions, you use the method of Example 2.3 to obtain the formula. The formula is $\mathbf{Fe_3(PO_4)_2}$.

b Titanium(IV) oxide is composed of titanium(IV) ions, Ti^{4+}, and oxide ions, O^{2-}. The formula of titanium(IV) oxide is $\mathbf{TiO_2}$.

Answer Check Whenever you have to write the formula of a compound from its name, make sure that the formula you have written will result in the correct name.

Exercise 2.6 A compound has the name thallium(III) nitrate. What is its formula? The symbol of thallium is Tl.

■ See Problems 2.79 and 2.80.

CONCEPT CHECK 2.6

Which of the following are polyatomic ions?

SO_3 SO_4^{2-} NO_2^- NO_2 I_2 I_3^-

Which of the following is a false statement regarding polyatomic ions?

a Polyatomic ions must contain two or more atoms in their chemical formula.
b Polyatomic ions must have either a positive or negative charge.
c Polyatomic ions must have at least two different elements chemically bonded together.
d Polyatomic ions can be combinations of metals and nonmetals.
e Polyatomic ions can be oxoanions.

Binary Molecular Compounds

A **binary compound** is *a compound composed of only two elements.* Binary compounds composed of a metal and a nonmetal are usually ionic and are named as ionic compounds, as we have just discussed. (For example, NaCl, $MgBr_2$, and Al_2N_3 are all binary ionic compounds.) Binary compounds composed of two nonmetals or metalloids are usually molecular and are named using a prefix system. Examples of binary molecular compounds are H_2O, NH_3, and CCl_4. Using this prefix system, you name the two elements using the order given by the formula of the compound.

Order of Elements in the Formula The order of elements in the formula of a binary molecular compound is established by convention. By this convention, the nonmetal or metalloid occurring first in the following sequence is written first in the formula of the compound.

Element	B	Si C	Sb As P N	H	Te Se S	I Br Cl	O	F
Group	IIIA	IVA	VA		VIA	VIIA		

You can reproduce this order easily if you have a periodic table available. You arrange the nonmetals and metalloids in the order of the groups of the periodic table, listing the elements from the bottom of the group upward. Then you place H between Groups VA and VIA, and move O so that it is just before F.

This order places the nonmetals and metalloids approximately in order of increasing nonmetallic character. Thus, in the formula, the first element is the more metallic and the second element is the more nonmetallic, similar to the situation with binary ionic compounds. For example, the compound whose molecule contains three fluorine atoms and one nitrogen atom is written NF_3, not F_3N.

Rules for Naming Binary Molecular Compounds Now let us look at the rules for naming binary molecular compounds by the *prefix system.*

1. The name of the compound usually has the elements in the order given in the formula.
2. You name the first element using the exact element name.
3. You name the second element by writing the stem name of the element with the suffix *-ide* (as if the element occurred as the anion).
4. You add a prefix, derived from the Greek, to each element name to denote the subscript of the element in the formula. (The Greek prefixes are listed in Table 2.7.) Generally, the prefix *mono-* is not used, unless it is needed to distinguish two compounds of the same two elements.

Table 2.7 Greek Prefixes for Naming Compounds

Number	Prefix
1	mono-
2	di-
3	tri-
4	tetra-
5	penta-
6	hexa-
7	hepta-
8	octa-
9	nona-
10	deca-

Consider N_2O_3. This is a binary molecular compound, as you would predict because N and O are nonmetals. According to Rule 1, you name it after the elements, N before O, following the order of elements in the formula. By Rule 2, the N is named exactly as the element (nitrogen), and by Rule 3, the O is named as the anion (oxide). The compound is a nitrogen oxide. Finally, you add prefixes to denote the subscripts in the formula (Rule 4). As Table 2.7 shows, the prefix for two is *di-*, and the prefix for three is *tri-*. The name of the compound is <u>di</u>nitrogen <u>tri</u>oxide.

Here are some examples to illustrate how the prefix *mono-* is used. There is only one compound of hydrogen and chlorine: HCl. It is called hydrogen chloride, not monohydrogen monochloride, since the prefix *mono-* is not generally written. On the other hand, there are two common compounds of carbon and oxygen: CO and CO_2. They are both carbon oxides. To distinguish one from the other, we name them carbon monoxide and carbon dioxide, respectively.

Here are some other examples of prefix names for binary molecular compounds.

SF_4	sulfur tetrafluoride	SF_6	sulfur hexafluoride
ClO_2	chlorine dioxide	Cl_2O_7	dichlorine hept(a)oxide

The final vowel in a prefix is often dropped before a vowel in a stem name, for ease in pronunciation. Tetraoxide, for instance, becomes tetroxide.

A few compounds have older, well-established names. The compound H_2S would be named dihydrogen sulfide by the prefix system, but is commonly called hydrogen sulfide. NO is still called nitric oxide, although its prefix name is nitrogen monoxide. The names water and ammonia are used for H_2O and NH_3, respectively.

Example 2.6 Naming a Binary Compound from Its Formula

Gaining Mastery Toolbox

Critical Concept 2.6
There are rules that can be applied to naming binary molecular compounds. These rules are specific to binary molecular compounds and do not apply to ionic compounds. Molecular compounds are combinations of nonmetal atoms.

Solution Essentials:
- Greek prefixes
- Binary molecular compound naming rules
- Naming order of elements
- Binary compound
- Periodic table
- Nonmetal

Name the following compounds: a N_2O_4, b P_4O_6.

Problem Strategy First, you need to determine the type of compound that you are dealing with: molecular or ionic. In this case the compound contains only nonmetals, so it is molecular. Molecular compounds require that you use the Greek prefixes (Table 2.7), naming the elements in the same order as in the formulas.

Solution

a N_2O_4 contains two nitrogen atoms and four oxygen atoms. Consulting Table 2.7 reveals that the corresponding prefixes are *di-* and *tetra-*, so the name is **dinitrogen tetroxide.**

b P_4O_6 contains four phosphorus atoms and six oxygen atoms. Consulting Table 2.7 reveals that the corresponding prefixes are *tetra-* and *hexa-*, so the name is **tetraphosphorus hexoxide.**

Answer Check Check to see whether the name leads back to the correct formula.

Exercise 2.7 Name the following compounds: a Cl_2O_6, b PCl_3, c PCl_5.

■ See Problems 2.83 and 2.84.

Example 2.7 Writing the Formula from the Name of a Binary Compound

Gaining Mastery Toolbox

Critical Concept 2.7
Greek prefixes are used to determine the correct formula of binary molecular compounds. The elements in the formula are written in the same order as they are presented in the name.

Solution Essentials:
- Greek prefixes
- Binary molecular compound naming rules
- Element symbols
- Periodic table
- Nonmetal

Give the formulas of the following compounds: a disulfur dichloride, b tetraphosphorus trisulfide.

Problem Strategy As always, first determine whether you are working with a molecular or ionic compound. Because these compounds contain only nonmetals, they must be molecular. Given that they are molecular, you will need to use the Greek prefixes in Table 2.7 to give you the corresponding names for each subscript.

Solution You change the names of the elements to symbols and translate the prefixes to subscripts. The formulas are a S_2Cl_2 and b P_4S_3.

Answer Check Check to see whether the formula you have written will result in the correct name.

Exercise 2.8 Give formulas for the following compounds: a carbon disulfide, b sulfur trioxide.

■ See Problems 2.85 and 2.86.

Example 2.8 Naming a Binary Chemical Compound from Its Molecular Model

Gaining Mastery Toolbox

Critical Concept 2.8
Models of molecular compounds depict the number and identity of each element in a single molecule. The molecular formula can be written directly from the information contained in the molecular model.

Solution Essentials:
- Element symbols
- Naming order of elements
- Example 2.6
- Molecular models

Name the chemical compounds shown in the margin at the top of the next page.

Problem Strategy After recognizing that these are molecular compounds, start by writing down the elements using subscripts to denote how many atoms of each element occur in the compound. Next, determine which element should be written first in the chemical formula. (The more metallic element is written first.)

(continued)

(*continued*)

Solution

a The elements in the compound are O and N. There are two O atoms for each N atom, so the chemical formula could be O_2N. Applying the rule for writing the more metallic element first, the formula is rearranged to yield the correct formula, NO_2. Using Greek prefixes and naming the elements in the order in which they appear, we get the correct chemical formula: **nitrogen dioxide.**

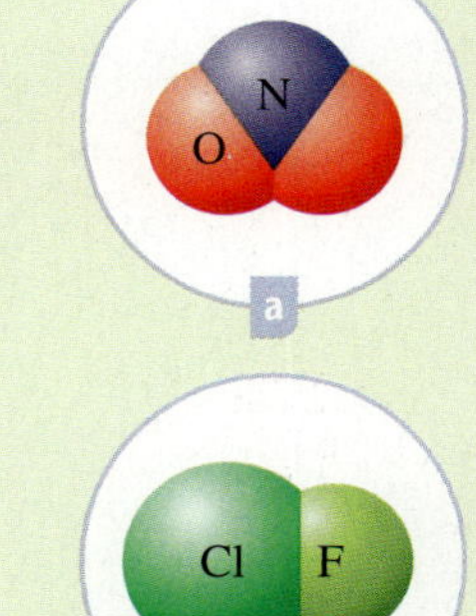

b Following the same procedure as part a, we get **chlorine monofluoride.**

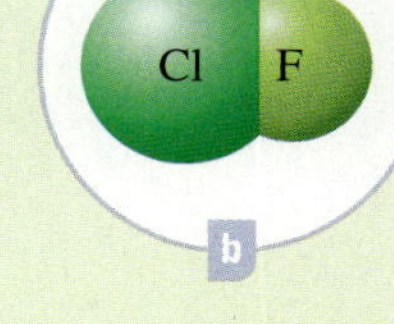

Answer Check Make sure that the answer reflects the correct number of each of the elements given in the molecular structure.

Exercise 2.9 Using the molecular models, name the following chemical compounds:

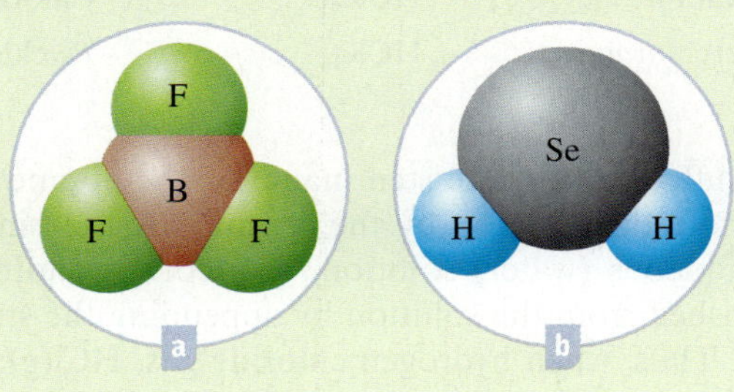

■ See Problems 2.87 and 2.88.

Acids and Corresponding Anions

Acids are an important class of compounds, and we will look closely at them in the next chapter. Here we want merely to see how we name these compounds and how they are related to the anions that we encounter in some ionic compounds. For our present purposes, an *acid* is a molecular compound that yields hydrogen ions, H^+, and an anion for each acid molecule when the acid dissolves in water. An example is nitric acid, HNO_3. The HNO_3 molecule yields one H^+ ion and one nitrate ion, NO_3^-, in aqueous (water) solution.

Nitric acid, HNO_3, is an oxoacid, or oxyacid (Figure 2.26). An **oxoacid** is *an acid containing hydrogen, oxygen, and another element (often called the central atom)*. In water the oxoacid molecule yields one or more hydrogen ions, H^+, and an oxoanion. The names of the oxoacids are related to the names of the corresponding oxoanions. If you know the name of the oxoanion, you can obtain the name of the corresponding acid by replacing the suffix as follows:

Anion Suffix	*Acid Suffix*
-ate	*-ic*
-ite	*-ous*

The following diagram illustrates how to apply this information to naming oxoacids.

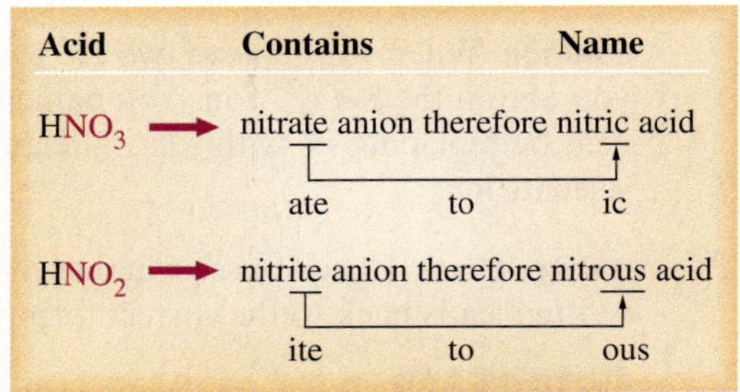

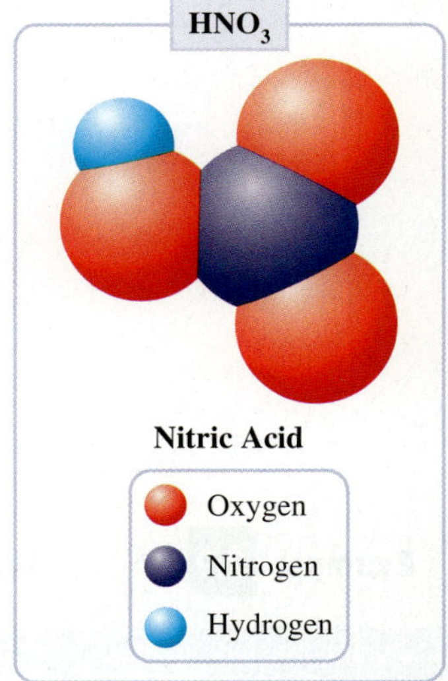

Figure 2.26 ▲
Molecular model of nitric acid

Table 2.8 lists some oxoanions and their corresponding oxoacids.

Some binary compounds of hydrogen and nonmetals yield acidic solutions when dissolved in water. These *solutions* are named like compounds by using the

Table 2.8 Some Oxoanions and Their Corresponding Oxoacids

Oxoanion		Oxoacid	
CO_3^{2-}	Carbon*ate ion*	H_2CO_3	Carbon*ic acid*
NO_2^-	Nitr*ite ion*	HNO_2	Nitr*ous acid*
NO_3^-	Nitr*ate ion*	HNO_3	Nitr*ic acid*
PO_4^{3-}	Phosph*ate ion*	H_3PO_4	Phosphor*ic acid*
SO_3^{2-}	Sulf*ite ion*	H_2SO_3	Sulfur*ous acid*
SO_4^{2-}	Sulf*ate ion*	H_2SO_4	Sulfur*ic acid*
ClO^-	*Hypo*chlor*ite ion*	$HClO$	*Hypo*chlor*ous acid*
ClO_2^-	Chlor*ite ion*	$HClO_2$	Chlor*ous acid*
ClO_3^-	Chlor*ate ion*	$HClO_3$	Chlor*ic acid*
ClO_4^-	*Per*chlor*ate ion*	$HClO_4$	*Per*chlor*ic acid*

prefix *hydro-* and the suffix *-ic* with the stem name of the nonmetal, followed by the word *acid.* We denote the solution by the formula of the binary compound followed by (*aq*) for aqueous (water) solution. The corresponding binary compound can be distinguished from the solution by appending the state of the compound to the formula. Thus, when hydrogen chloride gas, HCl(*g*), is dissolved in water, it forms hydrochloric acid, HCl(*aq*).

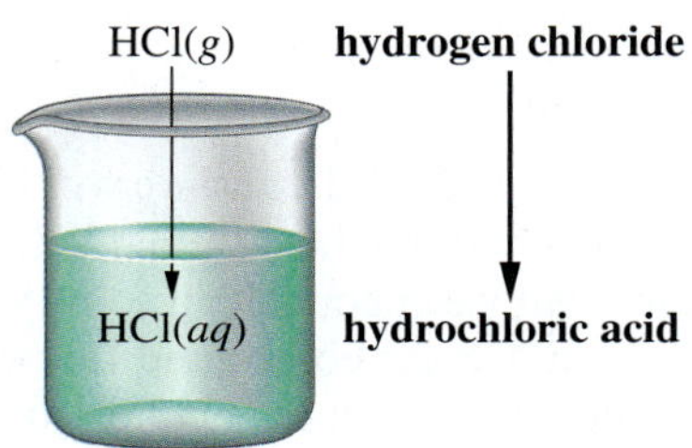

Here are some other examples:

Binary Compound	***Acid Solution***
HBr(*g*), hydrogen bromide	*hydro*brom*ic acid,* HBr(*aq*)
HF(*g*), hydrogen fluoride	*hydro*fluor*ic acid,* HF(*aq*)

Example 2.9 Writing the Name and Formula of an Anion from the Acid

Gaining Mastery Toolbox

Critical Concept 2.9
There are rules that can be applied to naming acids. Depending on the type of acid, there are two sets of rules: one for oxoacids and one for acids that are binary compounds of hydrogen and a nonmetal.

Solution Essentials:
- Recognition of the chemical formula of an acid
- Oxoacid
- Naming oxoacid from oxoanion name
- Polyatomic anion names

Selenium has an oxoacid, H_2SeO_4, called selenic acid. What is the formula and name of the corresponding anion?

Problem Strategy Knowing that H_2SeO_4 is an oxyacid, we must first determine the oxoanion name from the examples provided in Table 2.8. From the oxoanion name you determine whether a prefix (*per-* or *hypo-*) is necessary and whether the proper suffix is *-ite* or *-ate*. You then name the anion.

Solution When you remove two H^+ ions from H_2SeO_4, you obtain the $\mathbf{SeO_4^{2-}}$ ion. You name the ion from the acid by replacing *-ic* with *-ate*. The anion is called the **selenate ion.**

Answer Check Check to see whether the name you have written leads back to the correct formula.

Exercise 2.10 What are the name and formula of the anion corresponding to perbromic acid, $HBrO_4$?

See Problems 2.89 and 2.90.

Hydrates

A **hydrate** is *a compound that contains water molecules weakly bound in its crystals.* These substances are often obtained by evaporating an aqueous solution of the compound. Consider copper(II) sulfate. When an aqueous solution of this substance is evaporated, blue crystals form in which each formula unit of copper(II) sulfate, $CuSO_4$, is associated with five molecules of water. The formula of the hydrate is written $CuSO_4{\cdot}5H_2O$, where a centered dot separates $CuSO_4$ and $5H_2O$. When the blue crystals of the hydrate are heated, the water is driven off, leaving behind white crystals of copper(II) sulfate without associated water, a substance called *anhydrous* copper(II) sulfate (see Figure 2.27).

Hydrates are named from the anhydrous compound, followed by the word *hydrate* with a prefix to indicate the number of water molecules per formula unit of the compound. For example, $CuSO_4{\cdot}5H_2O$ is known as copper(II) sulfate pentahydrate.

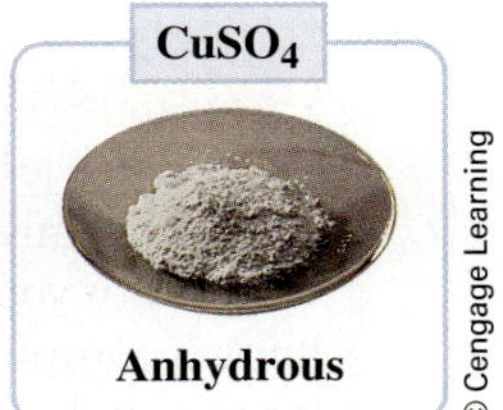

Figure 2.27 ◀ **Copper(II) sulfate** The hydrate $CuSO_4{\cdot}5H_2O$ is blue; the anhydrous compound, $CuSO_4$, is white.

Example 2.10 Naming a Hydrate from Its Formula

Gaining Mastery Toolbox

Critical Concept 2.10
Hydrates are ionic compounds that contain water molecules as part of the chemical formula. Naming hydrates requires that you first apply the rules of naming ionic compounds and then account for the number of water molecules in the chemical formula.

Solution Essentials:
- Recognition of the chemical formula of a hydrate
- Rules for naming hydrates
- Rules for naming ionic compounds
- Greek prefixes

Epsom salts has the formula $MgSO_4{\cdot}7H_2O$. What is the chemical name of the substance?

Problem Strategy You need to identify the type of compound that you are dealing with, ionic or molecular. Because this compound contains both metals and nonmetals, it is ionic. However, this compound has the additional feature that it contains water molecules. Therefore, you need to apply ionic naming rules and then use Greek prefixes (Table 2.7) to indicate the number of water molecules that are part of the formula. The word *hydrate* will be used to represent water in the name.

Solution $MgSO_4$ is magnesium sulfate. $MgSO_4{\cdot}7H_2O$ is **magnesium sulfate heptahydrate.**

Answer Check Make sure that the name leads back to the formula that you started with.

Exercise 2.11 Washing soda has the formula $Na_2CO_3{\cdot}10H_2O$. What is the chemical name of this substance?

■ See Problems 2.91 and 2.92.

Example 2.11 Writing the Formula from the Name of a Hydrate

Gaining Mastery Toolbox

Critical Concept 2.11
A hydrate formula has two parts: the formula of the ionic compound and the number of water molecules bound to the ionic compound. A centered dot is used to separate the two parts, and Greek prefixes are used to indicate the number of water molecules.

Solution Essentials:
- Rules for naming hydrates
- Rules for naming ionic compounds
- Greek prefixes

The mineral gypsum has the chemical name calcium sulfate dihydrate. What is the chemical formula of this substance?

Problem Strategy You need to identify the type of compound that you are dealing with, ionic or molecular. Because this compound contains both metals and nonmetals, it is ionic. However, this compound has the additional feature that it contains water molecules. Therefore, we need to apply ionic naming rules and then use Greek prefixes (Table 2.7) to indicate the number of water molecules that are part of the formula. The word *hydrate* will be used to represent water in the name.

Solution Calcium sulfate is composed of calcium ions (Ca^{2+}) and sulfate ions (SO_4^{2-}), so the formula of the anhydrous compound is $CaSO_4$. Since the mineral is a *di*hydrate, the formula of the compound is $\mathbf{CaSO_4{\cdot}2H_2O}$.

(continued)

(continued)

Answer Check Make sure that the formula will result in the correct name that you started with.

Exercise 2.12 Photographers' hypo, used to fix negatives during the development process, is sodium thiosulfate pentahydrate. What is the chemical formula of this compound?

■ See Problems 2.93 and 2.94.

CONCEPT CHECK 2.7

You take a job with the U.S. Environmental Protection Agency (EPA) inspecting college chemistry laboratories. On the first day of the job while inspecting Generic University you encounter a bottle with the formula Al_2Q_3 that was used as an unknown compound in an experiment. Before you send the compound off to the EPA lab for analysis, you want to narrow down the possibilities for element Q. What are the likely (real element) candidates for element Q?

Chemical Reactions: Equations

An important feature of atomic theory is its explanation of a chemical reaction as a rearrangement of the atoms of substances. The reaction of sodium metal with chlorine gas described in the chapter opening involves the rearrangement of the atoms of sodium and chlorine to give the new combination of atoms in sodium chloride. Such a rearrangement of atoms is conveniently represented by a chemical equation, which uses chemical formulas.

2.9 Writing Chemical Equations

A **chemical equation** is *the symbolic representation of a chemical reaction in terms of chemical formulas.* For example, the burning of sodium in chlorine to produce sodium chloride is written

$$2Na + Cl_2 \longrightarrow 2NaCl$$

The formulas on the left side of an equation (before the arrow) represent the reactants; a **reactant** is *a starting substance in a chemical reaction.* The arrow means "react to form" or "yield." The formulas on the right side represent the products; a **product** is *a substance that results from a reaction.* Note the *coefficient* of 2 in front of the formula NaCl. Coefficients in front of the formula give the relative number of molecules or formula units involved in the reaction. When no coefficient is written, it is understood to be 1.

In many cases, it is useful to indicate the states or phases of the substances in an equation. You do this by placing appropriate labels indicating the phases within parentheses following the formulas of the substances. You use the following phase labels:

$$(g) = \text{gas},\ (l) = \text{liquid},\ (s) = \text{solid},\ (aq) = \text{aqueous (water) solution}$$

When you use these labels, the previous equation becomes

$$2Na(s) + Cl_2(g) \longrightarrow 2NaCl(s)$$

You can also indicate in an equation the conditions under which a reaction takes place. If the reactants are heated to make the reaction go, you can indicate this by putting the symbol Δ (capital Greek delta) over the arrow. For example, the equation

$$2NaNO_3(s) \xrightarrow{\Delta} 2NaNO_2(s) + O_2(g)$$

indicates that solid sodium nitrate, $NaNO_3$, decomposes when heated to give solid sodium nitrite, $NaNO_2$, and oxygen gas, O_2.

When an aqueous solution of hydrogen peroxide, H_2O_2, comes in contact with platinum metal, Pt, the hydrogen peroxide decomposes into water and oxygen gas. The platinum acts as a *catalyst,* a substance that speeds up a reaction without undergoing any net change itself. You write the equation for this reaction as follows. The catalyst, Pt, is written over the arrow. ▶

$$2H_2O_2(aq) \xrightarrow{Pt} 2H_2O(l) + O_2(g)$$

You also can combine the symbol Δ with the formula of a catalyst, placing one above the arrow and the other below the arrow.

Platinum is a silvery-white metal used for jewelry. It is also valuable as a catalyst for many reactions, including those that occur in the catalytic converters of automobiles.

2.10 Balancing Chemical Equations

When the coefficients in a chemical equation are correctly given, the numbers of atoms of each element are equal on both sides of the arrow. The equation is then said to be *balanced.* That a chemical equation should be balanced follows from atomic theory. A chemical reaction involves simply a recombination of the atoms; none are destroyed and none are created. Consider the burning of natural gas, which is composed mostly of methane, CH_4. Using atomic theory, you describe this as the chemical reaction of one molecule of methane, CH_4, with two molecules of oxygen, O_2, to form one molecule of carbon dioxide, CO_2, and two molecules of water, H_2O.

CH_4	+	$2O_2$	$\longrightarrow$	CO_2	+	$2H_2O$
one molecule of methane	+	two molecules of oxygen	react to form	one molecule of carbon dioxide	+	two molecules of water

Figure 2.28 represents the reaction in terms of molecular models.

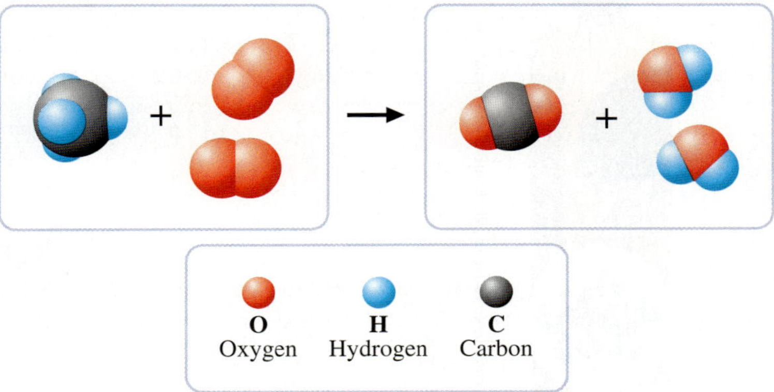

Figure 2.28 ◀

Representation of the reaction of methane with oxygen Molecular models represent the reaction of CH_4 with O_2 to give CO_2 and H_2O.

Before you can write a balanced chemical equation for a reaction, you must determine *by experiment* those substances that are reactants and those that are products. You must also determine the formulas of each substance. Once you know these things, you can write the balanced chemical equation.

As an example, consider the burning of propane gas (Figure 2.29). By experiment, you determine that propane reacts with oxygen in air to give carbon dioxide and water. You also determine from experiment that the formulas of propane, oxygen, carbon dioxide, and water are C_3H_8, O_2, CO_2, and H_2O, respectively. Then you can write

$$C_3H_8 + O_2 \longrightarrow CO_2 + H_2O$$

This equation is not balanced because the coefficients that give the relative number of molecules involved have not yet been determined. *To balance the equation, you select coefficients that will make the numbers of atoms of each element equal on both sides of the equation.*

Because there are three carbon atoms on the left side of the equation (C_3H_8), you must have three carbon atoms on the right. This is achieved by writing 3 for the coefficient of CO_2.

$$\underline{1}C_3H_8 + O_2 \longrightarrow \underline{3}CO_2 + H_2O$$

(We have written the coefficient for C_3H_8 to show that it is now determined; normally when the coefficient is 1, it is omitted.) Similarly, there are eight hydrogen atoms on the left (in C_3H_8), so you write 4 for the coefficient of H_2O.

$$\underline{1}C_3H_8 + O_2 \longrightarrow \underline{3}CO_2 + \underline{4}H_2O$$

The coefficients on the right are now determined, and there are ten oxygen atoms on this side of the equation: 6 from the three CO_2 molecules and 4 from the four

Figure 2.29 ▶

The burning of propane gas

Propane gas, C_3H_8, from the tank burns by reacting with oxygen, O_2, in air to give carbon dioxide, CO_2, and water, H_2O.

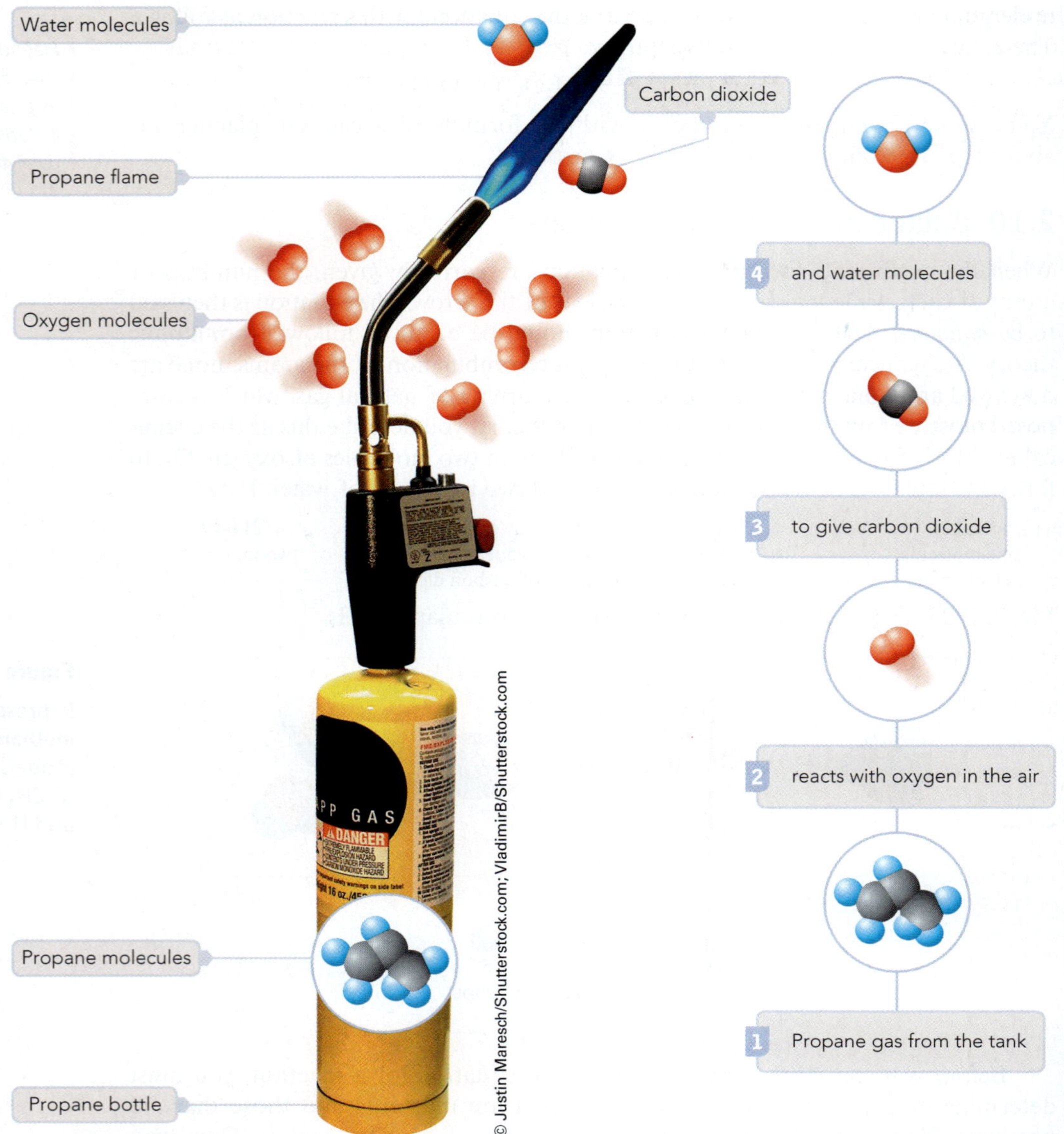

H_2O molecules. You now write 5 for the coefficient of O_2. We drop the coefficient 1 from C_3H_8 and have the balanced equation.

$$C_3H_8 + 5O_2 \longrightarrow 3CO_2 + 4H_2O$$

It is a good idea to check your work by counting the number of atoms of each element on the left side and then on the right side of the equation.

The combustion of propane can be equally well represented by

$$2C_3H_8 + 10O_2 \longrightarrow 6CO_2 + 8H_2O$$

in which the coefficients of the previous equation have been doubled. Usually, however, it is preferable to *write the coefficients so that they are the smallest whole numbers possible.* When we refer to a balanced equation, we will assume that this is the case unless otherwise specified.

The method we have outlined—called *balancing by inspection*—is essentially a trial-and-error method. ◀ It can be made easier, however, by observing the following rule:

Balancing equations this way is relatively quick for all but the most complicated equations. A special method for such equations is described in Chapter 19.

Balance first the atoms for elements that occur in only one substance on each side of the equation.

The usefulness of this rule is illustrated in the next example. Remember that in any formula, such as $Fe_2(SO_4)_3$, a subscript to the right of the parentheses multiplies *each* subscript within the parentheses. Thus, $Fe_2(SO_4)_3$ represents 4×3 oxygen atoms. Remember also that you cannot change a subscript in any formula; only the coefficients can be altered to balance an equation.

OWL Interactive Example 2.12 Balancing Simple Equations

Gaining Mastery Toolbox

Critical Concept 2.12
When a chemical reaction occurs, elements are neither created nor destroyed; they are rearranged. A balanced chemical equation is one that correctly represents the substances (elements and compounds) before and after a reaction. When the equation is balanced, the starting numbers of atoms of elements (reactants) equal those after the reaction (products).

Solution Essentials:

- Balancing chemical equations by inspection
- Balanced chemical equation
- Chemical equation
- Reactants
- Products

Balance the following equations.

a $H_3PO_3 \longrightarrow H_3PO_4 + PH_3$ **b** $Ca + H_2O \longrightarrow Ca(OH)_2 + H_2$

c $Fe_2(SO_4)_3 + NH_3 + H_2O \longrightarrow Fe(OH)_3 + (NH_4)_2SO_4$

Problem Strategy Look at the chemical equation for atoms of elements that occur in only one substance on each side of the equation. Begin by balancing the equation in one of these atoms. Use the number of atoms on the left side of the arrow as the coefficient of the substance containing that element on the right side, and vice versa. After balancing one element in an equation, the rest become easier.

Solution

a Oxygen occurs in just one of the products (H_3PO_4). It is therefore easiest to balance O atoms first. To do this, note that H_3PO_3 has three O atoms; use 3 as the coefficient of H_3PO_4 on the right side. There are four O atoms in H_3PO_4 on the right side; use 4 as the coefficient of H_3PO_3 on the left side.

$$\underline{4}H_3PO_3 \longrightarrow \underline{3}H_3PO_4 + PH_3$$

This equation is now also balanced in P and H atoms. Thus, the balanced equation is

$$\mathbf{4H_3PO_3 \longrightarrow 3H_3PO_4 + PH_3}$$

b The equation is balanced in Ca atoms as it stands.

$$\underline{1}Ca + H_2O \longrightarrow \underline{1}Ca(OH)_2 + H_2$$

O atoms occur in only one reactant and in only one product, so they are balanced next.

$$\underline{1}Ca + \underline{2}H_2O \longrightarrow \underline{1}Ca(OH)_2 + H_2$$

The equation is now also balanced in H atoms. Thus, the answer is

$$\mathbf{Ca + 2H_2O \longrightarrow Ca(OH)_2 + H_2}$$

c To balance the equation in Fe atoms, you write

$$\underline{1}Fe_2(SO_4)_3 + NH_3 + H_2O \longrightarrow \underline{2}Fe(OH)_3 + (NH_4)_2SO_4$$

You balance the S atoms by placing the coefficient 3 for $(NH_4)_2SO_4$.

$$\underline{1}Fe_2(SO_4)_3 + \underline{6}NH_3 + H_2O \longrightarrow \underline{2}Fe(OH)_3 + \underline{3}(NH_4)_2SO_4$$

Now you balance the N atoms.

$$\underline{1}Fe_2(SO_4)_3 + \underline{6}NH_3 + H_2O \longrightarrow \underline{2}Fe(OH)_3 + \underline{3}(NH_4)_2SO_4$$

To balance the O atoms, you first count the number of O atoms on the right (18). Then you count the number of O atoms in substances on the left with known coefficients. There are 12 O's in $Fe_2(SO_4)_3$; hence, the number of remaining O's (in H_2O) must be $18 - 12 = 6$.

$$\underline{1}Fe_2(SO_4)_3 + \underline{6}NH_3 + \underline{6}H_2O \longrightarrow \underline{2}Fe(OH)_3 + \underline{3}(NH_4)_2SO_4$$

Finally, note that the equation is now balanced in H atoms. The answer is

$$\mathbf{Fe_2(SO_4)_3 + 6NH_3 + 6H_2O \longrightarrow 2Fe(OH)_3 + 3(NH_4)_2SO_4}$$

(*continued*)

(continued)

Answer Check As a final step when balancing chemical equations, always count the number of atoms of each element on both sides of the equation to make certain that they are equal.

Exercise 2.13 Find the coefficients that balance the following equations.

a $O_2 + PCl_3 \longrightarrow POCl_3$

b $P_4 + N_2O \longrightarrow P_4O_6 + N_2$

c $As_2S_3 + O_2 \longrightarrow As_2O_3 + SO_2$

d $Ca_3(PO_4)_2 + H_3PO_4 \longrightarrow Ca(H_2PO_4)_2$

See Problems 2.97 and 2.98.

A Checklist for Review

OWL and Go Chemistry Sign in at www.cengage.com/owl to:

- View tutorials and simulations, develop problem-solving skills, and complete online homework assigned by your professor.
- For quick review and exam prep, download Go Chemistry mini lecture modules from OWL or purchase them at www.cengagebrain.com.

Summary of Facts and Concepts

Atomic theory is central to chemistry. According to this theory, all matter is composed of small particles, or *atoms.* Each *element* is composed of the same kind of atom, and a *compound* is composed of two or more elements chemically combined in fixed proportions. A *chemical reaction* consists of the rearrangement of the atoms present in the reacting substances to give new chemical combinations present in the substances formed by the reaction. Although Dalton considered atoms to be the ultimate particles of matter, we now know that atoms themselves have structure. An atom has a *nucleus* and *electrons.* These atomic particles were discovered by J. J. Thomson and Ernest Rutherford.

Thomson established that cathode rays consist of negatively charged particles, or electrons, that are constituents of all atoms. Thomson measured the mass-to-charge ratio of the electron, and later Millikan measured its charge. From these measurements, the electron was found to be more than 1800 times lighter than the lightest atom.

Rutherford proposed the nuclear model of the atom to account for the results of experiments in which alpha particles were scattered from metal foils. According to this model, the atom consists of a central core, or nucleus, around which the electrons exist. The nucleus has most of the mass of the atom and consists of *protons* (with a positive charge) and *neutrons* (with no charge). Each chemically distinct atom has a nucleus with a specific number of protons (*atomic number*), and around the nucleus in the neutral atom are an equal number of electrons. The number of protons plus neutrons in a nucleus equals the *mass number.* Atoms whose nuclei have the same number of protons but different numbers of neutrons are called *isotopes.*

The *atomic mass* can be calculated from the isotopic masses and *fractional abundances* of the isotopes in a naturally occurring element. These data can be determined by the use of a mass spectrometer, which separates ions according to their mass-to-charge ratios.

The elements can be arranged in rows and columns by atomic number to form the *periodic table.* Elements in a given *group* (column) have similar properties. (A *period* is a row in the periodic table.) Elements on the left and at the center of the table are *metals;* those on the right are *nonmetals.*

A *chemical formula* is a notation used to convey the relative proportions of the atoms of the different elements in a substance. If the substance is molecular, the formula gives the precise number of each kind of atom in the molecule. If the substance is ionic, the formula also gives the relative number of different ions in the compound.

Chemical nomenclature is the systematic naming of compounds based on their formulas or structures. Rules are given for naming *ionic compounds, binary molecular compounds, acids,* and *hydrates.*

A chemical reaction occurs when the atoms in substances rearrange and combine into new substances. We represent a reaction by a *chemical equation,* writing a chemical formula for each reactant and product. The coefficients in the equation indicate the relative numbers of reactant and product molecules or formula units. Once the reactants and products and their formulas have been determined by experiment, we determine the coefficients by *balancing* the numbers of each kind of atom on both sides of the equation.

Learning Objectives	Important Terms
2.1 Atomic Theory of Matter	
▪ List the postulates of atomic theory. ▪ Define *element, compound,* and *chemical reaction* in the context of these postulates. ▪ Recognize the atomic symbols of the elements. ▪ Explain the significance of the law of multiple proportions.	**atomic theory** **atom** **element** **compound** **chemical reaction** **atomic symbol** **law of multiple proportions**
2.2 The Structure of the Atom	
▪ Describe Thomson's experiment in which he discovered the electron. ▪ Describe Rutherford's experiment that led to the nuclear model of the atom.	**nucleus** **electron**
2.3 Nuclear Structure; Isotopes	
▪ Name and describe the nuclear particles making up the nucleus of the atom. ▪ Define *atomic number, mass number,* and *nuclide.* ▪ Write the nuclide symbol for a given nuclide. ▪ Define and provide examples of *isotopes* of an element. ▪ Write the nuclide symbol of an element. Example 2.1	**proton** **atomic number (Z)** **neutron** **mass number (A)** **nuclide** **isotope**
2.4 Atomic Masses	
▪ Define *atomic mass unit* and *atomic mass.* ▪ Describe how a *mass spectrometer* can be used to determine the *fractional abundance* of the isotopes of an element. ▪ Determine the atomic mass of an element from the isotopic masses and fractional abundances. Example 2.2	**atomic mass unit (amu)** **atomic mass** **fractional abundance**
2.5 Periodic Table of the Elements	
▪ Identify periods and groups on the periodic table. ▪ Find the *main-group* and *transition* elements on the periodic table. ▪ Locate the *alkali metal* and *halogen* groups on the periodic table. ▪ Recognize the portions of the periodic table that contain the *metals, nonmetals,* and *metalloids* (*semimetals*).	**periodic table** **period (of periodic table)** **group (of periodic table)** **metal** **nonmetal** **metalloid (semimetal)**
2.6 Chemical Formulas; Molecular and Ionic Substances	
▪ Determine when the *chemical formula* of a compound represents a *molecule.* ▪ Determine whether a chemical formula is also a *molecular formula.* ▪ Define *ion, cation,* and *anion.* ▪ Classify compounds as *ionic* or *molecular.* ▪ Define and provide examples for the term *formula unit.* ▪ Specify the charge on all substances, ionic and molecular. ▪ Write an ionic formula, given the ions. Example 2.3	**chemical formula** **molecule** **molecular formula** **polymer** **monomer** **ion** **anion** **cation** **ionic compound** **formula unit**

2.7 Organic Compounds	
■ List the attributes of molecular substances that make them *organic compounds.* ■ Explain what makes a molecule a *hydrocarbon.* ■ Recognize some *functional groups* of organic molecules.	**organic compound** **hydrocarbon** **functional group**
2.8 Naming Simple Compounds	
■ Recognize *inorganic compounds.* ■ Learn the rules for predicting the charges of *monatomic ions* in ionic compounds. ■ Apply the rules for naming monatomic ions. ■ Learn the names and charges of common *polyatomic ions.* ■ Name an ionic compound from its formula. Example 2.4 ■ Write the formula of an ionic compound from its name. Example 2.5 ■ Determine the order of elements in a *binary (molecular) compound.* ■ Learn the rules for naming binary molecular compounds, including the Greek prefixes. ■ Name a binary compound from its formula. Example 2.6 ■ Write the formula of a binary compound from its name. Example 2.7 ■ Name a binary molecular compound from its molecular model. Example 2.8 ■ Recognize molecular compounds that are acids. ■ Determine whether an acid is an *oxoacid.* ■ Learn the approach for naming binary acids and oxoacids. ■ Write the name and formula of an anion from the acid. Example 2.9 ■ Recognize compounds that are *hydrates.* ■ Learn the rules for naming hydrates. ■ Name a hydrate from its formula. Example 2.10 ■ Write the formula of a hydrate from its name. Example 2.11	**chemical nomenclature** **inorganic compound** **monatomic ion** **polyatomic ion** **binary compound** **oxoacid** **hydrate**
2.9 Writing Chemical Equations	
■ Identify the *reactants* and *products* in a chemical equation. ■ Write chemical equations using appropriate phase labels, symbols of reaction conditions, and the presence of a catalyst.	**chemical equation** **reactant** **product**
2.10 Balancing Chemical Equations	
■ Determine if a chemical reaction is balanced. ■ Master the techniques for balancing chemical equations. Example 2.12	

Questions and Problems

OWL Interactive versions of these problems may be assigned in OWL.

Self-Assessment and Review Questions

Key: These questions test your understanding of the ideas you worked with in the chapter. These problems vary in difficulty and often can be used for the basis of discussion.

2.1 Describe atomic theory and discuss how it explains the great variety of different substances. How does it explain chemical reactions?

2.2 Two compounds of iron and chlorine, A and B, contain 1.270 g and 1.904 g of chlorine, respectively, for each gram of iron. Show that these amounts are in the ratio 2:3. Is this consistent with the law of multiple proportions? Explain.

2.3 Explain the operation of a cathode-ray tube. Describe the deflection of cathode rays by electrically charged plates placed within the cathode-ray tube. What does this imply about cathode rays?

2.4 Explain Millikan's oil-drop experiment.

2.5 Describe the nuclear model of the atom. How does this model explain the results of alpha-particle scattering from metal foils?

2.6 What are the different kinds of particles in the atom's nucleus? Compare their properties with each other and with those of an electron.

2.7 Describe how protons and neutrons were discovered to be constituents of nuclei.

2.8 Oxygen consists of three different __________, each having eight protons but different numbers of neutrons.

2.9 Describe how Dalton obtained relative atomic masses.

2.10 Briefly explain how a mass spectrometer works. What kinds of information does one obtain from the instrument?

2.11 Define the term *atomic mass.* Why might the values of atomic masses on a planet elsewhere in the universe be different from those on earth?

2.12 What is the name of the element in Group IVA and Period 5?

2.13 Cite some properties that are characteristic of a metal.

2.14 Ethane consists of molecules with two atoms of carbon and six atoms of hydrogen. Write the molecular formula for ethane.

2.15 What is the difference between a molecular formula and a structural formula?

2.16 What is the fundamental difference between an organic substance and an inorganic substance? Write chemical formulas of three inorganic molecules that contain carbon.

2.17 Give an example of a binary compound that is ionic. Give an example of a binary compound that is molecular.

2.18 Which of the following models represent a(n):

a element
b compound
c mixture
d ionic solid
e gas made up of an element and a compound
f mixture of elements
g solid element
h solid
i liquid

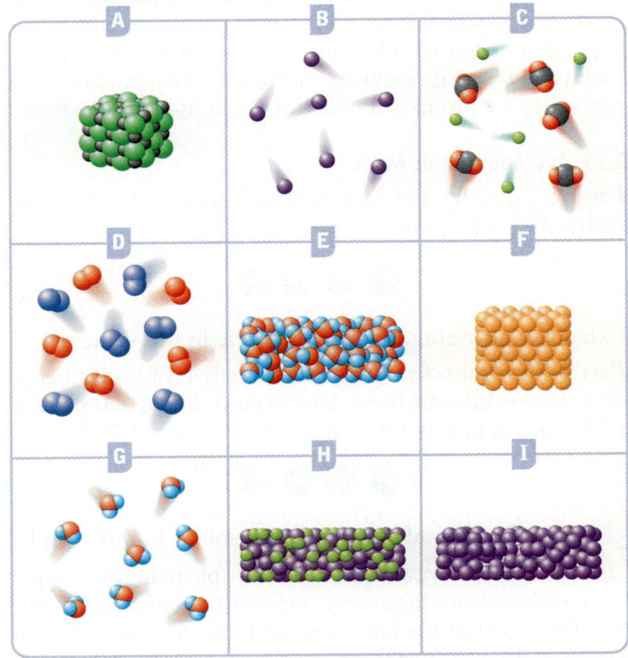

2.19 The compounds CuCl and $CuCl_2$ were formerly called cuprous chloride and cupric chloride, respectively. What are their names using the Stock system of nomenclature? What are the advantages of the Stock system of nomenclature over the former one?

2.20 Explain what is meant by the term *balanced chemical equation.*

2.21 How many protons, neutrons, and electrons are in $^{119}Sn^{2+}$?

a 50 p, 69 n, 48 e^-
b 50 p, 69 n, 50 e^-
c 119 p, 50 n, 69 e^-
d 69 p, 50 n, 69 e
e 50 p, 119 n, 52 e^-

2.22 The atomic mass of Ga is 69.72 amu. There are only two naturally occurring isotopes of gallium: ^{69}Ga, with a mass of 69.0 amu, and ^{71}Ga, with a mass of 71.0 amu. The natural abundance of the ^{69}Ga isotope is approximately:

a 15% b 30% c 50% d 65% d 80%

2.23 In which of the following are the name and formula correctly paired?

a sodium sulfite: Na_2S
b calcium carbonate: $Ca(CO_3)_2$
c magnesium hydroxide: $Mg(OH)_2$
d nitrite: NO_2
e iron (III) oxide: FeO

2.24 A chunk of an unidentified element (let's call it element "X") is reacted with sulfur to form an ionic compound with the chemical formula X_2S. Which of the following elements is the most likely identity of X?

a Mg b Li c Al d C e Cl

Concept Explorations

Key: Concept explorations are comprehensive problems that provide a framework that will enable you to explore and learn many of the critical concepts and ideas in each chapter. If you master the concepts associated with these explorations, you will have a better understanding of many important chemistry ideas and will be more successful in solving all types of chemistry problems. These problems are well suited for group work and for use as in-class activities.

2.25 Average Atomic Mass

Part 1: Consider the four identical spheres below, each with a mass of 2.00 g.

Calculate the average mass of a sphere in this sample.

Part 2: Now consider a sample that consists of four spheres, each with a different mass: blue mass is 2.00 g, red mass is 1.75 g, green mass is 3.00 g, and yellow mass is 1.25 g.

a Calculate the average mass of a sphere in this sample.

b How does the average mass for a sphere in this sample compare with the average mass of the sample that consisted just of the blue spheres? How can such different samples have their averages turn out the way they did?

Part 3: Consider two jars. One jar contains 100 blue spheres, and the other jar contains 25 each of red, blue, green, and yellow colors mixed together.

a If you were to remove 50 blue spheres from the jar containing just the blue spheres, what would be the total mass of spheres left in the jar? (Note that the masses of the spheres are given in Part 2.)

b If you were to remove 50 spheres from the jar containing the mixture (assume you get a representative distribution of colors), what would be the total mass of spheres left in the jar?

c In the case of the mixture of spheres, does the average mass of the spheres *necessarily* represent the mass of an individual sphere in the sample?

d If you had 80.0 grams of spheres from the blue sample, how many spheres would you have?

e If you had 60.0 grams of spheres from the mixed-color sample, how many spheres would you have? What assumption did you make about your sample when performing this calculation?

Part 4: Consider a sample that consists of three green spheres and one blue sphere. The green mass is 3.00 g, and the blue mass is 1.00 g.

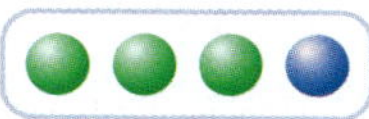

a Calculate the *fractional abundance* of each sphere in the sample.

b Use the fractional abundance to calculate the average mass of the spheres in this sample.

c How are the ideas developed in this Concept Exploration related to the atomic masses of the elements?

2.26 Model of the Atom

Consider the following depictions of two atoms, which have been greatly enlarged so you can see the subatomic particles.

a How many protons are present in atom A?

b What is the significance of the number of protons depicted in atom A or any atom?

c Can you identify the real element represented by the drawing of atom A? If so, what element does it represent?

d What is the charge on atom A? Explain how you arrived at your answer.

e Write the nuclide symbol of atom A.

f Write the atomic symbol and the atomic number of atom B.

g What is the mass number of atom B? How does this mass number compare with that of atom A?

h What is the charge on atom B?

i Write the nuclide symbol of atom B.

j Draw pictures like those above of ${}^{6}_{3}Li^{+}$ and ${}^{6}_{3}Li^{-}$ atoms. What are the mass number and atomic number of each of these atoms?

k Consider the two atoms depicted in this problem and the two that you just drew. What is the total number of lithium isotopes depicted? How did you make your decision?

l Is the mass number of an isotope of an atom equal to the mass of the isotope of the atom? Be sure to explain your answer.

Conceptual Problems

Key: These problems are designed to check your understanding of the concepts associated with some of the main topics presented in each chapter. A strong conceptual understanding of chemistry is the foundation for both applying chemical knowledge and solving chemical problems. These problems vary in level of difficulty and often can be used as a basis for group discussion.

2.27 One of the early models of the atom proposed that atoms were wispy balls of positive charge with the electrons evenly distributed throughout. What would you expect to observe if you conducted Rutherford's experiment and the atom had this structure?

2.28 A friend is trying to balance the following equation:

$$N_2 + H_2 \longrightarrow NH_3$$

He presents you with his version of the "balanced" equation:

$$N + H_3 \longrightarrow NH_3$$

You immediately recognize that he has committed a serious error; however, he argues that there is nothing wrong, since the equation is balanced. What reason can you give to convince him that his "method" of balancing the equation is flawed?

2.29 Given that the periodic table is an organizational scheme for the elements, what might be some other logical ways in which to group the elements that would provide meaningful chemical information in a periodic table of your own devising?

2.30 You discover a new set of polyatomic anions that has the newly discovered element "X" combined with oxygen. Since you made the discovery, you get to choose the names for these new polyatomic ions. In developing the names, you want to pay attention to convention. A colleague in your lab has come up with a name that she really likes for one of the ions. How would you name the following in a way consistent with her name?

Formula	*Name*
XO_4^{2-}	
XO_3^{2-}	
XO_2^{2-}	excite
XO^{2-}	

2.31 You have the mythical metal element "X" that can exist as X^+, X^{2+}, and X^{5+} ions.

a What would be the chemical formula for compounds formed from the combination of each of the X ions and SO_4^{2-}?

b If the name of the element X is exy, what would be the names of each of the compounds from part a of this problem?

2.32 Match the molecular model with the correct chemical formula: CH_3OH, NH_3, KCl, H_2O.

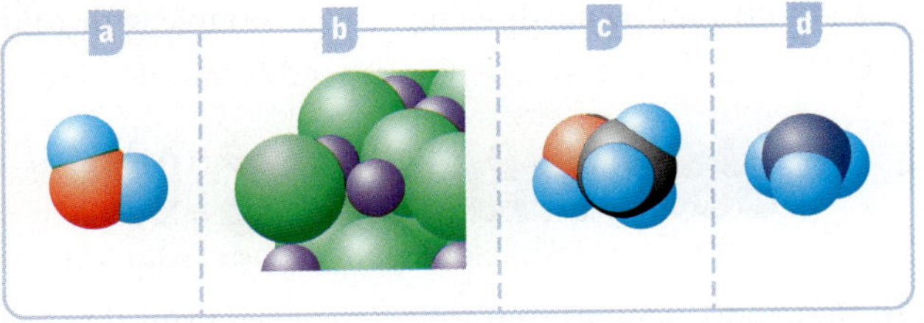

2.33 Consider a hypothetical case in which the charge on a proton is twice that of an electron. Using this hypothetical case, and the fact that atoms maintain a charge of 0, how many protons, neutrons, and electrons would a potassium-39 atom contain?

2.34 Currently, the atomic mass unit (amu) is based on being exactly one-twelfth the mass of a carbon-12 atom and is equal to 1.66×10^{-27} kg.

a If the amu were based on sodium-23 with a mass equal to exactly 1/23 of the mass of a sodium-23 atom, would the mass of the amu be different?

b If the new mass of the amu based on sodium-23 is 1.67×10^{-27} kg, how would the mass of a hydrogen atom, in amu, compare with the current mass of a hydrogen atom in amu?

2.35 For each of the following chemical reactions, write the correct element and/or compound symbols, formulas, and coefficients needed to produce complete, balanced equations. In all cases, the reactants are elements and the products are binary compounds.

a ______ + ______ $\longrightarrow$ LiCl

b Na + S $\longrightarrow$

c Al + I_2 $\longrightarrow$

d Ba + N_2 $\longrightarrow$

e ______ + ______ $\longrightarrow$ V_3P_5

2.36 You perform a chemical reaction using the hypothetical elements A and B. These elements are represented by their molecular models shown below:

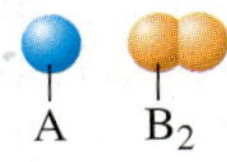

The product of the reaction represented by molecular models is

a Using the molecular models and the boxes, present a balanced chemical equation for the reaction of elements A and B.

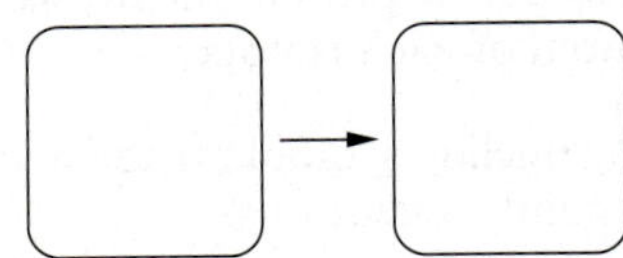

b Using the symbols A and B_2 for the chemical reaction, write a balanced chemical equation.

c What are some real-element possibilities for element B?

Practice Problems

Key: These problems are for practice in applying problem-solving skills. They are divided by topic, and some are keyed to exercises (see the ends of the exercises). The problems are arranged in matching pairs; the odd-numbered problem of each pair is listed first, and its answer is given in the back of the book.

2.37 What is the name of the element represented by each of the following atomic symbols?

a Ar   b Zn   c Ag   d Mg

2.38 For each atomic symbol, give the name of the element.

a Be   b Ag   c Si   d C

2.39 Give the atomic symbol for each of the following elements.

a potassium   b sulfur   c iron   d manganese

2.40 Give the atomic symbol for each of the following elements.

a copper   b calcium   c mercury   d tin

Electrons, Protons, and Neutrons

2.41 A student has determined the mass-to-charge ratio for an electron to be 5.64×10^{-12} kg/C. In another experiment, using Millikan's oil-drop apparatus, he found the charge on the electron to be 1.605×10^{-19} C. What would be the mass of the electron, according to these data?

2.42 The mass-to-charge ratio for the positive ion F^+ is 1.97×10^{-7} kg/C. Using the value of 1.602×10^{-19} C for the charge on the ion, calculate the mass of the fluorine atom. (The mass of the electron is negligible compared with that of the ion, so the ion mass is essentially the atomic mass.)

2.43 The following table gives the number of protons and neutrons in the nuclei of various atoms. Which atom is the isotope of atom A? Which atom has the same mass number as atom A?

	Protons	*Neutrons*
Atom A	18	19
Atom B	16	19
Atom C	18	18
Atom D	17	20

2.44 The following table gives the number of protons and neutrons in the nuclei of various atoms. Which atom is the isotope of atom A? Which atom has the same mass number as atom A?

	Protons	*Neutrons*
Atom A	32	39
Atom B	33	38
Atom C	38	50
Atom D	32	38

2.45 Naturally occurring chlorine is a mixture of the isotopes Cl-35 and Cl-37. How many protons and how many neutrons are there in each isotope? How many electrons are there in the neutral atoms?

2.46 Naturally occurring nitrogen is a mixture of ^{14}N and ^{15}N. Give the number of protons, neutrons, and electrons in the neutral atom of each isotope.

2.47 What is the nuclide symbol for the nucleus that contains 14 protons and 14 neutrons?

2.48 An atom contains 34 protons and 45 neutrons. What is the nuclide symbol for the nucleus?

Atomic Masses

2.49 Ammonia is a gas with a characteristic pungent odor. It is sold as a water solution for use in household cleaning. The gas is a compound of nitrogen and hydrogen in the atomic ratio 1 : 3. A sample of ammonia contains 7.933 g N and 1.712 g H. What is the atomic mass of N relative to H?

2.50 Hydrogen sulfide is a gas with the odor of rotten eggs. The gas can sometimes be detected in automobile exhaust. It is a compound of hydrogen and sulfur in the atomic ratio 2 : 1. A sample of hydrogen sulfide contains 0.587 g H and 9.330 g S. What is the atomic mass of S relative to H?

2.51 Calculate the atomic mass of an element with two naturally occurring isotopes, from the following data:

Isotope	*Isotopic Mass (amu)*	*Fractional Abundance*
X-63	62.930	0.6909
X-65	64.928	0.3091

What is the identity of element X?

2.52 An element has two naturally occurring isotopes with the following masses and abundances:

Isotopic Mass (amu)	*Fractional Abundance*
49.9472	2.500×10^{-3}
50.9440	0.9975

What is the atomic mass of this element? What is the identity of the element?

2.53 An element has three naturally occurring isotopes with the following masses and abundances:

Isotopic Mass (amu)	*Fractional Abundance*
38.964	0.9326
39.964	1.000×10^{-4}
40.962	0.0673

Calculate the atomic mass of this element. What is the identity of the element?

2.54 An element has three naturally occurring isotopes with the following masses and abundances:

Isotopic Mass (amu)	*Fractional Abundance*
27.977	0.9221
28.976	0.0470
29.974	0.0309

Calculate the atomic mass of this element. What is the identity of the element?

2.55 While traveling to a distant universe, you discover the hypothetical element "X." You obtain a representative sample of the element and discover that it is made up of two isotopes, X-23 and X-25. To help your science team calculate the atomic mass of the substance, you send the following drawing of your sample with your report.

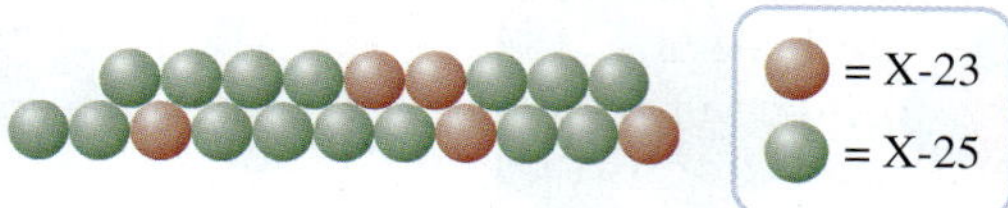

In the report, you also inform the science team that the brown atoms are X-23, which have an isotopic mass of 23.02 amu, and the green atoms are X-25, which have an isotopic mass of 25.147 amu. What is the atomic mass of element X?

2.56 While roaming a parallel universe, you discover the hypothetical element "Z." You obtain a representative sample of the element and discover that it is made up of two isotopes, Z-47 and Z-51. To help your science team calculate the atomic mass of the substance, you send the following drawing of your sample with your report.

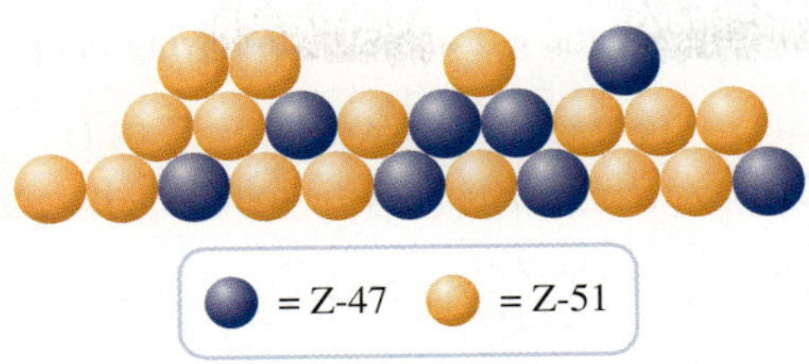

In the report, you also inform the science team that the blue atoms are Z-47, which have an isotopic mass of 47.621 amu, and the orange atoms are Z-51, which have an isotopic mass of 51.217 amu. What is the atomic mass of element Z?

Periodic Table

2.57 Identify the group and period for each of the following. Refer to the periodic table (Figure 2.15 or inside front cover). Label each as a metal, nonmetal, or metalloid.
a C b Po c Cr d Mg e B

2.58 Refer to the periodic table (Figure 2.15 or inside front cover) and obtain the group and period for each of the following elements. Also determine whether the element is a metal, nonmetal, or metalloid.
a V b Rb c B d I e He

2.59 Refer to the periodic table (Figure 2.15 or inside front cover) and answer the following questions.
a What Group VIA element is a metalloid?
b What is the Group IIIA element in Period 3?

2.60 Refer to the periodic table (Figure 2.15 or inside front cover) and answer the following questions.
a What Group VA element is a metal?
b What is the Group IIA element in Period 2?

2.61 Give one example (atomic symbol and name) for each of the following.
a a main-group (representative) element in the second period
b a halogen
c a transition element in the fourth period
d a lanthanide element

2.62 Give one example (atomic symbol and name) for each of the following.
a a transition element in the fourth period
b an alkali metal
c a main-group (representative) element in the third period
d an actinide element

Molecular and Ionic Substances

2.63 The normal form of the element sulfur is a brittle, yellow solid. This is a molecular substance, S_8. If this solid is vaporized, it first forms S_8 molecules; but at high temperature, S_2 molecules are formed. How do the molecules of the solid sulfur and of the hot vapor differ? How are the molecules alike?

2.64 White phosphorus is available in sticks, which have a waxy appearance. This is a molecular substance, P_4. When this solid is vaporized, it first forms P_4 molecules; but at high temperature, P_2 molecules are formed. How do the molecules of white phosphorus and those of the hot vapor differ? How are the molecules alike?

2.65 A 1.50-g sample of nitrous oxide (an anesthetic, sometimes called laughing gas) contains 2.05×10^{22} N_2O molecules. How many nitrogen atoms are in this sample? How many nitrogen atoms are in 44.0 g of nitrous oxide?

2.66 Nitric acid is composed of HNO_3 molecules. A sample weighing 4.50 g contains 4.30×10^{22} HNO_3 molecules. How many nitrogen atoms are in this sample? How many oxygen atoms are in 61.0 g of nitric acid?

2.67 A sample of ammonia, NH_3, contains 3.3×10^{21} hydrogen atoms. How many NH_3 molecules are in this sample?

2.68 A sample of ethanol (ethyl alcohol), C_2H_5OH, contains 4.2×10^{23} hydrogen atoms. How many C_2H_5OH molecules are in this sample?

2.69 Give the molecular formula for each of the following structural formulas.

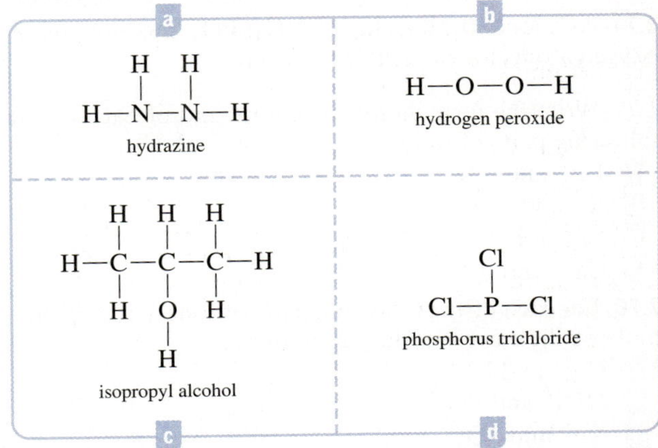

2.70 What molecular formula corresponds to each of the following structural formulas?

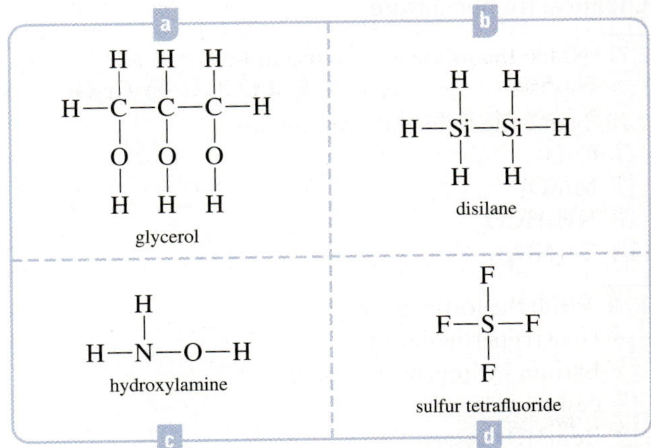

2.71 Write the molecular formula for each of the following compounds represented by molecular models.

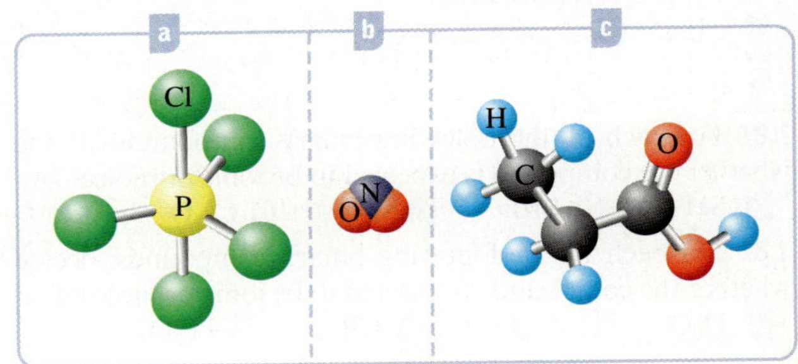

2.72 Write the molecular formula for each of the following compounds represented by molecular models.

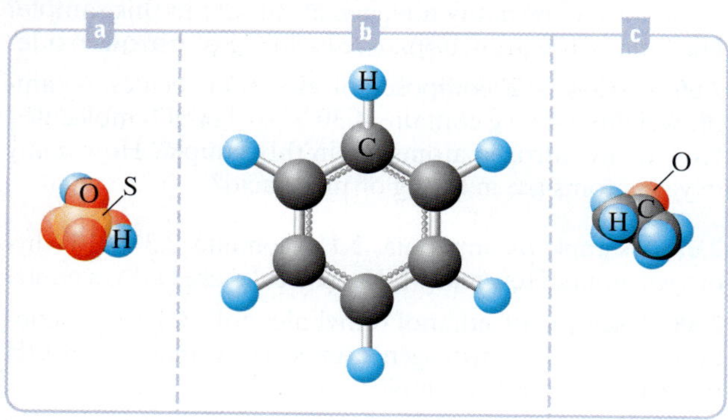

2.73 Iron(II) nitrate has the formula $Fe(NO_3)_2$. What is the ratio of iron atoms to oxygen atoms in this compound?

2.74 Ammonium phosphate, $(NH_4)_3PO_4$, has how many oxygen atoms for each nitrogen atom?

2.75 Write the formula for the compound of each of the following pairs of ions.

a. Fe^{3+} and CN^-
b. K^+ and CO_3^{2-}
c. Li^+ and N^{3-}
d. Ca^{2+} and P^{3-}

2.76 For each of the following pairs of ions, write the formula of the corresponding compound.

a. Co^{2+} and N^{3-}
b. NH_4^+ and PO_4^{3-}
c. Na^+ and SO_3^{2-}
d. Fe^{3+} and OH^-

Chemical Nomenclature

2.77 Name the following compounds.

a. Na_2SO_4 b. Na_3N c. CuCl d. Cr_2O_3

2.78 Name the following compounds.

a. CaO
b. Mn_2O_3
c. NH_4HCO_3
d. $Cu(NO_3)_2$

2.79 Write the formulas of:

a. lead(II) permanganate
b. barium hydrogen carbonate
c. cesium sulfide
d. iron(III) acetate

2.80 Write the formulas of:

a. sodium thiosulfate
b. copper(II) hydroxide
c. calcium hydrogen carbonate
d. chromium(III) phosphide

2.81 For each of the following binary compounds, decide whether the compound is expected to be ionic or molecular.

a. SeF_4 b. LiBr c. SiF_4 d. Cs_2O

2.82 For each of the following binary compounds, decide whether the compound is expected to be ionic or molecular.

a. H_2O b. As_4O_6 c. ClF_3 d. Fe_2O_3

2.83 Give systematic names to the following binary compounds.

a. N_2O b. P_4O_{10} c. $AsCl_3$ d. Cl_2O_7

2.84 Give systematic names to the following binary compounds.

a. N_2F_2 b. CF_4 c. N_2O_5 d. As_4O_6

2.85 Write the formulas of the following compounds.

a. nitrogen tribromide
b. xenon hexafluoride
c. carbon monoxide
d. dichlorine pentoxide

2.86 Write the formulas of the following compounds.

a. diphosphorus pentoxide
b. nitrogen dioxide
c. dinitrogen tetrafluoride
d. boron trifluoride

2.87 Write the systematic name for each of the following compounds represented by a molecular model.

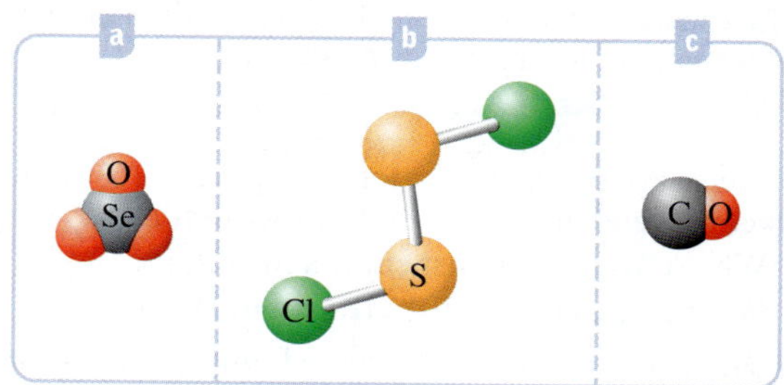

2.88 Write the systematic name for each of the following molecules represented by a molecular model.

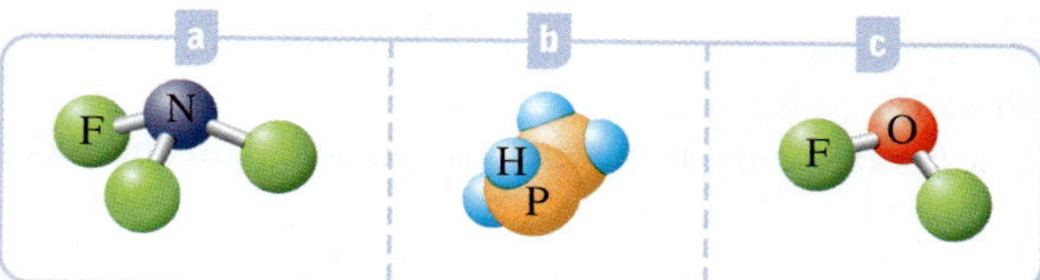

2.89 Give the name and formula of the acid corresponding to each of the following oxoanions.

a. sulfite ion, SO_3^{2-}
b. hyponitrite ion, $N_2O_2^{2-}$
c. disulfite ion, $S_2O_5^{2-}$
d. arsenate ion, AsO_4^{3-}

2.90 Give the name and formula of the acid corresponding to each of the following oxoanions.

a. selenite ion, SeO_3^{2-}
b. chlorite ion, ClO_2^-
c. hypoiodite ion, IO^-
d. nitrate ion, NO_3^-

2.91 Glauber's salt has the formula $Na_2SO_4 \cdot 10H_2O$. What is the chemical name of this substance?

2.92 Emerald-green crystals of the substance $NiSO_4 \cdot 6H_2O$ are used in nickel plating. What is the chemical name of this compound?

2.93 Iron(II) sulfate heptahydrate is a blue-green, crystalline compound used to prepare other iron compounds. What is the formula of iron(II) sulfate heptahydrate?

2.94 Cobalt(II) chloride hexahydrate has a pink color. It loses water on heating and changes to a blue-colored compound. What is the formula of cobalt(II) chloride hexahydrate?

Chemical Equations

2.95 For the balanced chemical equation $Pb(NO_3)_2 + K_2CO_3 \longrightarrow PbCO_3 + 2KNO_3$, how many oxygen atoms are on the right side?

2.96 In the equation $2PbS + O_2 \longrightarrow 2PbO + 2SO_2$, how many oxygen atoms are there on the right side? Is the equation balanced as written?

2.97 Balance the following equations.

a. $Sn + NaOH \longrightarrow Na_2SnO_2 + H_2$
b. $Al + Fe_3O_4 \longrightarrow Al_2O_3 + Fe$
c. $CH_3OH + O_2 \longrightarrow CO_2 + H_2O$
d. $P_4O_{10} + H_2O \longrightarrow H_3PO_4$
e. $PCl_5 + H_2O \longrightarrow H_3PO_4 + HCl$

2.98 Balance the following equations.

a. $Ca_3(PO_4)_2 + H_3PO_4 \longrightarrow Ca(H_2PO_4)_2$
b. $MnO_2 + HCl \longrightarrow MnCl_2 + Cl_2 + H_2O$
c. $Na_2S_2O_3 + I_2 \longrightarrow NaI + Na_2S_4O_6$
d. $Al_4C_3 + H_2O \longrightarrow Al(OH)_3 + CH_4$
e. $NO_2 + H_2O \longrightarrow HNO_3 + NO$

2.99 Solid calcium phosphate and aqueous sulfuric acid solution react to give calcium sulfate, which comes out of the solution as a solid. The other product is phosphoric acid, which remains in solution. Write a balanced equation for the reaction using complete formulas for the compounds with phase labels.

2.100 Solid sodium metal reacts with water, giving a solution of sodium hydroxide and releasing hydrogen gas. Write a balanced equation for the reaction using complete formulas for the compounds with phase labels.

2.101 An aqueous solution of ammonium chloride and barium hydroxide is heated, and the compounds react to give off ammonia gas. Barium chloride solution and water are also products. Write a balanced equation for the reaction using complete formulas for the compounds with phase labels; indicate that the reactants are heated.

2.102 Lead metal is produced by heating solid lead(II) sulfide with solid lead(II) sulfate, resulting in liquid lead and sulfur dioxide gas. Write a balanced equation for the reaction using complete formulas for the compounds with phase labels; indicate that the reactants are heated.

General Problems

Key: These problems provide more practice but are not divided by topic or keyed to exercises. Each section ends with essay questions, each of which is color coded to refer to the *A Chemist Looks at Frontiers* (purple) chapter essay on which it is based. Odd-numbered problems and the even-numbered problems that follow are similar; answers to all odd-numbered problems except the essay questions are given in the back of the book.

2.103 Two samples of different compounds of nitrogen and oxygen have the following compositions. Show that the compounds follow the law of multiple proportions. What is the ratio of oxygen in the two compounds for a fixed amount of nitrogen?

	Amount N	*Amount O*
Compound A	1.206 g	2.755 g
Compound B	1.651 g	4.714 g

2.104 Two samples of different compounds of sulfur and oxygen have the following compositions. Show that the compounds follow the law of multiple proportions. What is the ratio of oxygen in the two compounds for a fixed amount of sulfur?

	Amount S	*Amount O*
Compound A	1.210 g	1.811 g
Compound B	1.783 g	1.779 g

2.105 In a series of oil-drop experiments, the charges measured on the oil drops were -3.20×10^{-19} C, -6.40×10^{-19} C, -9.60×10^{-19} C, and -1.12×10^{-18} C. What is the smallest difference in charge between any two drops? If this is assumed to be the charge on the electron, how many excess electrons are there on each drop?

2.106 In a hypothetical universe, an oil-drop experiment gave the following measurements of charges on oil drops: -5.55×10^{-19} C, -9.25×10^{-19} C, -1.11×10^{-18} C, and -1.48×10^{-18} C. Assume that the smallest difference in charge equals the unit of negative charge in this universe. What is the value of this unit of charge? How many units of excess negative charge are there on each oil drop?

2.107 Compounds of europium, Eu, are used to make color television screens. The europium nucleus has a charge of +63. How many electrons are there in the neutral atom? in the Eu^{3+} ion?

2.108 Cesium, Cs, is used in photoelectric cells ("electric eyes"). The cesium nucleus has a charge of +55. What is the number of electrons in the neutral atom? in the Cs^+ ion?

2.109 A nucleus of mass number 81 contains 46 neutrons. An atomic ion of this element has 36 electrons in it. Write the symbol for this atomic ion (give the symbol for the nucleus and give the ionic charge as a right superscript).

2.110 One isotope of a metallic element has mass number 74 and has 51 neutrons in the nucleus. An atomic ion has 18 electrons. Write the symbol for this ion (give the symbol for the nucleus and give the ionic charge as a right superscript).

2.111 Obtain the fractional abundances for the two naturally occurring isotopes of copper. The masses of the isotopes are ${}^{63}_{29}Cu$, 62.9298 amu; ${}^{65}_{29}Cu$, 64.9278 amu. The atomic mass is 63.546 amu.

2.112 Silver has two naturally occurring isotopes, one of mass 106.91 amu and the other of mass 108.90 amu. Find the fractional abundances for these two isotopes. The atomic mass is 107.87 amu.

2.113 Identify the following elements, giving their name and atomic symbol.
- a a nonmetal that is normally a liquid
- b a normally gaseous element in Group IA
- c a transition element in Group VB, Period 5
- d the halogen in Period 2

2.114 Identify the following elements, giving their name and atomic symbol.
- a a normally liquid element in Group VIIA
- b a metal that is normally a liquid
- c a main-group element in Group IIIA, Period 3
- d the alkali metal in Period 4

2.115 Give the names of the following ions.
a Cr^{3+} b Pb^{4+} c Ti^{2+} d Cu^{2+}

2.116 Give the names of the following ions.
a Mn^{2+} b Ni^{2+} c Co^{2+} d Fe^{3+}

2.117 Write formulas for all the ionic compounds that can be formed by combinations of these ions: Na^+, Co^{2+}, SO_4^{2-}, and Cl^-.

2.118 Write formulas for all the ionic compounds that can be formed by combinations of these ions: Mg^{2+}, Cr^{3+}, S^{2-}, and NO^{3-}.

2.119 Name the following compounds.
a $Sn_3(PO_4)_2$ b NH_4NO_2
c $Mg(OH)_2$ d $NiSO_3$

2.120 Name the following compounds.
a $Cu(NO_2)_2$ b $(NH_4)_3P$
c K_2SO_3 d Hg_3N_2

2.121 Give the formulas for the following compounds.
- a mercury(I) sulfide
- b cobalt(III) sulfite
- c ammonium dichromate
- d aluminum fluoride

2.122 Give the formulas for the following compounds.
- a hydrogen peroxide
- b magnesium phosphate
- c lead(IV) phosphide
- d calcium carbonate

2.123 Name the following molecular compounds.
a $AsBr_3$ b H_2Te c P_2O_5 d SiO_2

2.124 Name the following molecular compounds.
a ClF_4 b CS_2 c PF_3 d SF_6

2.125 Balance the following equations.
- a $C_2H_6 + O_2 \longrightarrow CO_2 + H_2O$
- b $P_4O_6 + H_2O \longrightarrow H_3PO_3$
- c $KClO_3 \longrightarrow KCl + KClO_4$
- d $(NH_4)_2SO_4 + NaOH \longrightarrow NH_3 + H_2O + Na_2SO_4$
- e $NBr_3 + NaOH \longrightarrow N_2 + NaBr + HOBr$

2.126 Balance the following equations.
- a $NaOH + H_2CO_3 \longrightarrow Na_2CO_3 + H_2O$
- b $SiCl_4 + H_2O \longrightarrow SiO_2 + HCl$
- c $Ca_3(PO_4)_2 + C \longrightarrow Ca_3P_2 + CO$
- d $H_2S + O_2 \longrightarrow SO_2 + H_2O$
- e $N_2O_5 \longrightarrow NO_2 + O_2$

2.127 A monatomic ion has a charge of +2. The nucleus of the ion has a mass number of 62. The number of neutrons in the nucleus is 1.21 times that of the number of protons. How many electrons are in the ion? What is the name of the element?

2.128 A monatomic ion has a charge of +1. The nucleus of the ion has a mass number of 85. The number of neutrons in the nucleus is 1.30 times that of the number of protons. How many electrons are in the ion? What is the name of the element?

2.129 Natural carbon, which has an atomic mass of 12.011 amu, consists of carbon-12 and carbon-13 isotopes. Given that the mass of carbon-13 is 13.00335 amu, what would be the average atomic mass (in amu) of a carbon sample prepared by mixing equal numbers of carbon atoms from a sample of natural carbon and a sample of pure carbon-13?

2.130 Natural chlorine, which has an atomic mass of 35.4527 amu, consists of chlorine-35 and chlorine-37 isotopes. Given that the mass of chlorine-35 is 34.96885 amu, what is the average atomic mass (in amu) of a chlorine sample prepared by mixing equal numbers of chlorine atoms from a sample of natural chlorine and a sample of pure chlorine-35?

■ **2.131** Describe the island of stability. What nuclide is predicted to be most stable?

■ **2.132** Write the equation for the nuclear reaction in which element 112 was first produced.

Strategy Problems

Key: As noted earlier, all of the practice and general problems are matched pairs. This section is a selection of problems that are not in matched-pair format. These challenging problems require that you employ many of the concepts and strategies that were developed in the chapter. In some cases, you will have to integrate several concepts and operational skills in order to solve the problem successfully.

2.133 Correct any mistakes in the naming of the following compounds or ions.

SO_3: sulfite

NO_2: nitrate

PO_4^{3-}: phosphite ion

N_2: nitride

$Mg(OH)_2$: maganese dihydroxide

2.134 An unknown metal (let's call it "M") is reacted with sulfur to produce a compound with the chemical formula MS. What is the charge on the metal in the compound MS? Name a metal that could be metal M.

2.135 Aluminum is reacted with oxygen to form a binary compound. Write the balanced chemical reaction and write the name of the compound formed by the reaction.

2.136 Ammonia gas reacts with molecular oxygen gas to form nitrogen monoxide gas and liquid water. Write the complete balanced reaction with all proper state symbols.

2.137 A hypothetical element X is found to have an atomic mass of 37.45 amu. Element X has only two isotopes, X-37 and X-38. The X-37 isotope has a fractional abundance of 0.7721 and an isotopic mass of 37.24. What is the isotopic mass of the other isotope?

2.138 A monotomic ion has a charge of +3. The nucleus of the ion has a mass number of 27. The number of neutrons in the nucleus is equal to the number of electrons in a S^{2+} ion. Identify the element and indicate the number of protons, neutrons, and electrons.

2.139 A small crystal of $CaCl_2$ that weighs 0.12 g contains 6.5×10^{20} formula units of $CaCl_2$. What is the total number of ions (cations and anions) that make up this crystal?

2.140 Write the formulas and names for all the ionic compounds that can form by combinations of the following ions: Mg^{2+}, Cr^{3+}, the carbonate anion, and the nitride anion.

2.141 Name the following compounds:

SO_3

HNO_2

Mg_3N_3

$HI(aq)$

$Cu_3(PO_4)_2$

$CuSO_4{\cdot}5H_2O$

2.142 The IO_3^- anion is called iodate. There are three related ions: IO^-, IO_2^-, and IO_4^-. Using what you have learned about similar groups of anions, write the name for each of the following compounds:

HIO_3

$NaIO_4$

$Mg(IO_2)_2$

$Fe(IO_2)_3$

2.143 Write out in words the following chemical equations:

a $PbCl_2(aq) + Na_2S(aq) \longrightarrow PbS(s) + 2NaCl(aq)$

b $H_2O(l) + SO_3(g) \longrightarrow H_2SO_4(aq)$

c $C(s) + O_2(g) \longrightarrow CO_2(g)$

d $H_2(g) + I_2(g) \longrightarrow 2HI(g)$

2.144 From the following written description, write the balanced chemical equation for the reaction including state symbols. A diatomic gaseous molecule that contains 17 protons per atom is reacted with a solid element that has an atomic number of 19 to yield an ionic compound.

2.145 An 18-mL sample of water contains 6.0×10^{23} water molecules. Assuming that this sample consists only of the Hydrogen-1 and Oxygen-16 isotopes, calculate the number of neutrons, protons, and electrons in this water sample.

2.146 Name the following compounds:

a $HCl(g)$

b $HBr(aq)$

c $HF(g)$

d $HNO_3(aq)$

2.147 During nuclear decay a ^{238}U atom can break apart into a helium-4 atom and one other atom. Assuming that no subatomic particles are destroyed during this decay process, what is the other element produced?

2.148 Write the balanced chemical equation for the reaction of an aqueous oxacid that contains the phosphate anion with solid magnesium hydroxide. The reaction produces liquid dihydrogen monoxide and a solid ionic compound formed from the combination of magnesium cations and phosphate anions.

Cumulative-Skills Problems

Key: The problems under this heading combine skills introduced in the previous chapter with those given in the current one.

2.149 There are 2.619×10^{22} atoms in 1.000 g of sodium. Assume that sodium atoms are spheres of radius 1.86 Å and that they are lined up side by side. How many miles in length is the line of sodium atoms?

2.150 There are 1.699×10^{22} atoms in 1.000 g of chlorine. Assume that chlorine atoms are spheres of radius 0.99 Å and that they are lined up side by side in a 0.5-g sample. How many miles in length is the line of chlorine atoms in the sample?

2.151 A sample of green crystals of nickel(II) sulfate heptahydrate was heated carefully to produce the bluish green nickel(II) sulfate hexahydrate. What are the formulas of the hydrates? If 8.753 g of the heptahydrate produces 8.192 g of the hexahydrate, how many grams of anhydrous nickel(II) sulfate could be obtained?

2.152 Cobalt(II) sulfate heptahydrate has pink-colored crystals. When heated carefully, it produces cobalt(II) sulfate monohydrate, which has red crystals. What are the formulas of these hydrates? If 3.548 g of the heptahydrate yields 2.184 g of the monohydrate, how many grams of the anhydrous cobalt(II) sulfate could be obtained?

2.153 A sample of metallic element X, weighing 3.177 g, combines with 0.6015 L of O_2 gas (at normal pressure and 20.0°C) to form the metal oxide with the formula XO. If the density of O_2 gas under these conditions is 1.330 g/L, what is the mass of this oxygen? The atomic mass of oxygen is 15.9994 amu. What is the atomic mass of X? What is the identity of X?

2.154 A sample of metallic element X, weighing 4.315 g, combines with 0.4810 L of Cl_2 gas (at normal pressure and 20.0°C) to form the metal chloride with the formula XCl. If the density of Cl_2 gas under these conditions is 2.948 g/L, what is the mass of the chlorine? The atomic mass of chlorine is 35.453 amu. What is the atomic mass of X? What is the identity of X?

The **formula mass** (FM) of a substance is *the sum of the atomic masses of all atoms in a formula unit of the compound,* whether molecular or not. Sodium chloride, with the formula unit NaCl, has a formula mass of 58.44 amu (22.99 amu from Na plus 35.45 amu from Cl). NaCl is ionic, so strictly speaking the expression "molecular mass of NaCl" has no meaning. On the other hand, the molecular mass and the formula mass calculated from the molecular formula of a substance are identical. ◀

Some chemists use the term molecular mass *in a less strict sense for ionic as well as molecular compounds.*

Example 3.1 Calculating the Formula Mass from a Formula

Gaining Mastery Toolbox

Critical Concept 3.1
The formula mass of a compound is based on the number and type of each element in the compound. The formula mass is calculated by summing the masses of all atoms in the chemical formula.

Solution Essentials:
- Formula mass
- Atomic mass
- Atomic mass unit
- Periodic table
- Rules for significant figures and rounding

Calculate the formula mass of each of the following to three significant figures, using a table of atomic masses (AM): **a** chloroform, $CHCl_3$; **b** iron(III) sulfate, $Fe_2(SO_4)_3$.

Problem Strategy Identify the number and type of atoms in the chemical formula. Use the periodic table to obtain the atomic mass of each of the elements present in the compounds. Taking into account the number of each atom present in the formula, sum the masses. If atoms in the formula are enclosed within parentheses as in part b, the number of each element within the parentheses should be multiplied by the subscript that follows the final, or closing, parenthesis.

Solution

a The calculation is

$$\begin{aligned} 1 \times \text{AM of C} &= && 12.0 \text{ amu} \\ 1 \times \text{AM of H} &= && 1.0 \text{ amu} \\ 3 \times \text{AM of Cl} &= 3 \times 35.45 \text{ amu} &=& \ 106.4 \text{ amu} \\ \text{FM of } CHCl_3 &= && 119.4 \text{ amu} \end{aligned}$$

The answer rounded to three significant figures is **119 amu.**

b The calculation is

$$\begin{aligned} 2 \times \text{AM of Fe} &= 2 \times 55.8 \text{ amu} &=& \ 111.6 \text{ amu} \\ 3 \times \text{AM of S} &= 3 \times 32.1 \text{ amu} &=& \ 96.3 \text{ amu} \\ 3 \times 4 \times \text{AM of O} &= 12 \times 16.00 \text{ amu} &=& \ 192.0 \text{ amu} \\ \text{FM of } Fe_2(SO_4)_3 &= && 399.9 \text{ amu} \end{aligned}$$

The answer rounded to three significant figures is **4.00×10^2 amu.**

Answer Check The most common error when calculating formula masses is not correctly accounting for those atoms that are enclosed in parentheses.

Exercise 3.1 Calculate the formula masses of the following compounds, using a table of atomic masses. Give the answers to three significant figures. **a** nitrogen dioxide, NO_2; **b** glucose, $C_6H_{12}O_6$; **c** sodium hydroxide, NaOH; **d** magnesium hydroxide, $Mg(OH)_2$.

■ See Problems 3.27 and 3.28.

Example 3.2 Calculating the Formula Mass from Molecular Models

Gaining Mastery Toolbox

Critical Concept 3.2
A molecular model depicts the number and type of each atom present in the molecular compound. Molecular models can be used to determine the chemical formula of a molecular compound.

Solution Essentials:
- Formula mass
- Atomic mass
- Atomic mass unit
- Periodic table
- Molecular model
- Rules for significant figures and rounding

For the following two compounds, write the molecular formula and calculate the formula mass to four significant figures:

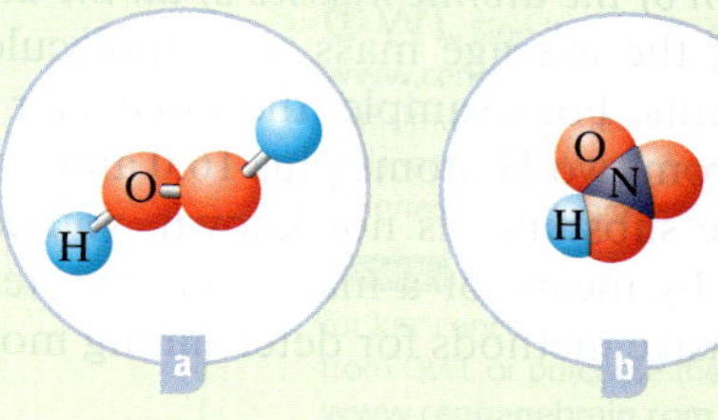

(continued)

(*continued*)

Problem Strategy In order to calculate the formula mass, you need to have the total number of each type of atom present in the structure, so look at the structure and keep a tally of all the atoms present.

Solution

a This molecular model is of a molecule that is composed of two O and two H atoms. For inorganic compounds, the elements in a chemical formula are written in order such that the most metallic element is listed first. (Even though H is not metallic, it is positioned in the periodic table in such a way that it is considered to be a metal when writing formulas.) Hence, the chemical formula is $\mathbf{H_2O_2}$. Using the same approach as Example 3.1, calculating the formula mass yields **34.02 amu.**

b This molecular model represents a molecule made up of one N atom, three O atoms, and one H atom. The chemical formula is then $\mathbf{HNO_3}$. The formula mass is **63.01 amu.**

Answer Check Always make sure that your answer is reasonable. If you obtain a molecular mass of more than 200 amu for a simple molecular compound (this is possible but not typical), it is advisable to check your work closely.

Exercise 3.2 For the following compounds, write the molecular formula and calculate the formula mass to three significant figures.

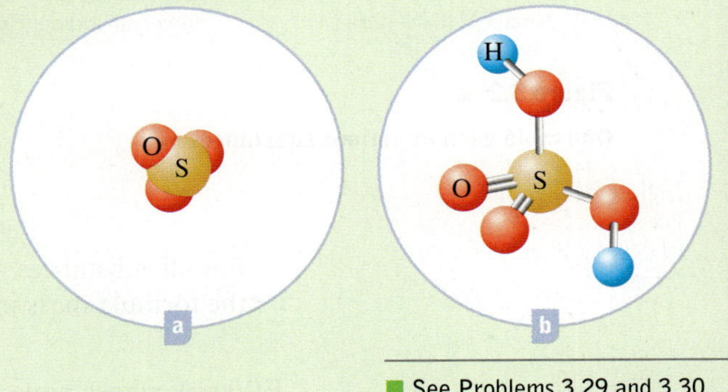

■ See Problems 3.29 and 3.30.

3.2 The Mole Concept

When we prepare a compound industrially or even study a reaction in the laboratory, we deal with tremendous numbers of molecules or ions. Suppose you wish to prepare acetic acid, starting from 10.0 g of ethanol. This small sample (less than 3 teaspoonsful) contains 1.31×10^{23} molecules, a truly staggering number. Imagine a device that counts molecules at the rate of one million per second. It would take more than four billion years—nearly the age of the earth—for this device to count that many molecules! Chemists have adopted the *mole concept* as a convenient way to deal with the enormous numbers of molecules or ions in the samples they work with.

Definition of Mole and Molar Mass

A **mole** (symbol **mol**) is defined as *the quantity of a given substance that contains as many molecules or formula units as the number of atoms in exactly 12 g of carbon-12.* One mole of ethanol, for example, contains the same number of ethanol molecules as there are carbon atoms in 12 g of carbon-12.

The number of atoms in a 12-g sample of carbon-12 is called **Avogadro's number** (to which we give the symbol N_A). Recent measurements of this number give the value 6.0221367×10^{23}, which to three significant figures is 6.02×10^{23}.

A mole of a substance contains Avogadro's number (6.02×10^{23}) of molecules (or formula units). The term *mole,* like a dozen or a gross, thus refers to a particular number of things. A dozen eggs equals 12 eggs, a gross of pencils equals 144 pencils, and a mole of ethanol equals 6.02×10^{23} ethanol molecules.

In using the term *mole* for ionic substances, we mean the number of formula units of the substance. For example, a mole of sodium carbonate, Na_2CO_3, is a quantity containing 6.02×10^{23} Na_2CO_3 units. But each formula unit of Na_2CO_3 contains two Na^+ ions and one CO_3^{2-} ion. Therefore, a mole of Na_2CO_3 also contains $2 \times 6.02 \times 10^{23}$ Na^+ ions and $1 \times 6.02 \times 10^{23}$ CO_3^{2-} ions. ▶

Sodium carbonate, Na_2CO_3, is a white, crystalline solid known commercially as soda ash. Large amounts of soda ash are used in the manufacture of glass. The hydrated compound, $Na_2CO_3 \cdot 10H_2O$, is known as washing soda.

When using the term *mole,* it is important to specify the formula of the unit to avoid any misunderstanding. For example, a mole of oxygen atoms (with the formula O) contains 6.02×10^{23} O atoms. A mole of oxygen molecules (formula O_2) contains 6.02×10^{23} O_2 molecules—that is, $2 \times 6.02 \times 10^{23}$ O atoms.

The **molar mass** of a substance is *the mass of one mole of the substance.* Carbon-12 has a molar mass of exactly 12 g/mol, by definition.

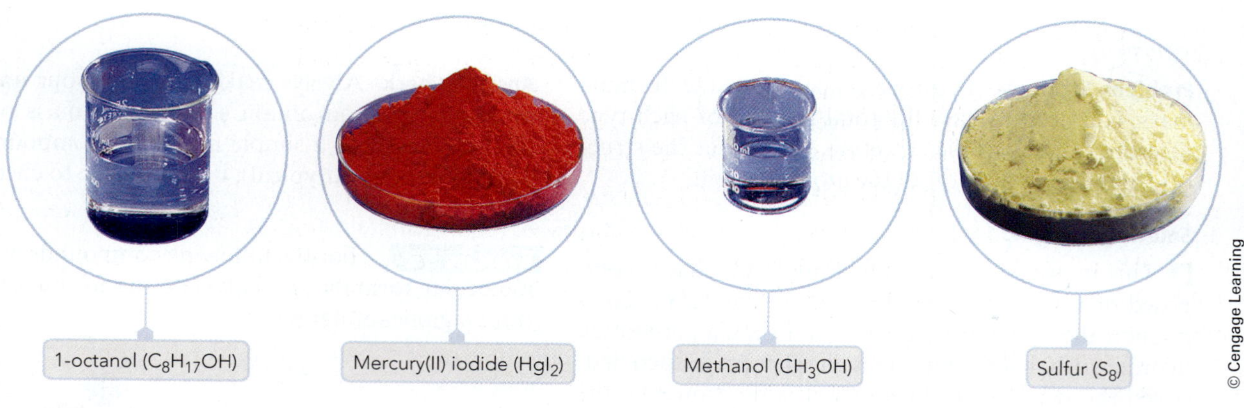

Figure 3.2 ▲

One mole each of various substances

For all substances, the molar mass in grams per mole is numerically equal to the formula mass in atomic mass units.

Ethanol, whose molecular formula is C_2H_6O (frequently written as the condensed structural formula C_2H_5OH), has a molecular mass of 46.1 amu and a molar mass of 46.1 g/mol. Figure 3.2 shows molar amounts of different substances.

Example 3.3 Calculating the Mass of an Atom or Molecule

Gaining Mastery Toolbox

Critical Concept 3.3

The mass of one mole (6.022×10^{23} units) of a substance is the molar mass of a substance. The units of molar mass are grams per mole (g/mol). The molar mass of each element can be found on the periodic table and the inside back cover of the text.

Solution Essentials:

- Mole (mol)
- Avogadro's number (6.022×10^{23})
- Molar mass
- Periodic table
- Rules for significant figures and rounding

a What is the mass in grams of a chlorine atom, Cl?
b What is the mass in grams of a hydrogen chloride molecule, HCl?

Problem Strategy In order to solve this type of problem, we need to consider using molar mass and the relationship between the number of atoms or molecules and the molar mass.

Solution

a The atomic mass of Cl is 35.5 amu, so the molar mass of Cl is 35.5 g/mol. Dividing 35.5 g (per mole) by 6.02×10^{23} (Avogadro's number) gives the mass of one atom.

$$\text{Mass of a Cl atom} = \frac{35.5 \text{ g}}{6.02 \times 10^{23}} = \mathbf{5.90 \times 10^{-23} \text{ g}}$$

b The molecular mass of HCl equals the AM of H plus the AM of Cl, or 1.01 amu + 35.5 amu = 36.5 amu. Therefore, 1 mol HCl contains 36.5 g HCl and

$$\text{Mass of an HCl molecule} = \frac{36.5 \text{ g}}{6.02 \times 10^{23}} = \mathbf{6.06 \times 10^{-23} \text{ g}}$$

Answer Check As you know, individual atoms and molecules are incredibly small. Therefore, whenever you are asked to calculate the mass of a few atoms or molecules, you should expect a very small mass.

Exercise 3.3 **a** What is the mass in grams of a calcium atom, Ca? **b** What is the mass in grams of an ethanol molecule, C_2H_5OH?

■ See Problems 3.33, 3.34, 3.35, and 3.36.

Mole Calculations

Now that you know how to find the mass of one mole of a substance, there are two important questions to ask. First, how much does a given number of moles of a substance weigh? Second, how many moles of a given formula unit does a given mass of substance contain? Both questions are easily answered using dimensional analysis, or the conversion-factor method.

Alternatively, because the molar mass is the mass per mole, you can relate mass and moles by means of the formula

Molar mass = mass/moles

To illustrate, consider the conversion of grams of ethanol, C_2H_5OH, to moles of ethanol. The molar mass of ethanol is 46.1 g/mol, so we write

$$1 \text{ mol } C_2H_5OH = 46.1 \text{ g } C_2H_5OH$$

Thus, the factor converting grams of ethanol to moles of ethanol is 1 mol C_2H_5OH/46.1 g C_2H_5OH. To convert moles of ethanol to grams of ethanol, we simply invert the conversion factor (46.1 g C_2H_5OH/1 mol C_2H_5OH). Note that the unit you are converting *from* is on the bottom of the conversion factor; the unit you are converting *to* is on the top.

Again, suppose you are going to prepare acetic acid from 10.0 g of ethanol, C_2H_5OH. How many moles of C_2H_5OH is this? You convert 10.0 g C_2H_5OH to moles C_2H_5OH by multiplying by the appropriate conversion factor.

$$10.0 \text{ g } \cancel{C_2H_5OH} \times \frac{1 \text{ mol } C_2H_5OH}{46.1 \text{ g } \cancel{C_2H_5OH}} = 0.217 \text{ mol } C_2H_5OH$$

Diagramming this solution (recall that the ⇒ indicates a conversion factor).

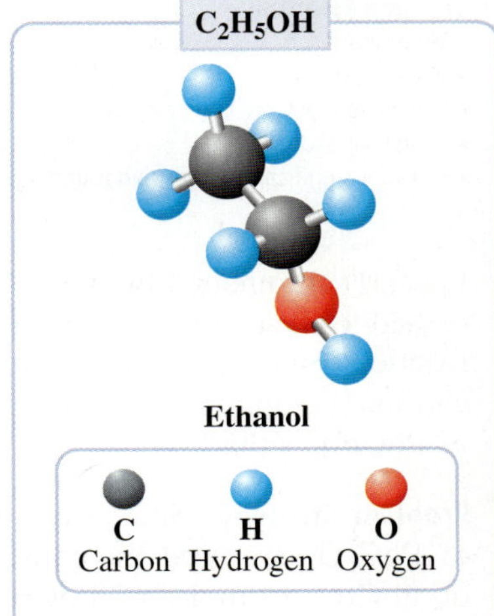

The following examples further illustrate this conversion-factor technique.

Example 3.4 Converting Moles of Substance to Grams

Gaining Mastery Toolbox

Critical Concept 3.4
Molar mass is a conversion factor. This conversion factor can be used with dimensional analysis to convert from grams of substance to moles of substance.

Solution Essentials:
- Molar mass
- Mole (mol)
- Conversion factor
- Dimensional analysis
- Rules for significant figures and rounding

Zinc iodide, ZnI_2, can be prepared by the direct combination of elements (Figure 3.3). A chemist determines from the amounts of elements that 0.0654 mol ZnI_2 can form. How many grams of zinc iodide is this?

Problem Strategy Use the molar mass to write the factor that converts from mol ZnI_2 to g ZnI_2. Note that the unit you are converting from (mol ZnI_2) is on the bottom of the conversion factor, and the unit you are converting to (g ZnI_2) is on the top.

The problem strategy is diagrammed as

Solution The molar mass of ZnI_2 is 319 g/mol. (The formula mass is 319 amu, which is obtained by summing the atomic masses in the formula.) Therefore,

$$0.0654 \cancel{\text{mol } ZnI_2} \times \frac{319 \text{ g } ZnI_2}{1 \cancel{\text{mol } ZnI_2}} = \mathbf{20.9 \text{ g } ZnI_2}$$

Answer Check Whenever you solve a problem of this type, be sure to write all units, making certain that they will cancel. This "built-in" feature of dimensional analysis ensures that you are correctly using the conversion factors.

Exercise 3.4 Hydrogen peroxide, H_2O_2, is a colorless liquid. A concentrated solution of it is used as a source of oxygen for rocket propellant fuels. Dilute aqueous solutions are used as a bleach. Analysis of a solution shows that it contains 0.909 mol H_2O_2 in 1.00 L of solution. What is the mass of hydrogen peroxide in this volume of solution?

Figure 3.3 **Reaction of zinc and iodine** Heat from the reaction of the elements causes some iodine to vaporize (violet vapor).

See Problems 3.37, 3.38, 3.39, and 3.40.

OWL Interactive Example 3.5 Converting Grams of Substance to Moles

Gaining Mastery Toolbox

Critical Concept 3.5
Molar mass is a conversion factor. This conversion factor can be used with dimensional analysis to convert from moles of substance to mass of substance.

Solution Essentials:
- Molar mass
- Mole (mol)
- Conversion factor
- Dimensional analysis
- Rules for significant figures and rounding

Lead(II) chromate, $PbCrO_4$, is a yellow paint pigment (called chrome yellow) prepared by a precipitation reaction (Figure 3.4). In a preparation, 45.6 g of lead(II) chromate is obtained as a precipitate. How many moles of $PbCrO_4$ is this?

Problem Strategy Since we are starting with a mass of $PbCrO_4$, we need the conversion factor for grams of $PbCrO_4$ to moles of $PbCrO_4$. The molar mass of $PbCrO_4$ will provide this information.

The problem strategy is diagrammed as

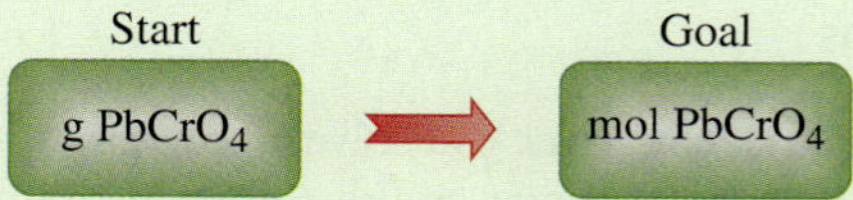

Solution The molar mass of $PbCrO_4$ is 323 g/mol. That is,

$$1 \text{ mol } PbCrO_4 = 323 \text{ g } PbCrO_4$$

Therefore,

$$45.6 \text{ g } \cancel{PbCrO_4} \times \frac{1 \text{ mol } PbCrO_4}{323 \text{ g } \cancel{PbCrO_4}} = \mathbf{0.141 \text{ mol } PbCrO_4}$$

Answer Check Note that the given amount of material in this problem (45.6 g $PbCrO_4$) is much less than its molar mass (323 g/mol). Therefore, we would expect the number of moles of $PbCrO_4$ to be much less than 1, which is the case here. Quick, alert comparisons such as this can be very valuable in checking for calculation errors.

Exercise 3.5 Nitric acid, HNO_3, is a colorless, corrosive liquid used in the manufacture of nitrogen fertilizers and explosives. In an experiment to develop new explosives for mining operations, a sample containing 28.5 g of nitric acid was poured into a beaker. How many moles of HNO_3 are there in this sample of nitric acid?

Figure 3.4 ▲

Preparation of lead(II) chromate When lead(II) nitrate solution (colorless) is added to potassium chromate solution (clear yellow), bright yellow solid lead(II) chromate forms (giving a cloudlike formation of fine crystals).

■ See Problems 3.41 and 3.42.

Example 3.6 Calculating the Number of Molecules in a Given Mass

Gaining Mastery Toolbox

Critical Concept 3.6
Avogadro's number is a conversion factor. Whenever you are required to know "how many" atoms, elements, molecules, etc., use the conversion factor 1 mol = 6.022×10^{23}.

Solution Essentials:
- Molar mass
- Avogadro's number (6.022×10^{23})
- Mole (mol)
- Conversion factor
- Dimensional analysis
- Rules for significant figures and rounding

How many molecules are there in a 3.46-g sample of hydrogen chloride, HCl?

Problem Strategy The number of molecules in a sample is related to moles of compound (1 mol HCl = 6.02×10^{23} HCl molecules). Therefore, if you first convert grams HCl to moles, then you can convert moles to number of molecules.

This problem strategy that requires two conversion factors is diagrammed as

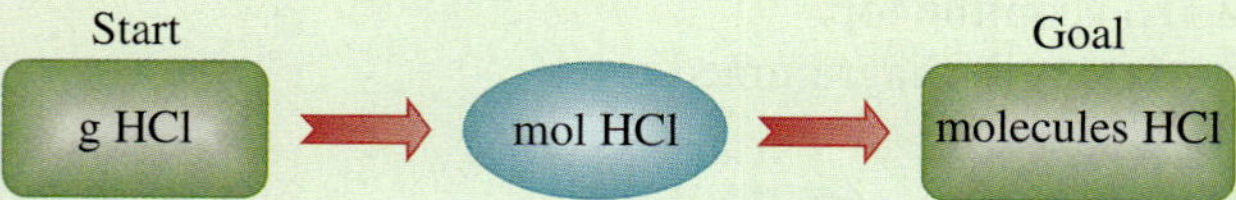

Solution Here is the calculation:

$$3.46 \text{ g } \cancel{HCl} \times \frac{1 \cancel{\text{ mol } HCl}}{36.5 \text{ g } \cancel{HCl}} \times \frac{6.02 \times 10^{23} \text{ HCl molecules}}{1 \cancel{\text{ mol } HCl}}$$
$$= \mathbf{5.71 \times 10^{22} \text{ HCl molecules}}$$

(continued)

(continued)

Note how the units in the numerator of a factor are canceled by the units in the denominator of the following factor.

Answer Check A very common mistake made when solving this type of problem is to use an incorrect conversion factor, such as 1 mol HCl = 6.02×10^{23} g HCl. If this statement were true, the mass of HCl contained in a mole would be far more than the mass of all the matter in the Milky Way galaxy! Therefore, if you end up with a gigantic mass for an answer, or get stuck on how to use the quantity 6.02×10^{23} as a conversion factor, improper unit assignment is a likely culprit.

Exercise 3.6 Hydrogen cyanide, HCN, is a volatile, colorless liquid with the odor of certain fruit pits (such as peach and cherry pits). The compound is highly poisonous. How many molecules are there in 56 mg HCN, the average toxic dose?

■ See Problems 3.45, 3.46, 3.47, and 3.48.

CONCEPT CHECK 3.1

You have 1.5 moles of tricycles.

a How many moles of seats do you have?

b How many moles of tires do you have?

c How could you use parts a and b as an analogy to teach a friend about the number of moles of OH^- ions in 1.5 moles of $Mg(OH)_2$?

Determining Chemical Formulas

When a chemist has discovered a new compound, the first question to answer is, What is the formula? To answer, you begin by analyzing the compound to determine amounts of the elements for a given amount of compound. This is conveniently expressed as **percentage composition**—that is, as *the mass percentages of each element in the compound.* You then determine the formula from this percentage composition. If the compound is a molecular substance, you must also find the molecular mass of the compound in order to determine the molecular formula.

The next section describes the calculation of mass percentages. Then, in two following sections, we describe how to determine a chemical formula.

3.3 Mass Percentages from the Formula

Suppose that A is a part of something—that is, part of a whole. It could be an element in a compound or one substance in a mixture. We define the **mass percentage** of *A as the parts of* A *per hundred parts of the total, by mass.* That is,

$$\text{Mass } \% A = \frac{\text{mass of } A \text{ in the whole}}{\text{mass of the whole}} \times 100\%$$

You can look at the mass percentage of A as the number of grams of A in 100 g of the whole.

The next example will provide practice with the concept of mass percentage. In this example we will start with a compound (formaldehyde, CH_2O) whose formula is given and obtain the percentage composition.

Example 3.7 Calculating the Percentage Composition from the Formula

Gaining Mastery Toolbox

Critical Concept 3.7
You assume that you have one mole of compound. Making this assumption, you then know the total mass of the compound (molar mass), and from the chemical formula, the number of moles of each element in the compound.

Solution Essentials:
- Percentage composition
- Mass percentage
- Molar mass
- Rules for significant figures and rounding

Formaldehyde, CH_2O, is a toxic gas with a pungent odor. Large quantities are consumed in the manufacture of plastics (Figure 3.5), and a water solution of the compound is used to preserve biological specimens. Calculate the mass percentages of the elements in formaldehyde (give answers to three significant figures).

Problem Strategy To calculate mass percentage, you need the mass of an element in a given mass of compound. You can get this information by interpreting the formula in molar terms and then converting moles to masses, using a table of atomic masses. Thus, 1 mol CH_2O has a mass of 30.0 g and contains 1 mol C (12.0 g), 2 mol H (2 × 1.01 g), and 1 mol O (16.0 g). You divide each mass of element by the molar mass, then multiply by 100, to obtain the mass percentage.

Solution Here are the calculations:

$$\%\ \mathbf{C} = \frac{12.0\ \text{g}}{30.0\ \text{g}} \times 100\% = \mathbf{40.0\%}$$

$$\%\ \mathbf{H} = \frac{2 \times 1.01\ \text{g}}{30.0\ \text{g}} \times 100\% = \mathbf{6.73\%}$$

You can calculate the percentage of O in the same way, but it can also be found by subtracting the percentages of C and H from 100%:

$$\%\ \mathbf{O} = 100\% - (40.0\% + 6.73\%) = \mathbf{53.3\%}$$

The clear solution contains formaldehyde, CH_2O, and resorcinol, $C_6H_4(OH)_2$. The formation of the plastic is started by adding several drops of potassium hydroxide.

The red plastic has formed in the beaker.

Figure 3.5 ▲
Preparing resorcinol-formaldehyde plastic

Answer Check A typical mistake when working a mass percentage problem is to forget to account for the number of moles of each element given in the chemical formula. An example would be to answer this question as though the formula were CHO instead of CH_2O.

Exercise 3.7 Ammonium nitrate, NH_4NO_3, which is prepared from nitric acid, is used as a nitrogen fertilizer. Calculate the mass percentages of the elements in ammonium nitrate (to three significant figures).

■ See Problems 3.57, 3.58, 3.59, and 3.60.

Example 3.8 Calculating the Mass of an Element in a Given Mass of Compound

Gaining Mastery Toolbox

Critical Concept 3.8
The percentage composition of an element is the *mass* percentage of the element. The mass percentage of an element can be used to calculate the mass of that element in a sample.

Solution Essentials:
- Percentage composition
- Mass percentage
- Rules for significant figures and rounding

How many grams of carbon are there in 83.5 g of formaldehyde, CH_2O? Use the percentage composition obtained in the previous example (40.0% C, 6.73% H, 53.3% O).

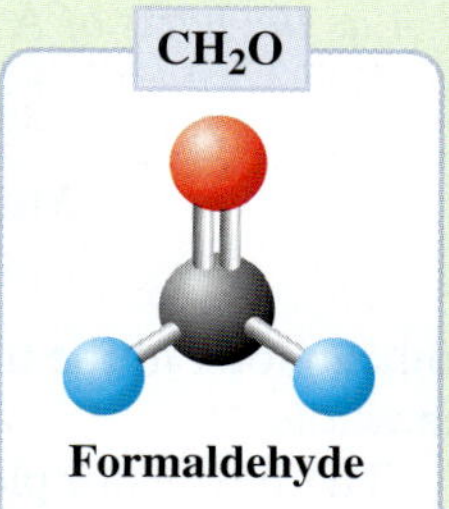

Problem Strategy Solving this type of problem requires that

(continued)

(continued)

you first calculate the mass percentage of the element of interest, carbon in this case (as noted in the problem, this was done in Example 3.7). Multiplying of this percentage, expressed as a decimal, times the mass of formaldehyde in the sample, will yield the mass of carbon present.

Solution CH_2O is 40.0% C, so the mass of carbon in 83.5 g CH_2O is

$$83.5 \text{ g} \times 0.400 = \mathbf{33.4 \text{ g}}$$

Answer Check Make sure that you have converted the percent to a decimal prior to multiplication. An answer that is more than the starting mass of material is an indication that you probably made this mistake.

Exercise 3.8 How many grams of nitrogen, N, are there in a fertilizer containing 48.5 g of ammonium nitrate and no other nitrogen-containing compound? See Exercise 3.7 for the percentage composition of NH_4NO_3.

■ See Problems 3.61 and 3.62.

3.4 Elemental Analysis: Percentages of Carbon, Hydrogen, and Oxygen

Suppose you have a newly discovered compound whose formula you wish to determine. The first step is to obtain its percentage composition. As an example, consider the determination of the percentages of carbon, hydrogen, and oxygen in compounds containing only these three elements. The basic idea is this: You burn a sample of the compound of known mass and get CO_2 and H_2O. Next you relate the masses of CO_2 and H_2O to the masses of carbon and hydrogen. Then you calculate the mass percentages of C and H. You find the mass percentage of O by subtracting the mass percentages of C and H from 100.

Figure 3.6 shows an apparatus used to find the amount of carbon and hydrogen in a compound. The compound is burned in a stream of oxygen gas. The vapor of the compound and its combustion products pass over copper pellets coated with copper(II) oxide, CuO, which supplies additional oxygen and ensures that the compound is completely burned. As a result of the combustion, every mole of carbon (C) in the compound ends up as a mole of carbon dioxide (CO_2), and every mole of hydrogen (H) ends up as one-half mole of water (H_2O). The water is collected by a drying agent, a substance that has a strong affinity for water. The carbon dioxide is collected by chemical reaction with sodium hydroxide, NaOH.

Figure 3.6 ▼

Combustion method for determining the percentages of carbon and hydrogen in a compound

Vapor of the compound burns in O_2 in the presence of CuO pellets, giving CO_2 and H_2O. The water vapor is collected by a drying agent, and CO_2 combines with the sodium hydroxide. Amounts of CO_2 and H_2O are obtained by weighing the U-tubes before and after combustion.

Copper(II) oxide (CuO) pellets

Oxygen in

Excess oxygen out

Furnace

Drying agent (traps H_2O)

Sodium hydroxide (traps CO_2)

The compound is placed in the sample dish and is heated by the furnace.

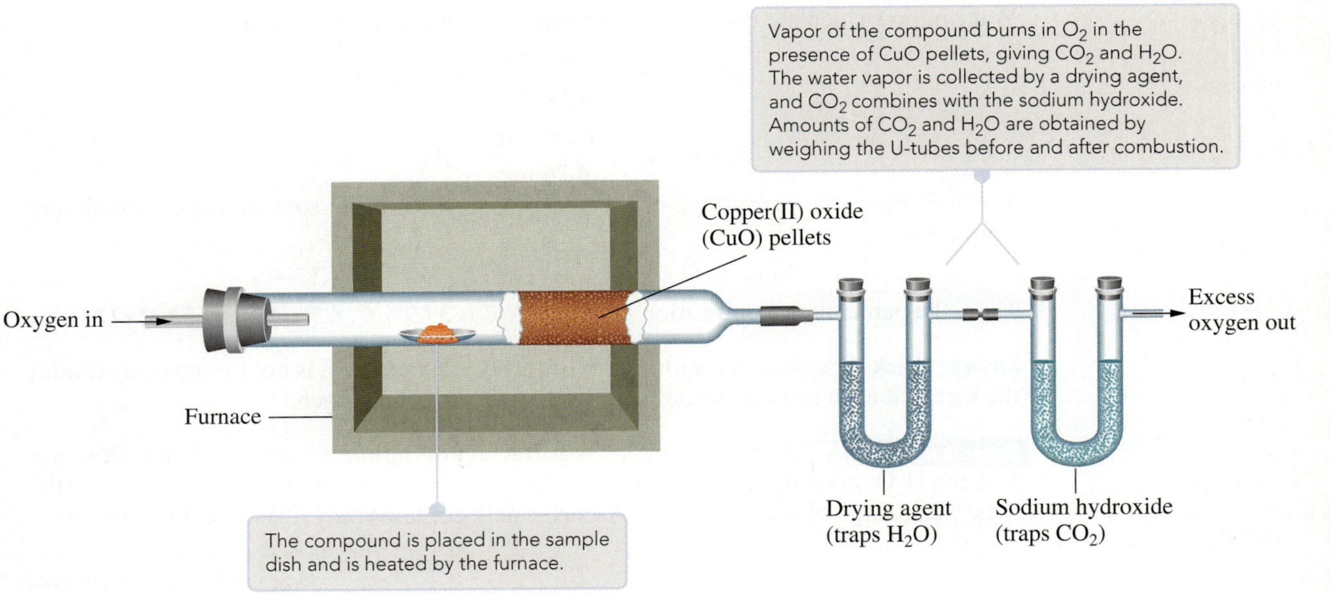

Sodium hydroxide reacts with carbon dioxide according to the following equations:

$NaOH + CO_2 \longrightarrow NaHCO_3$

$2NaOH + CO_2 \longrightarrow Na_2CO_3 + H_2O$

◀ By weighing the U-tubes containing the drying agent and the sodium hydroxide before and after combustion, it is possible to determine the masses of water and carbon dioxide produced. From these data, you can calculate the percentage composition of the compound.

The chapter opened with a discussion of acetic acid. The next example shows how to determine the percentage composition of this substance from combustion data. We will use this percentage composition later in Example 3.12 to determine the formula of acetic acid.

OWL Interactive Example 3.9 Calculating the Percentages of C and H by Combustion

Gaining Mastery Toolbox

Critical Concept 3.9

The mass percentages of carbon and hydrogen can be calculated from the masses of CO_2 and H_2O. During the reaction *all* of the carbon atoms in the sample end up in the CO_2 and *all* of the hydrogen atoms end up in the H_2O. Therefore, the mass of carbon in the CO_2 is equal to the mass of carbon in the sample, and the mass of hydrogen in the H_2O is equal to the mass of hydrogen in the sample. We *cannot* make the statement that the mass of oxygen in the H_2O and CO_2 is equal to the mass of oxygen in the sample because oxygen is both in the sample and is a reactant.

Solution Essentials:

- Elemental analysis
- Percentage composition
- Mass percentage
- Molar mass
- Dimensional analysis
- Rules for significant figures and rounding

Acetic acid contains only C, H, and O. A 4.24-mg sample of acetic acid is completely burned. It gives 6.21 mg of carbon dioxide and 2.54 mg of water. What is the mass percentage of each element in acetic acid?

Problem Strategy Note that in the products of the combustion, all of the carbon from the sample ends up in the CO_2, all of the hydrogen ends up in the H_2O, and the oxygen is in both compounds. Because of this, you should concentrate first on the carbon and hydrogen and worry about the oxygen last. If we can determine the mass of carbon and hydrogen in the original sample, we should then be able to determine the percentage of each of these elements present in the compound. Once we know the mass percentages of carbon and hydrogen in the original compound, the remaining mass percentage (100% total) must be due to oxygen. Let's start by determining the mass of carbon that was originally contained in the compound. You first convert the mass of CO_2 to moles of CO_2. Then you convert this to moles of C, noting that 1 mol C produces 1 mol CO_2. Finally, you convert to mass of C. Similarly, for hydrogen, you convert the mass of H_2O to mol H_2O, then to mol H, and finally to mass of H. (Remember that 1 mol H_2O produces 2 mol H.) Once you have the masses of C and H, you can calculate the mass percentages. Subtract from 100% to get % O.

Solution Following is the calculation of grams C:

$$6.21 \times 10^{-3}\ \cancel{g\ CO_2} \times \frac{1\ \cancel{mol\ CO_2}}{44.0\ \cancel{g\ CO_2}} \times \frac{1\ \cancel{mol\ C}}{1\ \cancel{mol\ CO_2}} \times \frac{12.0\ g\ C}{1\ \cancel{mol\ C}} = 1.69 \times 10^{-3}\ g\ C\ (\text{or } 1.69\ mg\ C)$$

For hydrogen, you note that 1 mol H_2O yields 2 mol H, so you write

$$2.54 \times 10^{-3}\ \cancel{g\ H_2O} \times \frac{1\ \cancel{mol\ H_2O}}{18.0\ \cancel{g\ H_2O}} \times \frac{2\ \cancel{mol\ H}}{1\ \cancel{mol\ H_2O}} \times \frac{1.01\ g\ H}{1\ \cancel{mol\ H}} = 2.85 \times 10^{-4}\ g\ H(\text{ or } 0.285\ mg\ H)$$

You can now calculate the mass percentages of C and H in acetic acid.

$$\text{Mass \% C} = \frac{1.69\ mg}{4.24\ mg} \times 100\% = 39.9\%$$

$$\text{Mass \% H} = \frac{0.285\ mg}{4.24\ mg} \times 100\% = 6.72\%$$

You find the mass percentage of oxygen by subtracting the sum of these percentages from 100%:

$$\text{Mass \% O} = 100\% - (39.9\% + 6.72\%) = 53.4\%$$

Thus, the percentage composition of acetic acid is **39.9% C, 6.7% H, and 53.4% O.**

Answer Check The most common error for this type of problem is not taking into account the fact that each mole of water contains two moles of hydrogen.

Exercise 3.9 A 3.87-mg sample of ascorbic acid (vitamin C) gives 5.80 mg CO_2 and 1.58 mg H_2O on combustion. What is the percentage composition of this compound (the mass percentage of each element)? Ascorbic acid contains only C, H, and O.

■ See Problems 3.63 and 3.64.

CONCEPT CHECK 3.2

You perform combustion analysis on a compound that contains only C and H.

a Considering the fact that the combustion products CO_2 and H_2O are colorless, how can you tell if some of the product got trapped in the CuO pellets (see Figure 3.6)?

b Would your calculated results of mass percentage of C and H be affected if some of the combustion products got trapped in the CuO pellets? If your answer is yes, how might your results differ from the expected values for the compound?

3.5 Determining Formulas

The percentage composition of a compound leads directly to its empirical formula. An **empirical formula** (or **simplest formula**) for a compound is *the formula of a substance written with the smallest integer (whole number) subscripts.* For most ionic substances, the empirical formula is the formula of the compound. ▶ This is often not the case for molecular substances. For example, hydrogen peroxide has the molecular formula H_2O_2. The molecular formula, you may recall, tells you the precise number of atoms of different elements in a molecule of the substance. The empirical formula, however, merely tells you the ratio of numbers of atoms in the compound. The empirical formula of hydrogen peroxide (H_2O_2) is HO (Figure 3.7).

The formula of sodium peroxide, an ionic compound of Na^+ and O_2^{2-}, is Na_2O_2. Its empirical formula is NaO.

Compounds with different molecular formulas can have the same empirical formula, and such substances will have the same percentage composition. An example is acetylene, C_2H_2, and benzene, C_6H_6. Acetylene is a gas used as a fuel and also in welding. Benzene, in contrast, is a liquid that is used in the manufacture of plastics and is a component of gasoline. Table 3.1 illustrates how these two compounds, with the same empirical formula, but different molecular formulas, also have different chemical structures. Because the empirical formulas of acetylene and benzene are the same, they have the same percentage composition: 92.3% C and 7.7% H, by mass.

To obtain the molecular formula of a substance, you need two pieces of information: (1) as in the previous section, the percentage composition, from which the empirical formula can be determined; and (2) the molecular mass. The molecular mass allows you to choose the correct multiple of the empirical formula for the molecular formula. We will illustrate these steps in the next three examples.

Empirical Formula from the Composition

The empirical formula of a compound shows the ratios of numbers of atoms in the compound. You can find this formula from the composition of the compound by converting masses of the elements to moles. The next two examples show these calculations in detail.

© Helen Sessions/Alamy

Figure 3.7 ▲

Molecular model of hydrogen peroxide (H_2O_2) Hydrogen peroxide has the empirical formula HO and the molecular formula H_2O_2. In order to arrive at the correct structure of molecular compounds like H_2O_2 shown here, chemists must have the molecular formula.

Table 1.3 Molecular Models of Two Compounds That Have the Empirical Formula CH

Although benzene and acetylene have the same empirical formula, they do not have the same molecular formula or structure.

Compound	Empirical Formula	Molecular Formula	Molecular Model
Acetylene	CH	C_2H_2	
Benzene	CH	C_6H_6	H C

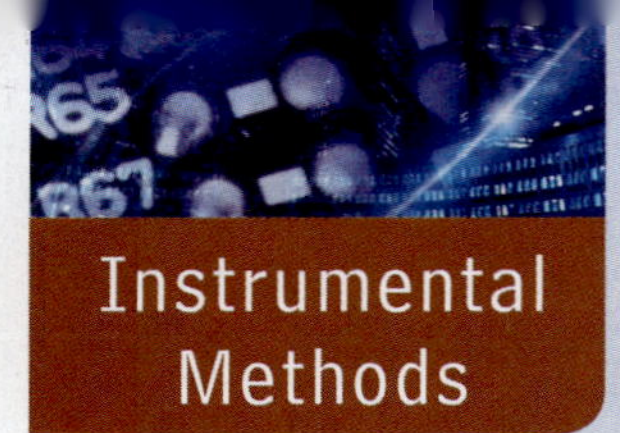

Instrumental Methods

Mass Spectrometry and Molecular Formula

Some very sophisticated instruments have become indispensable to modern chemical research. One such instrument is the **mass spectrometer,** which measures the masses of positive ions produced from an extremely small sample of a substance, displaying the data as a *mass spectrum.* Figure 3.8 shows the mass spectrum of methylene chloride, CH_2Cl_2. Each of the positive ions produced in the spectrometer from the substance is shown as a bar or line (called a *peak*) at the mass of the ion, in amu. (Strictly speaking, each peak appears at the mass-to-charge ratio of the ion, but the ion charge is usually one.) The height of each peak is proportional to the relative abundance of the ion. The mass spectrum of a substance can be used to identify it or, if the substance is a new one, to obtain its molecular formula. The spectrum may also give information about the structure of the molecule.

Figure 2.11 on page 51 shows a simplified diagram of a mass spectrometer, used there to separate atomic neon into its three isotopic ions. Neon gas (which consists of neon atoms) enters the spectrometer, where the atoms are ionized by bombarding them with a beam of high-energy electrons. (An electron from the beam strikes a neon atom, knocking out one of neon's electrons.) The positive ions are accelerated by a negative plate into the field of a magnet, which bends the path of each ion depending on its mass-to-charge ratio. The less massive ions are bent more strongly than the heavier ones for the same charge. In most modern spectrometers, the ions fall on an electronic detector, whose current is recorded and displayed as the mass spectrum on a computer screen and printed on chart paper.

The spectrum of a molecular substance is more complicated than that of an atom. Consider methylene chloride, CH_2Cl_2. When its molecule is struck by an electron of sufficient energy, it may lose one of its own electrons (now giving two free electrons), leaving behind a positive ion:

$$CH_2Cl_2 + e^- \longrightarrow CH_2Cl_2^+ + 2e^-$$

The ion gains a great deal of energy from the collision of the molecule with an electron, and it may lose this energy by breaking into smaller pieces. For example:

$$CH_2Cl_2^+ \longrightarrow CH_2Cl^+ + Cl$$

Thus, the original molecule, even one as simple as CH_2Cl_2, can give rise to a number of ions. In addition, each ion may consist of atoms having more than one natural isotope, so that each ion may occur in the spectrum as a group of several peaks. Note that the mass spectrum of methylene chloride shown in Figure 3.8 displays four prominent peaks and many smaller ones. A larger molecule can give an even more complicated mass spectrum, oftentimes yielding a "fingerprint" useful in identification. Only methylene chloride has exactly the spectrum shown in Figure 3.8. Thus, by comparing the mass spectrum of an unknown substance with those in a catalog of mass spectra of known compounds, you can determine its identity. The more information you have about the compound, the shorter the search through the catalog of spectra.

The mass spectrum itself contains a wealth of information about molecular structure. Some experience is needed to analyze the spectrum of a compound, but you can get an idea of how it is done by looking at Figure 3.8. Suppose you do not know the identity of the compound. The most intense peaks at the greatest mass usually correspond to the ion from the original molecule and give you the molecular mass. Thus, you would expect the peaks at mass 84 and mass 86 to be from the original molecular ion, so the molecular mass is approximately 84 to 86.

Example 3.10 Determining the Empirical Formula from Masses of Elements (Binary Compound)

Gaining Mastery Toolbox

Critical Concept 3.10

The mole ratio of elements in the compound is the same ratio as the element subscripts in the empirical formula. The subscripts of the compound must be written as whole (integer) numbers. In some cases the empirical formula is the same as the molecular formula.

Solution Essentials:

- Empirical formula
- Molar mass
- Dimensional analysis
- Rules for significant figures and rounding

A compound of nitrogen and oxygen is analyzed, and a sample weighing 1.587 g is found to contain 0.483 g N and 1.104 g O. What is the empirical formula of the compound?

Problem Strategy This compound has the formula N_xO_y, where x and y are whole numbers that we need to determine. Masses of N and O are given in the problem. If we convert the masses of each of these elements to moles, they will be proportional to the subscripts in the empirical formula. The formula must be the smallest integer ratio of the elements. To obtain the smallest integers from the moles, we divide each by the smallest one. If the results are all whole numbers, they

(continued)

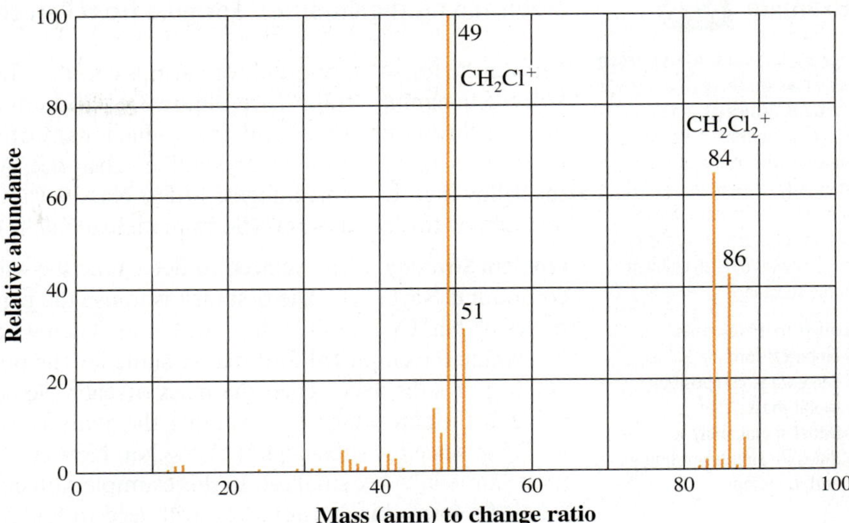

Figure 3.8 ▶

Mass spectrum of methylene chloride, CH_2Cl_2 The lines (*peaks*) at higher mass correspond to the ions $CH_2Cl_2^+$ with different isotopes, produced by the loss of a single electron from the original CH_2Cl_2 molecule. (The spectrum is from the NIST Chemistry WebBook, http://webbook.nist.gov/chemistry/.)

An elemental analysis is also possible. The two most intense peaks in Figure 3.8 are at 84 amu and 49 amu. They differ by 35 amu. Perhaps the original molecule that gives the peak at 84 amu contains a chlorine-35 atom, which is lost to give the peak at 49 amu. If this is true, you should expect another peak at mass 86, corresponding to the original molecular ion with chlorine-37 in place of a chlorine-35 atom. This is indeed what you see.

The relative heights of the peaks, which depend on the natural abundances of atoms, are also important, because they give you additional information about the elements present. The relative heights can also tell you how many atoms of a given element are in the original molecule. Naturally occurring chlorine is 75.8% chlorine-35 and 24.2% chlorine-37. If the original molecule contained only one Cl atom, the peaks at 84 amu and 86 amu would be in the ratio 0.758 : 0.242. That is, the peak at 86 amu would be about one-third the height of the one at 84 amu. In fact, the relative height is twice this value. This means that the molecular ion contains two chlorine atoms, because the chance that any such ion contains one chlorine-37 atom is then twice as great. Thus, simply by comparing relative peak heights, you can confirm the presence of particular elements.

The spectrum in Figure 3.8 shows the peaks at nearest whole-number masses. There are mass spectrometers, however, that can obtain the masses of moderate-sized molecules to perhaps four decimal places. With this high resolution, you can in principle obtain the molecular formula of a substance. For example, the CH_2Cl_2 molecule composed of the most abundant isotopes has a whole-number molecular mass of 84 amu (the largest peak at the right in Figure 3.8), but its mass to four decimals is 83.9534 amu. There are several other molecules with whole-number masses of 84 amu. One of these is hexene, C_6H_{12} (the species composed of the more abundant isotopes); but, its more exact mass is 84.0939 amu. In general, molecules with different molecular formulas will have different masses, so you can turn this problem around. Given a molecular mass, you can use special computer programs to obtain the molecular formula.

■ See Problems 3.115 and 3.116.

(continued)

will be the subscripts in the formula. (Otherwise, you will need to multiply by some factor, as illustrated in Example 3.11.)

Solution You convert the masses to moles:

$$0.483\ \cancel{\text{g N}} \times \frac{1\ \text{mol N}}{14.0\ \cancel{\text{g N}}} = 0.0345\ \text{mol N}$$

$$1.104\ \cancel{\text{g O}} \times \frac{1\ \text{mol O}}{16.00\ \cancel{\text{g O}}} = 0.06900\ \text{mol O}$$

In order to obtain the smallest integers, you divide each mole number by the smaller one (0.0345 mol). For N, you get 1.00; for O, you get 2.00. Thus, the ratio of the number of N atoms to the number of O atoms is 1 to 2. Hence, the empirical formula is $\mathbf{NO_2}$.

Answer Check Keep in mind that the empirical formula determined by this type of calculation is not necessarily the molecular formula of the compound. The only information that it provides is the smallest whole-number ratio of the elements.

Exercise 3.10 A sample of compound weighing 83.5 g contains 33.4 g of sulfur. The rest is oxygen. What is the empirical formula?

■ See Problems 3.65 and 3.66.

Example 3.11 Determining the Empirical Formula from Percentage Composition (General)

Gaining Mastery Toolbox

Critical Concept 3.11
You assume that you are starting with 100 g of substance. By making this assumption, it simplifies the process of determining the relative masses of each element in the substance.

Solution Essentials:
- Empirical formula
- Percentage composition
- Molar mass
- Dimensional analysis
- Rules for significant figures and rounding

Chromium forms compounds of various colors. (The word *chromium* comes from the Greek *khroma,* meaning "color"; see Figure 3.9.) Sodium dichromate is the most important commercial chromium compound, from which many other chromium compounds are manufactured. It is a bright orange, crystalline substance. An analysis of sodium dichromate gives the following mass percentages: 17.5% Na, 39.7% Cr, and 42.8% O. What is the empirical formula of this compound? (Sodium dichromate is ionic, so it has no molecular formula.)

Problem Strategy Here we need to determine the whole-number values for x, y, and z in the compound $Na_xCr_yO_z$. The first task is to use the percent composition data to determine the moles of Na, Cr, and O in the compound. Then we can use mole ratios of these elements to determine the empirical formula. Assume for the purposes of this calculation that you have 100.0 g of substance. Then the mass of each element in the sample equals the numerical value of the percentage. For example, the quantity of sodium in 100 g of sodium dichromate is 17.5 g, since the substance is 17.5% Na. Now convert the masses to moles and divide each mole number by the smallest. In this example, you do not obtain a series of integers, or whole numbers, from this division. You will need to find a whole-number factor to multiply these results by to obtain integers. Normally this factor will be 2 or 3, though it might be larger.

Solution Of the 100.0 g of sodium dichromate, 17.5 g is Na, 39.7 g is Cr, and 42.8 g is O. You convert these amounts to moles.

$$17.5\ \cancel{\text{g Na}} \times \frac{1\ \text{mol Na}}{23.0\ \cancel{\text{g Na}}} = 0.761\ \text{mol Na}$$

$$39.7\ \cancel{\text{g Cr}} \times \frac{1\ \text{mol Cr}}{52.0\ \cancel{\text{g Cr}}} = 0.763\ \text{mol Cr}$$

$$42.8\ \cancel{\text{g O}} \times \frac{1\ \text{mol O}}{16.0\ \cancel{\text{g O}}} = 2.68\ \text{mol O}$$

Now you divide all the mole numbers by the smallest one.

$$\text{For Na: } \frac{0.761\ \cancel{\text{mol}}}{0.761\ \cancel{\text{mol}}} = 1.00$$

$$\text{For Cr: } \frac{0.763\ \cancel{\text{mol}}}{0.761\ \cancel{\text{mol}}} = 1.00$$

$$\text{For O: } \frac{2.68\ \cancel{\text{mol}}}{0.761\ \cancel{\text{mol}}} = 3.52$$

You must decide whether these numbers are integers (whole numbers), within experimental error. If you round off the last digit, which is subject to experimental error, you get $Na_{1.0}Cr_{1.0}O_{3.5}$. In this case, the subscripts are not all integers. However, they can be made into integers by multiplying each one by 2; you get $Na_{2.0}Cr_{2.0}O_{7.0}$. Thus, the empirical formula is $\mathbf{Na_2Cr_2O_7}$.

Answer Check If you are working this type of problem and are not arriving at an answer that is readily converted to integers, it might be that you did not use enough significant figures throughout the calculation and/or that you rounded intermediate calculations.

Exercise 3.11 Benzoic acid is a white, crystalline powder used as a food preservative. The compound contains 68.8% C, 5.0% H, and 26.2% O, by mass. What is its empirical formula?

Figure 3.9 ▼
Chromium compounds of different colors

Potassium chromate, K_2CrO_4
Chromium(VI) oxide, CrO_3
Chromium(III) sulfate, $Cr_2(SO_4)_3$
Chromium metal
Potassium dichromate, $K_2Cr_2O_7$
chromium(III) oxide, Cr_2O_3

■ See Problems 3.67, 3.68, 3.69, and 3.70.

Molecular Formula from Empirical Formula

The molecular formula of a compound is a multiple of its empirical formula. For example, the molecular formula of acetylene, C_2H_2, is equivalent to $(CH)_2$, and the molecular formula of benzene, C_6H_6, is equivalent to $(CH)_6$. Therefore, the molecular mass is some multiple of the empirical formula mass, which is obtained by summing the atomic masses of the atoms in the empirical formula. For any molecular compound, you can write

$$\text{Molecular mass} = n \times \text{empirical formula mass}$$

where n is the number of empirical formula units in the molecule. You get the molecular formula by multiplying the subscripts of the empirical formula by n, which you calculate from the equation

$$n = \frac{\text{molecular mass}}{\text{empirical formula mass}}$$

Once you determine the empirical formula for a compound, you can calculate its empirical formula mass. If you have an experimental determination of its molecular mass, you can calculate n and then the molecular formula. The next example illustrates how you use percentage composition and molecular mass to determine the molecular formula of acetic acid.

Example 3.12 Determining the Molecular Formula from Percentage Composition and Molecular Mass

Gaining Mastery Toolbox

Critical Concept 3.12
The molecular mass of a compound is a multiple of the empirical formula mass. The multiple is an integer and can be used to determine the molecular formula of the compound from the empirical formula.

Solution Essentials:
- Empirical formula
- Empirical formula mass
- Molecular mass
- Percentage composition
- Dimensional analysis
- Rules for significant figures and rounding

In Example 3.9, we found the percentage composition of acetic acid to be 39.9% C, 6.7% H, and 53.4% O. Determine the empirical formula. The molecular mass of acetic acid was determined by experiment to be 60.0 amu. What is its molecular formula?

Problem Strategy This problem employs the same general strategy as Example 3.11, with a few additional steps. In this case the compound is composed of C, H, and O. After determining the empirical formula of the compound, you calculate the empirical formula mass. You then determine how many times more massive the molecular compound is than the empirical compound by dividing the molecular mass by the empirical formula mass. You now have a factor by which the subscripts of the empirical formula need to be multiplied to arrive at the molecular formula.

Solution A sample of 100.0 g of acetic acid contains 39.9 g C, 6.7 g H, and 53.4 g O. Converting these masses to moles gives 3.33 mol C, 6.6 mol H, and 3.34 mol O. Dividing the mole numbers by the smallest one gives 1.00 for C, 2.0 for H, and 1.00 for O. **The empirical formula of acetic acid is CH_2O.** (You may have noted that the percentage composition of acetic acid is, within experimental error, the same as that of formaldehyde—see Example 3.7—so they must have the same empirical formula.) The empirical formula mass is 30.0 amu. Dividing the empirical formula mass into the molecular mass gives the number by which the subscripts in CH_2O must be multiplied.

$$n = \frac{\text{molecular mass}}{\text{empirical formula mass}} = \frac{60.0 \text{ amu}}{30.0 \text{ amu}} = 2.00$$

The molecular formula of acetic acid is $(CH_2O)_2$, or $C_2H_4O_2$. (A molecular model of acetic acid is shown in Figure 3.10.)

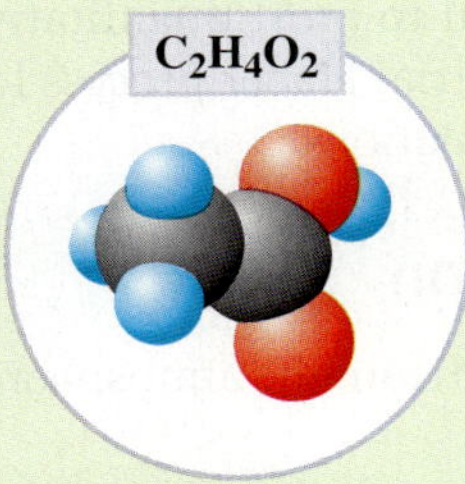

Figure 3.10 ▲ **Molecular model of acetic acid**

Answer Check Calculate the molecular mass of your answer. It should agree with the molecular mass of the compound given in the problem.

Exercise 3.12 The percentage composition of acetaldehyde is 54.5% C, 9.2% H, and 36.3% O, and its molecular mass is 44 amu. Obtain the molecular formula of acetaldehyde.

■ See Problems 3.73, 3.74, 3.75, and 3.76.

The condensed structural formula for acetaldehyde is CH_3CHO. Its structural formula is

```
   H  O
   |  ||
H—C—C—H
   |
   H
```

Determining this structural formula requires additional information.

The formula of acetic acid is often written $HC_2H_3O_2$ to indicate that one of the hydrogen atoms is acidic (lost easily), while the other three are not (Figure 3.10). Now that you know the formulas of acetic acid and acetaldehyde (determined from the data in Exercise 3.12 to be C_2H_4O), you can write the equations for the preparation of acetic acid described in the chapter opener. The first step consists of reacting ethanol with oxygen to obtain acetaldehyde and water. If you write the chemical equation and balance it, you obtain

$$\underset{\text{ethanol}}{2C_2H_5OH} + O_2 \longrightarrow \underset{\text{acetaldehyde}}{2C_2H_4O} + 2H_2O$$

In practice, the reaction is carried out with the reactants as gases (gas phase) at about 400°C using silver as a catalyst.

The second step consists of reacting acetaldehyde with oxygen to obtain acetic acid. Acetaldehyde liquid is mixed with a catalyst—manganese(II) acetate—and air is bubbled through it. The balanced equation is

$$\underset{\text{acetaldehyde}}{2C_2H_4O} + O_2 \longrightarrow \underset{\text{acetic acid}}{2HC_2H_3O_2}$$

Once you have this balanced equation, you are in a position to answer quantitative questions such as, How much acetic acid can you obtain from a 10.0-g sample of acetaldehyde? You will see how to answer such questions in the next sections.

CONCEPT CHECK 3.3

A friend has some questions about empirical formulas and molecular formulas. You can assume that he is good at performing the calculations.

a For a problem that asked him to determine the empirical formula, he came up with the answer $C_2H_8O_2$. Is this a possible answer to the problem? If not, what guidance would you offer your friend?

b For another problem he came up with the answer $C_{1.5}H_4$ as the empirical formula. Is this answer correct? Once again, if it isn't correct, what could you do to help your friend?

c Since you have been a big help, your friend asks one more question. He completed a problem of the same type as Example 3.12. His answers indicate that the compound had an empirical formula of C_3H_8O and the molecular formula C_3H_8O. Is this result possible?

Stoichiometry: Quantitative Relations in Chemical Reactions

In Chapter 2, we described a chemical equation as a representation of what occurs when molecules react. We will now study chemical equations more closely to answer questions about the stoichiometry of reactions. **Stoichiometry** (pronounced "stoy-key-om'-e-tree") is *the calculation of the quantities of reactants and products involved in a chemical reaction.* It is based on the chemical equation and on the relationship between mass and moles. Such calculations are fundamental to most quantitative work in chemistry. In the next sections, we will use the industrial Haber process for the production of ammonia to illustrate stoichiometric calculations.

3.6 Molar Interpretation of a Chemical Equation

In the Haber process for producing ammonia, NH_3, nitrogen (from the atmosphere) reacts with hydrogen at high temperature and pressure.

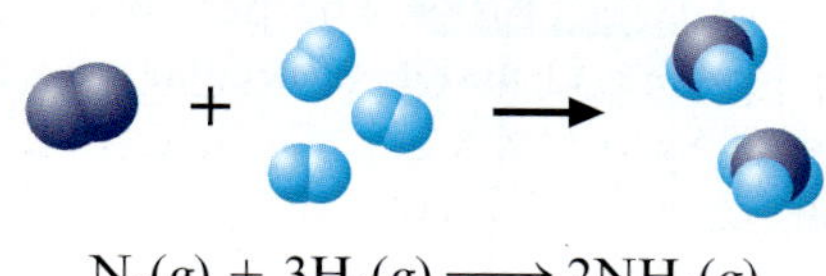

$$N_2(g) + 3H_2(g) \longrightarrow 2NH_3(g)$$

Hydrogen is usually obtained from natural gas or petroleum and so is relatively expensive. For this reason, the price of hydrogen partly determines the price of ammonia. Thus, an important question to answer is, How much hydrogen is required to give a particular quantity of ammonia? For example, how much hydrogen would be needed to produce one ton (907 kg) of ammonia? Similar kinds of questions arise throughout chemical research and industry.

To answer such quantitative questions, you must first look at the balanced chemical equation: one N_2 molecule and three H_2 molecules react to produce two NH_3 molecules (see models above). A similar statement involving multiples of these numbers of molecules is also correct. For example, 6.02×10^{23} N_2 molecules react with $3 \times 6.02 \times 10^{23}$ H_2 molecules, giving $2 \times 6.02 \times 10^{23}$ NH_3 molecules. This last statement can be put in molar terminology: one mole of N_2 reacts with three moles of H_2 to give two moles of NH_3.

You may interpret a chemical equation either in terms of numbers of molecules (or ions or formula units) or in terms of numbers of moles, depending on your needs.

Because moles can be converted to mass, you can also give a mass interpretation of a chemical equation. The molar masses of N_2, H_2, and NH_3 are 28.0, 2.02, and 17.0 g/mol, respectively. Therefore, 28.0 g of N_2 reacts with 3×2.02 g of H_2 to yield 2×17.0 g of NH_3.

We summarize these three interpretations as follows:

$$
\begin{array}{lclcll}
N_2 & + & 3H_2 & \longrightarrow & 2NH_3 & \\
1\text{ molecule } N_2 & + & 3\text{ molecules } H_2 & \longrightarrow & 2\text{ molecules } NH_3 & \text{(molecular interpretation)} \\
1\text{ mol } N_2 & + & 3\text{ mol } H_2 & \longrightarrow & 2\text{ mol } NH_3 & \text{(molar interpretation)} \\
28.0\text{ g } N_2 & + & 3 \times 2.02\text{ g } H_2 & \longrightarrow & 2 \times 17.0\text{ g } NH_3 & \text{(mass interpretation)}
\end{array}
$$

Suppose you ask how many grams of atmospheric nitrogen will react with 6.06 g (3×2.02 g) of hydrogen. You see from the last equation that the answer is 28.0 g N_2. We formulated this question for one mole of atmospheric nitrogen. Recalling the question posed earlier, you may ask how much hydrogen (in kg) is needed to yield 907 kg of ammonia in the Haber process. The solution to this problem depends on the fact that *the number of moles involved in a reaction is proportional to the coefficients in the balanced chemical equation.* In the next section, we will describe a procedure for solving such problems.

Exercise 3.13 In an industrial process, hydrogen chloride, HCl, is prepared by burning hydrogen gas, H_2, in an atmosphere of chlorine, Cl_2. Write the chemical equation for the reaction. Below the equation, give the molecular, molar, and mass interpretations.

■ See Problems 3.77 and 3.78.

3.7 Amounts of Substances in a Chemical Reaction

You see from the preceding discussion that a balanced chemical equation relates the amounts of substances in a reaction. The coefficients in the equation can be given a molar interpretation, and using this interpretation you can, for example, calculate the moles of product obtained from any given moles of reactant. Also, you can extend this type of calculation to answer questions about masses of reactants and products.

Again consider the Haber process for producing ammonia gas. Suppose you have a mixture of H_2 and N_2, and 4.8 mol H_2 in this mixture reacts with N_2 to produce NH_3. How many moles of NH_3 can you produce from this quantity of H_2?

In asking questions like this, you assume that the mixture contains a sufficient quantity of the other reactant (N_2, in this case).

The balanced chemical equation for the Haber process tells you that 3 mol H_2 produce 2 mol NH_3. You can express this as a conversion factor:

This conversion factor simply expresses the fact that the mole ratio of NH_3 to H_2 in the reaction is 2 to 3.

$$\underbrace{\frac{2 \text{ mol NH}_3}{3 \text{ mol H}_2}}_{\substack{\text{Converts from} \\ \text{mol H}_2 \text{ to mol NH}_3}}$$

Multiplying any quantity of H_2 by this conversion factor mathematically converts that quantity of H_2 to the quantity of NH_3 as specified by the balanced chemical equation. Note that you write the conversion factor with the quantity you are converting *from* on the bottom (3 mol H_2), and the quantity you are converting *to* on the top (2 mol NH_3).

To calculate the quantity of NH_3 produced from 4.8 mol H_2, you write 4.8 mol H_2 and multiply this by the preceding conversion factor:

$$4.8 \cancel{\text{mol H}_2} \times \frac{2 \text{ mol NH}_3}{3 \cancel{\text{mol H}_2}} = 3.2 \text{ mol NH}_3$$

Note how the unit mol H_2 cancels to give the answer in mol NH_3.

In this calculation, you converted moles of reactant to moles of product.

$$\text{Moles reactant} \longrightarrow \text{moles product}$$

It is just as easy to calculate the moles of reactant needed to obtain the specified moles of product. You mathematically convert moles of product to moles of reactant.

$$\text{Moles product} \longrightarrow \text{moles reactant}$$

To set up the conversion factor, you refer to the balanced chemical equation and place the quantity you are converting from on the bottom and the quantity you are converting to on the top.

$$\underbrace{\frac{3 \text{ mol H}_2}{2 \text{ mol NH}_3}}_{\substack{\text{Converts from} \\ \text{mol NH}_3 \text{ to mol H}_2}}$$

Now consider the problem asked in the previous section: How much hydrogen (in kg) is needed to yield 907 kg of ammonia by the Haber process? The balanced chemical equation directly relates moles of substances, not masses. Therefore, you must first convert the mass of ammonia to moles of ammonia, then convert moles of ammonia to moles of hydrogen. Finally, you convert moles of hydrogen to mass of hydrogen.

$$\text{Mass NH}_3 \longrightarrow \text{mol NH}_3 \longrightarrow \text{mol H}_2 \longrightarrow \text{mass H}_2$$

You learned how to convert mass to moles, and vice versa, in Section 3.2.

The calculation to convert 907 kg NH_3, or 9.07×10^5 g NH_3, to mol NH_3 is as follows:

$$9.07 \times 10^5 \cancel{\text{g NH}_3} \times \underbrace{\frac{1 \text{ mol NH}_3}{17.0 \cancel{\text{g NH}_3}}}_{\substack{\text{Converts from} \\ \text{g NH}_3 \text{ to mol NH}_3}} = 5.34 \times 10^4 \text{ mol NH}_3$$

Now you convert from moles NH_3 to moles H_2.

$$5.34 \times 10^4 \cancel{\text{mol NH}_3} \times \underbrace{\frac{3 \text{ mol H}_2}{2 \cancel{\text{mol NH}_3}}}_{\substack{\text{Converts from} \\ \text{mol NH}_3 \text{ to mol H}_2}} = 8.01 \times 10^4 \text{ mol H}_2$$

Finally, you convert moles H_2 to grams H_2.

$$8.01 \times 10^4 \text{ mol } H_2 \times \underbrace{\frac{2.02 \text{ g } H_2}{1 \text{ mol } H_2}}_{\substack{\text{Converts from} \\ \text{mol } H_2 \text{ to g } H_2}} = 1.62 \times 10^5 \text{ g } H_2$$

The result says that to produce 907 kg NH_3, you need 1.62×10^5 g H_2, or 162 kg H_2.

Once you feel comfortable with the individual conversions, you can do this type of calculation in a single step by multiplying successively by conversion factors, as follows:

$$9.07 \times 10^5 \text{ g } NH_3 \times \frac{1 \text{ mol } NH_3}{17.0 \text{ g } NH_3} \times \frac{3 \text{ mol } H_2}{2 \text{ mol } NH_3} \times \frac{2.02 \text{ g } H_2}{1 \text{ mol } H_2} =$$

$$1.62 \times 10^5 \text{ g } H_2 \text{ (or 162 kg } H_2)$$

Note that the unit in the denominator of each conversion factor cancels the unit in the numerator of the preceding factor. Figure 3.11 illustrates this calculation diagrammatically.

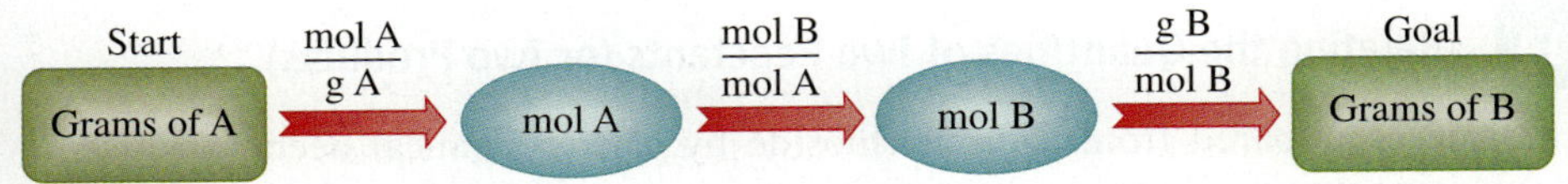

Figure 3.11 ◀ **Steps in a stoichiometric calculation** You convert the mass of substance A in a reaction to moles of substance A, then to moles of another substance B, and finally to mass of substance B.

The following two examples illustrate additional variations of this type of calculation.

OWL Interactive Example 3.13 Relating the Quantity of Reactant to Quantity of Product

Gaining Mastery Toolbox

Critical Concept 3.13
The coefficients in a balanced chemical equation represent molar relationships between the reactants and products. These relationships can be used as conversion factors when performing calculations.

Solution Essentials:
- Molar interpretation of a balanced chemical equation.
- Molar mass
- Balanced chemical equation
- Dimensional analysis
- Rules for significant figures and rounding

Hematite, Fe_2O_3, is an important ore of iron; see Figure 3.12. (An ore is a natural substance from which the metal can be profitably obtained.) The free metal is obtained by reacting hematite with carbon monoxide, CO, in a blast furnace. Carbon monoxide is formed in the furnace by partial combustion of carbon. The reaction is

$$Fe_2O_3(s) + 3CO(g) \longrightarrow 2Fe(s) + 3CO_2(g)$$

How many grams of iron can be produced from 1.00 kg Fe_2O_3?

Problem Strategy This calculation involves the conversion of a quantity of Fe_2O_3 to a quantity of Fe. In performing this calculation, we assume that there is enough CO for the complete reaction of the Fe_2O_3. An essential feature of this type of calculation is that you must use the information from the balanced chemical equation to convert from the moles of a given substance to the moles of another substance. Therefore, you first convert the mass of Fe_2O_3 (1.00 kg Fe_2O_3 = 1.0×10^3 g Fe_2O_3) to moles of Fe_2O_3. Then, using the relationship from the balanced chemical equation, you convert moles of Fe_2O_3 to moles of Fe. Finally, you convert the moles of Fe to grams of Fe.

The problem strategy is diagrammed as

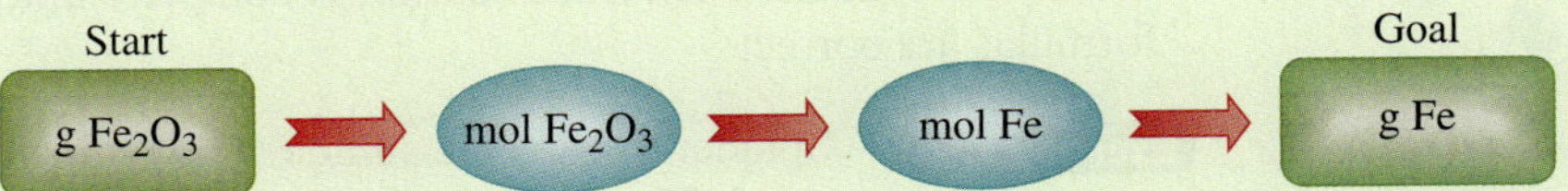

Solution The calculation is as follows:

$$1.00 \times 10^3 \text{ g } Fe_2O_3 \times \frac{1 \text{ mol } Fe_2O_3}{160 \text{ g } Fe_2O_3} \times \frac{2 \text{ mol Fe}}{1 \text{ mol } Fe_2O_3} \times \frac{55.8 \text{ g Fe}}{1 \text{ mol Fe}} = \mathbf{698 \text{ g Fe}}$$

(continued)

(continued)

© Cengage Learning

Figure 3.12 ▲

Hematite The name of this iron mineral stems from the Greek word for blood, which alludes to the color of certain forms of the mineral.

Answer Check When calculating the mass of product produced in a chemical reaction such as this, keep in mind that the total mass of reactants must equal the total mass of products. Because of this, you should be suspicious of any calculation where you find that a few grams of reactant produces a large mass of product.

Exercise 3.14 Sodium is a soft, reactive metal that instantly reacts with water to give hydrogen gas and a solution of sodium hydroxide, NaOH. How many grams of sodium metal are needed to give 7.81 g of hydrogen by this reaction? (Remember to write the balanced equation first.)

■ See Problems 3.83, 3.84, 3.85, and 3.86.

OWL Interactive Example 3.14 Relating the Quantities of Two Reactants (or Two Products)

Gaining Mastery Toolbox

Critical Concept 3.14
The coefficients in a balanced chemical equation represent molar relationships between any two substances in the chemical equation. In addition to the conversion factors between products and reactants, there are conversion factors relating reactants to other reactants and products to other products.

Solution Essentials:
- Molar interpretation of a balanced chemical equation.
- Molar mass
- Balanced chemical equation
- Dimensional analysis
- Rules for significant figures and rounding

Today chlorine is prepared from sodium chloride by electrochemical decomposition. Formerly chlorine was produced by heating hydrochloric acid with pyrolusite (manganese dioxide, MnO_2), a common manganese ore. Small amounts of chlorine may be prepared in the laboratory by the same reaction (see Figure 3.13):

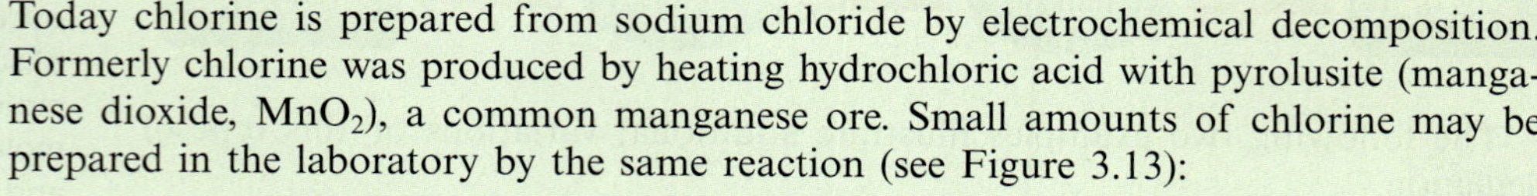

$$4HCl(aq) + MnO_2(s) \longrightarrow 2H_2O(l) + MnCl_2(aq) + Cl_2(g)$$

How many grams of HCl react with 5.00 g of manganese dioxide, according to this equation?

Problem Strategy When starting a problem like this, determine and note which quantities in the balanced chemical equation are being related. Here you are looking at the two reactants, and the relationship is that 4 mol HCl reacts with 1 mol MnO_2. Answer this problem following the same strategy as outlined in Example 3.13, only this time use the relationship between the HCl and MnO_2.

The problem strategy is diagrammed as

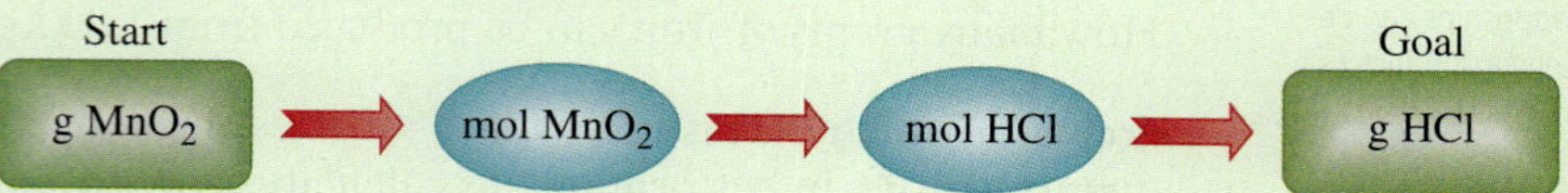

Solution You write what is given (5.00 g MnO_2) and convert this to moles, then to moles of what is desired (mol HCl). Finally you convert this to mass (g HCl). The calculation is:

$$5.00\ \text{g}\ \cancel{MnO_2} \times \frac{1\ \cancel{\text{mol}\ MnO_2}}{86.9\ \cancel{\text{g}\ MnO_2}} \times \frac{4\ \cancel{\text{mol HCl}}}{1\ \cancel{\text{mol}\ MnO_2}} \times \frac{36.5\ \text{g HCl}}{1\ \cancel{\text{mol HCl}}} = \mathbf{8.40\ g\ HCl}$$

Answer Check To avoid incorrect answers, before starting any stoichiometry problem, make sure that the chemical equation is complete and balanced and that the chemical formulas are correct.

Exercise 3.15 Sphalerite is a zinc sulfide (ZnS) mineral and an important commercial source of zinc metal. The first step in the processing of the ore consists of heating the sulfide with oxygen to give zinc oxide, ZnO, and sulfur dioxide, SO_2. How many kilograms of oxygen gas combine with 5.00×10^3 g of zinc sulfide in this reaction? (You must first write the balanced chemical equation.)

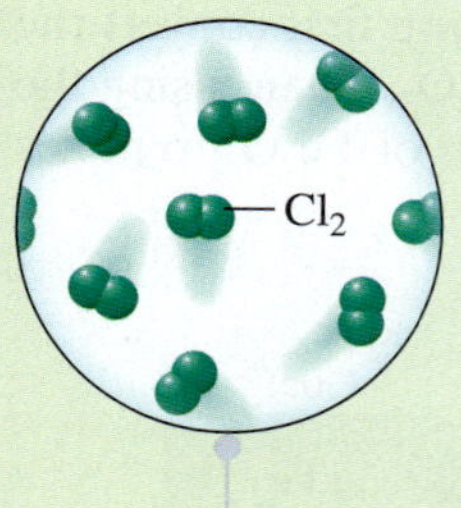

© Cengage Learning

Figure 3.13 ◀

Preparation of chlorine Concentrated hydrochloric acid was added to manganese dioxide in the beaker. Note the formation of yellowish green gas (chlorine), which is depicted by molecular models.

■ See Problems 3.87 and 3.88.

Exercise 3.16 The British chemist Joseph Priestley prepared oxygen in 1774 by heating mercury(II) oxide, HgO. Mercury metal is the other product. If 6.47 g of oxygen is collected, how many grams of mercury metal are also produced?

■ See Problems 3.89 and 3.90.

CONCEPT CHECK 3.4

The main reaction of a charcoal grill is $C(s) + O_2(g) \longrightarrow CO_2(g)$. Which of the statements below are incorrect? Why?

a 1 atom of carbon reacts with 1 molecule of oxygen to produce 1 molecule of CO_2.

b 1 g of C reacts with 1 g of O_2 to produce 2 grams of CO_2.

c 1 g of C reacts with 0.5 g of O_2 to produce 1 g of CO_2.

d 12 g of C reacts with 32 g of O_2 to produce 44 g of CO_2.

e 1 mol of C reacts with 1 mol of O_2 to produce 1 mol of CO_2.

f 1 mol of C reacts with 0.5 mol of O_2 to produce 1 mol of CO_2.

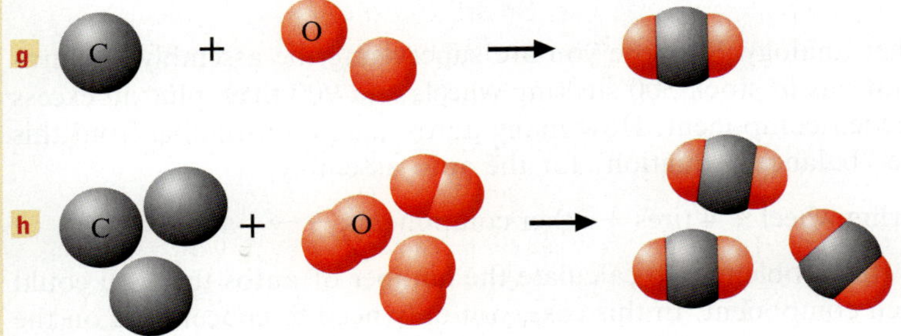

3.8 Limiting Reactant; Theoretical and Percentage Yields

Often reactants are added to a reaction vessel in amounts different from the molar proportions given by the chemical equation. In such cases, only one of the reactants may be completely consumed at the end of the reaction, whereas some amounts of other reactants will remain unreacted. The **limiting reactant** (or **limiting reagent**) is *the reactant that is entirely consumed when a reaction goes to completion.* A reactant that is not completely consumed is often referred to as an *excess reactant.* Once one of the reactants is used up, the reaction stops. This means that:

> The moles of product are always determined by the starting moles of limiting reactant.

A couple of analogies may help you understand the limiting reactant problem. Suppose you want to make some cheese sandwiches. Each is made from two slices of bread and a slice of cheese. Let's write that in the form of a chemical equation:

$$2 \text{ slices bread} + 1 \text{ slice cheese} \longrightarrow 1 \text{ sandwich}$$

You look in the kitchen and see that you have six slices of bread and two slices of cheese. The six slices of bread would be enough to make three sandwiches if you had enough cheese. The two slices of cheese would be enough to make two sandwiches if you had enough bread. How many sandwiches can you make?

Figure 3.14 ▶

Limiting reactant analogy using cheese sandwiches Start with six slices of bread, two slices of cheese, and the sandwich-making equation. Even though you have extra bread, you are limited to making two sandwiches by the amount of cheese you have on hand. Cheese is the limiting reactant.

What you will find is that once you have made two sandwiches, you will be out of cheese (Figure 3.14). Cheese is the "limiting reactant" in the language of chemistry. If you look at each "reactant," and ask how much "product" you can make from it, the reactant that limits your amount of product is called the limiting reactant.

Here is another analogy. Suppose you are supervising the assembly of automobiles. Your plant has in stock 300 steering wheels and 900 tires, plus an excess of every other needed component. How many autos can you assemble from this stock? Here is the "balanced equation" for the auto assembly:

$$1 \text{ steering wheel} + 4 \text{ tires} + \text{other components} \longrightarrow 1 \text{ auto}$$

One way to solve this problem is to calculate the number of autos that you could assemble from each component. In this case, you only need to concentrate on the effects of the steering wheels and tires since production will not be limited by the other, excess components. Looking at the equation, you can determine that from 300 steering wheels, you could assemble 300 autos; from 900 tires you could assemble 900 ÷ 4 = 225 autos.

How many autos can you produce, 225 or 300? Note that by the time you have assembled 225 autos, you will have exhausted your stock of tires, so no more autos can be assembled. Tires are the "limiting reactant," since they limit the number of automobiles that you can actually assemble (225).

You can set up these calculations in the same way for a chemical reaction. First, you calculate the numbers of autos you could assemble from the number of steering wheels and from the number of tires available, and then you compare the results. From the balanced equation, you can see that one steering wheel is equivalent to one auto, and four tires are equivalent to one auto. Therefore:

$$300 \text{ \sout{steering wheels}} \times \frac{1 \text{ auto}}{1 \text{ \sout{steering wheel}}} = 300 \text{ autos}$$

$$900 \text{ \sout{tires}} \times \frac{1 \text{ auto}}{4 \text{ \sout{tires}}} = 225 \text{ autos}$$

Comparing the numbers of autos produced (225 autos versus 300 autos), you conclude that tires are the limiting component. (The component producing the fewer autos is the limiting component.) Note that apart from the units (or factor labels), the calculation is identical with what was done previously.

Now consider a chemical reaction, the burning of hydrogen in oxygen.

$$2H_2(g) + O_2(g) \longrightarrow 2H_2O(g)$$

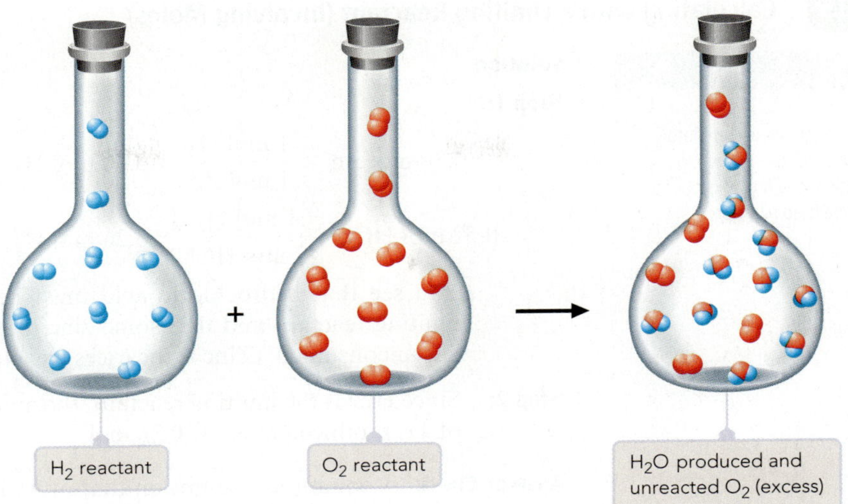

Figure 3.15 ◀ **Molecular view of $2H_2 + O_2 \longrightarrow 2H_2O$ reaction with H_2 as the limiting reactant** When an equal number of moles of H_2 and O_2 are reacted according to the equation $2H_2 + O_2 \longrightarrow 2H_2O$, all of the H_2 completely reacts, whereas only half of the O_2 is consumed. In this case, the H_2 is the limiting reactant and the O_2 is the excess reactant.

Suppose you put 1 mol H_2 and 1 mol O_2 into a reaction vessel. How many moles of H_2O will be produced? First, you note that 2 mol H_2 produces 2 mol H_2O and that 1 mol O_2 produces 2 mol H_2O. Now you calculate the moles of H_2O that you could produce from each quantity of reactant, assuming that there is sufficient other reactant.

$$1\ \cancel{\text{mol H}_2} \times \frac{2\ \text{mol H}_2\text{O}}{2\ \cancel{\text{mol H}_2}} = 1\ \text{mol H}_2\text{O}$$

$$1\ \cancel{\text{mol O}_2} \times \frac{2\ \text{mol H}_2\text{O}}{1\ \cancel{\text{mol O}_2}} = 2\ \text{mol H}_2\text{O}$$

Comparing these results, you see that hydrogen, H_2, yields the least amount of product, so it must be the limiting reactant. By the time 1 mol H_2O is produced, all of the hydrogen is used up; the reaction stops. Oxygen, O_2, is the excess reactant. Figure 3.15 depicts this reaction from a molecular viewpoint.

We can summarize the limiting-reactant problem as follows. Suppose you are given the amounts of reactants added to a vessel, and you wish to calculate the amount of product obtained when the reaction is complete. Unless you know that the reactants have been added in the molar proportions given by the chemical equation, the problem is twofold: (1) you must first identify the limiting reactant; (2) you then calculate the amount of product from the amount of limiting reactant. The next examples illustrate the steps.

CONCEPT CHECK 3.5

Solid ReO_3 is a material that is an extremely good conductor of electricity. It can be formed by the chemical reaction $Re(s) + 3Re_2O_7(s) \longrightarrow 7ReO_3(s)$. If 3 mol of Re and 3 mol of Re_2O_7 are allowed to react, 7 mol of ReO_3 are produced. Which of the following statements is the best reason that 7 mol of ReO_3 are produced during the chemical reaction?

a According to the balanced chemical equation, 7 mol of ReO_3 will always be formed regardless of the starting amounts of the reactants.

b Rhenium is the limiting reactant, so 7 mol of product will form.

c When 3 mol of Re_2O_7 undergoes reaction, 7 mol of ReO_3 is produced.

d Re_2O_7 is the excess reactant, and therefore 7 mol of ReO_3 are produced.

e The excess rhenium completely reacts producing 7 mol of ReO_3.

OWL Interactive Example 3.15 Calculating with a Limiting Reactant (Involving Moles)

Gaining Mastery Toolbox

Critical Concept 3.15
During a chemical reaction, after any one of the reactants is completely consumed, no additional products are formed. In order to calculate the amount of product formed during a chemical reaction, one must first identify the reactant—the limiting reactant— that would be used up prior to the other reactants.

Solution Essentials:
- Limiting reactant
- Molar interpretation of a balanced chemical equation.
- Molar mass
- Balanced chemical equation
- Dimensional analysis
- Rules for significant figures and rounding

Zinc metal reacts with hydrochloric acid by the following reaction:

$$Zn(s) + 2HCl(aq) \longrightarrow ZnCl_2(aq) + H_2(g)$$

If 0.30 mol Zn is added to hydrochloric acid containing 0.52 mol HCl, how many moles of H_2 are produced?

Problem Strategy

Step 1: Which is the limiting reactant? To answer this, using the relationship from the balanced chemical equation, you take each reactant in turn and ask how much product (H_2) would be obtained if each were totally consumed. The reactant that gives the smaller amount of product is the limiting reactant. (Remember how you obtained the limiting component in the auto-assembly analogy.)

Step 2: You obtain the amount of product actually obtained from the amount of limiting reactant.

Solution

Step 1:

$$0.30\ \cancel{\text{mol Zn}} \times \frac{1\ \text{mol H}_2}{1\ \cancel{\text{mol Zn}}} = 0.30\ \text{mol H}_2$$

$$0.52\ \cancel{\text{mol HCl}} \times \frac{1\ \text{mol H}_2}{2\ \cancel{\text{mol HCl}}} = 0.26\ \text{mol H}_2$$

You see that hydrochloric acid must be the limiting reactant and that some zinc must be left unconsumed. (Zinc is the excess reactant.)

Step 2: Since HCl is the limiting reactant, the amount of H_2 produced must be **0.26 mol.**

Answer Check A common assumption that often leads to errors in solving a problem of this type is to assume that whichever reactant is present in the least quantity (in mass or moles of material) is automatically the limiting reactant. Note how that assumption would have led to an incorrect answer in this problem. Always be sure to account for the stoichiometry of the balanced chemical equation.

Exercise 3.17 Aluminum chloride, $AlCl_3$, is used as a catalyst in various industrial reactions. It is prepared from hydrogen chloride gas and aluminum metal shavings.

$$2Al(s) + 6HCl(g) \longrightarrow 2AlCl_3(s) + 3H_2(g)$$

Suppose a reaction vessel contains 0.15 mol Al and 0.35 mol HCl. How many moles of $AlCl_3$ can be prepared from this mixture?

See Problems 3.91 and 3.92.

OWL Interactive Example 3.16 Calculating with a Limiting Reactant (Involving Masses)

Gaining Mastery Toolbox

Critical Concept 3.16
The amount (mol) of each reactant decreases during a chemical reaction. After the reaction is complete, the limiting reactant is completely consumed, and the other reactants are partially consumed.

Solution Essentials:
- Limiting reactant
- Molar interpretation of a balanced chemical equation.
- Molar mass
- Balanced chemical equation
- Dimensional analysis
- Rules for significant figures and rounding

In a process for producing acetic acid, oxygen gas is bubbled into acetaldehyde, CH_3CHO, containing manganese(II) acetate (catalyst) under pressure at 60°C.

$$2CH_3CHO(l) + O_2(g) \longrightarrow 2HC_2H_3O_2(l)$$

In a laboratory test of this reaction, 20.0 g CH_3CHO and 10.0 g O_2 were put into a reaction vessel. **a** How many grams of acetic acid can be produced by this reaction from these amounts of reactants? **b** How many grams of the excess reactant remain after the reaction is complete?

Problem Strategy

a This part is similar to the preceding example, but now you must convert grams of each reactant (acetaldehyde and oxygen) to moles of product (acetic acid). From these results, you decide which is the limiting reactant and the moles of product obtained, which you convert to grams of product.

b In order to calculate the amount of excess reactant remaining after the reaction is complete, you need to know the identity of the excess reactant and how much of this

(continued)

(continued)

excess reactant was needed for the reaction. The result from Step 1 provides the identity of the limiting reactant, so you will know which reactant was in excess. From the moles of product produced by the limiting reactant, you calculate the grams of the excess reactant needed in the reaction. You now know how much of this excess reactant was consumed, so subtracting the amount consumed from the starting amount will yield the amount of excess reactant that remains.

Solution

a How much acetic acid is produced?

Step 1: To determine which reactant is limiting, you convert grams of each reactant (20.0 g CH_3CHO and 10.0 g O_2) to moles of product, $HC_2H_3O_2$. Acetaldehyde has a molar mass of 44.1 g/mol, and oxygen has a molar mass of 32.0 g/mol.

$$20.0\ \text{g}\ \cancel{CH_3CHO} \times \frac{1\ \cancel{\text{mol}\ CH_3CHO}}{44.1\ \cancel{\text{g}\ CH_3CHO}} \times \frac{2\ \text{mol}\ HC_2H_3O_2}{2\ \cancel{\text{mol}\ CH_3CHO}} = 0.454\ \text{mol}\ HC_2H_3O_2$$

$$10.0\ \cancel{\text{g}\ O_2} \times \frac{1\ \cancel{\text{mol}\ O_2}}{32.0\ \cancel{\text{g}\ O_2}} \times \frac{2\ \text{mol}\ HC_2H_3O_2}{1\ \cancel{\text{mol}\ O_2}} = 0.625\ \text{mol}\ HC_2H_3O_2$$

Thus, acetaldehyde, CH_3CHO, is the limiting reactant, so 0.454 mol $HC_2H_3O_2$ was produced.

Step 2: You convert 0.454 mol $HC_2H_3O_2$ to grams of $HC_2H_3O_2$.

$$0.454\ \cancel{\text{mol}\ HC_2H_3O_2} \times \frac{60.1\ \text{g}\ HC_2H_3O_2}{1\ \cancel{\text{mol}\ HC_2H_3O_2}} = \mathbf{27.3\ g\ HC_2H_3O_2}$$

© Cengage Learning

Figure 3.16 ▲

Reaction of zinc and sulfur When a hot nail is stuck into a pile of zinc and sulfur, a fiery reaction occurs and zinc sulfide forms.

b How much of the excess reactant (oxygen) was left over? You convert the moles of acetic acid to grams of oxygen (the quantity of oxygen needed to produce this amount of acetic acid).

$$0.454\ \cancel{\text{mol}\ HC_2H_3O_2} \times \frac{1\ \cancel{\text{mol}\ O_2}}{2\ \cancel{\text{mol}\ HC_2H_3O_2}} \times \frac{32.0\ \text{g}\ O_2}{1\ \cancel{\text{mol}\ O_2}} = 7.26\ \text{g}\ O_2$$

You started with 10.0 g O_2, so the quantity remaining is

$$(10.0 - 7.26)\ \text{g}\ O_2 = \mathbf{2.7\ g\ O_2} \qquad \text{(mass remaining)}$$

Answer Check Whenever you are confronted with a stoichiometry problem you should always determine if you are going to have to solve a limiting reactant problem like this one, or a problem like Example 3.13 that involves a single reactant and one reactant in excess. A good rule of thumb is that when two or more reactant quantities are specified, you should approach the problem as was done here.

Exercise 3.18 In an experiment, 7.36 g of zinc was heated with 6.45 g of sulfur (Figure 3.16). Assume that these substances react according to the equation

$$8Zn + S_8 \longrightarrow 8ZnS$$

What amount of zinc sulfide was produced?

■ See Problems 3.93 and 3.94.

The **theoretical yield** of product is *the maximum amount of product that can be obtained by a reaction from given amounts of reactants.* It is the amount that you calculate from the stoichiometry based on the limiting reactant. In Example 3.16, the theoretical yield of acetic acid is 27.3 g. In practice, the *actual yield* of a product may be much less for several possible reasons. First, some product may be lost during the process of separating it from the final reaction mixture. Second, there may be other, competing reactions that occur simultaneously with the reactant on which the theoretical yield is based. Finally, many reactions appear to stop before they reach completion; they give mixtures of reactants and products.

Such reactions reach chemical equilibrium. We will discuss equilibrium quantitatively in Chapter 14.

It is important to know the actual yield from a reaction in order to make economic decisions about a preparation method. The reactants for a given method may not be too costly per kilogram, but if the actual yield is very low, the final cost can be very high. The **percentage yield** of product is *the actual yield (experimentally determined) expressed as a percentage of the theoretical yield (calculated).*

$$\text{Pecentage yield} = \frac{\text{actual yield}}{\text{theoretical yield}} \times 100\%$$

To illustrate the calculation of percentage yield, recall that the theoretical yield of acetic acid calculated in Example 3.16 was 27.3 g. If the actual yield of acetic acid obtained in an experiment, using the amounts of reactants given in Example 3.16, is 23.8 g, then

$$\text{Percentage yield of } HC_2H_3O_2 = \frac{23.8\text{ g}}{27.3\text{ g}} \times 100\% = 87.2\%$$

Exercise 3.19 New industrial plants for acetic acid react liquid methanol with carbon monoxide in the presence of a catalyst.

$$CH_3OH(l) + CO(g) \longrightarrow HC_2H_3O_2(l)$$

In an experiment, 15.0 g of methanol and 10.0 g of carbon monoxide were placed in a reaction vessel. What is the theoretical yield of acetic acid? If the actual yield is 19.1 g, what is the percentage yield?

■ See Problems 3.97 and 3.98.

CONCEPT CHECK 3.6

You perform the hypothetical reaction of an element, $X_2(g)$, with another element, $Y(g)$, to produce $XY(g)$.

a Write the balanced chemical equation for the reaction.

b If X_2 and Y were mixed in the quantities shown in the container on the left below and allowed to react, which of the three options is the correct representation of the contents of the container after the reaction has occurred?

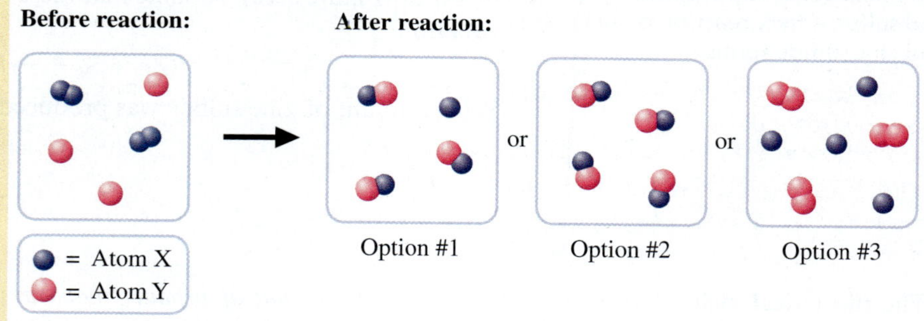

c Using the information presented in part b, identify the limiting reactant.

A Checklist for Review

OWL and Go Chemistry Sign in at www.cengage.com/owl to:

- View tutorials and simulations, develop problem-solving skills, and complete online homework assigned by your professor.
- For quick review and exam prep, download Go Chemistry mini lecture modules from OWL or purchase them at www.cengagebrain.com.

Summary of Facts and Concepts

A *formula mass* equals the sum of the atomic masses of the atoms in the formula of a compound. If the formula corresponds to that of a molecule, this sum of atomic masses equals the *molecular mass* of the compound. The mass of *Avogadro's number* (6.02×10^{23}) of formula units—that is, the mass of one *mole* of substance—equals the mass in grams that corresponds to the numerical value of the formula mass in amu. This mass is called the *molar mass.*

The *empirical formula (simplest formula)* of a compound is obtained from the *percentage composition* of the substance, which is expressed as *mass percentages* of the elements. To calculate the empirical formula, you convert mass percentages to ratios of moles, which, when expressed in smallest whole numbers, give the subscripts in the formula. A molecular formula is a multiple of the empirical formula; this multiple is determined from the experimental value of the molecular mass.

A chemical equation may be interpreted in terms of moles of reactants and products, as well as in terms of molecules. Using this *molar interpretation,* you can convert from the mass of one substance in a chemical equation to the mass of another. The maximum amount of product from a reaction is determined by the *limiting reactant,* the reactant that is completely used up; the other reactants are in excess.

Learning Objectives	Important Terms
3.1 Molecular Mass and Formula Mass	
▪ Define the terms *molecular mass* and *formula mass* of a substance. ▪ Calculate the formula mass from a formula. Example 3.1 ▪ Calculate the formula mass from molecular models. Example 3.2	**molecular mass** **formula mass**
3.2 The Mole Concept	
▪ Define the quantity called the *mole.* ▪ Learn *Avogadro's number.* ▪ Understand how the *molar mass* is related to the formula mass of a substance. ▪ Calculate the mass of atoms and molecules. Example 3.3 ▪ Perform calculations using the mole. ▪ Convert from moles of substance to grams of substance. Example 3.4 ▪ Convert from grams of substance to moles of substance. Example 3.5 ▪ Calculate the number of molecules in a given mass of substance. Example 3.6	**mole (mol)** **Avogadro's number (N_A)** **molar mass**
3.3 Mass Percentages from the Formula	
▪ Define *mass percentage.* ▪ Calculate the percentage composition of the elements in a compound. Example 3.7 ▪ Calculate the mass of an element in a given mass of compound. Example 3.8	**percentage composition** **mass percentage**
3.4 Elemental Analysis: Percentage of Carbon, Hydrogen, and Oxygen	
▪ Describe how C, H, and O combustion analysis is performed. ▪ Calculate the percentage of C, H, and O from combustion data. Example 3.9	

3.5 Determining Formulas	
▪ Define *empirical formula.* ▪ Determine the empirical formula of a binary compound from the masses of its elements. Example 3.10 ▪ Determine the empirical formula from the percentage composition. Example 3.11 ▪ Understand the relationship between the molecular mass of a substance and its *empirical formula mass.* ▪ Determine the molecular formula from the percentage composition and molecular mass. Example 3.12	**empirical (simplest) formula**
3.6 Molar Interpretation of a Chemical Equation	
▪ Relate the coefficients in a balanced chemical equation to the number of molecules or moles (*molar interpretation*).	**stoichiometry**
3.7 Amounts of Substances in a Chemical Reaction	
▪ Use the coefficients in a chemical reaction to perform calculations. ▪ Relate the quantities of reactant to the quantity of product. Example 3.13 ▪ Relate the quantities of two reactants or two products. Example 3.14	
3.8 Limiting Reactant; Theoretical and Percentage Yields	
▪ Understand how a *limiting reactant* or *limiting reagent* determines the moles of product formed during a chemical reaction and how much *excess reactant* remains. ▪ Calculate with a limiting reactant involving moles. Example 3.15 ▪ Calculate with a limiting reactant involving masses. Example 3.16 ▪ Define and calculate the *theoretical yield* of chemical reactions. ▪ Determine the *percentage yield* of a chemical reaction.	**limiting reactant (reagent)** **theoretical yield** **percentage yield**

Key Equations

$$\text{Mass\% } A = \frac{\text{mass of } A \text{ in the whole}}{\text{mass of the whole}} \times 100\%$$

$$\text{Percentage yield} = \frac{\text{actual yield}}{\text{theoretical yield}} \times 100\%$$

$$n = \frac{\text{molecular mass}}{\text{empirical formula mass}}$$

Questions and Problems

OWL Interactive versions of these problems may be assigned in OWL.

Self-Assessment and Review Questions

Key: These questions test your understanding of the ideas you worked with in the chapter. These problems vary in difficulty and often can be used for the basis of discussion.

3.1 What is the difference between a formula mass and a molecular mass? Could a given substance have both a formula mass and a molecular mass?

3.2 Describe in words how to obtain the formula mass of a compound from the formula.

3.3 One mole of N_2 contains how many N_2 molecules? How many N atoms are there in one mole of N_2? One mole of iron(III) sulfate, $Fe_2(SO_4)_3$, contains how many moles of SO_4^{2-} ions? How many moles of O atoms?

3.4 Explain what is involved in determining the composition of a compound of C, H, and O by combustion.

3.5 Explain what is involved in obtaining the empirical formula from the percentage composition.

3.6 A substance has the molecular formula $C_6H_{12}O_2$. What is its empirical formula?

3.7 Hydrogen peroxide has the empirical formula HO and an empirical formula weight of 17.0 amu. If the molecular mass is 34.0 amu, what is the molecular formula?

3.8 Describe in words the meaning of the equation

$$CH_4 + 2O_2 \longrightarrow CO_2 + 2H_2O$$

using a molecular, a molar, and then a mass interpretation.

3.9 Explain how a chemical equation can be used to relate the masses of different substances involved in a reaction.

3.10 What is a limiting reactant in a reaction mixture? Explain how it determines the amount of product.

3.11 Come up with some examples of limiting reactants that use the concept but don't involve chemical reactions.

3.12 Explain why it is impossible to have a theoretical yield of more than 100%.

3.13 How many grams of NH_3 will have the same number of molecules as 15.0 g of C_6H_6?

a 3.27
b 1.92
c 15.0
d 17.0
e 14.2

3.14 Which of the following has the largest number of molecules?

a 1 g of benzene, C_6H_6
b 1 g of formaldehyde, CH_2O
c 1 g of TNT, $C_7H_5N_3O_6$
d 1 g of naphthalene, $C_{10}H_8$
e 1 g of glucose, $C_6H_{12}O_6$

3.15 How many atoms are present in 123 g of magnesium cyanide?

a 9.7×10^{23}
b 2.91×10^{24}
c 2.83×10^{28}
d 4.85×10^{24}
e 5.65×10^{27}

3.16 When 2.56 g of a compound containing only carbon, hydrogen, and oxygen is burned completely, 3.84 g of CO_2 and 1.05 g of H_2O are produced. What is the empirical formula of the compound?

a $C_3H_4O_3$
b $C_5H_6O_4$
c $C_5H_6O_5$
d $C_4H_4O_3$
e $C_4H_6O_3$

Concept Explorations

Key: Concept explorations are comprehensive problems that provide a framework that will enable you to explore and learn many of the critical concepts and ideas in each chapter. If you master the concepts associated with these explorations, you will have a better understanding of many important chemistry ideas and will be more successful in solving all types of chemistry problems. These problems are well suited for group work and for use as in-class activities.

3.17 Moles and Molar Mass

Part 1: The mole provides a convenient package where we can make a connection between the mass of a substance and the number (count) of that substance. This is a familiar concept if you have ever bought pieces of bulk hard candy, where you purchase candy by mass rather than count. Typically, there is a scale provided for weighing the candy. For example, a notice placed above the candy bin might read, "For the candy in the bin below, there are 500 pieces of candy per kg." Using this conversion factor, perform the following calculations.

a How many candies would you have if you had 0.2 kg?
b If you had 10 dozen candies, what would be their mass?
c What is the mass of one candy piece?
d What is the mass of 2.0 moles of candies?

Part 2: The periodic table provides information about each element that serves somewhat the same purpose as the label on the nail bin described in Part 1, only in this case, the mass (molar mass) of each element is the number of grams of the element that contain 6.02×10^{23} atoms or molecules of the element. As you are aware, the quantity 6.02×10^{23} is called the mole.

a If you had 0.2 kg of helium, how many helium atoms would you have?
b If you had 10 dozen helium atoms, what would be their mass?
c What is the mass of one helium atom?
d What is the mass of 2.0 moles of helium atoms?

Part 3: Say there is a newly defined "package" called the binkle. One binkle is defined as being exactly 3×10^{12}.

a If you had 1.0 kg of nails and 1.0 kg of helium atoms, would you expect them to have the same number of binkles? Using complete sentences, explain your answer.
b If you had 3.5 binkles of nails and 3.5 binkles of helium atoms, which quantity would have more (count) and which would have more mass? Using complete sentences, explain your answers.
c Which would contain more atoms, 3.5 g of helium or 3.5 g of lithium? Using complete sentences, explain your answer.

3.18 Moles Within Moles and Molar Mass

Part 1:

a How many hydrogen and oxygen atoms are present in 1 molecule of H_2O?
b How many moles of hydrogen and oxygen atoms are present in 1 mol H_2O?
c What are the masses of hydrogen and oxygen in 1.0 mol H_2O?
d What is the mass of 1.0 mol H_2O?

Part 2: Two hypothetical ionic compounds are discovered with the chemical formulas XCl_2 and YCl_2, where X and Y represent symbols of the imaginary elements. Chemical analysis of the two compounds reveals that 0.25 mol XCl_2 has a mass of 100.0 g and 0.50 mol YCl_2 has a mass of 125.0 g.

a What are the molar masses of XCl_2 and YCl_2?

b If you had 1.0-mol samples of XCl_2 and YCl_2, how would the number of chloride ions compare?

c If you had 1.0-mol samples of XCl_2 and YCl_2, how would the masses of elements X and Y compare?

d What is the mass of chloride ions present in 1.0 mol XCl_2 and 1.0 mol YCl_2?

e What are the molar masses of elements X and Y?

f How many moles of X ions and chloride ions would be present in a 200.0-g sample of XCl_2?

g How many grams of Y ions would be present in a 250.0-g sample of YCl_2?

h What would be the molar mass of the compound YBr_3?

Part 3: A minute sample of $AlCl_3$ is analyzed for chlorine. The analysis reveals that there are 12 chloride ions present in the sample. How many aluminum ions must be present in the sample?

a What is the total mass of $AlCl_3$ in this sample?

b How many moles of $AlCl_3$ are in this sample?

Conceptual Problems

Key: These problems are designed to check your understanding of the concepts associated with some of the main topics presented in each chapter. A strong conceptual understanding of chemistry is the foundation for both applying chemical knowledge and solving chemical problems. These problems vary in level of difficulty and often can be used as a basis for group discussion.

3.19 You react nitrogen and hydrogen in a container to produce ammonia, $NH_3(g)$. The following figure depicts the contents of the container after the reaction is complete.

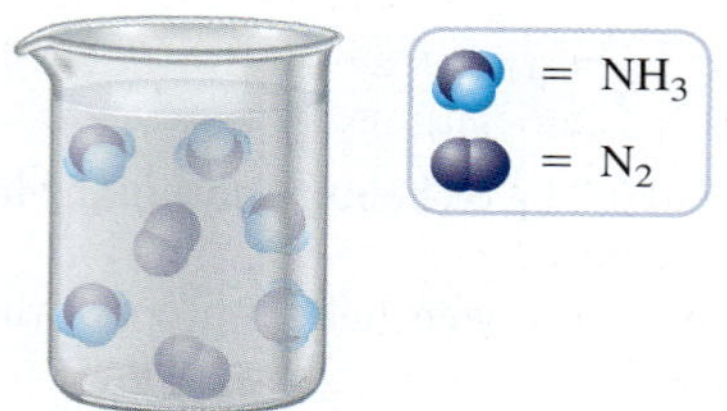

a Write a balanced chemical equation for the reaction.

b What is the limiting reactant?

c How many molecules of the limiting reactant would you need to add to the container in order to have a complete reaction (convert all reactants to products)?

3.20 Propane, C_3H_8, is the fuel of choice in a gas barbecue. When burning, the balanced equation is

$$C_3H_8 + 5O_2 \longrightarrow 3CO_2 + 4H_2O$$

a What is the limiting reactant in cooking with a gas grill?

b If the grill will not light and you know that you have an ample flow of propane to the burner, what is the limiting reactant?

c When using a gas grill you can sometimes turn the gas up to the point at which the flame becomes yellow and smokey. In terms of the chemical reaction, what is happening?

3.21 A critical point to master in becoming proficient at solving problems is evaluating whether or not your answer is reasonable. A friend asks you to look over her homework to see if she has done the calculations correctly. Shown below are descriptions of some of her answers. Without using your calculator or doing calculations on paper, see if you can judge the answers below as being reasonable or ones that will require her to go back and work the problems again.

a 0.33 mol of an element has a mass of 1.0×10^{-3} g.

b The mass of one molecule of water is 1.80×10^{-10} g.

c There are 3.01×10^{23} atoms of Na in 0.500 mol of Na.

d The molar mass of CO_2 is 44.0 kg/mol.

3.22 An exciting, and often loud, chemical demonstration involves the simple reaction of hydrogen gas and oxygen gas to produce water vapor:

$$2H_2(g) + O_2(g) \longrightarrow 2H_2O(g)$$

The reaction is carried out in soap bubbles or balloons that are filled with the reactant gases. We get the reaction to proceed by igniting the bubbles or balloons. The more H_2O that is formed during the reaction, the bigger the bang. Explain the following observations.

a A bubble containing just H_2 makes a quiet "fffft" sound when ignited.

b When a bubble containing equal amounts of H_2 and O_2 is ignited, a sizable bang results.

c When a bubble containing a ratio of 2 to 1 in the amounts of H_2 and O_2 is ignited, the loudest bang results.

d When a bubble containing just O_2 is ignited, virtually no sound is made.

3.23 High cost and limited availability of a reactant often dictate which reactant is limiting in a particular process. Identify the limiting reactant when the reactions below are run, and come up with a reason to support your decision.

a Burning charcoal on a grill:

$$C(s) + O_2(g) \longrightarrow CO_2(g)$$

b Burning a chunk of Mg in water:

$$Mg(s) + 2H_2O(l) \longrightarrow Mg(OH)_2(aq) + H_2(g)$$

c The Haber process of ammonia production:

$$3H_2(g) + N_2(g) \longrightarrow 2NH_3(g)$$

3.24 A few hydrogen and oxygen molecules are introduced into a container in the quantities depicted in the following drawing. The gases are then ignited by a spark, causing them to react and form H_2O.

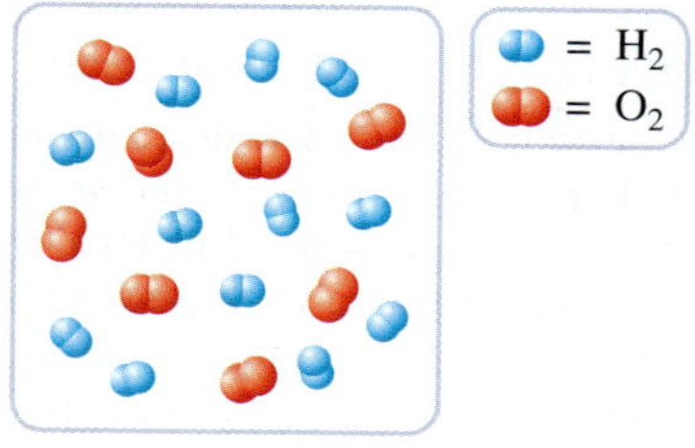

a What is the maximum number of water molecules that can be formed in the chemical reaction?

b Draw a molecular level representation of the container's contents after the chemical reaction.

3.25 A friend asks if you would be willing to check several homework problems to see if she is on the right track. Following are the problems and her proposed solutions. When you identify the problem with her work, make the appropriate correction.

a Calculate the number of moles of calcium in 27.0 g of Ca.

$$27.0 \text{ g Ca} \times \frac{1 \text{ mol Ca}}{6.022 \times 10^{23} \text{ g Ca}} = ?$$

b Calculate the number of potassium ions in 2.5 mol of K_2SO_4.

$$2.5 \text{ mol } K_2SO_4 \times \frac{1 \text{ mol } K^+ \text{ions}}{1 \text{ mol } K_2SO_4} \times \frac{6.022 \times 10^{23}\ K^+ \text{ions}}{1 \text{ mol } K^+ \text{ ions}} = ?$$

c Sodium reacts with water according to the following chemical equation.

$$2Na + 2H_2O \longrightarrow H_2 + 2NaOH$$

Assuming complete reaction, calculate the number of moles of water required to react with 0.50 mol of Na.

$$0.50 \text{ mol Na} \times \frac{1 \text{ mol } H_2O}{2 \text{ mol Na}} = ?$$

3.26 A friend is doing his chemistry homework and is working with the following chemical reaction.

$$2C_2H_2(g) + 5O_2(g) \longrightarrow 4CO_2(g) + 2H_2O(g)$$

He tells you that if he reacts 2 moles of C_2H_2 with 4 moles of O_2, then the C_2H_2 is the limiting reactant since there are fewer moles of C_2H_2 than O_2.

a How would you explain to him where he went wrong with his reasoning (what concept is he missing)?

b After providing your friend with the explanation from part a, he still doesn't believe you because he *had* a homework problem where 2 moles of calcium were reacted with 4 moles of sulfur and he needed to determine the limiting reactant. The reaction is

$$Ca(s) + S(s) \longrightarrow CaS(s)$$

He obtained the correct answer, Ca, by reasoning that since there were fewer moles of calcium reacting, calcium *had* to be the limiting reactant. How would you explain his reasoning flaw and why he got "lucky" in choosing the answer that he did?

Practice Problems

Key: These problems are for practice in applying problem-solving skills. They are divided by topic, and some are keyed to exercises (see the ends of the exercises). The problems are arranged in matching pairs; the odd-numbered problem of each pair is listed first, and its answer is given in the back of the book.

Formula Masses and Mole Calculations

3.27 Find the formula masses of the following substances to three significant figures.

a methanol, CH_3OH

b nitrogen trioxide, NO_3

c potassium carbonate, K_2CO_3

d nickel(II) phosphate, $Ni_3(PO_4)_2$

3.28 Find the formula masses of the following substances to three significant figures.

a sulfuric acid, H_2SO_4

b phosphorus pentachloride, PCl_5

c ammonium chloride, NH_4Cl

d calcium hydroxide, $Ca(OH)_2$

3.29 Calculate the formula mass of the following molecules to three significant figures.

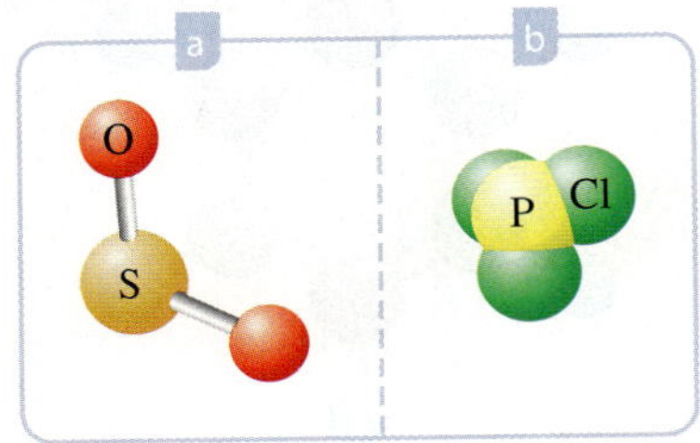

3.30 Calculate the formula mass of the following molecules to three significant figures.

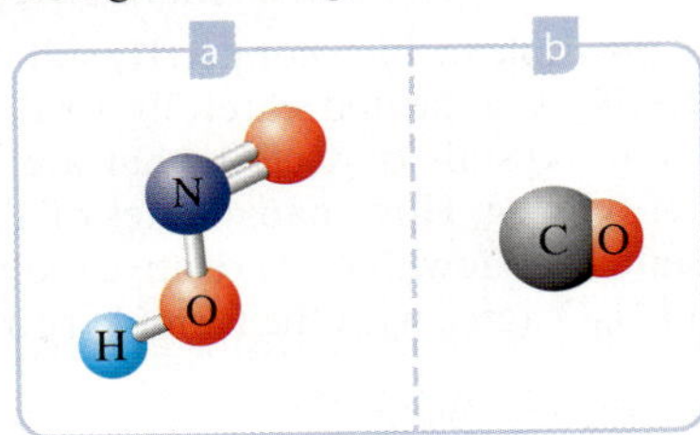

3.31 Ammonium nitrate, NH_4NO_3, is used as a nitrogen fertilizer and in explosives. What is the molar mass of NH_4NO_3?

3.32 Phosphoric acid, H_3PO_4, is used to make phosphate fertilizers and detergents and is also used in carbonated beverages. What is the molar mass of H_3PO_4?

3.33 Calculate the mass (in grams) of each of the following species.

a Na atom

b N atom

c CH_3Cl molecule

d $Hg(NO_3)_2$ formula unit

3.34 Calculate the mass (in grams) of each of the following species.

a Br atom

b Te atom

c PBr_3 molecule

d $Fe(OH)_2$ formula unit

3.35 Diethyl ether, $(C_2H_5)_2O$, commonly known as ether, is used as an anesthetic. What is the mass in grams of a molecule of diethyl ether?

3.36 Glycerol, $C_3H_8O_3$, is used as a moistening agent for candy and is the starting material for nitroglycerin. Calculate the mass of a glycerol molecule in grams.

3.37 Calculate the mass in grams of the following.
a 0.15 mol Na
b 0.594 mol S
c 2.78 mol CH_2Cl_2
d 38 mol $(NH_4)_2S$

3.38 Calculate the mass in grams of the following.
a 0.331 mol Fe
b 2.1 mol F
b 0.034 mol CO_2
d 1.89 mol K_2CrO_4

3.39 Boric acid, H_3BO_3, is a mild antiseptic and is often used as an eyewash. A sample contains 0.543 mol H_3BO_3. What is the mass of boric acid in the sample?

3.40 Carbon disulfide, CS_2, is a colorless, highly flammable liquid used in the manufacture of rayon and cellophane. A sample contains 0.0116 mol CS_2. Calculate the mass of carbon disulfide in the sample.

3.41 Obtain the moles of substance in the following.
a 2.86 g C
b 7.05 g Cl_2
c 76 g C_4H_{10}
d 26.2 g $Al_2(CO_3)_3$

3.42 Obtain the moles of substance in the following.
a 2.57 g As
b 7.83 g S_8
c 33.8 g N_2H_4
d 227 g $Al_2(SO_4)_3$

3.43 Calcium sulfate, $CaSO_4$, is a white, crystalline powder. Gypsum is a mineral, or natural substance, that is a hydrate of calcium sulfate. A 1.000-g sample of gypsum contains 0.791 g $CaSO_4$. How many moles of $CaSO_4$ are there in this sample? Assuming that the rest of the sample is water, how many moles of H_2O are there in the sample? Show that the result is consistent with the formula $CaSO_4{\cdot}2H_2O$.

3.44 A 1.547-g sample of blue copper(II) sulfate pentahydrate, $CuSO_4{\cdot}5H_2O$, is heated carefully to drive off the water. The white crystals of $CuSO_4$ that are left behind have a mass of 0.989 g. How many moles of H_2O were in the original sample? Show that the relative molar amounts of $CuSO_4$ and H_2O agree with the formula of the hydrate.

3.45 Calculate the following.
a number of atoms in 8.21 g Li
b number of atoms in 32.0 g Br_2
c number of molecules in 45 g NH_3
d number of formula units in 201 g $PbCrO_4$
e number of SO_4^{2-} ions in 14.3 g $Cr_2(SO_4)_3$

3.46 Calculate the following.
a number of atoms in 25.7 g Al
b number of atoms in 5.66 g I_2
c number of molecules in 14.9 g N_2O_5
d number of formula units in 2.99 g $NaClO_4$
e number of Ca^{2+} ions in 4.71 g $Ca_3(PO_4)_2$

3.47 Carbon tetrachloride is a colorless liquid used in the manufacture of fluorocarbons and as an industrial solvent. How many molecules are there in 7.58 mg of carbon tetrachloride?

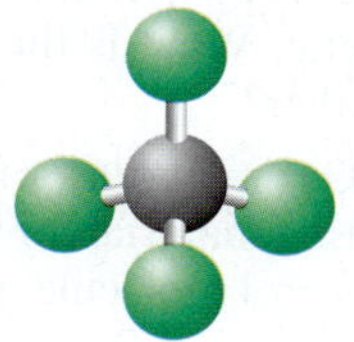

3.48 Chlorine trifluoride is a colorless, reactive gas used in nuclear fuel reprocessing. How many molecules are there in a 8.55-mg sample of chlorine trifluoride?

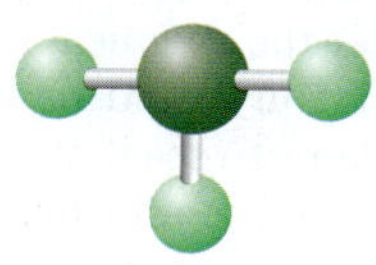

Mass Percentage

3.49 A 1.836-g sample of coal contains 1.584 g C. Calculate the mass percentage of C in the coal.

3.50 A 6.01-g aqueous solution of isopropyl alcohol contains 3.67 g of isopropyl alcohol. What is the mass percentage of isopropyl alcohol in the solution?

3.51 Phosphorus oxychloride is the starting compound for preparing substances used as flame retardants for plastics. An 8.53-mg sample of phosphorus oxychloride contains 1.72 mg of phosphorus. What is the mass percentage of phosphorus in the compound?

3.52 Ethyl mercaptan is an odorous substance added to natural gas to make leaks easily detectable. A sample of ethyl mercaptan weighing 3.52 mg contains 1.64 mg of sulfur. What is the mass percentage of sulfur in the substance?

3.53 A fertilizer is advertised as containing 14.0% nitrogen (by mass). How much nitrogen is there in 4.15 kg of fertilizer?

3.54 Seawater contains 0.0065% (by mass) of bromine. How many grams of bromine are there in 2.50 L of seawater? The density of seawater is 1.025 g/cm^3.

3.55 A sample of an alloy of aluminum contains 0.0898 mol Al and 0.0381 mol Mg. What are the mass percentages of Al and Mg in the alloy?

3.56 A sample of gas mixture from a neon sign contains 0.0856 mol Ne and 0.0254 mol Kr. What are the mass percentages of Ne and Kr in the gas mixture?

Chemical Formulas

3.57 Calculate the percentage composition for each of the following compounds (three significant figures).
a CO
b CO_2
c NaH_2PO_4
d $Co(NO_3)_2$

3.58 Calculate the percentage composition for each of the following compounds (three significant figures).
a NO_2
b H_2N_2
c $KClO_4$
d $Mg(NO_3)_2$

3.59 Calculate the mass percentage of each element in toluene, represented by the following molecular model.

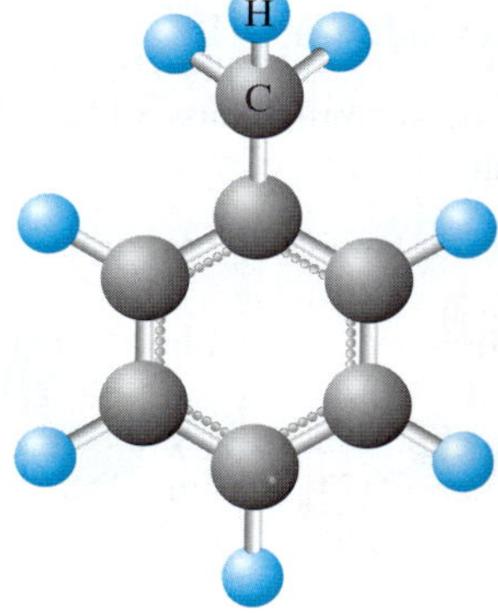

3.60 Calculate the mass percentage of each element in 2-propanol, represented by the following molecular model.

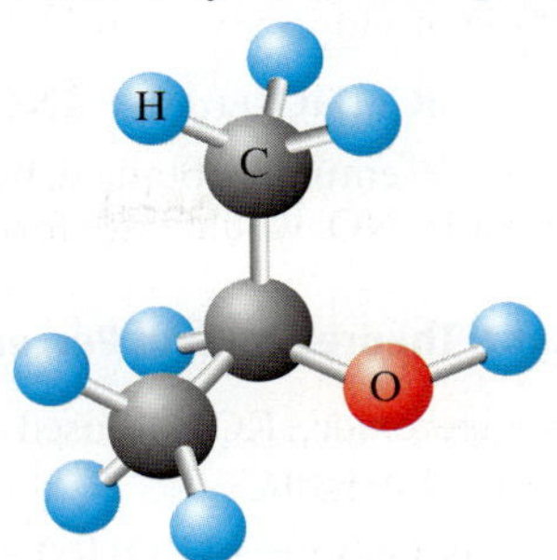

3.61 Which contains more carbon, 6.01 g of glucose, $C_6H_{12}O_6$, or 5.85 g of ethanol, C_2H_6O?

3.62 Which contains more sulfur, 40.8 g of calcium sulfate, $CaSO_4$, or 38.8 g of sodium sulfite, Na_2SO_3?

3.63 Ethylene glycol is used as an automobile antifreeze and in the manufacture of polyester fibers. The name glycol stems from the sweet taste of this poisonous compound. Combustion of 6.38 mg of ethylene glycol gives 9.06 mg CO_2 and 5.58 mg H_2O. The compound contains only C, H, and O. What are the mass percentages of the elements in ethylene glycol?

3.64 Phenol, commonly known as carbolic acid, was used by Joseph Lister as an antiseptic for surgery in 1865. Its principal use today is in the manufacture of phenolic resins and plastics. Combustion of 5.23 mg of phenol yields 14.67 mg CO_2 and 3.01 mg H_2O. Phenol contains only C, H, and O. What is the percentage of each element in this substance?

3.65 An oxide of osmium (symbol Os) is a pale yellow solid. If 2.89 g of the compound contains 2.16 g of osmium, what is its empirical formula?

3.66 An oxide of tungsten (symbol W) is a bright yellow solid. If 5.34 g of the compound contains 4.23 g of tungsten, what is its empirical formula?

3.67 Potassium manganate is a dark green, crystalline substance whose composition is 39.6% K, 27.9% Mn, and 32.5% O, by mass. What is its empirical formula?

3.68 Hydroquinone, used as a photographic developer, is 65.4% C, 5.5% H, and 29.1% O, by mass. What is the empirical formula of hydroquinone?

3.69 Acrylic acid, used in the manufacture of acrylic plastics, has the composition 50.0% C, 5.6% H, and 44.4% O. What is its empirical formula?

3.70 Malonic acid is used in the manufacture of barbiturates (sleeping pills). The composition of the acid is 34.6% C, 3.9% H, and 61.5% O. What is malonic acid's empirical formula?

3.71 Two compounds have the same composition: 92.25% C and 7.75% H.

a Obtain the empirical formula corresponding to this composition.

b One of the compounds has a molecular mass of 52.03 amu; the other, of 78.05 amu. Obtain the molecular formulas of both compounds.

3.72 Two compounds have the same composition: 85.62% C and 14.38% H.

a Obtain the empirical formula corresponding to this composition.

b One of the compounds has a molecular mass of 28.03 amu; the other, of 56.06 amu. Obtain the molecular formulas of both compounds.

3.73 Putrescine, a substance produced by decaying animals, has the empirical formula C_2H_6N. Several determinations of molecular mass give values in the range of 87 to 90 amu. Find the molecular formula of putrescine.

3.74 Compounds of boron with hydrogen are called boranes. One of these boranes has the empirical formula BH_3 and a molecular mass of 28 amu. What is its molecular formula?

3.75 Oxalic acid is a toxic substance used by laundries to remove rust stains. Its composition is 26.7% C, 2.2% H, and 71.1% O (by mass), and its molecular mass is 90 amu. What is its molecular formula?

3.76 Adipic acid is used in the manufacture of nylon. The composition of the acid is 49.3% C, 6.9% H, and 43.8% O (by mass), and the molecular mass is 146 amu. What is the molecular formula?

Stoichiometry: Quantitative Relations in Reactions

3.77 Ethylene, C_2H_4, burns in oxygen to give carbon dioxide, CO_2, and water. Write the equation for the reaction, giving molecular, molar, and mass interpretations below the equation.

3.78 Hydrogen sulfide gas, H_2S, burns in oxygen to give sulfur dioxide, SO_2, and water. Write the equation for the reaction, giving molecular, molar, and mass interpretations below the equation.

3.79 Butane, C_4H_{10}, burns with the oxygen in air to give carbon dioxide and water.

$$2C_4H_{10}(g) + 13O_2(g) \longrightarrow 8CO_2(g) + 10H_2O(g)$$

What is the amount (in moles) of carbon dioxide produced from 0.41 mol C_4H_{10}?

3.80 Ethanol, C_2H_5OH, burns with the oxygen in air to give carbon dioxide and water.

$$C_2H_5OH(l) + 3O_2(g) \longrightarrow 2CO_2(g) + 3H_2O(l)$$

What is the amount (in moles) of water produced from 0.77 mol C_2H_5OH?

3.81 Iron in the form of fine wire burns in oxygen to form iron(III) oxide.

$$4Fe(s) + 3O_2(g) \longrightarrow 2Fe_2O_3(s)$$

How many moles of O_2 are needed to produce 3.91 mol Fe_2O_3?

3.82 Nickel(II) chloride reacts with sodium phosphate to precipitate nickel(II) phosphate.

$$3NiCl_2(aq) + 2Na_3PO_4(aq) \longrightarrow Ni_3(PO_4)_2(s) + 6NaCl(aq)$$

How many moles of nickel(II) chloride are needed to produce 0.715 mol nickel(II) phosphate?

3.83 Nitric acid, HNO_3, is manufactured by the Ostwald process, in which nitrogen dioxide, NO_2, reacts with water.

$$3NO_2(g) + H_2O(l) \longrightarrow 2HNO_3(aq) + NO(g)$$

How many grams of nitrogen dioxide are required in this reaction to produce 7.50 g HNO_3?

3.84 White phosphorus, P_4, is prepared by fusing calcium phosphate, $Ca_3(PO_4)_2$, with carbon, C, and sand, SiO_2, in an electric furnace.

$$2Ca_3(PO_4)_2(s) + 6SiO_2(s) + 10C(s) \longrightarrow P_4(g) + 6CaSiO_3(l) + 10CO(g)$$

How many grams of calcium phosphate are required to give 30.0 g of phosphorus?

3.85 Tungsten metal, W, is used to make incandescent bulb filaments. The metal is produced from the yellow tungsten(VI) oxide, WO_3, by reaction with hydrogen.

$$WO_3(s) + 3H_2(g) \longrightarrow W(s) + 3H_2O(g)$$

How many grams of tungsten can be obtained from 4.81 kg of hydrogen with excess tungsten(VI) oxide?

3.86 Acrylonitrile, C_3H_3N, is the starting material for the production of a kind of synthetic fiber (acrylics). It can be made from propylene, C_3H_6, by reaction with nitric oxide, NO.

$$4C_3H_6(g) + 6NO(g) \longrightarrow 4C_3H_3N(g) + 6H_2O(g) + N_2(g)$$

How many grams of acrylonitrile are obtained from 452 kg of propylene and excess NO?

3.87 The following reaction, depicted using molecular models, is used to make carbon tetrachloride, CCl_4, a solvent and starting material for the manufacture of fluorocarbon refrigerants and aerosol propellants.

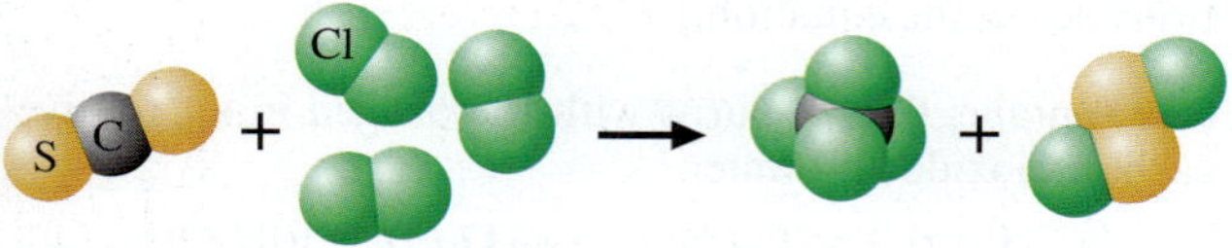

Calculate the number of grams of carbon disulfide, CS_2, needed for a laboratory-scale reaction with 62.7 g of chlorine, Cl_2.

3.88 Using the following reaction (depicted using molecular models), large quantities of ammonia are burned in the presence of a platinum catalyst to give nitric oxide, as the first step in the preparation of nitric acid.

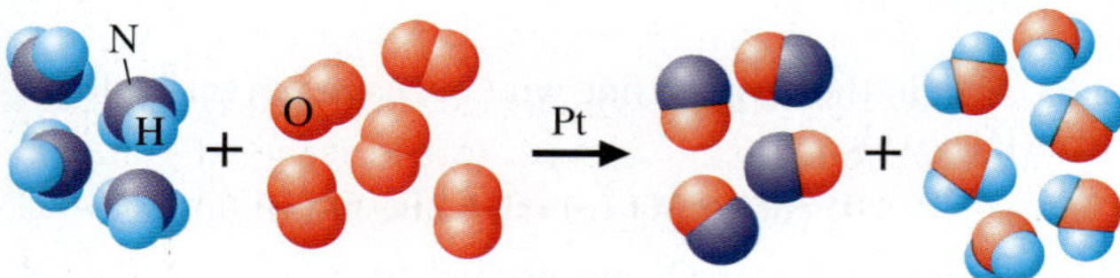

Suppose a vessel contains 5.5 g of NH_3. How many grams of O_2 are needed for a complete reaction?

3.89 When dinitrogen pentoxide, N_2O_5, a white solid, is heated, it decomposes to nitrogen dioxide and oxygen.

$$2N_2O_5(s) \xrightarrow{\Delta} 4NO_2(g) + O_2(g)$$

If a sample of N_2O_5 produces 1.381 g O_2, how many grams of NO_2 are formed?

3.90 Copper metal reacts with nitric acid. Assume that the reaction is

$$3Cu(s) + 8HNO_3(aq) \longrightarrow 3Cu(NO_3)_2(aq) + 2NO(g) + 4H_2O(l)$$

If 5.58 g $Cu(NO_3)_2$ is eventually obtained, how many grams of nitrogen monoxide, NO, would have formed?

Limiting Reactant; Theoretical and Percentage Yields

3.91 Potassium superoxide, KO_2, is used in rebreathing gas masks to generate oxygen.

$$4KO_2(s) + 2H_2O(l) \longrightarrow 4KOH(s) + 3O_2(g)$$

If a reaction vessel contains 0.25 mol KO_2 and 0.15 mol H_2O, what is the limiting reactant? How many moles of oxygen can be produced?

3.92 Solutions of sodium hypochlorite, NaClO, are sold as a bleach (such as Clorox). They are prepared by the reaction of chlorine with sodium hydroxide.

$$2NaOH(aq) + Cl_2(g) \longrightarrow NaCl(aq) + NaClO(aq) + H_2O(l)$$

If you have 1.44 mol of NaOH in solution and 1.47 mol of Cl_2 gas available to react, which is the limiting reactant? How many moles of NaClO(*aq*) could be obtained?

3.93 Methanol, CH_3OH, is prepared industrially from the gas-phase catalytic balanced reaction that has been depicted here using molecular models.

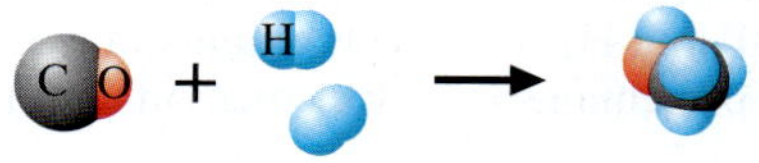

In a laboratory test, a reaction vessel was filled with 35.4 g CO and 10.2 g H_2. How many grams of methanol would be produced in a complete reaction? Which reactant remains unconsumed at the end of the reaction? How many grams of it remain?

3.94 Carbon disulfide, CS_2, burns in oxygen. Complete combustion gives the balanced reaction that has been depicted here using molecular models.

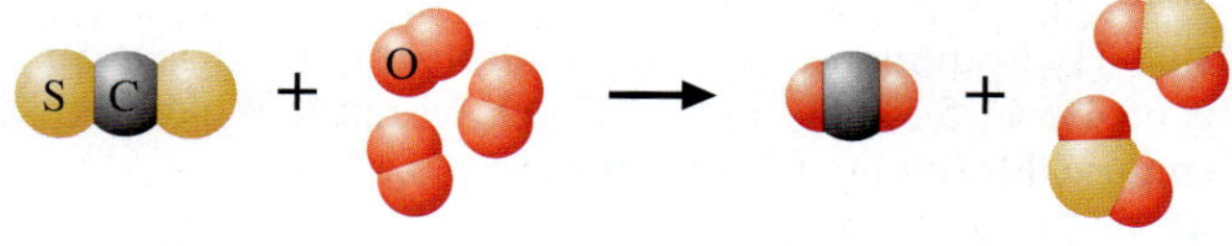

Calculate the grams of sulfur dioxide, SO_2, produced when a mixture of 35.0 g of carbon disulfide and 35.0 g of oxygen reacts. Which reactant remains unconsumed at the end of the combustion? How many grams remain?

3.95 Titanium, which is used to make airplane engines and frames, can be obtained from titanium(IV) chloride, which in turn is obtained from titanium(IV) oxide by the following process:

$$3TiO_2(s) + 4C(s) + 6Cl_2(g) \longrightarrow 3TiCl_4(g) + 2CO_2(g) + 2CO(g)$$

A vessel contains 4.15 g TiO_2, 5.67 g C, and 6.78 g Cl_2. Suppose the reaction goes to completion as written. How many grams of titanium(IV) chloride can be produced?

3.96 Hydrogen cyanide, HCN, is prepared from ammonia, air, and natural gas (CH_4) by the following process:

$$2NH_3(g) + 3O_2(g) + 2CH_4(g) \xrightarrow{Pt} 2HCN(g) + 6H_2O(g)$$

Hydrogen cyanide is used to prepare sodium cyanide, which is used in part to obtain gold from gold-containing rock. If a reaction vessel contains 11.5 g NH_3, 12.0 g O_2, and 10.5 g CH_4, what is the maximum mass in grams of hydrogen cyanide that could be made, assuming the reaction goes to completion as written?

3.97 Aspirin (acetylsalicylic acid) is prepared by heating salicylic acid, $C_7H_6O_3$, with acetic anhydride, $C_4H_6O_3$. The other product is acetic acid, $C_2H_4O_2$.

$$C_7H_6O_3 + C_4H_6O_3 \longrightarrow C_9H_8O_4 + C_2H_4O_2$$

What is the theoretical yield (in grams) of aspirin, $C_9H_8O_4$, when 2.00 g of salicylic acid is heated with 4.00 g of acetic anhydride? If the actual yield of aspirin is 1.86 g, what is the percentage yield?

3.98 Methyl salicylate (oil of wintergreen) is prepared by heating salicylic acid, $C_7H_6O_3$, with methanol, CH_3OH.

$$C_7H_6O_3 + CH_3OH \longrightarrow C_8H_8O_3 + H_2O$$

In an experiment, 2.50 g of salicylic acid is reacted with 10.31 g of methanol. The yield of methyl salicylate, $C_8H_8O_3$, is 1.27 g. What is the percentage yield?

General Problems

Key: These problems provide more practice but are not divided by topic or keyed to exercises. Each section ends with essay questions, each of which is color coded to refer to the *Instrumental Methods* (brown) chapter essay on which it is based. Odd-numbered problems and the even-numbered problems that follow are similar; answers to all odd-numbered problems except the essay questions are given in the back of the book.

3.99 Caffeine, the stimulant in coffee and tea, has the molecular formula $C_8H_{10}N_4O_2$. Calculate the mass percentage of each element in the substance. Give the answers to three significant figures.

3.100 Morphine, a narcotic substance obtained from opium, has the molecular formula $C_{17}H_{19}NO_3$. What is the mass percentage of each element in morphine (to three significant figures)?

3.101 A moth repellent, *para*-dichlorobenzene, has the composition 49.1% C, 2.7% H, and 48.2% Cl. Its molecular mass is 147 amu. What is its molecular formula?

3.102 Sorbic acid is added to food as a mold inhibitor. Its composition is 64.3% C, 7.2% H, and 28.5% O, and its molecular mass is 112 amu. What is its molecular formula?

3.103 Thiophene is a liquid compound of the elements C, H, and S. A sample of thiophene weighing 7.96 mg was burned in oxygen, giving 16.65 mg CO_2. Another sample was subjected to a series of reactions that transformed all of the sulfur in the compound to barium sulfate. If 4.31 mg of thiophene gave 11.96 mg of barium sulfate, what is the empirical formula of thiophene? Its molecular mass is 84 amu. What is its molecular formula?

3.104 Aniline, a starting compound for urethane plastic foams, consists of C, H, and N. Combustion of such compounds yields CO_2, H_2O, and N_2 as products. If the combustion of 9.71 mg of aniline yields 6.63 mg H_2O and 1.46 mg N_2, what is its empirical formula? The molecular mass of aniline is 93 amu. What is its molecular formula?

3.105 A sample of limestone (containing calcium carbonate, $CaCO_3$) weighing 438 mg is treated with oxalic acid, $H_2C_2O_4$, to give calcium oxalate, CaC_2O_4.

$$CaCO_3(s) + H_2C_2O_4(aq) \longrightarrow CaC_2O_4(s) + H_2O(l) + CO_2(g)$$

The mass of the calcium oxalate produced is 469 mg. What is the mass percentage of calcium carbonate in this limestone?

3.106 A titanium ore contains rutile (TiO_2) plus some iron oxide and silica. When it is heated with carbon in the presence of chlorine, titanium tetrachloride, $TiCl_4$, is formed.

$$TiO_2(s) + C(s) + 2Cl_2(g) \longrightarrow TiCl_4(g) + CO_2(g)$$

Titanium tetrachloride, a liquid, can be distilled from the mixture. If 35.4 g of titanium tetrachloride is recovered from 18.1 g of crude ore, what is the mass percentage of TiO_2 in the ore (assuming all TiO_2 reacts)?

3.107 Ethylene oxide, C_2H_4O, is made by the oxidation of ethylene, C_2H_4.

$$2C_2H_4(g) + O_2(g) \longrightarrow 2C_2H_4O(g)$$

Ethylene oxide is used to make ethylene glycol for automobile antifreeze. In a pilot study, 10.6 g of ethylene gave 9.91 g of ethylene oxide. What is the percentage yield of ethylene oxide?

3.108 Nitrobenzene, $C_6H_5NO_2$, an important raw material for the dye industry, is prepared from benzene, C_6H_6, and nitric acid, HNO_3.

$$C_6H_6(l) + HNO_3(l) \longrightarrow C_6H_5NO_2(l) + H_2O(l)$$

When 21.6 g of benzene and an excess of HNO_3 are used, what is the theoretical yield of nitrobenzene? If 30.0 g of nitrobenzene is recovered, what is the percentage yield?

3.109 Zinc metal can be obtained from zinc oxide, ZnO, by reaction at high temperature with carbon monoxide, CO.

$$ZnO(s) + CO(g) \longrightarrow Zn(s) + CO_2(g)$$

The carbon monoxide is obtained from carbon.

$$2C(s) + O_2(g) \longrightarrow 2CO(g)$$

What is the maximum amount of zinc that can be obtained from 75.0 g of zinc oxide and 50.0 g of carbon?

3.110 Hydrogen cyanide, HCN, can be made by a two-step process. First, ammonia is reacted with O_2 to give nitric oxide, NO.

$$4NH_3(g) + 5O_2(g) \longrightarrow 4NO(g) + 6H_2O(g)$$

Then nitric oxide is reacted with methane, CH_4.

$$2NO(g) + 2CH_4(g) \longrightarrow 2HCN(g) + 2H_2O(g) + H_2(g)$$

When 24.2 g of ammonia and 25.1 g of methane are used, how many grams of hydrogen cyanide can be produced?

3.111 Calcium carbide, CaC_2, used to produce acetylene, C_2H_2, is prepared by heating calcium oxide, CaO, and carbon, C, to high temperature.

$$CaO(s) + 3C(s) \longrightarrow CaC_2(s) + CO(g)$$

If a mixture contains 2.60 kg of each reactant, how many grams of calcium carbide can be prepared?

3.112 A mixture consisting of 11.9 g of calcium fluoride, CaF_2, and 12.1 g of sulfuric acid, H_2SO_4, is heated to drive off hydrogen fluoride, HF.

$$CaF_2(s) + H_2SO_4(l) \longrightarrow 2HF(g) + CaSO_4(s)$$

What is the maximum number of grams of hydrogen fluoride that can be obtained?

3.113 Alloys, or metallic mixtures, of mercury with another metal are called amalgams. Sodium in sodium amalgam reacts with water. (Mercury does not.)

$$2Na(s) + 2H_2O(l) \longrightarrow 2NaOH(aq) + H_2(g)$$

If a 15.23-g sample of sodium amalgam evolves 0.108 g of hydrogen, what is the percentage of sodium in the amalgam?

3.114 A sample of sandstone consists of silica, SiO_2, and calcite, $CaCO_3$. When the sandstone is heated, calcium carbonate, $CaCO_3$, decomposes into calcium oxide, CaO, and carbon dioxide.

$$CaCO_3(s) \longrightarrow CaO(s) + CO_2(g)$$

What is the percentage of silica in the sandstone if 18.7 mg of the rock yields 3.95 mg of carbon dioxide?

■ **3.115** What type of information can you obtain from a compound using a mass spectrometer?

■ **3.116** Why is the mass spectrum of a molecule much more complicated than that of an atom?

Strategy Problems

Key: As noted earlier, all of the practice and general problems are matched pairs. This section is a selection of problems that are not in matched-pair format. These challenging problems require that you employ many of the concepts and strategies that were developed in the chapter. In some cases, you will have to integrate several concepts and operational skills in order to solve the problem successfully.

3.117 Exactly 4.0 g of hydrogen gas combines with 32 g of oxygen gas according to the following reaction.

$$2H_2 + O_2 \longrightarrow 2H_2O$$

a. How many hydrogen molecules are required to completely react with 48 oxygen molecules?
b. If you conducted the reaction, and it produced 5.0 mol H_2O, how many moles of both O_2 and H_2 did you start with?
c. If you started with 37.5 g O_2, how many grams of H_2 did you start with to have a complete reaction?
d. How many grams of O_2 and H_2 were reacted to produce 30.0 g H_2O?

3.118 Aluminum metal reacts with iron(III) oxide to produce aluminum oxide and iron metal.

a. How many moles of Fe_2O_3 are required to completely react with 44 g Al?
b. How many moles of Fe are produced by the reaction of 3.14 mol Fe_2O_3 and 99.1 g Al?
c. How many atoms of Al are required to produce 7.0 g Fe?

3.119 Consider the equation

$$2KOH + H_2SO_4 \longrightarrow K_2SO_4 + 2H_2O$$

a. If 25 g H_2SO_4 is reacted with 7.7 g KOH, how many grams of K_2SO_4 are produced?
b. For part a of this problem, identify the limiting reactant and calculate the mass of excess reactant that remains after the reaction is completed.
c. Calculate the theoretical yield of the reaction. How many grams of material would you expect to obtain if the reaction has a 67.1% yield?

3.120 You perform a combustion analysis on a 255 mg sample of a substance that contains only C, H, and O, and you find that 561 mg CO_2 is produced, along with 306 mg H_2O.

a. If the substance contains only C, H, and O, what is the empirical formula?
b. If the molar mass of the compound is 180 g/mol, what is the molecular formula of the compound?

3.121 When ammonia and oxygen are reacted, they produce nitric oxide and water. When 8.5 g of ammonia is allowed to react with an excess of O_2, the reaction produces 12.0 g of nitrogen monoxide. What is the percentage yield of the reaction?

3.122 A 3.0-L sample of paint that has a density of 4.65 g/mL is found to contain 33.1 g $Pb_3N_2(s)$. How many grams of lead were in the paint sample?

3.123 A 12.1-g sample of Na_2SO_3 is mixed with a 14.6-g sample of $MgSO_4$. What is the total mass of oxygen present in the mixture?

3.124 Potassium superoxide, KO_2, is employed in a self-contained breathing apparatus used by emergency personnel as a source of oxygen. The reaction is

$$4KO_2(s) + 2H_2O(l) \longrightarrow 4KOH(s) + 3O_2(g)$$

If a self-contained breathing apparatus is charged with 750 g KO_2 and then is used to produce 188 g of oxygen, was all of the KO_2 consumed in this reaction? If the KO_2 wasn't all consumed, how much is left over and what mass of additional O_2 could be produced?

3.125 Calcium carbonate is a common ingredient in stomach antacids. If an antacid tablet has 68.4 mg of calcium carbonate, how many moles of calcium carbonate are there in 175 tablets?

3.126 While cleaning out your closet, you find a jar labeled "2.21 moles lead nitrite." Since Stock convention was not used, you do not know the oxidation number of the lead. You weigh the contents, and find a mass of 6.61×10^5 mg. What is the percentage composition of nitrite?

3.127 Sulfuric acid can be produced by the following sequence of reactions.

$$S_8 + 8O_2 \longrightarrow 8SO_2$$
$$2SO_2 + O_2 \longrightarrow 2SO_3$$
$$SO_3 + H_2O \longrightarrow H_2SO_4$$

You find a source of sulfur that is 65.0% sulfur by mass. How many grams of this starting material would be required to produce 87.0 g of sulfuric acid?

3.128 Copper reacts with nitric acid according to the following reaction.

$$3Cu(s) + 8HNO_3(aq) \longrightarrow 3Cu(NO_3)_2(aq) + 2NO(g) + 4H_2O(l)$$

If 2.40 g of Cu is added to a container with 2.00 mL of concentrated nitric acid (70% by mass HNO_3; density = 1.42 g/cm^3), what mass of nitrogen monoxide gas will be produced?

3.129 A sample of methane gas, $CH_4(g)$, is reacted with oxygen gas to produce carbon dioxide and water. If 20.0 L of methane (density = 1.82 kg/m^3) and 30.0 L of oxygen gas (density = 1.31 kg/m^3) are placed into a container and allowed to react, how many kg of carbon dioxide will be produced by the reaction?

3.130 A sample containing only boron and fluorine was decomposed yielding 4.75 mg of boron and 17.5 mL of fluorine gas (density = 1.43 g/L). What is the empirical formula of the sample compound?

3.131 Sodium azide, NaN_3, undergoes the reaction $NaN_3(s) \longrightarrow 2Na(s) + 3N_2(g)$. Because this reaction is very fast and produces nitrogen gas, NaN_3 is used to inflate airplane escape chutes. Sodium azide can be produced through two reaction steps.

$$2Na(s) + 2NH_3(g) \longrightarrow 2NaNH_2(s) + H_2(g)$$
$$2NaNH_2(s) + N_2O(g) \longrightarrow NaN_3(s) + NaOH(s) + NH_3(g)$$

Starting with 1.0 kg of Na, 6.0 kg of NH_3, and 1.0 kg of N_2O, what is the maximum mass (kg) of sodium azide that can be produced?

3.132 Dinotrogen monoxide, commonly known as laughing gas, can be obtained by cautiously warming ammonium nitrate according to the equation

$$NH_4NO_3(s) \longrightarrow N_2O(g) + 2H_2O(g)$$

If the reaction has a 75% yield, what mass of ammonium nitrate must be used to produce 685 mg of N_2O?

Cumulative-Skills Problems

Key: The problems under this heading combine skills introduced in previous chapters with those given in the current one.

3.133 A 0.500-g mixture of Cu_2O and CuO contains 0.425 g Cu. What is the mass of CuO in the mixture?

3.134 A mixture of Fe_2O_3 and FeO was found to contain 72.00% Fe by mass. What is the mass of Fe_2O_3 in 0.750 g of this mixture?

3.135 Hemoglobin is the oxygen-carrying molecule of red blood cells, consisting of a protein and a nonprotein substance. The nonprotein substance is called heme. A sample of heme weighing 35.2 mg contains 3.19 mg of iron. If a heme molecule contains one atom of iron, what is the molecular mass of heme?

3.136 Penicillin V was treated chemically to convert sulfur to barium sulfate, $BaSO_4$. An 8.19-mg sample of penicillin V gave 5.46 mg $BaSO_4$. What is the percentage of sulfur in penicillin V? If there is one sulfur atom in the molecule, what is the molecular mass?

3.137 A 3.41-g sample of a metallic element, M, reacts completely with 0.0158 mol of a gas, X_2, to form 4.52 g MX. What are the identities of M and X?

3.138 1.92 g M^{2+} ion reacts with 0.158 mol X^- ion to produce a compound, MX_2, which is 86.8% X by mass. What are the identities of M^+ and X^-?

3.139 An alloy of iron (54.7%), nickel (45.0%), and manganese (0.3%) has a density of 8.17 g/cm^3. How many iron atoms are there in a block of alloy measuring 10.0 cm × 20.0 cm × 15.0 cm?

3.140 An alloy of iron (71.0%), cobalt (12.0%), and molybdenum (17.0%) has a density of 8.20 g/cm^3. How many cobalt atoms are there in a cylinder with a radius of 2.50 cm and a length of 10.0 cm?

4 Chemical Reactions

© Cengage Learning; Siede Preis/Getty Images

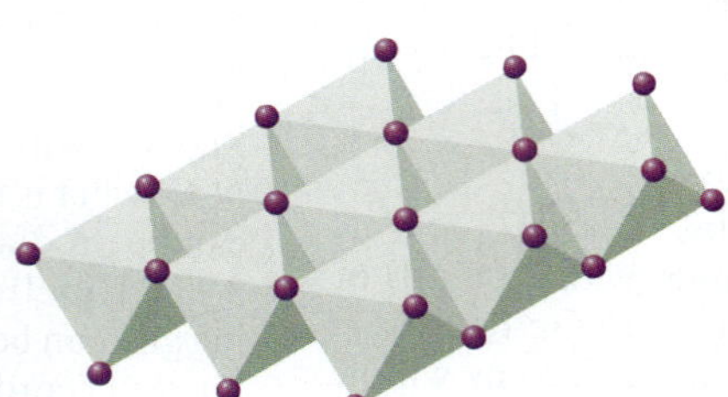

Adding lead(II) nitrate to potassium iodide gives a yellow precipitate of lead(II) iodide. Precipitation reactions are important in biological organisms—for example, in the production of bone (calcium phosphate) and seashell (calcium carbonate).

Contents and Concepts

Ions in Aqueous Solution

Explore how molecular and ionic substances behave when they dissolve in water to form *solutions.*

Types of Chemical Reactions

Investigate several important types of reactions that typically occur in aqueous solution: *precipitation reactions, acid–base reactions,* and *oxidation–reduction reactions.*

Working with Solutions

Now that we have looked at how substances behave in solution, it is time to quantitatively describe these solutions using *concentration.*

Quantitative Analysis

Using chemical reactions in aqueous solution, determine the amount of substance or species present in materials

OWL Sign in to OWL at **www.cengage.com/owl** to view tutorials and simulations, develop problem-solving skills, and complete online homework assigned by your professor.

go Chemistry Download mini lecture videos for key concept review and exam prep from OWL or purchase them from **www.cengagebrain.com**.

Chemical reactions are the heart of chemistry. Some reactions, such as those accompanying a forest fire or the explosion of dynamite, are quite dramatic. Others are much less obvious, although all chemical reactions must involve detectable change. A chemical reaction involves a change from reactant substances to product substances, and the product substances will have physical and chemical properties different from those of the reactants.

Figure 4.1 shows an experimenter adding a colorless solution of potassium iodide, KI, to a colorless solution of lead(II) nitrate, $Pb(NO_3)_2$. What you see is the formation of a cloud of bright yellow crystals where the two solutions have come into contact, clear evidence of a chemical reaction. The bright yellow crystals are lead(II) iodide, PbI_2, one of the reaction products. We call a solid that forms during a chemical reaction in solution a *precipitate;* the reaction is a precipitation reaction.

Figure 4.1 ▲

Reaction of potassium iodide solution and lead(II) nitrate solution The reactant solutions are colorless, but one of the products, lead(II) iodide, forms as a yellow precipitate.

In this chapter, we will discuss the major types of chemical reactions, including precipitation reactions. Some of the most important reactions we will describe involve ions in aqueous (water) solution. Therefore, we will first look at these ions and see how we represent by chemical equations the reactions involving ions in aqueous solution.

Some questions we will answer are: What is the evidence for ions in solution? How do we write chemical equations for reactions involving ions? How can we classify and describe the many reactions we observe so that we can begin to understand them? What is the quantitative description of solutions and reactions in solution?

Ions in Aqueous Solution

You probably have heard that you should not operate electrical equipment while standing in water. And you may have read a murder mystery in which the victim was electrocuted when an electrical appliance "accidentally" fell into his or her bath water. Actually, if the water were truly pure, the person would be safe from electrocution, because pure water is a nonconductor of electricity. Bath water, or water as it flows from the faucet, however, is a *solution* of water with small amounts of dissolved substances in it, and these dissolved substances make the solution an electrical conductor. This allows an electric current to flow from an electrical appliance to the human body. Let us look at the nature of such solutions.

4.1 Ionic Theory of Solutions and Solubility Rules

Chemists began studying the electrical behavior of substances in the early nineteenth century, and they knew that you could make pure water electrically conducting by dissolving certain substances in it. In 1884, the young Swedish chemist Svante Arrhenius proposed the *ionic theory of solutions* to account for this conductivity. He said that certain substances produce freely moving ions when they dissolve in water, and these ions conduct an electric current in an aqueous solution. ▶

Arrhenius submitted his ionic theory as part of his doctoral dissertation to the faculty at Uppsala, Sweden, in 1884. It was not well received and he barely passed. In 1903, however, he was awarded the Nobel Prize in chemistry for this theory.

Suppose you dissolve sodium chloride, NaCl, in water. From our discussion in Section 2.6, you may remember that sodium chloride is an ionic solid consisting of sodium ions, Na^+, and chloride ions, Cl^-, held in a regular, fixed array. When you dissolve solid sodium chloride in water, the Na^+ and Cl^- ions go into solution as freely moving ions. Now suppose you dip electric wires that are connected to the poles of a battery into a solution of sodium chloride. The wire that connects to the positive pole of the battery attracts the negatively charged chloride ions in solution, because of their opposite charges. Similarly, the wire connected to the negative pole of the battery attracts the positively charged sodium ions in

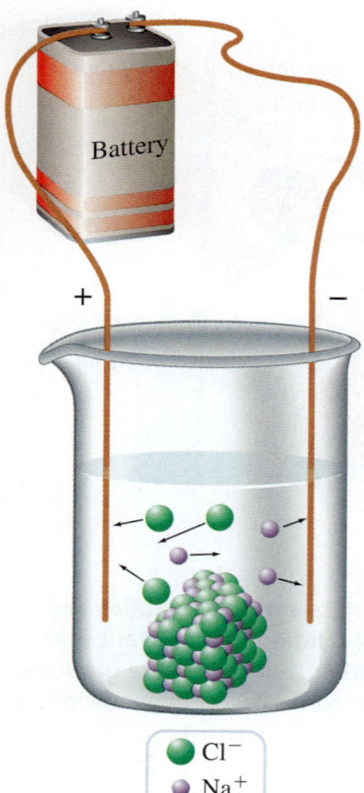

Figure 4.2 ▲

Motion of ions in solution Ions are in fixed positions in a crystal. During the solution process, however, ions leave the crystal and become freely moving. Note that Na^+ ions (small gray spheres) are attracted to the negative wire, whereas Cl^- ions (large green spheres) are attracted to the positive wire.

solution (Figure 4.2). Thus, the ions in the solution begin to move, and these moving charges form the electric current in the solution. (In a wire, it is moving electrons that constitute the electric current.)

Now consider pure water. Water consists of molecules, each of which is electrically neutral. Since each molecule carries no net electric charge, it carries no overall electric charge when it moves. Thus, pure water is a nonconductor of electricity.

In summary, although water is itself nonconducting, it has the ability to dissolve various substances, some of which go into solution as freely moving ions. An aqueous solution of ions is electrically conducting.

Electrolytes and Nonelectrolytes

We can divide the substances that dissolve in water into two broad classes, electrolytes and nonelectrolytes. An **electrolyte** is *a substance that dissolves in water to give an electrically conducting solution.* Sodium chloride, table salt, is an example of an electrolyte. When most ionic substances dissolve in water, ions that were in fixed sites in the crystalline solid go into the surrounding aqueous solution, where they are free to move about. The resulting solution is conducting because the moving ions form an electric current. Thus, in general, *ionic solids that dissolve in water are electrolytes.*

Not all electrolytes are ionic substances. Certain molecular substances dissolve in water to form ions. The resulting solution is electrically conducting, and so we say that the molecular substance is an electrolyte. An example is hydrogen chloride gas, $HCl(g)$, which is a molecular substance. Hydrogen chloride gas dissolves in water, giving $HCl(aq)$, which in turn produces hydrogen ions, H^+, and chloride ions, Cl^-, in aqueous solution. (The solution of H^+ and Cl^- ions is called hydrochloric acid.)

$$HCl(aq) \xrightarrow{H_2O} H^+(aq) + Cl^-(aq)$$

We will look more closely at molecular electrolytes, such as HCl, at the end of this section.

A **nonelectrolyte** is *a substance that dissolves in water to give a nonconducting or very poorly conducting solution.* A common example is sucrose, $C_{12}H_{22}O_{11}$, which is ordinary table sugar. Another example is methanol, CH_3OH, a compound used in car window washer solution. Both of these are molecular substances. The solution process occurs because molecules of the substance mix with molecules of water. Molecules are electrically neutral and cannot carry an electric current, so the solution is electrically nonconducting.

Observing the Electrical Conductivity of a Solution

Figure 4.3 shows a simple apparatus that allows you to observe the ability of a solution to conduct an electric current. The apparatus has two electrodes; here they are flat metal plates, dipping into the solution in a beaker. One electrode connects directly to a battery through a wire. The other electrode connects by a wire to a light bulb that connects with another wire to the other side of the battery. For an electric current to flow from the battery, there must be a complete circuit, which allows the current to flow from the positive pole of the battery through the circuit to the negative pole of the battery. To have a complete circuit, the solution in the beaker must conduct electricity, as the wires do. If the solution is conducting, the circuit is complete and the bulb lights. If the solution is nonconducting, the circuit is incomplete and the bulb does not light.

The beaker shown on the left side of Figure 4.3 contains pure water. Because the bulb is not lit, we conclude that pure water is a nonconductor (or very poor conductor) of electricity, which is what we expect from our earlier discussion. The beaker shown on the right side of Figure 4.3 contains a solution of sodium chloride

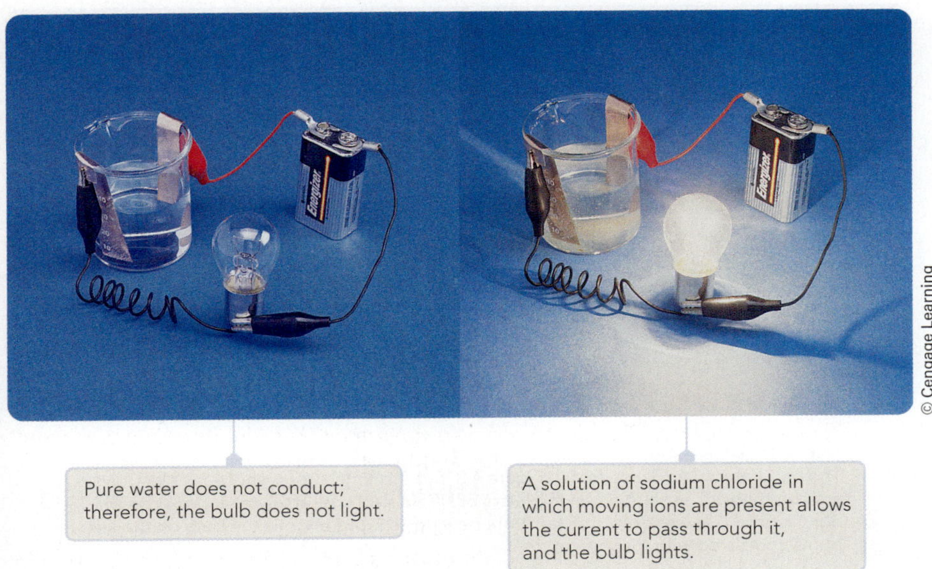

Figure 4.3 ◀

Testing the electrical conductivity of a solution

in water. In this case, the bulb burns brightly, showing that the solution is a very good conductor of electricity, due to the movement of ions in the solution.

How brightly the bulb lights tells you whether the solution is a very good conductor (contains a "strong" electrolyte) or only a moderately good conductor (contains a "weak" electrolyte). The solution of sodium chloride burns brightly, so we conclude that sodium chloride is a strong electrolyte. Let us look more closely at such strong and weak electrolytes.

Strong and Weak Electrolytes

When electrolytes dissolve in water they produce ions, but to varying extents. A **strong electrolyte** is *an electrolyte that exists in solution almost entirely as ions.* Most ionic solids that dissolve in water do so by going into the solution almost completely as ions, so they are strong electrolytes. An example is sodium chloride. We can represent the dissolution of sodium chloride in water by the following equation:

$$NaCl(s) \xrightarrow{H_2O} Na^+(aq) + Cl^-(aq)$$

A **weak electrolyte** is *an electrolyte that dissolves in water to give a relatively small percentage of ions.* These are generally molecular substances. Ammonia, NH_3, is an example. Pure ammonia is a gas that readily dissolves in water and goes into solution as ammonia molecules, $NH_3(aq)$. When you buy "ammonia" in the grocery store, you are buying an aqueous solution of ammonia. Ammonia molecules react with water to form ammonium ions, NH_4^+, and hydroxide ions, OH^-.

$$NH_3(aq) + H_2O(l) \longrightarrow NH_4^+(aq) + OH^-(aq)$$

However, these ions, $NH_4^+ + OH^-$, react with each other to give back ammonia molecules and water molecules.

$$NH_4^+(aq) + OH^-(aq) \longrightarrow NH_3(aq) + H_2O(l)$$

Both reactions, the original one and its reverse, occur constantly and simultaneously. We denote this situation by writing a single equation with a double arrow:

$$NH_3(aq) + H_2O(l) \rightleftharpoons NH_4^+(aq) + OH^-(aq)$$

As a result of this forward and reverse reaction, just a small percentage of the NH_3 molecules (about 3%) have reacted at any given moment to form ions. Thus, ammonia is a weak electrolyte. The electrical conductivities of a strong and a weak electrolyte are contrasted in Figure 4.4.

Figure 4.4

Comparing strong and weak electrolytes

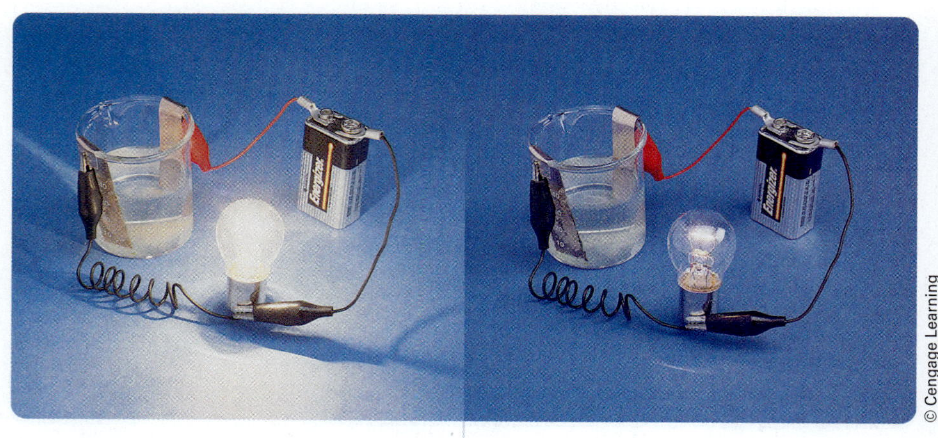

The apparatus is similar to that in Figure 4.3, but this time strong and weak electrolytes are compared. The solution on the left is of HCl (a strong electrolyte) and that on the right is of NH_3 (a weak electrolyte). Note how much more brightly the bulb on the left burns compared with that on the right.

Most soluble molecular substances are either nonelectrolytes or weak electrolytes. An exception is hydrogen chloride gas, HCl(*g*), which dissolves in water to produce hydrogen ions and chloride ions. We represent its reaction with water by an equation with a single arrow:

$$HCl(aq) \longrightarrow H^+(aq) + Cl^-(aq)$$

Since hydrogen chloride dissolves to give almost entirely ions, hydrogen chloride (or hydrochloric acid) is a strong electrolyte.

Solubility Rules

It is clear from the preceding discussion that substances vary widely in their *solubility,* or ability to dissolve, in water. Some compounds, such as sodium chloride and ethyl alcohol (CH_3CH_2OH), dissolve readily and are said to be *soluble.* Others, such as calcium carbonate (which occurs naturally as limestone and marble) and benzene (C_6H_6), have quite limited solubilities and are thus said to be *insoluble.*

Soluble ionic compounds form solutions that contain many ions and therefore are strong electrolytes. To predict the solubility of ionic compounds, chemists have developed solubility rules. Table 4.1 lists eight solubility rules for ionic compounds. These rules apply to most of the common ionic compounds that we will discuss in this course. Example 4.1 illustrates how to use the rules.

CONCEPT CHECK 4.1

Which of the following would you expect to be strong electrolytes when placed in water?

NH_4Cl $MgBr_2$ H_2O HCl $Ca_3(PO_4)_2$ CH_3OH

Which of the statements below support the answers you chose above? (Pick as many as apply.)

a Ionic compounds always produce a significant number of ions in aqueous solution.

b A molecular compound can be either a weak electrolyte or a nonelectrolyte.

c Soluble ionic compounds are strong electrolytes.

d Water conducts electricity so that whenever water is present, we would expect a strong electrolyte.

e Some molecular substances completely ionize in aqueous solution.

Table 4.1 Solubility Rules for Ionic Compounds

Rule	Applies to	Statement	Exceptions
1	Li^+, Na^+, K^+, NH_4^+	Group IA and ammonium compounds are soluble.	—
2	$C_2H_3O_2^-$, NO_3^-	Acetates and nitrates are soluble.	—
3	Cl^-, Br^-, I^-	Most chlorides, bromides, and iodides are soluble.	AgCl, Hg_2Cl_2, $PbCl_2$, AgBr, $HgBr_2$, Hg_2Br_2, $PbBr_2$, AgI, HgI_2, Hg_2I_2, PbI_2
4	SO_4^{2-}	Most sulfates are soluble.	$CaSO_4$, $SrSO_4$, $BaSO_4$, Ag_2SO_4, Hg_2SO_4, $PbSO_4$
5	CO_3^{2-}	Most carbonates are insoluble.	Group IA carbonates, $(NH_4)_2CO_3$
6	PO_4^{3-}	Most phosphates are insoluble.	Group IA phosphates, $(NH_4)_3PO_4$
7	S^{2-}	Most sulfides are insoluble.	Group IA sulfides, $(NH_4)_2S$
8	OH^-	Most hydroxides are insoluble.	Group IA hydroxides, $Ca(OH)_2$, $Sr(OH)_2$, $Ba(OH)_2$

Example 4.1 Using the Solubility Rules

Gaining Mastery Toolbox

Critical Concept 4.1
The solubility rules are used to predict if an ionic compound will dissolve in water. Soluble ionic compounds are strong electrolytes; therefore, they exist as ions in solution. Insoluble ionic compounds do not appreciably dissolve in water.

Solution Essentials:
- Solubility rules
- Ionic compound
- Ions

Determine whether the following compounds are soluble or insoluble in water.

a Hg_2Cl_2 b KI c lead(II) nitrate

Problem Strategy You should refer to Table 4.1 for the solubility rules of ionic compounds to see which compounds are soluble in water. The soluble ionic compounds are those that readily dissolve in water. The insoluble compounds hardly dissolve at all.

Solution

a According to Rule 3 in Table 4.1, most compounds that contain chloride, Cl^-, are soluble. However, Hg_2Cl_2 is listed as one of the exceptions to this rule, so it does not dissolve in water. Therefore, **Hg_2Cl_2 is not soluble in water**.

b According to both Rule 1, Group IA compounds are soluble, and Rule 3, most iodides are soluble, KI is expected to be soluble. Therefore, **KI is soluble in water**.

c According to Rule 2, compounds containing nitrates, NO_3^-, are soluble. Since there are no exceptions to this rule, **lead(II) nitrate, $Pb(NO_3)_2$, is soluble in water**.

Answer Check Most incorrect determinations of solubility occur when the exceptions are overlooked, so always be sure to check the exceptions column of Table 4.1.

Exercise 4.1 Determine whether the following compounds are soluble or insoluble in water.

a NaBr b $Ba(OH)_2$ c calcium carbonate

■ See Problems 4.29 and 4.30.

Let us summarize the main points in this section. Compounds that dissolve in water are soluble; those that dissolve little, or not at all, are insoluble. Soluble substances are either electrolytes or nonelectrolytes. Nonelectrolytes form nonconducting aqueous solutions because they dissolve completely as molecules. Electrolytes form electrically conducting solutions in water because they dissolve to give ions in solution. Electrolytes can be strong or weak. Almost all soluble ionic substances are strong electrolytes. Soluble molecular substances usually are nonelectrolytes or weak electrolytes; the latter solution consists primarily of molecules, but has a small percentage of ions. Ammonia, NH_3, is an example of a molecular substance that is a weak electrolyte. A few molecular substances (such as HCl) dissolve almost entirely as ions in the solution and are therefore strong electrolytes. The solubility rules can be used to predict the solubility of ionic compounds in water.

CONCEPT CHECK 4.2

$LiI(s)$ and $CH_3OH(l)$ are introduced into separate beakers containing water. Using the drawings shown here, label each beaker with the appropriate compound and indicate whether you would expect each substance to be a strong electrolyte, a weak electrolyte, or a nonelectrolyte.

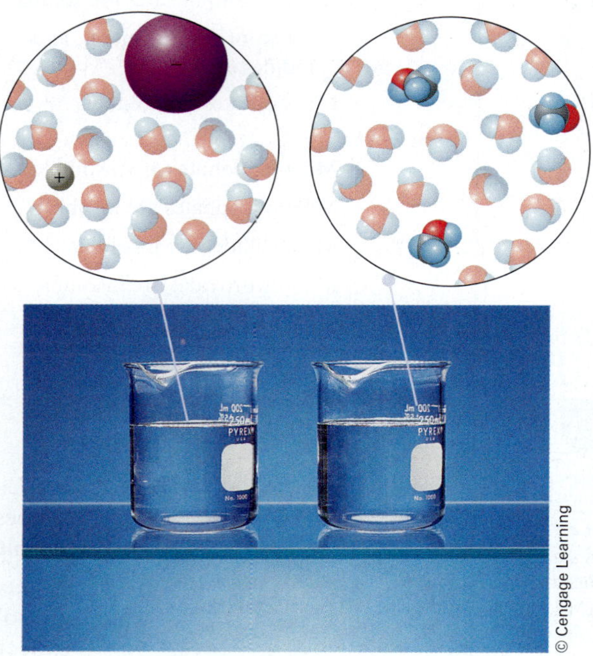

4.2 Molecular and Ionic Equations

We use chemical equations to help us describe chemical reactions. For a reaction involving ions, we have a choice of chemical equations, depending on the kind of information we want to convey. We can represent such a reaction by a *molecular equation,* a *complete ionic equation,* or a *net ionic equation.*

To illustrate these different kinds of equations, consider the preparation of precipitated calcium carbonate, $CaCO_3$. This white, fine, powdery compound is used as a paper filler to brighten and retain ink, as an antacid (as in the trade-named Tums), and as a mild abrasive in toothpastes. One way to prepare this compound is to react calcium hydroxide, $Ca(OH)_2$, with sodium carbonate, Na_2CO_3. Let us look at the different ways to write the equation for this reaction.

Molecular Equations

You could write the equation for this reaction as follows:

$$Ca(OH)_2(aq) + Na_2CO_3(aq) \longrightarrow CaCO_3(s) + 2NaOH(aq)$$

We call this a **molecular equation**, which is *a chemical equation in which the reactants and products are written as if they were molecular substances, even though they may actually exist in solution as ions.* The molecular equation is useful because it is explicit about what the reactant solutions are and what products you obtain. The equation says that you add aqueous solutions of calcium hydroxide and sodium carbonate to the reaction vessel. As a result of the reaction, the insoluble, white calcium carbonate solid forms in the solution; that is, calcium carbonate precipitates. After you remove the precipitate, you are left with a solution of sodium hydroxide. The molecular equation closely describes what you actually do in the laboratory or in an industrial process.

Complete Ionic Equations

Although a molecular equation is useful in describing the actual reactant and product substances, it does not tell you what is happening at the level of ions. That is, it does not give you an ionic-theory interpretation of the reaction. Because this kind of information is useful, you often need to rewrite the molecular equation as an ionic equation.

Again, consider the reaction of calcium hydroxide, $Ca(OH)_2$, and sodium carbonate, Na_2CO_3. Both are soluble ionic substances and therefore strong electrolytes; when they dissolve in water, they go into solution as ions. Each formula unit of $Ca(OH)_2$ forms one Ca^{2+} ion and two OH^- ions in solution. If you want to emphasize that the solution contains freely moving ions, it would be better to write $Ca^{2+}(aq) + 2OH^-(aq)$ in place of $Ca(OH)_2(aq)$. Similarly, each formula unit of Na_2CO_3 forms two Na^+ ions and one CO_3^{2-} ion in solution, and you would emphasize this by writing $2Na^+(aq) + CO_3^{2-}(aq)$ in place of $Na_2CO_3(aq)$. The reactant side of the equation becomes

$$Ca^{2+}(aq) + 2OH^-(aq) + 2Na^+(aq) + CO_3^{2-}(aq) \longrightarrow$$

Thus, the reaction mixture begins as a solution of four different kinds of ions.

Now let us look at the product side of the equation. One product is the precipitate $CaCO_3(s)$. According to the solubility rules, this is an insoluble ionic compound, so it will exist in water as a solid. We leave the formula as $CaCO_3(s)$ to convey this information in the equation. On the other hand, NaOH is a soluble ionic substance and therefore a strong electrolyte; it dissolves in aqueous solution to give the freely moving ions, which we denote by writing $Na^+(aq) + OH^-(aq)$. The complete equation is

$$Ca^{2+}(aq) + 2OH^-(aq) + 2Na^+(aq) + CO_3^{2-}(aq) \longrightarrow$$
$$CaCO_3(s) + 2Na^+(aq) + 2OH^-(aq)$$

The purpose of such a *complete ionic equation* is to represent each substance by its predominant form in the reaction mixture. For example, if the substance is a soluble ionic compound, it probably dissolves as individual ions (so it is a strong electrolyte). In a complete ionic equation, you represent the compound as separate ions. If the substance is a weak electrolyte, it is present in solution primarily as molecules, so you represent it by its molecular formula. If the substance is an insoluble ionic compound, you represent it by the formula of the compound, not by the formulas of the separate ions in solution.

Thus, a **complete ionic equation** is *a chemical equation in which strong electrolytes (such as soluble ionic compounds) are written as separate ions in the solution.* You represent other reactants and products by the formulas of the compounds, indicating any soluble substance by (*aq*) after its formula and any insoluble solid substance by (*s*) after its formula.

Net Ionic Equations

In the complete ionic equation representing the reaction of calcium hydroxide and sodium carbonate, some ions (OH^- and Na^+) appear on both sides of the equation. This means that nothing happens to these ions as the reaction occurs. They are called spectator ions. A **spectator ion** is *an ion in an ionic equation that does not take part in the reaction.* You can cancel such ions from both sides to express the essential reaction that occurs.

$$Ca^{2+}(aq) + \cancel{2OH^-(aq)} + \cancel{2Na^+(aq)} + CO_3^{2-}(aq) \longrightarrow$$
$$CaCO_3(s) + \cancel{2Na^+(aq)} + \cancel{2OH^-(aq)}$$

The resulting equation is

$$Ca^{2+}(aq) + CO_3^{2-}(aq) \longrightarrow CaCO_3(s)$$

This is the **net ionic equation,** *an ionic equation from which spectator ions have been canceled.* It shows that the reaction that actually occurs at the ionic level is between calcium ions and carbonate ions to form solid calcium carbonate.

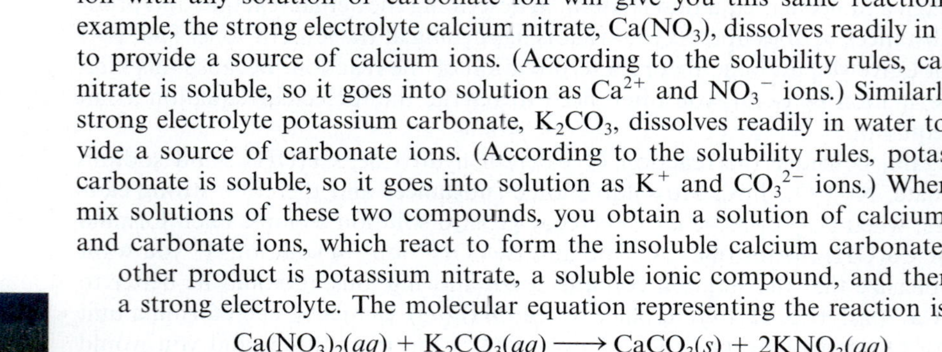

From the net ionic equation, you can see that mixing any solution of calcium ion with any solution of carbonate ion will give you this same reaction. For example, the strong electrolyte calcium nitrate, $Ca(NO_3)$, dissolves readily in water to provide a source of calcium ions. (According to the solubility rules, calcium nitrate is soluble, so it goes into solution as Ca^{2+} and NO_3^- ions.) Similarly, the strong electrolyte potassium carbonate, K_2CO_3, dissolves readily in water to provide a source of carbonate ions. (According to the solubility rules, potassium carbonate is soluble, so it goes into solution as K^+ and CO_3^{2-} ions.) When you mix solutions of these two compounds, you obtain a solution of calcium ions and carbonate ions, which react to form the insoluble calcium carbonate. The other product is potassium nitrate, a soluble ionic compound, and therefore a strong electrolyte. The molecular equation representing the reaction is

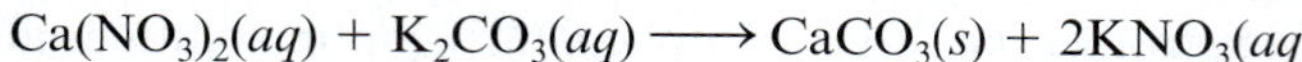

$$Ca(NO_3)_2(aq) + K_2CO_3(aq) \longrightarrow CaCO_3(s) + 2KNO_3(aq)$$

Figure 4.5 ▲

Limestone formations It is believed that most limestone formed as a precipitate of calcium carbonate (and other carbonates) from seawater. The photograph shows limestone formations at Bryce Point, Bryce Canyon National Park, Utah. More than 60 million years ago, this area was covered by seawater.

© Royalty-Free/Corbis

You obtain the complete ionic equation from this molecular equation by rewriting each of the soluble ionic compounds as ions, but retaining the formula for the precipitate $CaCO_3(s)$:

$$Ca^{2+}(aq) + \cancel{2NO_3^-(aq)} + \cancel{2K^+(aq)} + CO_3^{2-}(aq) \longrightarrow CaCO_3(s) + \cancel{2K^+(aq)} + \cancel{2NO_3^-(aq)}$$

The net ionic equation is

$$Ca^{2+}(aq) + CO_3^{2-}(aq) \longrightarrow CaCO_3(s)$$

Note that the net ionic equation is identical to the one obtained from the reaction of $Ca(OH)_2$ and Na_2CO_3. The essential reaction is the same whether you mix solutions of calcium hydroxide and sodium carbonate or solutions of calcium nitrate and potassium carbonate.

The value of the net ionic equation is its generality. For example, seawater contains Ca^{2+} and CO_3^{2-} ions from various sources. Whatever the sources of these ions, you expect them to react to form a precipitate of calcium carbonate. In seawater, this precipitate results in sediments of calcium carbonate, which eventually form limestone (Figure 4.5).

Example 4.2 Writing Net Ionic Equations

Gaining Mastery Toolbox

Critical Concept 4.2

The net ionic equation depicts the reaction that is occurring. In the net ionic equation, soluble ionic compounds are represented as aqueous ions, whereas nonelectrolytes and weak electrolytes are written with their molecular formula (not as ions). You expect a reaction to occur when a nonelectrolyte, weak electrolyte, or insoluble solid (precipitate) is formed as a product. If no reaction occurs, then there will be no net ionic equation.

Solution Essentials:

- Net ionic equation
- Spectator ion
- Complete ionic equation
- Molecular equation
- Solubility rules
- Nonelectrolyte
- Weak electrolyte

Write a net ionic equation for each of the following molecular equations.

a $2HClO_4(aq) + Ca(OH)_2(aq) \longrightarrow Ca(ClO_4)_2(aq) + 2H_2O(l)$
Perchloric acid, $HClO_4$, is a strong electrolyte, forming H^+ and ClO_4^- ions in solution. $Ca(ClO_4)_2$ is a soluble ionic compound.

b $HC_2H_3O_2(aq) + NaOH(aq) \longrightarrow NaC_2H_3O_2(aq) + H_2O(l)$
Acetic acid, $HC_2H_3O_2$, is a molecular substance and a weak electrolyte.

Problem Strategy You will need to convert the molecular equation to the complete ionic equation, then cancel spectator ions to obtain the net ionic equation. For each ionic compound in the reaction, use the solubility rules to determine if the compound will be soluble (in the solution as ions) or insoluble (present as an undissolved solid). For the complete ionic equation, represent all of the strong electrolytes by their separate ions in solution, adding (*aq*) after the formula of each. Retain the formulas of the other compounds. An ionic compound should have (*aq*) after its formula if it is soluble or (*s*) if it is insoluble.

Solution

a According to the solubility rules presented in Table 4.1 and the problem statement, $Ca(OH)_2$ and $Ca(ClO_4)_2$ are soluble ionic compounds, so they are strong electrolytes. The problem statement notes that $HClO_4$ is also a strong electrolyte. You write each strong electrolyte in the form of separate ions. Water, H_2O, is a nonelectrolyte (or very

(*continued*)

(continued)

weak electrolyte), so you retain its molecular formula. The complete ionic equation is

$$2H^+(aq) + \cancel{2ClO_4^-(aq)} + \cancel{Ca^{2+}(aq)} + 2OH^-(aq) \longrightarrow \cancel{Ca^{2+}(aq)} + \cancel{2ClO_4^-(aq)} + 2H_2O(l)$$

After canceling spectator ions and dividing by 2, you get the following net ionic equation:

$$\mathbf{H^+(aq) + OH^-(aq) \longrightarrow H_2O(l)}$$

b According to the solubility rules, NaOH and $NaC_2H_3O_2$ are soluble ionic compounds, so they are strong electrolytes. The problem statement notes that $HC_2H_3O_2$ is a weak electrolyte, which you write by its molecular formula. Water, H_2O, is a nonelectrolyte, so you retain its molecular formula also. The complete ionic equation is

$$HC_2H_3O_2(aq) + \cancel{Na^+(aq)} + OH^-(aq) \longrightarrow \cancel{Na^+(aq)} + C_2H_3O_2^-(aq) + H_2O(l)$$

and the net ionic equation is

$$\mathbf{HC_2H_3O_2(aq) + OH^-(aq) \longrightarrow C_2H_3O_2^-(aq) + H_2O(l)}$$

Answer Check If both of the reactants are strong electrolytes and a reaction occurs, your correct net ionic equation should not have any ions as products of the reaction. In this case, because the reaction is with a weak electrolyte, ions in the net ionic equation are possible.

Exercise 4.2 Write complete ionic and net ionic equations for each of the following molecular equations.

a $2HNO_3(aq) + Mg(OH)_2(s) \longrightarrow 2H_2O(l) + Mg(NO_3)_2(aq)$
Nitric acid, HNO_3, is a strong electrolyte.

b $Pb(NO_3)_2(aq) + Na_2SO_4(aq) \longrightarrow PbSO_4(s) + 2NaNO_3(aq)$

See Problems 4.33 and 4.34.

Types of Chemical Reactions

Among the several million known substances, many millions of chemical reactions are possible. Beginning students are often bewildered by the possibilities. How can I know whether two substances will react when they are mixed? How can I predict the products? Although it is not possible to give completely general answers to these questions, it is possible to make sense of chemical reactions. Most of the reactions we will study belong to one of three types:

1. Precipitation reactions. In these reactions, you mix solutions of two ionic substances, and a solid ionic substance (a precipitate) forms.
2. Acid–base reactions. An acid substance reacts with a substance called a base. Such reactions involve the transfer of a proton between reactants.
3. Oxidation–reduction reactions. These involve the transfer of electrons between reactants.

We will look at each of these types of reactions. By the time you finish this chapter, you should feel much more comfortable with the descriptions of chemical reactions that you will encounter in this course.

4.3 Precipitation Reactions

In the previous section, we used a precipitation reaction to illustrate how to convert a molecular equation to an ionic equation. A precipitation reaction occurs in aqueous solution because one product is insoluble. A **precipitate** is *an insoluble solid compound formed during a chemical reaction in solution.* To predict whether a precipitate will form when you mix two solutions of ionic compounds, you need to know whether any of the potential products that might form are insoluble. This is another application of the solubility rules (Section 4.1).

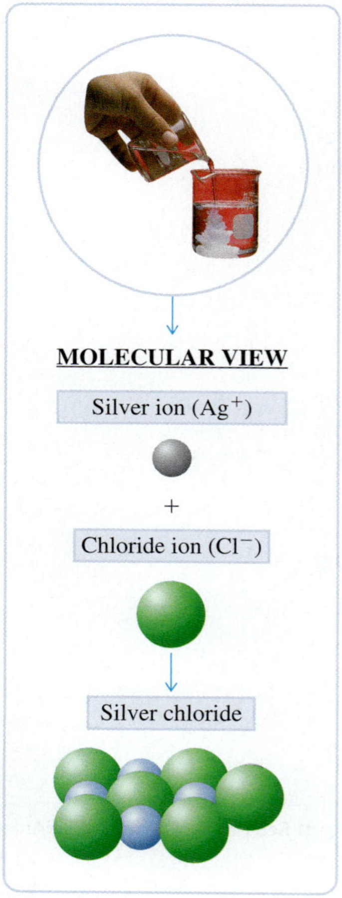

Predicting Precipitation Reactions

Now let us see how you would go about predicting whether a precipitation reaction will occur. Suppose you mix together solutions of magnesium chloride, $MgCl_2$, and silver nitrate, $AgNO_3$. You can write the potential reactants as follows:

$$MgCl_2 + AgNO_3 \longrightarrow$$

How can you tell if a reaction will occur, and if it does, what products to expect?

When you write a precipitation reaction as a molecular equation, the reaction has the form of an exchange reaction. An **exchange** (or **metathesis**) **reaction** is *a reaction between compounds that, when written as a molecular equation, appears to involve the exchange of parts between the two reactants.* In a precipitation reaction, the anions exchange between the two cations (or vice versa).

Let us momentarily assume that a reaction does occur between magnesium chloride and silver nitrate. If you exchange the anions, you get silver chloride and magnesium nitrate. Once you figure out the formulas of these potential products, you can write the molecular equation. The formulas are AgCl and $Mg(NO_3)_2$. (If you do not recall how to write the formula of an ionic compound, given the ions, turn back to Example 2.3.) The balanced equation, assuming there *is* a reaction, is

$$MgCl_2 + 2AgNO_3 \longrightarrow 2AgCl + Mg(NO_3)_2$$

Let us verify that $MgCl_2$ and $AgNO_3$ are soluble and then check the solubilities of the products. Rule 3 in Table 4.1 says that chlorides are soluble, with certain exceptions, which do not include magnesium chloride. Thus, we predict that magnesium chloride is soluble. Rule 2 indicates that nitrates are soluble, so $AgNO_3$ is soluble as well. The potential products are silver chloride and magnesium nitrate. According to Rule 3, silver chloride is one of the exceptions to the general solubility of chlorides. Therefore, we predict that the silver chloride is insoluble. Magnesium nitrate is soluble according to Rule 2.

Now we can append the appropriate phase labels to the compounds in the preceding equation.

$$MgCl_2(aq) + 2AgNO_3(aq) \longrightarrow 2AgCl(s) + Mg(NO_3)_2(aq)$$

We predict that reaction occurs because silver chloride is insoluble, and precipitates from the reaction mixture. Figure 4.6 shows the formation of the white silver chloride from this reaction and depicts the net ionic equation using molecular models. If you separate the precipitate from the solution by pouring it through filter paper, the solution that passes through (the filtrate) contains magnesium nitrate, which you could obtain by evaporating the water. The molecular equation is a summary of the actual reactants and products in the reaction. ◀

MOLECULAR VIEW

Silver ion (Ag^+)

+

Chloride ion (Cl^-)

Silver chloride

Figure 4.6 ▲

Reaction of magnesium chloride and silver nitrate Magnesium chloride solution is added to a beaker of silver nitrate solution. A white precipitate of silver chloride forms.

The reactants $MgCl_2$ and $AgNO_3$ must be added in correct amounts; otherwise, the excess reactant will remain along with the product $Mg(NO_3)_2$.

To see the reaction that occurs on an ionic level, you need to rewrite the molecular equation as a net ionic equation. You first write the strong electrolytes (here soluble ionic compounds) in the form of ions, leaving the formula of the precipitate unchanged.

$$\cancel{Mg^{2+}(aq)} + 2Cl^-(aq) + 2Ag^+(aq) + \cancel{2NO_3^-(aq)} \longrightarrow 2AgCl(s) + \cancel{Mg^{2+}(aq)} + \cancel{2NO_3^-(aq)}$$

After canceling spectator ions and reducing the coefficients to the smallest whole numbers, you obtain the net ionic equation:

$$Ag^+(aq) + Cl^-(aq) \longrightarrow AgCl(s)$$

This equation represents the essential reaction that occurs: Ag^+ ions and Cl^- ions in aqueous solution react to form solid silver chloride.

If silver chloride were soluble, a reaction would not have occurred. When you first mix solutions of $MgCl_2(aq)$ and $AgNO_3(aq)$, you obtain a solution of four ions: $Mg^{2+}(aq)$, $Cl^-(aq)$, $Ag^+(aq)$, and $NO_3^-(aq)$. If no precipitate formed, you would end up simply with a solution of these four ions. However, because Ag^+ and Cl^- react to give the precipitate AgCl, the effect is to remove these ions from the reaction mixture as an insoluble compound and leave behind a solution of $Mg(NO_3)_2(aq)$.

OWL Interactive Example 4.3 Deciding Whether a Precipitation Reaction Occurs

Gaining Mastery Toolbox

Critical Concept 4.3
Use the solubility rules to predict if a precipitate will form during the reaction. Whenever one of the products of an aqueous reaction between two soluble salts is an insoluble compound, then a precipitation reaction will occur. Otherwise, there will be no reaction, and we predict that the solution will consist only of dissolved ions.

Solution Essentials
- Exchange (metathesis) reaction
- Precipitate
- Net ionic equation
- Spectator ion
- Complete ionic equation
- Molecular equation
- Solubility rules

For each of the following, decide whether a precipitation reaction occurs. If it does, write the balanced molecular equation and then the net ionic equation. If no reaction occurs, write the compounds followed by an arrow and then *NR* (no reaction).

a Aqueous solutions of sodium chloride and iron(II) nitrate are mixed.
b Aqueous solutions of aluminum sulfate and sodium hydroxide are mixed.

Problem Strategy When aqueous solutions containing soluble salts are mixed, a precipitation reaction occurs if an insoluble compound (precipitate) forms. Start by writing the formulas of the compounds that are mixed. (If you have trouble with this, see Examples 2.3 and 2.4.) Then, assuming momentarily that the compounds do react, write the exchange reaction. Make sure that you write the correct formulas of the products. Using the solubility rules, append phase labels to each formula in the equation consulting the solubility rules. If one of the products forms an insoluble compound (precipitate), reaction occurs; otherwise, no reaction occurs.

Solution

a The formulas of the compounds are NaCl and $Fe(NO_3)_2$. Exchanging anions, you get sodium nitrate, $NaNO_3$, and iron(II) chloride, $FeCl_2$. The equation for the exchange reaction is

$$NaCl + Fe(NO_3)_2 \longrightarrow NaNO_3 + FeCl_2 \qquad \text{(not balanced)}$$

Referring to Table 4.1, note that NaCl and $NaNO_3$ are soluble (Rule 1). Also, iron(II) nitrate is soluble (Rule 2), and iron(II) chloride is soluble. (Rule 3 says that chlorides are soluble with some exceptions, none of which include $FeCl_2$.) Since there is no precipitate, no reaction occurs. You obtain simply an aqueous solution of the four different ions (Na^+, Cl^-, Fe^{2+}, and NO_3^-). For the answer, we write

$$\mathbf{NaCl(\mathit{aq}) + Fe(NO_3)_2(\mathit{aq}) \longrightarrow NR}$$

b The formulas of the compounds are $Al_2(SO_4)_3$ and NaOH. Exchanging anions, you get aluminum hydroxide, $Al(OH)_3$, and sodium sulfate, Na_2SO_4. The equation for the exchange reaction is

$$Al_2(SO_4)_3 + NaOH \longrightarrow Al(OH)_3 + Na_2SO_4 \qquad \text{(not balanced)}$$

From Table 4.1, you see that $Al_2(SO_4)_3$ is soluble (Rule 4), NaOH and Na_2SO_4 are soluble (Rule 1), and $Al(OH)_3$ is insoluble (Rule 8). Thus, aluminum hydroxide precipitates. The balanced molecular equation with phase labels is

$$\mathbf{Al_2(SO_4)_3(\mathit{aq}) + 6NaOH(\mathit{aq}) \longrightarrow 2Al(OH)_3(\mathit{s}) + 3Na_2SO_4(\mathit{aq})}$$

To get the net ionic equation, you write the strong electrolytes (here, soluble ionic compounds) as ions in aqueous solution and cancel spectator ions.

$$2Al^{3+}(aq) + \cancel{3SO_4^{2-}(aq)} + \cancel{6Na^+(aq)} + 6OH^-(aq) \longrightarrow 2Al(OH)_3(s) + \cancel{6Na^+(aq)} + \cancel{3SO_4^{2-}(aq)}$$

The net ionic equation is

$$\mathbf{Al^{3+}(\mathit{aq}) + 3OH^-(\mathit{aq}) \longrightarrow Al(OH)_3(\mathit{s})}$$

Thus, aluminum ion reacts with hydroxide ion to precipitate aluminum hydroxide.

(*continued*)

(continued)

Answer Check One of the most critical aspects of getting this type of problem right is making certain that you are using the correct chemical formulas for the reactants and products.

Exercise 4.3 You mix aqueous solutions of sodium iodide and lead(II) acetate. If a reaction occurs, write the balanced molecular equation and the net ionic equation. If no reaction occurs, indicate this by writing the formulas of the compounds and an arrow followed by *NR.*

See Problems 4.37, 4.38, 4.39, and 4.40.

CONCEPT CHECK 4.3

Your lab partner tells you that she mixed two solutions that contain ions. You analyze the solution and find that it contains the ions and precipitate shown in the beaker.

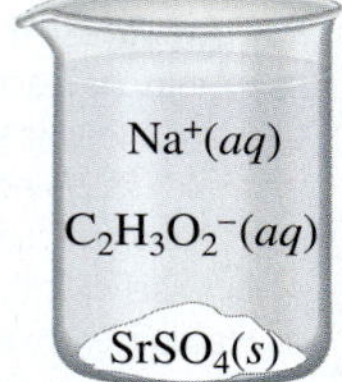

a Write the molecular equation for the reaction.

b Write the complete ionic equation for the reaction.

c Write the net ionic equation for the reaction.

4.4 Acid–Base Reactions

Acids and bases are some of the most important electrolytes. You can recognize acids and bases by some simple properties. Acids have a sour taste. Solutions of bases, on the other hand, have a bitter taste and a soapy feel. (Of course, you should *never* taste laboratory chemicals.) Some examples of acids are acetic acid, present in vinegar; citric acid, a constituent of lemon juice; and hydrochloric acid, found in the digestive fluid of the stomach. An example of a base is aqueous ammonia, often used as a household cleaner. Table 4.2 lists further examples; see also Figure 4.7.

Another simple property of acids and bases is their ability to cause color changes in certain dyes. An **acid–base indicator** is *a dye used to distinguish between acidic and basic solutions by means of the color changes it undergoes in these solutions.* Such dyes are common in natural materials. The amber color of tea, for example, is lightened by the addition of lemon juice (citric acid). Red cabbage

Table 4.2 Common Acids and Bases

Name	Formula	Remarks
Acids		
Acetic acid	$HC_2H_3O_2$	Found in vinegar
Acetylsalicylic acid	$HC_9H_7O_4$	Aspirin
Ascorbic acid	$H_2C_6H_6O_6$	Vitamin C
Citric acid	$H_3C_6H_5O_7$	Found in lemon juice
Hydrochloric acid	HCl	Found in gastric juice (digestive fluid in stomach)
Sulfuric acid	H_2SO_4	Battery acid
Bases		
Ammonia	NH_3	Aqueous solution used as a household cleaner
Calcium hydroxide	$Ca(OH)_2$	Slaked lime (used in mortar for building construction)
Magnesium hydroxide	$Mg(OH)_2$	Milk of magnesia (antacid and laxative)
Sodium hydroxide	$NaOH$	Drain cleaners, oven cleaners

juice changes to green and then yellow when a base is added (Figure 4.8). The green and yellow colors change back to red when an acid is added. Litmus is a common laboratory acid–base indicator. This dye, produced from certain species of lichens, turns red in acidic solution and blue in basic solution. Phenolphthalein (fee' nol thay' leen), another laboratory acid–base indicator, is colorless in acidic solution and pink in basic solution.

An important chemical characteristic of acids and bases is the way they react with one another. To understand these acid–base reactions, we need to have precise definitions of the terms *acid* and *base.*

Definitions of Acid and Base

When Arrhenius developed his ionic theory of solutions, he also gave the classic definitions of acids and bases. According to Arrhenius, an **acid** is *a substance that produces hydrogen ions, H^+, when it dissolves in water.* An example is nitric acid, HNO_3, a molecular substance that dissolves in water to give H^+ and NO_3^-.

$$HNO_3(aq) \xrightarrow{H_2O} H^+(aq) + NO_3^-(aq)$$

An Arrhenius **base** is *a substance that produces hydroxide ions, OH^-, when it dissolves in water.* For example, sodium hydroxide is a base.

$$NaOH(s) \xrightarrow{H_2O} Na^+(aq) + OH^-(aq)$$

The molecular substance ammonia, NH_3, is also a base because it yields hydroxide ions when it reacts with water. ▶

$$NH_3(aq) + H_2O(l) \rightleftharpoons NH_4^+(aq) + OH^-(aq)$$

Although the Arrhenius concept of acids and bases is useful, it is somewhat limited. For example, it tends to single out the OH^- ion as the source of base character, when other ions or molecules can play a similar role. In 1923, Johannes N. Brønsted and Thomas M. Lowry independently noted that many reactions involve nothing more than the transfer of a proton (H^+) between reactants, and they realized that they could use this idea to expand the definitions of acids and bases

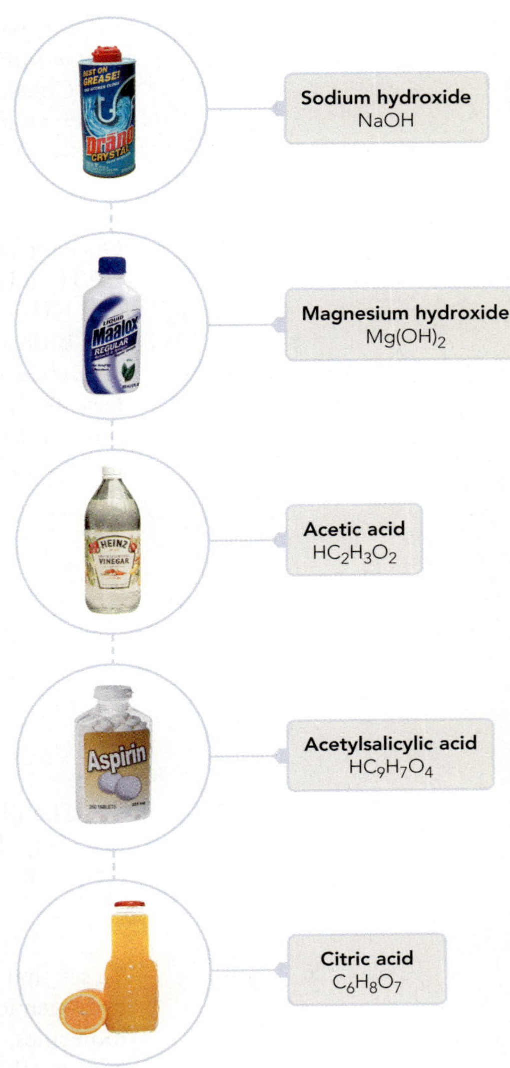

Figure 4.7 ▲

Household acids and bases Shown are a variety of household products that are either acids or bases.

Solutions of ammonia are sometimes called ammonium hydroxide and given the formula NH_4OH(aq), based on analogy with NaOH. No NH_4OH molecule or compound has ever been found, however.

Preparation of red cabbage juice. The beaker (green solution) contains red cabbage juice and sodium bicarbonate (baking soda), which is a basic solution. The flask in the center contains red cabbage juice in an acidic solution.

Red cabbage juice has been added from a pipet to solutions in the beakers. These solutions vary in acidity from highly acidic on the left to highly basic on the right.

Figure 4.8 ◀

Red cabbage juice as an acid–base indicator

to describe a large class of chemical reactions. In this view, acid–base reactions are *proton-transfer reactions.*

Consider the preceding reaction. It involves the transfer of a proton from the water molecule, H_2O, to the ammonia molecule, NH_3.

$$\overset{H^+}{\curvearrowleft}$$
$$NH_3(aq) + H_2O(l) \rightleftharpoons NH_4^+(aq) + OH^-(aq)$$

Once the proton, H^+, has left the H_2O molecule, it leaves behind an OH^- ion. When the H^+ adds to NH_3, an NH_4^+ ion results. H_2O is said to *donate* a proton to NH_3, and NH_3 is said to *accept* a proton from H_2O.

Brønsted and Lowry defined an **acid** as *the species (molecule or ion) that donates a proton to another species in a proton-transfer reaction.* They defined a **base** as *the species (molecule or ion) that accepts a proton in a proton-transfer reaction.* In the reaction of ammonia with water, the H_2O molecule is the acid, because it donates a proton. The NH_3 molecule is a base, because it accepts a proton.

$$\overset{H^+}{\curvearrowleft}$$
$$\underset{\text{base}}{NH_3(aq)} + \underset{\text{acid}}{H_2O(l)} \rightleftharpoons NH_4^+(aq) + OH^-(aq)$$

The dissolution of nitric acid, HNO_3, in water is actually a proton-transfer reaction, although the following equation, which we used as an illustration of an Arrhenius acid, does not spell that out.

$$HNO_3(aq) \xrightarrow{H_2O} H^+(aq) + NO_3^-(aq)$$

To see that this is a proton-transfer reaction, we need to clarify the structure of the hydrogen ion, $H^+(aq)$. This ion consists of a proton (H^+) in association with water molecules, which is what (aq) means. This is not a weak association, however, because the proton (or hydrogen nucleus) would be expected to attract electrons strongly to itself. In fact, the $H^+(aq)$ ion might be better thought of as a proton chemically bonded to a water molecule to give the H_3O^+ ion, with other water molecules less strongly associated with this ion, which we represent by the phase-labeled formula $H_3O^+(aq)$. Written in this form, we usually call this the *hydronium ion.* It is important to understand that the hydrogen ion, $H^+(aq)$, and the hydronium ion, $H_3O^+(aq)$, represent precisely the same physical ion. For simplicity, we often write the formula for this ion as $H^+(aq)$, but when we want to be explicit about the proton-transfer aspect of a reaction, we write the formula as $H_3O^+(aq)$. ◀

The H^+(aq) ion has variable structure in solution. In solids, there is evidence for $[H(H_2O)_n]^+$, where n is 1, 2, 3, 4, or 6.

Now let us rewrite the preceding equation by replacing $H^+(aq)$ by $H_3O^+(aq)$. To maintain a balanced equation, we will also need to add $H_2O(l)$ to the left side.

$$HNO_3(aq) + H_2O(l) \longrightarrow NO_3^-(aq) + H_3O^+(aq)$$

Note that the reaction involves simply a transfer of a proton (H^+) from HNO_3 to H_2O:

$$\overset{H^+}{\curvearrowright}$$
$$\underset{\text{acid}}{HNO_3(aq)} + \underset{\text{base}}{H_2O(l)} \longrightarrow NO_3^-(aq) + H_3O^+(aq)$$

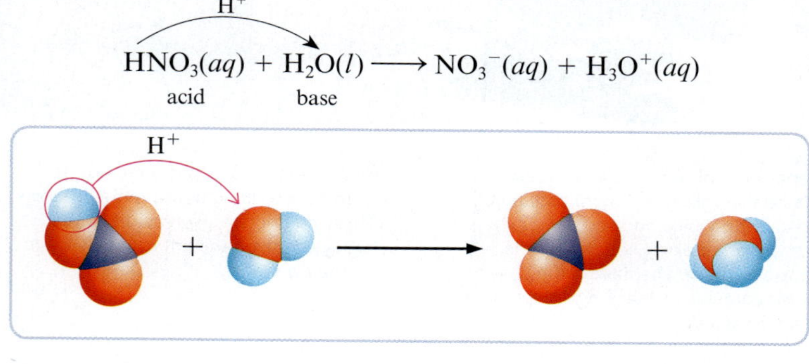

The HNO_3 molecule is an acid (proton donor) and H_2O is a base (proton acceptor). Note that H_2O may function as an acid or a base, depending on the other reactant.

The Arrhenius definitions and those of Brønsted and Lowry are essentially equivalent for aqueous solutions, although their points of view are different. For instance, sodium hydroxide and ammonia are bases in the Arrhenius view because they increase the percentage of OH^- ion in the aqueous solution. They are bases in the Brønsted–Lowry view because they provide species (OH^- from the strong electrolyte sodium hydroxide and NH_3 from ammonia) that can accept protons. ▶

We will discuss the Brønsted–Lowry concept of acids and bases more thoroughly in Chapter 15.

Strong and Weak Acids and Bases

Acids and bases are classified as strong or weak, depending on whether they are strong or weak electrolytes. A **strong acid** is *an acid that ionizes completely in water; it is a strong electrolyte.* Hydrochloric acid, $HCl(aq)$, and nitric acid, $HNO_3(aq)$, are examples of strong acids. Using the hydronium ion notation, we write the respective equations as follows:

$$HCl(aq) + H_2O(l) \longrightarrow H_3O^+(aq) + Cl^-(aq)$$

$$HNO_3(aq) + H_2O(l) \longrightarrow H_3O^+(aq) + NO_3^-(aq)$$

Table 4.3 lists six common strong acids. Most of the other acids we will discuss are weak acids.

Table 4.3 Common Strong Acids and Bases

Strong Acids	Strong Bases
$HClO_4$	LiOH
H_2SO_4	NaOH
HI	KOH
HBr	$Ca(OH)_2$
HCl	$Sr(OH)_2$
HNO_3	$Ba(OH)_2$

A **weak acid** is *an acid that only partly ionizes in water; it is a weak electrolyte.* Examples of weak acids are hydrocyanic acid, $HCN(aq)$, and hydrofluoric acid, $HF(aq)$. These molecules react with water to produce a small percentage of ions in solution.

$$HCN(aq) + H_2O(l) \rightleftharpoons H_3O^+(aq) + CN^-(aq)$$

$$HF(aq) + H_2O\,(l) \rightleftharpoons H_3O^+(aq) + F^-(aq)$$

Figure 4.9 presents molecular views of the strong acid $HCl(aq)$ and the weak acid $HF(aq)$. Note how both the strong and weak acid undergo the same reaction with water (proton donors); however, in the case of the weak acid such as HF, only a small portion of the acid molecules actually undergoes reaction, leaving the majority of the acid molecules unreacted. On the other hand, strong acids like HCl and HNO_3 undergo complete reaction with water. The result is that weak acids can produce little H_3O^+ in aqueous solution, whereas strong acids can produce large amounts. Hence, when chemists refer to a weak acid, they are often thinking about a substance that produces a limited amount of $H_3O^+(aq)$, whereas when chemists refer to a strong acid, they are thinking about a substance that can produce large amounts of $H_3O^+(aq)$.

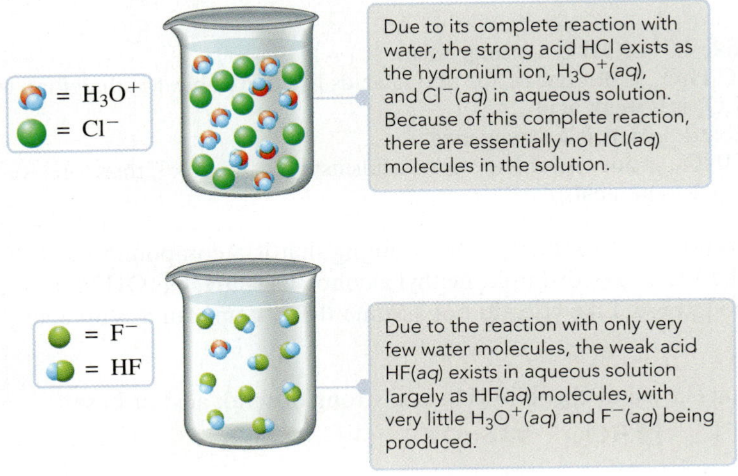

Figure 4.9 ◀ **Molecular views comparing the strong acid HCl and the weak acid HF in water (H_2O molecules are omitted for clarity)**

A **strong base** is *a base that is present in aqueous solution entirely as ions, one of which is* OH^-*; it is a strong electrolyte.* The ionic compound sodium hydroxide, NaOH, is an example of a strong base. It dissolves in water as Na^+ and OH^-.

$$NaOH(s) \xrightarrow{H_2O} Na^+(aq) + OH^-(aq)$$

The hydroxides of Groups IA and IIA elements, except for beryllium hydroxide, are strong bases (see Table 4.3).

A **weak base** is *a base that is only partly ionized in water; it is a weak electrolyte.* Ammonia, NH_3, is an example.

$$NH_3(aq) + H_2O(l) \rightleftharpoons NH_4^+(aq) + OH^-(aq)$$

You will find it important to be able to identify an acid or base as strong or weak. When you write an ionic equation, you represent strong acids and bases by the ions they form and weak acids and bases by the formulas of the compounds. The next example will give you some practice identifying acids and bases as strong or weak.

In many cases, when chemists are thinking about strong versus weak bases in aqueous solution, they consider a weak base to be a substance that can produce only limited amounts of OH^-*(aq) and a strong base to be a substance that can produce large amounts of* OH^-*(aq).*

CONCEPT CHECK 4.4

Complete and balance the two chemical equations.

$$HClO_4(aq) + H_2O(l) \longrightarrow$$

$$HBr(aq) + H_2O(l) \longrightarrow$$

Which of the following statements is true regarding the solutions formed after the two chemical reactions have occurred?

- **a** The HBr is partially ionized in water.
- **b** The $HClO_4$ is completely ionized in water.
- **c** In the solution of $HClO_4$, you would expect to find very few $HClO_4(aq)$ molecules relative to $ClO_4^-(aq)$ ions.
- **d** In the HBr solution there are significant quantities of $HBr(aq)$ molecules.
- **e** None of the other statements (a–d) is true.

Example 4.4 Classifying Acids and Bases as Strong or Weak

Gaining Mastery Toolbox

Critical Concept 4.4
Assume that if an acid or base is not listed in Table 4.3 ("Common Strong Acids and Bases"), it is a weak acid or a weak base. This assumption works very well for most acids and bases typically encountered in the laboratory.

Solution Essentials:
- Weak acid
- Weak base
- Strong acid
- Strong base

Identify each of the following compounds as a strong or weak acid or base:

a LiOH **b** $HC_2H_3O_2$ **c** HBr **d** HNO_2

Problem Strategy Refer to Table 4.3 for the common strong acids and bases. You can assume that other acids and bases are weak.

Solution

a As noted in Table 4.3, **LiOH is a strong base.**

b Acetic acid, $HC_2H_3O_2$, is not one of the strong acids listed in Table 4.3; therefore, we assume **$HC_2H_3O_2$ is a weak acid.**

c As noted in Table 4.3, **HBr is a strong acid.**

d Nitrous acid, HNO_2, is not one of the strong acids listed in Table 4.3; therefore, we assume **HNO_2 is a weak acid.**

Answer Check Do not fall into the trap of assuming that if a compound has OH in the formula, it must be a base. For example, methyl alcohol, CH_3OH, has OH in the formula, but it is not a strong base. Likewise, do not assume that a compound is an acid just because it has H.

Exercise 4.4 Label each of the following as a strong or weak acid or base:

a H_3PO_4 **b** HClO **c** $HClO_4$ **d** $Sr(OH)_2$

■ See Problems 4.41 and 4.42.

Neutralization Reactions

One of the chemical properties of acids and bases is that they neutralize one another. A **neutralization reaction** is *a reaction of an acid and a base that results in an ionic compound and possibly water.* When a base is added to an acid solution, the acid is said to be neutralized. *The ionic compound that is a product of a neutralization reaction* is called a **salt**. Most ionic compounds other than hydroxides and oxides are salts. Salts can be obtained from neutralization reactions such as

$$\underset{\text{acid}}{2HCl(aq)} + \underset{\text{base}}{Ca(OH)_2(aq)} \longrightarrow \underset{\text{salt}}{CaCl_2(aq)} + 2H_2O(l)$$

$$\underset{\text{acid}}{HCN(aq)} + \underset{\text{base}}{KOH(aq)} \longrightarrow \underset{\text{salt}}{KCN(aq)} + H_2O(l)$$

The salt formed in a neutralization reaction consists of cations obtained from the base and anions obtained from the acid. In the first example, the base is $Ca(OH)_2$, which supplies Ca^{2+} cations; the acid is HCl, which supplies Cl^- anions. The salt contains Ca^{2+} and Cl^- ions ($CaCl_2$).

We wrote these reactions as molecular equations. Written in this form, the equations make explicit the reactant compounds and the salts produced. However, to discuss the essential reactions that occur, you need to write the net ionic equations. You will see that the net ionic equation for each acid–base reaction involves the transfer of a proton.

The first reaction involves a strong acid, HCl(*aq*), and a strong base, $Ca(OH)_2(aq)$. Both $Ca(OH)_2$ and the product $CaCl_2$, being soluble ionic compounds, are strong electrolytes. Writing these strong electrolytes in the form of ions gives the following complete ionic equation:

$$2H^+(aq) + \cancel{2Cl^-(aq)} + \cancel{Ca^{2+}(aq)} + 2OH^-(aq) \longrightarrow \cancel{Ca^{2+}(aq)} + \cancel{2Cl^-(aq)} + 2H_2O(l)$$

Canceling spectator ions and dividing by 2 gives the net ionic equation:

$$\overset{H^+}{\curvearrowright}$$
$$H^+(aq) + OH^-(aq) \longrightarrow H_2O(l)$$

Note the transfer of a proton from the hydrogen ion (or hydronium ion, H_3O^+) to the hydroxide ion.

The second reaction involves HCN(*aq*), a weak acid, and KOH(*aq*), a strong base; the product is KCN, a strong electrolyte. The net ionic equation is

$$\overset{H^+}{\curvearrowright}$$
$$HCN(aq) + OH^-(aq) \longrightarrow CN^-(aq) + H_2O(l)$$

Note the proton transfer, characteristic of an acid–base reaction.

In each of these examples, hydroxide ions latch strongly onto protons to form water. Because water is a very stable substance, it effectively provides the driving force of the reaction.

Although water is one of the products in most neutralization reactions, the reaction of an acid with the base ammonia provides a prominent exception. Consider the reaction of sulfuric acid with ammonia:

$$\underset{\text{acid}}{H_2SO_4(aq)} + \underset{\text{base}}{2NH_3(aq)} \longrightarrow \underset{\text{salt}}{(NH_4)_2SO_4(aq)}$$

The net ionic equation is

$$\overset{H^+}{\curvearrowright}$$
$$H^+(aq) + NH_3(aq) \longrightarrow NH_4^+(aq)$$

Note the proton transfer, a hallmark of an acid–base reaction. In this case, NH_3 molecules latch onto protons and form relatively stable NH_4^+ ions.

Example 4.5 Writing an Equation for a Neutralization

Gaining Mastery Toolbox

Critical Concept 4.5
One product of a neutralization reaction will be a salt. The salt is an ionic compound formed from the acid anions and the base cations. Water is often an additional product of neutralization reactions.

Solution Essentials:
- Neutralization reaction
- Salt
- Strong and weak acid
- Net ionic equation
- Spectator ion
- Strong and weak electrolyte

Write the molecular equation and then the net ionic equation for the neutralization of nitrous acid, HNO_2, by sodium hydroxide, NaOH, both in aqueous solution. Use an arrow with H^+ over it to show the proton transfer.

Problem Strategy Because this is a neutralization reaction, you know that the reactants are an acid and a base and that there will be a salt and possibly water as a product. Therefore, start by writing the acid and base reactants and the salt product. Recall that the salt consists of cations from the base and anions from the acid. Note also that water is frequently a product; if it is, you will not be able to balance the equation without it. Once you have the balanced molecular equation, write the complete ionic equation. Do that by representing any strong electrolytes by the formulas of their ions. Finally, write the net ionic equation by canceling any spectator ions, and from it note the proton transfer.

Solution The salt consists of the cation from the base (Na^+) and the anion from the acid (NO_2^-); its formula is $NaNO_2$. You will need to add H_2O as a product to complete and balance the molecular equation:

$$\mathbf{HNO_2(\mathit{aq}) + NaOH(\mathit{aq}) \longrightarrow NaNO_2(\mathit{aq}) + H_2O(\mathit{l})} \qquad \textbf{(molecular equation)}$$

Note that NaOH (a strong base) and $NaNO_2$ (a soluble ionic substance) are strong electrolytes; HNO_2 is a weak electrolyte (it is not one of the strong acids in Table 4.3). You write both NaOH and $NaNO_2$ in the form of ions. The complete ionic equation is

$$HNO_2(aq) + \cancel{Na^+(aq)} + OH^-(aq) \longrightarrow \cancel{Na^+(aq)} + NO_2^-(aq) + H_2O(l)$$

The net ionic equation is

$$HNO_2(aq) + OH^-(aq) \longrightarrow NO_2^-(aq) + H_2O(l)$$

Note that a proton is transferred from HNO_2 to the OH^- ion to yield the products:

$$\mathbf{\overset{H^+}{\overbrace{HNO_2(\mathit{aq}) + OH^-}}(\mathit{aq}) \longrightarrow NO_2^-(\mathit{aq}) + H_2O(\mathit{l})} \qquad \textbf{(net ionic equation)}$$

Answer Check Remember that correct answers include phase labels and that ions must have correct charges.

Exercise 4.5 Write the molecular equation and the net ionic equation for the neutralization of hydrocyanic acid, HCN, by lithium hydroxide, LiOH, both in aqueous solution.

■ See Problems 4.43, 4.44, 4.45, and 4.46.

Acids such as HCl and HNO_3 that have only one acidic hydrogen atom per acid molecule are called *monoprotic* acids. A **polyprotic acid** is *an acid that yields two or more acidic hydrogens per molecule.* Phosphoric acid is an example of a triprotic acid. By reacting this acid with different amounts of a base, you can obtain a series of salts:

$$H_3PO_4(aq) + NaOH(aq) \longrightarrow NaH_2PO_4(aq) + H_2O(l)$$
$$H_3PO_4(aq) + 2NaOH(aq) \longrightarrow Na_2HPO_4(aq) + 2H_2O(l)$$
$$H_3PO_4(aq) + 3NaOH(aq) \longrightarrow Na_3PO_4(aq) + 3H_2O(l)$$

Salts such as NaH_2PO_4 and Na_2HPO_4 that have acidic hydrogen atoms and can undergo neutralization with bases are called *acid salts.*

Exercise 4.6 Write molecular and net ionic equations for the successive neutralizations of each of the acidic hydrogens of sulfuric acid with potassium hydroxide. (That is, write equations for the reaction of sulfuric acid with KOH to give the acid salt and for the reaction of the acid salt with more KOH to give potassium sulfate.)

■ See Problems 4.47, 4.48, 4.49, and 4.50.

Acid–Base Reactions with Gas Formation

Certain salts, notably carbonates, sulfites, and sulfides, react with acids to form a gaseous product. Consider the reaction of sodium carbonate with hydrochloric acid. The molecular equation for the reaction is

$$\underset{\text{carbonate}}{Na_2CO_3(aq)} + \underset{\text{acid}}{2HCl(aq)} \longrightarrow \underset{\text{salt}}{2NaCl(aq)} + H_2O(l) + CO_2(g)$$

Here, a carbonate (sodium carbonate) reacts with an acid (hydrochloric acid) to produce a salt (sodium chloride), water, and carbon dioxide gas. A similar reaction is shown in Figure 4.10, in which baking soda (sodium hydrogen carbonate) reacts with the acetic acid in vinegar. Note the bubbles of carbon dioxide gas that evolve during the reaction. This reaction of a carbonate with an acid is the basis of a simple test for carbonate minerals. When you treat a carbonate mineral or rock, such as limestone, with hydrochloric acid, the material fizzes, as bubbles of odorless carbon dioxide form.

It is useful to consider the preceding reaction as an exchange, or metathesis, reaction. Recall that in an exchange reaction, you obtain the products from the reactants by exchanging the anions (or the cations) between the two reactants. In this case, we interchange the carbonate ion with the chloride ion and we obtain the following:

$$Na_2CO_3(aq) + 2HCl(aq) \longrightarrow 2NaCl(aq) + H_2CO_3(aq)$$

The last product shown in the equation is carbonic acid, H_2CO_3. (You obtain the formula of carbonic acid by noting that you need to associate two H^+ ions with the carbonate ion, CO_3^{2-}, to obtain a neutral compound.) Carbonic acid is unstable and decomposes to water and carbon dioxide gas. The overall result is the equation we wrote earlier:

$$Na_2CO_3(aq) + 2HCl(aq) \longrightarrow 2NaCl(aq) + \underbrace{H_2O(l) + CO_2(g)}_{H_2CO_3(aq)}$$

The net ionic equation for this reaction is

$$CO_3^{2-}(aq) + 2H^+(aq) \longrightarrow H_2O(l) + CO_2(g)$$

The carbonate ion reacts with hydrogen ion from the acid. If you write the hydrogen ion as H_3O^+ (hydronium ion), you can see also that the reaction involves a proton transfer:

$$CO_3^{2-}(aq) + 2H_3O^+(aq) \longrightarrow H_2CO_3(aq) + 2H_2O(l) \longrightarrow 3H_2O(l) + CO_2(g)$$

(arrow labeled $2H^+$ from $2H_3O^+$ to CO_3^{2-})

From the broader perspective of the Brønsted–Lowry view, this is an acid–base reaction.

Sulfites behave similarly to carbonates. When a sulfite, such as sodium sulfite (Na_2SO_3), reacts with an acid, sulfur dioxide (SO_2) is a product. Sulfides, such as sodium sulfide (Na_2S), react with acids to produce hydrogen sulfide gas. The reaction can be viewed as a simple exchange reaction. The reactions are summarized in Table 4.4. ▶

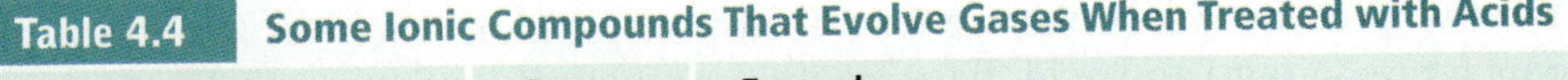

Table 4.4 Some Ionic Compounds That Evolve Gases When Treated with Acids

Ionic Compound	Gas	Example
Carbonate (CO_3^{2-})	CO_2	$Na_2CO_3 + 2HCl \longrightarrow 2NaCl + H_2O + CO_2$
Sulfite (SO_3^{2-})	SO_2	$Na_2SO_3 + 2HCl \longrightarrow 2NaCl + H_2O + SO_2$
Sulfide (S^{2-})	H_2S	$Na_2S + H_2SO_4 \longrightarrow Na_2SO_4 + H_2S$

+

=

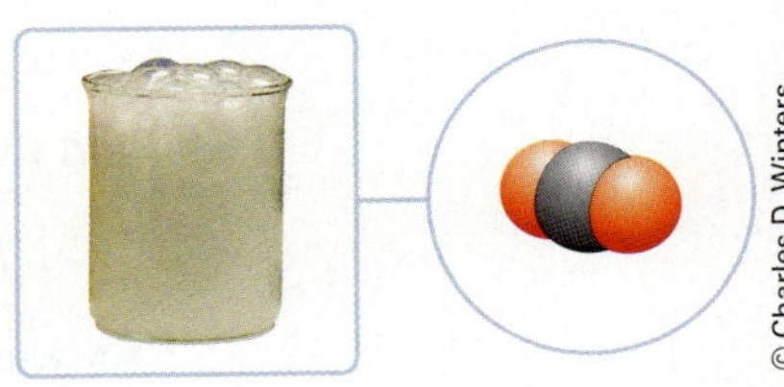

Figure 4.10 ▲

Reaction of a carbonate with an acid Baking soda (sodium hydrogen carbonate) reacts with acetic acid in vinegar to evolve bubbles of carbon dioxide.

CO_3^{2-}, S^{2-}, and SO_3^{2-} are bases. Thus the net ionic equations involve a basic anion reacting with H^+, an acid.

OWL Interactive Example 4.6 Writing an Equation for a Reaction with Gas Formation

Gaining Mastery Toolbox

Critical Concept 4.6
Ionic compounds that contain the carbonate, sulfite, or sulfide anions will produce a gas when reacted with an acid. In addition to the gas, these reactions will produce a salt.

Solution Essentials:
- Acid
- Salt
- Exchange reaction
- Solubility rules
- Carbonate, sulfite, and sulfide anions

Write the molecular equation and the net ionic equation for the reaction of zinc sulfide with hydrochloric acid.

Problem Strategy Start by writing the reactants, using the solubility rules to determine the phase labels. Next think about the ways in which this reaction can occur; first try to write it as an exchange reaction. Examine the products to see whether an insoluble compound forms and/or there is gas formation; either indicates a reaction.

Solution First write the reactants, noting from Table 4.1 that most sulfides (except those of the group IA metals and $(NH_4)_2S$) are insoluble.

$$ZnS(s) + HCl(aq) \longrightarrow$$

If you look at this as an exchange reaction, the products are $ZnCl_2$ and H_2S. Since chlorides are soluble (with some exceptions, not including $ZnCl_2$), we write the following balanced equation.

$$\mathbf{ZnS(s) + 2HCl(aq) \longrightarrow ZnCl_2(aq) + H_2S(g)}$$

The complete ionic equation is

$$ZnS(s) + 2H^+(aq) + \cancel{2Cl^-(aq)} \longrightarrow Zn^{2+}(aq) + \cancel{2Cl^-(aq)} + H_2S(g)$$

and the net ionic equation is

$$\mathbf{ZnS(s) + 2H^+(aq) \longrightarrow Zn^{2+}(aq) + H_2S(g)}$$

Answer Check When you are trying to solve problems of this type, if starting with an exchange reaction doesn't yield the desired results, use the examples in Table 4.4 as a guide.

Exercise 4.7 Write the molecular equation and the net ionic equation for the reaction of calcium carbonate with nitric acid.

■ See Problems 4.51, 4.52, 4.53, and 4.54.

CONCEPT CHECK 4.5

At times, we want to generalize the formula of certain important chemical substances; acids and bases fall into this category. Given the following reactions, try to identify the acids, bases, and some examples of what the general symbols (M and A^-) represent.

a $MOH(s) \longrightarrow M^+(aq) + OH^-(aq)$

b $HA(aq) + H_2O(l) \rightleftharpoons H_3O^+(aq) + A^-(aq)$

c $H_2A(aq) + H_2O(l) \rightleftharpoons H_3O^+(aq) + HA^-(aq)$

d For parts a through c, provide real examples for M and A.

4.5 Oxidation–Reduction Reactions

In the two preceding sections, we described precipitation reactions (reactions producing a precipitate) and acid–base reactions (reactions involving proton transfer). Here we discuss the third major class of reactions, oxidation–reduction reactions, which are reactions involving a transfer of electrons from one species to another.

As a simple example of an oxidation–reduction reaction, let us look at what happens when you dip an iron nail into a blue solution of copper(II) sulfate

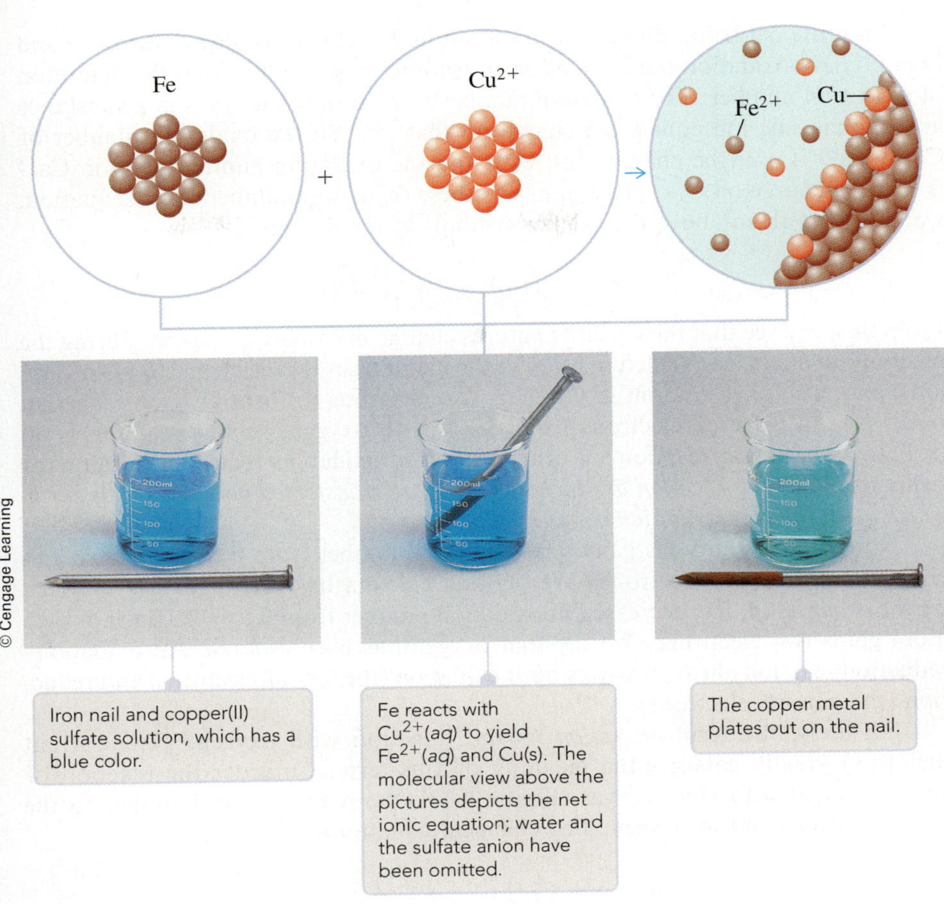

Figure 4.11 ◀
Reaction of iron with Cu^{2+}(aq)

(Figure 4.11). What you see is that the iron nail becomes coated with a reddish-brown tinge of metallic copper. The molecular equation for this reaction is

$$Fe(s) + CuSO_4(aq) \longrightarrow FeSO_4(aq) + Cu(s)$$

The net ionic equation is

$$Fe(s) + Cu^{2+}(aq) \longrightarrow Fe^{2+}(aq) + Cu(s)$$

The electron-transfer aspect of the reaction is apparent from this equation. Note that each iron atom in the metal loses two electrons to form an iron(II) ion, and each copper(II) ion gains two electrons to form a copper atom in the metal. The net effect is that two electrons are transferred from each iron atom in the metal to each copper(II) ion.

The concept of *oxidation numbers* was developed as a simple way of keeping track of electrons in a reaction. Using oxidation numbers, you can determine whether electrons have been transferred from one atom to another. If electrons have been transferred, an oxidation–reduction reaction has occurred.

Oxidation Numbers

We define the **oxidation number** (or **oxidation state**) of an atom in a substance as *the actual charge of the atom if it exists as a monatomic ion, or a hypothetical charge assigned to the atom in the substance by simple rules.* An oxidation–reduction reaction is one in which one or more atoms change oxidation number, implying that there has been a transfer of electrons.

Consider the combustion of calcium metal in oxygen gas (Figure 4.12).

$$2Ca(s) + O_2(g) \longrightarrow 2CaO(s)$$

This is an oxidation–reduction reaction. To see this, you assign oxidation numbers to the atoms in the equation and then note that the atoms change oxidation number during the reaction.

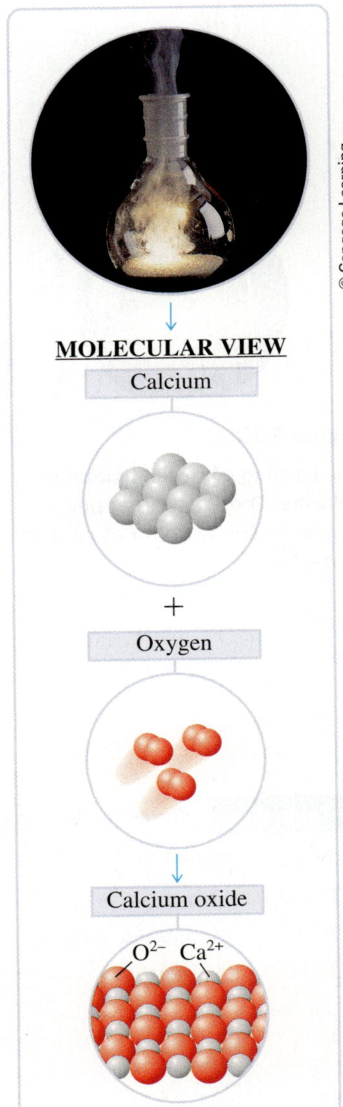

Figure 4.12 ▲
The burning of calcium metal in oxygen The burning calcium emits a red-orange flame.

Since the oxidation number of an atom in an element is always zero, Ca and O in O_2 have oxidation numbers of zero. Another rule follows from the definition of oxidation number: The oxidation number of an atom that exists in a substance as a monatomic ion equals the charge on that ion. So the oxidation number of Ca in CaO is +2 (the charge on Ca^{2+}), and the oxidation number of O in CaO is −2 (the charge on O^{2-}). To emphasize these oxidation numbers in an equation, we will write them above the atomic symbols in the formulas.

$$2\overset{0}{\text{Ca}}(s) + \overset{0}{\text{O}}_2(g) \longrightarrow 2\overset{+2}{\text{Ca}}\overset{-2}{\text{O}}(s)$$

From this, you see that the Ca and O atoms change in oxidation number during the reaction. In effect, each calcium atom in the metal loses two electrons to form Ca^{2+} ions, and each oxygen atom in O_2 gains two electrons to form O^{2-} ions. The net result is a transfer of electrons from calcium to oxygen, so this reaction is an oxidation–reduction reaction. In other words, an **oxidation–reduction reaction** (or **redox reaction**) is *a reaction in which electrons are transferred between species or in which atoms change oxidation number.*

Note that calcium has gained in oxidation number from 0 to +2. (Each calcium atom loses two electrons.) We say that calcium has been *oxidized.* Oxygen, on the other hand, has decreased in oxidation number from 0 to −2. (Each oxygen atom gains two electrons.) We say that oxygen has been *reduced.* An oxidation–reduction reaction always involves both oxidation (the loss of electrons) and reduction (the gain of electrons).

Formerly, the term *oxidation* meant "reaction with oxygen." The current definition greatly enlarges the meaning of this term. Consider the reaction of calcium metal with chlorine gas (Figure 4.13); the reaction looks similar to the burning of calcium in oxygen. The chemical equation is

$$\overset{0}{\text{Ca}}(s) + \overset{0}{\text{Cl}}_2(g) \longrightarrow \overset{+2}{\text{Ca}}\overset{-1}{\text{Cl}}_2(s)$$

Figure 4.13 ▲

The burning of calcium metal in chlorine The reaction appears similar to the burning of calcium in oxygen.

In this reaction, the calcium atom is oxidized, because it increases in oxidation number (from 0 to +2, as in the previous equation). Chlorine is reduced; it decreases in oxidation number from 0 to −1. This is clearly an oxidation–reduction reaction that does not involve oxygen.

Oxidation-Number Rules

So far, we have used two rules for obtaining oxidation numbers: (1) the oxidation number of an atom in an element is zero, and (2) the oxidation number of an atom in a monatomic ion equals the charge on the ion. These and several other rules for assigning oxidation numbers are given in Table 4.5.

Table 4.5 Rules for Assigning Oxidation Numbers

Rule	Applies to	Statement
1	Elements	The oxidation number of an atom in an element is zero.
2	Monatomic ions	The oxidation number of an atom in a monatomic ion equals the charge on the ion.
3	Oxygen	The oxidation number of oxygen is −2 in most of its compounds. (An exception is O in H_2O_2 and other peroxides, where the oxidation number is −1.)
4	Hydrogen	The oxidation number of hydrogen is +1 in most of its compounds. (The oxidation number of hydrogen is −1 in binary compounds with a metal, such as CaH_2.)
5	Halogens	The oxidation number of fluorine is −1 in all of its compounds. Each of the other halogens (Cl, Br, I) has an oxidation number of −1 in binary compounds, except when the other element is another halogen above it in the periodic table or the other element is oxygen.
6	Compounds and ions	The sum of the oxidation numbers of the atoms in a compound is zero. The sum of the oxidation numbers of the atoms in a polyatomic ion equals the charge on the ion.

In molecular substances, we use these rules to give the *approximate* charges on the atoms. Consider the molecule SO_2. Oxygen atoms tend to attract electrons, pulling them from other atoms (sulfur in the case of SO_2). As a result, an oxygen atom in SO_2 takes on a negative charge relative to the sulfur atom. The magnitude of the charge on an oxygen atom in a molecule is not a full -2 charge as in the O^{2-} ion. However, it is convenient to assign an oxidation number of -2 to oxygen in SO_2 (and in most other compounds of oxygen) to help us express the approximate charge distribution in the molecule. Rule 3 in Table 4.5 says that an oxygen atom has an oxidation number of -2 in most of its compounds.

Rules 4 and 5 are similar in that they tell you what to expect for the oxidation number of certain elements in their compounds. Rule 4, for instance, says that hydrogen has an oxidation number of $+1$ in most of its compounds.

Rule 6 states that the sum of the oxidation numbers of the atoms in a compound is zero. This rule follows from the interpretation of oxidation numbers as (hypothetical) charges on the atoms. Because any compound is electrically neutral, the sum of the charges on its atoms must be zero. This rule is easily extended to ions: the sum of the oxidation numbers (hypothetical charges) of the atoms in a polyatomic ion equals the charge on the ion.

You can use Rule 6 to obtain the oxidation number of one atom in a compound or ion, if you know the oxidation numbers of the other atoms in the compound or ion. Consider the SO_2 molecule. According to Rule 6,

$$(\text{Oxidation number of S}) + 2 \times (\text{oxidation number of O}) = 0$$

or

$$(\text{Oxidation number of S}) + 2 \times (-2) = 0.$$

Therefore,

$$\text{Oxidation number of S (in } SO_2) = -2 \times (-2) = +4$$

The next example illustrates how to use the rules in Table 4.5 to assign oxidation numbers.

Example 4.7 **Assigning Oxidation Numbers**

Gaining Mastery Toolbox

Critical Concept 4.7
The sum of all the oxidation numbers in a compound or polyatomic ion is equal to the overall charge on the compound or ion. This concept is very helpful in assigning the oxidation number of elements that are not covered by the rules (Table 4.5).

Solution Essentials:
- Oxidation number
- Rules for assigning oxidation numbers (Table 4.5)

Use the rules from Table 4.5 to obtain the oxidation number of the chlorine atom in each of the following: **a** $HClO_4$ (perchloric acid), **b** ClO_3^- (chlorate ion).

Problem Strategy We need to apply the rules in Table 4.5. Rule 6 is the best place to start. Therefore, in each case, write the expression for the sum of the oxidation numbers, equating this to zero for a compound or to the charge for an ion. Now, use Rules 2 to 5 to substitute oxidation numbers for particular atoms, such as -2 for oxygen and $+1$ for hydrogen, and solve for the unknown oxidation number (Cl in this example).

Solution

a For perchloric acid, Rule 6 gives the equation

$$(\text{Oxidation number of H}) + (\text{oxidation number of Cl}) + 4 \times (\text{oxidation number of O}) = 0$$

Using Rules 3 and 4, you obtain

$$(+1) + (\text{oxidation number of Cl}) + 4 \times (-2) = 0$$

(continued)

(continued)

Therefore,

$$\text{Oxidation number of Cl (in } HClO_4) = -(+1) - 4 \times (-2) = \mathbf{+7}$$

b For the chlorate ion, Rule 6 gives the equation

$$(\text{Oxidation number of Cl}) + 3 \times (\text{oxidation number of O}) = -1$$

Using Rule 3, you obtain

$$(\text{Oxidation number of Cl}) + 3 \times (-2) = -1$$

Therefore,

$$\text{Oxidation number of Cl (in } ClO_3^-) = -1 - 3 \times (-2) = \mathbf{+5}$$

Answer Check Because most compounds do not have elements with very large positive or very large negative oxidation numbers, you should always be on the alert for a *possible* assignment mistake when you find oxidation states greater than +6 or less than −4. (From this example you see that a +7 oxidation state is possible; however, it occurs in only a limited number of cases.)

Exercise 4.8 Obtain the oxidation numbers of the atoms in each of the following: **a** potassium dichromate, $K_2Cr_2O_7$, **b** permanganate ion, MnO_4^-.

See Problems 4.55, 4.56, 4.57, and 4.58.

Describing Oxidation–Reduction Reactions

We use special terminology to describe oxidation–reduction reactions. To illustrate this, we will look again at the reaction of iron with copper(II) sulfate. The net ionic equation is

$$\overset{0}{Fe}(s) + \overset{+2}{Cu^{2+}}(aq) \longrightarrow \overset{+2}{Fe^{2+}}(aq) + \overset{0}{Cu}(s)$$

We can write this reaction in terms of two half-reactions. A **half-reaction** is *one of two parts of an oxidation–reduction reaction, one part of which involves a loss of electrons (or increase of oxidation number) and the other a gain of electrons (or decrease of oxidation number).* The half-reactions for the preceding equation are

$$\overset{0}{Fe}(s) \longrightarrow \overset{+2}{Fe^{2+}}(aq) + 2e^- \quad \text{(electrons lost by Fe)}$$

$$\overset{+2}{Cu^{2+}}(aq) + 2e^- \longrightarrow \overset{0}{Cu}(s) \quad \text{(electrons gained by } Cu^{2+})$$

Oxidation is *the half-reaction in which there is a loss of electrons by a species (or an increase of oxidation number of an atom).* **Reduction** is *the half-reaction in which there is a gain of electrons by a species (or a decrease in the oxidation number of an atom).* Thus, the equation $Fe(s) \longrightarrow Fe^{2+}(aq) + 2e^-$ represents the oxidation half-reaction, and the equation $Cu^{2+}(aq) + 2e^- \longrightarrow Cu(s)$ represents the reduction half-reaction.

Recall that a species that is *oxidized* loses electrons (or contains an atom that increases in oxidation number) and a species that is *reduced* gains electrons (or contains an atom that decreases in oxidation number). An **oxidizing agent** is *a species that oxidizes another species; it is itself reduced.* Similarly, a **reducing agent** is *a species that reduces another species; it is itself oxidized.* In our example reaction, the copper(II) ion is the oxidizing agent, whereas iron metal is the reducing agent.

The relationships among these terms are shown in the following diagram for the reaction of iron with copper(II) ion.

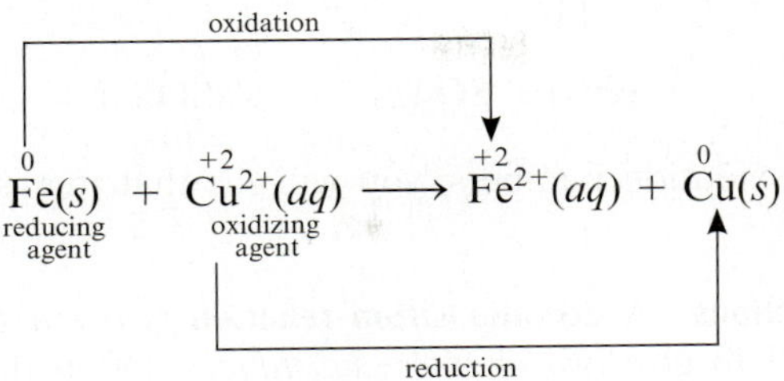

Some Common Oxidation–Reduction Reactions

Many oxidation–reduction reactions can be described as one of the following:

1. Combination reaction
2. Decomposition reaction
3. Displacement reaction
4. Combustion reaction

We will describe examples of each of these in this section.

Combination Reactions A **combination reaction** is *a reaction in which two substances combine to form a third substance.* Note that not all combination reactions are oxidation–reduction reactions. However, the simplest cases are those in which two elements react to form a compound; these are clearly oxidation–reduction reactions. In Chapter 2, we discussed the reaction of sodium metal and chlorine gas, which is a redox reaction (Figure 4.14).

$$2Na(s) + Cl_2(g) \longrightarrow 2NaCl(s)$$

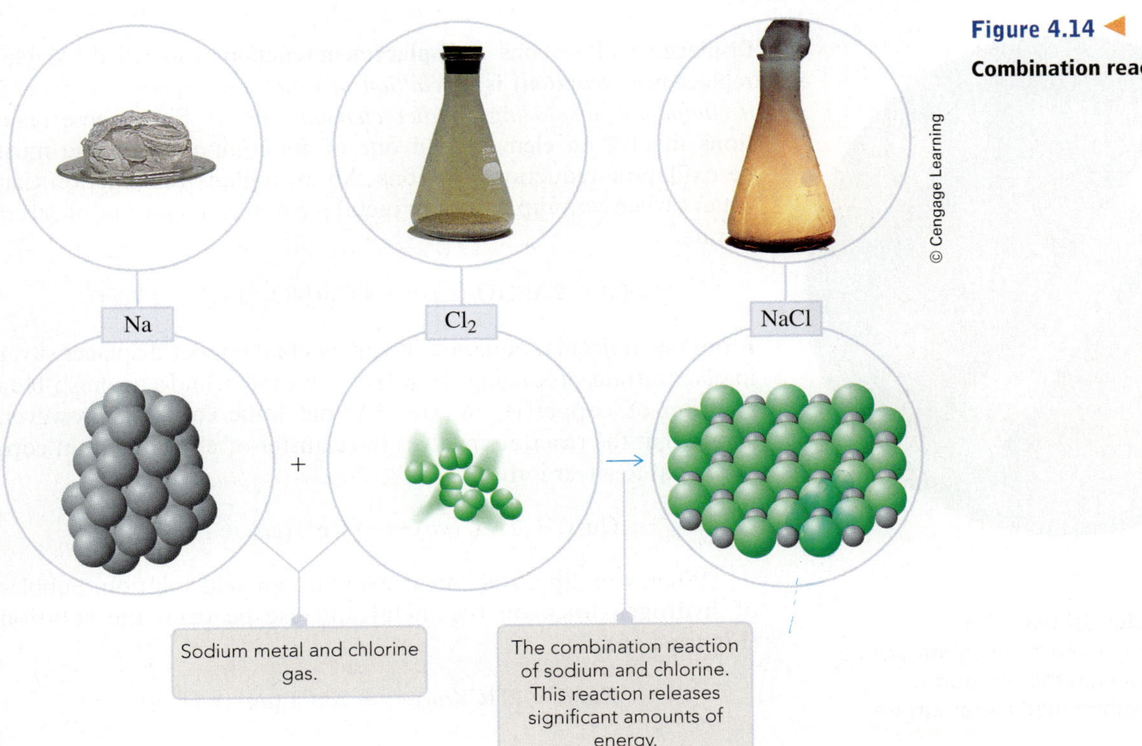

Figure 4.14 Combination reaction

Figure 4.15 ▲

Decomposition reaction The decomposition reaction of mercury(II) oxide into its elements, mercury and oxygen.

Antimony and chlorine also combine in a fiery reaction.

$$2Sb + 3Cl_2 \longrightarrow 2SbCl_3$$

Some combination reactions involve compounds as reactants and are not oxidation–reduction reactions. For example,

$$CaO(s) + SO_2(g) \longrightarrow CaSO_3(s)$$

(If you check the oxidation numbers, you will see that this is not an oxidation–reduction reaction.)

Decomposition Reactions A **decomposition reaction** is *a reaction in which a single compound reacts to give two or more substances.* Often these reactions occur when the temperature is raised. In Chapter 1, we described the decomposition of mercury(II) oxide into its elements when the compound is heated (Figure 4.15). This is an oxidation–reduction reaction.

$$2HgO(s) \xrightarrow{\Delta} 2Hg(l) + O_2(g)$$

Another example is the preparation of oxygen by heating potassium chlorate with manganese(IV) oxide as a catalyst.

$$2KClO_3(s) \xrightarrow[MnO_2]{\Delta} 2KCl(s) + 3O_2(g)$$

In this reaction, a compound decomposes into another compound and an element; it also is an oxidation–reduction reaction.

Not all decomposition reactions are of the oxidation–reduction type. For example, calcium carbonate at high temperatures decomposes into calcium oxide and carbon dioxide.

$$CaCO_3(s) \xrightarrow{\Delta} CaO(s) + CO_2(g)$$

Is there a change in oxidation numbers? If not, this confirms that this is not an oxidation–reduction reaction.

Figure 4.16 ▲

Displacement reaction Displacement reaction of zinc metal and hydrochloric acid. Hydrogen gas formed in the reaction bubbles from the submerged metal surface.

Displacement Reactions A **displacement reaction** (also called a **single-replacement reaction**) is *a reaction in which an element reacts with a compound, displacing another element from it.* Since these reactions involve an element and one of its compounds, these must be oxidation–reduction reactions. An example is the reaction that occurs when you dip a copper metal strip into a solution of silver nitrate.

$$Cu(s) + 2AgNO_3(aq) \longrightarrow Cu(NO_3)_2(aq) + 2Ag(s)$$

From the molecular equation, it appears that copper displaces silver in silver nitrate, producing crystals of silver metal and a greenish-blue solution of copper(II) nitrate. The net ionic equation, however, shows that the reaction involves the transfer of electrons from copper metal to silver ion:

$$Cu(s) + 2Ag^+(aq) \longrightarrow Cu^{2+}(aq) + 2Ag(s)$$

When you dip a zinc metal strip into an acid solution, bubbles of hydrogen form on the metal and escape from the solution (Figure 4.16).

$$Zn(s) + 2HCl(aq) \longrightarrow ZnCl_2(aq) + H_2(g)$$

Table 4.6 Activity Series of the Elements

React vigorously with acidic solutions and water to give H_2	Li K Ba Ca Na
React with acids to give H_2	Mg Al Zn Cr Fe Cd Co Ni Sn Pb
	H_2
Do not react with acids to give H_2*	Cu Hg Ag Au

*Cu, Hg, and Ag react with HNO_3 but do not produce H_2. In these reactions, the metal is oxidized to the metal ion, and NO_3^- ion is reduced to NO_2 or other nitrogen species.

Zinc displaces hydrogen in the acid, producing zinc chloride solution and hydrogen gas. The net ionic equation is

$$Zn(s) + 2H^+(aq) \rightarrow Zn^{2+}(aq) + H_2(g)$$

Whether a reaction occurs between a given element and a monatomic ion depends on the relative ease with which the two species gain or lose electrons. Table 4.6 shows the activity series of the elements, a listing of the elements in decreasing order of their ease of losing electrons during reactions in aqueous solution. The metals listed at the top are the strongest reducing agents (they lose electrons easily); those at the bottom, the weakest. A free element reacts with the monatomic ion of another element if the free element is above the other element in the activity series. The highlighted elements react slowly with liquid water, but readily with steam, to give H_2.

Consider this reaction:

$$2K(s) + 2H^+(aq) \longrightarrow 2K^+(aq) + H_2(g)$$

You would expect this reaction to proceed as written, because potassium metal (K) is well above hydrogen in the activity series. In fact, potassium metal reacts violently with water, which contains only a very small percentage of H^+ ions. Imagine the reaction of potassium metal with a strong acid like HCl!

Combustion Reactions A **combustion reaction** is *a reaction in which a substance reacts with oxygen, usually with the rapid release of heat to produce a flame.* The products include one or more oxides. Oxygen changes oxidation number from 0 to -2, so combustions are oxidation–reduction reactions.

Organic compounds usually burn in oxygen or air to yield carbon dioxide. If the compound contains hydrogen (as most do), water is also a product. For instance, butane (C_4H_{10}) burns in air as follows:

$$2C_4H_{10}(g) + 13O_2(g) \longrightarrow 8CO_2(g) + 10H_2O(g)$$

Many metals burn in air as well. Although chunks of iron do not burn readily in air, iron wool, which consists of fine strands of iron, does (Figure 4.17). The increased surface area of the metal in iron wool allows oxygen from air to react quickly with it.

$$4Fe(s) + 3O_2(g) \longrightarrow 2Fe_2O_3(s)$$

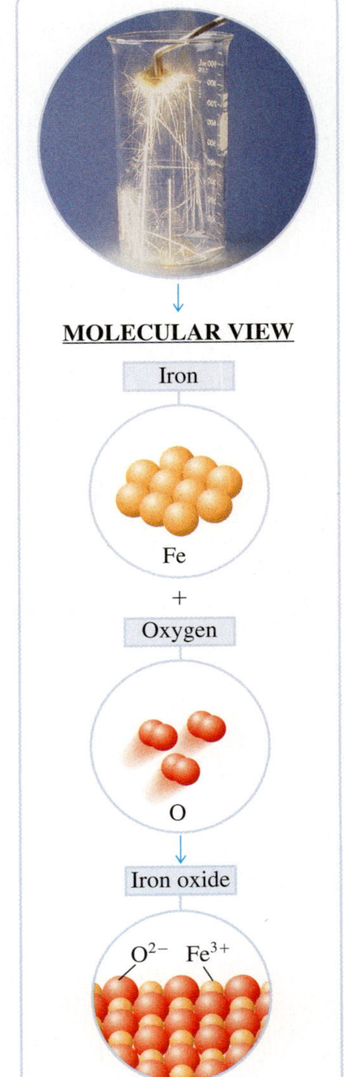

Figure 4.17 ▲

The combustion of iron wool Iron reacts with oxygen in the air to produce iron(III) oxide, Fe_2O_3. The reaction is similar to the rusting of iron but is much faster.

4.6 Balancing Simple Oxidation–Reduction Equations

Oxidation–reduction reactions can often be quite difficult to balance. ▶ Some are so complex in fact that chemists have written computer programs to accomplish the

Balancing equations was discussed in Section 2.10.

task. In this section, we will develop a method for balancing simple oxidation–reduction reactions that can later be generalized for far more complex reactions. One of the advantages of using this technique for even simple reactions is that you focus on what makes oxidation–reduction reactions different from other reaction types. See Chapter 19 for a comprehensive treatment of this topic.

At first glance, the equation representing the reaction of zinc metal with silver(I) ions in solution might appear to be balanced.

$$Zn(s) + Ag^{+}(aq) \longrightarrow Zn^{2+}(aq) + Ag(s)$$

However, because a balanced chemical equation must have a charge balance as well as a mass balance, this equation is not balanced: it has a total charge of +1 for the reactants and +2 for the products. Let us apply the half-reaction method for balancing this equation.

Half-Reaction Method Applied to Simple Oxidation–Reduction Equations

The *half-reaction method* consists of first separating the equation into two half-reactions, one for oxidation, the other for reduction. You balance each half-reaction, then combine them to obtain a balanced oxidation–reduction reaction. Here is an illustration of the process. First we identify the species being oxidized and reduced and assign the appropriate oxidation states.

$$\overset{0}{Zn}(s) + \overset{+1}{Ag^{+}}(aq) \longrightarrow \overset{+2}{Zn^{2+}}(aq) + \overset{0}{Ag}(s)$$

Next, write the half-reactions in an unbalanced form.

$$Zn \longrightarrow Zn^{2+} \qquad \text{(oxidation)}$$

$$Ag^{+} \longrightarrow Ag \qquad \text{(reduction)}$$

Next, balance the charge in each equation by adding electrons to the more positive side to create balanced half-reactions. Following this procedure, the balanced half-reactions are:

$$Zn \longrightarrow Zn^{2+} + 2e^{-} \qquad \text{(oxidation half-reaction)}$$

$$Ag^{+} + e^{-} \longrightarrow Ag \qquad \text{(reduction half-reaction)}$$

Note that the number of electrons that Zn loses during the oxidation process (two) exceeds the number of electrons gained by Ag^{+} during the reduction (one). Since, according to the reduction half-reaction, each Ag^{+} is capable of gaining only one electron, we need to double the amount of Ag^{+} in order for it to accept all of the electrons produced by Zn during oxidation. To meet this goal and obtain the balanced oxidation–reduction reaction, we multiply each half-reaction by a factor (integer) so that when we add them together, the electrons cancel. We multiply the first equation by 1 (the number of electrons in the second half-reaction) and multiply the second equation by 2 (the number of electrons in the first half-reaction).

$$1 \times (Zn \longrightarrow Zn^{2+} + 2e^{-})$$

$$Zn + 2Ag^{+} + 2e^{-} -$$

The electrons cancel, which finally yields the balanced oxidation–reduction equation:

$$Zn(s) + 2Ag^{+}(aq) \longrightarrow Zn^{2+}(aq) + 2Ag(s)$$

Example 4.8 further illustrates this technique.

Example 4.8 **Balancing Simple Oxidation–Reduction Reactions by the Half-Reaction Method**

Gaining Mastery Toolbox

Critical Concept 4.8
A balanced oxidation-reduction reaction has both mass balance and charge balance. Using the half-reaction method of balancing oxidation-reduction reactions assures that there is charge balance.

Solution Essentials:
- Half-reaction (oxidation and reduction)
- Oxidation
- Reduction
- Oxidation-reduction reaction
- Rules for assigning oxidation numbers (Table 4.5)

Consider a more difficult problem, the combination (oxidation–reduction) reaction of magnesium metal and nitrogen gas:

$$Mg(s) + N_2(g) \longrightarrow Mg_3N_2(s)$$

Apply the half-reaction method to balance this equation.

Problem Strategy Start by identifying the species undergoing oxidation and reduction and assigning oxidation numbers. Then write the two balanced half-reactions, keeping in mind that you add the electrons to the more positive side of the reaction. Next, multiply each of the half-reactions by a whole number that will cancel the electrons on each side of the equation. Finally, add the half-reactions together to yield the balanced equation.

Solution Identify the oxidation states of the elements:

$$\overset{0}{Mg}(s) + \overset{0}{N_2}(g) \longrightarrow \overset{+2}{Mg_3}\overset{-3}{N_2}(s)$$

In this problem, a molecular compound, nitrogen (N_2), is undergoing reduction. When a species undergoing reduction or oxidation is a molecule, write the formula of the molecule (do not split it up) in the half-reaction. Also, make sure that both the mass and the charge are balanced. (Note the $6e^-$ required to balance the charge due to the $2N^{3-}$.)

$$Mg \longrightarrow Mg^{2+} + 2e^- \quad \text{(balanced oxidation half-reaction)}$$

$$N_2 + 6e^- \longrightarrow 2N^{3-} \quad \text{(balanced reduction half-reaction)}$$

We now need to multiply each half-reaction by a factor that will cancel the electrons.

$$3 \times (Mg \longrightarrow Mg^{2+} + 2e^-)$$
$$1 \times (N_2 + 6e^- \longrightarrow 2N^{3-})$$
$$\overline{3Mg + N_2 + \cancel{6e^-} \longrightarrow 3Mg^{2+} + 2N^{3-} + \cancel{6e^-}}$$

Therefore, the balanced combination (oxidation–reduction) reaction is

$$\mathbf{3Mg + N_2 \longrightarrow 3Mg^{2+} + 2N^{3-}}$$

From inspecting the coefficients in this reaction, looking at the original equation, and knowing that Mg^{2+} and N^{3-} will combine to form an ionic compound (Mg_3N_2), we can rewrite the equation in the following form:

$$3Mg(s) + N_2(g) \longrightarrow Mg_3N_2(s)$$

Answer Check Do a final check for balance by counting the number of atoms of each element on both sides of the equation.

Exercise 4.9 Use the half-reaction method to balance the equation $Ca(s) + Cl_2(g) \longrightarrow CaCl_2(s)$.

■ See Problems 4.65 and 4.66.

Working with Solutions

The majority of the chemical reactions discussed in this chapter take place in solution. This is because the reaction between two solid reactants often proceeds very slowly or not at all. In a solid, the molecules or ions in a crystal tend to occupy approximately fixed positions, so the chance of two molecules or ions coming together to react is small. In liquid solutions, reactant molecules are free to move throughout the liquid; therefore, reaction is much faster. When you run reactions in liquid solutions, it is convenient to dispense the amounts of reactants by measuring out volumes of reactant solutions. In the next two sections, we will discuss calculations involved in making up solutions, and in Section 4.10 we will describe stoichiometric calculations involving such solutions.

4.7 Molar Concentration

When we dissolve a substance in a liquid, we call the substance the *solute* and the liquid the *solvent*. Consider ammonia solutions. Ammonia gas dissolves readily in water, and aqueous ammonia solutions are often used in the laboratory. In such solutions, ammonia gas is the solute and water is the solvent.

The general term *concentration* refers to the quantity of solute in a standard quantity of solution. Qualitatively, we say that a solution is *dilute* when the solute concentration is low and *concentrated* when the solute concentration is high. Usually these terms are used in a comparative sense and do not refer to a specific concentration. We say that one solution is more dilute, or less concentrated, than another. However, for commercially available solutions, the term *concentrated* refers to the maximum, or near maximum, concentration available. For example, concentrated aqueous ammonia contains about 28% NH_3 by mass.

In this example, we expressed the concentration quantitatively by giving the mass percentage of solute—that is, the mass of solute in 100 g of solution. However, we need a unit of concentration that is convenient for dispensing reactants in solution, such as one that specifies moles of solute per solution volume.

Molar concentration, or **molarity (*M*)**, is defined as *the moles of solute dissolved in one liter (cubic decimeter) of solution.*

This is the last instance where dimensional analysis and the rules for significant figures and rounding will be listed in the Solution Essentials. Any time that you perform a calculation and report an answer, you must consider the rules for rounding and significant figures. Dimensional analysis is a powerful tool that is applicable to most problem solving in which molar quantities are involved, as they are in this example.

$$\text{Molarity } (M) = \frac{\text{moles of solute}}{\text{liters of solution}}$$

An aqueous solution that is 0.15 *M* NH_3 (read this as "0.15 molar NH_3") contains 0.15 mol NH_3 per liter of solution. If you want to prepare a solution that is, for example, 0.200 *M* $CuSO_4$, you place 0.200 mol $CuSO_4$ in a 1.000-L volumetric flask, or a proportional amount in a flask of a different size (Figure 4.18). You add a small quantity of water to dissolve the $CuSO_4$. Then you fill the flask with additional water to the mark on the neck and mix the solution. The following example shows how to calculate the molarity of a solution given the mass of solute and the volume of solution.

Example 4.9 Calculating Molarity from Mass and Volume

Gaining Mastery Toolbox

Critical Concept 4.9
Molarity is a common way for chemists to express solution concentration. Molarity is a ratio of the moles of dissolved substance (solute) to the volume of solution in liters: mol/L. The concentration of a solution is always independent of the amount of solution. 

Solution Essentials:
- Molarity (molar concentration)
- Mole
- Molar mass
- Dimensional analysis
- Rules for significant figures and rounding

A sample of $NaNO_3$ weighing 0.38 g is placed in a 50.0 mL volumetric flask. The flask is then filled with water to the mark on the neck, dissolving the solid. What is the molarity of the resulting solution?

Problem Strategy To calculate the molarity, you need the moles of solute. Therefore, you first convert grams $NaNO_3$ to moles. The molarity equals the moles of solute divided by the liters of solution.

(*continued*)

0.0500 mol $CuSO_4 \cdot 5H_2O$ (12.48 g) is weighed on a platform balance.

The copper(II) sulfate pentahydrate is transferred carefully to the volumetric flask.

Water is added to bring the solution level to the mark on the neck of the 250-mL volumetric flask. The molarity is 0.0500 mol/0.250 L = 0.200 *M*.

Figure 4.18 ▲

Preparing a 0.200 *M* $CuSO_4$ solution

(continued)

Solution You find that 0.38 g $NaNO_3$ is $4.\underline{4}7 \times 10^{-3}$ mol $NaNO_3$; the last significant figure is underlined. The volume of solution is 50.0 mL, or 50.0×10^{-3} L, so the molarity is

$$\text{Molarity} = \frac{4.\underline{4}7 \times 10^{-3}\ \text{mol NaNO}_3}{50.0 \times 10^{-3}\ \text{L soln}} = \mathbf{0.089\ \mathit{M}\ NaNO_3}$$

Answer Check Although very dilute solutions are possible, there is a limit as to how concentrated solutions can be. Therefore, any answer that leads to solution concentrations that are in excess of 20 *M* should be suspect.

Exercise 4.10 A sample of sodium chloride, NaCl, weighing 0.0678 g is placed in a 25.0-mL volumetric flask. Enough water is added to dissolve the NaCl, and then the flask is filled to the mark with water and carefully shaken to mix the contents. What is the molarity of the resulting solution?

■ See Problems 4.67, 4.68, 4.69, and 4.70.

The advantage of molarity as a concentration unit is that the amount of solute is related to the volume of solution. Rather than having to weigh out a specified mass of substance, you can instead measure out a definite volume of solution of the substance, which is usually easier. As the following example illustrates, molarity can be used as a factor for converting from moles of solute to liters of solution, and vice versa.

Example 4.10 Using Molarity as a Conversion Factor

Gaining Mastery Toolbox

Critical Concept 4.10
Molarity is a conversion factor. As a conversion factor it can be used to convert between moles of solute and the volume of solution or from volume of solution to moles of solute.

Solution Essentials:
- Molarity (molar concentration)
- Mole
- Molar mass
- Conversion factor

An experiment calls for the addition to a reaction vessel of 0.184 g of sodium hydroxide, NaOH, in aqueous solution. How many milliliters of 0.150 *M* NaOH should be added?

Problem Strategy Because molarity relates moles of solute to volume of solution, you first need to convert grams of NaOH to moles of NaOH. Then, you can convert moles NaOH to liters of solution, using the molarity as a conversion factor. Here, 0.150 *M* means that 1 L of solution contains 0.150 moles of solute, so the conversion factor is

$$\underbrace{\frac{1\ \text{L soln}}{0.150\ \text{mol NaOH}}}_{\substack{\text{Converts} \\ \text{mol NaOH to L soln}}}$$

(continued)

(*continued*)

Solution Here is the calculation. (The molar mass of NaOH is 40.0 g/mol.)

$$0.184\ \cancel{\text{g NaOH}} \times \frac{1\ \cancel{\text{mol NaOH}}}{40.0\ \cancel{\text{g NaOH}}} \times \frac{1\ \text{L soln}}{0.150\ \cancel{\text{mol NaOH}}} = 3.07 \times 10^{-2}\ \text{L soln (or 30.7 mL)}$$

You need to add **30.7 mL** of 0.150 *M* NaOH solution to the reaction vessel.

Answer Check Note that the mass of NaOH required for the experiment is relatively small. Because of this, you should not expect that a large volume of the 0.150 *M* NaOH solution will be required.

Exercise 4.11 How many milliliters of 0.163 *M* NaCl are required to give 0.0958 g of sodium chloride?

■ See Problems 4.71, 4.72, 4.73, and 4.74.

Exercise 4.12 How many moles of sodium chloride should be put in a 50.0-mL volumetric flask to give a 0.15 *M* NaCl solution when the flask is filled to the mark with water? How many grams of NaCl is this?

■ See Problems 4.75, 4.76, 4.77, and 4.78.

4.8 Diluting Solutions

Commercially available aqueous ammonia (28.0% NH_3) is 14.8 *M* NH_3. Suppose, however, that you want a solution that is 1.00 *M* NH_3. You need to dilute the concentrated solution with a definite quantity of water. For this purpose, you must know the relationship between the molarity of the solution before dilution (the *initial molarity*) and that after dilution (the *final molarity*).

To obtain this relationship, first recall the equation defining molarity:

$$\text{Molarity} = \frac{\text{moles of solute}}{\text{liters of solution}}$$

You can rearrange this to give

$$\text{Moles of solute} = \text{molarity} \times \text{liters of solution}$$

The product of molarity and the volume (in liters) gives the moles of solute in the solution. Writing M_i for the initial molar concentration and V_i for the initial volume of solution, you get

$$\text{Moles of solute} = M_i \times V_i$$

When the solution is diluted by adding more water, the concentration and volume change to M_f (the final molar concentration) and V_f (the final volume), and the moles of solute equals

$$\text{Moles of solute} = M_f \times V_f$$

Because the moles of solute has not changed during the dilution (Figure 4.19),

$$M_i \times V_i = M_f \times V_f$$

(Note: You can use any volume units, but both V_i and V_f must be in the same units.)

The next example illustrates how you can use this formula to find the volume of a concentrated solution needed to prepare a given volume of dilute solution.

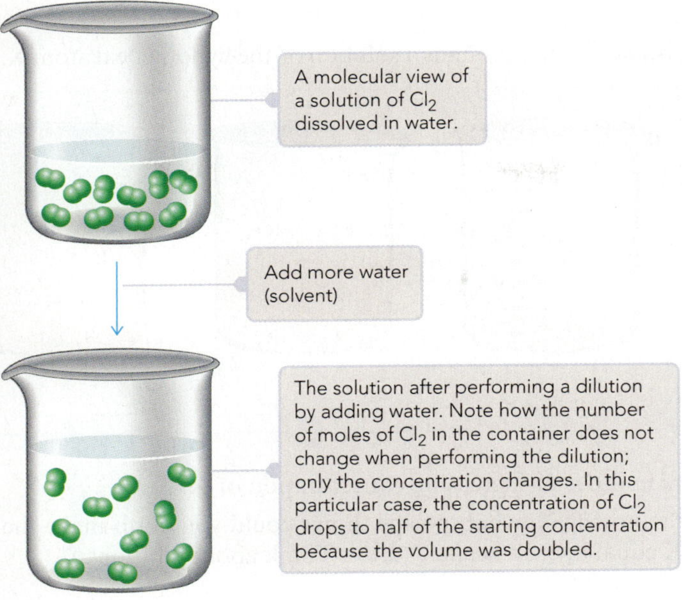

Figure 4.19 ▲

Molecular view of the dilution process

Figure 4.20 ▲

Preparing a solution by diluting a concentrated one A volume of concentrated ammonia, similar to the one in the beaker, has been added to the volumetric flask. Here water is being added from the plastic squeeze bottle to bring the volume up to the mark on the flask.

Example 4.11 Diluting a Solution

Gaining Mastery Toolbox

Critical Concept 4.11

The product of a solution's molar concentration and the volume of the solution gives the moles of solute in a solution. This relationship can be expressed mathematically as $(M_{solute})(V_{solute}) = mol_{solute}$. This relationship can be applied to derive the dilution equation ($M_i \times V_i = M_f \times V_f$).

Solution Essentials:

- Dilution
- Dilution equation ($M_i \times V_i = M_f \times V_f$)
- Molarity

You are given a solution of 14.8 M NH_3. How many milliliters of this solution do you require to give 100.0 mL of 1.00 M NH_3 when diluted (Figure 4.20)?

Problem Strategy Because this problem is a dilution, you can use the dilution formula. In this case you want to know the initial volume (V_i) of the solution that you are going to dilute; all of the other quantities needed for the formula are given in the problem.

Solution You know the final volume (100.0 mL), final concentration (1.00 M), and initial concentration (14.8 M). You write the dilution formula and rearrange it to give the initial volume.

$$M_iV_i = M_fV_f$$

$$V_i = \frac{M_fV_f}{M_i}$$

Now you substitute the known values into the right side of the equation.

$$V_i = \frac{1.00\ \cancel{M} \times 100.0\ \text{mL}}{14.8\ \cancel{M}} = \mathbf{6.76\ mL}$$

Answer Check When you perform a dilution, the volume of the more concentrated solution should always be less than the volume of the final solution, as it is in the photo. Related to this concept is the fact that the initial concentration of a solution is always greater than the final concentration after dilution. These two concepts will enable you to check the reasonableness of your calculated quantities whenever you use the dilution equation.

Exercise 4.13 You have a solution that is 1.5 M H_2SO_4 (sulfuric acid). How many milliliters of this acid do you need to prepare 100.0 mL of 0.18 M H_2SO_4?

■ See Problems 4.79 and 4.80.

CONCEPT CHECK 4.6

Consider the following beakers. Each contains a solution of the hypothetical atom X.

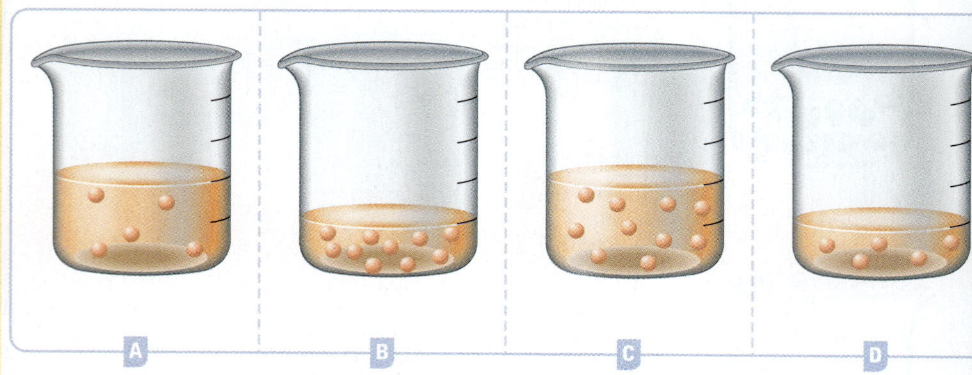

a Arrange the beakers in order of increasing concentration of X.

b Without adding or removing X, what specific things could you do to make the concentrations of X equal in each beaker? (*Hint:* Think about dilutions.)

Quantitative Analysis

Analytical chemistry deals with the determination of composition of materials—that is, the analysis of materials. The materials that one might analyze include air, water, food, hair, body fluids, pharmaceutical preparations, and so forth. The analysis of materials is divided into qualitative and quantitative analysis. *Qualitative analysis* involves the identification of substances or species present in a material. For instance, you might determine that a sample of water contains lead(II) ion. **Quantitative analysis**, which we will discuss in the last sections of this chapter, involves *the determination of the amount of a substance or species present in a material.* In a quantitative analysis, you might determine that the amount of lead(II) ion in a sample of water is 0.067 mg/L.

4.9 Gravimetric Analysis

Gravimetric analysis is *a type of quantitative analysis in which the amount of a species in a material is determined by converting the species to a product that can be isolated completely and weighed.* Precipitation reactions are frequently used in gravimetric analyses. In these reactions, you determine the amount of an ionic species by precipitating it from solution. The precipitate, or solid formed in the reaction, is then filtered from the solution, dried, and weighed. The advantages of a gravimetric analysis are its simplicity (at least in theory) and its accuracy. The chief disadvantage is that it requires meticulous, time-consuming work. Because of this, whenever possible, chemists use modern instrumental methods.

As an example of a gravimetric analysis, consider the problem of determining the amount of lead in a sample of drinking water. Lead, if it occurs in the water, probably exists as the lead(II) ion, Pb^{2+}. Lead(II) sulfate is a very insoluble compound of lead(II) ion. When sodium sulfate, Na_2SO_4, is added to a solution containing Pb^{2+}, lead(II) sulfate precipitates (that is, $PbSO_4$ comes out of the solution as a fine, crystalline solid). If you assume that the lead is present in solution as lead(II) nitrate, you can write the following equation for the reaction:

$$Na_2SO_4(aq) + Pb(NO_3)_2(aq) \longrightarrow 2NaNO_3(aq) + PbSO_4(s)$$

You can separate the white precipitate of lead(II) sulfate from the solution by filtration. Then you dry and weigh the precipitate. Figure 4.21 shows the laboratory setup used in a similar analysis.

© Cengage Learning

Figure 4.21 ◀

Gravimetric analysis for barium ion

a A solution of potassium chromate (yellow) is poured down a stirring rod into a solution containing an unknown amount of barium ion, Ba^{2+}. The yellow precipitate that forms is barium chromate, $BaCrO_4$.

b The solution is filtered by pouring it into a crucible containing a porous glass partition. Afterward, the crucible is heated to dry the barium chromate. By weighing the crucible before and afterward, you can determine the mass of precipitate.

OWL Interactive Example 4.12 Determining the Amount of a Species by Gravimetric Analysis

Gaining Mastery Toolbox

Critical Concept 4.12
The precipitate will contain the species (element, ion, etc.) of interest. If you know the mass and chemical formula of the precipitate, you can determine the mass of each element in the precipitate as outlined in Example 3.7.

Solution Essentials:

- Mass percentage
- Molar mass
- Example 3.7

A 1.000-L sample of polluted water was analyzed for lead(II) ion, Pb^{2+}, by adding an excess of sodium sulfate to it. The mass of lead(II) sulfate that precipitated was 229.8 mg. What is the mass of lead in a liter of the water? Give the answer as milligrams of lead per liter of solution.

Problem Strategy Because an excess of sodium sulfate was added to the solution, you can expect that all of the lead is precipitated as lead(II) sulfate, $PbSO_4$. If you determine the percentage of lead in the $PbSO_4$ precipitate, you can calculate the quantity of lead in the water sample.

Solution Following Example 3.7, you obtain the mass percentage of Pb in $PbSO_4$ by dividing the molar mass of Pb by the molar mass of $PbSO_4$, then multiplying by 100%:

$$\% \text{ Pb} = \frac{207.2 \text{ g/mol}}{303.3 \text{ g/mol}} \times 100\% = 68.32\%$$

Therefore, the 1.000-L sample of water contains

$$\text{Amount Pb in sample} = 229.8 \text{ mg PbSO}_4 \times 0.6832 = 157.0 \text{ mg Pb}$$

The water sample contains **157.0 mg Pb per liter**.

Answer Check Check to make sure that the mass of the element of interest, 157.0 mg Pb in this case, is less than the total mass of the precipitate, 229.8 mg in this case. If you do not find this to be true, you have made an error.

Exercise 4.14 You are given a sample of limestone, which is mostly $CaCO_3$, to determine the mass percentage of Ca in the rock. You dissolve the limestone in hydrochloric acid, which gives a solution of calcium chloride. Then you precipitate the calcium ion in solution by adding sodium oxalate, $Na_2C_2O_4$. The precipitate is calcium oxalate, CaC_2O_4. You find that a sample of limestone weighing 128.3 mg gives 140.2 mg of CaC_2O_4. What is the mass percentage of calcium in the limestone?

■ See Problems 4.83 and 4.84.

4.10 Volumetric Analysis

As you saw earlier, you can use molarity as a conversion factor, and in this way you can calculate the volume of solution that is equivalent to a given mass of solute (see Example 4.10). This means that you can replace mass measurements in solution reactions by volume measurements. In the next example, we look at the volumes of solutions involved in a given reaction.

OWL Interactive Example 4.13 Calculating the Volume of Reactant Solution Needed

Gaining Mastery Toolbox

Critical Concept 4.13
You need to start with a balanced chemical equation. From the balanced chemical equation, you can determine the molar relationship (conversion factor) between the reacting quantities. This conversion factor is critical for correctly solving this type of problem. Do *not* use the dilution equation ($M_i \times V_i = M_f \times V_f$) to solve problems that involve chemical reactions in solution.

Solution Essentials:
- Molarity (molar concentration)
- Molar interpretation of a balanced chemical equation.
- Stoichiometry

Consider the reaction of sulfuric acid, H_2SO_4, with sodium hydroxide, NaOH.

$$H_2SO_4(aq) + 2NaOH(aq) \longrightarrow 2H_2O(l) + Na_2SO_4(aq)$$

Suppose a beaker contains 35.0 mL of 0.175 M H_2SO_4. How many milliliters of 0.250 M NaOH must be added to react completely with the sulfuric acid?

Problem Strategy This is a stoichiometry problem that involves the reaction of sulfuric acid and sodium hydroxide. You have a known volume and concentration of H_2SO_4 in the beaker, and you want to determine what volume of a known concentration of NaOH is required for complete reaction. If you can determine the number of moles of H_2SO_4 that are contained in the beaker, you can then use the balanced chemical reaction to determine the number of moles of NaOH required to react completely with the H_2SO_4. Finally, you can use the concentration of the NaOH to determine the required volume of NaOH. Following this strategy, you convert from 35.0 mL (or 35.0×10^{-3} L) H_2SO_4 solution to moles H_2SO_4 (using the molarity of H_2SO_4), then to moles NaOH (from the chemical equation). Finally, you convert this to volume of NaOH solution (using the molarity of NaOH).

The problem strategy is diagrammed as

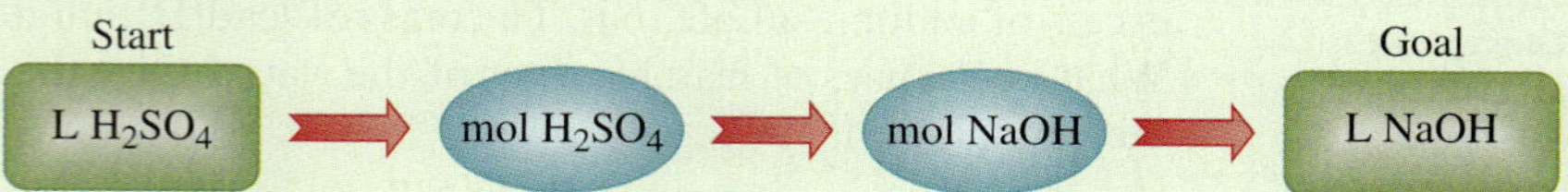

Solution The calculation is as follows:

$$35.0 \times 10^{-3}\ \cancel{\text{L } H_2SO_4}\text{ soln} \times \frac{0.175\ \cancel{\text{mol } H_2SO_4}}{1\ \cancel{\text{L } H_2SO_4 \text{ soln}}} \times \frac{2\ \cancel{\text{mol NaOH}}}{1\ \cancel{\text{mol } H_2SO_4}} \times$$

$$\frac{1 \text{ L NaOH soln}}{0.250\ \cancel{\text{mol NaOH}}} = 4.90 \times 10^{-2} \text{ L NaOH soln (or 49.0 mL NaOH soln)}$$

Thus, 35.0 mL of 0.175 M sulfuric acid solution reacts with exactly **49.0 mL** of 0.250 M sodium hydroxide solution.

Answer Check Whenever you perform a titration calculation, be sure that you have taken into account the stoichiometry of the reaction between the acid and base (use the balanced chemical equation). In this case, two moles of NaOH are required to neutralize each mole of acid. Furthermore, when performing titration calculations, do not be tempted to apply the dilution equation to solve the problem. If you were to take such an approach here, you would arrive at an incorrect result since the dilution equation fails to take into account the stoichiometry of the reaction.

Exercise 4.15 Nickel sulfate, $NiSO_4$, reacts with sodium phosphate, Na_3PO_4, to give a pale yellow-green precipitate of nickel phosphate, $Ni_3(PO_4)_2$, and a solution of sodium sulfate, Na_2SO_4.

$$3NiSO_4(aq) + 2Na_3PO_4(aq) \longrightarrow Ni_3(PO_4)_2(s) + 3Na_2SO_4(aq)$$

How many milliliters of 0.375 M $NiSO_4$ will react with 45.7 mL of 0.265 M Na_3PO_4?

■ See Problems 4.89 and 4.90.

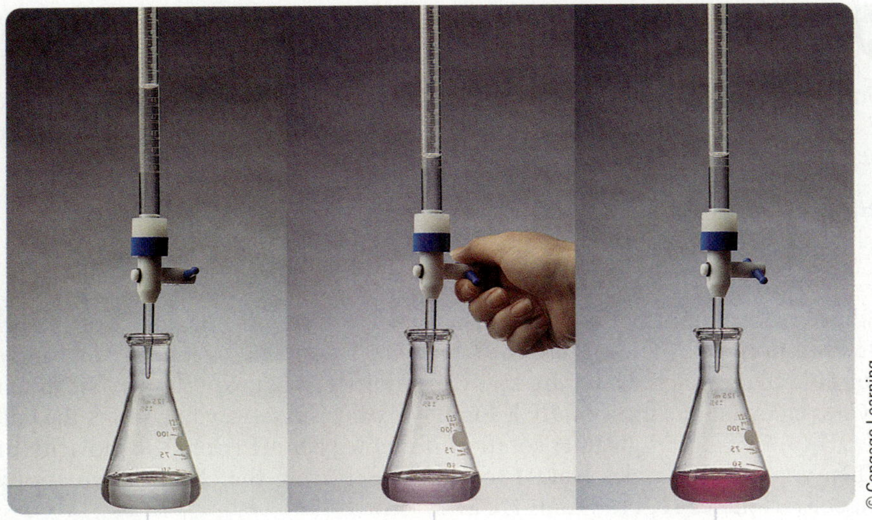

Figure 4.22 ▲

Titration of an unknown amount of HCl with NaOH

An important method for determining the amount of a particular substance is based on measuring the volume of reactant solution. Suppose substance *A* reacts in solution with substance *B*. If you know the volume and concentration of a solution of *B* that just reacts with substance *A* in a sample, you can determine the amount of *A*. **Titration** is *a procedure for determining the amount of substance* A *by adding a carefully measured volume of a solution with known concentration of* B *until the reaction of* A *and* B *is just complete.* **Volumetric analysis** is *a method of analysis based on titration.*

Figure 4.22 shows a flask containing hydrochloric acid with an unknown amount of HCl being titrated with sodium hydroxide solution, NaOH, of known molarity. The reaction is

$$NaOH(aq) + HCl(aq) \longrightarrow NaCl(aq) + H_2O(l)$$

To the HCl solution are added a few drops of phenolphthalein indicator. ▶ Phenolphthalein is colorless in the hydrochloric acid but turns pink at the completion of the reaction of NaOH with HCl. Sodium hydroxide with a concentration of 0.207 *M* is contained in a *buret,* a glass tube graduated to measure the volume of liquid delivered from the stopcock. The solution in the buret is added to the HCl in the flask until the phenolphthalein just changes from colorless to pink. At this point, the reaction is complete and the volume of NaOH that reacts with the HCl is read from the buret. This volume is then used to obtain the mass of HCl in the original solution.

An indicator is a substance that undergoes a color change when a reaction approaches completion. See Section 4.4.

OWL Interactive Example 4.14 Calculating the Quantity of Substance in a Titrated Solution

Gaining Mastery Toolbox

Critical Concept 4.14
You need to start with a balanced chemical equation for the acid-base neutralization reaction. Just as in Example 4.13, a balanced chemical equation is essential for correctly solving the problem. In most instances the titration calculations will be based on acid-base neutralization reactions.

Solution Essentials:
- Molarity
- Acid-base neutralization reaction
- Molar interpretation of a balanced chemical equation.
- Stoichimetry

A flask contains a solution with an unknown amount of HCl. This solution is titrated with 0.207 *M* NaOH. It takes 4.47 mL of the NaOH solution to complete the reaction. What is the mass of the HCl?

Problem Strategy First, in order to determine the stoichiometry of the reaction, you start by writing the balanced chemical equation.

$$NaOH(aq) + HCl(aq) \longrightarrow NaCl(aq) + H_2O(l)$$

If you use the numerical information from the problem to determine the moles of NaOH added to the solution, you then can use the stoichiometry of the reaction to determine the moles of HCl that reacted. Once you know the moles of HCl, you can use the molar mass of HCl to calculate the mass of HCl. Employing this strategy, you convert the volume of NaOH (4.47×10^{-3} L NaOH solution) to moles NaOH (from the molarity of NaOH). Then you convert moles NaOH to moles HCl (from the chemical equation). Finally, you convert moles HCl to grams HCl.

The problem strategy is diagrammed as

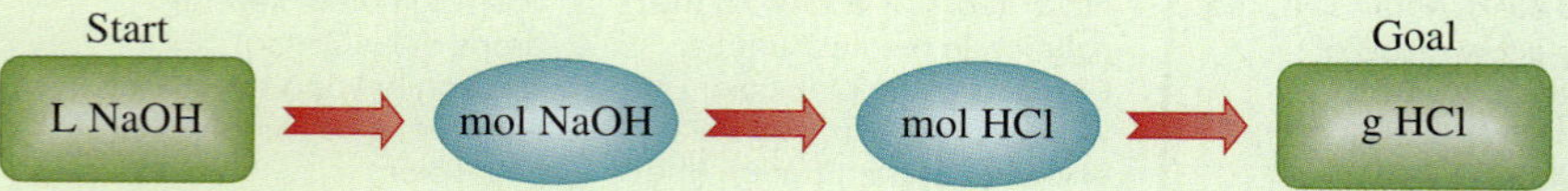

Solution The calculation is as follows:

$$4.47 \times 10^{-3}\cancel{\text{L NaOH soln}} \times \frac{0.207\ \cancel{\text{mol NaOH}}}{1\ \cancel{\text{L NaOH soln}}} \times \frac{1\ \cancel{\text{mol HCl}}}{1\ \cancel{\text{mol NaOH}}} \times \frac{36.5\text{ g HCl}}{1\ \cancel{\text{mol HCl}}} = \mathbf{0.0338\ g\ HCl}$$

Answer Check Before you perform the titration calculations, always write down the balanced chemical equation.

Exercise 4.16 A 5.00-g sample of vinegar is titrated with 0.108 *M* NaOH. If the vinegar requires 39.1 mL of the NaOH solution for complete reaction, what is the mass percentage of acetic acid, $HC_2H_3O_2$, in the vinegar? The reaction is

$$HC_2H_3O_2(aq) + NaOH(aq) \longrightarrow NaC_2H_3O_2(aq) + H_2O(l)$$

■ See Problems 4.91 and 4.92.

CONCEPT CHECK 4.7

Consider three flasks, each containing 0.10 mol of acid. You need to learn something about the acids in each of the flasks, so you perform titration using a NaOH solution. Here are the results of the experiment:

Flask A	10 mL of NaOH required for neutralization
Flask B	20 mL of NaOH required for neutralization
Flask C	30 mL of NaOH required for neutralization

a What have you learned about each of these acids from performing the experiment?

b Could you use the results of this experiment to determine the concentration of the NaOH? If not, what assumption about the molecular formulas of the acids would allow you to make the concentration determination?

A Checklist for Review

OWL and Go Chemistry Sign in at www.cengage.com/owl to:

- View tutorials and simulations, develop problem-solving skills, and complete online homework assigned by your professor.
- For quick review and exam prep, download Go Chemistry mini lecture modules from OWL or purchase them at www.cengagebrain.com.

Summary of Facts and Concepts

Reactions often involve ions in aqueous solution. Many of the compounds in such reactions are *electrolytes,* which are substances that dissolve in water to give ions. Electrolytes that exist in solution almost entirely as ions are called *strong electrolytes.* Electrolytes that dissolve in water to give a relatively small percentage of ions are called *weak electrolytes.* The *solubility rules* can be used to predict the extent to which an ionic compound will dissolve in water. Most soluble ionic compounds are strong electrolytes.

We can represent a reaction involving ions in one of three different ways, depending on what information we want to convey. A *molecular equation* is one in which substances are written as if they were molecular, even though they are ionic. This equation closely describes what you actually do in the laboratory. However, this equation does not describe what is happening at the level of ions and molecules. For that purpose, we rewrite the molecular equation as a *complete ionic equation* by replacing the formulas for strong electrolytes by their ion formulas. If you cancel *spectator ions* from the complete ionic equation, you obtain the *net ionic equation.*

Most of the important reactions we consider in this course can be divided into three major classes: (1) precipitation reactions, (2) acid–base reactions, and (3) oxidation–reduction reactions. A *precipitation reaction* occurs in aqueous solution because one product is insoluble. You can decide whether two ionic compounds will result in a precipitation reaction, if you know from solubility rules that one of the potential products is insoluble.

Acids are substances that yield hydrogen ions in aqueous solution or donate protons. *Bases* are substances that yield hydroxide ions in aqueous solution or accept protons. These *acid–base reactions* are proton-transfer reactions. In this chapter, we covered neutralization reactions (reactions of acids and bases to yield salts) and reactions of certain salts with acids to yield a gas.

Oxidation–reduction reactions are reactions involving a transfer of electrons from one species to another or a change in the oxidation number of atoms. The concept of *oxidation numbers* helps us describe this type of reaction. The atom that increases in oxidation number is said to undergo *oxidation;* the atom that decreases in oxidation number is said to undergo *reduction.* Oxidation and reduction must occur together in a reaction. Many oxidation–reduction reactions fall into the following categories: combination reactions, decomposition reactions, displacement reactions, and combustion reactions. Oxidation–reduction reactions can be balanced by the *half-reaction method.*

Molar concentration, or *molarity,* is the moles of solute in a liter of solution. Knowing the molarity allows you to calculate the amount of solute in any volume of solution. Because the moles of solute are constant during the *dilution of a solution,* you can determine to what volume to dilute a concentrated solution to give one of desired molarity.

Quantitative analysis involves the determination of the amount of a species in a material. In *gravimetric analysis,* you determine the amount of a species by converting it to a product you can weigh. In *volumetric analysis,* you determine the amount of a species by titration. *Titration* is a method of chemical analysis in which you measure the volume of solution of known molarity that reacts with a compound of unknown amount. You determine the amount of the compound from this volume of solution.

Learning Objectives	Important Terms
4.1 Ionic Theory of Solutions and Solubility Rules	
■ Describe how an ionic substance can form ions in aqueous solution. ■ Explain how an *electrolyte* makes a solution electrically conductive. ■ Give examples of substances that are electrolytes. ■ Define *nonelectrolyte,* and provide an example of a molecular substance that is a nonelectrolyte. ■ Compare the properties of solutions that contain *strong electrolytes* and *weak electrolytes.* ■ Learn the *solubility rules* for ionic compounds. ■ Use the solubility rules. Example 4.1	**electrolyte** **nonelectrolyte** **strong electrolyte** **weak electrolyte**
4.2 Molecular and Ionic Equations	
■ Write the *molecular equation* of a chemical reaction. ■ From the molecular equation of both strong electrolytes and weak electrolytes, determine the *complete ionic equation.* ■ From the complete ionic equation, write the *net ionic equation.* ■ Write net ionic equations. Example 4.2	**molecular equation** **complete ionic equation** **spectator ion** **net ionic equation**

4.3 Precipitation Reactions	
▪ Recognize *precipitation* (*exchange*) reactions. ▪ Write molecular, complete ionic, and net ionic equations for precipitation reactions. ▪ Decide whether a precipitation reaction will occur. Example 4.3 ▪ Determine the product of a precipitation reaction.	**precipitate** **exchange (metathesis) reaction**
4.4 Acid–Base Reactions	
▪ Understand how an *acid–base* indicator is used to determine whether a solution is acidic or basic. ▪ Define *Arrhenius acid* and *Arrhenius base.* ▪ Write the chemical equation of an Arrhenius base or acid in aqueous solution. ▪ Define *Brønsted–Lowry acid* and *Brønsted–Lowry base.* ▪ Write the chemical equation of a Brønsted–Lowry base or acid in aqueous solution. ▪ Write the chemical equation of an acid in aqueous solution using a *hydronium ion.* ▪ Learn the common *strong acids* and *strong bases.* ▪ Distinguish between a strong acid and a *weak acid* and the solutions they form. ▪ Distinguish between a strong base and a *weak base* and the solutions they form. ▪ Classify acids and bases as strong or weak. Example 4.4 ▪ Recognize *neutralization reactions.* ▪ Write an equation for a neutralization reaction. Example 4.5 ▪ Write the reactions for a *polyprotic acid* in aqueous solution. ▪ Recognize acid–base reactions that lead to gas formation. ▪ Write an equation for a reaction with gas formation. Example 4.6	**acid–base indicator** **acid (Arrhenius)** **base (Arrhenius)** **acid (Brønsted–Lowry)** **base (Brønsted–Lowry)** **strong acid** **weak acid** **strong base** **weak base** **neutralization reaction** **salt** **polyprotic acid**
4.5 Oxidation–Reduction Reactions	
▪ Define *oxidation–reduction* reaction. ▪ Learn the oxidation-number rules. ▪ Assign oxidation numbers. Example 4.7 ▪ Write the *half-reactions* of an oxidation–reduction reaction. ▪ Determine the species undergoing *oxidation* and *reduction.* ▪ Recognize *combination reactions, decomposition reactions, displacement reactions,* and *combustion reactions.* ▪ Use the activity series to predict when displacement reactions will occur.	**oxidation number (oxidation state)** **oxidation–reduction reaction (redox reaction)** **half-reaction** **oxidation** **reduction** **oxidizing agent** **reducing agent** **combination reaction** **decomposition reaction** **displacement reaction (single-replacement reaction)** **combustion reaction**
4.6 Balancing Simple Oxidation–Reduction Equations	
▪ Balance simple oxidation–reduction reactions by the half-reaction method. Example 4.8	
4.7 Molar Concentration	
▪ Define *molarity* or *molar concentration* of a solution. ▪ Calculate the molarity from mass and volume. Example 4.9 ▪ Use molarity as a conversion factor. Example 4.10	**molar concentration (molarity) (*M*)**

4.8 Diluting Solutions	
■ Describe what happens to the concentration of a solution when it is diluted. ■ Perform calculations associated with dilution. ■ Diluting a solution. **Example 4.11**	
4.9 Gravimetric Analysis	
■ Determine the amount of a species by *gravimetric analysis.* **Example 4.12**	**quantitative analysis** **gravimetric analysis**
4.10 Volumetric Analysis	
■ Calculate the volume of reactant solution needed to perform a reaction. **Example 4.13** ■ Understand how to perform a *titration.* ■ Calculate the quantity of substance in a titrated solution. **Example 4.14**	**titration** **volumetric analysis**

Key Equations

$$\text{Molarity } (M) = \frac{\text{moles of solute}}{\text{liters of solution}}$$

$$M_i \times V_i = M_f \times V_f$$

Questions and Problems

OWL Interactive versions of these problems may be assigned in OWL.

Self-Assessment and Review Questions

Key: These questions test your understanding of the ideas you worked with in the chapter. These problems vary in difficulty and often can be used for the basis of discussion.

4.1 Explain why some electrolyte solutions are strongly conducting, whereas others are weakly conducting.

4.2 Define the terms *strong electrolyte* and *weak electrolyte.* Give an example of each.

4.3 Explain the terms *soluble* and *insoluble.* Use the solubility rules to write the formula of an insoluble ionic compound.

4.4 What are the advantages and disadvantages of using a molecular equation to represent an ionic reaction?

4.5 What is a *spectator ion*? Illustrate with a complete ionic reaction.

4.6 What is a net ionic equation? What is the value in using a net ionic equation? Give an example.

4.7 What are the major types of chemical reactions? Give a brief description and an example of each.

4.8 Describe in words how you would prepare pure crystalline AgCl and $NaNO_3$ from solid $AgNO_3$ and solid NaCl.

4.9 Give an example of a neutralization reaction. Label the acid, base, and salt.

4.10 Give an example of a polyprotic acid and write equations for the successive neutralizations of the acidic hydrogen atoms of the acid molecule to produce a series of salts.

4.11 Why must oxidation and reduction occur together in a reaction?

4.12 Give an example of a displacement reaction. What is the oxidizing agent? What is the reducing agent?

4.13 Why is the product of molar concentration and volume constant for a dilution problem?

4.14 Describe how the amount of sodium hydroxide in a mixture can be determined by titration with hydrochloric acid of known molarity.

4.15 What is the net ionic equation for the following molecular equation?

$$HF(aq) + KOH(aq) \longrightarrow KF(aq) + H_2O(l)$$

Hydrofluoric acid, HF, is a molecular substance and a weak electrolyte.

a $H^+(aq) + OH^-(aq) \longrightarrow H_2O(l)$

b $H^+(aq) + KOH(aq) \longrightarrow K^+(aq) + H_2O(l)$

c $HF(aq) + KOH(aq) \longrightarrow K^+(aq) + F^-(aq)$

d $HF(aq) + K^+(aq) + OH^-(aq) \longrightarrow KF(aq) + H_2O(l)$

e $HF(aq) + OH^-(aq) \longrightarrow F^-(aq) + H_2O(l)$

4.16 An aqueous sodium hydroxide solution mixed with an aqueous magnesium nitrate solution yields which of the following products?

a magnesium hydroxide(*aq*)

b magnesium dihydroxide(*s*)

c magnesium hydroxide(*s*)

d dimagnesium hydroxide(*s*)

e sodium nitrate(*l*)

4.17 Which of the following compounds would produce the highest concentration of Cl^- ions when 0.10 mol of each is placed in separate beakers containing equal volumes of water?

a $NaCl$ b $PbCl_2$ c $HClO_4$
d $MgCl_2$ e HCl

4.18 In an aqueous 0.10 *M* HNO_2 solution (HNO_2 is a weak electrolyte), which of the following would you expect to find in the highest concentration?

a H_3O^+ b NO_2^- c H^+
d HNO_2 e OH^-

Concept Explorations

Key: Concept explorations are comprehensive problems that provide a framework that will enable you to explore and learn many of the critical concepts and ideas in each chapter. If you master the concepts associated with these explorations, you will have a better understanding of many important chemistry ideas and will be more successful in solving all types of chemistry problems. These problems are well suited for group work and for use as in-class activities.

4.19 The Behavior of Substances in Water

Part 1:

a Ammonia, NH_3, is a weak electrolyte. It forms ions in solution by *reacting* with water molecules to form the ammonium ion and hydroxide ion. Write the balanced chemical reaction for this process, including state symbols.

b From everyday experience you are probably aware that table sugar (sucrose), $C_{12}H_{22}O_{11}$, is soluble in water. When sucrose dissolves in water, it doesn't form ions through any reaction with water. It just *dissolves* without forming ions, so it is a nonelectrolyte. Write the chemical equation for the dissolving of sucrose in water.

c Both NH_3 and $C_{12}H_{22}O_{11}$ are soluble molecular compounds, yet they behave differently in aqueous solution. Briefly explain why one is a weak electrolyte and the other is a nonelectrolyte.

d Hydrochloric acid, HCl, is a molecular compound that is a strong electrolyte. Write the chemical reaction of HCl with water.

e Compare the ammonia reaction with that of hydrochloric acid. Why are both of these substances considered electrolytes?

f Explain why HCl is a strong electrolyte and ammonia is a weak electrolyte.

g Classify each of the following substances as either ionic or molecular.

- KCl
- NH_3
- CO_2
- $MgBr_2$
- HCl
- $Ca(OH)_2$
- PbS
- $HC_2H_3O_2$

h For those compounds above that you classified as ionic, use the solubility rules to determine which are soluble.

i The majority of ionic substances are solids at room temperature. Describe what you would observe if you placed a soluble ionic compound and an insoluble ionic compound in separate beakers of water.

j Write the chemical equation(s), including state symbols, for what happens when each soluble ionic compound that you identified above is placed in water. Are these substances reacting with water when they are added to water?

k How would you classify the soluble ionic compounds: strong electrolyte, weak electrolyte, or nonelectrolyte? Explain your answer.

l Sodium chloride, NaCl, is a strong electrolyte, as is hydroiodic acid, HI. Write the chemical equations for what happens when these substances are added to water.

m Are NaCl and HI strong electrolytes because they have similar behavior in aqueous solution? If not, describe, using words and equations, the different chemical process that takes place in each case.

Part 2: You have two hypothetical molecular compounds, AX and AY. AX is a strong electrolyte and AY is a weak electrolyte. The compounds undergo the following chemical reactions when added to water.

$$AX(aq) + H_2O(l) \longrightarrow AH_2O^+(aq) + X^-(aq)$$

$$AY(aq) + H_2O(l) \longrightarrow AH_2O^+(aq) + Y^-(aq)$$

a Explain how the relative amounts of $AX(aq)$ and $AY(aq)$ would compare if you had a beaker of water with AX and a beaker of water with AY.

b How would the relative amounts of $X^-(aq)$ and $Y^-(aq)$ in the two beakers compare? Be sure to explain your answer.

4.20 Working with Concentration (Molarity Concepts)

Note: You should be able to answer all of the following questions without using a calculator.

Part 1:

a Both NaCl and $MgCl_2$ are soluble ionic compounds. Write the balanced chemical equations for these two substances dissolving in water.

b Consider the pictures below. These pictures represent 1.0-L solutions of 1.0 *M* NaCl(*aq*) and 1.0 *M* $MgCl_2(aq)$. The representations of the ions in solution are the correct relative amounts. Water molecules have been omitted for clarity. Correctly label each of the beakers, provide a key to help identify the ions, and give a brief explanation of how you made your assignments.

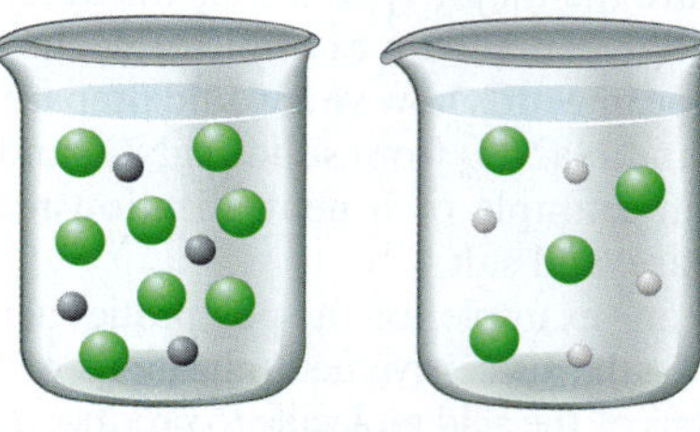

Keeping in mind that the pictures represent the relative amounts of ions in the solution and that the numerical information about these solutions is presented above, answer the following questions c through f.

c How many moles of NaCl and $MgCl_2$ are in each beaker?

d How many moles of chloride ions are in each beaker? How did you arrive at this answer?

e What is the concentration of chloride ions in each beaker? Without using mathematical equations, briefly explain how you obtained your answer.

f Explain how it is that the concentrations of chloride ions in these beakers are different even though the concentrations of each substance (compound) are the same.

Part 2: Say you were to dump out half of the $MgCl_2$ solution from the beaker above.

a What would be the concentration of the $MgCl_2(aq)$ and of the chloride ions in the remaining solution?

b How many moles of the $MgCl_2$ and of the chloride ions would remain in the beaker?

c Explain why the concentration of $MgCl_2(aq)$ *would not* change, whereas the number of moles of $MgCl_2$ *would* change when solution was removed from the beaker. As part of your answer, you are encouraged to use pictures.

Part 3: Consider the beaker containing 1.0 L of the 1.0 *M* NaCl(*aq*) solution. You now add 1.0 L of water to this beaker.

a What is the concentration of this NaCl(*aq*) solution?

b How many moles of NaCl are present in the 2.0 L of NaCl(*aq*) solution?

c Explain why the concentration of NaCl(*aq*) *does* change with the addition of water, whereas the number of moles *does not* change. Here again, you are encouraged to use pictures to help answer the question.

Conceptual Problems

Key: These problems are designed to check your understanding of the concepts associated with some of the main topics presented in each chapter. A strong conceptual understanding of chemistry is the foundation for both applying chemical knowledge and solving chemical problems. These problems vary in level of difficulty and often can be used as a basis for group discussion.

4.21 You need to perform gravimetric analysis of a water sample in order to determine the amount of Ag^+ present.

a List three aqueous solutions that would be suitable for mixing with the sample to perform the analysis.

b Would adding $KNO_3(aq)$ allow you to perform the analysis?

c Assume you have performed the analysis and the silver solid that formed is moderately soluble. How might this affect your analysis results?

4.22 In this problem you need to draw two pictures of solutions in beakers at different points in time. Time zero ($t = 0$) will be the *hypothetical* instant at which the reactants dissolve in the solution (*if* they dissolve) *before* they react. Time after mixing ($t > 0$) will be the time required to allow sufficient interaction of the materials. For now, we assume that insoluble solids have no ions in solution and do not worry about representing the stoichiometric amounts of the dissolved ions. Here is an example: Solid NaCl and solid $AgNO_3$ are added to a beaker containing 250 mL of water.

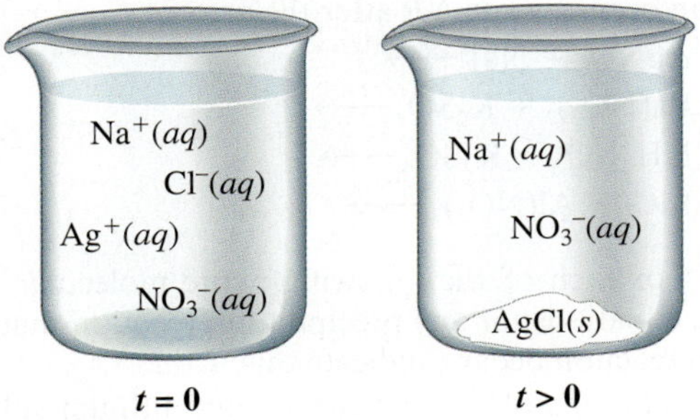

Note that we are not showing the H_2O, and we are representing only the ions and solids in solution. Using the same conditions as the example (adding the solids to H_2O), draw pictures of the following:

a Solid lead(II) nitrate and solid ammonium chloride at $t = 0$ and $t > 0$

b FeS(*s*) and $NaNO_3(s)$ at $t = 0$ and $t > 0$

c Solid lithium iodide and solid sodium carbonate at $t = 0$ and $t > 0$

4.23 You come across a beaker that contains water, aqueous ammonium acetate, and a precipitate of calcium phosphate.

a Write the balanced molecular equation for a reaction between two solutions containing ions that could produce this solution.

b Write the complete ionic equation for the reaction in part a.

c Write the net ionic equation for the reaction in part a.

4.24 Three acid samples are prepared for titration by 0.01 *M* NaOH:

1. Sample 1 is prepared by dissolving 0.01 mol of HCl in 50 mL of water.
2. Sample 2 is prepared by dissolving 0.01 mol of HCl in 60 mL of water.
3. Sample 3 is prepared by dissolving 0.01 mol of HCl in 70 mL of water.

a Without performing a formal calculation, compare the concentrations of the three acid samples (rank them from highest to lowest).

b When the titration is performed, which sample, if any, will require the largest volume of the 0.01 *M* NaOH for neutralization?

4.25 Would you expect a precipitation reaction between an ionic compound that is an electrolyte and an ionic compound that is a nonelectrolyte? Justify your answer.

4.26 Equal quantities of the hypothetical strong acid HX, weak acid HA, and weak base BZ are added to separate beakers of water, producing the solutions depicted in the drawings. In the drawings, the relative amounts of each substance present in the solution (neglecting the water) are shown. Identify the acid or base that was used to produce each of the solutions (HX, HA, or BZ).

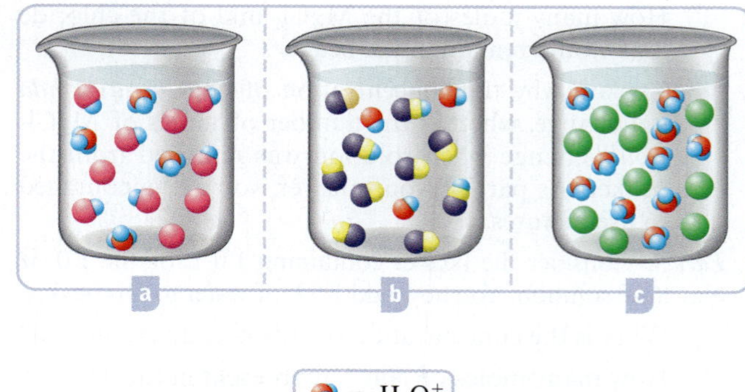

4.27 Try and answer the following questions without using a calculator.

a. A solution is made by mixing 1.0 L of 0.5 *M* NaCl and 0.5 L of 1.0 *M* $CaCl_2$. Which ion is at the highest concentration in the solution?

b. Another solution is made by mixing 0.50 L of 1.0 *M* KBr and 0.50 L of 1.0 *M* K_3PO_4. What is the concentration of each ion in the solution?

4.28 If one mole of the following compounds were each placed into separate beakers containing the same amount of water, rank the $Cl^-(aq)$ concentrations from highest to lowest (some may be equivalent): KCl, $AlCl_3$, $PbCl_2$, NaCl, HCl, NH_3, KOH, and HCN.

Practice Problems

Key: These problems are for practice in applying problem-solving skills. They are divided by topic, and some are keyed to exercises (see the ends of the exercises). The problems are arranged in matching pairs; the odd-numbered problem of each pair is listed first, and its answer is given in the back of the book.

Solubility Rules

4.29 Using solubility rules, predict the solubility in water of the following ionic compounds.

a. PbS
b. $AgNO_3$
c. Na_2CO_3
d. CaI_2

4.30 Using solubility rules, predict the solubility in water of the following ionic compounds.

a. $Al(OH)_3$
b. Ca_3N_2
c. NH_4Cl
d. KOH

4.31 Using solubility rules, decide whether the following ionic solids are soluble or insoluble in water. If they are soluble, write the chemical equation for dissolving in water and indicate what ions you would expect to be present in solution.

a. AgBr
b. Li_2SO_4
c. $Ca_3(PO_4)_2$
d. Na_2CO_3

4.32 Using solubility rules, decide whether the following ionic solids are soluble or insoluble in water. If they are soluble, write the chemical equation for dissolving in water and indicate what ions you would expect to be present in solution.

a. $(NH_4)_2SO_4$
b. $BaCO_3$
c. $Pb(NO_3)_2$
d. $Ca(OH)_2$

Ionic Equations

4.33 Write net ionic equations for the following molecular equations. HBr is a strong electrolyte.

a. $HBr(aq) + KOH(aq) \longrightarrow KBr(aq) + H_2O(l)$
b. $AgNO_3(aq) + NaBr(aq) \longrightarrow AgBr(s) + NaNO_3(aq)$
c. $K_2S(aq) + 2HBr(aq) \longrightarrow 2KBr(aq) + H_2S(g)$
d. $NaOH(aq) + NH_4Br(aq) \longrightarrow$ $NaBr(aq) + NH_3(g) + H_2O(l)$

4.34 Write net ionic equations for the following molecular equations. HBr is a strong electrolyte.

a. $HBr(aq) + NH_3(aq) \longrightarrow NH_4Br(aq)$
b. $2HCl(aq) + Ba(OH)_2(aq) \longrightarrow 2H_2O(l) + BaCl_2(aq)$
c. $Pb(NO_3)_2(aq) + 2NaBr(aq) \longrightarrow$ $PbBr_2(s) + 2NaNO_3(aq)$
d. $MgCO_3(s) + H_2SO_4(aq) \longrightarrow$ $MgSO_4(aq) + H_2O(l) + CO_2(g)$

4.35 Lead(II) nitrate solution and sodium sulfate solution are mixed. Crystals of lead(II) sulfate come out of solution, leaving a solution of sodium nitrate. Write the molecular equation and the net ionic equation for the reaction.

4.36 Lithium carbonate solution reacts with aqueous hydrobromic acid to give a solution of lithium bromide, carbon dioxide gas, and water. Write the molecular equation and the net ionic equation for the reaction.

Precipitation

4.37 Write the molecular equation and the net ionic equation for each of the following aqueous reactions. If no reaction occurs, write *NR* after the arrow.

a. $FeSO_4 + NaCl \longrightarrow$
b. $Na_2CO_3 + MgBr_2 \longrightarrow$
c. $MgSO_4 + NaOH \longrightarrow$
d. $NiCl_2 + NaBr \longrightarrow$

4.38 Write the molecular equation and the net ionic equation for each of the following aqueous reactions. If no reaction occurs, write *NR* after the arrow.

a. $AgNO_3 + NaI \longrightarrow$
b. $Ba(NO_3)_2 + K_2SO_4 \longrightarrow$
c. $NH_4NO_3 + K_2SO_4 \longrightarrow$
d. $LiCl + Al(NO_3)_3 \longrightarrow$

4.39 For each of the following, write molecular and net ionic equations for any precipitation reaction that occurs. If no reaction occurs, indicate this.

a. Solutions of barium nitrate and lithium sulfate are mixed.

b. Solutions of sodium bromide and calcium nitrate are mixed.

c Solutions of aluminum sulfate and sodium hydroxide are mixed.

d Solutions of calcium bromide and sodium phosphate are mixed.

4.40 For each of the following, write molecular and net ionic equations for any precipitation reaction that occurs. If no reaction occurs, indicate this.

a Zinc chloride and sodium sulfide are dissolved in water.

b Sodium sulfide and calcium chloride are dissolved in water.

c Magnesium sulfate and potassium bromide are dissolved in water.

d Magnesium sulfate and potassium carbonate are dissolved in water.

Strong and Weak Acids and Bases

4.41 Classify each of the following as a strong or weak acid or base.

a HF b KOH c $HClO_4$ d HIO

4.42 Classify each of the following as a strong or weak acid or base.

a NH_3 b HCNO c $Mg(OH)_2$ d $HClO_3$

Neutralization Reactions

4.43 Complete and balance each of the following molecular equations (in aqueous solution); include phase labels. Then, for each, write the net ionic equation.

a $NaOH + HNO_3 \longrightarrow$

b $HCl + Ba(OH)_2 \longrightarrow$

c $HC_2H_3O_2 + Ca(OH)_2 \longrightarrow$

d $NH_3 + HNO_3 \longrightarrow$

4.44 Complete and balance each of the following molecular equations (in aqueous solution); include phase labels. Then, for each, write the net ionic equation.

a $Al(OH)_3 + HCl \longrightarrow$

b $HClO + Sr(OH)_2 \longrightarrow$

c $Ba(OH)_2 + HC_2H_3O_2 \longrightarrow$

d $H_2SO_4 + KOH \longrightarrow$

4.45 For each of the following, write the molecular equation, including phase labels. Then write the net ionic equation. Note that the salts formed in these reactions are soluble.

a the neutralization of hydrobromic acid with calcium hydroxide solution

b the reaction of solid aluminum hydroxide with nitric acid

c the reaction of aqueous hydrogen cyanide with calcium hydroxide solution

d the neutralization of lithium hydroxide solution by aqueous hydrogen cyanide

4.46 For each of the following, write the molecular equation, including phase labels. Then write the net ionic equation. Note that the salts formed in these reactions are soluble.

a the neutralization of lithium hydroxide solution by aqueous chloric acid

b the reaction of barium hydroxide solution and aqueous nitrous acid

c the reaction of sodium hydroxide solution and aqueous nitrous acid

d the neutralization of aqueous hydrogen cyanide by aqueous strontium hydroxide

4.47 Complete the right side of each of the following molecular equations. Then write the net ionic equations. Assume all salts formed are soluble. Acid salts are possible.

a $2KOH(aq) + H_3PO_4(aq) \longrightarrow$

b $3H_2SO_4(aq) + 2Al(OH)_3(s) \longrightarrow$

c $2HC_2H_3O_2(aq) + Ca(OH)_2(aq) \longrightarrow$

d $H_2SO_3(aq) + NaOH(aq) \longrightarrow$

4.48 Complete the right side of each of the following molecular equations. Then write the net ionic equations. Assume all salts formed are soluble. Acid salts are possible.

a $Ca(OH)_2(aq) + 2H_2SO_4(aq) \longrightarrow$

b $2H_3PO_4(aq) + Ca(OH)_2(aq) \longrightarrow$

c $NaOH(aq) + H_2SO_4(aq) \longrightarrow$

d $Sr(OH)_2(aq) + 2H_2CO_3(aq) \longrightarrow$

4.49 Write molecular and net ionic equations for the successive neutralizations of each acidic hydrogen of sulfurous acid by aqueous calcium hydroxide. $CaSO_3$ is insoluble; the acid salt is soluble.

4.50 Write molecular and net ionic equations for the successive neutralizations of each acidic hydrogen of phosphoric acid by calcium hydroxide solution. $Ca_3(PO_4)_2$ is insoluble; assume that the acid salts are soluble.

Reactions Evolving a Gas

4.51 The following reactions occur in aqueous solution. Complete and balance the molecular equations using phase labels. Then write the net ionic equations.

a $CaS + HBr \longrightarrow$ b $MgCO_3 + HNO_3 \longrightarrow$

c $K_2SO_3 + H_2SO_4 \longrightarrow$

4.52 The following reactions occur in aqueous solution. Complete and balance the molecular equations using phase labels. Then write the net ionic equations.

a $BaCO_3 + HNO_3 \longrightarrow$

b $K_2S + HCl \longrightarrow$

c $CaSO_3(s) + HI \longrightarrow$

4.53 Write the molecular equation and the net ionic equation for the reaction of solid iron(II) sulfide and hydrochloric acid. Add phase labels.

4.54 Write the molecular equation and the net ionic equation for the reaction of solid barium carbonate and hydrogen bromide in aqueous solution. Add phase labels.

Oxidation Numbers

4.55 Obtain the oxidation number for the element noted in each of the following.

a Ga in Ga_2O_3 b Nb in NbO_2

c Br in $KBrO_4$ d Mn in K_2MnO_4

4.56 Obtain the oxidation number for the element noted in each of the following.

a Cr in CrO_3 b Hg in Hg_2Cl_2

c Ga in $Ga(OH)_3$ d P in Na_3PO_4

4.57 Obtain the oxidation number for the element noted in each of the following.

a N in NH_2^- b I in IO_3^-
c H in H_2 d Cl in $HClO_4$

4.58 Obtain the oxidation number for the element noted in each of the following.

a N in N_2 b Cr in CrO_4^{2-}
c Zn in $Zn(OH)_4^{2-}$ d As in $H_2AsO_3^-$

4.59 Determine the oxidation numbers of all the elements in each of the following compounds. (*Hint:* Look at the ions present.)

a $Mn(ClO_2)_2$ b $Fe_2(CrO_4)_3$
c $HgCr_2O_7$ d $Co_3(PO_4)_2$

4.60 Determine the oxidation numbers of all the elements in each of the following compounds. (*Hint:* Look at the ions present.)

a $Hg_2(ClO_4)_2$ b $Cr_2(SO_4)_3$
c $CoSeO_4$ d $Pb(OH)_2$

Describing Oxidation–Reduction Reactions

4.61 In the following reactions, label the oxidizing agent and the reducing agent.

a $P_4(s) + 5O_2(g) \longrightarrow P_4O_{10}(s)$
b $Co(s) + Cl_2(g) \longrightarrow CoCl_2(s)$

4.62 In the following reactions, label the oxidizing agent and the reducing agent.

a $ZnO(s) + C(s) \longrightarrow Zn(g) + CO(g)$
b $8Fe(s) + S_8(s) \longrightarrow 8FeS(s)$

4.63 In the following reactions, label the oxidizing agent and the reducing agent.

a $2Al(s) + 3F_2(g) \longrightarrow 2AlF_3(s)$
b $Hg^{2+}(aq) + NO_2^-(aq) + H_2O(l) \longrightarrow Hg(s) + 2H^+(aq) + NO_3^-(aq)$

4.64 In the following reactions, label the oxidizing agent and the reducing agent.

a $Fe_2O_3(s) + 3CO(g) \longrightarrow 2Fe(s) + 3CO_2(g)$
b $PbS(s) + 4H_2O_2(aq) \longrightarrow PbSO_4(s) + 4H_2O(l)$

Balancing Oxidation–Reduction Reactions

4.65 Balance the following oxidation–reduction reactions by the half-reaction method.

a $CuCl_2(aq) + Al(s) \longrightarrow AlCl_3(aq) + Cu(s)$
b $Cr^{3+}(aq) + Zn(s) \longrightarrow Cr(s) + Zn^{2+}(aq)$

4.66 Balance the following oxidation–reduction reactions by the half-reaction method.

a $FeI_3(aq) + Mg(s) \longrightarrow Fe(s) + MgI_2(aq)$
b $H_2(g) + Ag^+(aq) \longrightarrow Ag(s) + H^+(aq)$

Molarity

4.67 A sample of 0.0512 mol of iron(III) chloride, $FeCl_3$, was dissolved in water to give 25.0 mL of solution. What is the molarity of the solution?

4.68 A 50.0-mL volume of $AgNO_3$ solution contains 0.0345 mol $AgNO_3$ (silver nitrate). What is the molarity of the solution?

4.69 An aqueous solution is made from 0.798 g of potassium permanganate, $KMnO_4$. If the volume of solution is 50.0 mL, what is the molarity of $KMnO_4$ in the solution?

4.70 A sample of oxalic acid, $H_2C_2O_4$, weighing 1.200 g is placed in a 100.0-mL volumetric flask, which is then filled to the mark with water. What is the molarity of the solution?

4.71 What volume of 0.120 *M* $CuSO_4$ is required to give 0.150 mol of copper(II) sulfate, $CuSO_4$?

4.72 How many milliliters of 0.126 *M* $HClO_4$ (perchloric acid) are required to give 0.150 mol $HClO_4$?

4.73 An experiment calls for 0.0353 g of potassium hydroxide, KOH. How many milliliters of 0.0176 *M* KOH are required?

4.74 What is the volume (in milliliters) of 0.100 *M* H_2SO_4 (sulfuric acid) containing 0.949 g H_2SO_4?

4.75 Heme, obtained from red blood cells, binds oxygen, O_2. How many moles of heme are there in 150 mL of 0.0019 *M* heme solution?

4.76 Insulin is a hormone that controls the use of glucose in the body. How many moles of insulin are required to make up 28 mL of 0.0048 *M* insulin solution?

4.77 How many grams of sodium dichromate, $Na_2Cr_2O_7$, should be added to a 100.0-mL volumetric flask to prepare 0.025 *M* $Na_2Cr_2O_7$ when the flask is filled to the mark with water?

4.78 Describe how you would prepare 2.50×10^2 mL of 0.50 *M* Na_2SO_4. What mass (in grams) of sodium sulfate, Na_2SO_4, is needed?

4.79 You wish to prepare 0.12 *M* HNO_3 from a stock solution of nitric acid that is 15.8 *M*. How many milliliters of the stock solution do you require to make up 1.00 L of 0.12 *M* HNO_3?

4.80 A chemist wants to prepare 0.75 *M* HCl. Commercial hydrochloric acid is 12.4 *M*. How many milliliters of the commercial acid does the chemist require to make up 1.50 L of the dilute acid?

4.81 A 4.00 g sample of KCl is dissolved in 10.0 mL of water. The resulting solution is then added to 60.0 mL of a 0.500 *M* $CaCl_2(aq)$ solution. Assuming that the volumes are additive, calculate the concentrations of each ion present in the final solution.

4.82 Calculate the concentrations of each ion present in a solution that results from mixing 50.0 mL of a 0.20 *M* $NaClO_3(aq)$ solution with 25.0 mL of a 0.20 *M* $Na_2SO_4(aq)$ solution. Assume that the volumes are additive.

Gravimetric Analysis

4.83 A chemist added an excess of sodium sulfate to a solution of a soluble barium compound to precipitate all of the barium ion as barium sulfate, $BaSO_4$. How many grams of barium ion are in a 458-mg sample of the barium compound if a solution of the sample gave 513 mg $BaSO_4$ precipitate? What is the mass percentage of barium in the compound?

4.84 A soluble iodide was dissolved in water. Then an excess of silver nitrate, $AgNO_3$, was added to precipitate all of the iodide ion as silver iodide, AgI. If 1.545 g of the soluble iodide gave 2.185 g of silver iodide, how many grams of iodine are in the sample of soluble iodide? What is the mass percentage of iodine, I, in the compound?

4.85 Copper has compounds with copper(I) ion or copper(II) ion. A compound of copper and chlorine was treated with a solution of silver nitrate, $AgNO_3$, to convert the chloride ion in the compound to a precipitate of AgCl. A 59.40-mg sample of the copper compound gave 86.00 mg AgCl.

a Calculate the percentage of chlorine in the copper compound.

b Decide whether the formula of the compound is CuCl or $CuCl_2$.

4.86 Gold has compounds containing gold(I) ion or gold(III) ion. A compound of gold and chlorine was treated with a solution of silver nitrate, $AgNO_3$, to convert the chloride ion in the compound to a precipitate of AgCl. A 162.7-mg sample of the gold compound gave 100.3 mg AgCl.

a Calculate the percentage of the chlorine in the gold compound.

b Decide whether the formula of the compound is AuCl or $AuCl_3$.

4.87 A compound of iron and chlorine is soluble in water. An excess of silver nitrate was added to precipitate the chloride ion as silver chloride. If a 134.8-mg sample of the compound gave 304.8 mg AgCl, what is the formula of the compound?

4.88 A 1.345-g sample of a compound of barium and oxygen was dissolved in hydrochloric acid to give a solution of barium ion, which was then precipitated with an excess of potassium chromate to give 2.012 g of barium chromate, $BaCrO_4$. What is the formula of the compound?

Volumetric Analysis

4.89 What volume of 0.250 *M* HNO_3 (nitric acid) reacts with 44.8 mL of 0.150 *M* Na_2CO_3 (sodium carbonate) in the following reaction?

$$2HNO_3(aq) + Na_2CO_3(aq) \longrightarrow 2NaNO_3(aq) + H_2O(l) + CO_2(g)$$

4.90 A flask contains 49.8 mL of 0.150 *M* $Ca(OH)_2$ (calcium hydroxide). How many milliliters of 0.500 *M* Na_2CO_3 (sodium carbonate) are required to react completely with the calcium hydroxide in the following reaction?

$$Na_2CO_3(aq) + Ca(OH)_2(aq) \longrightarrow CaCO_3(s) + 2NaOH(aq)$$

4.91 How many milliliters of 0.150 *M* H_2SO_4 (sulfuric acid) are required to react with 8.20 g of sodium hydrogen carbonate, $NaHCO_3$, according to the following equation?

$$H_2SO_4(aq) + 2NaHCO_3(aq) \longrightarrow Na_2SO_4(aq) + 2H_2O(l) + 2CO_2(g)$$

4.92 How many milliliters of 0.250 *M* $KMnO_4$ are needed to react with 3.55 g of iron(II) sulfate, $FeSO_4$? The reaction is as follows:

$$10FeSO_4(aq) + 2KMnO_4(aq) + 8H_2SO_4(aq) \longrightarrow 5Fe_2(SO_4)_3(aq) + 2MnSO_4(aq) + K_2SO_4(aq) + 8H_2O(l)$$

4.93 A solution of hydrogen peroxide, H_2O_2, is titrated with a solution of potassium permanganate, $KMnO_4$. The reaction is

$$5H_2O_2(aq) + 2KMnO_4(aq) + 3H_2SO_4(aq) \longrightarrow 5O_2(g) + 2MnSO_4(aq) + K_2SO_4(aq) + 8H_2O(l)$$

It requires 51.7 mL of 0.145 *M* $KMnO_4$ to titrate 20.0 g of the solution of hydrogen peroxide. What is the mass percentage of H_2O_2 in the solution?

4.94 A 3.75-g sample of iron ore is transformed to a solution of iron(II) sulfate, $FeSO_4$, and this solution is titrated with 0.150 *M* $K_2Cr_2O_7$ (potassium dichromate). If it requires 43.7 mL of potassium dichromate solution to titrate the iron(II) sulfate solution, what is the percentage of iron in the ore? The reaction is

$$6FeSO_4(aq) + K_2Cr_2O_7(aq) + 7H_2SO_4(aq) \longrightarrow 3Fe_2(SO_4)_3(aq) + Cr_2(SO_4)_3(aq) + 7H_2O(l) + K_2SO_4(aq)$$

General Problems

Key: These problems provide more practice but are not divided by topic or keyed to exercises. Each section ends with essay questions. Odd-numbered problems and the even-numbered problems that follow are similar; answers to all odd-numbered problems except the essay questions are given in the back of the book.

4.95 Magnesium metal reacts with hydrobromic acid to produce hydrogen gas and a solution of magnesium bromide. Write the molecular equation for this reaction. Then write the corresponding net ionic equation.

4.96 Aluminum metal reacts with perchloric acid to produce hydrogen gas and a solution of aluminum perchlorate. Write the molecular equation for this reaction. Then write the corresponding net ionic equation.

4.97 Nickel(II) sulfate solution reacts with sodium hydroxide solution to produce a precipitate of nickel(II) hydroxide and a solution of sodium sulfate. Write the molecular equation for this reaction. Then write the corresponding net ionic equation.

4.98 Potassium sulfate solution reacts with barium bromide solution to produce a precipitate of barium sulfate and a solution of potassium bromide. Write the molecular equation for this reaction. Then write the corresponding net ionic equation.

4.99 Decide whether a reaction occurs for each of the following. If it does not, write *NR* after the arrow. If it does, write the balanced molecular equation; then write the net ionic equation.

a $LiOH + HCN \longrightarrow$
b $Li_2CO_3 + HNO_3 \longrightarrow$
c $LiCl + AgNO_3 \longrightarrow$
d $NaCl + MgSO_4 \longrightarrow$

4.100 Decide whether a reaction occurs for each of the following. If it does not, write *NR* after the arrow. If it does, write the balanced molecular equation; then write the net ionic equation.

a $Al(OH)_3 + HNO_3 \longrightarrow$
b $NaBr + HClO_4 \longrightarrow$
c $CaCl_2 + NaNO_3 \longrightarrow$
d $MgSO_4 + Ba(NO_3)_2 \longrightarrow$

4.101 Complete and balance each of the following molecular equations, including phase labels, if a reaction occurs. Then write the net ionic equation. If no reaction occurs, write *NR* after the arrow.

a $Sr(OH)_2 + HC_2H_3O_2 \longrightarrow$
b $NH_4I + CsCl \longrightarrow$
c $NaNO_3 + CsCl \longrightarrow$
d $NH_4I + AgNO_3 \longrightarrow$

4.102 Complete and balance each of the following molecular equations, including phase labels, if a reaction occurs. Then write the net ionic equation. If no reaction occurs, write *NR* after the arrow.

a $HClO_4 + BaCO_3 \longrightarrow$
b $H_2CO_3 + Sr(OH)_2 \longrightarrow$
c $K_3PO_4 + MgCl_2 \longrightarrow$
d $FeSO_4 + MgCl_2 \longrightarrow$

4.103 Describe in words how you would do each of the following preparations. Then give the molecular equation for each preparation.

a $CuCl_2(s)$ from $CuSO_4(s)$
b $Ca(C_2H_3O_2)_2(s)$ from $CaCO_3(s)$
c $NaNO_3(s)$ from $Na_2SO_3(s)$
d $MgCl_2(s)$ from $Mg(OH)_2(s)$

4.104 Describe in words how you would do each of the following preparations. Then give the molecular equation for each preparation.

a $MgCl_2(s)$ from $MgCO_3(s)$
b $NaNO_3(s)$ from $NaCl(s)$
c $Al(OH)_3(s)$ from $Al(NO_3)_3(s)$
d $HCl(aq)$ from $H_2SO_4(aq)$

4.105 Classify each of the following reactions as a combination reaction, decomposition reaction, displacement reaction, or combustion reaction.

a When they are heated, ammonium dichromate crystals, $(NH_4)_2Cr_2O_7$, decompose to give nitrogen, water vapor, and solid chromium(III) oxide, Cr_2O_3.
b When aqueous ammonium nitrite, NH_4NO_2, is heated, it gives nitrogen and water vapor.
c When gaseous ammonia, NH_3, reacts with hydrogen chloride gas, HCl, fine crystals of ammonium chloride, NH_4Cl, are formed.
d Aluminum added to an aqueous solution of sulfuric acid, H_2SO_4, forms a solution of aluminum sulfate, $Al_2(SO_4)_3$. Hydrogen gas is released.

4.106 Classify each of the following reactions as a combination reaction, decomposition reaction, displacement reaction, or combustion reaction.

a When solid calcium oxide, CaO, is exposed to gaseous sulfur trioxide, SO_3, solid calcium sulfate, $CaSO_4$, is formed.
b Calcium metal (solid) reacts with water to produce a solution of calcium hydroxide, $Ca(OH)_2$, and hydrogen gas.
c When solid sodium hydrogen sulfite, $NaHSO_3$, is heated, solid sodium sulfite, Na_2SO_3, sulfur dioxide gas, SO_2, and water vapor are formed.
d Magnesium reacts with bromine to give magnesium bromide, $MgBr_2$.

4.107 Consider the reaction of all pairs of the following compounds in water solution: $Ba(OH)_2$, $Pb(NO_3)_2$, H_2SO_4, $NaNO_3$, $MgSO_4$.

a Which pair (or pairs) forms one insoluble compound and one soluble compound (not water)?
b Which pair (or pairs) forms two insoluble compounds?
c Which pair (or pairs) forms one insoluble compound and water?

4.108 Consider the reaction of all pairs of the following compounds in water solution: $Sr(OH)_2$, $AgNO_3$, H_3PO_4, KNO_3, $CuSO_4$.

a Which pair (or pairs) forms one insoluble compound and one soluble compound (not water)?
b Which pair (or pairs) forms two insoluble compounds?
c Which pair (or pairs) forms one insoluble compound and water?

4.109 An aqueous solution contains 5.00 g of calcium chloride, $CaCl_2$, per liter. What is the molarity of $CaCl_2$? When calcium chloride dissolves in water, the calcium ions, Ca^{2+}, and chloride ions, Cl^-, in the crystal go into the solution. What is the molarity of each ion in the solution?

4.110 An aqueous solution contains 3.75 g of iron(III) sulfate, $Fe_2(SO_4)_3$, per liter. What is the molarity of $Fe_2(SO_4)_3$? When the compound dissolves in water, the Fe^{3+} ions and SO_4^{2-} ions in the crystal go into the solution. What is the molar concentration of each ion in the solution?

4.111 A stock solution of potassium dichromate, $K_2Cr_2O_7$, is made by dissolving 89.3 g of the compound in 1.00 L of solution. How many milliliters of this solution are required to prepare 1.00 L of 0.100 *M* $K_2Cr_2O_7$?

4.112 A 71.2-g sample of oxalic acid, $H_2C_2O_4$, was dissolved in 1.00 L of solution. How would you prepare 2.50 L of 0.150 *M* $H_2C_2O_4$ from this solution?

4.113 A solution contains 6.00% (by mass) NaBr (sodium bromide). The density of the solution is 1.046 g/cm^3. What is the molarity of NaBr?

4.114 An aqueous solution contains 3.75% NH_3 (ammonia) by mass. The density of the aqueous ammonia is 0.979 g/mL. What is the molarity of NH_3 in the solution?

4.115 A barium mineral was dissolved in hydrochloric acid to give a solution of barium ion. An excess of potassium sulfate was added to 50.0 mL of the solution, and 1.128 g of barium sulfate precipitate formed. Assume that the original solution was barium chloride. What was the molarity of $BaCl_2$ in this solution?

4.116 Bone was dissolved in hydrochloric acid, giving 50.0 mL of solution containing calcium chloride, $CaCl_2$. To precipitate the calcium ion from the resulting solution, an excess of potassium oxalate was added. The precipitate of calcium oxalate, CaC_2O_4, weighed 1.437 g. What was the molarity of $CaCl_2$ in the solution?

4.117 You have a sample of a rat poison whose active ingredient is thallium(I) sulfate. You analyze this sample for the mass percentage of active ingredient by adding potassium iodide to precipitate yellow thallium(I) iodide. If the sample of rat poison weighed 759.0 mg and you obtained 212.2 mg of the dry precipitate, what is the mass percentage of the thallium(I) sulfate in the rat poison?

4.118 An antacid tablet has calcium carbonate as the active ingredient; other ingredients include a starch binder. You dissolve the tablet in hydrochloric acid and filter off insoluble material. You add potassium oxalate to the filtrate (containing calcium ion) to precipitate calcium oxalate. If a tablet weighing 0.750 g gave 0.629 g of calcium oxalate, what is the mass percentage of active ingredient in the tablet?

4.119 A sample of $CuSO_4 \cdot 5H_2O$ was heated to 110°C, where it lost water and gave another hydrate of copper(II) ion that contains 32.50% Cu. A 98.77-mg sample of this new hydrate gave 116.66 mg of barium sulfate precipitate when treated with a barium nitrate solution. What is the formula of the new hydrate?

4.120 A sample of $CuSO_4 \cdot 5H_2O$ was heated to 100°C, where it lost water and gave another hydrate of copper(II) ion that contained 29.76% Cu. An 85.42-mg sample of this new hydrate gave 93.33 mg of barium sulfate precipitate when treated with a barium nitrate solution. What is the formula of the new hydrate?

4.121 A water-soluble compound of gold and chlorine is treated with silver nitrate to convert the chlorine completely to silver chloride, AgCl. In an experiment, 328 mg of the compound gave 464 mg of silver chloride. Calculate the percentage of Cl in the compound. What is its empirical formula?

4.122 A solution of scandium chloride was treated with silver nitrate. The chlorine in the scandium compound was converted to silver chloride, AgCl. A 58.9-mg sample of scandium chloride gave 167.4 mg of silver chloride. What are the mass percentages of Sc and Cl in scandium chloride? What is its empirical formula?

4.123 A 0.608-g sample of fertilizer contained nitrogen as ammonium sulfate, $(NH_4)_2SO_4$. It was analyzed for nitrogen by heating with sodium hydroxide.

$$(NH_4)_2SO_4(s) + 2NaOH(aq) \longrightarrow Na_2SO_4(aq) + 2H_2O(l) + 2NH_3(g)$$

The ammonia was collected in 46.3 mL of 0.213 *M* HCl (hydrochloric acid), with which it reacted.

$$NH_3(g) + HCl(aq) \longrightarrow NH_4Cl(aq)$$

This solution was titrated for excess hydrochloric acid with 44.3 mL of 0.128 *M* NaOH.

$$NaOH(aq) + HCl(aq) \longrightarrow NaCl(aq) + H_2O(l)$$

What is the percentage of nitrogen in the fertilizer?

4.124 An antacid tablet contains sodium hydrogen carbonate, $NaHCO_3$, and inert ingredients. A 0.500-g sample of powdered tablet was mixed with 50.0 mL of 0.190 M HCl (hydrochloric acid). The mixture was allowed to stand until it reacted.

$$NaHCO_3(s) + HCl(aq) \longrightarrow NaCl(aq) + H_2O(l) + CO_2(g)$$

The excess hydrochloric acid was titrated with 45.3 mL of 0.128 *M* NaOH (sodium hydroxide).

$$HCl(aq) + NaOH(aq) \longrightarrow NaCl(aq) + H_2O(l)$$

What is the percentage of sodium hydrogen carbonate in the antacid?

Strategy Problems

Key: As noted earlier, all of the practice and general problems are matched pairs. This section is a selection of problems that are not in matched-pair format. These challenging problems require that you employ many of the concepts and strategies that were developed in the chapter. In some cases, you will have to integrate several concepts and operational skills in order to solve the problem successfully.

4.125 You order a glass of juice in a restaurant, only to discover that it is warm and too sweet. The sugar concentration of the juice is 3.47 *M*, but you would like it reduced to a concentration of 1.78 *M*. How many grams of ice should you add to 100 mL of juice, knowing that only a third of the ice will melt before you take the first sip? (The density of water is 1.00 g/mL.)

4.126 If 45.1 mL of a solution containing 8.30 g of silver nitrate is added to 31.3 mL of 0.511 *M* sodium carbonate solution, calculate the molarity of silver ion in the resulting solution. (Assume volumes are additive.)

4.127 If 38.2 mL of 0.248 *M* aluminum sulfate solution is diluted with deionized water to a total volume of 0.639 L, how many grams of aluminum ion are present in the diluted solution?

4.128 An aluminum nitrate solution is labeled 0.256 *M*. If 31.6 mL of this solution is diluted to a total of 65.1 mL, calculate the molarity of nitrate ion in the resulting solution.

4.129 Zinc acetate is sometimes prescribed by physicians for the treatment of Wilson's disease, which is a genetically caused condition wherein copper accumulates to toxic levels in the body. If you were to analyze a sample of zinc acetate and find that it contains 3.33×10^{23} acetate ions, how many grams of zinc acetate must be present in the sample?

4.130 Arsenic acid, H_3AsO_4, is a poisonous acid that has been used in the treatment of wood to prevent insect damage. Arsenic acid has three acidic protons. Say you take a 25.00-mL sample of arsenic acid and prepare it for titration with NaOH by adding 25.00 mL of water. The complete neutralization of this solution requires the addition of 53.07 mL of 0.6441 *M* NaOH solution. Write the balanced chemical reaction for the titration, and calculate the molarity of the arsenic acid sample.

4.131 When the following equation is balanced by the half-reaction method using the smallest set of whole-number stoichiometric coefficients possible, how many electrons are canceled when the two half-reactions are added together?

$$K(s) + N_2(g) \longrightarrow K_3N(s)$$

4.132 Identify each of the following reactions as being a neutralization, precipitation, or oxidation–reduction reaction.

a. $Fe_2O_3(s) + 3CO(g) \longrightarrow 2Fe(s) + 3CO_2(g)$

b. $Na_2SO_4(aq) + Hg(NO_3)_2(aq) \longrightarrow HgSO_4(s) + 2NaNO_3(aq)$

c. $CsOH(aq) + HClO_4(aq) \longrightarrow Cs^+(aq) + 2H_2O(l) + ClO_4^-(aq)$

d. $Mg(NO_3)_2(g) + Na_2S(aq) \longrightarrow MgS(s) + 2NaNO_3(aq)$

4.133 A 414-mL sample of 0.196 *M* $MgBr_2$ solution is prepared in a large flask. A 43.0-mL portion of the solution is then placed into an empty 100.0-mL beaker. What is the concentration of the solution in the beaker?

4.134 Three 1.0-g samples of $PbCl_2$, KCl, and $CaCl_2$ are placed in separate 500-mL beakers. In each case, enough 25°C water is added to bring the total volume of the mixture to 250 mL. Each of the mixtures is then stirred for five minutes. Which of the mixtures will have the highest concentration of chloride (Cl^-) ion?

4.135 A 25-mL sample of 0.50 *M* NaOH is combined with a 75-mL sample of 0.50 *M* NaOH. What is the concentration of the resulting NaOH solution?

4.136 What is the molarity of pure water with a density of 1.00 g/mL?

4.137 Nitric acid can be reacted with zinc according to the following chemical equation.

$$4HNO_3(aq) + Zn(s) \longrightarrow Zn(NO_3)_2(aq) + 2H_2O(l) + 2NO_2(g)$$

If 3.75 g of Zn is added to 175 mL of 0.500 *M* HNO_3, what mass of NO_2 would be produced by the chemical reaction?

4.138 How many grams of precipitate are formed if 175 mL of a 0.750 *M* aluminum sulfate solution and 375 mL of a 1.15 *M* sodium hydroxide solution are mixed together?

4.139 You are asked to prepare 0.250 L of a solution that is 0.500 *M* in nitrate ion. Your only source of nitrate ion is a bottle of 1.00 *M* calcium nitrate. What volume (mL) of the calcium nitrate solution must you use?

4.140 Potassium hydrogen phthalate (abbreviated as KHP) has the molecular formula $KHC_8H_4O_4$ and a molar mass of 204.22 g/mol. KHP has one acidic hydrogen. A solid sample of KHP is dissolved in 50 mL of water and titrated to the equivalence point with 22.90 mL of a 0.5010 *M* NaOH solution. How many grams of KHP were used in the titration?

Cumulative-Skills Problems

Key: The problems under this heading combine skills introduced in previous chapters with those given in the current one.

4.141 Lead(II) nitrate reacts with cesium sulfate in an aqueous precipitation reaction. What are the formulas of lead(II) nitrate and cesium sulfate? Write the molecular equation and net ionic equation for the reaction. What are the names of the products? Give the molecular equation for another reaction that produces the same precipitate.

4.142 Silver nitrate reacts with strontium chloride in an aqueous precipitation reaction. What are the formulas of silver nitrate and strontium chloride? Write the molecular equation and net ionic equation for the reaction. What are the names of the products? Give the molecular equation for another reaction that produces the same precipitate.

4.143 Elemental bromine is the source of bromine compounds. The element is produced from certain brine solutions that occur naturally. These brines are essentially solutions of calcium bromide that, when treated with chlorine gas, yield bromine in a displacement reaction. What are the molecular equation and net ionic equation for the reaction? A solution containing 40.0 g of calcium bromide requires 14.2 g of chlorine to react completely with it, and 22.2 g of calcium chloride is produced in addition to whatever bromine is obtained. How many grams of calcium bromide are required to produce 10.0 pounds of bromine?

4.144 Barium carbonate is the source of barium compounds. It is produced in an aqueous precipitation reaction from barium sulfide and sodium carbonate. (Barium sulfide is a soluble compound obtained by heating the mineral barite, which is barium sulfate, with carbon.) What are the molecular equation and net ionic equation for the precipitation reaction? A solution containing 33.9 g of barium sulfide requires 21.2 g of sodium carbonate to react completely with it, and 15.6 g of sodium sulfide is produced in addition to whatever barium carbonate is obtained. How many grams of barium sulfide are required to produce 5.00 tons of barium carbonate? (One ton equals 2000 pounds.)

4.145 Mercury(II) nitrate is treated with hydrogen sulfide, H_2S, forming a precipitate and a solution. Write the molecular equation and the net ionic equation for the reaction. An acid is formed; is it strong or weak? Name each of the products. If 81.15 g of mercury(II) nitrate and 8.52 g of hydrogen sulfide are mixed in 550.0 g of water to form 58.16 g of precipitate, what is the mass of the solution after the reaction?

4.146 Mercury(II) nitrate is treated with hydrogen sulfide, H_2S, forming a precipitate and a solution. Write the molecular equation and the net ionic equation for the reaction. An acid is formed; is it strong or weak? Name each of the products. If 65.65 g of mercury(II) nitrate and 4.26 g of hydrogen sulfide are mixed in 395.0 g of water to form 54.16 g of precipitate, what is the mass of the solution after the reaction?

4.147 Iron forms a sulfide with the approximate formula Fe_7S_8. Assume that the oxidation state of sulfur is −2 and that iron atoms exist in both +2 and +3 oxidation states. What is the ratio of Fe(II) atoms to Fe(III) atoms in this compound?

4.148 A transition metal X forms an oxide of formula X_2O_3. It is found that only 50% of X atoms in this compound are in the +3 oxidation state. The only other stable oxidation states of X are +2 and +5. What percentage of X atoms are in the +2 oxidation state in this compound?

4.149 What volume of a solution of ethanol, C_2H_6O, that is 94.0% ethanol by mass contains 0.200 mol C_2H_6O? The density of the solution is 0.807 g/mL.

4.150 What volume of a solution of ethylene glycol, $C_2H_6O_2$, that is 56.0% ethylene glycol by mass contains 0.350 mol $C_2H_6O_2$? The density of the solution is 1.072 g/mL.

4.151 A 10.0-mL sample of potassium iodide solution was analyzed by adding an excess of silver nitrate solution to produce silver iodide crystals, which were filtered from the solution.

$$KI(aq) + AgNO_3(aq) \longrightarrow KNO_3(aq) + AgI(s)$$

If 2.183 g of silver iodide was obtained, what was the molarity of the original KI solution?

4.152 A 25.0-mL sample of sodium sulfate solution was analyzed by adding an excess of barium chloride solution to produce barium sulfate crystals, which were filtered from the solution.

$$Na_2SO_4(aq) + BaCl_2(aq) \longrightarrow 2NaCl(aq) + BaSO_4(s)$$

If 5.719 g of barium sulfate was obtained, what was the molarity of the original Na_2SO_4 solution?

4.153 A metal, M, was converted to the sulfate, $M_2(SO_4)_3$. Then a solution of the sulfate was treated with barium chloride to give barium sulfate crystals, which were filtered off.

$$M_2(SO_4)_3(aq) + 3BaCl_2(aq) \longrightarrow 2MCl_3(aq) + 3BaSO_4(s)$$

If 1.200 g of the metal gave 6.026 g of barium sulfate, what is the atomic weight of the metal? What is the metal?

4.154 A metal, M, was converted to the chloride MCl_2. Then a solution of the chloride was treated with silver nitrate to give silver chloride crystals, which were filtered from the solution.

$$MCl_2(aq) + 2AgNO_3(aq) \longrightarrow M(NO_3)_2(aq) + 2AgCl(s)$$

If 2.434 g of the metal gave 7.964 g of silver chloride, what is the atomic weight of the metal? What is the metal?

4.155 Phosphoric acid is prepared by dissolving phosphorus(V) oxide, P_4O_{10}, in water. What is the balanced equation for this reaction? How many grams of P_4O_{10} are required to make 1.50 L of aqueous solution containing 5.00% phosphoric acid by mass? The density of the solution is 1.025 g/mL.

4.156 Iron(III) chloride can be prepared by reacting iron metal with chlorine. What is the balanced equation for this reaction? How many grams of iron are required to make 3.00 L of aqueous solution containing 9.00% iron(III) chloride by mass? The density of the solution is 1.067 g/mL.

4.157 An alloy of aluminum and magnesium was treated with sodium hydroxide solution, in which only aluminum reacts.

$$2Al(s) + 2NaOH(aq) + 6H_2O(l) \longrightarrow 2NaAl(OH)_4(aq) + 3H_2(g)$$

If a sample of alloy weighing 1.118 g gave 0.1068 g of hydrogen, what is the percentage of aluminum in the alloy?

4.158 An alloy of iron and carbon was treated with sulfuric acid, in which only iron reacts.

$$2Fe(s) + 3H_2SO_4(aq) \longrightarrow Fe_2(SO_4)_3(aq) + 3H_2(g)$$

If a sample of alloy weighing 2.358 g gave 0.1067 g of hydrogen, what is the percentage of iron in the alloy?

4.159 Determine the volume of sulfuric acid solution needed to prepare 37.4 g of aluminum sulfate, $Al_2(SO_4)_3$, by the reaction

$$2Al(s) + 3H_2SO_4(aq) \longrightarrow Al_2(SO_4)_3(aq) + 3H_2(g)$$

The sulfuric acid solution, whose density is 1.104 g/mL, contains 15.0% H_2SO_4 by mass.

4.160 Determine the volume of sodium hydroxide solution needed to prepare 26.2 g sodium phosphate, Na_3PO_4, by the reaction

$$3NaOH(aq) + H_3PO_4(aq) \longrightarrow Na_3PO_4(aq) + 3H_2O(l)$$

The sodium hydroxide solution, whose density is 1.133 g/mL, contains 12.0% NaOH by mass.

4.161 The active ingredients of an antacid tablet contained only magnesium hydroxide and aluminum hydroxide. Complete neutralization of a sample of the active ingredients required 48.5 mL of 0.187 *M* hydrochloric acid. The chloride salts from this neutralization were obtained by evaporation of the filtrate from the titration; they weighed 0.4200 g. What was the percentage by mass of magnesium hydroxide in the active ingredients of the antacid tablet?

4.162 The active ingredients in an antacid tablet contained only calcium carbonate and magnesium carbonate. Complete reaction of a sample of the active ingredients required 39.20 mL of 0.08750 *M* hydrochloric acid. The chloride salts from the reaction were obtained by evaporation of the filtrate from this titration; they weighed 0.1900 g. What was the percentage by mass of the calcium carbonate in the active ingredients of the antacid tablet?

7 Quantum Theory of the Atom

Colored flames from several metallic compounds; the visible light is emitted from the metal atoms. The yellow emission from sodium atoms is used in sodium street lamps.

Contents and Concepts

Light Waves, Photons, and the Bohr Theory

To understand the formation of chemical bonds, you need to know something about the electronic structure of atoms. Because light gives us information about this structure, we begin by discussing the nature of light. Then we look at the Bohr theory of the simplest atom, hydrogen.

Quantum Mechanics and Quantum Numbers

The Bohr theory firmly establishes the concept of energy levels but fails to account for the details of atomic structure. Here we discuss some basic notions of quantum mechanics, which is the theory currently applied to extremely small particles, such as electrons in atoms.

According to Rutherford's model (Section 2.2), an atom consists of a nucleus many times smaller than the atom itself, with electrons occupying the remaining space. How are the electrons distributed in this space? Or, we might ask, what are the electrons doing in the atom?

The answer was to come from an unexpected area: the study of colored flames. When metallic compounds burn in a flame, they emit bright colors (Figure 7.1). The spectacular colors of fireworks are due to the burning of metal compounds. Lithium and strontium compounds give a deep red color; barium compounds, a green color; and copper compounds, a bluish green color.

Figure 7.1 ▲

Flame tests of Groups IA and IIA elements A wire loop containing a sample of a metallic compound is placed in a flame. *Left to right:* flames of lithium (red), sodium (yellow), strontium (red), and calcium (orange).

Although the red flames of lithium and strontium appear similar, the light from each can be resolved (separated) by means of a prism into distinctly different colors. This resolution easily distinguishes the two elements. A prism disperses the colors of white light just as small raindrops spread the colors of sunlight into a rainbow or spectrum. But the light from a flame, when passed through a prism, reveals something other than a rainbow. Instead of the continuous range of color from red to yellow to violet, the spectrum from a strontium flame, for example, shows a cluster of red lines and blue lines against a black background. The spectrum of lithium is different, showing a red line, a yellow line, and two blue lines against a black background. (See Figure 7.2.)

Each element, in fact, has a characteristic line spectrum because of the emission of light from atoms in the hot gas. The spectra can be used to identify elements. How is it that each atom emits particular colors of light? What does a line spectrum tell us about the structure of an atom? If you know something about the structures of atoms, can you explain the formation of ions and molecules? We will answer these questions in this and the next few chapters.

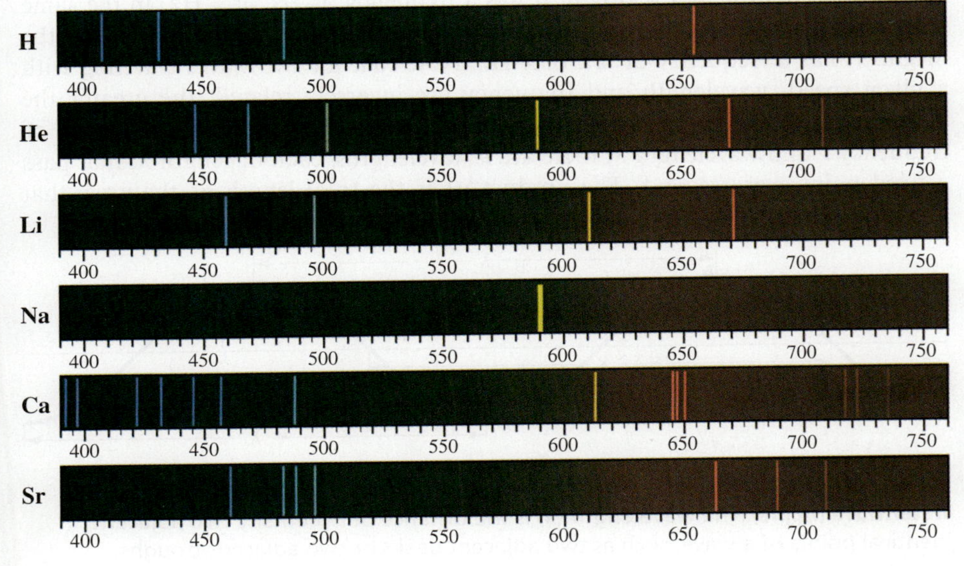

Figure 7.2 ◀

Emission (line) spectra of some elements The lines correspond to visible light emitted by atoms. (Wavelengths of lines are given in nanometers.)

Light Waves, Photons, and the Bohr Theory

In Chapter 2 we looked at the *basic structure* of atoms and we introduced the concept of a chemical bond. To understand the formation of a chemical bond between atoms, however, you need to know something about the *electronic structure* of atoms. The present theory of the electronic structure of atoms started with an explanation of the colored light produced in hot gases and flames. Before we can discuss this, we need to describe the nature of light.

7.1 The Wave Nature of Light

If you drop a stone into one end of a quiet pond, the impact of the stone with the water starts an up-and-down motion of the water surface. This up-and-down motion travels outward from where the stone hit; it is a familiar example of a wave. A *wave* is a continuously repeating change or oscillation in matter or in a physical field. Light is also a wave. It consists of oscillations in electric and magnetic fields that can travel through space. Visible light, x rays, and radio waves are all forms of *electromagnetic radiation.*

You characterize a wave by its wavelength and frequency. The **wavelength,** denoted by the Greek letter λ (lambda), is *the distance between any two adjacent identical points of a wave.* Thus, the wavelength is the distance between two adjacent peaks or troughs of a wave. Figure 7.3 shows a cross section of a water wave at a given moment, with the wavelength (λ) identified. Radio waves have wavelengths from approximately 100 mm to several hundred meters. Visible light has much shorter wavelengths, about 10^{-6} m. Wavelengths of visible light are often given in nanometers (1 nm = 10^{-9} m). For example, light of wavelength 5.55×10^{-7} m, the greenish yellow light to which the human eye is most sensitive, equals 555 nm.

The **frequency** of a wave is *the number of wavelengths of that wave that pass a fixed point in one unit of time* (usually one second). For example, imagine you are anchored in a small boat on a pond when a stone is dropped into the water. Waves travel outward from this point and move past your boat. The number of wavelengths that pass you in one second is the frequency of that wave. Frequency is denoted by the Greek letter ν (nu, pronounced "new"). The unit of frequency is /s, or s^{-1}, also called the *hertz* (Hz).

The wavelength and frequency of a wave are related to each other. Figure 7.4 shows two waves, each traveling from left to right at the same speed; that is, each wave moves the same total length in 1 s. The top wave, however, has a wavelength twice that of the bottom wave. In 1 s, two complete wavelengths of the top wave move left to right from the origin. It has a frequency of 2/s, or 2 Hz. In the same time, four complete wavelengths of the bottom wave move left to right from the origin. It has a frequency of 4/s, or 4 Hz. Note that for two waves traveling with a given speed, wavelength and frequency are inversely related: the greater the wavelength, the lower the frequency, and vice versa. In general, with a wave of frequency ν and wavelength λ, there are ν wavelengths, each of length λ, that pass a fixed point every second. The product $\nu\lambda$ is the total length of the wave that

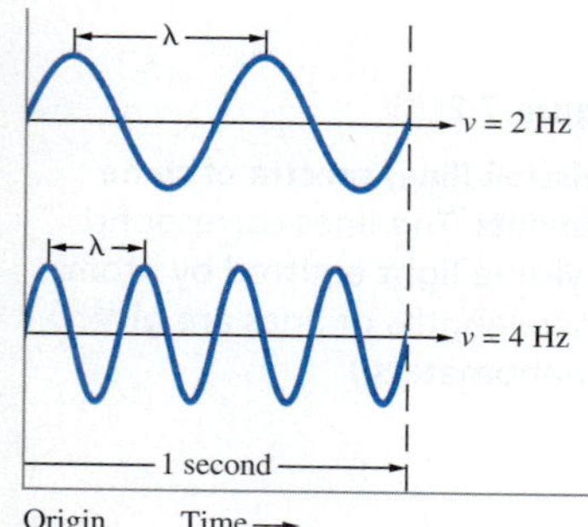

Figure 7.4 ▲

Relation between wavelength and frequency Both waves are traveling at the same speed. The top wave has a wavelength twice that of the bottom wave. The bottom wave, however, has twice the frequency of the top wave.

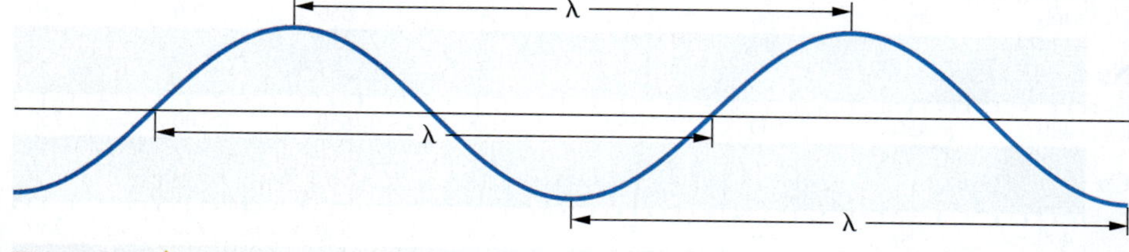

Figure 7.3 ▲

Water wave (ripple) The wavelength (λ) is the distance between any two adjacent identical points of a wave, such as two adjacent peaks or two adjacent troughs.

has passed the point in 1s. This length of wave per second is the speed of the wave. For light of speed c,

$$c = \nu\lambda$$

The speed of light waves in a vacuum is a constant and is independent of wavelength or frequency. This speed is 3.00×10^8 m/s, which is the value for c that we use in the following examples. ▶

To understand how fast the speed of light is, it might help to realize that it takes only 2.6 s for radar waves (which travel at the speed of light) to leave Earth, bounce off the moon, and return—a total distance of 478,000 miles.

Example 7.1 **Obtaining the Wavelength of Light from Its Frequency**

Gaining Mastery Toolbox

Critical Concept 7.1
A wave is characterized by its wavelength, λ, and its frequency, ν, which are related by the speed of the wave (which for light waves is denoted by the symbol c). You need to be able to manipulate this relationship to solve for wavelength in terms of frequency.

Solution Essentials:
- Equation relating frequency and wavelength: $c = \nu\lambda$

What is the wavelength of the yellow sodium emission, which has a frequency of 5.09×10^{14}/s?

Problem Strategy Note that frequency and wavelength are related (their product equals the speed of light). Write this as an equation, and then rearrange it to give the wavelength on the left and the frequency and speed of light on the right.

Solution The frequency and wavelength are related by the formula $c = \nu\lambda$. You rearrange this formula to give

$$\lambda = \frac{c}{\nu}$$

in which c is the speed of light (3.00×10^8 m/s). Substituting yields

$$\lambda = \frac{3.00 \times 10^8 \text{ m/}\cancel{\text{s}}}{5.09 \times 10^{14}/\cancel{\text{s}}} = 5.89 \times 10^{-7} \text{ m, or } \mathbf{589\ nm}$$

Answer Check Make sure you use the same units for c and ν (we used SI units in the solution). Check that the units cancel on the right to give the units of wavelength. As a check on your arithmetic, note that the wavelength for the answer should be in the visible range (400 nm to 750 nm).

Exercise 7.1 The frequency of the strong red line in the spectrum of potassium is 3.91×10^{14}/s. What is the wavelength of this light in nanometers?

■ See Problems 7.35 and 7.36.

OWL InteractiveExample 7.2 **Obtaining the Frequency of Light from Its Wavelength**

Gaining Mastery Toolbox

Critical Concept 7.2
A wave is characterized by its wavelength, λ, and its frequency, ν, which are related by the speed of the wave (which for light waves is denoted by the symbol c). You need to be able to manipulate this relationship to solve for frequency in terms of wavelength.

Solution Essentials:
- Equation relating frequency and wavelength: $c = \nu\lambda$

What is the frequency of violet light with a wavelength of 408 nm?

Problem Strategy Rearrange the equation relating frequency and wavelength so that the frequency is on the left and the wavelength and speed of light are on the right.

Solution You rearrange the equation relating frequency and wavelength to give

$$\nu = \frac{c}{\lambda}$$

Substituting for λ (408 nm = 408×10^{-9} m) gives

$$\nu = \frac{3.00 \times 10^8 \cancel{\text{m}}/\text{s}}{408 \times 10^{-9} \cancel{\text{m}}} = \mathbf{7.35 \times 10^{14}/s}$$

(continued)

(*continued*)

In 1854, Kirchhoff found that each element has a unique spectrum. Later, Bunsen and Kirchhoff developed the prism spectroscope and used it to confirm their discovery of two new elements, cesium (in 1860) and rubidium (in 1861).

Answer Check Use the same units for wavelength and speed of light; then check that the units cancel after substituting to give the units of frequency. We have used SI units here.

Exercise 7.2 The element cesium was discovered in 1860 by Robert Bunsen and Gustav Kirchhoff, who found two bright blue lines in the spectrum of a substance isolated from a mineral water. One of the spectral lines of cesium has a wavelength of 456 nm. What is its frequency?

See Problems 7.37 and 7.38.

The range of frequencies or wavelengths of electromagnetic radiation is called the **electromagnetic spectrum,** shown in Figure 7.5. Visible light extends from the violet end of the spectrum, which has a wavelength of about 400 nm, to the red end, with a wavelength of less than 800 nm. Beyond these extremes, electromagnetic radiation is not visible to the human eye. Infrared radiation has wavelengths greater than 800 nm (greater than the wavelength of red light), and ultraviolet radiation has wavelengths less than 400 nm (less than the wavelength of violet light).

CONCEPT CHECK 7.1

Laser light of a specific frequency falls on a crystal that converts a portion of this light into light with double the original frequency. How is the wavelength of this frequency-doubled light related to the wavelength of the original laser light? Suppose the original laser light was red. In which region of the spectrum would the frequency-doubled light be? (If this is in the visible region, what color is the light?)

Figure 7.5

The electromagnetic spectrum Divisions between regions are not defined precisely.

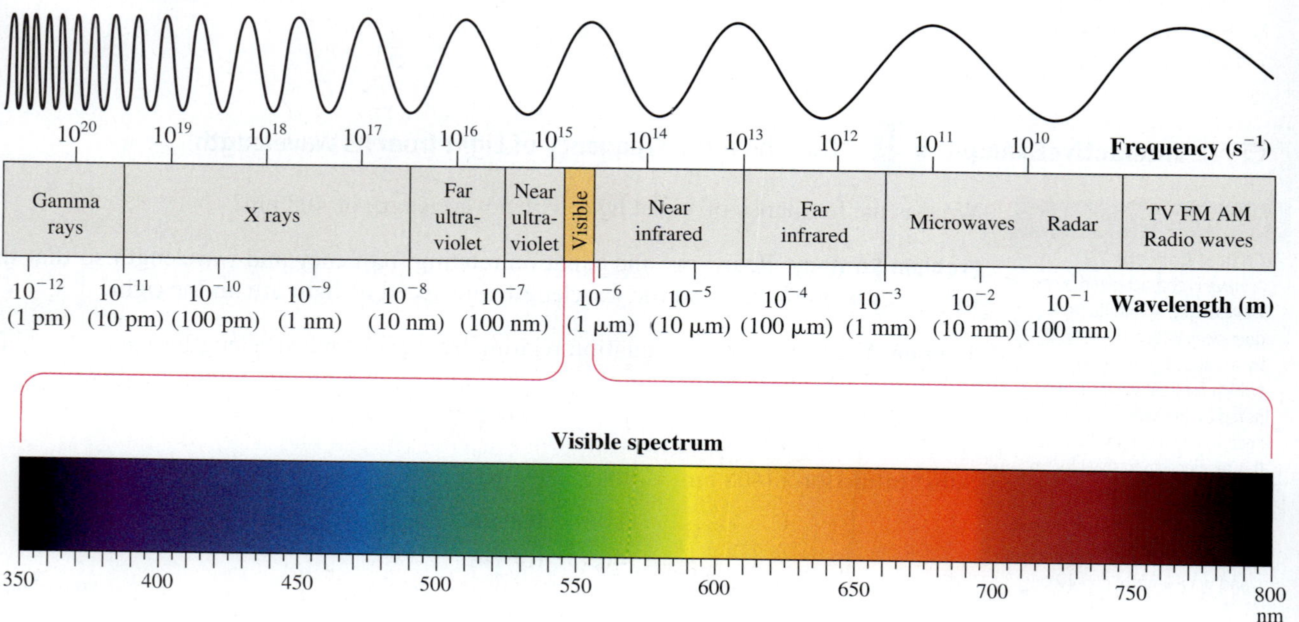

7.2 Quantum Effects and Photons

Isaac Newton, who studied the properties of light in the seventeenth century, believed that light consisted of a beam of particles. In 1801, however, British physicist Thomas Young showed that light, like waves, could be diffracted. *Diffraction* is a property of waves in which the waves spread out when they encounter an obstruction or small hole about the size of the wavelength. You can observe diffraction by viewing a light source through a hole—for example, a streetlight through a mesh curtain. The image of the streetlight is blurred by diffraction.

By the early part of the twentieth century, the wave theory of light appeared to be well entrenched. But in 1905 the German physicist Albert Einstein (1879–1955; emigrated to the United States in 1933) discovered that he could explain a phenomenon known as the *photoelectric effect* by postulating that light had both wave and particle properties. Einstein based this idea on the work of the German physicist Max Planck (1858–1947).

Planck's Quantization of Energy

In 1900 Max Planck found a theoretical formula that exactly describes the intensity of light of various frequencies emitted by a hot solid at different temperatures. ▶ Earlier, others had shown experimentally that the light of maximum intensity from a hot solid varies in a definite way with temperature. A solid glows red at 750°C, then white as the temperature increases to 1200°C. At the lower temperature, chiefly red light is emitted. As the temperature increases, more yellow and blue light become mixed with the red, giving white light. ▶

Max Planck was professor of physics at the University of Berlin when he did this research. He received the Nobel Prize in physics for it in 1918.

Radiation emitted from the human body and warm objects is mostly infrared, which is detected by burglar alarms, military night-vision scopes, and similar equipment.

According to Planck, the atoms of the solid oscillate, or vibrate, with a definite frequency ν, depending on the solid. But in order to reproduce the results of experiments on glowing solids, he found it necessary to accept a strange idea. An atom could have only certain energies of vibration, E, those allowed by the formula

$$E = nh\nu, \qquad n = 1, 2, 3, \ldots$$

where h is a constant, now called **Planck's constant,** *a physical constant relating energy and frequency, having the value* 6.63×10^{-34} *J·s.* The value of n must be 1 or 2 or some other whole number. Thus, the only energies a vibrating atom can have are $h\nu$, $2h\nu$, $3h\nu$, and so forth.

The numbers symbolized by n are called *quantum numbers*. The vibrational energies of the atoms are said to be *quantized;* that is, the possible energies are limited to certain values.

The quantization of energy seems contradicted by everyday experience. Consider the potential energy of an object, such as a tennis ball. Its potential energy depends on its height above the surface of the earth: the greater the upward height, the greater the potential energy. (Recall the discussion of potential energy in Section 6.1.) We have no problem in placing the tennis ball at any height, so it can have any energy. Imagine, however, that you could only place the tennis ball on the steps of a stairway. In that case, you could only put the tennis ball on one of the steps, so the potential energy of the tennis ball could have only certain values; its energy would be quantized. Of course, this restriction of the tennis ball is artificial; in fact, a tennis ball can have a range of energies, not just particular values. As we will see, quantum effects depend on the mass of the object: the smaller the mass, the more likely you will see quantum effects. Atoms, and particularly electrons, have small enough masses to exhibit quantization of energy; tennis balls do not.

Photoelectric Effect

Planck himself was uneasy with the quantization assumption and tried unsuccessfully to eliminate it from his theory. Albert Einstein, on the other hand, boldly extended Planck's work to include the structure of light itself. Einstein reasoned that if a vibrating atom changed energy, say from $3h\nu$ to $2h\nu$, it would decrease in energy by $h\nu$, and this energy would be emitted as a bit (or quantum) of light

Albert Einstein obtained his Ph.D. in 1905. In the same year he published five groundbreaking research papers: one on the photoelectric effect (for which he received the Nobel Prize in physics in 1921); two on special relativity; a paper on the determination of molecular size; and one on Brownian motion (which led to experiments that tested kinetic-molecular theory and ended remaining doubts about the existence of atoms and molecules).

energy. He therefore postulated that light consists of quanta (now called **photons**), or *particles of electromagnetic energy, with energy E proportional to the observed frequency of the light:* ◀

$$E = h\nu$$

In 1905 Einstein used this photon concept to explain the photoelectric effect.

The **photoelectric effect** is *the ejection of electrons from the surface of a metal or from another material when light shines on it* (see Figure 7.6). Electrons are ejected, however, only when the frequency of light exceeds a certain *threshold value* characteristic of the particular metal. For example, although violet light will cause potassium metal to eject electrons, no amount of red light (which has a lower frequency) has any effect.

To explain this dependence of the photoelectric effect on the frequency, Einstein assumed that an electron is ejected from a metal when it is struck by a single photon. Therefore, this photon must have at least enough energy to remove the electron from the attractive forces of the metal. No matter how many photons strike the metal, if no single one has sufficient energy, an electron cannot be ejected. A photon of red light has insufficient energy to remove an electron from potassium. But a photon corresponding to the threshold frequency has just enough energy, and at higher frequencies it has more than enough energy. When the photon hits the metal, its energy $h\nu$ is taken up by the electron. The photon ceases to exist as a particle; it is said to be *absorbed.*

The minimum energy that a photon can have and still eject an electron from a material in the photoelectric effect is characteristic of the material; this energy is called its photoelectric *work function*. The work function is frequently expressed in electron volts, eV. (An electron volt is the energy imparted to an electron when it is accelerated by an electric potential of one volt; 1 eV equals 1.602×10^{-19} J.) For example, the work function of potassium metal is 2.30 eV. If a photon has energy greater than the work function, the excess energy appears as kinetic energy of the ejected electron.

The wave and particle pictures of light should be regarded as complementary views of the same physical entity. This is called the *wave–particle duality* of light. The equation $E = h\nu$ displays this duality; E is the energy of a light particle or photon, and ν is the frequency of the associated wave. Neither the wave nor the particle view alone is a complete description of light.

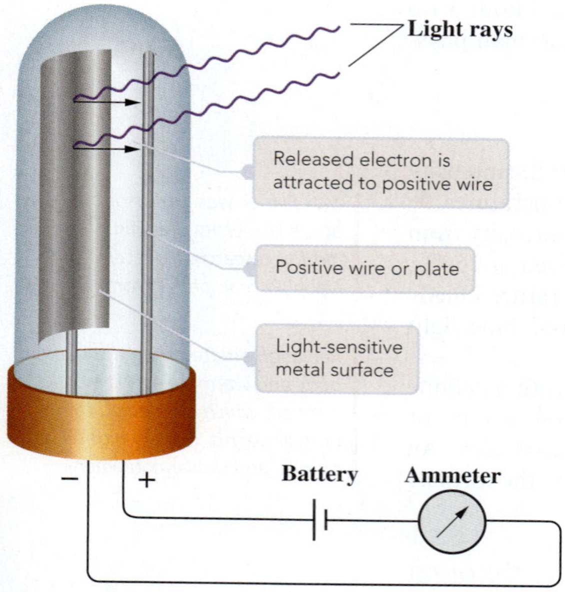

Figure 7.6 ▲

The photoelectric effect Light shines on a metal surface, knocking out electrons. The metal surface is contained in an evacuated tube, which allows the ejected electrons to be accelerated to a positively charged plate. As long as light of sufficient frequency shines on the metal, free electrons are produced and a current flows through the tube. (The current is measured by an ammeter.) When the light is turned off, the current stops flowing.

OWL Interactive Example 7.3 Calculating the Energy of a Photon

Gaining Mastery Toolbox

Critical Concept 7.3
The energy of a photon, E, is related to Planck's constant, h, times the frequency of the corresponding light wave, ν.

Solution Essentials:
- Equation relating the photon energy to frequency: $E = h\nu$
- Equation relating frequency and wavelength: $c = \nu\lambda$

The red spectral line of lithium occurs at 671 nm (6.71×10^{-7} m). Calculate the energy of one photon of this light.

Problem Strategy Note that the energy of a photon is related to its corresponding frequency. Therefore, you will first need to obtain this frequency from the wavelength of the spectral line.

Solution The frequency of this light is

$$\nu = \frac{c}{\lambda} = \frac{3.00 \times 10^{8} \cancel{\text{m}}/\text{s}}{6.71 \times 10^{-7} \cancel{\text{m}}} = 4.47 \times 10^{14}/\text{s}$$

(continued)

A CHEMIST Looks at...

Daily Life

Zapping Hamburger with Gamma Rays

In 1993 in Seattle, Washington, four children died and hundreds of people became sick from food poisoning when they ate undercooked hamburgers containing a dangerous strain of a normally harmless bacterium, *Escherichia coli*. A similar bout of food poisoning from hamburgers containing the dangerous strain of *E. coli* occurred in the summer of 1997; 17 people became ill. This time the tainted hamburger was traced to a meat processor in Nebraska, which immediately recalled 25 million pounds of ground beef. Within months, the U.S. Food and Drug Administration approved the irradiation of red meat by gamma rays to kill harmful bacteria. Gamma rays are a form of electromagnetic radiation similar to x rays, but the photons of gamma rays have higher energy.

The idea of using high-energy radiation to disinfect foods is not new (Figure 7.7). It has been known for about 50 years that high-energy radiation kills bacteria and molds in foods, prolonging their shelf life. Gamma rays kill organisms by breaking up the DNA molecules within their cells. This DNA holds the information for producing vital cell proteins.

Currently, the gamma rays used in food irradiation come from the radioactive decay of cobalt-60. It is perhaps this association with radioactivity that has slowed the acceptance of irradiation as a food disinfectant. However, food is not made radioactive by irradiating it with gamma rays. The two major concerns about food irradiation—whether a food's nutritional value might be significantly diminished and whether decomposition products might cause cancer—appear to be without foundation. Irradiation of foods with gamma rays may soon join pasteurization as a way to protect our food supply from harmful bacteria.

Figure 7.7 ▲

Gamma irradiation of foods Here, fresh produce has been irradiated to reduce spoilage and to ensure that the produce does not contain harmful bacteria. Now, meat can also be irradiated.

■ See Problems 7.91 and 7.92.

(continued)

Hence, the energy of one photon is

$$E = h\nu = 6.63 \times 10^{-34}\,\text{J}\cdot\cancel{\text{s}} \times 4.47 \times 10^{14}/\cancel{\text{s}} = \mathbf{2.96 \times 10^{-19}\,J}$$

Answer Check Be sure to use the same system of units for the wavelength, the speed of light, and Planck's constant. Check that units cancel to give the proper units for energy in the final answer.

Exercise 7.3 The following are representative wavelengths in the infrared, ultraviolet, and x-ray regions of the electromagnetic spectrum, respectively: 1.0×10^{-6} m, 1.0×10^{-8} m, and 1.0×10^{-10} m. What is the energy of a photon of each radiation? Which has the greatest amount of energy per photon? Which has the least?

■ See Problems 7.43, 7.44, 7.45, and 7.46.

Figure 7.8 ▲

Niels Bohr (1885–1962) After Bohr developed his quantum theory of the hydrogen atom, he used his ideas to explain the periodic behavior of the elements. Later, when the new quantum mechanics was discovered by Schrödinger and Heisenberg, Bohr spent much of his time developing its philosophical basis. He received the Nobel Prize in physics in 1922.

7.3 The Bohr Theory of the Hydrogen Atom

According to Rutherford's nuclear model, the atom consists of a nucleus with most of the mass of the atom and a positive charge, around which move enough electrons to make the atom electrically neutral. But this model, offered in 1911, posed a dilemma. Using the then-current theory, one could show that an electrically charged particle (such as an electron) that revolves around a center would continuously lose energy as electromagnetic radiation. As an electron in an atom lost energy, it would spiral into the nucleus (in about 10^{-10} s, according to available theory). The stability of the atom could not be explained.

A solution to this theoretical dilemma was found in 1913 by Niels Bohr (1885–1962), a Danish physicist, who at the time was working with Rutherford (see Figure 7.8). Using the work of Planck and Einstein, Bohr applied a new theory to the simplest atom, hydrogen. Before we look at Bohr's theory, we need to consider the line spectra of atoms, which we looked at briefly in the chapter opening.

Atomic Line Spectra

As described in the previous section, a heated solid emits light. A heated tungsten filament in an ordinary lightbulb is a typical example. With a prism we can spread out the light from a bulb to give a **continuous spectrum**—that is, *a spectrum containing light of all wavelengths,* like that of a rainbow (see Figure 7.9). The light emitted by a heated gas, however, yields different results. Rather than a continuous spectrum, with all colors of the rainbow, we obtain a **line spectrum**—*a spectrum showing only certain colors or specific wavelengths of light.* When the light from a hydrogen gas discharge tube (in which an electrical discharge heats hydrogen gas) is separated into its components by a prism, it gives a spectrum of lines, each line corresponding to light of a given wavelength. The light produced in the discharge tube is emitted by hydrogen atoms. Figure 7.2 shows the line spectrum of the hydrogen atom, as well as line spectra of other atoms.

The line spectrum of the hydrogen atom is especially simple. In the visible region, it consists of only four lines (a red, a blue-green, a blue, and a violet), although others appear in the infrared and ultraviolet regions. In 1885 J. J. Balmer showed that the wavelengths λ in the visible spectrum of hydrogen could be reproduced by a simple formula:

$$\frac{1}{\lambda} = 1.097 \times 10^7/\text{m}\left(\frac{1}{2^2} - \frac{1}{n^2}\right)$$

Figure 7.9 ▶

Dispersion of white light by a prism White light, entering at the left, strikes a prism, which disperses the light into a continuous spectrum of wavelengths.

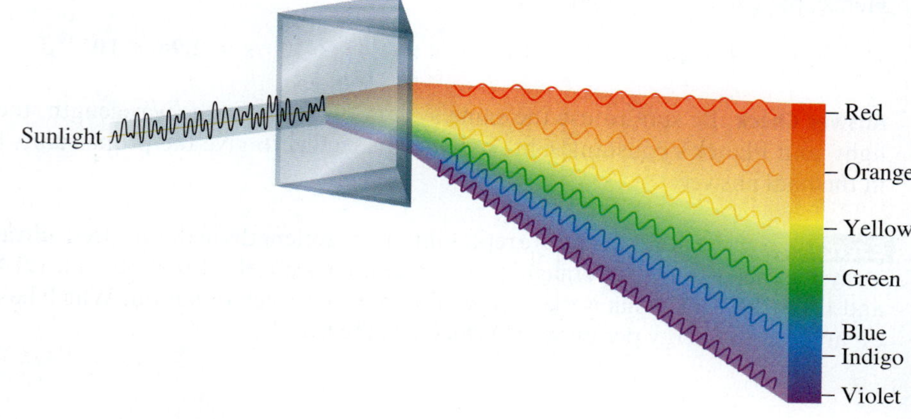

Here n is some whole number (integer) greater than 2. By substituting $n = 3$, for example, and calculating $1/\lambda$ and then λ, one finds $\lambda = 6.56 \times 10^{-7}$ m, or 656 nm, a wavelength corresponding to red light. The wavelengths of the other lines in the hydrogen atom visible spectrum are obtained by successively substituting $n = 4$, $n = 5$, and $n = 6$.

Bohr's Postulates

Bohr set down the following postulates to account for (1) the stability of the hydrogen atom (that the atom exists and its electron does not continuously radiate energy and spiral into the nucleus) and (2) the line spectrum of the atom.

1. Energy-level Postulate An electron can have only *specific energy values in an atom,* which are called its **energy levels.** Therefore, the atom itself can have only specific total energy values.

Bohr borrowed the idea of quantization of energy from Planck. Bohr, however, devised a rule for this quantization that could be applied to the motion of an electron in an atom. From this he derived the following formula for the energy levels of the electron in the hydrogen atom:

The energies have negative values because the energy of the separated nucleus and electron is taken to be zero. As the nucleus and electron come together to form a stable state of the atom, energy is released and the energy becomes less than zero, or negative.

$$E = -\frac{R_H}{n^2} \quad n = 1, 2, 3, \ldots \infty \quad \text{(for H atom)}$$

where R_H is the *Rydberg constant* (expressed in energy units) with the value 2.179×10^{-18} J. Different values of the possible energies of the electron are obtained by putting in different values of n, which can have only the integral values 1, 2, 3, and so forth (up to infinity, ∞). Here n is called the *principal quantum number*. The diagram in Figure 7.10 shows the energy levels of the electron in the H atom. ▶

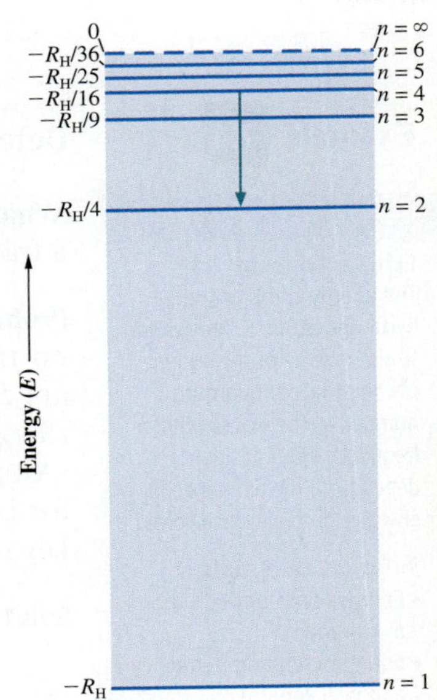

Figure 7.10 ▲

Energy-level diagram for the electron in the hydrogen atom Energy is plotted on the vertical axis (in fractional multiples of R_H). The arrow represents an electron transition (discussed in Postulate 2) from level $n = 4$ to level $n = 2$. Light of wavelength 486 nm (blue-green) is emitted. (See Example 7.4 for the calculation of this wavelength.)

2. Transitions Between Energy Levels An electron in an atom can change energy only by going from one energy level to another energy level. By so doing, the electron undergoes a *transition.*

By using only these two postulates, Bohr was able to explain Balmer's formula for the wavelengths in the spectrum of the hydrogen atom. According to Bohr, the emission of light from an atom occurs as follows. An electron in a higher energy level (initial energy level, E_i) undergoes a transition to a lower energy level (final energy level, E_f) (see Figure 7.10). In this process, the electron loses energy, which is emitted as a photon. Balmer's formula follows by this reasoning: The energy lost by the hydrogen atom is $\Delta E = E_f - E_i$. If we write n_i for the principal quantum number of the initial energy level, and n_f for the principal quantum number of the final energy level, then from Postulate 1,

$$E_i = -\frac{R_H}{n_i^2} \quad \text{and} \quad E_f = -\frac{R_H}{n_f^2}$$

so

$$\Delta E = \left(-\frac{R_H}{n_f^2}\right) - \left(-\frac{R_H}{n_i^2}\right) = -R_H\left(\frac{1}{n_f^2} - \frac{1}{n_i^2}\right)$$

For example, if the electron undergoes a transition from $n_i = 4$ to $n_f = 2$,

$$\Delta E = -2.179 \times 10^{-18}\,\text{J}\left(\frac{1}{2^2} - \frac{1}{4^2}\right) = -4.086 \times 10^{-19}\,\text{J}$$

The sign of the energy change is negative. This means that energy equal to 4.086×10^{-19} J is lost by the atom in the form of a photon.

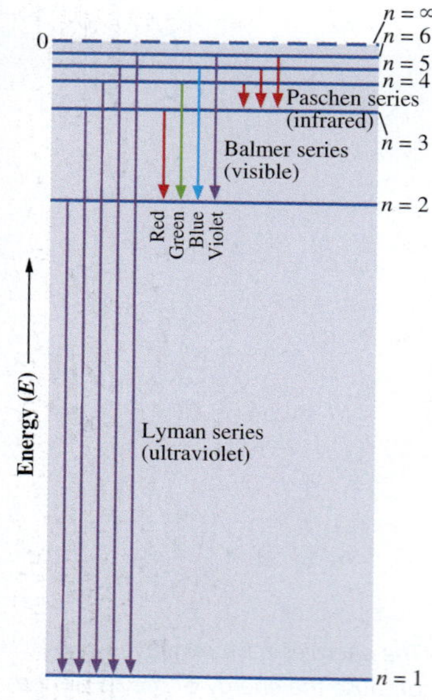

Figure 7.11

Transitions of the electron in the hydrogen atom The diagram shows the Lyman, Balmer, and Paschen series of transitions that occur for $n_f = 1$, 2, and 3, respectively.

In general, the energy of the emitted photon, $h\nu$, equals the positive energy lost by the atom ($-\Delta E$):

$$\text{Energy of emitted photon} = h\nu = -\Delta E = -(E_f - E_i)$$

That is,

$$h\nu = R_H\left(\frac{1}{n_f^2} - \frac{1}{n_i^2}\right)$$

Recalling that $\nu = c/\lambda$, you can rewrite this as

$$\frac{1}{\lambda} = \frac{R_H}{hc}\left(\frac{1}{n_f^2} - \frac{1}{n_i^2}\right)$$

By substituting $R_H = 2.179 \times 10^{-18}$ J, $h = 6.626 \times 10^{-34}$ J·s, and $c = 2.998 \times 10^8$ m/s, you find that $R_H/hc = 1.097 \times 10^7$/m, which is the constant given in the Balmer formula. In Balmer's formula, the quantum number n_f is 2. This means that Balmer's formula gives wavelengths that occur when electrons in H atoms undergo transitions from energy levels $n_i > 2$ to level $n_f = 2$. If you change n_f to other integers, you obtain different series of lines (or wavelengths) for the spectrum of the H atom (see Figure 7.11).

Example 7.4 Determining the Wavelength or Frequency of a Hydrogen Atom Transition

Gaining Mastery Toolbox

Critical Concept 7.4
The energy levels of the hydrogen atom, E, are related to the inverse of the square of the principal quantum number, n. The wavelength or frequency of a transition depends on the difference in energies of the levels involved.

Solution Essentials:
- Energy levels of the H atom: $E = -R_H/n^2$
- Equation relating frequency and wavelength: $c = \nu\lambda$

What is the wavelength of light emitted when the electron in a hydrogen atom undergoes a transition from energy level $n = 4$ to level $n = 2$?

Problem Strategy Remember that the wavelength or frequency of a transition depends on the difference in energies of the levels involved. For an H atom, the energy levels are $E = -R_H/n^2$. You calculate the difference in energy for the two levels *to obtain the energy of the photon.* Then, calculate the frequency and wavelength of the emitted light. (Although you could do this problem by "plugging into" the previously derived equation for $1/\lambda$, the method followed here requires you to remember only a minimum number of key formulas.)

Solution From the formula for the energy levels, you know that

$$E_i = \frac{-R_H}{4^2} = \frac{-R_H}{16} \quad \text{and} \quad E_f = \frac{-R_H}{2^2} = \frac{-R_H}{4}$$

You subtract the lower value from the higher value, to get a positive result. (The energy of the photon is positive. If your result is negative, it means you calculated $E_f - E_i$, rather than $E_i - E_f$. Simply reverse the subtraction to obtain a positive result.) Because this result equals the energy of the photon, you equate it to $h\nu$:

$$\left(\frac{-R_H}{16}\right) - \left(\frac{-R_H}{4}\right) = \frac{-4R_H + 16R_H}{64} = \frac{-R_H + 4R_H}{16} = \frac{3R_H}{16} = h\nu$$

The frequency of the light emitted is

$$\nu = \frac{3R_H}{16h} = \frac{3}{16} \times \frac{2.179 \times 10^{-18}\ \cancel{J}}{6.626 \times 10^{-34}\ \cancel{J}\cdot s} = 6.17 \times 10^{14}/s$$

(continued)

(continued)

Since $\lambda = c/\nu$,

$$\lambda = \frac{3.00 \times 10^8 \text{ m/}\cancel{s}}{6.17 \times 10^{14}/\cancel{s}} = 4.86 \times 10^{-7} \text{ m, or } \mathbf{486 \text{ nm}}$$

The color is blue-green (see Figure 7.10).

Answer Check Balmer lines (lines ending with $n = 2$) are in the visible region of the spectrum. Wavelengths of visible light are in the range of 400 nm to about 800 nm. Note that the answer lies in this range.

Exercise 7.4 Calculate the wavelength of light emitted from the hydrogen atom when the electron undergoes a transition from level $n = 3$ to level $n = 1$.

■ See Problems 7.49, 7.50, 7.51, and 7.52.

According to Bohr's theory, the *emission* of light from an atom occurs when an electron undergoes a transition from an upper energy level to a lower one. To complete the explanation, we need to describe how the electron gets into the upper level prior to emission. Normally, the electron in a hydrogen atom exists in its lowest, or $n = 1$, level. To get into a higher energy level, the electron must gain energy, or be *excited*. One way this can happen is through the collision of two hydrogen atoms. During this collision, some of the kinetic energy of one atom can be gained by the electron of another atom, thereby boosting, or exciting, the electron from the $n = 1$ level to a higher energy level. The excitation of atoms and the subsequent emission of light are most likely to occur in a hot gas, where atoms have large kinetic energies.

Bohr's theory explains not only the emission but also the *absorption* of light. When an electron in the hydrogen atom undergoes a transition from $n = 3$ to $n = 2$, a photon of red light (wavelength 656 nm) is emitted. When red light of wavelength 656 nm shines on a hydrogen atom in the $n = 2$ level, a photon can be absorbed. If the photon is absorbed, the energy is gained by the electron, which undergoes a transition to the $n = 3$ level. (This is the reverse of the emission process we just discussed.) Materials that have a color, such as dyed textiles and painted walls, appear colored because of the absorption of light. For example, when white light falls on a substance that absorbs red light, the color components that are not absorbed, the yellow and blue light, are reflected. The substance appears blue-green.

Postulates 1 and 2 hold for atoms other than hydrogen, except that the energy levels cannot be obtained by a simple formula. However, if you know the wavelength of the emitted light, you can relate it to ν and then to the difference in energy levels of the atom. The energy levels of atoms have been experimentally determined in this way.

Exercise 7.5 What is the difference in energy levels of the sodium atom if emitted light has a wavelength of 589 nm?

■ See Problems 7.55 and 7.56.

CONCEPT CHECK 7.2

An atom has a line spectrum consisting of a red line and a blue line. Assume that each line corresponds to a transition between two adjacent energy levels. Sketch an energy-level diagram with three energy levels that might explain this line spectrum, indicating the transitions on this diagram. Consider the transition from the highest energy level on this diagram to the lowest energy level. How would you describe the color or region of the spectrum corresponding to this transition?

A **CHEMIST** Looks at...

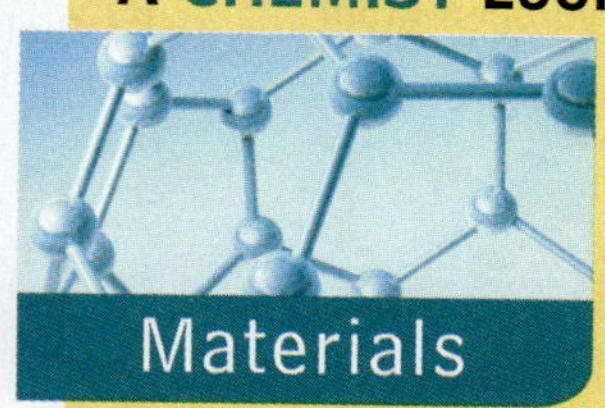

Lasers and CD and DVD Players

Lasers are sources of intense, highly directed beams of *monochromatic* light—light of very narrow wavelength range. The word *laser* is an acronym meaning **l**ight **a**mplification by **s**timulated **e**mission of **r**adiation. Many different kinds of lasers now exist, but the general principle of a laser can be understood by looking at the ruby laser, the first type constructed (in 1960).

Ruby is aluminum oxide containing a small concentration of chromium(III) ions, Cr^{3+}, in place of some aluminum ions. The electron transitions in a ruby laser are those of Cr^{3+} ions. Figure 7.12 shows an energy-level diagram of this ion in ruby. When you shine light of wavelength 545 nm on a ruby crystal, the light is absorbed and Cr^{3+} undergoes a transition from level 1 to level 3. Most of the ions in level 3 then undergo *radiationless transitions* to level 2. (In these transitions, the ions lose energy as heat to the crystal, rather than by emitting photons.) When an intense green light at 545 nm is flashed on a ruby crystal, Cr^{3+} ions of the ruby end up in level 2. Like all excited states, level 2 spontaneously emits photons, going to the ground state. But whereas most excited states quickly (within 10^{-8} s) emit photons, level 2 has a much longer lifetime (a fraction of a millisecond), resulting in a buildup of excited Cr^{3+} ions in this level. If these accumulated excited ions can be triggered to emit all at once, or nearly so, an intense emission of monochromatic light at 694 nm will occur.

The process of *stimulated emission* is ideal for this triggering. When a photon corresponding to 694 nm encounters a Cr^{3+} ion in level 2, it stimulates the ion to undergo the transition from level 2 to level 1. The ion emits a photon corresponding to exactly the same wavelength as the original photon. In place of just one photon, there are now two photons, the original one and the one obtained by stimulated emission. The net effect is to increase the intensity of the light at this wavelength. Thus, a weak light at 694 nm can be amplified by stimulated emission of the excited ruby.

Figure 7.13 shows a sketch of a ruby laser. A flash lamp emits green light (at 545 nm) that is absorbed by the ruby, building up a concentration of Cr^{3+} in level 2. A few of these excited ions spontaneously emit photons corresponding to 694 nm (red light), and these photons then stimulate other ions to emit, which in turn stimulate more ions to emit, and so forth, resulting in a pulse of laser light at 694 nm.

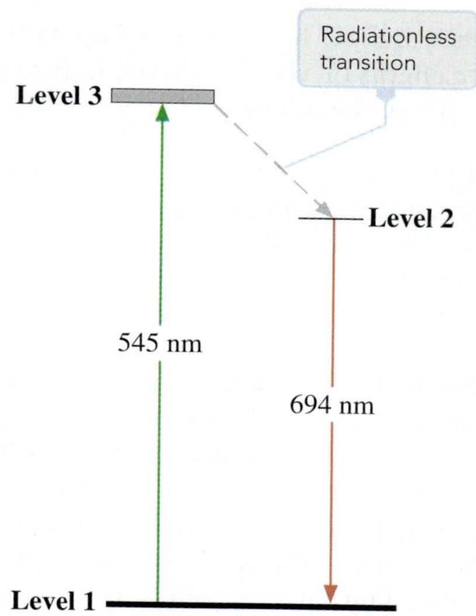

Figure 7.12 ▲

Energy levels of the chromium(III) ion in ruby The energy levels have been numbered by increasing energy from 1 to 3. (Level 3 is broadened in the solid state.) When chromium(III) ions absorb the 545-nm portion of the light from a flash lamp, electrons in the ions first undergo transitions to level 3 and then radiationless transitions to level 2. (The energy is lost as heat.) Electrons accumulate in level 2 until more exist in level 2 than in level 1. Then a photon from a spontaneous emission from level 2 to level 1 can stimulate the emission from other atoms in level 2. These photons move back and forth between reflective surfaces, stimulating additional emissions and forming a laser pulse at 694 nm.

Laser light is *coherent*. This means that the waves forming the beam are all in phase; that is, the waves have their maxima and minima at the same points in space and time. The property of coherence of a laser beam is used in CDs (compact discs) and DVDs. Music, video, or other information is encoded on the discs in the form of pits, or indentations, on a spiral track, starting at the center of the disc. Figure 7.14 shows how a CD works; DVDs work similarly but with shorter wavelengths of laser light.

When the disc is played, a small laser beam scans the track and is reflected back to a detector. Light reflected from an indentation is out of phase with light from the laser and interferes with it. Because of this interference, the reflected beam is diminished in intensity and gives a diminished detector signal. Fluctuations in the signal are then converted to sound, video, or other information. The difference between CDs and DVDs is in the wavelength of laser light used. A shorter wavelength allows the indentations on a disc to be smaller so that more information can be placed on a disc of the same size. A CD disc uses laser light at 780 nm (in the near infrared), whereas a DVD disc uses light at 650 nm (red). A Blu-ray disc uses laser light at 405 nm, which is in the violet (although commonly called a "blue" laser).

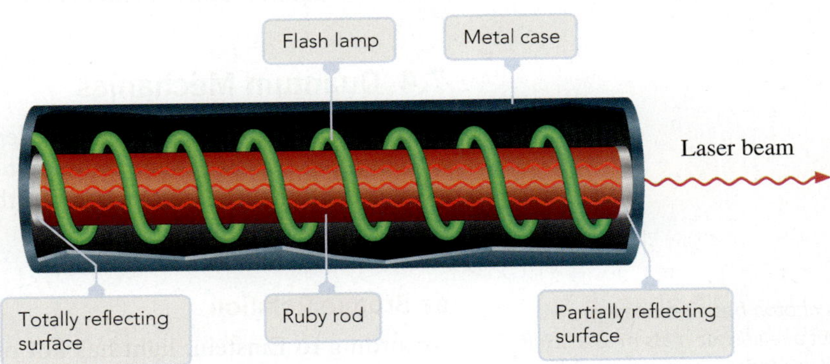

Figure 7.13

A ruby laser A flash lamp encircles a ruby rod. Light of 545 nm (green) from the flash "pumps" electrons from level 1 to level 3, then level 2, from which stimulated emission forms a laser pulse at 694 nm (red). The stimulated emission bounces back and forth between reflective surfaces at the ends of the ruby rod, building up a coherent laser beam. One end has a partially reflective surface to allow the laser beam to exit from the ruby.

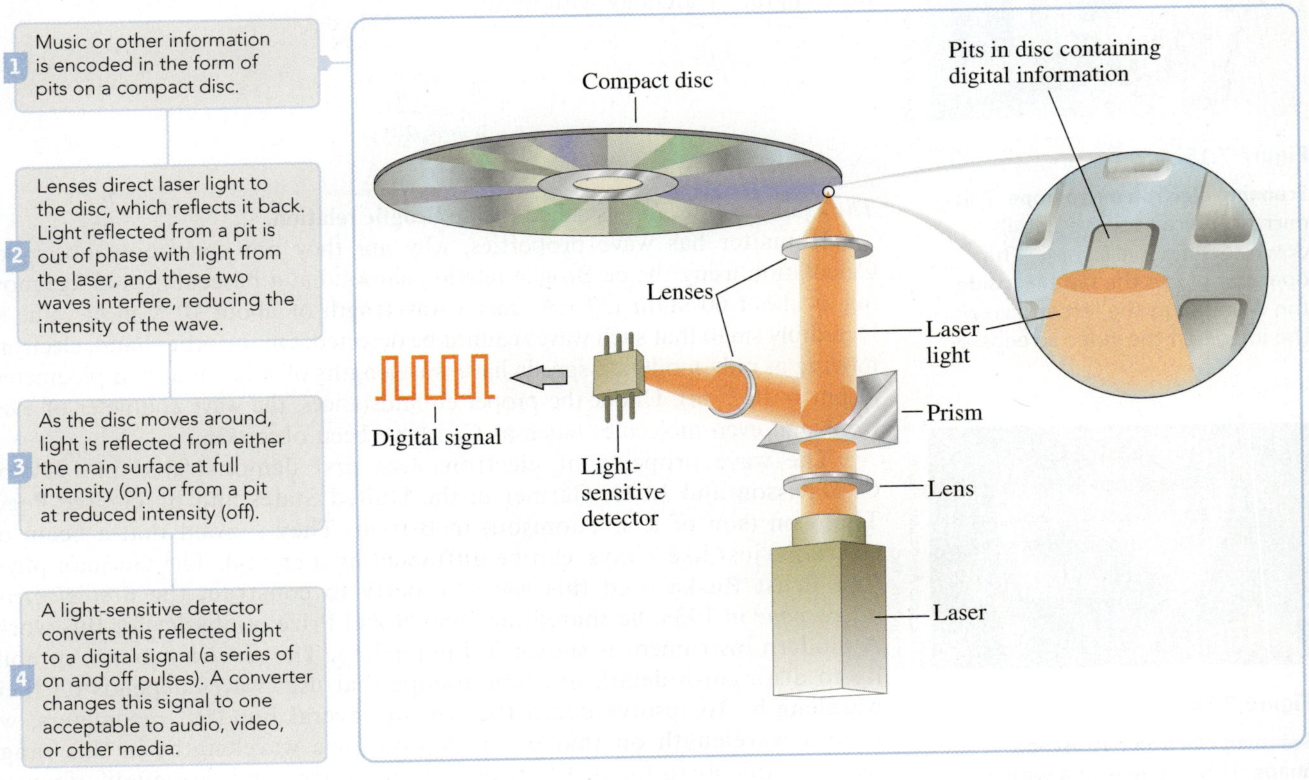

1 Music or other information is encoded in the form of pits on a compact disc.

2 Lenses direct laser light to the disc, which reflects it back. Light reflected from a pit is out of phase with light from the laser, and these two waves interfere, reducing the intensity of the wave.

3 As the disc moves around, light is reflected from either the main surface at full intensity (on) or from a pit at reduced intensity (off).

4 A light-sensitive detector converts this reflected light to a digital signal (a series of on and off pulses). A converter changes this signal to one acceptable to audio, video, or other media.

Compact disc
Pits in disc containing digital information
Lenses
Laser light
Digital signal
Prism
Light-sensitive detector
Lens
Laser

Figure 7.14

How a compact disc player works

See Problems 7.93 and 7.94.

Quantum Mechanics and Quantum Numbers

Bohr's theory firmly established the concept of atomic energy levels. It was unsuccessful, however, in accounting for the details of atomic structure and in predicting energy levels for atoms other than hydrogen. Further understanding of atomic structure required other theoretical developments.

7.4 Quantum Mechanics

Current ideas about atomic structure depend on the principles of *quantum mechanics,* a theory that applies to submicroscopic (that is, extremely small) particles of matter, such as electrons. The development of this theory was stimulated by the discovery of the de Broglie relation.

de Broglie Relation

According to Einstein, light has not only wave properties, which we characterize by frequency and wavelength, but also particle properties. For example, a particle of light, the photon, has a definite energy $E = h\nu$. One can also show that the photon has momentum. (The momentum of a particle is the product of its mass and speed.) This momentum, mc, is related to the wavelength of the light: $mc = h/\lambda$ or $\lambda = h/mc$. ◀

A photon has a rest mass of zero but a relativistic mass m as a result of its motion. Einstein's equation $E = mc^2$ relates this relativistic mass to the energy of the photon, which also equals $h\nu$. Therefore, $mc^2 = h\nu$, or $mc = h\nu/c = h/\lambda$.

In 1923 the French physicist Louis de Broglie reasoned that if light (considered as a wave) exhibits particle aspects, then perhaps particles of matter show characteristics of waves under the proper circumstances. He therefore postulated that a particle of matter of mass m and speed v has an associated wavelength, by analogy with light:

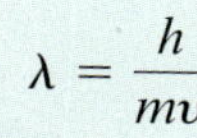

$$\lambda = \frac{h}{mv}$$

The equation $\lambda = h/mv$ is called the **de Broglie relation.**

If matter has wave properties, why are they not commonly observed? Calculation using the de Broglie relation shows that a baseball (0.145 kg) moving at about 60 mi/hr (27 m/s) has a wavelength of about 10^{-34} m, a value so incredibly small that such waves cannot be detected. On the other hand, electrons moving at only moderate speeds have wavelengths of a few hundred picometers (1 pm = 10^{-12} m). Under the proper circumstances, the wave character of electrons and even molecules (such as C_{60}) have been observed.

The wave property of electrons was first demonstrated in 1927 by C. Davisson and L. H. Germer in the United States and by George Paget Thomson (son of J. J. Thomson) in Britain. They showed that a beam of electrons, just like x rays, can be diffracted by a crystal. The German physicist Ernst Ruska used this wave property to construct the first *electron microscope* in 1933; he shared the 1986 Nobel Prize in physics for this work. A modern instrument is shown in Figure 7.15. The *resolving power,* or ability to distinguish detail, of a microscope that uses waves depends on their wavelength. To resolve detail the size of several hundred picometers, we need a wavelength on that order. X rays have wavelengths in this range but are difficult to focus. Electrons, on the other hand, are readily focused with electric and magnetic fields. Figure 7.16 shows a photograph taken with an electron microscope, displaying the detail that is possible with this instrument.

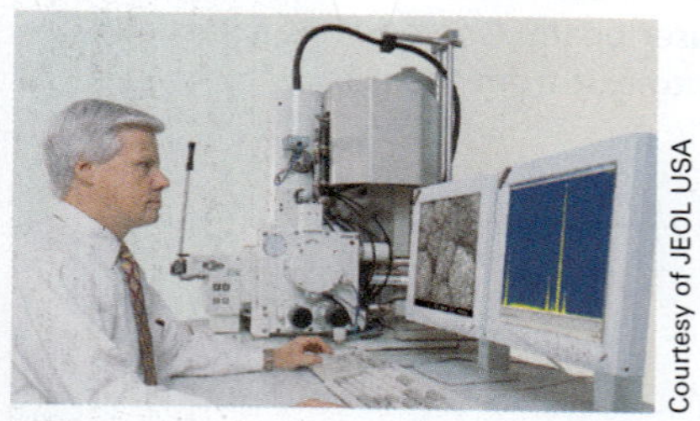

Figure 7.15 ▲

Scanning electron microscope This microscope can resolve details down to 3 nm in a sample. The operator places the sample inside the chamber at the left and views the image on the video screens.

Figure 7.16 ▲

Scanning electron microscope image This image is of a wasp's head. Color has been added by computer for contrast in discerning different parts of the image.

Example 7.5 Calculate the Wavelength of a Moving Particle

Gaining Mastery Toolbox

Critical Concept 7.5
A particle has an associated wavelength, λ, that depends inversely on the mass, m, and speed of the particle, v.

Solution Essentials:
- DeBroglie relation: $\lambda = h/mv$

a Calculate the wavelength (in meters) of the wave associated with a 1.00-kg mass moving at 1.00 km/hr. **b** What is the wavelength (in picometers) associated with an electron, whose mass is 9.11×10^{-31} kg, traveling at a speed of 4.19×10^6 m/s? (This speed can be attained by an electron accelerated between two charged plates differing by 50.0 volts; voltages in the kilovolt range are used in electron microscopes.)

Problem Strategy The questions ask for the wavelength associated with a moving mass, which is given by de Broglie's relation, $\lambda = h/mv$. Be careful to use consistent units in substituting into this equation. If you use SI units, then in part a you will need to convert km/hr to m/s. In part b, you will need to convert the answer in meters to picometers.

Solution

a A speed v of 1.00 km/hr equals

$$1.00 \frac{\text{km}}{\text{hr}} \times \frac{1 \text{hr}}{3600 \text{ s}} \times \frac{10^3 \text{ m}}{1 \text{ km}} = 0.278 \text{ m/s}$$

Substituting quantities (all expressed in SI units for consistency), you get

$$\lambda = \frac{h}{mv} = \frac{6.63 \times 10^{-34} \text{ J}\cdot\text{s}}{1.00 \text{ kg} \times 0.278 \text{ m/s}} = \mathbf{2.38 \times 10^{-33} \text{ m}}$$

b $$\lambda = \frac{h}{mv} = \frac{6.63 \times 10^{-34} \text{ J}\cdot\text{s}}{9.11 \times 10^{-31} \text{ kg} \times 4.19 \times 10^6 \text{ m/s}} = 1.74 \times 10^{-10} \text{ m} = \mathbf{174 \text{ pm}}$$

Answer Check Make sure that you are using the correct equation for the de Broglie relation; units should cancel to give units of length (for wavelength). Check that the final units are those asked for in the problem statement.

Exercise 7.6 Calculate the wavelength (in picometers) associated with an electron traveling at a speed of 2.19×10^6 m/s.

■ See Problems 7.57 and 7.58.

CONCEPT CHECK 7.3

A proton is approximately 2000 times heavier than an electron. How would the speeds of these particles compare if their corresponding wavelengths were about equal?

Wave Functions

De Broglie's relation applies quantitatively only to particles in a force-free environment. It cannot be applied directly to an electron in an atom, where the electron is subject to the attractive force of the nucleus. But in 1926 Erwin Schrödinger, guided by de Broglie's work, devised a theory that could be used to find the wave properties of electrons in atoms and molecules. *The branch of physics that mathematically describes the wave properties of submicroscopic particles* is called **quantum mechanics** or **wave mechanics.** ▶

Schrödinger received the Nobel Prize in physics in 1933 for his wave formulation of quantum mechanics. Werner Heisenberg won the Nobel Prize the previous year for his matrix-algebra formulation of quantum mechanics. The two formulations yield identical results.

Without going into the mathematics of quantum mechanics here, we will discuss some of the most important conclusions of the theory. In particular, quantum mechanics alters the way we think about the motion of particles.

Our usual concept of motion comes from what we see in the everyday world. We might, for instance, visually follow a ball that has been thrown. The path of the ball is given by its position and velocity (or momentum) at various times. We are therefore conditioned to think in terms of a continuous path for moving objects. In Bohr's theory, the electron was thought of as moving about, or orbiting, the nucleus in the way the earth orbits the sun. Quantum mechanics vastly changes this view of motion. We can no longer think of an electron as having a precise orbit in an atom. To describe such an orbit, we would have to know the exact position of the electron at various times and exactly how long it would take it to travel to a nearby position in the orbit. That is, at any moment we would have to know not only the precise position but also the precise momentum (mass times speed) of the electron.

In 1927 Werner Heisenberg showed from quantum mechanics that it is impossible to know simultaneously, with absolute precision, both the position and the momentum of a particle such as an electron. Heisenberg's **uncertainty principle** is *a relation that states that the product of the uncertainty in position and the uncertainty in momentum of a particle can be no smaller than Planck's constant divided by* 4π. Thus, letting Δx be the uncertainty in the x coordinate of the particle and letting Δp_x be the uncertainty in the momentum in the x direction, we have

$$(\Delta x)(\Delta p_x) \geq \frac{h}{4\pi}$$

There are similar relations in the y and z directions. The uncertainty principle says that the more precisely you know the position (the smaller Δx, Δy, and Δz), the less well you know the momentum of the particle (the larger Δp_x, Δp_y, and Δp_z). In other words, if you know very well where a particle is, you cannot know where it is going!

The uncertainty principle is significant only for particles of very small mass such as electrons. You can see this by noting that the momentum equals mass times velocity, so $p_x = mv_x$. The preceding relation becomes

$$(\Delta x)(\Delta v_x) \geq \frac{h}{4\pi m}$$

For dust particles and baseballs, where m is relatively large, the term on the right becomes nearly zero, and the uncertainties of position and velocity are quite small. The path of a baseball has meaning. For electrons, however, the uncertainties in position and momentum are normally quite large relative to atomic scale dimensions. We cannot describe the electron in an atom as moving in a definite orbit.

Although quantum mechanics does not allow us to describe the electron in the hydrogen atom as moving in an orbit, it does allow us to make *statistical* statements about where we would find the electron if we were to look for it. For example, we can obtain the *probability* of finding an electron at a certain point in a hydrogen atom. Although we cannot say that an electron will definitely be at a particular position at a given time, we can say that the electron is likely (or not likely) to be at this position.

Information about a particle in a given energy level (such as an electron in an atom) is contained in a mathematical expression called a *wave function,* denoted by the Greek letter psi, ψ. The wave function is obtained by solving an equation of quantum mechanics (Schrödinger's equation). Its square, ψ^2, gives the probability of finding the particle within a region of space.

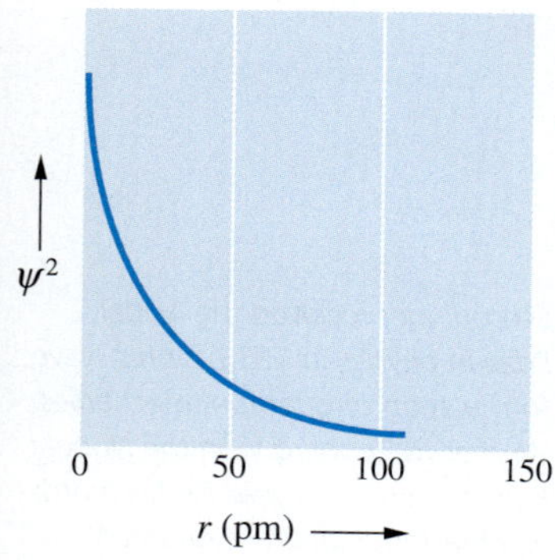

Figure 7.17

Plot of ψ^2 for the lowest energy level of the hydrogen atom The square of the wave function is plotted versus the distance, r, from the nucleus.

The wave function and its square, ψ^2, have values for all locations about a nucleus. Figure 7.17 shows values of ψ^2 for the electron in the lowest energy level of the hydrogen atom along a line starting from the nucleus. Note that ψ^2 is large near the nucleus ($r = 0$), indicating that the electron is most likely to be found in this region. The value of ψ^2 decreases rapidly as the distance from the nucleus increases, but ψ^2 never goes to exactly zero, although the probability does become extremely small at large distances from the nucleus. This means that an atom does not have a definite boundary, unlike in the Bohr model of the atom.

Figure 7.18 shows another view of this electron probability. The graph plots the probability of finding the electron in different spherical shells at particular

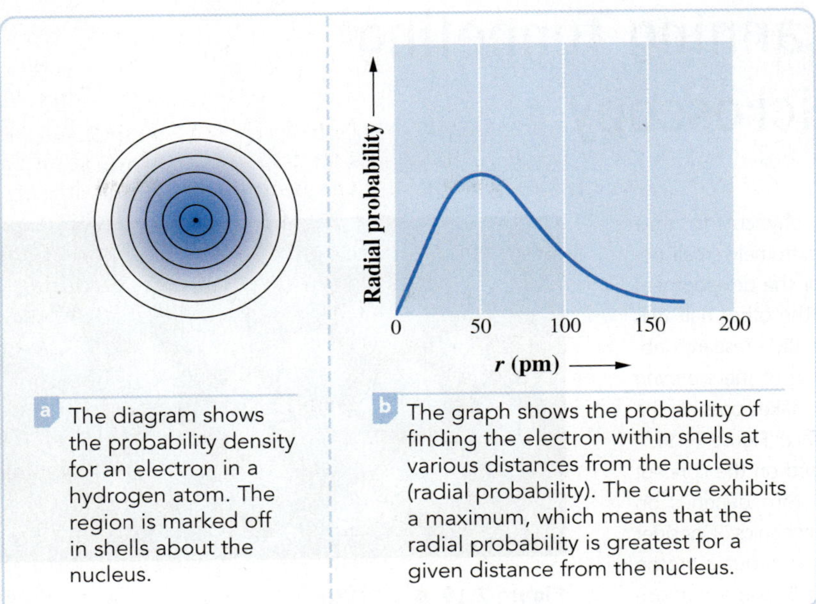

Figure 7.18 Probability of finding an electron in a spherical shell about the nucleus

distances from the nucleus, rather than the probability at a point. Even though the probability of finding the electron at a point near the nucleus is high, the volume of any shell there is small. Therefore, the probability of finding the electron *within a shell* is greatest at some distance from the nucleus. This distance just happens to equal the radius that Bohr calculated for an electron orbit in his model.

7.5 Quantum Numbers and Atomic Orbitals

According to quantum mechanics, each electron in an atom is described by four different quantum numbers, three of which (n, l, and m_l) specify the wave function that gives the probability of finding the electron at various points in space. *A wave function for an electron in an atom* is called an **atomic orbital.** An atomic orbital is pictured qualitatively by describing the region of space where there is high probability of finding the electrons. The atomic orbital so pictured has a definite shape. A fourth quantum number (m_s) refers to a magnetic property of electrons called *spin*. We first look at quantum numbers, then at atomic orbitals.

Three different quantum numbers are needed because there are three dimensions to space.

Quantum Numbers

The allowed values and general meaning of each of the four quantum numbers of an electron in an atom are as follows:

1. Principal Quantum Number (n) *This quantum number is the one on which the energy of an electron in an atom principally depends; it can have any positive value: 1, 2, 3, and so on.* The energy of an electron in an atom depends *principally* on n. The smaller n is, the lower the energy. In the case of the hydrogen atom or single-electron atomic ions, such as Li^{2+} and He^{+}, n is the only quantum number determining the energy (which is given by Bohr's formula, discussed in Section 7.3). For other atoms, the energy also depends to a slight extent on the l quantum number.

The *size* of an orbital also depends on n. The larger the value of n is, the larger the orbital. Orbitals of the same quantum state n are said to belong to the same *shell*. Shells are sometimes designated by the following letters:

Letter	*K*	*L*	*M*	*N* . . .
n	1	2	3	4 . . .

2. Angular Momentum Quantum Number (l) (Also Called *Azimuthal Quantum Number*) *This quantum number distinguishes orbitals of given n having different shapes; it can have any integer value from 0 to n − 1.* Within each shell of quantum number n, there

Instrumental Methods

Scanning Tunneling Microscopy

The 1986 Nobel Prize in physics went to three physicists for their work in developing microscopes for viewing extremely small objects. Half of the prize went to Ernst Ruska for the development of the electron microscope, described earlier. The other half was awarded to Gerd Binnig and Heinrich Rohrer, at IBM's research laboratory in Zurich, Switzerland, for their invention of the *scanning tunneling microscope* in 1981. This instrument makes possible the viewing of atoms and molecules on a solid surface (Figure 7.19).

The scanning tunneling microscope is based on the concept of quantum mechanical *tunneling,* which in turn depends on the probability interpretation of quantum mechanics. Consider a hydrogen atom, which consists of an electron about a proton (call it A), and imagine another proton (called B) some distance from the first. In classical terms, if we were to move the electron from the region of proton A to the region of proton B, energy would have to be supplied to remove the electron from the attractive field of proton A. Quantum mechanics, however, gives us a different picture. The probability of the electron in the hydrogen atom being at a location far from proton A, say near proton B, is very small but not zero. In effect, this means that the electron, which nominally belongs to proton A, may find itself near proton B without extra energy having been supplied. The electron is said to have *tunneled* from one atom to another.

The scanning tunneling microscope consists of a tungsten metal needle with an extremely fine point (the probe) placed close to the sample to be viewed (Figure 7.20). If the probe is close enough to the sample, electrons can tunnel from the probe to the sample. The probability for this can be increased by having a small voltage applied between the probe and sample. Electrons tunneling from the probe to the sample give rise to a measurable electric current. The magnitude of the current depends on the distance between the probe and the sample (as well as on the wave function of the atom in the sample). By adjusting this distance, the current can be maintained at a fixed value. As the probe scans the sample, it moves toward or away from the sample to maintain a fixed current; in effect, the probe follows the contours of the sample. (The probe is attached to the end of a *piezoelectric* rod, which undergoes very small changes in length when small voltages are applied to it.)

Researchers routinely use the scanning tunneling microscope to study the arrangement of atoms and molecules on surfaces. Lately, scientists have also used the microscope probe to move atoms about a surface in an effort to construct miniature devices.

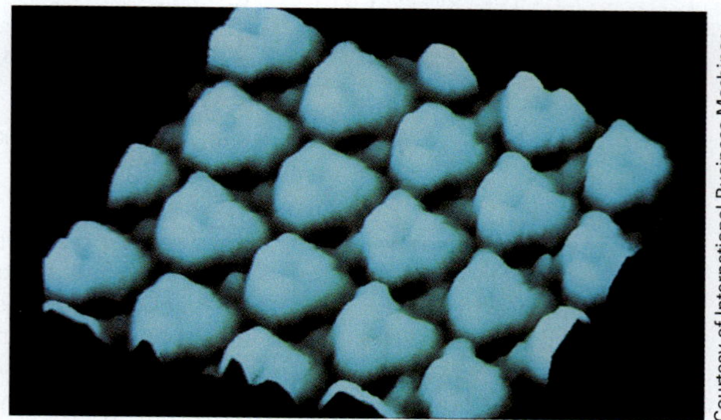

Figure 7.19 ▲

Scanning tunneling microscope image of benzene molecules on a metal surface Benzene molecules, C_6H_6, are arranged in a regular array on a rhodium metal surface.

are n different kinds of orbitals, each with a distinctive shape denoted by an l quantum number. For example, if an electron has a principal quantum number of 3, the possible values for l are 0, 1, and 2. Thus, within the M shell ($n = 3$), there are three kinds of orbitals, each having a different shape for the region where the electron is most likely to be found. These orbital shapes will be discussed later in this section.

Although the energy of an orbital is principally determined by the n quantum number, the energy also depends somewhat on the l quantum number (except for the H atom). For a given n, the energy of an orbital increases with l.

Orbitals of the same n but different l are said to belong to different *subshells* of a given shell. The different subshells are usually denoted by letters as follows:

Letter	*s*	*p*	*d*	*f*	*g* . . .
l	0	1	2	3	4 . . .

The rather odd choice of letter symbols for l quantum numbers survives from old spectroscopic terminology (describing the lines in a spectrum as sharp, principal, diffuse, and fundamental).

To denote a subshell within a particular shell, we write the value of the n quantum number for the shell, followed by the letter designation for the subshell. For example, $2p$ denotes a subshell with quantum numbers $n = 2$ and $l = 1$. ◀

Figure 7.21 shows 48 iron atoms that were arranged in a circle on a copper surface. Each of the iron atoms looks like a sharp mountain peak rising from a plain. Inside this "quantum corral" (as it has come to be called), you can see the wavelike distribution of electrons trapped within. Here is graphic proof of the wavelike nature of electrons!

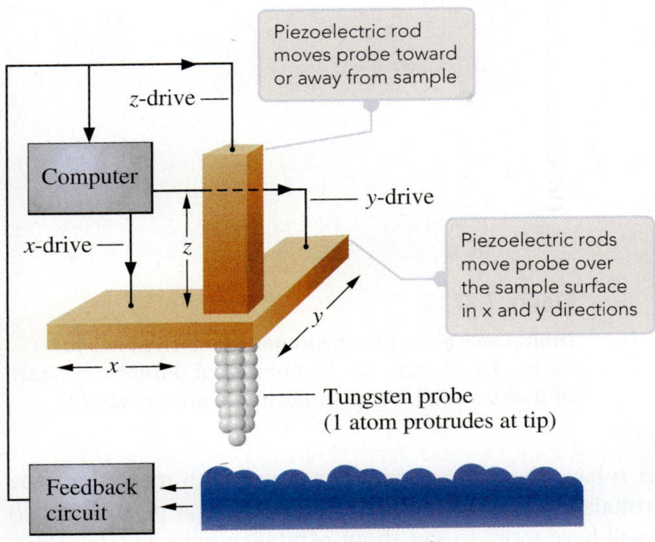

Figure 7.20 ▲

The scanning tunneling microscope A tunneling current flows between the probe and the sample when there is a small voltage between them. A feedback circuit, which provides this voltage, senses the current and varies the voltage on a piezoelectric rod (z-drive) in order to keep the distance constant between the probe and sample. A computer has been programmed to provide voltages to the x-drive and the y-drive to move the probe over the surface of the sample.

Figure 7.21 ▲

Quantum corral IBM scientists used a scanning tunneling microscope probe to arrange 48 iron atoms in a circle on a copper metal surface. The iron atoms appear as the tall peaks in the diagram. Note the wavelike ripple of electrons trapped within the circle of atoms (called *a quantum corral*). [IBM Research Division, Almaden Research Center; research done by Donald M. Eigler and coworkers.]

■ See Problems 7.95 and 7.96.

3. Magnetic Quantum Number (m_l) *This quantum number distinguishes orbitals of given n and l—that is, of given energy and shape but having a different orientation in space; the allowed values are the integers from $-l$ to $+l$.* For $l = 0$ (*s* subshell), the allowed m_l quantum number is 0 only; there is only one orbital in the *s* subshell. For $l = 1$ (*p* subshell), $m_l = -1$, 0, and $+1$; there are three different orbitals in the *p* subshell. The orbitals have the same shape but different orientations in space. In addition, all orbitals of a given subshell have the same energy. Note that there are $2l + 1$ orbitals in each subshell of quantum number l.

4. Spin Quantum Number (m_s) *This quantum number refers to the two possible orientations of the spin axis of an electron; possible values are $+\frac{1}{2}$ and $-\frac{1}{2}$.* An electron acts as though it were spinning on its axis like the earth. Such an electron spin would give rise to a circulating electric charge that would generate a magnetic field. In this way, an electron behaves like a small bar magnet, with a north and a south pole. ▶

Electron spin will be discussed further in Section 8.1.

Table 7.1 lists the permissible quantum numbers for all orbitals through the $n = 4$ shell. These values follow from the rules just given. Energies for the orbitals are shown in Figure 7.22 for the hydrogen atom. Note that all orbitals with the same

Table 7.1 Permissible Values of Quantum Numbers for Atomic Orbitals

n	l	m_l*	Subshell Notation	Number of Orbitals in the Subshell
1	0	0	1*s*	1
2	0	0	2*s*	1
2	1	−1, 0, +1	2*p*	3
3	0	0	3*s*	1
3	1	−1, 0, +1	3*p*	3
3	2	−2, −1, 0, +1, +2	3*d*	5
4	0	0	4*s*	1
4	1	−1, 0, +1	4*p*	3
4	2	−2, −1, 0, +1, +2	4*d*	5
4	3	−3, −2, −1, 0, +1, +2, +3	4*f*	7

*Any one of the m_l quantum numbers may be associated with the n and l quantum numbers on the same line.

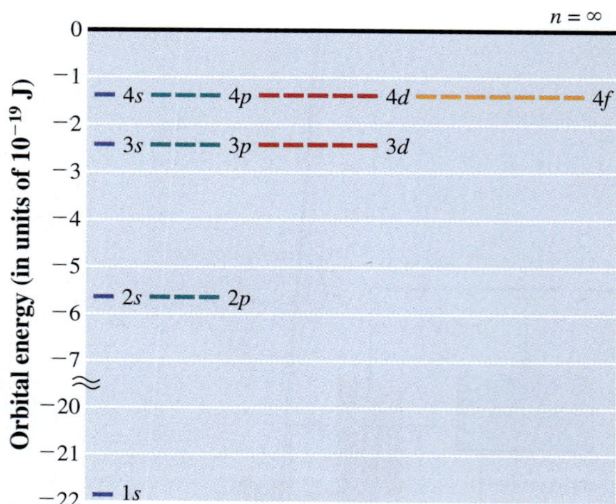

Figure 7.22 ▲

Orbital energies of the hydrogen atom The lines for each subshell indicate the number of different orbitals of that subshell. (Note break in the energy scale.)

principal quantum number n have the same energy. For atoms with more than one electron, however, only orbitals in the same subshell (denoted by a given n and l) have the same energy. We will have more to say about orbital energies in Chapter 8.

Example 7.6 Applying the Rules for Quantum Numbers

Gaining Mastery Toolbox

Critical Concept 7.6
An electron in an atom is described by four quantum numbers: the principal quantum number (n), the angular momentum quantum number (l), the magnetic quantum number (m_l), and the spin quantum number (m_s). You need to know the allowed values of each quantum number.

Solution Essentials:
Allowed values of the quantum numbers of an atom: n, l, m_l, and m_s.

State whether each of the following sets of quantum numbers is permissible for an electron in an atom. If a set is not permissible, explain why.

a $n = 1, l = 1, m_l = 0, m_s = +\frac{1}{2}$
b $n = 3, l = 1, m_l = -2, m_s = -\frac{1}{2}$
c $n = 2, l = 1, m_l = 0, m_s = +\frac{1}{2}$
d $n = 2, l = 0, m_l = 0, m_s = 1$

Problem Strategy Apply the rules for quantum numbers in order, first to n, then to l and m_l, and finally to m_s. A set of quantum numbers is impermissible if it disobeys any rule.

Solution

a **Not permissible.** The l quantum number is equal to n; it must be less than n.
b **Not permissible.** The magnitude of the m_l quantum number (that is, the m_l value, ignoring its sign) must not be greater than l.
c **Permissible.**
d **Not permissible.** The m_s quantum number can be only $+\frac{1}{2}$ or $-\frac{1}{2}$.

Answer Check Check that n is a positive integer (it cannot be zero). Also, check that l is a positive integer (but zero is allowed) and that m_l is an integer whose magnitude (its value except for sign) is equal to or less than l. The m_s quantum number can be only $+\frac{1}{2}$ or $-\frac{1}{2}$.

Exercise 7.7 Explain why each of the following sets of quantum numbers is not permissible for an orbital.

a $n = 0, l = 1, m_l = 0, m_s = +\frac{1}{2}$
b $n = 2, l = 3, m_l = 0, m_s = -\frac{1}{2}$
c $n = 3, l = 2, m_l = +3, m_s = +\frac{1}{2}$
d $n = 3, l = 2, m_l = +2, m_s = 0$

■ See Problems 7.69 and 7.70.

Atomic Orbital Shapes

An *s* orbital has a spherical shape, though specific details of the probability distribution depend on the value of *n*. Figure 7.23 shows cross-sectional representations of the probability distributions of a 1*s* and a 2*s* orbital. The color shading is darker where the electron is more likely to be found. In the case of a 1*s* orbital, the electron is most likely to be found near the nucleus. The shading becomes lighter as the distance from the nucleus increases, indicating that the electron is less likely to be found there.

The orbital does not abruptly end at some particular distance from the nucleus. An atom, therefore, has an indefinite extension, or "size." We can gauge the "size" of the orbital by means of the *99% contour*. The electron has a 99% probability of being found within the space of the 99% contour (the sphere indicated by the dashed line in the diagram).

A 2*s* orbital differs in detail from a 1*s* orbital. The electron in a 2*s* orbital is likely to be found in two regions, one near the nucleus and the other in a spherical shell about the nucleus. (The electron is most likely to be here.) The 99% contour shows that the 2*s* orbital is larger than the 1*s* orbital.

A cross-sectional diagram cannot portray the three-dimensional aspect of the 1*s* and 2*s* atomic orbitals. Figure 7.24 shows cutaway diagrams, which better illustrate this three-dimensionality.

There are three *p* orbitals in each *p* subshell. All *p* orbitals have the same basic shape (two lobes arranged along a straight line with the nucleus between the lobes) but differ in their orientations in space. Because the three orbitals are set at right angles to each other, we can show each one as oriented along a different coordinate axis (Figure 7.25). We denote these orbitals as $2p_x$, $2p_y$, and $2p_z$. A $2p_x$ orbital has its greatest electron probability along the *x*-axis, a $2p_y$ orbital along the *y*-axis, and a $2p_z$ orbital along the *z*-axis. Other *p* orbitals, such as 3*p*, have this same general shape, with differences in detail depending on *n*. We will discuss *s* and *p* orbital shapes again in Chapter 10 in reference to chemical bonding.

There are five *d* orbitals, which have more complicated shapes than do *s* and *p* orbitals. These are represented in Figure 7.26.

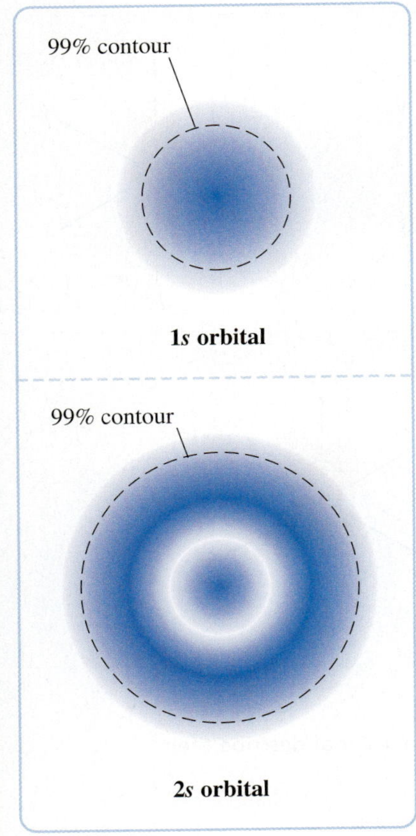

Figure 7.23 ◀

Cross-sectional representations of the probability distributions of *s* orbitals In a 1*s* orbital, the probability distribution is largest near the nucleus. In a 2*s* orbital, it is greatest in a spherical shell about the nucleus. Note the relative "size" of the orbitals, indicated by the 99% contours.

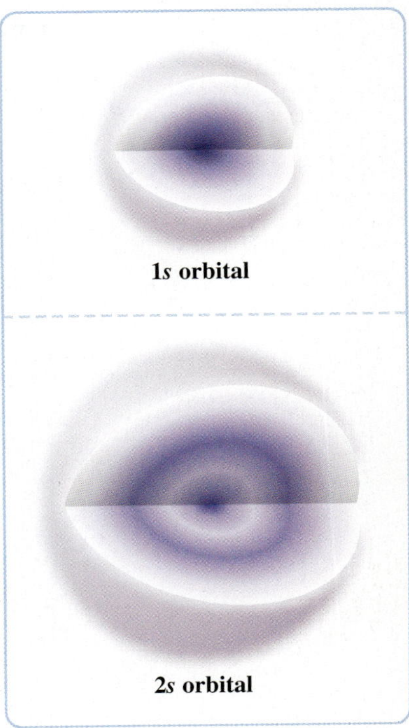

Figure 7.24 ▲

Cutaway diagrams showing the spherical shape of *s* orbitals In both diagrams, a segment of each orbital is cut away to reveal the electron distribution of the orbital.

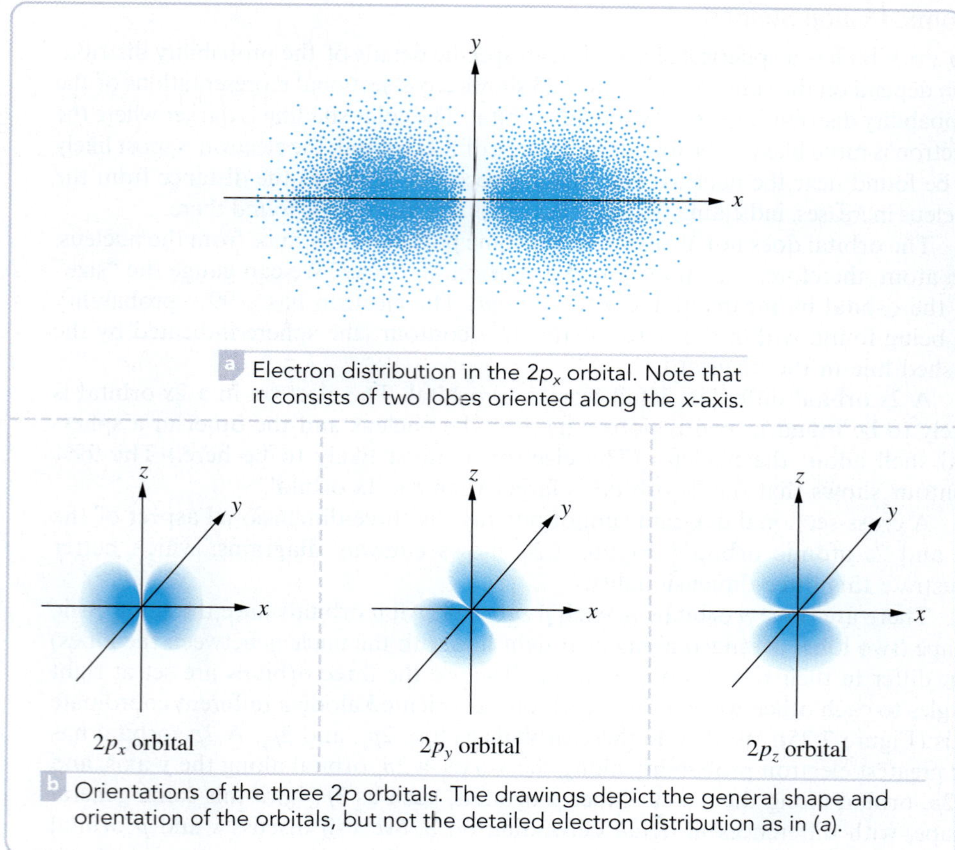

Figure 7.25
The 2p orbitals

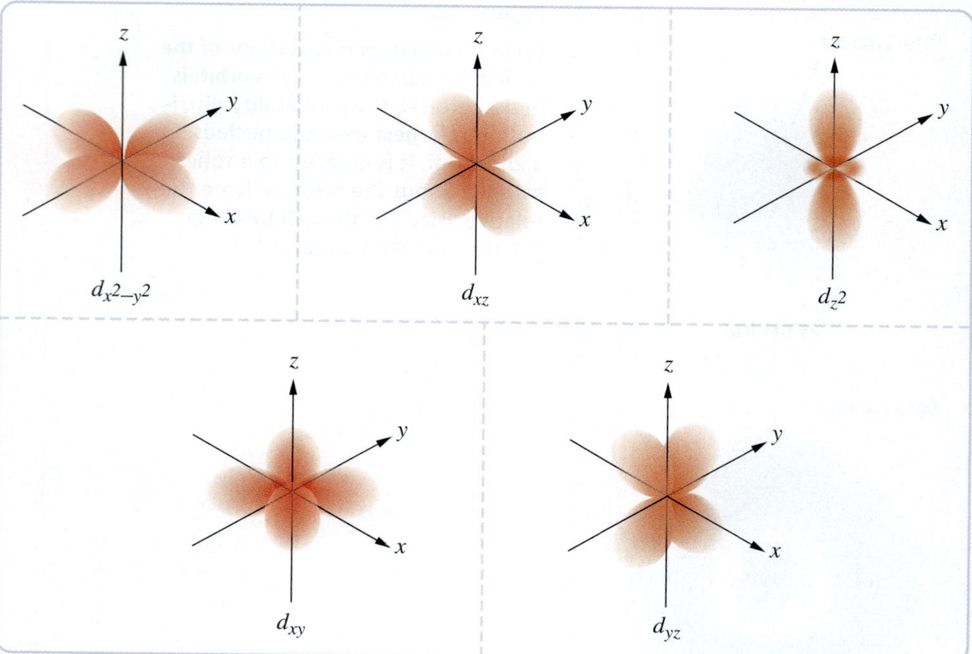

Figure 7.26
The five 3d orbitals These are labeled by subscripts, as in d_{xy}, that describe their mathematical characteristics.

A Checklist for Review

Summary of Facts and Concepts

One way to study the electronic structure of the atom is to analyze the electromagnetic radiation that is emitted from an atom. Electromagnetic radiation is characterized by its *wavelength* λ and frequency ν, and these quantities are related to the speed of light c ($c = \nu\lambda$).

Einstein showed that light consists of particles (*photons*), each of energy $E = h\nu$, where h is Planck's constant. According to Bohr, electrons in an atom have *energy levels,* and when an electron in a higher energy level drops (or undergoes a transition) to a lower energy level, a photon is emitted. The energy of the photon equals the difference in energy between the two levels.

Electrons and other particles of matter have both particle and wave properties. For a particle of mass m and speed v, the wavelength is related to momentum mv by the *de Broglie relation:* $\lambda = h/mv$. The wave properties of a particle are described by a *wave function,* from which we can get the probability of finding the particle in different regions of space.

Each electron in an atom is characterized by four different quantum numbers. The distribution of an electron in space—its *atomic orbital*—is characterized by three of these quantum numbers: the principal quantum number, the angular momentum quantum number, and the magnetic quantum number. The fourth quantum number (spin quantum number) describes the magnetism of the electron.

Learning Objectives	Important Terms
7.1 The Wave Nature of Light	
▪ Define the *wavelength* and *frequency* of a wave. ▪ Relate the wavelength, frequency, and speed of light. Examples 7.1 and 7.2 ▪ Describe the different regions of the electromagnetic spectrum.	**wavelength (λ)** **frequency (ν)** **electromagnetic spectrum**
7.2 Quantum Effects and Photons	
▪ State Planck's quantization of vibrational energy. ▪ Define *Planck's constant* and *photon.* ▪ Describe the photoelectric effect. ▪ Calculate the energy of a photon from its frequency or wavelength. Example 7.3	**Planck's constant** **photons** **photoelectric effect**
7.3 The Bohr Theory of the Hydrogen Atom	
▪ State the postulates of Bohr's theory of the hydrogen atom. ▪ Relate the energy of a photon to the associated energy levels of an atom. ▪ Determine the wavelength or frequency of a hydrogen atom transition. Example 7.4 ▪ Describe the difference between *emission* and *absorption* of light by an atom.	**continuous spectrum** **line spectrum** **energy levels**
7.4 Quantum Mechanics	
▪ State the de Broglie relation. ▪ Calculate the wavelength of a moving particle. Example 7.5 ▪ Define *quantum mechanics.* ▪ State Heisenberg's uncertainty principle. ▪ Relate the wave function for an electron to the probability of finding it at a location in space.	**de Broglie relation** **quantum (wave) mechanics** **uncertainty principle**

7.5 Quantum Numbers and Atomic Orbitals	
▪ Define *atomic orbital.* ▪ Define each of the quantum numbers for an atomic orbital. ▪ State the rules for the allowed values for each quantum number. ▪ Apply the rules for quantum numbers. **Example 7.6** ▪ Describe the shapes of *s*, *p*, and *d* orbitals.	**atomic orbital** **principal quantum number (n)** **angular momentum quantum number (l)** **magnetic quantum number (m_l)** **spin quantum number (m_s)**

Key Equations

$c = \nu\lambda$

$E = h\nu$

$E = -\frac{R_H}{n^2} \quad n = 1, 2, 3, \ldots \infty \quad \text{(for H atom)}$

Energy of emitted photon $= h\nu = -(E_f - E_i)$

$\lambda = \frac{h}{mv}$

Questions and Problems

OWL Interactive versions of these problems may be assigned in OWL.

Self-Assessment and Review Questions

Key: These questions test your understanding of the ideas you worked with in the chapter. These problems vary in difficulty and often can be used for the basis of discussion.

7.1 Give a brief wave description of light. What are two characteristics of light waves?

7.2 What is the mathematical relationship among the different characteristics of light waves? State the meaning of each of the terms in the equation.

7.3 Briefly describe the portions of the electromagnetic spectrum, starting with shortest wavelengths and going to longer wavelengths.

7.4 Planck originated the idea that energies can be quantized. What does the term *quantized* mean? What was Planck trying to explain when he was led to the concept of quantization of energy? Give the formula he arrived at and explain each of the terms in the formula.

7.5 In your own words, explain the photoelectric effect. How does the photon concept explain this effect?

7.6 Describe the wave–particle picture of light.

7.7 Give the equation that relates particle properties of light. Explain the meaning of each symbol in the equation.

7.8 Physical theory at the time Rutherford proposed his nuclear model of the atom was not able to explain how this model could give a stable atom. Explain the nature of this difficulty.

7.9 Explain the main features of Bohr's theory. Do these features solve the difficulty alluded to in Question 7.8?

7.10 Explain the process of emission of light by an atom.

7.11 Explain the process of absorption of light by an atom.

7.12 What is the evidence for electron waves? Give a practical application.

7.13 What kind of information does a wave function give about an electron in an atom?

7.14 The atom is sometimes said to be similar to a miniature planetary system, with electrons orbiting the nucleus. What does the uncertainty principle have to say about this view of the atom?

7.15 Bohr described the hydrogen atom as an electron orbiting a hydrogen nucleus. Although certain aspects of his theory are still valid, his theory agreed quantitatively with experiment only in the case of the hydrogen atom. In what way does quantum mechanics change Bohr's original picture of the hydrogen atom?

7.16 Give the possible values of a the principal quantum number, b the angular momentum quantum number, c the magnetic quantum number, and d the spin quantum number.

7.17 What is the notation for the subshell in which $n = 4$ and $l = 3$? How many orbitals are in this subshell?

7.18 What is the general shape of an *s* orbital? of a *p* orbital?

7.19 Which of the following statements about a hydrogen atom is *false*?

a An electron in the $n = 1$ level of the hydrogen atom is in its ground state.

b On average, an electron in the $n = 3$ level is farther from the nucleus than an electron in the $n = 2$ state.

c The wavelength of light emitted when the electron goes from the $n = 3$ level to the $n = 1$ level is the same as the wavelength of light absorbed when the electron goes from the $n = 1$ level to $n = 3$ level.

d An electron in the $n = 1$ level is higher in energy than an electron in the $n = 4$ level.

e Light of greater frequency is required for a transition from the $n = 1$ level to $n = 3$ level than is required for a transition from the $n = 2$ level to $n = 3$ level.

7.20 Which of the following statements is (are) true?

I. The product of wavelength and frequency of light is a constant.

II. As the energy of electromagnetic radiation increases, its frequency decreases.

III. As the wavelength of light increases, its frequency increases.

a I only
b II only
c III only
d I and III only
e II and III only

7.21 Of the following possible transitions of an electron in a hydrogen atom, which emits light of the highest energy?

a Transition from the $n = 1$ to the $n = 3$ level
b Transition from the $n = 1$ to the $n = 2$ level
c Transition from the $n = 3$ to the $n = 1$ level
d Transition from the $n = 2$ to the $n = 1$ level
e Transition from the $n = 5$ to the $n = 4$ level

7.22 What wavelength of electromagnetic radiation corresponds to a frequency of $3.46 \times 10^{13}\ s^{-1}$?

a 8.66×10^{-6} m
b 1.15×10^{5} m
c 7.65×10^{-29} m
d 9.10×10^{-6} m
e 8.99×10^{-6} m

Concept Explorations

Key: Concept explorations are comprehensive problems that provide a framework that will enable you to explore and learn many of the critical concepts and ideas in each chapter. If you master the concepts associated with these explorations, you will have a better understanding of many important chemistry ideas and will be more successful in solving all types of chemistry problems. These problems are well suited for group work and for use as in-class activities.

7.23 Light, Energy, and the Hydrogen Atom

a Which has the greater wavelength, blue light or red light?

b How do the frequencies of blue light and red light compare?

c How does the energy of blue light compare with that of red light?

d Does blue light have a greater speed than red light?

e How does the energy of three photons from a blue light source compare with the energy of one photon of blue light from the same source? How does the energy of two photons corresponding to a wavelength of 451 nm (blue light) compare with the energy of three photons corresponding to a wavelength of 704 nm (red light)?

f A hydrogen atom with an electron in its ground state interacts with a photon of light with a wavelength of 1.22×10^{-6} m. Could the electron make a transition from the ground state to a higher energy level? If it does make a transition, indicate which one. If no transition can occur, explain.

g If you have one mole of hydrogen atoms with their electrons in the $n = 1$ level, what is the minimum number of photons you would need to interact with these atoms in order to have all of their electrons promoted to the $n = 3$ level? What wavelength of light would you need to perform this experiment?

7.24 Investigating Energy Levels

Consider the hypothetical atom X that has one electron like the H atom but has different energy levels. The energies of an electron in an X atom are described by the equation

$$E = -\frac{R_H}{n^3}$$

where R_H is the same as for hydrogen (2.179×10^{-18} J). Answer the following questions, *without calculating energy values.*

a How would the ground-state energy levels of X and H compare?

b Would the energy of an electron in the $n = 2$ level of H be higher or lower than that of an electron in the $n = 2$ level of X? Explain your answer.

c How do the spacings of the energy levels of X and H compare?

d Which would involve the emission of a higher frequency of light, the transition of an electron in an H atom from the $n = 5$ to the $n = 3$ level or a similar transition in an X atom?

e Which atom, X or H, would require more energy to completely remove its electron?

f A photon corresponding to a particular frequency of blue light produces a transition from the $n = 2$ to the $n = 5$ level of a hydrogen atom. Could this photon produce the same transition ($n = 2$ to $n = 5$) in an atom of X? Explain.

Conceptual Problems

Key: These problems are designed to check your understanding of the concepts associated with some of the main topics presented in each chapter. A strong conceptual understanding of chemistry is the foundation for both applying chemical knowledge and solving chemical problems. These problems vary in level of difficulty and often can be used as a basis for group discussion.

7.25 Consider two beams of the same yellow light. Imagine that one beam has its wavelength doubled; the other has its frequency doubled. Which of these two beams is then in the ultraviolet region?

7.26 Some infrared radiation has a wavelength that is 1000 times larger than that of a certain visible light. This visible light has a frequency that is 1000 times smaller than that of some X radiation. How many times more energy is there in a photon of this X radiation than there is in a photon of the infrared radiation?

7.27 One photon of green light has less than twice the energy of two photons of red light. Consider two hypothetical experiments. In one experiment, potassium metal is exposed to one photon of green light; in another experiment, potassium metal is exposed to two photons of red light. In one of these experiments, no electrons are ejected by the photoelectric effect (no matter how many times this experiment is repeated). In the other experiment, at least one electron was observed to be ejected. What is the maximum number of electrons that could be ejected during this other experiment, one or two?

7.28 An atom in its ground state absorbs a photon (photon 1), then quickly emits another photon (photon 2). One of these photons corresponds to ultraviolet radiation, whereas the other one corresponds to red light. Explain what is happening. Which electromagnetic radiation, ultraviolet or red light, is associated with the emitted photon (photon 2)?

7.29 Three emission lines involving three energy levels in an atom occur at wavelengths x, $1.5x$, and $3.0x$ nanometers. Which wavelength corresponds to the transition from the highest to the lowest of the three energy levels?

7.30 An atom emits yellow light when an electron makes the transition from the $n = 5$ to the $n = 1$ level. In separate experiments, suppose you bombarded the $n = 1$ level of this atom with red light, yellow light (obtained from the previous emission), and blue light. In which experiment or experiments would the electron be promoted to the $n = 5$ level?

7.31 Which of the following particles has the longest wavelength?

a. an electron traveling at x meters per second
b. a proton traveling at x meters per second
c. a proton traveling at $2x$ meters per second

7.32 Imagine a world in which the rule for the l quantum number is that values start with 1 and go up to n. The rules for the n and m_l quantum numbers are unchanged from those of our world. Write the quantum numbers for the first two shells (i.e., $n = 1$ and $n = 2$).

7.33 Given the following energy level diagram for an atom that contains an electron in the $n = 3$ level, answer the following questions.

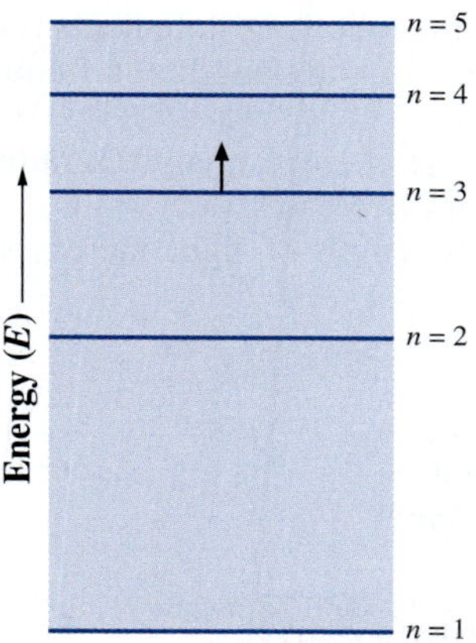

a. Which transition of the electron will emit light of the lowest frequency?
b. Using only those levels depicted in the diagram, which transition of the electron would require the highest-frequency light?
c. If the transition from the $n = 3$ level to the $n = 1$ level emits green light, what color light is absorbed when an electron makes the transition from the $n = 1$ to $n = 3$ level?

7.34 The following shapes each represent an orbital of an atom in a hypothetical universe. The small circle is the location of the nucleus in each orbital.

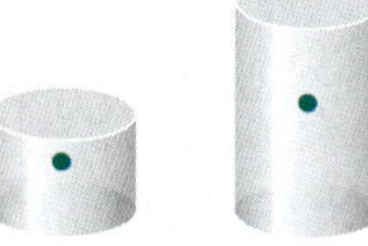

a. If you placed an electron in each orbital, which one would be higher in energy?
b. When an electron makes a transition from the orbital represented on the right to the orbital on the left, would you expect energy to be absorbed or released?
c. Draw a sketch of an orbital of the same type that would be higher in energy than either of the two pictured orbitals.

Practice Problems

Key: These problems are for practice in applying problem-solving skills. They are divided by topic, and some are keyed to exercises (see the ends of the exercises). The problems are arranged in matching pairs; the odd-numbered problem of each pair is listed first, and its answer is given in the back of the book.

Electromagnetic Waves

7.35 Radio waves in the AM region have frequencies in the range 530 to 1700 kilocycles per second (530 to 1700 kHz). Calculate the wavelength corresponding to a radio wave of frequency 1.365×10^6/s (that is, 1365 kHz).

7.36 Microwaves have frequencies in the range 10^9 to 10^{12}/s (cycles per second), equivalent to between 1 gigahertz and 1 terahertz. What is the wavelength of microwave radiation whose frequency is 1.395×10^{10}/s?

7.37 Light with a wavelength of 478 nm lies in the blue region of the visible spectrum. Calculate the frequency of this light.

7.38 Calculate the frequency associated with light of wavelength 434 nm. (This corresponds to one of the wavelengths of light emitted by the hydrogen atom.)

7.39 At its closest approach, Mars is 56 million km from Earth. How long would it take to send a radio message from a space probe of Mars to Earth when the planets are at this closest distance?

7.40 The space probe *Pioneer 11* was launched April 5, 1973, and reached Jupiter in December 1974, traveling a distance of 998 million km. How long did it take an electromagnetic signal to travel to Earth from *Pioneer 11* when it was near Jupiter?

7.41 The meter was defined in 1963 as the length equal to 1,650,763.73 wavelengths of the orange-red radiation emitted by the krypton-86 atom (the meter has since been redefined). What is the wavelength of this transition? What is the frequency?

7.42 The second is defined as the time it takes for 9,192,631,770 wavelengths of a certain transition of the cesium-133 atom to pass a fixed point. What is the frequency of this electromagnetic radiation? What is the wavelength?

Photons

7.43 What is the energy of a photon corresponding to radio waves of frequency 1.365×10^6/s?

7.44 What is the energy of a photon corresponding to microwave radiation of frequency 1.258×10^{10}/s?

7.45 The green line in the atomic spectrum of thallium has a wavelength of 535 nm. Calculate the energy of a photon of this light.

7.46 Molybdenum compounds give a yellowish-green flame test. The atomic emission responsible for this color has a wavelength of 551 nm. Obtain the energy of a single photon of this wavelength.

7.47 A particular transition of the rubidium atom emits light whose frequency is 3.84×10^{14} Hz. (Hz is the abbreviation for *hertz,* which is equivalent to the unit/s, or s^{-1}.) Is this light in the visible spectrum? If so, what is the color of the light? (See Figure 7.5.)

7.48 Selenium atoms have a particular transition that emits light of frequency 1.53×10^{15} Hz. (Hz is the abbreviation for *hertz,* which is equivalent to the unit/s, or s^{-1}.) Is this light in the visible spectrum? If so, what is the color of the light? (See Figure 7.5.)

Bohr Theory

7.49 An electron in a hydrogen atom in the level $n = 5$ undergoes a transition to level $n = 3$. What is the frequency of the emitted radiation?

7.50 Calculate the frequency of electromagnetic radiation emitted by the hydrogen atom in the electron transition from $n = 6$ to $n = 3$.

7.51 The first line of the Lyman series of the hydrogen atom emission results from a transition from the $n = 2$ level to the $n = 1$ level. What is the wavelength of the emitted photon? Using Figure 7.5, describe the region of the electromagnetic spectrum in which this emission lies.

7.52 What is the wavelength of the electromagnetic radiation emitted from a hydrogen atom when the electron undergoes the transition $n = 4$ to $n = 1$? In what region of the spectrum does this line occur? (See Figure 7.5.)

7.53 Calculate the shortest wavelength of the electromagnetic radiation emitted by the hydrogen atom in undergoing a transition from the $n = 6$ level.

7.54 Calculate the longest wavelength of the electromagnetic radiation emitted by the hydrogen atom in undergoing a transition from the $n = 6$ level.

7.55 What is the difference in energy between the two levels responsible for the violet emission line of the calcium atom at 422.7 nm?

7.56 What is the difference in energy between the two levels responsible for a red emission line of the strontium atom at 640.8 nm?

de Broglie Waves

Note: Masses of the electron, proton, and neutron are listed on the inside back cover of this book.

7.57 What is the wavelength of a neutron traveling at a speed of 4.15 km/s? (Neutrons of these speeds are obtained from a nuclear pile.)

7.58 What is the wavelength of a proton traveling at a speed of 6.21 km/s? What would be the region of the spectrum for electromagnetic radiation of this wavelength?

7.59 At what speed must an electron travel to have a wavelength of 10.0 pm?

7.60 At what speed must a neutron travel to have a wavelength of 10.6 pm?

7.61 What is the de Broglie wavelength of a 145-g baseball traveling at 30.0 m/s (67.1 mph)? Is the wavelength much smaller or much larger than the diameter of an atom (on the order of 100 pm)?

7.62 What is the de Broglie wavelength of an oxygen molecule, O_2, traveling at 535 m/s? Is the wavelength much smaller or much larger than the diameter of an atom (on the order of 100 pm)?

Atomic Orbitals

7.63 If the n quantum number of an atomic orbital is 4, what are the possible values of l? If the l quantum number is 3, what are the possible values of m_l?

7.64 The n quantum number of an atomic orbital is 5. What are the possible values of l? What are the possible values of m_l if the l quantum number is 4?

7.65 How many subshells are there in the M shell? How many orbitals are there in the d subshell?

7.66 How many subshells are there in the N shell? How many orbitals are there in the f subshell?

7.67 Give the notation (using letter designations for l) for the subshells denoted by the following quantum numbers.

a $n = 6, l = 2$
b $n = 5, l = 4$
c $n = 4, l = 3$
d $n = 6, l = 1$

7.68 Give the notation (using letter designations for l) for the subshells denoted by the following quantum numbers.

a $n = 3, l = 1$
b $n = 4, l = 2$
c $n = 4, l = 0$
d $n = 5, l = 3$

7.69 Explain why each of the following sets of quantum numbers would not be permissible for an electron, according to the rules for quantum numbers.

a $n = 1, l = 0, m_l = 0, m_s = +1$
b $n = 1, l = 3, m_l = +3, m_s = +\frac{1}{2}$
c $n = 3, l = 2, m_l = +3, m_s = -\frac{1}{2}$
d $n = 0, l = 1, m_l = 0, m_s = +\frac{1}{2}$
e $n = 2, l = 1, m_l = -1, m_s = +\frac{3}{2}$

7.70 State which of the following sets of quantum numbers would be possible and which impossible for an electron in an atom.

a $n = 2, l = 0, m_l = 0, m_s = +\frac{1}{2}$
b $n = 1, l = 1, m_l = 0, m_s = +\frac{1}{2}$
c $n = 0, l = 0, m_l = 0, m_s = -\frac{1}{2}$
d $n = 2, l = 1, m_l = -1, m_s = +\frac{1}{2}$
e $n = 2, l = 1, m_l = -2, m_s = +\frac{1}{2}$

General Problems

Key: These problems provide more practice but are not divided by topic or keyed to exercises. Each section ends with essay questions, each of which is color coded to refer to the *A Chemist Looks at Daily Life* (orange), *A Chemist Looks at Materials* (blue), and *Instrumental Methods* (brown) chapter essay on which it is based. Odd-numbered problems and the even-numbered problems that follow are similar; answers to all odd-numbered problems except the essay questions are given in the back of the book.

7.71 The blue line of the strontium atom emission has a wavelength of 461 nm. What is the frequency of this light? What is the energy of a photon of this light?

7.72 The barium atom has an emission with wavelength 554 nm (green). Calculate the frequency of this light and the energy of a photon of this light.

7.73 The energy of a photon is 4.10×10^{-19} J. What is the wavelength of the corresponding light? What is the color of this light?

7.74 The energy of a photon is 3.05×10^{-19} J. What is the wavelength of the corresponding light? What is the color of this light?

7.75 The photoelectric work function of a metal is the minimum energy needed to eject an electron by irradiating the metal with light. For calcium, this work function equals 4.34×10^{-19} J. What is the minimum frequency of light for the photoelectric effect in calcium?

7.76 The photoelectric work function for magnesium is 5.90×10^{-19} J. Calculate the minimum frequency of light required to eject electrons from magnesium.

7.77 Light of wavelength 345 nm shines on a piece of calcium metal. What is the speed of the ejected electron? (Light energy greater than that of the work function of calcium ends up as kinetic energy of the ejected electron. See Problem 7.75 for the definition of work function and its value for calcium.)

7.78 Light of wavelength 285 nm shines on a piece of magnesium metal. What is the speed of the ejected electron? (Light energy greater than that of the work function of magnesium ends up as kinetic energy of the ejected electron. See Problem 7.76 for the definition of work function and its value for magnesium.)

7.79 Calculate the wavelength of the Balmer line of the hydrogen spectrum in which the initial n quantum number is 5 and the final n quantum number is 2.

7.80 Calculate the wavelength of the Balmer line of the hydrogen spectrum in which the initial n quantum number is 3 and the final n quantum number is 2.

7.81 One of the lines in the Balmer series of the hydrogen atom emission spectrum is at 397 nm. It results from a transition from an upper energy level to $n = 2$. What is the principal quantum number of the upper level?

7.82 A line of the Lyman series of the hydrogen atom spectrum has the wavelength 9.50×10^{-8} m. It results from a transition from an upper energy level to $n = 1$. What is the principal quantum number of the upper level?

7.83 A hydrogen-like ion has a nucleus of charge $+Ze$ and a single electron outside this nucleus. The energy levels of these ions are $-Z^2R_H/n^2$ (where Z = atomic number). Calculate the wavelength of the transition from $n = 3$ to $n = 2$ for He^+, a hydrogen-like ion. In what region of the spectrum does this emission occur?

7.84 What is the wavelength of the transition from $n = 4$ to $n = 2$ for Li^{2+}? In what region of the spectrum does this emission occur? Li^{2+} is a hydrogen-like ion. Such an ion has a nucleus of charge $+Ze$ and a single electron outside this nucleus. The energy levels of the ion are $-Z^2R_H/n^2$, where Z is the atomic number.

7.85 An electron microscope employs a beam of electrons to obtain an image of an object. What energy must be imparted to each electron of the beam to obtain a wavelength of 10.0 pm? Obtain the energy in electron volts (eV) (1 eV = 1.602×10^{-19} J).

7.86 Neutrons are used to obtain images of the hydrogen atoms in molecules. What energy must be imparted to each neutron in a neutron beam to obtain a wavelength of 10.0 pm? Obtain the energy in electron volts (eV) (1 eV = 1.602×10^{-19} J).

7.87 What is the number of different orbitals in each of the following subshells?

a 3*d* b 4*f* c 4*p* d 5*s*

7.88 What is the number of different orbitals in each of the following subshells?

a 5*f* b 5*g* c 6*s* d 5*p*

7.89 List the possible subshells for the $n = 6$ shell.

7.90 List the possible subshells for the $n = 8$ shell.

■ **7.91** What are gamma rays? How does the gamma radiation of foods improve their shelf life?

■ **7.92** How can gamma rays that are used in food irradiation be produced? Does such irradiated food show any radioactivity?

■ **7.93** The word *laser* is an acronym meaning *l*ight *a*mplification by *s*timulated *e*mission of *r*adiation. What is the stimulated emission of radiation?

■ **7.94** Explain how lasers are used to "read" a compact disc.

■ **7.95** Explain the concept of quantum mechanical tunneling.

■ **7.96** Explain how the probe in a scanning tunneling microscope scans a sample on the surface of a metal.

Strategy Problems

Key: As noted earlier, all of the practice and general problems are matched pairs. This section is a selection of problems that are not in matched-pair format. These challenging problems require that you employ many of the concepts and strategies that were developed in the chapter. In some cases, you will have to integrate several concepts and operational skills in order to solve the problem successfully.

Note: See back cover for physical constants and conversion factors.

7.97 What wavelength of electromagnetic radiation corresponds to a frequency of $7.76 \times 10^9\ s^{-1}$? Note that Planck's constant is 6.63×10^{-34} J·s, and the speed of light is 3.00×10^8 m/s.

7.98 AM radio stations broadcast at frequencies between 530 kHz and 1700 kHz. (1 kHz = $10^3\ s^{-1}$.) For a station broadcasting at 1.69×10^3 kHz, what is the energy of this radio wave? Note that Planck's constant is 6.63×10^{-34} J·s, and the speed of light is 3.00×10^8 m/s.

7.99 A particular microwave oven delivers 800 watts. (A watt is a unit of power, which is the joules of energy delivered, or used, per second.) If the oven uses microwave radiation of wavelength 12.2 cm, how many photons of this radiation are required to heat 1.00 g of water 1.00°C, assuming that all of the photons are absorbed?

7.100 A ruby laser puts out a pulse of red light at a wavelength of 694 nm. (The ruby laser is described in the essay on p. 280.) If a pulse delivers 1.05×10^5 watts of power for 280 μs, how many photons are there in this pulse? (1 watt = 1 J/s.)

7.101 The retina of the eye contains two types of light-sensitive cells: rods (responsible for night vision) and cones (responsible for color vision). Rod cells are about a hundred times more sensitive to light than cone cells and are able to detect a single photon. Suppose a group of rod cells are radiated with a pulse of light having an energy equal to 1.60×10^{-16} J. If the wavelength of this light was 498 nm (the wavelength at which rod cells are most sensitive), how many photons are in this light pulse?

7.102 Ozone in the stratosphere absorbs ultraviolet light of wavelengths shorter than 320 nm, thus filtering out the most energetic radiation from sunlight. During this absorption, an ozone molecule absorbs a photon, which breaks an oxygen-oxygen bond, yielding an oxygen molecule and an oxygen atom:

$$O_3(g) + h\nu \longrightarrow O_2(g) + O(g)$$

(Here, $h\nu$ denotes a photon.) Suppose a flask of ozone is irradiated with a pulse of UV light of wavelength 275 nm. Assuming that each photon of this pulse that is absorbed breaks up one ozone molecule, calculate the energy absorbed per mole of O_2 produced, giving the answer in kJ/mol.

7.103 The photoelectric *work function* of a metal is the minimum energy required to eject an electron by shining light on the metal. The work function of calcium is 4.60×10^{-19} J. What is the longest wavelength of light (in nanometers) that can cause an electron to be ejected from calcium metal.

7.104 The photoelectric work function of potassium is 2.29 eV. A photon of energy greater than this ejects the electron with the excess as kinetic energy. Suppose light of wavelength 455 nm ejects an electron from the surface of potassium. What is the speed of the ejected electron?

7.105 The surface of a metal was illuminated with light whose wavelength is 389 nm. Upon illumination, the metal ejected an electron with a speed of 3.34×10^5 m/s. What is the photoelectric work function of the metal (in eV)? Assuming that the metal is an element, use values given in the *CRC Handbook of Chemistry and Physics* to discover the element.

7.106 Calculate the shortest wavelength of *visible* light (in nanometers) seen in the spectrum of the hydrogen atom. What are the principal quantum numbers for the levels in this transition? Does Figure 7.11 include all visible lines?

7.107 Light of wavelength 1.03×10^{-7} m is emitted when an electron in an excited level of a hydrogen atom undergoes a transition to the $n = 1$ level. What is the region of the spectrum of this light? What is the principal quantum number of the excited level?

7.108 A hydrogen atom in the ground state absorbs a photon whose wavelength is 95.0 nm. The resulting excited atom then emits a photon of 1282 nm. What are the regions of the electromagnetic spectrum for the radiations involved in these transitions? What is the principal quantum number of the final state resulting from the emission from the excited atom?

7.109 Hydrogen-like ions are atomic ions consisting of just one electron about a nucleus. The energy levels of such an ion are given by a formula similar to that of the hydrogen atom except that the Rydberg constant for a hydrogen-like ion of atomic number Z has a value of $2.179 \times 10^{-18}\ Z^2$. What is the shortest wavelength transition in the Li^{2+} ion?

7.110 It requires 799 kJ of energy to break one mole of carbon–oxygen double bonds in carbon dioxide. What wavelength of light does this correspond to per bond? Is there any transition in the hydrogen atom that has at least this quantity of energy in one photon?

7.111 The root-mean-square speed of an oxygen molecule, O_2, at 21°C is 479 m/s. Calculate the de Broglie wavelength for an O_2 molecule traveling at this speed. How does this wavelength compare with the approximate length of this molecule, which is about 242 pm? (For this comparison, state the wavelength as a percentage of the molecular length.)

7.112 An electron is accelerated through a potential difference of 15.6 kilovolts (giving the electron a kinetic energy of 15.6 keV). What is the associated wavelength of the electron in angstroms?

7.113 In X-ray fluorescence spectroscopy, a material can be analyzed for its constituent elements by radiating the material with short-wavelength X rays, which induce the atoms to emit longer-wavelength X rays characteristic of those atoms. Tungsten, for example, emits characteristic X rays of wavelength 0.1476 nm. If an electron has an equivalent wavelength, what is its kinetic energy?

7.114 For each of the following combinations of quantum numbers, make changes that produce an allowed combination. Count 3 for each change of n, 2 for each change of l, and 1 for each change of m_l. What is the lowest possible count that you can obtain?

a $n = 3, l = 0, m_l = -2$
b $n = 5, l = 5, m_l = 4$
c $n = 3, l = 3, m_l = -3$
d $n = 5, l = 6, m_l = 3$

7.115 The term *degeneracy* means the number of different quantum states of an atom or molecule having the same energy. For example, the degeneracy of the $n = 2$ level of the hydrogen atom is 4 (a 2*s* quantum state, and three different 2*p* states). What is the degeneracy of the $n = 5$ level?

7.116 In a hypothetical universe, the quantum numbers for an atomic orbital follow these rules:

n = any positive integer value from 2 to ∞
l = any positive integer value from 2 to $n + 1$
m_l = any integer from $-(l - 1)$ to $(l - 1)$

How many orbitals are possible for the $n = 2$ shell?

Cumulative-Skills Problems

Key: The problems under this heading combine skills introduced in previous chapters with those given in the current one.

7.117 The energy required to dissociate the Cl_2 molecule to Cl atoms is 239 kJ/mol Cl_2. If the dissociation of a Cl_2 molecule were accomplished by the absorption of a single photon whose energy was exactly the quantity required, what would be its wavelength (in meters)?

7.118 The energy required to dissociate the H_2 molecule to H atoms is 432 kJ/mol H_2. If the dissociation of an H_2 molecule were accomplished by the absorption of a single photon whose energy was exactly the quantity required, what would be its wavelength (in meters)?

7.119 A microwave oven heats by radiating food with microwave radiation, which is absorbed by the food and converted to heat. Suppose an oven's radiation wavelength is 12.5 cm. A container with 0.250 L of water was placed in the oven, and the temperature of the water rose from 20.0°C to 100.0°C. How many photons of this microwave radiation were required? Assume that all the energy from the radiation was used to raise the temperature of the water.

7.120 Warm objects emit electromagnetic radiation in the infrared region. Heat lamps employ this principle to generate infrared radiation. Water absorbs infrared radiation with wavelengths near 2.80 μm. Suppose this radiation is absorbed by the water and converted to heat. A 1.00-L sample of water absorbs infrared radiation, and its temperature increases from 20.0°C to 30.0°C. How many photons of this radiation are used to heat the water?

7.121 Light with a wavelength of 425 nm fell on a potassium surface, and electrons were ejected at a speed of 4.88×10^5 m/s. What energy was expended in removing an electron from the metal? Express the answer in joules (per electron) and in kilojoules per mole (of electrons).

7.122 Light with a wavelength of 405 nm fell on a strontium surface, and electrons were ejected. If the speed of an ejected electron is 3.36×10^5 m/s, what energy was expended in removing the electron from the metal? Express the answer in joules (per electron) and in kilojoules per mole (of electrons).

7.123 When an electron is accelerated by a voltage difference, the kinetic energy acquired by the electron equals the voltage times the charge on the electron. Thus, one volt imparts a kinetic energy of 1.602×10^{-19} volt-coulombs, which equals 1.602×10^{-19} J. What is the wavelength associated with electrons accelerated by 4.00×10^3 volts?

7.124 When an electron is accelerated by a voltage difference, the kinetic energy acquired by the electron equals the voltage times the charge on the electron. Thus, one volt imparts a kinetic energy of 1.602×10^{-19} volt-coulombs, or 1.602×10^{-19} J. What is the wavelength for electrons accelerated by 1.00×10^4 volts?

8

Electron Configurations and Periodicity

© Cengage Learning; Richard Megna/ Fundamental Photographs, NYC

Sodium metal reacts vigorously with water to produce hydrogen gas (which catches fire); other Group IA metals similarly react with water. The other product is sodium hydroxide, which is used in products such as oven cleaner and soap.

Contents and Concepts

Electronic Structure of Atoms

In the previous chapter, you learned that we characterize an atomic orbital by four quantum numbers: n, l, m_l, and m_s. In the first section, we look further at electron spin; then we discuss how electrons are distributed among the possible orbitals of an atom.

Periodicity of the Elements

You learned how the periodic table can be explained by the periodicity of the ground-state configurations of the elements. Now we will look at various aspects of the periodicity of the elements.

OWL Sign in to OWL at **www.cengage.com/owl** to view tutorials and simulations, develop problem-solving skills, and complete online homework assigned by your professor.

go Chemistry Download mini lecture videos for key concept review and exam prep from OWL or purchase them from **www.cengagebrain.com**.

Figure 8.1 ▲

Marie Sklodowska Curie (1867–1934), with Pierre Curie Marie Sklodowska Curie, born in Warsaw, Poland, began her doctoral work with Henri Becquerel soon after he discovered the spontaneous radiation emitted by uranium salts. She found this radiation to be an atomic property and coined the word *radioactivity* for it. In 1903 the Curies and Becquerel were awarded the Nobel Prize in physics for their discovery of radioactivity. Three years later, Pierre Curie was killed in a carriage accident. Marie Curie continued their work on radium and in 1911 was awarded the Nobel Prize in chemistry for the discovery of polonium and radium and the isolation of pure radium metal. This was the first time a scientist had received two Nobel awards. (Since then two others have been so honored.)

Marie Curie, a Polish-born French chemist, and her husband, Pierre, announced the discovery of radium in 1898 (Figure 8.1). They had separated a very radioactive mixture from pitchblende, an ore of uranium. This mixture was primarily a compound of barium. When the mixture was heated in a flame, however, it gave a new atomic line spectrum, in addition to the spectrum for barium. The Curies based their discovery of a new element on this finding. It took them four more years to obtain a pure compound of radium. Radium, like uranium, is a radioactive element. But in most of its chemical and physical properties, radium is similar to the nonradioactive element barium. It was this similarity that made the final separation of the new element so difficult.

Chemists had long known that groups of elements have similar properties. In 1869 Dmitri Mendeleev found that when the elements were arranged in a particular way, they fell into columns, with elements in the same column displaying similar properties. Thus, Mendeleev placed beryllium, calcium, strontium, and barium in one column. Now, with the Curies' discovery, radium was added to this column.

Mendeleev's arrangement of the elements, the *periodic table,* was originally based on the observed chemical and physical properties of the elements and their compounds. We now explain this arrangement in terms of the electronic structure of atoms. In this chapter we will look at this electronic structure and its relationship to the periodic table of elements.

Electronic Structure of Atoms

In Chapter 7 we found that an electron in an atom has four quantum numbers—n, l, m_l, and m_s—associated with it. The first three quantum numbers characterize the orbital that describes the region of space where an electron is most likely to be found; we say that the electron "occupies" this orbital. The spin quantum number, m_s, describes the spin orientation of an electron. In the first section, we will look further at electron spin; then we will discuss how electrons are distributed among the possible orbitals of an atom.

8.1 Electron Spin and the Pauli Exclusion Principle

Otto Stern and Walther Gerlach first observed electron spin magnetism in 1921. They directed a beam of silver atoms into the field of a specially designed magnet. The same experiment can be done with hydrogen atoms. The beam of hydrogen atoms is split into two by the magnetic field; half of the atoms are bent in one direction and half in the other (see Figure 8.2). The fact that the atoms are affected by the laboratory magnet shows that they themselves act as magnets.

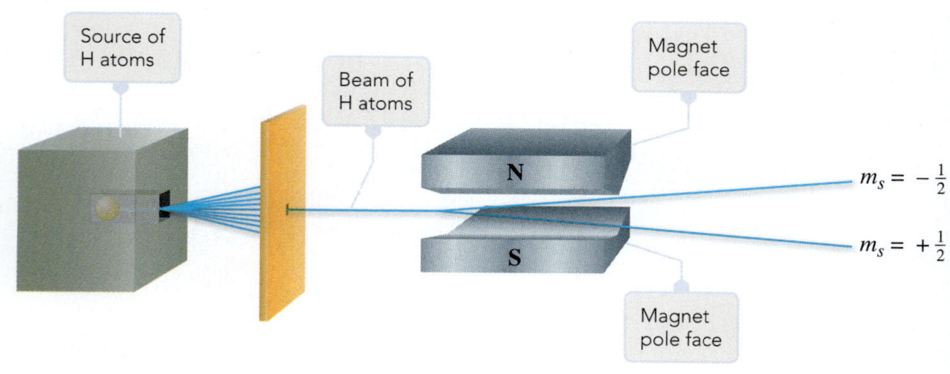

Figure 8.2 ▶

The Stern–Gerlach experiment The diagram shows the experiment using hydrogen atoms (simpler to interpret theoretically), although the original experiment employed silver atoms. A beam of hydrogen atoms (shown in blue) is split into two by a nonuniform magnetic field. One beam consists of atoms each with an electron having $m_s = +\frac{1}{2}$; the other beam consists of atoms each having an electron with $m_s = -\frac{1}{2}$.

The beam of hydrogen atoms is split into two because the electron in each atom behaves as a tiny magnet with only two possible orientations. In effect, the electron acts as though it were a ball of spinning charge (Figure 8.3), and, like a circulating electric charge, the electron would create a magnetic field. Electron spin, however, is subject to a quantum restriction on the possible directions of the spin axis. The resulting directions of spin magnetism correspond to spin quantum numbers $m_s = +\frac{1}{2}$ and $m_s = -\frac{1}{2}$. ▶

Protons and many nuclei also have spin. See the Instrumental Methods essay in this section.

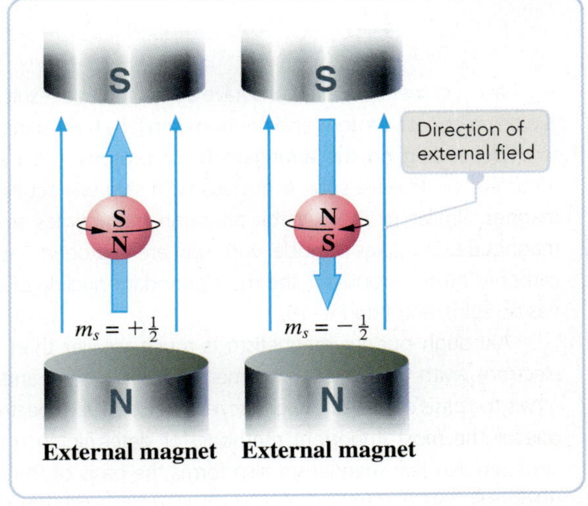

Figure 8.3 ▲

A representation of electron spin The two possible spin orientations are indicated by the models. By convention, the spin direction is given as shown by the large arrow on the spin axis. Electrons behave as tiny bar magnets, as shown in the figure.

Electron Configurations and Orbital Diagrams

We describe the electrons in an atom by the atom's electron configuration. An **electron configuration** of an atom is *a particular distribution of electrons among the available subshells.* A subshell, as we noted in the previous chapter, consists of a group of orbitals having the same n and l quantum numbers but different m_l values. Recall that we denote each subshell by its principal quantum number, n, followed by a letter standing for its l quantum number (s, p, d, f, etc.).

The notation for a configuration lists the subshell symbols one after the other, with a superscript giving the number of electrons in that subshell. For example, a configuration of the lithium atom (atomic number 3) with two electrons in the $1s$ subshell and one electron in the $2s$ subshell is written $1s^22s^1$.

Whereas the configuration gives the number of electrons in each subshell, we use *a diagram to show how the orbitals of a subshell are occupied by electrons.* It is called an **orbital diagram**. An orbital is represented by a circle. Each group of orbitals in a subshell is labeled by its subshell notation. An electron in an orbital is shown by an arrow; the arrow points up when $m_s = +\frac{1}{2}$ and down when $m_s = -\frac{1}{2}$. The orbital diagram

(↑↓) (↑↓) (↑)( )( )
$1s$ $2s$ $2p$

shows the electronic structure of an atom in which there are two electrons in the $1s$ subshell, or orbital (one electron with $m_s = +\frac{1}{2}$, the other with $m_s = -\frac{1}{2}$); two electrons in the $2s$ subshell ($m_s = +\frac{1}{2}, m_s = -\frac{1}{2}$); and one electron in the $2p$ subshell ($m_s = +\frac{1}{2}$). The electron configuration is $1s^22s^22p^1$.

Pauli Exclusion Principle

Not all of the conceivable arrangements of electrons among the orbitals of an atom are physically possible. The **Pauli exclusion principle**, which summarizes experimental observations, states that *no two electrons in an atom can have the same four quantum numbers.* If one electron in an atom has the quantum numbers $n = 1, l = 0, m_l = 0$, and $m_s = +\frac{1}{2}$, no other electron can have these same quantum numbers. In other words, you cannot place two electrons with the same value of m_s in a $1s$ orbital. The orbital diagram

$1s$

is not a possible arrangement of electrons.

Because there are only two possible values of m_s, an orbital can hold no more than two electrons—and then only if the two electrons have different spin quantum numbers. In an orbital diagram, an orbital with two electrons must be written

Instrumental Methods

Nuclear Magnetic Resonance (NMR)

You have just seen that electrons have a spin and as a result behave like tiny magnets. Protons and neutrons similarly have spins. Therefore, depending on the arrangement of protons and neutrons, a nucleus could have spin. A nucleus with spin will act like a bar magnet, similar to the electron although many times smaller in magnitude. Examples of nuclei with spin are hydrogen-1 (proton), carbon-13 (but carbon-12, the most abundant nuclide of carbon, has no spin), and fluorine-19.

Although nuclear magnetism is much smaller than that of electrons, with the correct equipment it is easily seen and in fact forms the basis of *nuclear magnetic resonance* (NMR) spectroscopy, one of the most important methods for determining molecular structure. Nuclear magnetism also forms the basis of the medical diagnostic tool *magnetic resonance imaging,* or MRI (Figure 8.4).

The essential features of NMR can be seen if you consider the proton. Like the electron, the proton has two spin states. In the absence of a magnetic field, these spin states have the same energy, but in the field of a strong magnet (external field), they have different energies. The state in which the proton magnetism is aligned with the external field, so the south pole of the proton magnet faces the north pole of the external magnet, will have lower energy. The state in which the proton magnet is turned 180°, with its south pole facing the south pole of the external magnet, will have higher energy. Now if a proton in the lower spin state is irradiated with electromagnetic waves of the proper frequency (in which the photon has energy equal to the difference in energy of the spin states), the proton can change to the higher spin state. The frequency absorbed by the proton depends on the magnitude of the magnetic field. For the magnets used in these instruments, the radiation lies in the radio-frequency range. Frequencies commonly used are 300 MHz and 900 MHz.

Figure 8.5 shows a diagram of an NMR spectrometer. It consists of a sample in the field of a variable electromagnet and near two coils, one a radio wave transmitter and the other a receiver coil perpendicular to the transmitter coil (so that the receiver will not pick up the signal from the transmitter). Suppose the transmitter radiates waves of 300 MHz. If the sample absorbs these radio waves, protons will undergo transitions from the lower to the higher spin state. Once protons are in the higher energy state, they tend to lose energy, going back to the lower spin state and radiating 300-MHz radio waves. Thus, the sample acts like a transmitter, but one with coils in various directions, so the signal can be detected by the receiver coil.

In general, the sample will not absorb at the chosen frequency. But you can change the energy difference between spin states, and therefore the frequency that is absorbed by the sample, by increasing or decreasing the magnitude of the external magnetic field using small coils on the magnet pole faces. In so doing you can, in effect, "tune" the sample, or bring it into "resonance" with the transmitter frequency. (Alternatively, many modern instruments vary the transmitter frequency to obtain resonance.)

Because each proton in a substance is surrounded by electrons that have their own magnetic fields, the magnetic environment of a proton depends to some extent on the bonding, or chemical, environment in which the proton is involved. So, the external magnetic field needed to bring a proton into resonance with the 300-MHz radiation, say, varies with the chemical (bonding) environment. A spectrum is produced by recording the magnetic field (or frequency) at which the protons in a molecule produce a resonance signal. Figure 8.6 shows a high-resolution NMR spectrum of ethanol.

Ethanol, whose molecular structure is

```
        H   H
        |   |
  H—O—C—C—H
        |   |
        H   H
```

has protons in three different chemical environments: a proton bonded to an oxygen atom (H—O—), the protons in a $—CH_2—$ group, and the protons in a $—CH_3$ group. Each of these three

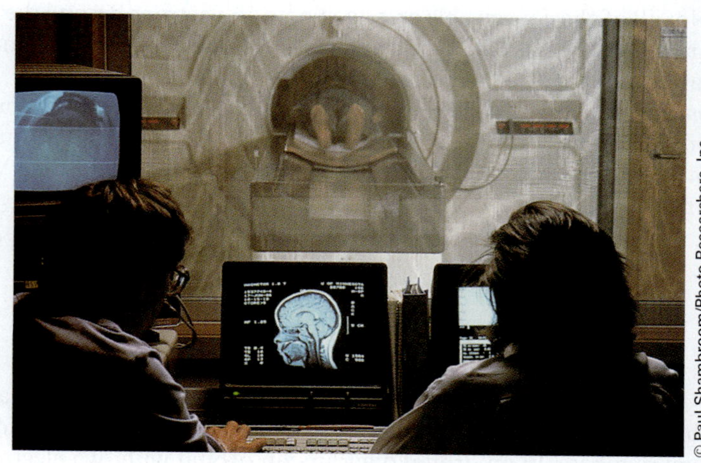

© Paul Shambroom/Photo Researchers, Inc.

Figure 8.4 ▶

Magnetic resonance imaging A patient's head is placed in a large magnet and subjected to a radio pulse. Proton spin transitions give rise to a radio wave emission that can be analyzed electronically and converted by computer to a two-dimensional image of a plane portion of the brain. Paul C. Lauterbur, a chemist at the University of Illinois (Urbana), and Sir Peter Mansfield, University of Nottingham (UK), won the Nobel Prize in Physiology or Medicine in 2003 for their work in the early 1970s in developing MRI.

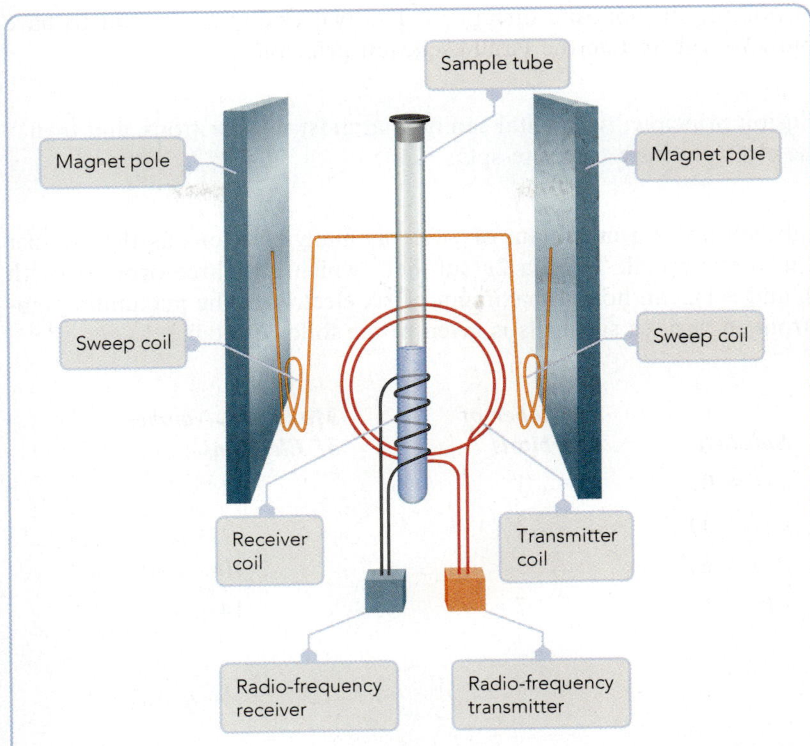

Figure 8.5 ◀

Nuclear magnetic resonance spectrometer A sample in the tube is in a magnetic field, which can be varied by changing the electric current in the *sweep* coils (shown in yellow). A radio-frequency transmitter radiates radio waves from a coil (shown in red) to the sample. The sample absorbs these waves if the radio frequency corresponds to the difference in energy of the sample's nuclear spin states. When the nuclear spins go back to the ground state, the sample emits radio waves, which are detected by the receiver coil (in black).

types of protons shows up as a peak or, as we will see, a group of peaks in the NMR spectrum. The position of the center of a group of peaks for a given proton relative to some standard is referred to as the *chemical shift* for that proton, because it depends on the chemical environment of the proton. By measuring the area under the peaks for a given type of proton (usually done electronically by the spectrometer), a researcher can discover the number of protons of that type, effectively giving the chemist structural information about the molecule.

As we noted earlier, a given proton may give rise to several peaks in a high-resolution spectrum. (At low resolution, these separate peaks may appear as only one for each proton.) For example, the protons for the —CH_2— group of ethanol give four closely spaced peaks. These peaks arise because of the interaction of the proton spins on this group with those protons on the adjacent —CH_3 group. The number of peaks (four) for the —CH_2— group is one more than the number of protons on the adjacent group (three).

NMR spectroscopy is one of the most important tools a chemist has in determining the identity of a substance and its molecular structure. The method is quick and can yield such information as the chemical bonding environment of the protons and the number of protons having a given environment. It can also give information about other nuclei, such as carbon-13.

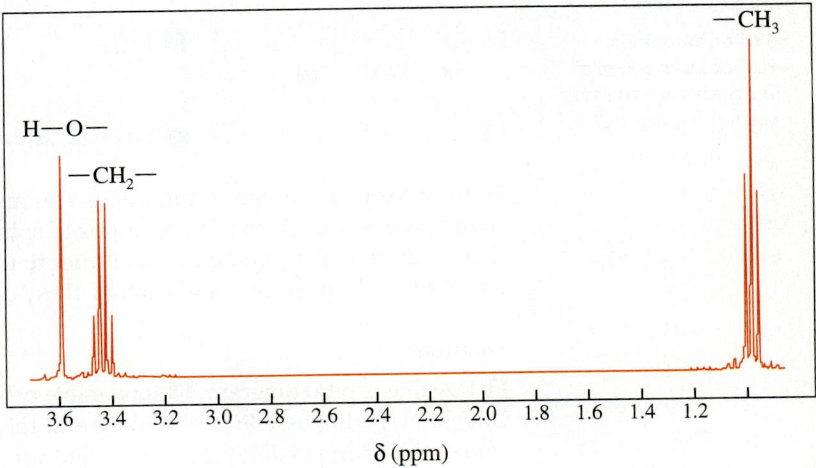

Figure 8.6 ▲

High-resolution NMR spectrum of ethanol, CH_3CH_2OH The protons of a given type occur as a peak or closely spaced group of peaks. The position of the center of such a group of peaks relative to some standard is referred to as the *chemical shift*. The peaks within a group arise from the interaction of the spins between protons of one type with those of another type.

■ See Problems 8.85 and 8.86.

with arrows pointing in opposite directions. The two electrons are said to have opposite spins. We can restate the Pauli exclusion principle:

Pauli exclusion principle: An orbital can hold at most two electrons, and then only if the electrons have opposite spins.

Each subshell holds a maximum of twice as many electrons as the number of orbitals in the subshell. Thus, a $2p$ subshell, which has three orbitals (with $m_l = -1$, 0, and +1), can hold a maximum of six electrons. The maximum number of electrons in various subshells is given in the following table.

Subshell	***Number of Orbitals***	***Maximum Number of Electrons***
s ($l = 0$)	1	2
p ($l = 1$)	3	6
d ($l = 2$)	5	10
f ($l = 3$)	7	14

Example 8.1 Applying the Pauli Exclusion Principle

Gaining Mastery Toolbox

Critical Concept 8.1
The Pauli exclusion principle states that an orbital can hold at most two electrons and then only if the electrons have opposite spin.

Solution Essentials:
- Pauli exclusion principle
- The number of orbitals in a subshell of given l is $2l + 1$

Which of the following orbital diagrams or electron configurations are possible and which are impossible, according to the Pauli exclusion principle? Explain.

a (↑↓) (↑↓) (↑)( )( )
$1s$ $2s$ $2p$

b (↑↓) (↑↓↑) ( )( )( )
$1s$ $2s$ $2p$

c (↑↓) (↑) (↑↑)( )( )
$1s$ $2s$ $2p$

d $1s^3 2s^1$

e $1s^2 2s^1 2p^7$

f $1s^2 2s^2 2p^6 3s^2 3p^6 3d^8 4s^2$

Problem Strategy In any orbital diagram, make sure that each orbital contains no more than two electrons, which have opposite spins. In any electron configuration, make sure that a subshell of given l contains no more than $2(2l + 1)$ electrons (there are $2l + 1$ different orbitals of given l, and each of these orbitals can hold two electrons).

Solution

a **Possible** orbital diagram. **b** **Impossible** orbital diagram; there are three electrons in the $2s$ orbital. **c** **Impossible** orbital diagram; there are two electrons in a $2p$ orbital with the same spin. **d** **Impossible** electron configuration; there are three electrons in the $1s$ subshell (one orbital). **e** **Impossible** electron configuration; there are seven electrons in the $2p$ subshell (which can hold only six electrons). **f** **Possible**. Note that the $3d$ subshell can hold as many as ten electrons.

Answer Check The overarching idea here is that each orbital in an atom can hold a maximum of only two electrons, and then only if the two electrons have opposite spins. For example, a p subshell has three orbitals and therefore holds a maximum of six electrons.

(continued)

(continued)

Exercise 8.1 Look at the following orbital diagrams and electron configurations. Which are possible and which are not, according to the Pauli exclusion principle? Explain.

a (↑) (↑) ( )( )( ) — 1s 2s 2p

b (↑) (↑) (↑↓)(↑↓)(↑↓) — 1s 2s 2p

c (↑↓) (↑↓) (↑↑)(↑↓)(↑↓) — 1s 2s 2p

d $1s^22s^22p^4$

e $1s^22s^42p^2$

f $1s^22s^22p^63s^23p^{10}3d^{10}$

■ See Problems 8.41, 8.42, 8.43, and 8.44.

8.2 Building-Up Principle and the Periodic Table

Every atom has an infinite number of possible electron configurations. The configuration associated with the lowest energy level of the atom corresponds to a quantum-mechanical state called the *ground state.* Other configurations correspond to *excited states,* associated with energy levels other than the lowest. For example, the ground state of the sodium atom is known from experiment to have the electron configuration $1s^22s^22p^63s^1$. The electron configuration $1s^22s^22p^63p^1$ represents an excited state of the sodium atom. ▶

The transition of the sodium atom from the excited state $1s^22s^22p^63p^1$ to the ground state $1s^22s^22p^63s^1$ is accompanied by the emission of yellow light at 589 nm. Excited states of an atom are needed to describe its spectrum.

The chemical properties of an atom are related primarily to the electron configuration of its ground state. Table 8.1 lists the experimentally determined

Table 8.1 Ground-State Electron Configurations of Atoms Z = 1 to 36*

Z	Element	Configuration	Z	Element	Configuration
1	H	$1s^1$	19	K	$1s^22s^22p^63s^23p^64s^1$
2	He	$1s^2$	20	Ca	$1s^22s^22p^63s^23p^64s^2$
3	Li	$1s^22s^1$	21	Sc	$1s^22s^22p^63s^23p^63d^14s^2$
4	Be	$1s^22s^2$	22	Ti	$1s^22s^22p^63s^23p^63d^24s^2$
5	B	$1s^22s^22p^1$	23	V	$1s^22s^22p^63s^23p^63d^34s^2$
6	C	$1s^22s^22p^2$	24	Cr	$1s^22s^22p^63s^23p^63d^54s^1$
7	N	$1s^22s^22p^3$	25	Mn	$1s^22s^22p^63s^23p^63d^54s^2$
8	O	$1s^22s^22p^4$	26	Fe	$1s^22s^22p^63s^23p^63d^64s^2$
9	F	$1s^22s^22p^5$	27	Co	$1s^22s^22p^63s^23p^63d^74s^2$
10	Ne	$1s^22s^22p^6$	28	Ni	$1s^22s^22p^63s^23p^63d^84s^2$
11	Na	$1s^22s^22p^63s^1$	29	Cu	$1s^22s^22p^63s^23p^63d^{10}4s^1$
12	Mg	$1s^22s^22p^63s^2$	30	Zn	$1s^22s^22p^63s^23p^63d^{10}4s^2$
13	Al	$1s^22s^22p^63s^23p^1$	31	Ga	$1s^22s^22p^63s^23p^63d^{10}4s^24p^1$
14	Si	$1s^22s^22p^63s^23p^2$	32	Ge	$1s^22s^22p^63s^23p^63d^{10}4s^24p^2$
15	P	$1s^22s^22p^63s^23p^3$	33	As	$1s^22s^22p^63s^23p^63d^{10}4s^24p^3$
16	S	$1s^22s^22p^63s^23p^4$	34	Se	$1s^22s^22p^63s^23p^63d^{10}4s^24p^4$
17	Cl	$1s^22s^22p^63s^23p^5$	35	Br	$1s^22s^22p^63s^23p^63d^{10}4s^24p^5$
18	Ar	$1s^22s^22p^63s^23p^6$	36	Kr	$1s^22s^22p^63s^23p^63d^{10}4s^24p^6$

*A complete table appears in Appendix D.

The quantum numbers and characteristics of orbitals were discussed in Section 7.5.

ground-state electron configurations of atoms $Z = 1$ to $Z = 36$. (A complete table appears in Appendix D.)

Building-Up Principle (Aufbau Principle)

Most of the configurations in Table 8.1 can be explained in terms of the **building-up principle** (or **Aufbau principle**), *a scheme used to reproduce the electron configurations of the ground states of atoms by successively filling subshells with electrons in a specific order (the building-up order).* Following this principle, you obtain the electron configuration of an atom by successively filling subshells in the following order: 1*s*, 2*s*, 2*p*, 3*s*, 3*p*, 4*s*, 3*d*, 4*p*, 5*s*, 4*d*, 5*p*, 6*s*, 4*f*, 5*d*, 6*p*, 7*s*, 5*f*. This order reproduces the experimentally determined electron configurations (with some exceptions, which we will discuss later). You need not memorize this order. As you will see, you can very easily obtain it from the periodic table. (You can also reproduce this order from the mnemonic diagram shown in Figure 8.7.)

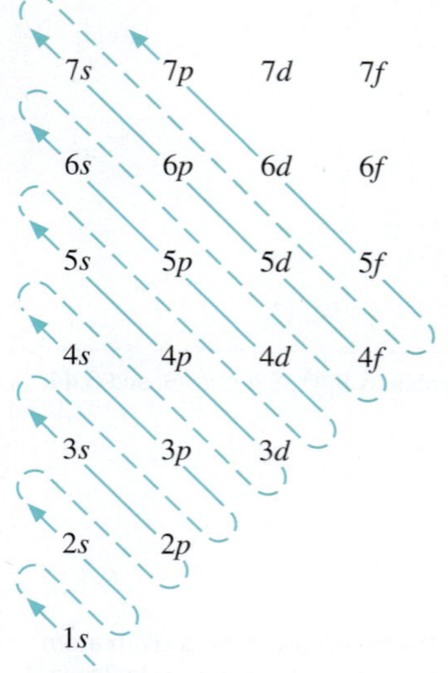

Figure 8.7 ▲

A mnemonic diagram for the building-up order (diagonal rule) You obtain this diagram by writing the subshell(s) in rows, each row having subshell(s) of given *n*. Within each row, you arrange the subshell(s) by increasing *l*. (You can stop after writing the *nf* subshell(s), since no known elements contain *g* or higher subshell(s).) Now, starting with the 1*s* subshell(s), draw a series of diagonals, as shown. The building-up order is the order in which these diagonals strike the subshell(s).

The building-up order corresponds for the most part to increasing energy of the subshells. You might expect this. By filling orbitals of lowest energy first, you usually get the lowest total energy (ground state) of the atom. Recall that the energy of an orbital depends only on the quantum numbers *n* and *l*. ◀ (The energy of the H atom, however, depends only on *n*.) Orbitals with the same *n* and *l* but different m_l—that is, different orbitals of the same subshell—have the same energy. The energy depends primarily on *n*, increasing with its value. For example, a 3*s* orbital has greater energy than a 2*s* orbital. Except for the H atom, the energies of orbitals with the same *n* increase with the *l* quantum number. A 3*p* orbital has slightly greater energy than a 3*s* orbital. The orbital of lowest energy is 1*s*; next higher are 2*s* and 2*p*, then 3*s* and 3*p*. The order of these subshells by energy, from 1*s* to 3*p*, follows the building-up order as listed earlier.

When subshells have nearly the same energy, however, the building-up order is not strictly determined by the order of their energies. The ground-state configurations, which we are trying to predict by the building-up order, are determined by the *total* energies of the atoms. The total energy of an atom depends not only on the energies of the subshells but also on the energies of interaction among the different subshells. It so happens that for all elements with $Z = 21$ or greater, the energy of the 3*d* subshell is lower than the energy of the 4*s* subshell (Figure 8.8), which is opposite to the building-up order. You need the building-up order to predict the electron configurations of the ground states of atoms.

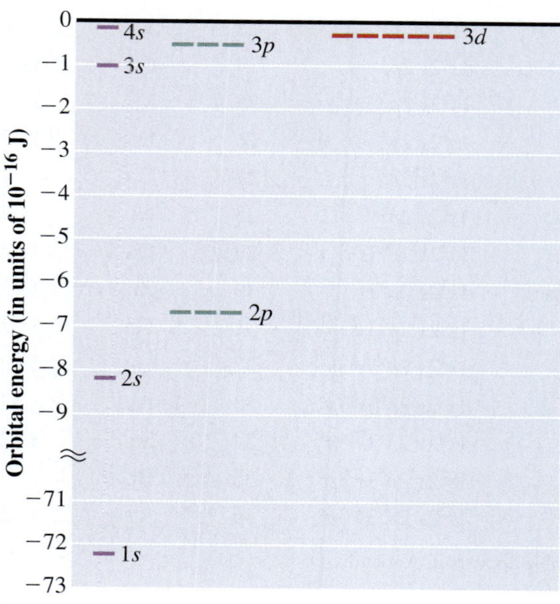

Figure 8.8 ▶

Orbital energies for the scandium atom (Z = 21) Note that in the scandium atom, unlike the hydrogen atom, the subshells for each *n* are spread apart in energy. Thus, the 2*p* energy is above the 2*s*. Similarly, the $n = 3$ subshells are spread to give the order $3s < 3p < 3d$. The 3*d* subshell energy is now just below the 4*s*. (Values for this figure were calculated from theory by Charlotte F. Fischer, Vanderbilt University.)

Now you can see how to reproduce the electron configurations of Table 8.1 using the building-up principle. Remember that the number of electrons in a neutral atom equals the atomic number Z. (The nuclear charge is $+Z$.) In the case of the simplest atom, hydrogen ($Z = 1$), you obtain the ground state by placing the single electron into the 1*s* orbital, giving the configuration $1s^1$ (this is read as "one-ess-one"). Now you go to helium ($Z = 2$). The first electron goes into the 1*s* orbital, as in hydrogen, followed by the second electron, because any orbital can hold two electrons. The configuration is $1s^2$. Filling the $n = 1$ shell creates a very stable configuration, and as a result, helium is chemically unreactive.

You continue this way through the elements, each time increasing Z by 1 and adding another electron. You obtain the configuration of an atom from that of the preceding element by adding an electron into the next available orbital, following the building-up order. In lithium ($Z = 3$), the first two electrons give the configuration $1s^2$, like helium, but the third electron goes into the next higher orbital in the building-up order, because the 1*s* orbital is now filled. This gives the configuration $1s^22s^1$. In beryllium ($Z = 4$), the fourth electron fills the 2*s* orbital, giving the configuration $1s^22s^2$.

Using the abbreviation [He] for $1s^2$, the configurations are

$Z = 3$	lithium	$1s^22s^1$	or	$[\text{He}]2s^1$
$Z = 4$	beryllium	$1s^22s^2$	or	$[\text{He}]2s^2$

With boron ($Z = 5$), the electrons begin filling the 2*p* subshell. You get

$Z = 5$	boron	$1s^22s^22p^1$	or	$[\text{He}]2s^22p^1$
$Z = 6$	carbon	$1s^22s^22p^2$	or	$[\text{He}]2s^22p^2$
⋮				
$Z = 10$	neon	$1s^22s^22p^6$	or	$[\text{He}]2s^22p^6$

Having filled the 2*p* subshell, you again find a particularly stable configuration. Neon is chemically unreactive as a result.

With sodium ($Z = 11$), the 3*s* orbital begins to fill. Using the abbreviation [Ne] for $1s^22s^22p^6$, you have

$Z = 11$	sodium	$1s^22s^22p^63s^1$	or	$[\text{Ne}]3s^1$
$Z = 12$	magnesium	$1s^22s^22p^63s^2$	or	$[\text{Ne}]3s^2$

Then the 3*p* subshell begins to fill.

$Z = 13$	aluminum	$1s^22s^22p^63s^23p^1$	or	$[\text{Ne}]3s^23p^1$
⋮				
$Z = 18$	argon	$1s^22s^22p^63s^23p^6$	or	$[\text{Ne}]3s^23p^6$

With the 3*p* subshell filled, a stable configuration has been attained; argon is an unreactive element.

Now the 4*s* orbital begins to fill. You get $[\text{Ar}]4s^1$ for potassium ($Z = 19$) and $[\text{Ar}]4s^2$ for calcium ($Z = 20$) ($[\text{Ar}] = 1s^22s^22p^63s^23p^6$). At this point the 3*d* subshell begins to fill. You get $[\text{Ar}]3d^14s^2$ for scandium ($Z = 21$), $[\text{Ar}]3d^24s^2$ for titanium ($Z = 22$), and $[\text{Ar}]3d^34s^2$ for vanadium ($Z = 23$). *Note that we have written the configurations with subshells arranged in order by shells.* This generally places the subshells in order by energy and puts the subshells involved in chemical reactions at the far right. ▶

Although the building-up order reproduces the ground-state electron configurations, it has no other significance. The order by n (and then by l within a given n), however, generally places the most easily ionized orbitals at the far right. For example, the electron configuration of Fe (in order by shells) is $1s^22s^22p^63s^23p^63d^64s^2$. The 4s electrons ionize first.

Let us skip to zinc ($Z = 30$). The 3*d* subshell has filled; the configuration is $[\text{Ar}]3d^{10}4s^2$. Now the 4*p* subshell begins to fill, starting with gallium ($Z = 31$), configuration $[\text{Ar}]3d^{10}4s^24p^1$, and ending with krypton ($Z = 36$), configuration $[\text{Ar}]3d^{10}4s^24p^6$.

Electron Configurations and the Periodic Table

By this time you can see a pattern develop among the ground-state electron configurations of the atoms. This pattern explains the periodic table, which was briefly described in Section 2.5. Consider helium, neon, argon, and krypton, elements in Group VIIIA of the periodic table. Neon, argon, and krypton have configurations in which a p subshell has just filled. (Helium has a filled $1s$ subshell; no $1p$ subshell is possible.)

helium	$1s^2$
neon	$1s^22s^22p^6$
argon	$1s^22s^22p^63s^23p^6$
krypton	$1s^22s^22p^63s^23p^63d^{10}4s^24p^6$

These elements are the first members of the group called *noble gases* because of their relative unreactivity.

Look now at the configurations of beryllium, magnesium, and calcium, members of the group of *alkaline earth metals* (Group IIA), which are similar, moderately reactive elements.

beryllium	$1s^22s^2$	or	$[He]2s^2$
magnesium	$1s^22s^22p^63s^2$	or	$[Ne]3s^2$
calcium	$1s^22s^22p^63s^23p^64s^2$	or	$[Ar]4s^2$

Each of these configurations consists of a **noble-gas core**, that is, *an inner-shell configuration corresponding to one of the noble gases,* plus two outer electrons with an ns^2 configuration.

The elements boron, aluminum, and gallium (Group IIIA) also have similarities. Their configurations are

boron	$1s^22s^22p^1$	or	$[He]2s^22p^1$
aluminium	$1s^22s^22p^63s^23p^1$	or	$[Ne]3s^23p^1$
gallium	$1s^22s^22p^63s^23p^63d^{10}4s^24p^1$	or	$[Ar]3d^{10}4s^24p^1$

Boron and aluminum have noble-gas cores plus three electrons with the configuration ns^2np^1. Gallium has an additional filled $3d$ subshell. *The noble-gas core together with* $(n - 1)d^{10}$ *electrons* is often referred to as a **pseudo-noble-gas core**, because these electrons usually are not involved in chemical reactions.

An electron in an atom outside the noble-gas or pseudo-noble-gas core is called a **valence electron**. Such electrons are primarily involved in chemical reactions, and similarities among the configurations of valence electrons (the *valence-shell configurations*) account for similarities of the chemical properties among groups of elements.

Figure 8.9 shows a periodic table with the valence-shell configurations included. Note the similarity in electron configuration within any group (column) of elements. This similarity explains what chemists since Mendeleev have known—the properties of elements in any group are similar.

The *main-group* (or *representative*) *elements* all have valence-shell configurations ns^anp^b, with some choice of a and b. (b could be equal to 0.) In other words, the outer s or p subshell is being filled. Similarly, in the *d-block transition elements* (often called simply *transition elements* or *transition metals*), a d subshell is being filled. In the *f-block transition elements* (or *inner transition elements*), an f subshell is being filled. (See Figure 8.9 or Appendix D for the configurations of these elements.)

Exceptions to the Building-Up Principle

As we have said, the building-up principle reproduces most of the ground-state configurations correctly. There are some exceptions, however, and chromium (Z = 24) is the first we encounter. The building-up principle predicts the configuration

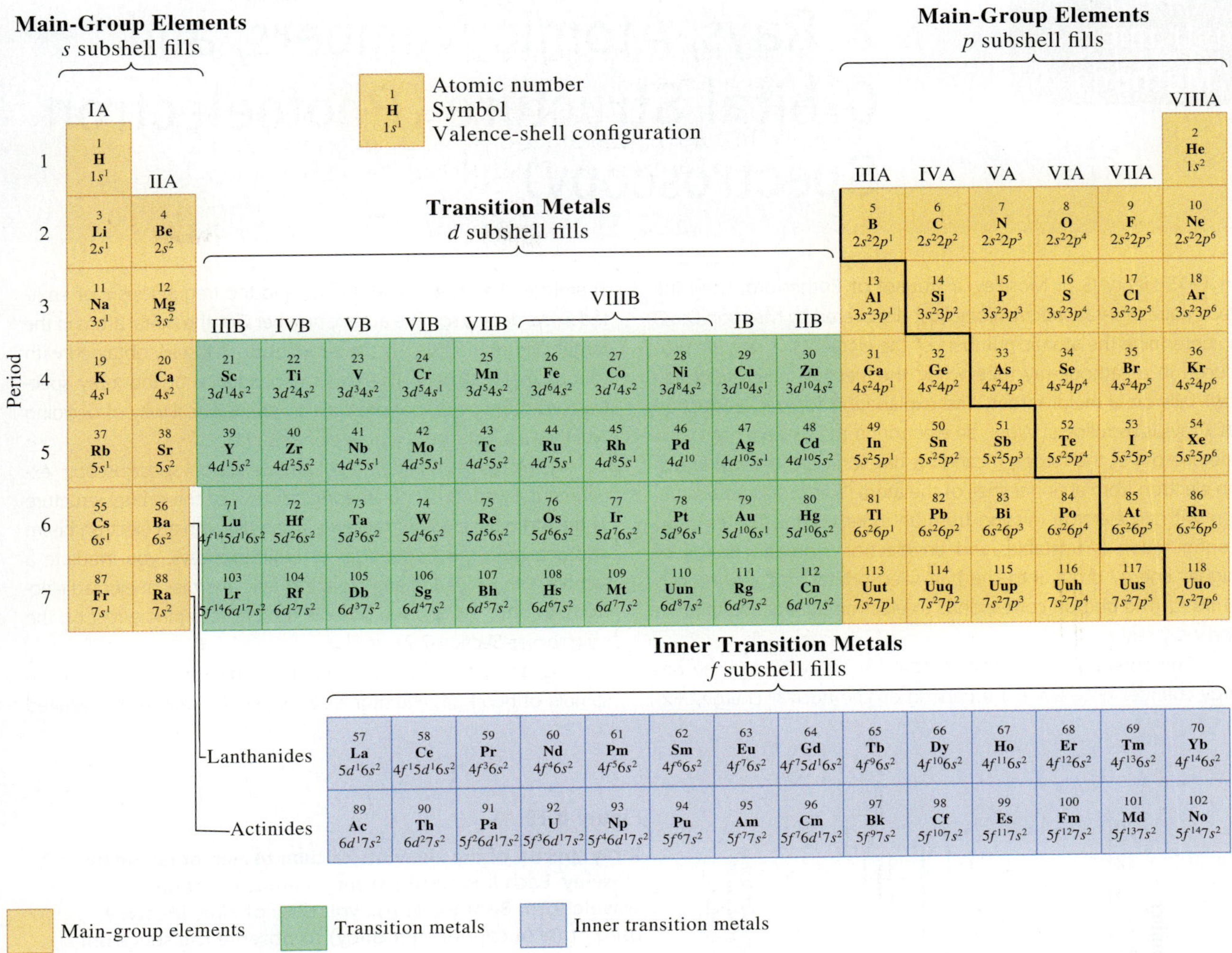

Figure 8.9 ▲

A periodic table This table shows the valence-shell configurations of the elements.

$[Ar]3d^44s^2$, though the correct one is found experimentally to be $[Ar]3d^54s^1$. These two configurations are actually very close in total energy because of the closeness in energies of the $3d$ and $4s$ orbitals (Figure 8.8). For that reason, small effects can influence which of the two configurations is actually lower in energy. Copper ($Z = 29$) is another exception to the building-up principle, which predicts the configuration $[Ar]3d^94s^2$, although experiment shows the ground-state configuration to be $[Ar]3d^{10}4s^1$.

We need not dwell on these exceptions beyond noting that they occur. The point to remember is that the configuration predicted by the building-up principle is very close in energy to the ground-state configuration (if it is not the ground state). Most of the qualitative conclusions regarding the chemistry of an element are not materially affected by arguing from the configuration given by the building-up principle. ▶

More exceptions occur among the heavier transition elements, where the outer subshells are very close together. We must concede that simplicity was not the uppermost concern in the construction of the universe!

CONCEPT CHECK 8.1

Imagine a world in which the Pauli principle is "No more than one electron can occupy an atomic orbital, irrespective of its spin." How many elements would there be in the second row of the periodic table, assuming that nothing else is different about this world?

Instrumental Methods

X Rays, Atomic Numbers, and Orbital Structure (Photoelectron Spectroscopy)

In 1913 Henry G. J. Moseley, a student of Rutherford, used the technique of *x-ray spectroscopy* (just discovered by Max von Laue) to determine the atomic numbers of the elements. X rays are produced in a cathode-ray tube when the electron beam (cathode ray) falls on a metal target. The explanation for the production of x rays is as follows: When an electron in the cathode ray hits a metal atom in the target, it can (if it has sufficient energy) knock an electron from an inner shell of the atom. This produces a metal ion with an electron missing from an inner orbital. The electron configuration is unstable, and an electron from an orbital of higher energy drops into the half-filled orbital and a photon is emitted. The photon corresponds to electromagnetic radiation in the x-ray region.

The energies of the inner orbitals of an atom and the energy changes between them depend on the nuclear charge, $+Z$. Therefore, the photon energies $h\nu$ and the frequencies ν of emitted x rays depend on the atomic number Z of the metal atom in the target. Figure 8.10 shows the x-ray spectra Moseley obtained with various metal targets. The direct dependence of the x-ray spectrum on atomic number provides an unequivocal way of deciding whether a substance is a pure element or not.

A related technique, *x-ray photoelectron spectroscopy,* experimentally confirms our theoretical view of the orbital structure of the atom. Instead of irradiating a sample with an electron beam and analyzing the frequencies of emitted x rays, you irradiate a sample with x rays and analyze the kinetic energies of ejected electrons. In other words, you observe the *photoelectric effect* on the sample (see Section 7.2).

As an example of photoelectron spectroscopy, consider a sample of neon gas (Ne atoms). Suppose the sample is irradiated

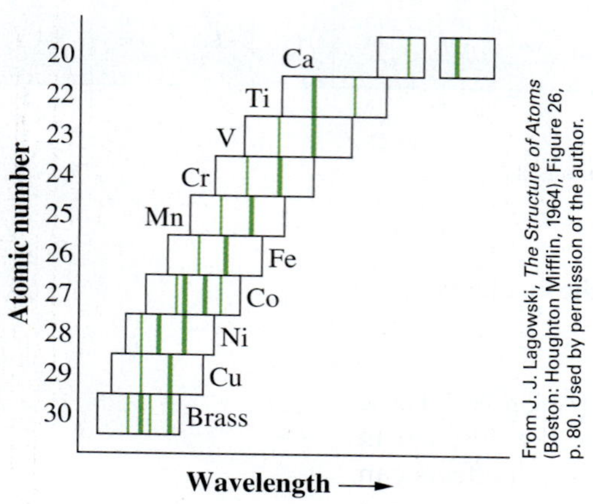

From J. J. Lagowski, *The Structure of Atoms* (Boston: Houghton Mifflin, 1964), Figure 26, p. 80. Used by permission of the author.

Figure 8.10

X-ray spectra of the elements calcium to zinc, obtained by Moseley Each line results from an emission of given wavelength. Because of the volatility of zinc, Moseley used brass (a copper–zinc alloy) to observe the spectrum of zinc. Note the copper lines in brass. Also note how the lines progress to the right (indicating increasing wavelength or decreasing energy difference) with decreasing atomic number.

8.3 Writing Electron Configurations Using the Periodic Table

To discuss bonding and the chemistry of the elements coherently, you must be able to reproduce the electron configurations with ease, following the building-up principle. All you need is some facility in recall of the building-up order of subshells.

One approach is to recall the structure of the periodic table. Because that structure is basic, it offers a sound way to remember the building-up order. There is a definite pattern to the order of filling of the subshells as you go through the with x rays of a specific frequency great enough to remove a 1*s* electron from the neon atom. Part of the energy of the x-ray photon, $h\nu$, is used to remove the electron from the atom. (This is the *ionization energy, I.E.*, for that electron.) The remaining energy appears as kinetic energy, E_k, of the ejected electron. From the law of conservation of energy, you can write

$$E_k = h\nu - I.E.$$

Because $h\nu$ is fixed, E_k will depend linearly on *I.E.*, the ionization energy.

If you look at the electrons ejected from neon, you find that they have kinetic energies related to the ionization energies from all possible orbitals (1*s*, 2*s*, and 2*p*) in the atom. When you scan the various kinetic energies of ejected electrons, you see a spectrum with peaks corresponding to the different occupied orbitals (see Figure 8.11*a*). These ionization energies are approximately equal to the positive values of the orbital energies (Figure 8.11*B*), so this spectrum provides direct experimental verification of the discrete energy levels associated with the electrons of the atom.

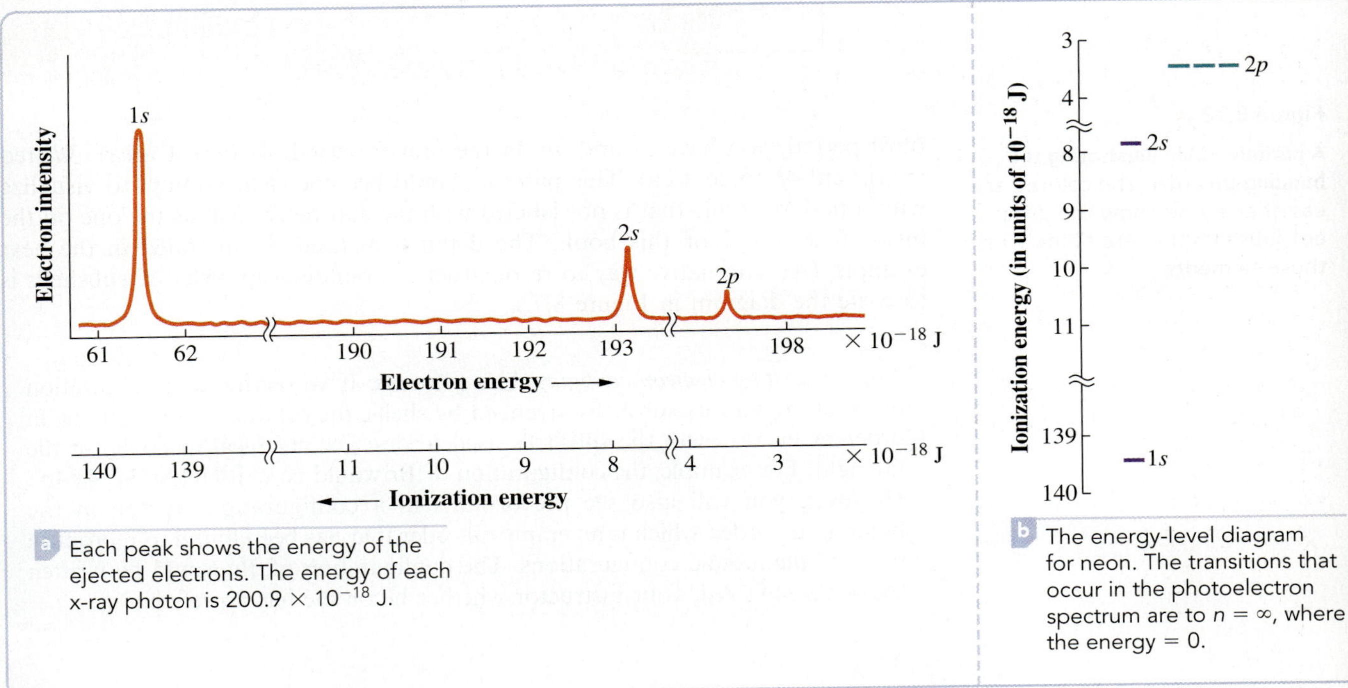

a Each peak shows the energy of the ejected electrons. The energy of each x-ray photon is 200.9×10^{-18} J.

b The energy-level diagram for neon. The transitions that occur in the photoelectron spectrum are to $n = \infty$, where the energy = 0.

Figure 8.11 ▲

X-ray photoelectron spectrum of neon

■ See Problems 8.87 and 8.88.

elements in the periodic table, and from this you can write the building-up order. Figure 8.12 shows a periodic table stressing this pattern. For example, in the violet-colored area, an *ns* subshell is being filled. In the blue-colored area, an *np* subshell is being filled. The value of *n* is obtained from the period (row) number. In the red area, an $(n - 1)d$ subshell is being filled.

You read the building-up order by starting with the first period, in which the 1*s* subshell is being filled. In the second period, you have 2*s* (violet area); then, staying in the same period but jumping across, you have 2*p* (blue area). In the

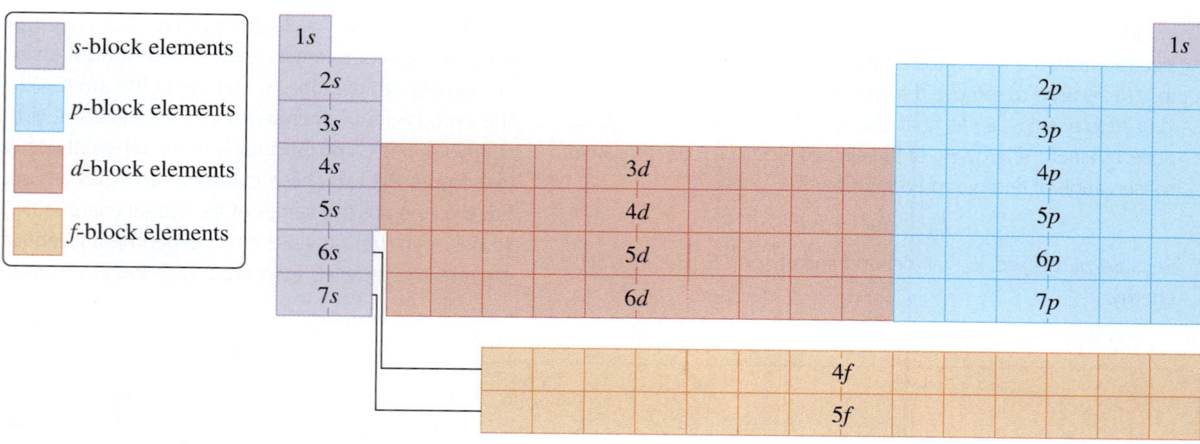

Figure 8.12 ▲

A periodic table illustrating the building-up order The colored areas of elements show the different subshells that are filling with those elements.

third period, you have 3*s* and 3*p*; in the fourth period, 4*s* (violet area), 3*d* (red area), and 4*p* (blue area). This pattern should become clear enough to visualize with a periodic table that is not labeled with the subshells, such as the one on the inside front cover of this book. The detailed method is illustrated in the next example. (An alternative way to reconstruct the building-up order of subshells is to write the diagram in Figure 8.7.)

Note on writing electron configurations of atoms: If you write the configuration of an atom with its subshells arranged by shells, the orbitals will usually be in order by energy, with the subshells used to describe chemical reactivity at the far right. For example, the configuration of Br would be written $[Ar]3d^{10}4s^24p^5$. However, you will also see the orbitals in a configuration written in the building-up order, which is an empirical order that has been found to reproduce most of the atomic configurations. The configuration of Br would be written $[Ar]4s^23d^{10}4p^5$. Ask your instructor whether he or she has a preference.

OWL Interactive Example 8.2 Determining the Configuration of an Atom Using the Building-Up Principle

Gaining Mastery Toolbox

Critical Concept 8.2

To write the electron configuration for the ground state of a neutral atom of atomic number *Z*, write the subshells in building-up order and then distribute *Z* electrons into these subshells, starting with the 1*s* subshell.

Solution Essentials:

- Building-up order of subshells
- Pauli exclusion principle
- The number of orbitals in a subshell of given *l* is $2l + 1$.

Use the building-up principle to obtain the configuration for the ground state of the gallium atom ($Z = 31$). Give the configuration in complete form (do not abbreviate for the core). What is the valence-shell configuration?

Problem Strategy Write the subshells in their building-up order. You can use any periodic table to do this. All you have to remember is the pattern in which subshells are filled as you progress through the table (Figure 8.12). Go through the periods, starting with hydrogen, writing the subshells that are being filled, and stopping with the element whose configuration you want (gallium in this example). Then you distribute the electrons (equal to the atomic number of the element) to the subshells. The valence-shell configuration includes only the subshells outside the noble-gas or pseudo-noble-gas core.

Solution From a periodic table, you get the following building-up order:

	1*s*	2*s* 2*p*	3*s* 3*p*	4*s* 3*d* 4*p*
Period:	first	second	third	fourth

Now you fill the subshells with electrons, remembering that you have a total of 31 electrons to distribute. You get

$$1s^22s^22p^63s^23p^64s^23d^{10}4p^1$$

Or, if you rearrange the subshells by shells, you write

$$\mathbf{1s^22s^22p^63s^23p^63d^{10}4s^24p^1}$$

The valence-shell configuration is $\mathbf{4s^24p^1}$.

(continued)

(continued)

Answer Check The principal quantum number of the valence shell must equal the period of the element. If your answer is not consistent with this, examine your work for error. Here, gallium is in period 4, and the principal quantum number of the valence shell is 4.

Exercise 8.2 Use the building-up principle to obtain the electron configuration for the ground state of the manganese atom ($Z = 25$).

■ See Problems 8.47, 8.48, 8.49, and 8.50.

In many cases, you need only the configuration of the outer electrons. You can determine this from the position of the element in the periodic table. Recall that the valence-shell configuration of a main-group element is ns^anp^b, where n, the principal quantum number of the outer shell, also equals the period number for the element. The total number of valence electrons, which equals $a + b$, can be obtained from the group number. For example, gallium is in Period 4, so $n = 4$. It is in Group IIIA, so the number of valence electrons is 3. This gives the valence-shell configuration $4s^24p^1$. The configuration of outer shells of a transition element is obtained in a similar fashion. The next example gives the details.

Example 8.3 Determining the Configuration of an Atom Using the Period and Group Numbers

Gaining Mastery Toolbox

Critical Concept 8.3
To write the valence-shell (or outer-shell) configuration of an element using the periodic table, determine the period and group number of the element, noting whether it is a main-group or transition element. The valence-shell configuration of a main-group element is ns^anp^b, where n equals the period of the element and $a + b$ the group number. (Distribute the electrons to the ns subshell, then to the np subshell.) The outer-shell configuration of a transition element is $(n - 1)d^{a-2}ns^2$, where a is the group number.

Solution Essentials:
- Determine the period and group number of an element
- Pauli exclusion principle
- The number of orbitals in a subshell of given l is $2l + 1$.

What are the configurations for the outer electrons of **a** tellurium, $Z = 52$, and **b** nickel, $Z = 28$?

Problem Strategy Find the period and the group number (the Roman numeral) of the element for the atom. Note whether it is a main-group or a transition element.

If the atom is that of a main-group element, the valence configuration is ns^anp^b, where n equals the period of the element and $a + b$ equals the group number (which equals the number of valence electrons). Distribute these electrons to the ns orbital; then distribute any remaining electrons to the np orbitals.

If the atom is that of a transition element, the outer-shell configuration is $(n - 1)d^{a-2}ns^2$, where n equals the period and a is the group number—except for Group VIIIB, which has three columns. For Group VIIIB, count 8, 9, then 10 for successive columns, which then equals a.

Solution

a You locate tellurium in a periodic table and find it to be in Period 5, Group VIA. Thus, it is a main-group element, and the outer subshells are $5s$ and $5p$. These subshells contain six electrons, because the group is VIA. The valence-shell configuration is $\mathbf{5s^25p^4}$. **b** Nickel is a Period 4 transition element, in which the general form of the outer-shell configuration is $3d^{a-2}4s^2$. To determine a, you note that it equals the Roman numeral group number up to iron (8). After that you count Co as 9 and Ni as 10. Hence, the outer-shell configuration is $\mathbf{3d^84s^2}$.

Answer Check Note whether the configuration does follow the general form, either for a main-group element or a transition element. Check that the principal quantum number, n, equals the period of the element. For a main-group element, the total number of valence electrons should equal the group number. For a transition element, the number of electrons in the outer shell equals a.

Exercise 8.3 Using the periodic table on the inside front cover, write the valence-shell configuration of arsenic (As).

■ See Problems 8.51, 8.52, 8.53, and 8.54.

Exercise 8.4 The lead atom has the ground-state configuration $[Xe]4f^{14}5d^{10}6s^26p^2$. Find the period and group for this element. From its position in the periodic table, would you classify lead as a main-group element, a transition element, or an inner transition element?

■ See Problems 8.55 and 8.56.

CONCEPT CHECK 8.2

Two elements in Period 3 are adjacent to one another in the periodic table. The ground-state atom of one element has only *s* electrons in its valence shell; the other one has at least one *p* electron in its valence shell. Identify the elements.

8.4 Orbital Diagrams of Atoms; Hund's Rule

In discussing the ground states of atoms, we have not yet described how the electrons are arranged within each subshell. There may be several different ways of arranging electrons in a particular configuration. Consider the carbon atom ($Z = 6$) with the ground-state configuration $1s^22s^22p^2$. Three possible arrangements are given in the following orbital diagrams.

Diagram 1: (↑↓) (↑↓) (↑)(↑)( )
1*s* 2*s* 2*p*

Diagram 2: (↑↓) (↑↓) (↑)(↓)( )
1*s* 2*s* 2*p*

Diagram 3: (↑↓) (↑↓) (↑↓)( )( )
1*s* 2*s* 2*p*

These orbital diagrams show different states of the carbon atom. Each state has a different energy and, as you will see, different magnetic characteristics.

Hund's Rule

In about 1927, Friedrich Hund discovered an empirical rule determining the lowest-energy arrangement of electrons in a subshell. **Hund's rule** states that *the lowest-energy arrangement of electrons in a subshell is obtained by putting electrons into separate orbitals of the subshell with the same spin before pairing electrons.* Let us see how this would apply to the carbon atom, whose ground-state configuration is $1s^22s^22p^2$. The first four electrons go into the 1*s* and 2*s* orbitals.

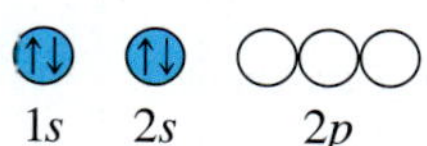

The next two electrons go into separate 2*p* orbitals, with both electrons having the same spin, following Hund's rule.

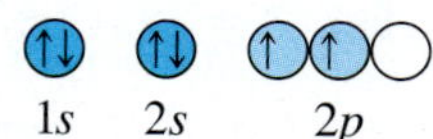

We see that the orbital diagram corresponding to the lowest energy is the one we previously labeled Diagram 1.

To apply Hund's rule to the oxygen atom, whose ground-state configuration is $1s^22s^22p^4$, we place the first seven electrons as follows:

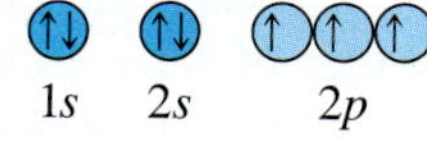

The last electron is paired with one of the 2*p* electrons to give a doubly occupied orbital. The orbital diagram for the ground state of the oxygen atom is

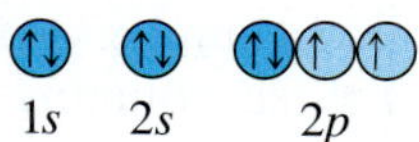

In the following example, Hund's rule is used to determine the orbital diagram for the ground state of a more complicated atom. Table 8.2 gives orbital diagrams for the ground states of the first ten elements.

Example 8.4 Applying Hund's Rule

Gaining Mastery Toolbox

Critical Concept 8.4
Hund's rule states that the lowest-energy arrangement of electrons in a subshell is obtained by putting electrons with the same spin into separate orbitals of the subshell before pairing electrons.

Solution Essentials:
- Hund's rule
- Building-up order of subshells
- Pauli exclusion principle
- The number of orbitals in a subshell of given l is $2l + 1$.

Write an orbital diagram for the ground state of the iron atom.

Problem Strategy First obtain the electron configuration, as described in Example 8.2. Then draw circles for the orbitals of each subshell. A filled subshell should have doubly occupied orbitals (two electrons with opposite spins). For a partially filled subshell, apply Hund's rule, putting electrons into separate orbitals with the same spin (either all up or all down) before pairing electrons.

Solution The electron configuration of the iron atom is $1s^2 2s^2 2p^6 3s^2 3p^6 3d^6 4s^2$. All the subshells except the $3d$ are filled. In placing the six electrons in the $3d$ subshell, you note that the first five go into separate $3d$ orbitals with their spin arrows in the same direction. The sixth electron must doubly occupy a $3d$ orbital. The orbital diagram is

(↑↓)	(↑↓)	(↑↓)(↑↓)(↑↓)	(↑↓)	(↑↓)(↑↓)(↑↓)	(↑↓)(↑)(↑)(↑)(↑)	(↑↓)
$1s$	$2s$	$2p$	$3s$	$3p$	$3d$	$4s$

You can write this diagram in abbreviated form using [Ar] for the argon-like core of the iron atom:

[Ar]	(↑↓)(↑)(↑)(↑)(↑)	(↑↓)
	3d	**4s**

Answer Check Check that you have started with the correct configuration for the atom. Then note that you have followed Hund's rule for a partially filled subshell.

Exercise 8.5 Write an orbital diagram for the ground state of the phosphorus atom ($Z = 15$). Write all orbitals.

■ See Problems 8.57 and 8.58.

Table 8.2 Orbital Diagrams for the Ground States of Atoms from $Z = 1$ to $Z = 10$

			Orbital Diagram		
Atom	**Z**	**Configuration**	**1s**	**2s**	**2p**
Hydrogen	1	$1s^1$	(↑)	( )	( )( )( )
Helium	2	$1s^2$	(↑↓)	( )	( )( )( )
Lithium	3	$1s^2 2s^1$	(↑↓)	(↑)	( )( )( )
Beryllium	4	$1s^2 2s^2$	(↑↓)	(↑↓)	( )( )( )
Boron	5	$1s^2 2s^2 2p^1$	(↑↓)	(↑↓)	(↑)( )( )
Carbon	6	$1s^2 2s^2 2p^2$	(↑↓)	(↑↓)	(↑)(↑)( )
Nitrogen	7	$1s^2 2s^2 2p^3$	(↑↓)	(↑↓)	(↑)(↑)(↑)
Oxygen	8	$1s^2 2s^2 2p^4$	(↑↓)	(↑↓)	(↑↓)(↑)(↑)
Fluorine	9	$1s^2 2s^2 2p^5$	(↑↓)	(↑↓)	(↑↓)(↑↓)(↑)
Neon	10	$1s^2 2s^2 2p^6$	(↑↓)	(↑↓)	(↑↓)(↑↓)(↑↓)

Magnetic Properties of Atoms

The magnetic properties of a substance can reveal certain information about the arrangement of electrons in an atom (or molecule). Although an electron in an atom behaves like a small magnet, the magnetic attractions from two electrons that are opposite in spin cancel each other. As a result, an atom that has only doubly occupied orbitals has no net spin magnetism. However, an atom with *unpaired* electrons—that is, with an excess of one kind of spin—does exhibit a net magnetism.

A CHEMIST Looks at...

Frontiers

Levitating Frogs and People

Have you ever seen a magic act in which a person rises above the stage in apparent defiance of gravity? Magicians have often included such acts of "levitation" as part of their performances. Of course, magical levitation uses tricks to deceive you. But levitation can be done, without tricks, using the known laws of electromagnetism.

Recently, researchers wanted to demonstrate that you can levitate almost anything if you have the proper magnetic field. They placed a frog within the magnetic field of a powerful electromagnet (which consists of an electric current flowing through a coil of wire). The frog floated in midair above the coil, at the point where the upward repulsion of the frog from the magnetic field just balanced the downward force of gravity (Figure 8.13).

We don't usually think of things such as frogs as being magnetic. In fact, most materials, including most substances in a biological organism, are diamagnetic. They contain pairs of electrons, in which the magnetism of each electron in a pair offsets the equal and opposite magnetism of the other electron. You might expect diamagnetic materials to be nonmagnetic because of this balance of opposite electron spins. But when you place a diamagnetic material in an external magnetic field, its electrons move so as to induce, or generate, a smaller magnetic field that is opposite in direction to the external field. This results in a repulsive force between the diamagnetic material and the external field. The repulsive force, though generally small, can be easily observed if the external magnet is large enough. Although, with a sufficiently large magnet, it would be possible to levitate a person as well as a frog, the researchers don't plan to do that.

Courtesy of A. Geim, University of Nijmegen

Figure 8.13 ▲

Levitation of a frog A frog placed in a powerful magnetic field, generated by a current flowing through a water-cooled coil of wire, appears to float in midair in defiance of gravity. The frog contains materials that are diamagnetic (composed of electron pairs), which are repelled by the magnetic field.

■ See Problems 8.89 and 8.90.

The magnetic properties of an atom can be observed. The most direct way is to determine whether the atomic substance is attracted to the field of a strong magnet. A **paramagnetic substance** is *a substance that is weakly attracted by a magnetic field, and this attraction is generally the result of unpaired electrons.* ◄ For example, sodium vapor has been found experimentally to be paramagnetic. The explanation is that the vapor consists primarily of sodium atoms, each containing an unpaired electron. (The configuration is $[Ne]3s^1$.) A **diamagnetic substance** is *a substance that is not attracted by a magnetic field or is very slightly repelled by such a field. This property generally means that the substance has only paired electrons.* Mercury vapor is found experimentally to be diamagnetic. The explanation is that mercury vapor consists of mercury atoms, with the electron configuration $[Xe]4f^{14}5d^{10}6s^2$, which has only paired electrons.

The strong, permanent magnetism seen in iron objects is called ferromagnetism *and is due to the cooperative alignment of electron spins in many iron atoms. Paramagnetism is a much weaker effect. Nevertheless, paramagnetic substances can be attracted to a strong magnet. Liquid oxygen is composed of paramagnetic O_2 molecules. When poured over a magnet, the liquid clings to the poles. (See Figure 10.29.)*

We expect the different orbital diagrams presented at the beginning of this section for the $1s^22s^22p^2$ configuration of the carbon atom to have different magnetic properties. Diagram 1, predicted by Hund's rule to be the ground state, would give a magnetic atom, whereas the other diagrams would not. If we could prepare a vapor of carbon atoms, it should be attracted to a magnet. (It should be paramagnetic.) It is difficult to prepare a vapor of free carbon atoms in sufficient concentration to observe a result. However, the visible–ultraviolet spectrum of carbon atoms can be obtained easily from dilute vapor. From an analysis of this spectrum, it is possible to show that the ground-state atom is magnetic, which is consistent with the prediction of Hund's rule.

Periodicity of the Elements

You have seen that the periodic table that Mendeleev discovered in 1869 can be explained by the periodicity of the ground-state electron configurations of the atoms. Now we will look at various aspects of the periodicity of the elements.

8.5 Mendeleev's Predictions from the Periodic Table

One of Mendeleev's periodic tables is reproduced in Figure 8.14. Though somewhat different from modern tables, it shows essentially the same arrangement. In this early form of the periodic table, within each column some elements were placed toward the left side and some toward the right. With some exceptions, the elements on a given side have similar properties. Mendeleev left spaces in his periodic table for what he felt were undiscovered elements. There are blank spaces in his row 5—for example, one directly under aluminum and another under silicon (looking at just the elements on the right side of the column). By writing the known elements in this row with their atomic weights, he could determine approximate values (between the known ones) for the missing elements (values in parentheses).

Cu	Zn	—	—	As	Se	Br
63 amu	65 amu	(68 amu)	(72 amu)	75 amu	78 amu	80 amu

The Group III element directly under aluminum Mendeleev called eka-aluminum, with the symbol Ea. (*Eka* is the Sanskrit word meaning "first"; thus eka-aluminum is the first element under aluminum.) The known Group III elements have oxides of the form R_2O_3, so Mendeleev predicted that eka-aluminum would have an oxide with the formula Ea_2O_3.

The physical properties of this undiscovered element could be predicted by comparing values for the neighboring known elements. For eka-aluminum Mendeleev predicted a density of 5.9 g/cm^3, a low *melting point* (the temperature at which a substance melts), and a high *boiling point* (the temperature at which a substance boils).

In 1874 the French chemist Paul-Émile Lecoq de Boisbaudran found two previously unidentified lines in the atomic spectrum of a sample of sphalerite (a zinc sulfide, ZnS, mineral). Realizing he was on the verge of a discovery, Lecoq de Boisbaudran quickly prepared a large batch of the zinc mineral, from which he isolated a gram of a new element. He called this new element gallium. The properties of gallium were remarkably close to those Mendeleev predicted for eka-aluminum.

Tabelle II.

Reihen	Gruppe I. — R^2O	Gruppe II. — RO	Gruppe III. — R^2O^3	Gruppe IV. RH^4 RO^2	Gruppe V. RH^3 R^2O^5	Gruppe VI. RH^2 RO^3	Gruppe VII. RH R^2O^7	Gruppe VIII. — RO^4
1	H=1							
2	Li=7	Be=9,4	B=11	C=12	N=14	O=16	F=19	
3	Na=23	Mg=24	Al=27,3	Si=28	P=31	S=32	Cl=35,5	
4	K=39	Ca=40	—=44	Ti=48	V=51	Cr=52	Mn=55	Fe=56, Co=59, Ni=59, Cu=63.
5	(Cu=63)	Zn=65	—=68	—=72	As=75	Se=78	Br=80	
6	Rb=85	Sr=87	?Yt=88	Zr=90	Nb=94	Mo=96	—=100	Ru=104, Rh=104, Pd=106, Ag=108.
7	(Ag=108)	Cd=112	In=113	Sn=118	Sb=122	Te=125	J=127	
8	Cs=133	Ba=137	?Di=138	?Ce=140	—	—	—	— — — —
9	(—)	—	—	—	—	—	—	
10	—	—	?Er=178	?La=180	Ta=182	W=184	—	Os=195, Ir=197, Pt=198, Au=199.
11	(Au=199)	Hg=200	Tl=204	Pb=207	Bi=208	—	—	
12	—	—	—	Th=231	—	U=240	—	— — — —

Figure 8.14 ◀ **Mendeleev's periodic table** This one was published in 1872.

Property	*Predicted for Eka-Aluminum*	*Found for Gallium*
Atomic weight	68 amu	69.7 amu
Formula of oxide	Ea_2O_3	Ga_2O_3
Density of the element	5.9 g/cm^3	5.91 g/cm^3
Melting point of the element	Low	30.1°C
Boiling point of the element	High	1983°C

The predictive power of Mendeleev's periodic table was demonstrated again when scandium (eka-boron) was discovered in 1879 and germanium (eka-silicon) in 1886. Both elements had properties remarkably like those predicted by Mendeleev. These early successes won acceptance for the organizational and predictive power of the periodic table.

8.6 Some Periodic Properties

The electron configurations of the atoms display a periodic variation with increasing atomic number (nuclear charge). As a result, the elements show periodic variations of physical and chemical behavior. The **periodic law** states that *when the elements are arranged by atomic number, their physical and chemical properties vary periodically.* In this section, we will look at three physical properties of an atom: atomic radius, ionization energy, and electron affinity. These three quantities, especially ionization energy and electron affinity, are important in discussions of chemical bonding (the subject of Chapter 9).

Atomic Radius

The definition of atomic radius is somewhat arbitrary; in fact, different definitions have been given. Because the statistical distribution of electrons does not abruptly end but merely decreases to smaller and smaller values as the distance from the nucleus increases, an atom has no definite boundary. You can see this in the plot of the electron distribution for the argon atom, shown in Figure 8.15. This distribution has been calculated from an approximate wavefunction obtained for the isolated atom. Note that the distribution shows a maximum at each shell of the atom and falls steadily to a negligible value after the maximum for the outer shell, which occurs at 71 pm.

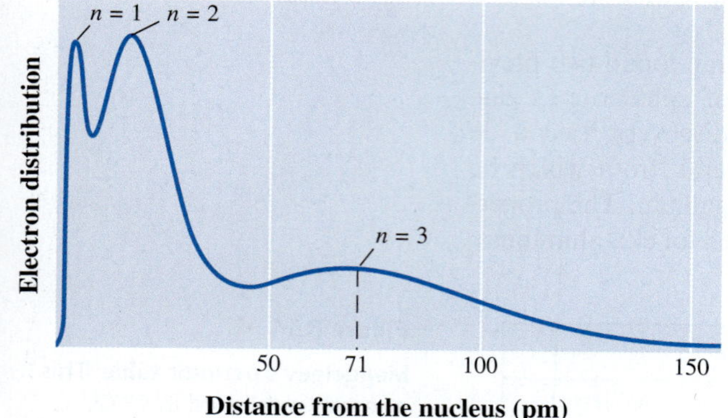

Figure 8.15 ▲

Electron distribution for the argon atom This is a radial distribution, showing the probability of finding an electron at a given distance from the nucleus. The distribution shows three maxima, for the $n = 1$, $n = 2$, and $n = 3$ shells. The outermost maximum occurs at 71 pm; then the distribution falls steadily, becoming negligibly small after several hundred picometers.

It seems reasonable, as one possibility, to define the **atomic radius** (calculated from theory for the isolated atom) as *the maximum in the radial distribution function of the outer shell of the atom.* From this definition, we would say that argon has a radius of 71 pm. Similar calculations have been done on most of the atoms in the periodic table. In Figure 8.16, we have graphically plotted the atomic radii from a particular set of calculations versus atomic number to display their periodic variations. Figure 8.17 shows the same data visually.

There are other definitions of atomic radii. As we will discuss in the next chapter on chemical bonding, the *covalent bond* is one type of chemical bond, and atomic radii have been defined in a way that reproduces the empirically determined covalent bond distances (the distances between atomic nuclei) in chemical compounds. By this definition, the **atomic radius** (in this case, often called the **covalent radius**) is *the value for that atom in a set of covalent radii assigned to atoms in such a way that the sum of the covalent radii of atoms* A *and* B *predicts the approximate* A—B *bond length.* Note that these atomic radii are from empirical data rather than calculated from wavefunctions, and as we noted, they refer to atoms in compounds and not to isolated atoms. We will discuss these covalent radii again in Section 9.10.

Although the definition of atomic radius is somewhat arbitrary, as long as we refer to a set of values obtained in a consistent manner, they can be useful.

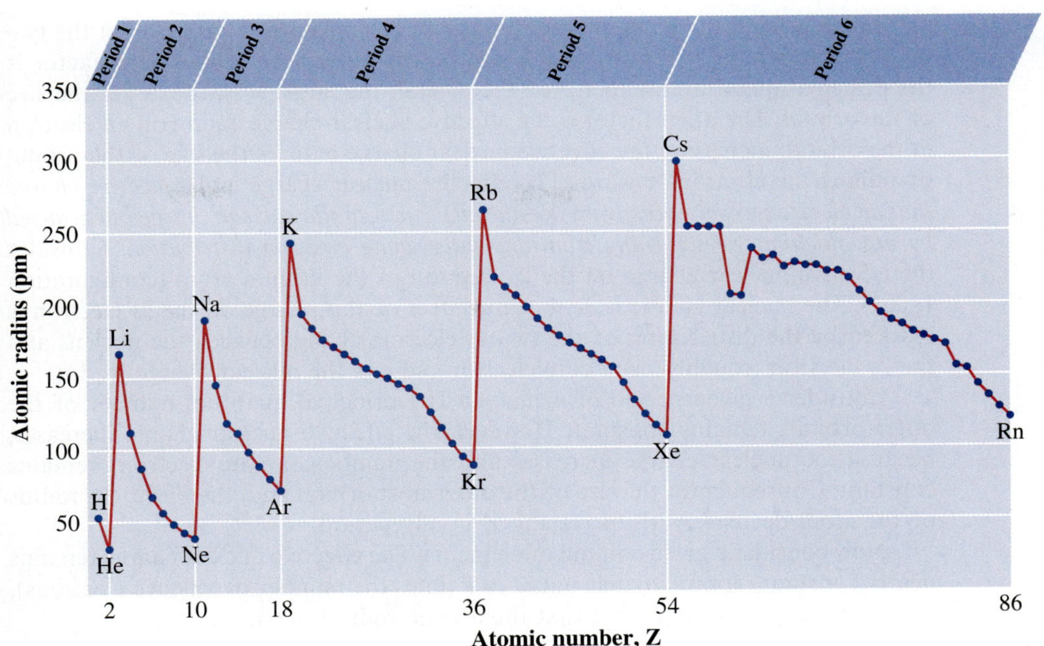

Figure 8.16 ◀ **Atomic radius (isolated atom) versus atomic number** These radii are calculated values for isolated atoms. Note that the curve is periodic (tends to repeat). Each period of elements begins with the Group IA atom, and the atomic radius tends to decrease until the Group VIIIA atom.

In particular, atomic radii in a given set follow general periodic trends, which can be seen in Figures 8.16 and 8.17. These trends are:

1. **Within each period (horizontal row)**, the atomic radius tends to decrease with increasing atomic number (nuclear charge). Thus, the largest atom in a period tends to be the Group IA atom and the smallest the noble-gas atom.
2. **Within each group (vertical column)**, the atomic radius tends to increase with the period number.

Note that the atomic radius increases greatly going from any noble-gas atom to the following Group IA atom, giving the curve in Figure 8.16 a saw-toothed appearance. We should stress that these are general trends, and you may see occasional exceptions from these trends in some lists of atomic radii.

Figure 8.17 ▼ **Representation of atomic radii (isolated atoms)** The atomic radii are in picometers. Note the trends within each period and each group.

PERIOD	IA	IIA	IIB	IVB	VB	VIB	VIIB	VIIIB	VIIIB	VIIIB	IB	IIB	IIIA	IVA	VA	VIA	VIIA	VIIIA
1	1 **H** 53																	2 **He** 31
2	3 **Li** 167	4 **Be** 112											5 **B** 87	6 **C** 67	7 **N** 56	8 **O** 48	9 **F** 42	10 **Ne** 38
3	11 **Na** 190	12 **Mg** 145											13 **Al** 118	14 **Si** 111	15 **P** 98	16 **S** 88	17 **Cl** 79	18 **Ar** 71
4	19 **K** 243	20 **Ca** 194	21 **Sc** 184	22 **Ti** 176	23 **V** 171	24 **Cr** 166	25 **Mn** 161	26 **Fe** 156	27 **Co** 152	28 **Ni** 149	29 **Cu** 145	30 **Zn** 142	31 **Ga** 136	32 **Ge** 125	33 **As** 114	34 **Se** 103	35 **Br** 94	36 **Kr** 88
5	37 **Rb** 265	38 **Sr** 219	39 **Y** 212	40 **Zr** 206	41 **Nb** 198	42 **Mo** 190	43 **Tc** 183	44 **Ru** 178	45 **Rh** 173	46 **Pd** 169	47 **Ag** 165	48 **Cd** 161	49 **In** 156	50 **Sn** 145	51 **Sb** 133	52 **Te** 123	53 **I** 115	54 **Xe** 108
6	55 **Cs** 298	56 **Ba** 253	71 **Lu** 217	72 **Hf** 208	73 **Ta** 200	74 **W** 193	75 **Re** 188	76 **Os** 185	77 **Ir** 180	78 **Pt** 177	79 **Au** 174	80 **Hg** 171	81 **Tl** 156	82 **Pb** 154	83 **Bi** 143	84 **Po** 135	85 **At** 127	86 **Rn** 120

Atomic radius increases ↓

Atomic radius decreases →

These general trends in atomic radius can be explained if you look at the two factors that primarily determine the size of the outermost orbital. One factor is the principal quantum number n of the orbital; the larger n is, the larger the size of the orbital. The other factor is the effective nuclear charge acting on an electron in the orbital; increasing the effective nuclear charge reduces the size of the orbital by pulling the electrons inward. The **effective nuclear charge** is *the positive charge that an electron experiences from the nucleus, equal to the nuclear charge but reduced by any shielding or screening from any intervening electron distribution.* Consider the effective nuclear charge on the $2s$ electron in the lithium atom (configuration $1s^2 2s^1$). The nuclear charge is $3e$, but the effect of this charge on the $2s$ electron is reduced by the distribution of the two $1s$ electrons lying between the nucleus and the $2s$ electron (roughly, each core electron reduces the nuclear charge by $1e$).

Consider a given period of elements. The principal quantum number of the outer orbitals remains constant. However, the effective nuclear charge increases, because the nuclear charge increases and the number of core electrons remains constant. Consequently, the size of the outermost orbital and, therefore, the radius of the atom decrease with increasing Z in any period.

Now consider a given column of elements. The effective nuclear charge remains nearly constant (approximately equal to e times the number of valence electrons), but n gets larger. You observe that the atomic radius increases.

OWL Interactive Example 8.5 Determining Relative Atomic Sizes from Periodic Trends

Gaining Mastery Toolbox

Critical Concept 8.5
The general trends in atomic radii are that they decrease left to right in a period and increase from top to bottom in a group.

Solution Essentials:
- General trends of atomic radii

Refer to a periodic table and use the trends noted for size of atomic radii to arrange the following in order of increasing atomic radius: Al, C, Si.

Problem Strategy The general trends in atomic radius are that it decreases left to right in a period and that it increases from top to bottom of a group. Compare any two elements in the same period, or in the same group, to decide which of the two is smaller. From this, decide the order of the elements.

Solution Note that C is above Si in Group IVA. Therefore, the radius of C is smaller than that of Si (the atomic radius increases going down a group of elements). Note that Al and Si are in the same period. Therefore, the radius of Si is smaller than that of Al (radius decreases with Z in a period). Hence the order of elements by increasing radius is **C, Si, Al.**

Answer Check After you have decided on the order of the atoms, check that the atomic radius does indeed increase, left to right.

Exercise 8.6 Using a periodic table, arrange the following in order of increasing atomic radius: Na, Be, Mg.

■ See Problems 8.61 and 8.62.

Ionization Energy

Ionization energies are also given in electron volts (eV). This is the amount of energy imparted to an electron when it is accelerated through an electrical potential of one volt. One electron volt is equivalent to 96.5 kJ/mol. (This is an easy conversion to remember; it is approximately one hundred kilojoules per mole.)

The **first ionization energy** (or **first ionization potential**) of an atom is *the minimum energy needed to remove the highest-energy (that is, the outermost) electron from the neutral atom in the gaseous state.* (When the unqualified term *ionization energy* is used, it generally means first ionization energy.) For the lithium atom, the first ionization energy is the energy needed for the following process (electron configurations are in parentheses):

$$\text{Li}(1s^2 2s^1) \longrightarrow \text{Li}^+(1s^2) + e^-$$

Values of this energy are often quoted for one mole of atoms (6.02×10^{23} atoms). Thus, the ionization energy of the lithium atom is 520 kJ/mol. ◀

Ionization energies display a periodic variation when plotted against atomic number, as Figure 8.18 shows. Within any period, values tend to increase with atomic number. Thus, the lowest values in a period are found for the Group IA elements *(alkali metals)*. It is characteristic of reactive metals such as these to

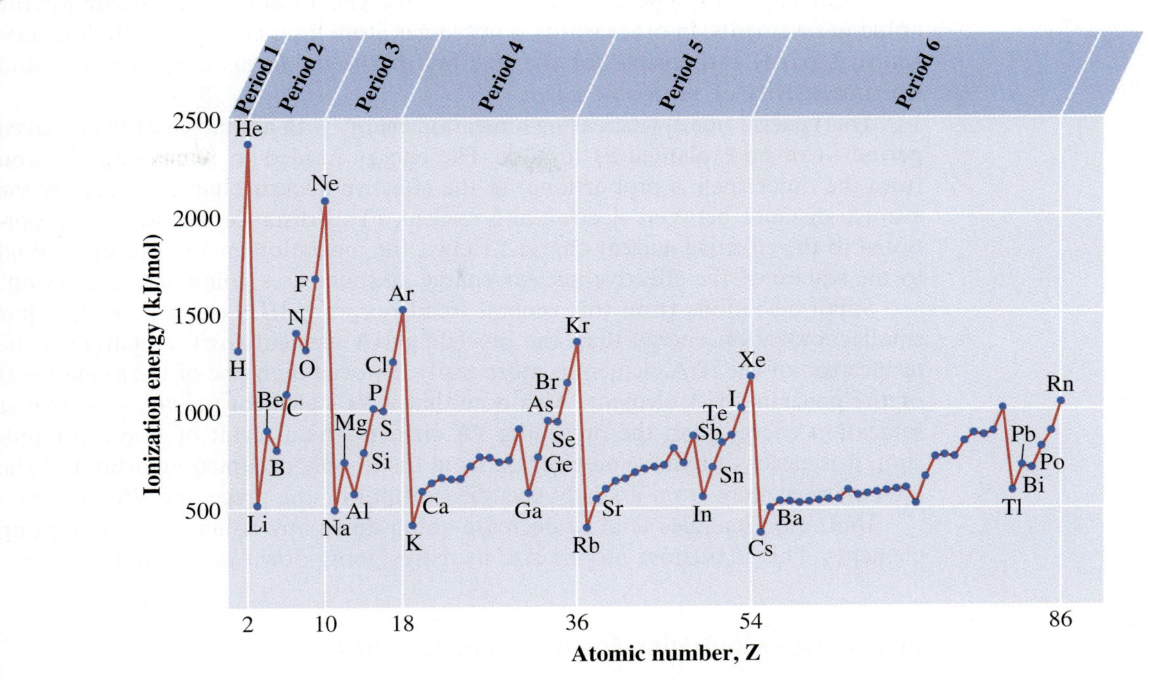

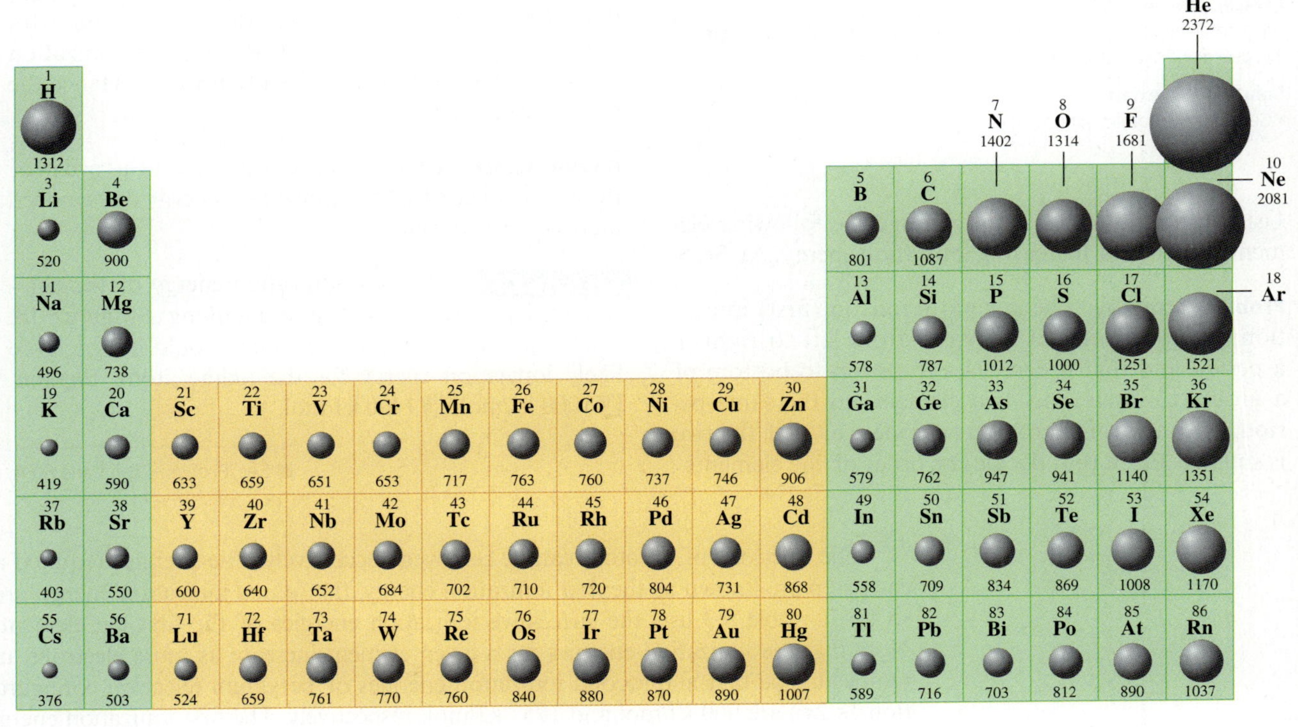

Figure 8.18 ▲

Ionization energy versus atomic number *Top:* Note that the values tend to increase within each period, except for small drops in ionization energy at the Group IIIA and VIA elements. Large drops occur when a new period begins. *Bottom:* Note the trends within each period. The size of the sphere for each atom indicates the relative magnitude of ionization energy. The values given are in kilojoules/mole.

lose electrons easily. The largest ionization energies in any period occur for the noble-gas elements. In other words, a noble-gas atom loses electrons with difficulty, which is partly responsible for the stability of the noble-gas configurations and the unreactivity of the noble gases.

This general trend—increasing ionization energy with atomic number in a given period—can be explained as follows: The energy needed to remove an electron from the outer shell is proportional to the effective nuclear charge divided by the average distance between electron and nucleus. (This distance is inversely proportional to the effective nuclear charge.) Hence, the ionization energy is proportional to the square of the effective nuclear charge and increases going across a period.

Small deviations from this general trend occur. A IIIA element (ns^2np^1) has smaller ionization energy than the preceding IIA element (ns^2). Apparently, the *np* electron of the IIIA element is more easily removed than one of the *ns* electrons of the preceding IIA element. Also note that a VIA element (ns^2np^4) has smaller ionization energy than the preceding VA element. As a result of electron repulsion, it is easier to remove an electron from the doubly occupied *np* orbital of the VIA element than from a singly occupied orbital of the preceding VA element.

Ionization energies tend to decrease going down any column of main-group elements. This is because atomic size increases going down the column.

Example 8.6 Determining Relative Ionization Energies from Periodic Trends

Gaining Mastery Toolbox

Critical Concept 8.6
The general trends in first ionization energy are that it increases left to right in a period and decreases from top to bottom in a group.

Solution Essentials:

- General trends of first ionization energy

Using a periodic table only, arrange the following elements in order of increasing ionization energy: Ar, Se, S.

Problem Strategy The general trends in (first) ionization energy are that it increases from left to right in a period and that it decreases from top to bottom of a group. Compare any two elements in the same period, or in the same group, to decide which of the two is smaller. From this, decide the order of the elements.

Solution Note that Se is below S in Group VIA. Therefore, the ionization energy of Se should be less than that of S. Also, S and Ar are in the same period, with Z increasing from S to Ar. Therefore, the ionization energy of S should be less than that of Ar. Hence the order is **Se, S, Ar**.

Answer Check After you have decided on the order of the atoms, check that the ionization energy does indeed increase, left to right.

Exercise 8.7 The first ionization energy of the chlorine atom is 1251 kJ/mol. Without looking at Figure 8.18, state which of the following values would be the more likely ionization energy for the iodine atom. Explain. **a** 1000 kJ/mol. **b** 1400 kJ/mol.

■ See Problems 8.63 and 8.64.

The electrons of an atom can be removed successively. The energies required at each step are known as the *first* ionization energy, the *second* ionization energy, and so forth. Table 8.3 lists the successive ionization energies of the first ten elements. Note that the ionization energies for a given element increase as more electrons are removed. The first and second ionization energies of beryllium (electron configuration $1s^22s^2$) are 900 kJ/mol and 1757 kJ/mol, respectively. The first ionization energy is the energy needed to remove a 2*s* electron from the Be atom. The second ionization energy is the energy needed to remove a 2*s* electron from the positive ion Be^+. Its value is greater than that of the first ionization energy because the electron is being removed from a positive ion, which strongly attracts the electron.

Note that there is a large jump in value from the second ionization energy of Be (1757 kJ/mol) to the third ionization energy (14,848 kJ/mol). The second ionization energy corresponds to removing a valence electron (from the 2*s* orbital), which is relatively easy. The third ionization energy corresponds to removing an electron from the core of the atom (from the 1*s* orbital)—that is, from a noble-gas configuration ($1s^2$). A vertical line in Table 8.3 separates the energies needed to

Table 8.3 Successive Ionization Energies of the First Ten Elements (kJ/mol)*

Element	First	Second	Third	Fourth	Fifth	Sixth	Seventh
H	1312						
He	2372	5250					
Li	520	7298	11,815				
Be	900	1757	14,848	21,006			
B	801	2427	3660	25,026	32,827		
C	1086	2353	4620	6223	37,831	47,277	
N	1402	2856	4578	7475	9445	53,267	64,360
O	1314	3388	5300	7469	10,990	13,326	71,330
F	1681	3374	6050	8408	11,023	15,164	17,868
Ne	2081	3952	6122	9371	12,177	15,238	19,999

*Ionization energies to the right of a vertical line correspond to removal of electrons from the core of the atom.

remove valence electrons from those needed to remove core electrons. For each element, a large increase in ionization energy occurs when this line is crossed. The large increase results from the fact that once the valence electrons are removed, a stable noble-gas configuration is obtained. Further ionizations become much more difficult. We will see in Chapter 9 that metal atoms often form compounds by losing valence electrons; core electrons are not significantly involved in the formation of compounds.

Electron Affinity

When a neutral atom in the gaseous state picks up an electron to form a stable negative ion, energy is released. For example, a chlorine atom can pick up an electron to give a chloride ion, Cl^-, and 349 kJ/mol of energy is released. We can write this process, including the electron configurations, as follows:

$$\text{Cl}([\text{Ne}]3s^2 3p^5) + e^- \longrightarrow \text{Cl}^-([\text{Ne}]3s^2 3p^6);\ \Delta E = -349 \text{ kJ/mol}$$

What is often measured is the energy of the *electron detachment* from the negative ion, which is essentially the ionization energy of the negative ion.

$$\text{Cl}^-([\text{Ne}]3s^2 3p^6) \longrightarrow \text{Cl}([\text{Ne}]3s^2 3p^5) + e^-;\ \Delta E = 349 \text{ kJ/mol}$$

The **electron affinity** of an atom is defined as *the energy required to remove an electron from the atom's negative ion (in its ground state)*. Or, if you wish, the electron affinity is the negative of the energy change obtained when the neutral atom picks up an electron. The quantity is positive if a stable negative ion forms. A large positive value means that the neutral atom has a strong affinity for an electron (picks up an electron easily). Table 8.4 displays the electron affinities of the main-group elements. ▶

It should be noted that electron affinities are also defined with the opposite sign. In this convention, the electron affinity of the chlorine atom is −349 kJ/mol. This is essentially the energy change (energy is released) in adding an electron to the chlorine atom, Cl.

Table 8.4 Electron Affinities of the Main-Group Elements (kJ/mol)

Period	IA	IIA	IIIA	IVA	VA	VIA	VIIA	VIIIA
1	**H** 73							**He** ≤0
2	**Li** 60	**Be** ≤0	**B** 27	**C** 122	**N** ≤0	**O** 141	**F** 328	**Ne** ≤0
3	**Na** 53	**Mg** ≤0	**Al** 44	**Si** 134	**P** 72	**S** 200	**Cl** 349	**Ar** ≤0
4	**K** 48	**Ca** 2	**Ga** 41	**Ge** 119	**As** 78	**Se** 195	**Br** 325	**Kr** ≤0
5	**Rb** 47	**Sr** 5	**In** 37	**Sn** 107	**Sb** 101	**Te** 190	**I** 295	**Xe** ≤0
6	**Cs** 46	**Ba** 14	**Tl** 36	**Pb** 35	**Bi** 91	**Po** 180	**At** 270	**Rn** ≤0

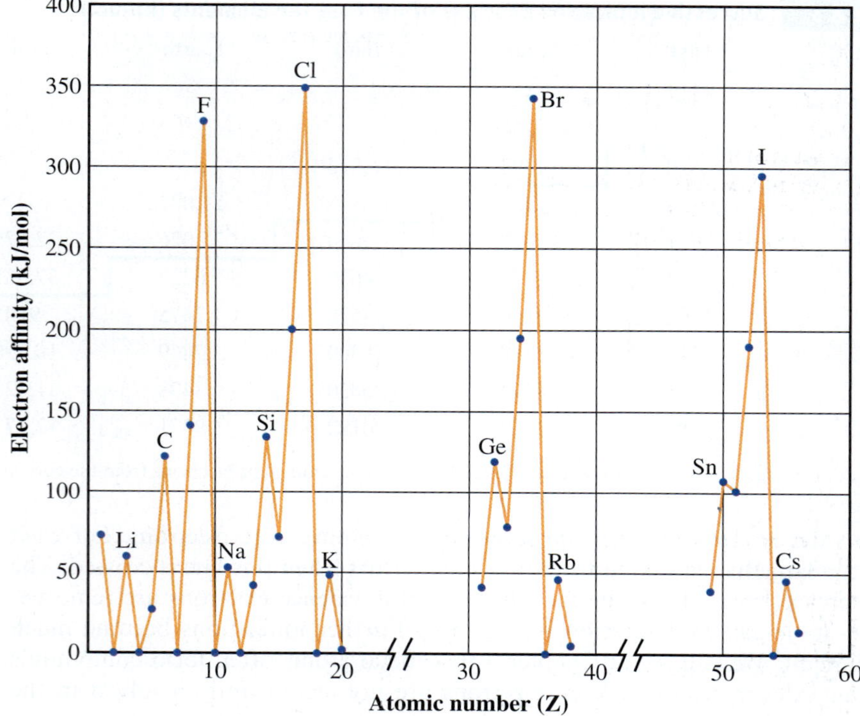

Figure 8.19 ▶

Electron affinity versus atomic number in the main-group elements Electron affinities in the main-group elements show a periodic variation when plotted against atomic number, although this variation is somewhat more complicated than that displayed by ionization energies. In a given period, the electron affinity rises from the Group IA element to the Group VIIA element but with sharp drops in the Group IIA and Group VA elements.

If we look at these main-group elements, we see that the electron affinities vary in going across any period in a fairly predictable fashion. See the graph in Figure 8.19. Broadly speaking, the general trend in any period of the main-group elements is an increase in electron affinities from left to right, although the IIA element and the VA element tend to have smaller electron affinities than the preceding element, and the Group VIIIA elements (noble gases) have zero or small negative values (indicating unstable negative ions). Note especially that the Group VIA and Group VIIA elements have the largest electron affinities of any of the other main-group elements. The stability of the negative ions of these elements explains why they have compounds with monatomic anions (such as F^-). While the trend in going across a period is rather clear, this is not the case in going down a column of elements; there is no simple trend.

Let's look more closely at these main-group elements. Consider those in Group IA. All of the Group IA elements have moderately positive electron affinities, as you might expect since the added electron just completes the outer *ns* subshell. When you add an electron to a potassium atom, for example, it goes into the 4*s* orbital to form a moderately stable negative ion, releasing energy and yielding a positive electron affinity (EA):

$$\text{K}([\text{Ar}]4s^1) + e^- \longrightarrow \text{K}^-([\text{Ar}]4s^2);\ \Delta E = -48 \text{ kJ/mol (EA} = 48 \text{ kJ/mol)}$$

The Group IIA elements have a filled outer *ns* subshell, so the added electron has to go into the next higher subshell, the *np* subshell. Thus, we expect the electron affinities of these elements to be smaller than those for the IA elements. Beryllium and magnesium appear to have values close to zero (no stable negative ions found), whereas the elements from calcium on have small positive values (weakly stable negative ions). For the calcium atom, we have

$$\text{Ca}([\text{Ar}]4s^2) + e^- \longrightarrow \text{Ca}^-([\text{Ar}]4s^24p^1);\ \Delta E = -2 \text{ kJ/mol (EA} = 2 \text{ kJ/mol)}$$

From the Group IIIA elements to the Group VIIA elements, the added electron goes into the outer *np* subshell, which already has one or more electrons in it. With the exception of the Group VA element, the electron affinity in any period tends to rise from the Group IIA element to the Group VIIA element. For example, in the Period 4 elements from gallium to bromine, we observe:

$$Ga([Ar]4s^2 4p^1) + e^- \longrightarrow Ga^-([Ar]4s^2 4p^2);\ \Delta E = -41\ \text{kJ/mol}\ (\text{EA} = 41\ \text{kJ/mol})$$
$$Ge([Ar]4s^2 4p^2) + e^- \longrightarrow Ge^-([Ar]4s^2 4p^3);\ \Delta E = -119\ \text{kJ/mol}\ (\text{EA} = 119\ \text{kJ/mol})$$
$$As([Ar]4s^2 4p^3) + e^- \longrightarrow As^-([Ar]4s^2 4p^4);\ \Delta E = -78\ \text{kJ/mol}\ (\text{EA} = 78\ \text{kJ/mol})$$
$$Se([Ar]4s^2 4p^4) + e^- \longrightarrow Se^-([Ar]4s^2 4p^5);\ \Delta E = -195\ \text{kJ/mol}\ (\text{EA} = 195\ \text{kJ/mol})$$
$$Br([Ar]4s^2 4p^5) + e^- \longrightarrow Br^-([Ar]4s^2 4p^6);\ \Delta E = -325\ \text{kJ/mol}\ (\text{EA} = 325\ \text{kJ/mol})$$

The drop in electron affinity of the Group VA element from the preceding element is easily explained. In the Group VA element (arsenic, As, in the above list), the added electron must pair up with one of the *np* electrons since all of the *np* orbitals have one electron, whereas in the preceding element the extra electron goes into an empty *np* orbital. This pairing of electrons in an orbital requires some energy, resulting in a smaller electron affinity for the Group VA element compared with the preceding IVA element.

No stable negative ions of the Group VIIIA elements (noble-gas elements) have been found. We might expect this since these atoms have filled outer shells so that the addition of an electron would have to go into the next higher shell. The electron affinities are therefore zero (or possibly negative), as we noted earlier.

Exercise 8.8 Without looking at Table 8.4 but using the general comments in this section, decide which has the larger electron affinity, C or F.

■ See Problems 8.65 and 8.66.

CONCEPT CHECK 8.3

Given the following information for element E, identify the element's group in the periodic table: The electron affinity of E is negative (that is, it does not form a stable negative ion). The first ionization energy of E is less than the second ionization energy, which in turn is very much less than its third ionization energy.

8.7 Periodicity in the Main-Group Elements

The chemical and physical properties of the main-group elements clearly display periodic character. For instance, as pointed out in Section 2.5, the metallic elements lie to the left of the "staircase" line in the periodic table (see inside front cover), nonmetals lie to the right, and the metalloids (with intermediate characteristics) lie along this line. So, as you move left to right in any row of the periodic table, the metallic character of the elements decreases. As you progress down a column, however, the elements tend to increase in metallic character.

These variations of metallic–nonmetallic character can be attributed in part to variations in the ionization energies of the corresponding atoms. Elements with low ionization energy tend to be metallic, whereas those with high ionization energy tend to be nonmetallic. As you saw in the previous section, ionization energy is a periodic property, so it is not surprising that the metallic–nonmetallic character of an element is similarly periodic.

The basic–acidic behavior of the oxides of the elements is a good indicator of the metallic–nonmetallic character of the elements. Oxides are classified as basic or acidic depending on their reactions with acids and bases. A **basic oxide** is *an oxide that reacts with acids.* Most metal oxides are basic. An **acidic oxide** is *an oxide that reacts with bases.* Most nonmetal oxides are acidic oxides. An **amphoteric oxide** is *an oxide that has both basic and acidic properties.*

In the following brief descriptions of the main-group elements, we will note the metallic–nonmetallic behavior of the elements, as well as the basic–acidic character of the oxides. Although elements in a given group are expected to be similar, the degree of similarity does vary among the groups. The alkali metals (Group IA) show marked similarities, as do the halogens (Group VIIA). On the other hand, the Group IVA elements range from a nonmetal (carbon) at the top of the column to a metal (lead) at the bottom. In either case, however, the changes from one element in a column to the next lower one are systematic, and the periodic table helps us to correlate these systematic changes.

Hydrogen ($1s^1$)

At very high pressures, however, hydrogen is believed to have metallic properties.

Although the electron configuration of hydrogen would seem to place the element in Group IA, its properties are quite different, and it seems best to consider this element as belonging in a group by itself. The element is a colorless gas composed of H_2 molecules. ◀

Group IA Elements, the Alkali Metals (ns^1)

The *alkali metals* are soft and reactive, with the reactivities increasing as you move down the column of elements. All of the metals react with water to produce hydrogen.

$$2Li(s) + 2H_2O(l) \longrightarrow 2LiOH(aq) + H_2(g)$$

The vigor of the reaction increases from lithium (moderate) to rubidium (violent). All of the alkali metals form basic oxides with the general formula R_2O.

Group IIA Elements, the Alkaline Earth Metals (ns^2)

The *alkaline earth metals* are also chemically reactive but much less so than the alkali metals. Reactivities increase going down the group. The alkaline earth metals form basic oxides with the general formula RO.

© Cengage Learning

Figure 8.20 ▲

Gallium This metal melts from the heat of a hand.

Group IIIA Elements (ns^2np^1)

Groups IA and IIA exhibit only slight increases in metallic character down a column, but with Group IIIA we see a significant increase. The first Group IIIA element, boron, is a metalloid. Other elements in this group—aluminum, gallium, indium, and thallium—are metals. (Gallium is a curious metal; it melts readily in the palm of the hand—see Figure 8.20.)

The oxides in this group have the general formula R_2O_3. Boron oxide, B_2O_3, is an acidic oxide; aluminum oxide, Al_2O_3, and gallium oxide, Ga_2O_3, are amphoteric oxides. The change in the oxides from acidic to amphoteric to basic is indicative of an increase in metallic character of the elements.

Group IVA Elements (ns^2np^2)

Bronze, one of the first alloys (metallic mixtures) used by humans, contains about 90% copper and 10% tin. Bronze melts at a lower temperature than copper but is much harder.

This group shows the most distinct change in metallic character. It begins with the nonmetal carbon, C, followed by the metalloids silicon, Si, and germanium, Ge, and then the metals tin, Sn, and lead, Pb. Both tin and lead were known to the ancients. ◀

All the elements in this group form oxides with the general formula RO_2, which progress from acidic to amphoteric. Carbon dioxide, CO_2, an acidic oxide, is a gas. (Carbon also forms the monoxide, CO.) Silicon dioxide, SiO_2, an acidic oxide, exists as quartz and white sand (particles of quartz). Germanium dioxide, GeO_2, is acidic, though less so than silicon dioxide. Tin dioxide, SnO_2, an amphoteric oxide, is found as the mineral cassiterite, the principal ore of tin. Lead(IV) oxide, PbO_2, is amphoteric. Lead has a more stable monoxide, PbO. Figure 8.21 shows oxides of some Group IVA elements.

© Cengage Learning

Figure 8.21 ▲

Oxides of some Group IVA elements Powdered lead(II) oxide (yellow), lead(IV) oxide (dark brown), tin dioxide (white), and crystalline silicon dioxide (clear quartz).

Group VA Elements (ns^2np^3)

The Group VA elements also show the distinct transition from nonmetal (nitrogen, N, and phosphorus, P) to metalloid (arsenic, As, and antimony, Sb) to metal (bismuth, Bi). Nitrogen occurs as a colorless, odorless, relatively unreactive gas with N_2 molecules; white phosphorus is a white, waxy solid with P_4 molecules. Gray arsenic is a brittle solid with metallic luster; antimony is a brittle solid with a silvery, metallic luster. Bismuth is a hard, lustrous metal with a pinkish tinge.

The Group VA elements form oxides with empirical formulas R_2O_3 and R_2O_5. In some cases, the molecular formulas are twice these formulas—that is, R_4O_6 and R_4O_{10}. Nitrogen has the acidic oxides N_2O_3 and N_2O_5, although it also has other, better known oxides, such as NO. Phosphorus has the acidic oxides P_4O_6 and P_4O_{10}. Arsenic has the acidic oxides As_2O_3 and As_2O_5; antimony has the amphoteric oxides Sb_2O_3 and Sb_2O_5; and bismuth has the basic oxide Bi_2O_3.

Figure 8.22 ▲

Burning sulfur The combustion of sulfur in air gives a blue flame. The product is primarily sulfur dioxide, detectable by its acrid odor.

Group VIA Elements, the Chalcogens (ns^2np^4)

These elements, the *chalcogens* (pronounced kal′-ke-jens), show the transition from nonmetal (oxygen, O, sulfur, S, and selenium, Se) to metalloid (tellurium, Te) to metal (polonium, Po). Oxygen occurs as a colorless, odorless gas with O_2 molecules. It also has an allotrope, ozone, with molecular formula O_3. Sulfur is a brittle, yellow solid with molecular formula S_8. Tellurium is a shiny gray, brittle solid; polonium is a silvery metal.

Sulfur, selenium, and tellurium form oxides with the formulas RO_2 and RO_3. (Sulfur burns in air to form sulfur dioxide; see Figure 8.22.) These oxides, except for TeO_2, are acidic; TeO_2 is amphoteric. Polonium has an oxide PoO_2, which is amphoteric, though more basic in character than TeO_2.

Group VIIA Elements, the Halogens (ns^2np^5)

The *halogens* are reactive nonmetals with the general molecular formula X_2, where X symbolizes a halogen. Fluorine, F_2, is a pale yellow gas; chlorine, Cl_2, a pale greenish yellow gas; bromine, Br_2, a reddish brown liquid; and iodine, I_2, a bluish black solid that has a violet vapor (see Figure 8.23). Little is known about the chemistry of astatine, At, because all isotopes are radioactive with very short half-lives. (The half-life of a radioactive isotope is the time it takes for half of the isotope to decay, or break down, to another element.) It might be expected to be a metalloid.

Each halogen forms several compounds with oxygen; these are generally unstable, acidic oxides.

Figure 8.23 ▲

The halogens From left to right, the flasks contain chlorine, bromine, and iodine. (Bromine and iodine were warmed to produce the vapors; bromine is normally a reddish brown liquid and iodine a bluish black solid.)

Group VIIIA Elements, the Noble Gases (ns^2np^6)

The Group VIIIA elements exist as gases consisting of uncombined atoms. For a long time these elements were thought to be chemically inert, because no compounds were known. Then, in the early 1960s, several compounds of xenon were prepared. Now compounds are also known for argon, krypton, and radon. These elements are known as the *noble gases* because of their relative unreactivity.

CONCEPT CHECK 8.4

A certain element is a metalloid that forms an acidic oxide with the formula R_2O_5. Identify the element.

A Checklist for Review

OWL and Go Chemistry Sign in at **www.cengage.com/owl** to:

- View tutorials and simulations, develop problem-solving skills, and complete online homework assigned by your professor.
- For quick review and exam prep, download Go Chemistry mini lecture modules from OWL or purchase them at **www.cengagebrain.com.**

Summary of Facts and Concepts

To understand the similarities that exist among the members of a group of elements, it is necessary to know the *electron configurations* for the ground states of atoms. Only those arrangements of electrons allowed by the *Pauli exclusion principle* are possible. The ground-state configuration of an atom represents the electron arrangement that has the lowest total energy. This arrangement can be reproduced by the *building-up principle* (Aufbau principle), where electrons fill the subshells in particular order (the building-up order) consistent with the Pauli exclusion principle. The arrangement of electrons in partially filled subshells is governed by *Hund's rule.*

Elements in the same group of the periodic table have similar valence-shell configurations. As a result, chemical and physical properties of the elements show periodic behavior. *Atomic radii,* for example, tend to decrease across any period (left to right) and increase down any group (top to bottom). *First ionization energies* tend to increase across a period and decrease down a group. *Electron affinities* of the Group VIA and Group VIIA elements have large values.

Learning Objectives	Important Terms
8.1 Electron Spin and the Pauli Exclusion Principle	
▪ Define *electron configuration* and *orbital diagram.* ▪ State the Pauli exclusion principle. ▪ Apply the Pauli exclusion principle. Example 8.1	**electron configuration** **orbital diagram** **Pauli exclusion principle**
8.2 Building-Up Principle and the Periodic Table	
▪ Define *building-up principle.* ▪ Define *noble-gas core, pseudo-noble-gas core,* and *valence electron.* ▪ Define *main-group element* and *(d-block* and *f-block) transition element.*	**building-up (Aufbau) principle** **noble-gas core** **pseudo-noble-gas core** **valence electron**
8.3 Writing Electron Configurations Using the Periodic Table	
▪ Determine the configuration of an atom using the building-up principle. Example 8.2 ▪ Determine the configuration of an atom using the period and group numbers. Example 8.3	
8.4 Orbital Diagrams of Atoms; Hund's Rule	
▪ State Hund's rule. ▪ Apply Hund's rule. Example 8.4 ▪ Define *paramagnetic substance* and *diamagnetic substance.*	**Hund's rule** **paramagnetic substance** **diamagnetic substance**
8.5 Mendeleev's Predictions from the Periodic Table	
▪ Describe how Mendeleev predicted the properties of undiscovered elements.	
8.6 Some Periodic Properties	
▪ State the periodic law. ▪ State the general periodic trends in size of atomic radii. ▪ Define *effective nuclear charge.* ▪ Determine relative atomic sizes from periodic trends. Example 8.5 ▪ State the general periodic trends in ionization energy. ▪ Define *first ionization energy.* ▪ Determine relative ionization energies from periodic trends. Example 8.6 ▪ Define *electron affinity.* ▪ State the broad general trend in electron affinity across any period.	**periodic law** **atomic radius** **effective nuclear charge** **first ionization energy (first ionization potential)** **electron affinity**
8.7 Periodicity in the Main-Group Elements	
▪ Define *basic oxide, acidic oxide,* and *amphoteric oxide.* ▪ State the main group corresponding to an *alkali metal,* an *alkaline earth metal,* a *chalcogen,* a *halogen,* and a *noble gas.* ▪ Describe the change in metallic/nonmetallic character (or reactivities) in going through any main group of elements.	**basic oxide** **acidic oxide** **amphoteric oxide**

Questions and Problems

OWL Interactive versions of these problems may be assigned in OWL.

Self-Assessment and Review Questions

Key: These questions test your understanding of the ideas you worked with in the chapter. These problems vary in difficulty and often can be used for the basis of discussion.

8.1 Describe the experiment of Stern and Gerlach. How are the results for the hydrogen atom explained?

8.2 Describe the model of electron spin given in the text. What are the restrictions on electron spin?

8.3 How does the Pauli exclusion principle limit the possible electron configurations of an atom?

8.4 What is the maximum number of electrons that can occupy a *g* subshell ($l = 4$)?

8.5 List the orbitals in order of increasing orbital energy up to and including 3*p* orbitals.

8.6 Define each of the following: noble-gas core, pseudo-noble-gas core, valence electron.

8.7 Give two different possible orbital diagrams for the $1s^22s^22p^4$ configuration of the oxygen atom, one of which should correspond to the ground state. Label the diagram for the ground state.

8.8 Define the terms *diamagnetic substance* and *paramagnetic substance.* Does the ground-state oxygen atom give a diamagnetic or a paramagnetic substance? Explain.

8.9 What kind of subshell is being filled in Groups IA and IIA? in Groups IIIA to VIIIA? in the transition elements? in the lanthanides and actinides?

8.10 How was Mendeleev able to predict the properties of gallium before it was discovered?

8.11 Describe the major trends that emerge when atomic radii are plotted against atomic number. Describe the trends observed when first ionization energies are plotted against atomic number.

8.12 What atom has the smallest radius among the alkaline earth elements?

8.13 What main group in the periodic table has elements with the largest electron affinities for each period? What group of elements has only unstable negative ions?

8.14 The ions Na^+ and Mg^{2+} occur in chemical compounds, but the ions Na^{2+} and Mg^{3+} do not. Explain.

8.15 Describe the major trends in metallic character observed in the periodic table of the elements.

8.16 Distinguish between an acidic and a basic oxide. Give examples of each.

8.17 What is the name of the alkali metal atom with valence-shell configuration $5s^1$?

8.18 What would you predict for the atomic number of the halogen in Period 7?

8.19 List the elements in Groups IIIA to VIA in the same order as in the periodic table. Label each element as a metal, a metalloid, or a nonmetal. Does each column of elements display the expected trend of increasing metallic characteristics?

8.20 For the list of elements you made for Question 8.19, note whether the oxides of each element are acidic, basic, or amphoteric.

8.21 Write an equation for the reaction of potassium metal with water.

8.22 From what is said in Section 8.7 about Group IIA elements, list some properties of barium.

8.23 Give the names and formulas of two oxides of carbon.

8.24 Match each description in the left column with the appropriate element in the right column.

a A waxy, white solid, normally stored under water	Sulfur
	Sodium
b A yellow solid that burns in air	White phosphorus
c A reddish brown liquid	Bromine
d A soft, light metal that reacts vigorously with water	

8.25 Phosphorus has ________ unpaired electrons.

a 1 b 2 c 3 d 4 e 5

8.26 Which of the following atoms, designated by their electron configurations, has the *highest* ionization energy?

a $[Ne]3s^23p^2$
b $[Ne]3s^23p^3$
c $[Ar]3d^{10}4s^24p^3$
d $[Kr]4d^{10}5s^25p^3$
e $[Xe]4f^{14}5d^{10}6s^26p^3$

8.27 When trying to remove electrons from Be, which of the following sets of ionization energy makes the most sense going from first to third ionization energy? Explain your answer.

a First IE 900 KJ/mol, second IE 1750 kJ/mol, third IE 15,000 kJ/mol
b First IE 1750 KJ/mol, second IE 900 kJ/mol, third IE 15,000 kJ/mol
c First IE 15,000 KJ/mol, second IE 1750 kJ/mol, third IE 900 kJ/mol
d First IE 900 KJ/mol, second IE 15,000 kJ/mol, third IE 22,000 kJ/mol
e First IE 900 KJ/mol, second IE 1750 kJ/mol, third IE 1850 kJ/mol

8.28 Consider the following orderings.

I. Al < Si < P < S
II. Be < Mg < Ca < Sr
III. I < Br < Cl < F

Which of these give(s) a correct trend in atomic size?

a I only
b II only
c III only
d I and II only
e II and III only

Concept Explorations

Key: Concept explorations are comprehensive problems that provide a framework that will enable you to explore and learn many of the critical concepts and ideas in each chapter. If you master the concepts associated with these explorations, you will have a better understanding of many important chemistry ideas and will be more successful in solving all types of chemistry problems. These problems are well suited for group work and for use as in-class activities.

8.29 Periodic Properties I

A hypothetical element, X, has the following ionization energy values:

First ionization energy: 900 kJ/mol
Second ionization energy: 1750 kJ/mol
Third ionization energy: 14,900 kJ/mol
Fourth ionization energy: 21,000 kJ/mol

Another element, Y, has the following ionization energy values:

First ionization energy: 1200 kJ/mol
Second ionization energy: 2500 kJ/mol
Third ionization energy: 19,900 kJ/mol
Fourth ionization energy: 26,000 kJ/mol

a. To what family of the periodic table would element X be most likely to belong? Explain?

b. What charge would you expect element X to have when it forms an ion?

c. If you were to place elements X and Y into the periodic table, would element Y be in the same period as element X? If not in the same period, where might they be relative to each other in the periodic table?

d. Would an atom of Y be smaller or larger than an atom of X? Explain your reasoning.

8.30 Periodic Properties II

Consider two hypothetical elements, X and Z. Element X has an electron affinity of 300 kJ/mol, and element Z has an electron affinity of 75 kJ/mol.

a. If you have an X^- ion and Z^- ion, from which ion would it require more energy to remove an electron? Explain your answer.

b. If elements X and Z are in the same period of the periodic table, which atom would you expect to have the greater atomic radius? Why?

c. Assuming that the elements are in the same period, which element would you expect to have the smaller first ionization energy?

d. Do the valence electrons in element Z feel a greater effective nuclear charge than those in element X? Explain how you arrived at your answer.

Conceptual Problems

Key: These problems are designed to check your understanding of the concepts associated with some of the main topics presented in each chapter. A strong conceptual understanding of chemistry is the foundation for both applying chemical knowledge and solving chemical problems. These problems vary in level of difficulty and often can be used as a basis for group discussion.

8.31 Suppose that the Pauli exclusion principle were "No more than two electrons can have the same four quantum numbers." What would be the electron configurations of the ground states for the first six elements of the periodic table, assuming that, except for the Pauli principle, the usual building-up principle held?

8.32 Imagine a world in which all quantum numbers, except the l quantum number, are as they are in the real world. In this imaginary world, l begins with 1 and goes up to n (the value of the principal quantum number). Assume that the orbitals fill in the order by n, then l; that is, the first orbital to fill is for $n = 1, l = 1$; the next orbital to fill is for $n = 2, l = 1$, and so forth. How many elements would there be in the first period of the periodic table?

8.33 Two elements in Period 5 are adjacent to one another in the periodic table. The ground-state atom of one element has only s electrons in its valence shell; the other has at least one d electron in an unfilled shell. Identify the elements.

8.34 Two elements are in the same column of the periodic table, one above the other. The ground-state atom of one element has two s electrons in its outer shell, and no d electrons anywhere in its configuration. The other element has d electrons in its configuration. Identify the elements.

8.35 You travel to an alternate universe where the atomic orbitals are different from those on earth, but all other aspects of the atoms are the same. In this universe, you find that the first (lowest energy) orbital is filled with three electrons and the second orbital can hold a maximum of nine electrons. You discover an element Z that has five electrons in its atom. Would you expect Z to be more likely to form a cation or an anion? Indicate a possible charge on this ion.

8.36 Would you expect to find an element having both a very large (positive) first ionization energy and an electron affinity that is also a large positive number? Explain.

8.37 Two elements are in the same group, one following the other. One is a metalloid; the other is a metal. Both form oxides of the formula RO_2. The first is acidic; the next is amphoteric. Identify the two elements.

8.38 A metalloid has an acidic oxide of the formula R_2O_3. The element has no oxide of the formula R_2O_5. What is the name of the element?

8.39 Given the following information, identify the group from the periodic table that contains elements that behave like main-group element "E."

i. The electron affinity of E is small.
ii. The ionization energy (IE) trend for element E is: first ionization energy $<$ second ionization energy $<<<$ third ionization energy.
iii. Samples of E are lustrous and are good electrical conductors.

8.40 A hypothetical element A has the following properties:

First ionization energy: 850 kJ/mol
Second ionization energy: 1700 kJ/mol
Third ionization energy: 13,999 kJ/mol

a. If you were to react element A with oxygen, what would be the chemical formula of the resulting compound?

b. Write the balanced chemical reaction of A reacting with oxygen to give the product from part a.

c. Would you expect the product of the chemical reaction of A with oxygen to be a basic, an acidic, or an amphoteric oxide?

Practice Problems

Key: These problems are for practice in applying problem-solving skills. They are divided by topic, and some are keyed to exercises (see the ends of the exercises). The problems are arranged in matching pairs; the odd-numbered problem of each pair is listed first, and its answer is given in the back of the book.

Pauli Exclusion Principle

8.41 Which of the following orbital diagrams are allowed by the Pauli exclusion principle? Explain how you arrived at this decision. Give the electron configuration for the allowed ones.

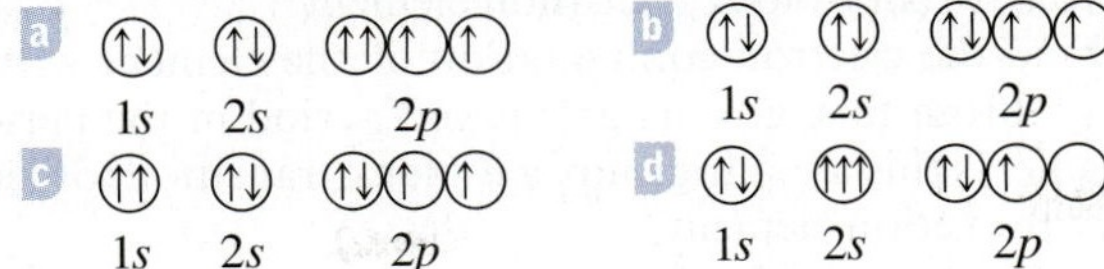

8.42 Which of the following orbital diagrams are allowed and which are not allowed by the Pauli exclusion principle? Explain. For those that are allowed, write the electron configuration.

a (↑↑) (↑↓) (↑↓)(↑)( )
1s 2s 2p

b (↑↓) (↑↓) (↑↓)(↑)(↑)
1s 2s 2p

c (↑↓) (↑↓↑) (↑)(↑)( )
1s 2s 2p

d (↑↓) (↑↓) (↑↓)(↓)(↑)
1s 2s 2p

8.43 Which of the following electron configurations are possible? Explain why the others are not.

a $1s^12s^22p^7$
b $1s^22s^22p^5$
c $1s^22s^22p^63s^33d^7$
d $1s^22s^22p^63s^23d^8$

8.44 Choose the electron configurations that are possible from among the following. Explain why the others are impossible.

a $1s^22s^12p^6$
b $1s^22s^22p^8$
c $1s^22s^22p^63s^23p^63d^7$
d $1s^22s^32p^63s^13d^9$

8.45 Write all of the possible orbital diagrams for the electron configuration $1s^22p^1$. (There are six different diagrams.)

8.46 Give the different orbital diagrams for the configuration $1s^12p^1$. (There are twelve different diagrams.)

Building-Up Principle and Hund's Rule

8.47 Give the electron configuration of the ground state of iodine, using the building-up principle.

8.48 Use the building-up principle to obtain the ground-state configuration of arsenic.

8.49 Use the building-up principle to obtain the electron configuration of the ground state of manganese.

8.50 Give the electron configuration of the ground state of nickel, using the building-up principle.

8.51 Bromine is a Group VIIA element in Period 4. Deduce the valence-shell configuration of bromine.

8.52 Bismuth is a Group VA element in Period 6. Write the valence-shell configuration of bismuth.

8.53 Zirconium is a Group IVB element in Period 5. What would you expect for the valence-shell configuration of zirconium?

8.54 Manganese is a Group VIIB element in Period 4. What would you expect for the configuration of outer electrons of manganese?

8.55 Thallium has the ground-state configuration $[Xe]4f^{14}5d^{10}6s^26p^1$. Give the group and period for this element. Classify it as a main-group, a *d*-transition, or an *f*-transition element.

8.56 The configuration for the ground state of iridium is $[Xe]4f^{14}5d^76s^2$. What are the group and period for this element? Is it a main-group, a *d*-transition, or an *f*-transition element?

8.57 Write the orbital diagram for the ground state of cobalt. The electron configuration is $[Ar]3d^74s^2$.

8.58 Write the orbital diagram for the ground state of samarium. The electron configuration is $[Xe]4f^66s^2$.

8.59 Write an orbital diagram for the ground state of the potassium atom. Is the atomic substance diamagnetic or paramagnetic?

8.60 Write an orbital diagram for the ground state of the zinc atom. Is the atomic substance diamagnetic or paramagnetic?

Periodic Trends

8.61 Order the following elements by increasing atomic radius according to what you expect from periodic trends: Se, S, As.

8.62 Using periodic trends, arrange the following elements in order of increasing atomic radius: F, S, Cl.

8.63 Using periodic trends, arrange the following elements by increasing ionization energy: Ar, Na, Cl, Al.

8.64 Arrange the following elements in order of increasing ionization energy: Mg, Ca, S. Do not look at Figure 8.18.

8.65 From what you know in a general way about electron affinities, state which member of each of the following pairs has the greater value: a As, Br b F, Li.

8.66 From what you know in a general way about electron affinities, state which member of each of the following pairs has the greater value: a Cl, S b Se, K.

8.67 If potassium chlorate has the formula $KClO_3$, what formula would you expect for lithium bromate?

8.68 Write the simplest formulas expected for two oxides of selenium.

General Problems

Key: These problems provide more practice but are not divided by topic or keyed to exercises. Each section ends with essay questions, each of which is color coded to refer to the *Instrumental Methods* (brown) and *A Chemist Looks at Frontiers* (purple) chapter essay on which it is based. Odd-numbered problems and the even-numbered problems that follow are similar; answers to all odd-numbered problems except the essay questions are given in the back of the book.

8.69 Write the complete ground-state electron configuration of the strontium atom, Sr, using the building-up principle.

8.70 Write the complete ground-state electron configuration of the tellurium atom, Te, using the building-up principle.

8.71 Obtain the valence-shell configuration of the polonium atom, Po, using the position of this atom in the periodic table.
8.72 Obtain the valence-shell configuration of the lead atom, Pb, using the position of this atom in the periodic table.

8.73 Write the orbital diagram for the ground state of the arsenic atom. Give all orbitals.
8.74 Write the orbital diagram for the ground state of the selenium atom. Give all orbitals.

8.75 For eka-lead, predict the electron configuration, whether the element is a metal or nonmetal, and the formula of an oxide.
8.76 For eka-bismuth, predict the electron configuration, whether the element is a metal or nonmetal, and the formula of an oxide.

8.77 From Figure 8.18, predict the first ionization energy of francium ($Z = 87$).
8.78 From Figure 8.18, predict the first ionization energy of ununseptium ($Z = 117$).

8.79 Write the orbital diagram corresponding to the ground state of Nb, whose configuration is $[Kr]4d^4 5s^1$.
8.80 Write the orbital diagram for the ground state of rhodium. The configuration is $[Kr]4d^8 5s^1$.

8.81 Match each set of characteristics on the left with an element in the column at the right.

a. reactive nonmetal; the atom has a large electron affinity	Sodium (Na)
b. A soft metal; the atom has a low ionization energy	Antimony (Sb)
c. A metalloid that forms an oxide of formula R_2O_3	Argon (Ar)
d. A chemically unreactive gas	Chlorine (Cl_2)

8.82 Match each element on the right with a set of characteristics on the left.

a. A reactive, pale yellow gas; the atom has a large electron affinity	Oxygen (O_2)
b. A soft metal that reacts with water to produce hydrogen	Gallium (Ga)
c. A metal that forms an oxide of formula R_2O_3	Barium (Ba)
d. A colorless gas; the atom has a moderately large negative electron affinity	Fluorine (F_2)

8.83 Find the electron configuration of the element with $Z = 23$. From this, give its group and period in the periodic table. Classify the element as a main-group, a *d*-block transition, or an *f*-block transition element.
8.84 Find the electron configuration of the element with $Z = 33$. From this, give its group and period in the periodic table. Is this a main-group, a *d*-block transition, or an *f*-block transition element?
■ **8.85** What property of an atom does nuclear magnetic resonance depend on? What frequency range of the electromagnetic spectrum does NMR use?
■ **8.86** Explain the origin of the "chemical shift" seen for a proton in a molecule.
■ **8.87** Explain how x rays are produced when an electron beam falls on a metal target.
■ **8.88** Describe how x-ray photoelectron spectroscopy is done. What kind of information is obtained from photoelectron spectroscopy?
■ **8.89** How is it possible for a diamagnetic material to be levitated by a magnetic field?
■ **8.90** Researchers have shown that it is possible to levitate diamagnetic materials. What animal did they use in their experiments to show this?

Strategy Problems

Key: As noted earlier, all of the practice and general problems are matched pairs. This section is a selection of problems that are not in matched-pair format. These challenging problems require that you employ many of the concepts and strategies that were developed in the chapter. In some cases, you will have to integrate several concepts and operational skills in order to solve the problem successfully.

8.91 The following are orbital diagrams for presumed ground-state atoms. Several, though, violate Pauli's principle or Hund's rule. Which of the these follow both Pauli's principle and Hund's rule, and which violate one or the other (state whether Pauli's principle or Hund's rule is violated)?

a. (↑↓) 1s (↑↓) 2s (↑)(↑)(↓) 2p
b. (↑↓) 1s (↑↑) 2s (↑↓)(↑↓)(↑↓) 2p
c. (↑↓) 1s (↑↓) 2s (↑↓)(↑↑)(↑↓) 2p (↑) 3s
d. [Ne] (↑↓) 3s (↑↓)(↑↓)(↑↓) 3p (↑) 4s
e. [Ne] (↑↓) 3s (↑↓)(↑↑)(↑↓) 3p (↑↓) 4s

8.92 As we noted at the beginning of Section 8.2, an atom has an infinite number of electron configurations, including not just the ground state but excited states as well. The following are excited states of neutral atoms. Identify the element and write the ground-state electron configuration for that atom.
a. $1s^2 2s^2 2p^6 3s^2 3p^6 3d^{10} 4s^1 4p^1$
b. $1s^1 2s^2 2p^6 3s^2 3p^6$
c. $1s^2 2s^2 2p^6 3s^2 3p^6 3d^1$
d. $1s^2 2s^2 2p^5 3s^2 3p^1$
e. $1s^2 2s^2 2p^6 3s^2 3p^6 3d^9 4s^2 4p^5$
8.93 A metallic element, M, reacts vigorously with water to form a solution of MOH. If M is in Period 5, what is the ground-state electron configuration of the atom?
8.94 A nonmetallic element, R, burns brightly in air to give the oxide R_4O_{10}. If R is in Period 3, what is the ground-state valence-shell configuration of the atom?
8.95 The ground-state electron configuration of an atom is $1s^2 2s^2 2p^6 3s^2 3p^4$. What is the valence-shell configuration of the atom in the same group, but in Period 5?
8.96 An atom of an element has the following ground-state configuration: $[Xe]4f^{14} 5d^9 6s^1$. Using this configuration, give the symbol of the element. Explain your reasoning.
8.97 An atom of an element has the following ground-state configuration: $[Kr]4d^{10} 5s^2 5p^b$. If the element is in Group VIA, what is the symbol of the element? What is *b*?

8.98 A main-group atom in Period 5 has the following orbital diagram for its valence shell (ground state): (↑↓) (↑)(↑)(↑). What is the symbol for the element? Write the formula of the oxide containing the most oxygen.

8.99 The following are electron configurations of neutral atoms, which may or may not be in ground states: $1s^22s^22p^53s^23p^2$ and $1s^22s^22p^63s^23p^5$. Write the formula of a compound of these two elements.

8.100 A neutral atom has the electron configuration $1s^22s^22p^63s^13p^2$. Is this the ground-state configuration? What is the period and column of the element? What is the name of the element? What is the formula of the oxide of this element?

8.101 A moderately reactive metallic element (whose melting point is 850°C) burns in a container of a greenish gas (also an element) to produce a white solid. The white solid dissolves in water to form a clear solution. When carbon dioxide gas, CO_2, is bubbled into this solution, it produces a white precipitate. Identify, by names and formulas, the substances referred to as the white solid and the white precipitate. Use a handbook of chemistry to obtain any data you may need.

8.102 A metallic element reacts vigorously with water, evolving hydrogen gas. An excited atom of this element has its outer electron in the $3p$ orbital. When this electron drops to its ground state in the $3s$ orbital, light is emitted of wavelength 589 nm. What is the identity of the element? Explain how you arrived at your answer. What is the color of the emitted light?

8.103 All elements up to atomic number 118 have been discovered. What electron configuration would you expect for the yet-to-be discovered element 120? Explain your answer.

8.104 The densities (in g/cm^3) of some elements in Period 5 that span indium, In, are as follows: Ag, 10.49; Cd, 8.65; Sn, 7.31; and Sb, 6.68. Plot the density of each element against its atomic number. Using the graph, estimate the density of indium. Compare your estimate to the value that you find in a handbook of chemistry or online.

8.105 Write the orbital diagram for the ground-state valence electrons of the main-group atom in Period 5 that has the smallest radius. Explain how you got this answer.

8.106 Using the periodic table, decide which of the following atoms you expect to have the largest radius: H, Sc, As, Br, V, K. Why?

8.107 An atom easily loses two electrons to form the ion R^{2+}. The element, which is in Period 6, forms the oxides RO and RO_2. Give the orbital diagram for the ground-state valence shell of the element that immediately follows R in the periodic table.

8.108 List three elements in order of increasing ionization energy. Two of the elements should be halogens (but not F), and the other should be Se.

8.109 The energy of dissociation of the Cl_2 molecule is 240 kJ/mol. Using the appropriate data from this chapter, calculate the energy obtained (or released) in adding electrons to the Cl_2 molecule to produce Cl^- ions. Is this process endothermic or exothermic?

8.110 The electron affinity of the lutetium atom (element 71) was measured using the technique of photoelectron spectroscopy with an infrared laser (the essay on p. 310 describes this instrumental method, using x rays). In this experiment, a beam of lutetium negative ions, Lu^-, was prepared and irradiated with a laser beam having a wavelength at 1064 nm. The energy supplied by a photon in this laser beam removes an electron from a negative ion, leaving the neutral atom. The energy needed to remove the electron from the negative ion to give the neutral atom (both in their ground states) is the electron affinity of lutetium. Any excess energy of the photon shows up as kinetic energy of the emitted electron. If the emitted electron in this experiment has a kinetic energy of 0.825 eV, what is the electron affinity of lutetium?

Cumulative-Skills Problems

Key: The problems under this heading combine skills introduced in previous chapters with those given in the current one.

8.111 A 2.50-g sample of barium reacted completely with water. What is the equation for the reaction? How many milliliters of dry H_2 evolved at 21°C and 748 mmHg?

8.112 A sample of cesium metal reacted completely with water, evolving 48.1 mL of dry H_2 at 19°C and 768 mmHg. What is the equation for the reaction? What was the mass of cesium in the sample?

8.113 What is the formula of radium oxide? What is the percentage of radium in this oxide?

8.114 What is the formula of hydrogen telluride? What is the percentage of tellurium in this compound?

8.115 How much energy would be required to ionize 5.00 mg of Na(g) atoms to Na^+(g) ions? The first ionization energy of Na atoms is 496 kJ/mol.

8.116 How much energy is evolved when 2.65 mg of Cl(g) atoms adds electrons to give Cl^-(g) ions?

8.117 Using the Bohr formula for the energy levels, calculate the energy required to raise the electron in a hydrogen atom from $n = 1$ to $n = \infty$. Express the result for 1 mol H atoms. Because the $n = \infty$ level corresponds to removal of the electron from the atom, this energy equals the ionization energy of the H atom.

8.118 Calculate the ionization energy of the He^+ ion in kJ/mol (this would be the second ionization energy of He). See Problem 8.117. The Bohr formula for the energy levels of an ion consisting of a nucleus of charge Z and a single electron is $-R_HZ^2/n^2$.

8.119 The lattice energy of an ionic solid such as NaCl is the enthalpy change $\Delta H°$ for the process in which the solid changes to ions. For example,

$$NaCl(s) \longrightarrow Na^+(g) + Cl^-(g) \quad \Delta H = 786 \text{ kJ/mol}$$

Assume that the ionization energy and electron affinity are ΔH values for the processes defined by those terms. The ionization energy of Na is 496 kJ/mol. Use this, the electron affinity from Table 8.4, and the lattice energy of NaCl to calculate ΔH for the following process:

$$Na(g) + Cl(g) \longrightarrow NaCl(s)$$

8.120 Calculate ΔH for the following process:

$$K(g) + Br(g) \longrightarrow KBr(s)$$

The lattice energy of KBr is 689 kJ/mol, and the ionization energy of K is 419 kJ/mol. The electron affinity of Br is given in Table 8.4. See Problem 8.119.

9 Ionic and Covalent Bonding

© Clyde H. Smith/Peter Arnold Images/Photolibrary; David Stoecklein/Corbis

The shape of snowflakes results from bonding (and intermolecular) forces in H_2O.

Contents and Concepts

Ionic Bonds

Molten salts and aqueous solutions of salts are electrically conducting. This conductivity results from the motion of ions in the liquids. This suggests the possibility that ions exist in certain solids, held together by the attraction of ions of opposite charge.

Covalent Bonds

Not all bonds can be ionic. Hydrogen, H_2, is a clear example, in which there is a strong bond between two like atoms. The bonding in the hydrogen molecule is covalent. A *covalent bond* forms between atoms by the sharing of a pair of electrons.

OWL Sign in to OWL at **www.cengage.com/owl** to view tutorials and simulations, develop problem-solving skills, and complete online homework assigned by your professor.

go Chemistry Download mini lecture videos for key concept review and exam prep from OWL or purchase them from **www.cengagebrain.com**.

The properties of a substance, such as sodium chloride (Figure 9.1), are determined in part by the chemical bonds that hold the atoms together. A *chemical bond* is a strong attractive force that exists between certain atoms in a substance. In Chapter 2 we described how sodium (a silvery metal) reacts with chlorine (a pale greenish yellow gas) to produce sodium chloride (table salt, a white solid). The substances in this reaction are quite different, as are their chemical bonds. Sodium chloride, NaCl, consists of Na^+ and Cl^- ions held in a regular arrangement, or crystal, by *ionic bonds.* Ionic bonding results from the attractive force of oppositely charged ions.

A second kind of chemical bond is a *covalent bond.* In a covalent bond, two atoms share valence electrons (outer-shell electrons), which are attracted to the positively charged cores of both atoms, thus linking them. For example, chlorine gas consists of Cl_2 molecules. You would not expect the two Cl atoms in each Cl_2 molecule to acquire the opposite charges required for ionic bonding. Rather, a covalent bond holds the two atoms together. This is consistent with the equal sharing of electrons that you would expect between identical atoms. In most molecules, the atoms are linked by covalent bonds.

Metallic bonding, seen in sodium and other metals, represents another important type of bonding. A crystal of sodium metal consists of a regular arrangement of sodium atoms. The valence electrons of these atoms move throughout the crystal, attracted to the positive cores of all Na^+ ions. This attraction holds the crystal together.

What determines the type of bonding in each substance? How do you describe the bonding in various substances? In this chapter we will look at some simple, useful concepts of bonding that can help us answer these questions. We will be concerned with ionic and covalent bonds in particular.

Figure 9.1

Sodium chloride crystals Natural crystals of sodium chloride mineral (halite).

Ionic Bonds

The first explanation of chemical bonding was suggested by the properties of *salts,* substances now known to be ionic. Salts are generally crystalline solids that melt at high temperatures. Sodium chloride, for example, melts at 801°C. A molten salt (the liquid after melting) conducts an electric current. A salt dissolved in water gives a solution that also conducts an electric current. The electrical conductivity of the molten salt and the salt solution results from the motion of ions in the liquids. This suggests the possibility that ions exist in certain solids, held together by the attraction of opposite charges.

9.1 Describing Ionic Bonds

An **ionic bond** is *a chemical bond formed by the electrostatic attraction between positive and negative ions.* The bond forms between two atoms when one or more electrons are transferred from the valence shell of one atom to the valence shell of the other. The atom that loses electrons becomes a *cation* (positive ion), and the atom that gains electrons becomes an *anion* (negative ion). Any given ion tends to attract as many neighboring ions of opposite charge as possible. When large numbers of ions gather together, they form an ionic solid. The solid normally has a regular, crystalline structure that allows for the maximum attraction of ions, given their particular sizes.

To understand why ionic bonding occurs, consider the transfer of a valence electron from a sodium atom (electron configuration $[Ne]3s^1$) to the valence shell of a chlorine atom ($[Ne]3s^23p^5$). You can represent the electron transfer by the following equation:

$$Na([Ne]3s^1) + Cl([Ne]3s^23p^5) \longrightarrow Na^+([Ne]) + Cl^-([Ne]3s^23p^6)$$

As a result of the electron transfer, ions are formed, each of which has a noble-gas configuration. The sodium atom has lost its 3*s* electron and has taken on the neon

Table 9.1 Lewis Electron-Dot Symbols for Atoms of the Second and Third Periods

Period	IA ns^1	IIA ns^2	IIIA ns^2np^1	IVA ns^2np^2	VA ns^2np^3	VIA ns^2np^4	VIIA ns^2np^5	VIIIA ns^2np^6
Second	Li·	·Be·	·Ḃ·	·Ċ·	:Ṅ·	:Ȯ·	:Ḟ̤·	:Ne:
Third	Na·	·Mg·	·Al̇·	·Ṡi·	:Ṗ·	:Ṡ·	:Cl·	:Ar:

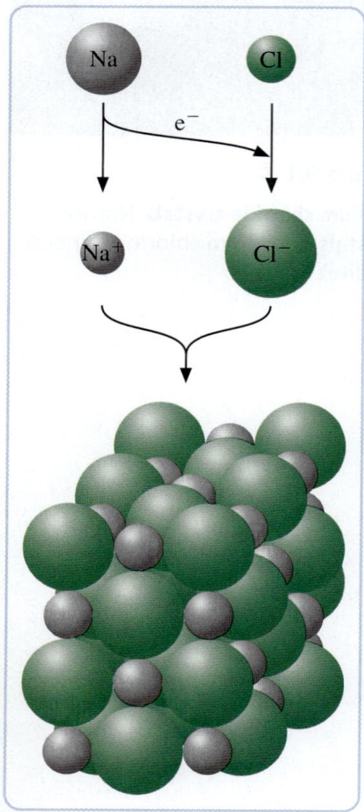

configuration, [Ne]. The chlorine atom has accepted the electron into its 3*p* subshell and has taken on the argon configuration, $[Ne]3s^23p^6$. Such noble-gas configurations and the corresponding ions are particularly stable. This stability of the ions accounts in part for the formation of the ionic solid NaCl. Once a cation or anion forms, it attracts ions of opposite charge. Within the sodium chloride crystal, NaCl, every Na^+ ion is surrounded by six Cl^- ions, and every Cl^- ion by six Na^+ ions. (Figure 2.19 shows the arrangement of ions in the NaCl crystal.)

Lewis Electron-Dot Symbols

You can simplify the preceding equation for the electron transfer between Na and Cl by writing Lewis electron-dot symbols for the atoms and monatomic ions. A **Lewis electron-dot symbol** is *a symbol in which the electrons in the valence shell of an atom or ion are represented by dots placed around the letter symbol of the element.* Table 9.1 lists Lewis symbols and corresponding valence-shell electron configurations for the atoms of the second and third periods. Note that dots are placed one to each side of a letter symbol until all four sides are occupied. Then the dots are written two to a side until all valence electrons are accounted for. The exact placement of the single dots is immaterial. For example, the single dot in the Lewis symbol for chlorine can be written on any one of the four sides. [This pairing of dots does not always correspond to the pairing of electrons in the ground state. Thus, you write ·Ḃ· for boron, rather than :B·, which more closely corresponds to the ground-state configuration $[He]2s^22p^1$. The first symbol better reflects boron's chemistry, in which each single electron (single dot) tends to be involved in bond formation.]

The equation representing the transfer of an electron from the sodium atom to the chlorine atom is

$$\text{Na}\cdot + \cdot\ddot{\underset{\cdot\cdot}{\text{Cl}}}: \longrightarrow \text{Na}^+ + [:\ddot{\underset{\cdot\cdot}{\text{Cl}}}:]^-$$

The noble-gas configurations of the ions are apparent from the symbols. No dots are shown for the cation. (All valence electrons have been removed, leaving the noble-gas core.) Eight dots are shown in brackets for the anion (noble-gas configuration ns^2np^6).

Example 9.1 Using Lewis Symbols to Represent Ionic Bond Formation

Gaining Mastery Toolbox

Critical Concept 9.1
To represent the formation of the ions of an ionic compound from the atoms using Lewis symbols, you show dots being transferred from the metal to nonmetal atoms to give the ions with noble-gas configurations.

Solution Essentials:
- Lewis electron-dot symbols
- Noble-gas configuration
- Formula of a binary ionic compound

Use Lewis electron-dot symbols to represent the transfer of electrons from magnesium to fluorine atoms to form ions with noble-gas configurations.

Problem Strategy Write down the Lewis symbols for the two atoms. Note how many electrons the metal atom should lose to assume a noble-gas configuration (all of the dots, representing valence electrons) and how many electrons the nonmetal atom should gain to assume such a configuration (enough to give eight dots about the atomic symbol). Represent the transfer of electrons between ions diagrammatically in the form of an equation.

(continued)

(*continued*)

Solution The Lewis symbols for the atoms are $:\ddot{\underset{..}{F}}\cdot$ and $\cdot Mg\cdot$ (see Table 9.1). The magnesium atom loses two electrons to assume a noble-gas configuration. But because a fluorine atom can accept only one electron to fill its valence shell, two fluorine atoms must take part in the electron transfer. We can represent this electron transfer as follows:

$$:\ddot{\underset{..}{F}}\cdot + \cdot Mg\cdot + \cdot\ddot{\underset{..}{F}}: \longrightarrow [:\ddot{\underset{..}{F}}:]^- + Mg^{2+} + [:\ddot{\underset{..}{F}}:]^-$$

Answer Check Check that you have the correct symbols for the atoms and that the symbols for the ions do have noble-gas configurations.

Exercise 9.1 Represent the transfer of electrons from magnesium to oxygen atoms to assume noble-gas configurations. Use Lewis electron-dot symbols.

■ See Problems 9.37 and 9.38.

Energy Involved in Ionic Bonding

You have seen in a qualitative way why a sodium atom and a chlorine atom might be expected to form an ionic bond. It is instructive, however, to look at the energy changes involved in ionic bond formation. From this analysis, you can gain further understanding of why certain atoms bond ionically and others do not.

If atoms come together and bond, there should be a net decrease in energy, because the bonded state should be more stable and therefore at a lower energy level. Consider again the formation of an ionic bond between a sodium atom and a chlorine atom. You can think of this as occurring in two steps: (1) An electron is transferred between the two separate atoms to give ions. (2) The ions then attract one another to form an ionic bond. In reality, the transfer of the electron and the formation of an ionic bond occur simultaneously, rather than in discrete steps, as the atoms approach one another. But the *net* quantity of energy involved is the same whether the steps occur one after the other or at the same time.

The first step requires removal of the 3*s* electron from the sodium atom and the addition of this electron to the valence shell of the chlorine atom. Removing the electron from the sodium atom requires energy (the first ionization energy of the sodium atom, which equals 496 kJ/mol). Adding the electron to the chlorine atom releases energy (equal to −349 kJ/mol, which is the negative of the electron affinity of the chlorine atom). ▶ The overall energy of this step is (496 – 349) kJ/mol, or 147 kJ/mol (Figure 9.2, Step 1). So, the process requires more energy to remove an electron from the sodium atom than is gained when the electron is added to the chlorine atom. The formation of ions from the atoms is not in itself energetically favorable.

Ionization energies and electron affinities of atoms were discussed in Section 8.6.

When positive and negative ions bond, however, more than enough energy is released to make the overall process favorable. What principally determines the

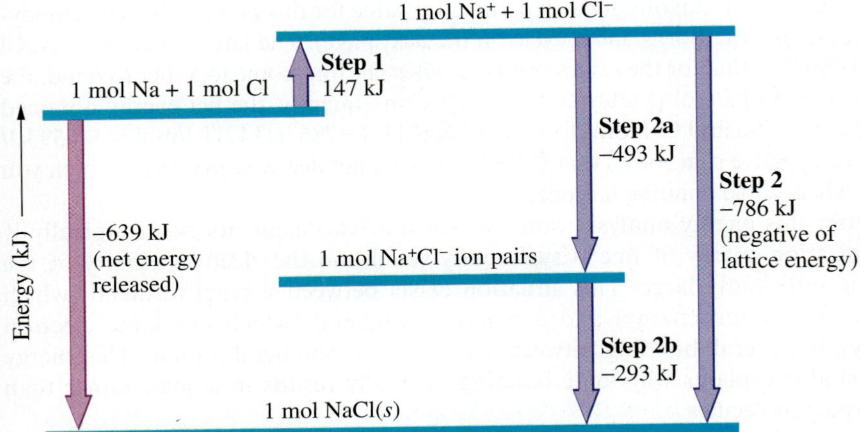

Figure 9.2 ◀

Energetics of ionic bonding The transfer of an electron from a Na atom to a Cl atom is not in itself energetically favorable; it requires 147 kJ/mol of energy (Step 1). However, 493 kJ of energy is released when these oppositely charged ions come together to form ion pairs (Step 2a). Moreover, additional energy (293 kJ) is released when these ion pairs form the solid NaCl crystal (Step 2b). The lattice energy released when one mole each of Na^+ and Cl^- ions react to produce NaCl(*s*) is 786 kJ/mol, and the overall process of NaCl formation is energetically favorable, releasing 639 kJ/mol if one starts with gaseous Na and Cl atoms.

energy released when ions bond is the attraction of oppositely charged ions. To see this, let's look first at the energy obtained when a Na^+ ion and a Cl^- ion come together to form an ion-pair molecule. We will estimate this energy by making the simplifying assumption that the ions are spheres, just touching, with the distance between the nuclei of the ions equal to this distance in the NaCl crystal. From experiment, this distance is known to be 282 pm, or 2.82×10^{-10} m. To calculate the energy obtained when the ion spheres come together, we use Coulomb's law.

Coulomb's law states that *the potential energy obtained in bringing two charges* Q_1 *and* Q_2*, initially far apart, up to a distance* r *apart is directly proportional to the product of the charges and inversely proportional to the distance between them:*

$$E = \frac{kQ_1Q_2}{r}$$

Here k is a physical constant, equal to 8.99×10^9 J·m/C^2 (C is the symbol for coulomb). The charge on Na^+ is $+e$ and that on Cl^- is $-e$, where e equals 1.602×10^{-19} C. Thus, our estimate of the energy of attraction of Na^+ and Cl^- ions to form an ion pair is

$$E = \frac{-(8.99 \times 10^9 \text{ J·m/C}^2) \times (1.602 \times 10^{-19} \text{ C})^2}{2.82 \times 10^{-10} \text{ m}} = -8.18 \times 10^{-19} \text{ J}$$

The minus sign means energy is released. This energy is for the formation of one ion pair. To express this for one mole of Na^+Cl^- ion pairs, we multiply by Avogadro's number, 6.02×10^{23}. We obtain -493 kJ/mol for the energy obtained when one mole of Na^+ and one mole of Cl^- come together to form Na^+Cl^- ion pairs (Figure 9.2, Step 2a). ◀

The energy value −493 kJ/mol is approximate, because of the simplifying assumption we made (that the ions are spheres, just touching).

What we see is that the formation of ion pairs from sodium and chlorine atoms is energetically favorable. The attraction of oppositely charged ions, however, does not stop with the bonding of pairs of ions. The maximum attraction of ions of opposite charge with the minimum repulsion of ions of the same charge is obtained with the formation of the crystalline solid. Then additional energy is released. The energy of this step (Figure 9.2, Step 2b) is most easily obtained as the difference between Figure 9.2, Step 2a, which we just calculated, and Figure 9.2, Step 2, the energy released when a crystalline solid forms from the ions. This is the negative of the *lattice energy* of NaCl. The additional energy (in going from ion pairs to the crystalline solid), Figure 9.2, Step 2b, equals -293 kJ/mol.

The **lattice energy** is *the change in energy that occurs when an ionic solid is separated into isolated ions in the gas phase.* For sodium chloride, the process is

$$NaCl(s) \longrightarrow Na^+(g) + Cl^-(g)$$

The distances between ions in the crystal are continuously enlarged until the ions are very far apart. You can obtain an experimental value for this process from thermodynamic data (see the Born-Haber cycle on the next page). The lattice energy for NaCl is 786 kJ/mol so that for the reverse process, when the ions come together to bond, the energy is -786 kJ/mol (Figure 9.2, Step 2). Consequently, the net energy obtained when gaseous Na and Cl atoms form solid NaCl is $(-786 + 147)$ kJ/mol $= -639$ kJ/mol. The negative sign shows that there has been a net decrease in energy, which you expect when stable bonding has occurred.

From this energy analysis, you can see that two elements bond ionically if the ionization energy of one is sufficiently small and the electron affinity of the other is sufficiently large. This situation exists between a reactive metal (which has low ionization energy) and a reactive nonmetal (which has large electron affinity). In general, bonding between a metal and a nonmetal is ionic. This energy analysis also explains why ionic bonding normally results in a solid rather than in ion-pair molecules.

Lattice Energies from the Born–Haber Cycle

The preceding energy analysis requires us to know the lattice energy of solid sodium chloride. Direct experimental determination of the lattice energy of an ionic solid is difficult. However, this quantity can be indirectly determined from experiment by means of a thermochemical "cycle" originated by Max Born and Fritz Haber in 1919 and now called the *Born–Haber cycle.* The reasoning is based on Hess's law.

To obtain the lattice energy of NaCl, you think of solid sodium chloride being formed from the elements by two different routes, as shown in Figure 9.3. In one route, NaCl(*s*) is formed directly from the elements, Na(*s*) and $\frac{1}{2}Cl_2(g)$. The enthalpy change for this is $\Delta H_f°$, which is given in Table 6.2 as −411 kJ per mole of NaCl. The second route consists of the following five steps, along with the enthalpy change for each. (To be precise, the ionization energy and electron affinity are energy changes, ΔE, and we should add small corrections to give the enthalpy changes, ΔH.)

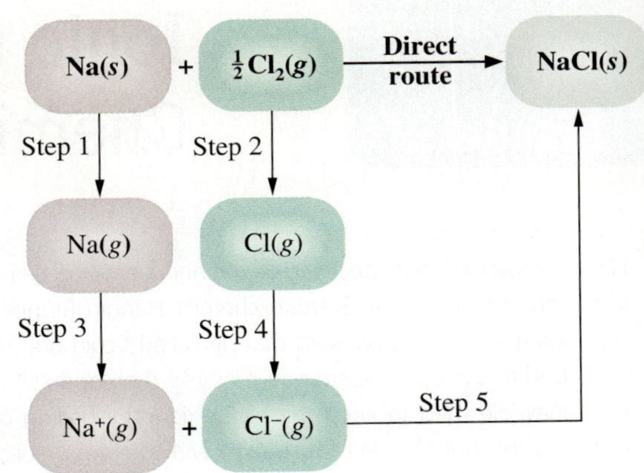

Figure 9.3 ▲

Born–Haber cycle for NaCl The formation of NaCl(s) from the elements is accomplished by two different routes. The direct route is the formation reaction (shown in boldface), and the enthalpy change is $\Delta H_f°$. The indirect route occurs in five steps.

1. *Sublimation of sodium.* Metallic sodium is vaporized to a gas of sodium atoms. (*Sublimation* is the transformation of a solid to a gas.) The enthalpy change for this process, measured experimentally, is 108 kJ per mole of sodium.
2. *Dissociation of chlorine.* Chlorine molecules are dissociated to atoms. The enthalpy change for this equals the Cl—Cl bond dissociation energy, which is 240 kJ per mole of bonds, or 120 kJ per mole of Cl atoms.
3. *Ionization of sodium.* Sodium atoms are ionized to Na^+ ions. The enthalpy change is essentially the ionization energy of atomic sodium, which equals 496 kJ per mole of Na.
4. *Formation of chloride ion.* The electrons from the ionization of sodium atoms are transferred to chlorine atoms. The enthalpy change for this is the negative of the electron affinity of atomic chlorine, equal to −349 kJ per mole of Cl atoms.
5. *Formation of NaCl(s) from ions.* The ions Na^+ and Cl^- formed in Steps 3 and 4 combine to give solid sodium chloride. Because this process is just the reverse of the one corresponding to the lattice energy (breaking the solid into ions), the enthalpy change is the negative of the lattice energy. If we let *U* be the lattice energy, the enthalpy change for Step 5 is −*U*.

Let us write these five steps and add them. We also add the corresponding enthalpy changes, following Hess's law.

$$
\begin{array}{lll}
\text{Na}(s) \longrightarrow \cancel{\text{Na}(g)} & \Delta H_1 = & 108 \text{ kJ} \\
\tfrac{1}{2}\text{Cl}_2(g) \longrightarrow \cancel{\text{Cl}(g)} & \Delta H_2 = & 120 \text{ kJ} \\
\cancel{\text{Na}(g)} \longrightarrow \cancel{\text{Na}^+(g)} + \cancel{e^-(g)} & \Delta H_3 = & 496 \text{ kJ} \\
\cancel{\text{Cl}(g)} + \cancel{e^-(g)} \longrightarrow \cancel{\text{Cl}^-(g)} & \Delta H_4 = & -349 \text{ kJ} \\
\cancel{\text{Na}^+(g)} + \cancel{\text{Cl}^-(g)} \longrightarrow \text{NaCl}(s) & \Delta H_5 = & -U \\
\hline
\text{Na}(s) + \tfrac{1}{2}\text{Cl}_2(g) \longrightarrow \text{NaCl}(s) & \Delta H_f° = & 375 \text{ kJ} - U
\end{array}
$$

In summing the equations, we have canceled terms that appear on both the left and right sides of the arrows. The final equation is simply the formation reaction for NaCl(*s*). Adding the enthalpy changes, we find that the enthalpy change for this formation reaction is 375 kJ − *U*. But the enthalpy of formation has been determined calorimetrically and equals −411 kJ. Equating these two values, we get

$$375 \text{ kJ} - U = -411 \text{ kJ}$$

Solving for *U* yields the lattice energy of NaCl:

$$U = (375 + 411) \text{ kJ} = 786 \text{ kJ}$$

A **CHEMIST** Looks at...

Ionic Liquids and Green Chemistry

New substances with peculiar or strange behavior have always intrigued chemists. The German chemist Hennig Brand in 1669 discovered a white, waxy solid that he called "cold fire," because it glowed in the dark—a strange property then and still captivating today. He tried to keep the recipe (lengthy boiling of putrid urine) secret, but people's fascination with the waxy solid, which was white phosphorus, proved irresistible.

Modern chemists continue to expand the boundaries of known materials. They have discovered plastics that conduct electricity like metals and materials that look solid but are so porous they are almost as light as air. Now chemists have produced room-temperature ionic liquids (Figure 9.4). Most of these are clear, well-behaved substances that look and pour much like water, and like water their strength is as solvents (liquids that dissolve other substances). In fact, they promise to be "super solvents." Given a material—an organic substance, a plastic, or even a rock—researchers believe you will be able to find an ionic liquid that is capable of dissolving it!

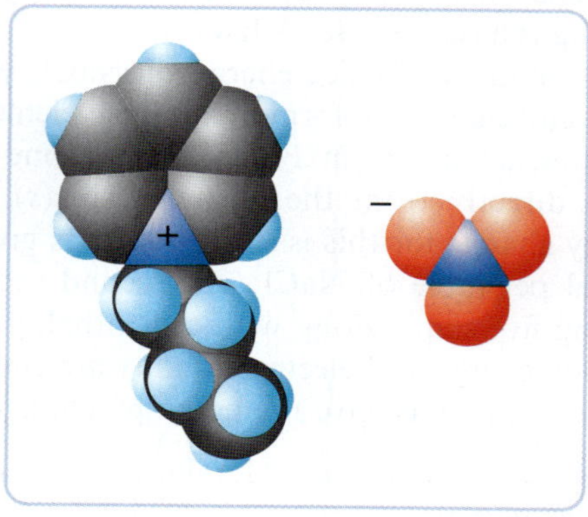

Figure 9.5 ▲

The ions composing an ionic liquid Shown here are space-filling models of the ions composing *N*-butylpyridinium nitrate. Note the bulky cation and small nitrate anion.

Compare room-temperature ionic liquids with their more prosaic cousins. Sodium chloride, a typical example, is solid at room temperature and doesn't melt until exposed to temperatures over 800°C. The molten liquid, though clear, is very corrosive (chemically reactive). The high melting point of sodium chloride is easily explained. It consists of small, spherical ions that pack closely together. Thus, the ions interact strongly, giving a solid with a high melting point. Room-temperature ionic liquids, by contrast, consist of large, nonspherical cations with various anions (Figure 9.5). It is the large, bulky cations that keep the ions from packing closely; the large distances between ions result in weak interactions, yielding a substance whose melting point is often well below room temperature.

Figure 9.4 ▲

Room-temperature ionic liquids A series of tubes containing luminescent ionic liquids, spelling out QUILL (Queen's University Ionic Liquid Laboratories Research Centre, Belfast. Ireland).

The demand for *green chemistry,* the commercial production of chemicals using environmentally sound methods, has spurred much of the research into ionic liquids. Many chemical processes use volatile organic solvents. These solvents are liquids that evaporate easily into the surrounding air, where they can contribute to air pollution. Organic solvents are often flammable, too. Ionic liquids are neither volatile nor flammable. In addition to these environmental rewards, however, ionic liquids appear to offer another bonus: The proper choice of ionic liquid may improve the yield and lower the costs of a chemical process.

■ See Problems 9.111 and 9.112.

Properties of Ionic Substances

Typically, ionic substances are high-melting solids. Sodium chloride, NaCl, ordinary salt, melts at 801°C, and magnesium oxide, MgO, a ceramic, melts at 2800°C. The explanation for the high melting points of these substances is simple.

Small, spherical cations and anions interact by strong bonds that essentially depend on the electrical force of attraction described by Coulomb's law. Large groups of such cations and anions attract one another with strong ionic bonds linking all of the ions together, forming a crystalline solid. When you heat any solid, the atoms or ions of that solid begin to vibrate; and, as you raise the temperature, these atoms or ions vibrate through larger and larger distances. At a high enough temperature, the atoms or ions of the solid may move sufficiently far apart that the solid melts to a liquid (or the solid may simply decompose to different substances). The temperature at which melting occurs depends on the strength of the interaction between the atoms or ions. Typical ionic solids require high temperatures for this process to occur because of the strong interactions between the ions.

Coulomb's law also explains why magnesium oxide, which is composed of ions having double charges (Mg^{2+} and O^{2-}), has such a high melting point compared with sodium chloride (Na^{+} and Cl^{-}). Coulombic interactions depend on the product of the ion charges. For NaCl, this is 1×1, whereas for MgO, this is 2×2, or 4 times greater. (Coulomb's law also depends on the distance between charges, or ions, which is about the same in NaCl and MgO.) The much larger Coulombic interaction in MgO requires a much greater temperature to initiate melting.

The liquid melt from an ionic solid consists of ions, and so the liquid conducts an electric current. If the ionic solid dissolves in a molecular liquid, such as water, the resulting solution consists of ions dispersed among molecules; the solution also conducts an electric current.

Recently, as the accompanying essay on ionic liquids describes, chemists have prepared ionic substances that behave atypically. Their unusual properties are the result of large, nonspherical cations that lead to especially weak ionic bonding. The melting points of these ionic substances are unusually low; often they are liquids at room temperature (that is, their melting points are below room temperature).

9.2 Electron Configurations of Ions

In the previous section, we described the formation of ions from atoms. Often you can understand what monatomic ions form by looking at the electron configurations of the atoms and deciding what configurations you would expect for the ions.

Ions of the Main-Group Elements

In Chapter 2, we listed the common monatomic ions of the main-group elements (Table 2.3). Most of the cations are obtained by removing all the valence electrons from the atoms of metallic elements. Once these atoms have lost their valence electrons, they have stable noble-gas or pseudo-noble-gas configurations. The stability of these configurations can be seen by looking at the successive ionization energies of some atoms. Table 9.2 lists the first through the fourth

Table 9.2 Ionization Energies of Na, Mg, and Al (in kJ/mol)*

	Successive Ionization Energies			
Element	**First**	**Second**	**Third**	**Fourth**
Na	496	4,562	6,910	9,543
Mg	738	1,451	7,733	10,542
Al	578	1,817	2,745	11,577

*Energies for the ionization of valence electrons lie to the left of the colored line.

ionization energies of Na, Mg, and Al. The energy needed to remove the first electron from the Na atom is only 496 kJ/mol (first ionization energy). But the energy required to remove another electron (the second ionization energy) is nearly ten times greater (4562 kJ/mol). The electron in this case must be taken from Na^+, which has a neon configuration. Magnesium and aluminum atoms are similar. Their valence electrons are easily removed, but the energy needed to take an electron from either of the ions that result (Mg^{2+} and Al^{3+}) is extremely high. That is why no compounds are found with ions having charges greater than the group number.

The loss of successive electrons from an atom requires increasingly more energy. Consequently, Group IIIA elements show less tendency to form ionic compounds than do Group IA and IIA elements, which primarily form ionic compounds. Boron, in fact, forms no compounds with B^{3+} ions. The bonding is normally covalent, a topic discussed later in this chapter. However, the tendency to form ions becomes greater going down any column of the periodic table because of decreasing ionization energy. The remaining elements of Group IIIA do form compounds containing 3+ ions.

There is also a tendency for Group IIIA to VA elements of higher periods, particularly Period 6, to form compounds with ions having a positive charge of two less than the group number. Thallium in IIIA, Period 6, has compounds with 1+ ions and compounds with 3+ ions. Ions with charge equal to the group number minus two are obtained when the *np* electrons of an atom are lost but the ns^2 electrons are retained. For example,

$$\mathrm{Tl}([\mathrm{Xe}]4f^{14}5d^{10}6s^26p^1) \longrightarrow \mathrm{Tl}^+([\mathrm{Xe}]4f^{14}5d^{10}6s^2) + e^-$$

Few compounds of 4+ ions are known because the energy required to form 4+ ions is so great. The first three elements of Group IVA—C, Si, and Ge—are nonmetals (or metalloids) and usually form covalent rather than ionic bonds. Tin and lead, however, the fourth and fifth elements of Group IVA, commonly form compounds with 2+ ions (ionic charge equal to the group number minus two). For example, tin forms tin(II) chloride, $SnCl_2$, which is an ionic compound. It also forms tin(IV) chloride, $SnCl_4$, but this is a covalent, not an ionic, compound. Bismuth, in Group VA, is a metallic element that forms compounds of Bi^{3+} (ionic charge equal to the group number minus two), where only the 6*p* electrons have been lost.

© Cengage Learning

Group VIA and Group VIIA elements, whose atoms have the largest electron affinities, would be expected to form monatomic ions by gaining electrons to give noble-gas or pseudo-noble-gas configurations. An atom of a Group VIIA element (valence-shell configuration ns^2np^5) picks up one electron to give a 1− anion (ns^2np^6); examples are F^- and Cl^-. (Hydrogen also forms compounds of the 1− ion, H^-. The hydride ion, H^-, has a $1s^2$ configuration, like the noble-gas atom helium.) An atom of a Group VIA element (valence-shell configuration ns^2np^4) picks up two electrons to give a 2− anion (ns^2np^6); examples are O^{2-} and S^{2-}. Although the electron affinity of nitrogen ($2s^22p^3$) is essentially zero, the N^{3-} ion ($2s^22p^6$) is stable in the presence of certain positive ions, including Li^+ and those of the alkaline earth elements. ◀

Interestingly, lithium metal reacts with nitrogen at room temperature to form a layer of lithium nitride, Li_3N, on the metal surface (see photo above).

To summarize, the common monatomic ions found in compounds of the main-group elements fall into three categories (see Table 2.3):

1. Cations of Groups IA to IIIA having noble-gas or pseudo-noble-gas configurations. The ion charges equal the group numbers.
2. Cations of Groups IIIA to VA having ns^2 electron configurations. The ion charges equal the group numbers minus two. Examples are Tl^+, Sn^{2+}, Pb^{2+}, and Bi^{3+}.
3. Anions of Groups VA to VIIA having noble-gas or pseudo-noble-gas configurations. The ion charges equal the group numbers minus eight.

Example 9.2 Writing the Electron Configuration and Lewis Symbol for a Main-Group Ion

Gaining Mastery Toolbox

Critical Concept 9.2
The common monatomic ions in the main-group elements fall into three categories: (1) cations of Groups IA to IIIA usually have ion charges equal to group numbers, (2) cations of Groups IIIA to VA may have ion charges equal to group numbers minus two, and (3) anions of Groups VA to VIIA have charges equal to the group numbers minus 8.

Solution Essentials:
- Categories of main-group monatomic ions
- Number of valence electrons in a main-group atom
- Electron configuration of an atom

Write the electron configuration and the Lewis symbol for N^{3-}.

Problem Strategy Recall the three categories of main-group monatomic ions: (1) cations of Groups IA to IIIA usually have ion charges equal to group numbers; (2) cations of Groups IIIA to VA may have ion charges equal to group numbers minus two—an example is Tl^+; and (3) anions of Groups VA to VIIA have charges equal to the group numbers minus eight. Write the electron configuration of the atom, and then subtract or add electrons to give the ion. Similarly, write the Lewis symbol of the atom, and subtract or add dots to give the ion.

Solution The electron configuration of the N atom is $[He]2s^22p^3$. By gaining three electrons, the atom assumes a 3− charge and the neon configuration $\mathbf{[He]2s^22p^6}$. The Lewis symbol is

$$[:\ddot{\underset{\cdot\cdot}{N}}:]^{3-}$$

Answer Check Check that the ion has a noble-gas or pseudo-noble-gas configuration and that the Lewis symbol has eight dots, or none, around the atomic symbol.

Exercise 9.2 Write the electron configuration and the Lewis symbol for Ca^{2+} and for S^{2-}.

■ See Problems 9.39 and 9.40.

Exercise 9.3 Write the electron configurations of Pb and Pb^{2+}.

■ See Problems 9.41 and 9.42.

Polyatomic Ions

Many ions, particularly anions, are polyatomic. Some common polyatomic ions are listed in Table 2.5. The atoms in these ions are held together by covalent bonds, which we will discuss in Section 9.4.

Transition-Metal Ions

Most transition elements form several cations of different charges. For example, iron has the cations Fe^{2+} and Fe^{3+}. Neither has a noble-gas configuration; that would require the energetically impossible loss of eight electrons from the neutral atom.

In forming ions in compounds, the atoms of transition elements generally lose the *ns* electrons first; then they may lose one or more $(n-1)d$ electrons. The 2+ ions are common for the transition elements and are obtained by the loss of the highest-energy *s* electrons from the atom. Many transition elements also form 3+ ions by losing one $(n-1)d$ electron in addition to the two *ns* electrons. Table 2.4 lists some common transition-metal ions. Many compounds of transition-metal ions are colored because of transitions involving *d* electrons, whereas the compounds of the main-group elements are usually colorless (Figure 9.6).

Figure 9.6 **Common transition-metal cations in aqueous solution** *Left to right*: Cr^{3+} (red-violet), Mn^{2+} (pale pink), Fe^{2+} (pale green), Fe^{3+} (pale yellow), Co^{2+} (pink), Ni^{2+} (green), Cu^{2+} (blue), and Zn^{2+} (colorless).

OWL Interactive Example 9.3 Writing Electron Configurations of Transition-Metal Ions

Gaining Mastery Toolbox

Critical Concept 9.3
The atoms of transition elements generally lose the *ns* electrons first; then they may lose one or more $(n - 1)d$ electrons.

Solution Essentials:
- Electron configuration of a transition-metal atom
- Building-up principle

Write the electron configurations of Fe^{2+} and Fe^{3+}.

Problem Strategy After obtaining the electron configuration of the atom (using the building-up principle as in Example 8.2 or Example 8.3), remove *ns* electrons, then $(n - 1)d$ electrons, until you have the correct positive charge on the ion.

Solution The electron configuration of the Fe atom ($Z = 26$) is $[Ar]3d^64s^2$. To obtain the configuration of Fe^{2+}, remove the $4s^2$ electrons. To obtain the configuration of Fe^{3+}, also remove a $3d$ electron. **The configuration of Fe^{2+} is $[Ar]3d^6$, and that of Fe^{3+} is $[Ar]3d^5$.**

Answer Check Check that the number of outer electrons in an ion is correct. The number of outer electrons equals the group number (but count the columns of Group VIIIB as 8, 9, and 10) minus the ion charge. For example, the configuration of Fe^{2+} is given as $[Ar]3d^6$. Iron is in the first column of Group VIIIB (count 8); the ion has two fewer electrons ($8 - 2 = 6$). This checks with the configuration as written.

Exercise 9.4 Write the electron configuration of Mn^{2+}.

See Problems 9.43 and 9.44.

CONCEPT CHECK 9.1

The following are electron configurations for some ions. Which ones would you expect to see in chemical compounds? State the concept or rule you used to decide for or against any ion.

a Fe^{2+} $[Ar]3d^44s^2$
b N^{2-} $[He]2s^22p^5$
c Zn^{2+} $[Ar]3d^{10}$
d Na^{2+} $[He]2s^22p^5$
e Ca^{2+} $[Ne]3s^23p^6$

9.3 Ionic Radii

A monatomic ion, like an atom, is a nucleus surrounded by a distribution of electrons. The **ionic radius** is *a measure of the size of the spherical region around the nucleus of an ion within which the electrons are most likely to be found.* As for an atomic radius, defining an ionic radius is somewhat arbitrary, because an electron distribution never abruptly ends. However, if we imagine ions to be spheres of definite size, we can obtain their radii from known distances between nuclei in crystals. (These distances can be determined accurately by observing how crystals diffract x rays.)

The study of crystal structure by x-ray diffraction is discussed in Chapter 11.

To understand how you compute ionic radii, consider the determination of the radius of an I^- ion in a lithium iodide (LiI) crystal. Figure 9.7 depicts a layer of ions in LiI. The distance between adjacent iodine nuclei equals twice the I^- radius. From x-ray diffraction experiments, the iodine–iodine distance is found to be 426 pm ($1 \text{ pm} = 1 \times 10^{-12}$ m). Therefore, the I^- radius in LiI is $\frac{1}{2} \times 426$ pm $= 213$ pm. Other crystals give approximately the same radius for the I^- ion. Table 9.3, which lists average values of ionic radii obtained from many compounds of the main-group elements, gives 216 pm for the I^- radius.

That you can find values of ionic radii that agree with the known structures of many crystals is strong evidence for the existence of ions in the solid state. Moreover, these values of ionic radii compare with atomic radii in ways that you might expect. For example, you expect a cation to be smaller and an anion to be larger than the corresponding atom (see Figure 9.8).

A cation formed when an atom loses all its valence electrons is smaller than the atom because it has one less shell of electrons. But even when only some of the valence electrons are lost from an atom, the ion is smaller. With fewer electrons in the valence orbitals, the electron–electron repulsion is initially less, so these

a A three-dimensional view of the crystal.

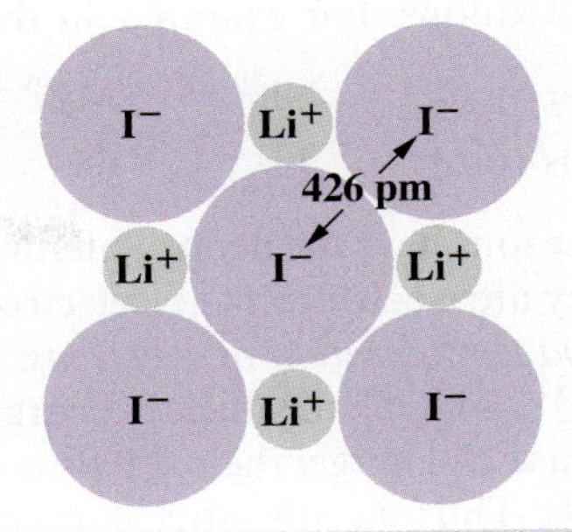

b Cross section through a layer of ions. Iodide ions are assumed to be spheres in contact with one another. The distance between iodine nuclei (426 pm) is determined experimentally. One-half of this distance (213 pm) equals the iodide ion radius.

Figure 9.7 

Determining the iodide ion radius in the lithium iodide (LiI) crystal

Table 9.3 Ionic Radii (in pm) of Some Main-Group Elements

Period	IA	IIA	IIIA	VIA	VIIA
2	Li^+	Be^{2+}		O^{2-}	F^-
	60	31		140	136
3	Na^+	Mg^{2+}	Al^{3+}	S^{2-}	Cl^-
	95	65	50	184	181
4	K^+	Ca^{2+}	Ga^{3+}	Se^{2-}	Br^-
	133	99	62	198	195
5	Rb^+	Sr^{2+}	In^{3+}	Te^{2-}	I^-
	148	113	81	221	216
6	Cs^+	Ba^{2+}	Tl^{3+}		
	169	135	95		

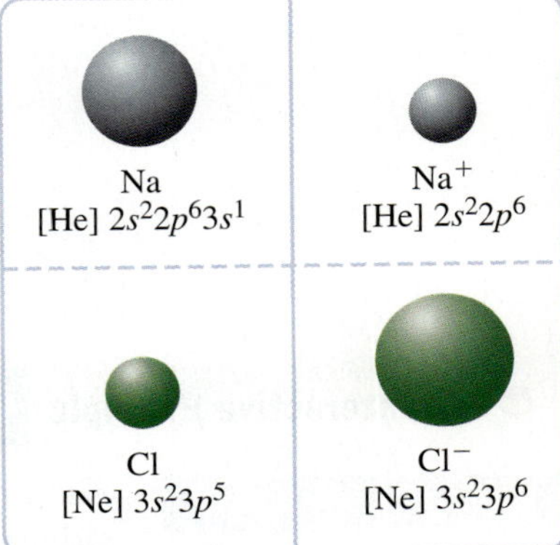

Figure 9.8

Comparison of atomic and ionic radii Note that the sodium atom loses its outer shell in forming the Na^+ ion. Thus, the cation is smaller than the atom. The Cl^- ion is larger than the Cl atom, because the same nuclear charge holds a greater number of electrons less strongly.

orbitals can shrink to increase the attraction of the electrons for the nucleus. Similarly, because an anion has more electrons than the atom, the electron–electron repulsion is greater, so the valence orbitals expand. Thus, the anion radius is larger than the atomic radius.

Exercise 9.5 Which has the larger radius, S or S^{2-}? Explain.

■ See Problems 9.45 and 9.46.

The ionic radii of the main-group elements shown in Table 9.3 follow a regular pattern, just as atomic radii do. *Ionic radii increase down any column because of the addition of electron shells.*

Exercise 9.6 Without looking at Table 9.3, arrange the following ions in order of increasing ionic radius: Sr^{2+}, Mg^{2+}, Ca^{2+}. (You may use a periodic table.)

■ See Problems 9.47 and 9.48.

The pattern across a period becomes clear if you look first at the cations and then at the anions. For example, in the third period we have

Cation	Na^+	Mg^{2+}	Al^{3+}
Radius (pm)	95	65	50

All of these ions have the neon configuration $1s^22s^22p^6$ but different nuclear charges; that is, they are isoelectronic. **Isoelectronic** *refers to different species having the same number and configuration of electrons.* To understand the decrease in radius from Na^+ to Al^{3+}, imagine the nuclear charge (atomic number) of Na^+ to increase. With each increase of charge, the orbitals contract due to the greater attractive force of the nucleus. Thus, in any isoelectronic sequence of atomic ions, the ionic radius decreases with increasing atomic number (just as it does for the atoms).

If you look at the anions in the third-period elements, you notice that they are much larger than the cations in the same period. This abrupt increase in ionic radius is due to the fact that the anions S^{2-} and Cl^- have configurations with one more shell of electrons than the cations. And because these anions also constitute an isoelectronic sequence (with argon configuration), the ionic radius decreases with atomic number (as with the atoms):

Anion	S^{2-}	Cl^-
Radius (pm)	184	181

Thus, *in general, across a period the cations decrease in radius. When you reach the anions, there is an abrupt increase in radius, and then the radius again decreases.*

OWL Interactive Example 9.4 Using Periodic Trends to Obtain Relative Ionic Radii

Gaining Mastery Toolbox

Critical Concept 9.4
In any series of isoelectronic ions, the ionic radius decreases with increasing atomic number. Thus, across a period, the cations decrease in radius; however, when you reach the anions in that period, you find an abrupt increase in radius, followed by a further decrease in radius.

Solution Essentials:
- Obtain valence-shell configuration using periodic table
- Building-up principle

Without looking at Table 9.3, arrange the following ions in order of decreasing ionic radius: F^-, Mg^{2+}, O^{2-}. (You may use a periodic table.)

Problem Strategy Note that in a series of isoelectronic ions, the ion radius decreases with an increase in nuclear charge (or atomic number).

Solution Note that F^-, Mg^{2+}, and O^{2-} are isoelectronic. If you arrange them by increasing nuclear charge, they will be in order of decreasing ionic radius. **The order is O^{2-}, F^-, Mg^{2+}.**

Answer Check Check that the ions are isoelectronic by writing their electron configurations.

Exercise 9.7 Without looking at Table 9.3, arrange the following ions in order of increasing ionic radius: Cl^-, Ca^{2+}, P^{3-}. (You may use a periodic table.)

■ See Problems 9.49 and 9.50.

Covalent Bonds

In the preceding sections we looked at ionic substances, which are typically high-melting solids. Many substances, however, are molecular—gases, liquids, or low-melting solids consisting of molecules (Figure 9.9). A molecule is a group of atoms, frequently nonmetal atoms, strongly linked by chemical bonds. Often the forces that hold atoms together in a molecular substance cannot be understood on the basis of the attraction of oppositely charged ions (the ionic model). An obvious example is the molecule H_2, in which the two H atoms are held together tightly and no ions are present. In 1916 Gilbert Newton Lewis proposed that the strong attractive force between two atoms in a molecule results from a **covalent bond,** *a chemical*

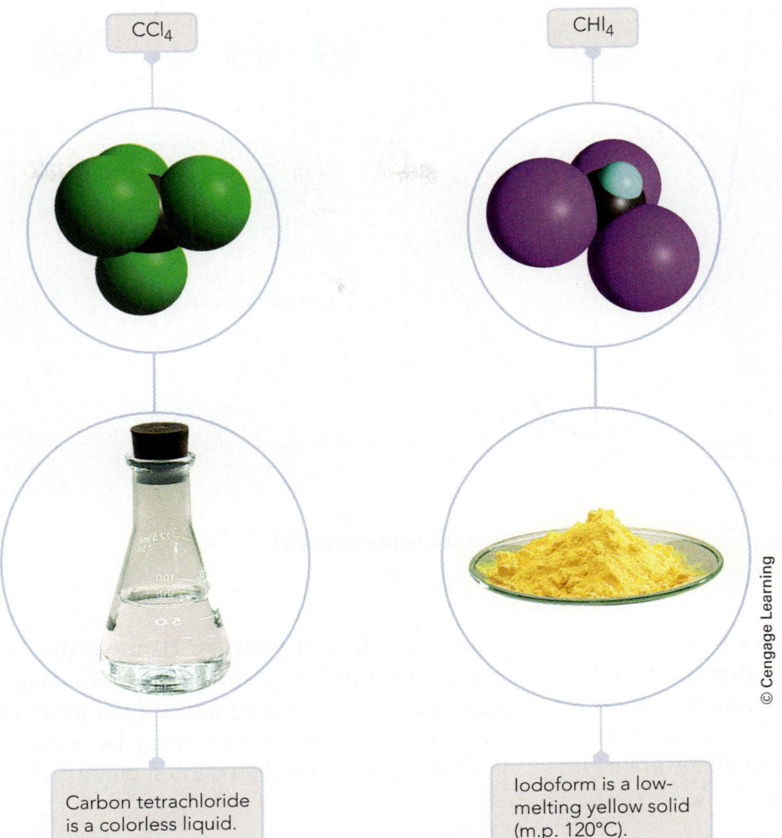

Figure 9.9 ◀

Two molecular substances

bond formed by the sharing of a pair of electrons between atoms. ▶ In 1926 Walter Heitler and Fritz London showed that the covalent bond in H_2 could be quantitatively explained by the newly discovered theory of quantum mechanics. We will discuss the descriptive aspects of covalent bonding in the following sections.

G. N. Lewis (1875–1946) was professor of chemistry at the University of California at Berkeley. Well known for work on chemical bonding, he was also noted for his research in molecular spectroscopy and thermodynamics (the study of heat involved in chemical and physical processes).

9.4 Describing Covalent Bonds

Consider the formation of a covalent bond between two H atoms to give the H_2 molecule. As the atoms approach one another, their 1*s* orbitals begin to overlap. Each electron can then occupy the space around both atoms. In other words, the two electrons can be shared by the atoms (Figure 9.10). The electrons are attracted simultaneously by the positive charges of the two hydrogen nuclei. This attraction that bonds the electrons to both nuclei is the force holding the atoms together. Although ions do not exist in H_2, the force that holds the atoms together can still be regarded as arising from the attraction of oppositely charged particles: nuclei and electrons.

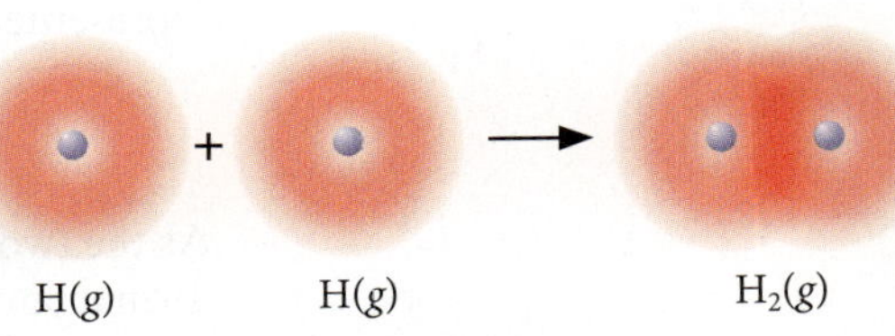

Figure 9.10 ▲

The electron probability distribution for the H_2 molecule The electron density (shown in red) occupies the space around both atoms.

It is interesting to see how the potential energy of the atoms changes as they approach and then bond. Figure 9.11 shows the potential energy of the atoms for various distances between nuclei. The potential energy of the atoms when they are some distance apart is indicated by a position on the potential-energy curve at the far right. As the atoms approach (moving from right to left on the potential-energy curve), the potential energy gets lower and lower. The decrease in energy is a reflection of the bonding of the atoms. Eventually, as the atoms get close enough, the repulsion of the positive charges on the nuclei becomes larger than the attraction of electrons for nuclei. In other words, the potential energy reaches a minimum value and then increases. The distance between nuclei at this minimum energy is called the *bond length* of H_2. It is the normal distance between nuclei in the molecule.

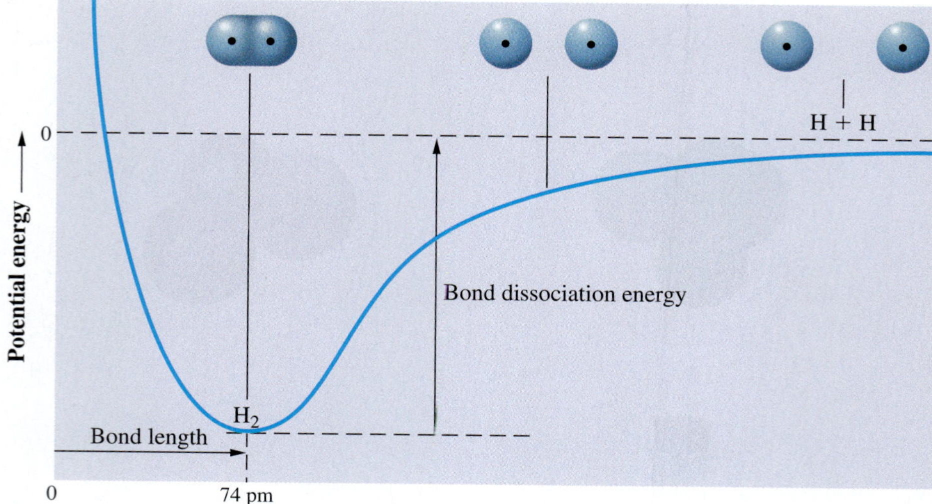

Figure 9.11 **Potential-energy curve for H_2** The stable molecule occurs at the bond distance corresponding to the minimum in the potential-energy curve.

Now imagine the reverse process. You start with the H_2 molecule, the atoms at their normal bond length apart (at the minimum of the potential-energy curve). To separate the atoms in the molecule, energy must be added (you move along the curve to the flat portion at the right). The energy that must be added is called the *bond dissociation energy*. The larger the bond dissociation energy, the stronger the bond.

Lewis Formulas

You can represent the formation of the covalent bond in H_2 from atoms as follows:

$$\mathrm{H\cdot} + \mathrm{\cdot H} \longrightarrow \mathrm{H:H}$$

This uses the Lewis electron-dot symbol for the hydrogen atoms and represents the covalent bond by a pair of dots. Recall that the two electrons from the covalent bond spend part of the time in the region of each atom. In this sense, each atom in H_2 has a helium configuration. We can draw a circle about each atom to emphasize this.

$$\mathrm{H:H}$$

The formation of a bond between H and Cl to give an HCl molecule can be represented in a similar way.

$$\mathrm{H\cdot} + \mathrm{\cdot\ddot{\underset{\cdot\cdot}{Cl}}:} \longrightarrow \mathrm{H:\ddot{\underset{\cdot\cdot}{Cl}}:}$$

As the two atoms approach each other, unpaired electrons on each atom pair up to form a covalent bond. The pair of electrons is shared by the two atoms. Each atom then acquires a noble-gas configuration of electrons, the H atom having two electrons about it (as in He), and the Cl atom having eight valence electrons about it (as in Ar).

A formula using dots to represent valence electrons is called a **Lewis electron-dot formula.** An electron pair represented by a pair of dots in such a formula is either a **bonding pair** *(an electron pair shared between two atoms)* or a **lone, or nonbonding, pair** *(an electron pair that remains on one atom and is not shared).* For example,

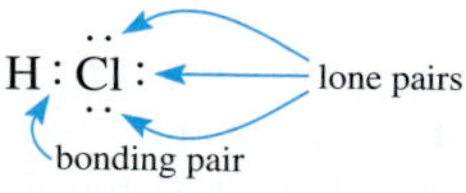

Bonding pairs are often represented by dashes rather than by pairs of dots.

Frequently, the number of covalent bonds formed by an atom equals the number of unpaired electrons shown in its Lewis symbol. Consider the formation of NH_3.

$$3\text{H}\cdot + \cdot\dot{\underset{\cdot}{\text{N}}}: \longrightarrow \text{H}:\overset{\text{H}}{\underset{\text{H}}{\ddot{\underset{\cdot\cdot}{\text{N}}}}}:$$

Each bond is formed between an unpaired electron on one atom and an unpaired electron on another atom. In many instances, the number of bonds formed by an atom in Groups IVA to VIIA equals the number of unpaired electrons, which is eight minus the group number. For example, a nitrogen atom (Group VA) forms $8 - 5 = 3$ covalent bonds. ▶

The numbers of unpaired electrons in the Lewis symbols for the atoms of elements in Groups IA and IIIA equal the group numbers. But except for the first elements of these groups, the atoms usually form ionic bonds.

Coordinate Covalent Bonds

When bonds form between atoms that both donate an electron, you have

$$\text{A}\cdot + \cdot\text{B} \longrightarrow \text{A}:\text{B}$$

However, it is possible for both electrons to come from the same atom. A **coordinate covalent bond** is *a bond formed when both electrons of the bond are donated by one atom:*

$$\text{A} + :\text{B} \longrightarrow \text{A}:\text{B}$$

A coordinate covalent bond is not essentially different from other covalent bonds; it involves the sharing of a pair of electrons between two atoms. An example is the formation of the ammonium ion, in which an electron pair on the N atom of NH_3 forms a bond with H^+.

$$\text{H}^+ + :\text{NH}_3 \longrightarrow \left[\text{H}:\overset{\text{H}}{\underset{\text{H}}{\ddot{\underset{\cdot\cdot}{\text{N}}}}}:\text{H}\right]^+$$

The new N—H bond is clearly identical to the other N—H bonds.

Octet Rule

In each of the molecules we have described so far, the atoms have obtained noble-gas configurations through the sharing of electrons. Atoms other than hydrogen have obtained eight electrons in their valence shells; hydrogen atoms have obtained two. *The tendency of atoms in molecules to have eight electrons in their valence shells (two for hydrogen atoms)* is known as the **octet rule.** Many of the molecules we will discuss follow the octet rule. Some do not.

Multiple Bonds

In the molecules we have described so far, each of the bonds has been a **single bond**—that is, *a covalent bond in which a single pair of electrons is shared by two atoms.* But it is possible for atoms to share two or more electron pairs. A **double bond** is *a covalent bond in which two pairs of electrons are shared by two atoms.* A **triple bond** is *a covalent bond in which three pairs of electrons are shared by two atoms.* As examples, consider ethylene, C_2H_4, and acetylene, C_2H_2. Their Lewis formulas are

$$\overset{\text{H}}{\underset{\text{H}}{}}\text{C}::\text{C}\overset{\text{H}}{\underset{\text{H}}{}} \quad \text{or} \quad \overset{\text{H}}{\underset{\text{H}}{}}\text{C}{=}\text{C}\overset{\text{H}}{\underset{\text{H}}{}}$$

Ethylene

$$\text{H}:\text{C}:::\text{C}:\text{H} \quad \text{or} \quad \text{H}{-}\text{C}{\equiv}\text{C}{-}\text{H}$$

Acetylene

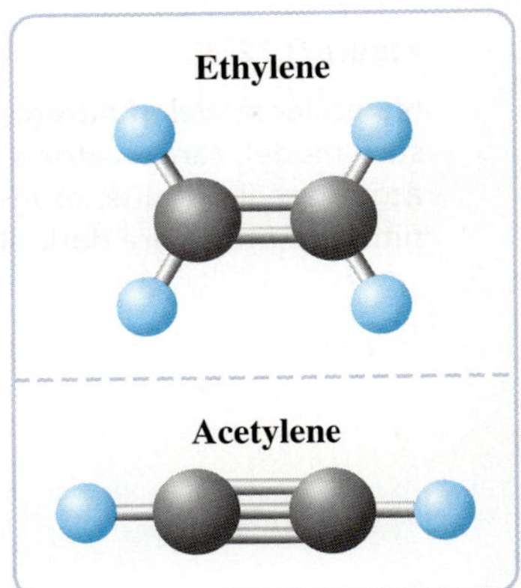

Note the octet of electrons on each C atom. Double bonds form primarily with C, N, O, and S atoms. Triple bonds form mostly to C and N atoms.

A **CHEMIST** Looks at...

Chemical Bonds in Nitroglycerin

Nitroglycerin gained a nasty reputation soon after its discovery in 1846. Unless kept cold, the pale yellow, oily liquid detonates from even the mildest vibration. In the French film *The Wages of Fear,* four men agree to drive two trucks loaded with nitroglycerin over mountainous roads from a remote village in South America. They are to drive the nitroglycerin to an out-of-control oil-well fire, where the explosive will be used to close off the well. On the way, they find the road obstructed by a huge boulder that has rolled across it, and they decide to blast the boulder out of their way using some of their explosive cargo. You see one of the men gingerly pouring nitroglycerin from a thermos bottle down a stick and into a hole in the rock. Beads of sweat form on his contorted face, while he tries desperately to suspend his breathing and any extra movement that might set off the nitroglycerin.

Here is the structure of nitroglycerin (also see the molecular model in Figure 9.12):

```
       H          H          H
       |          |          |
    H—C—————————C—————————C—H
       |          |          |
      :O:        :O:        :O:
       |          |          |
       N          N          N
      / \\       / \\       / \\
   :O.   .O:  :O.   .O:  :O.   .O:
```

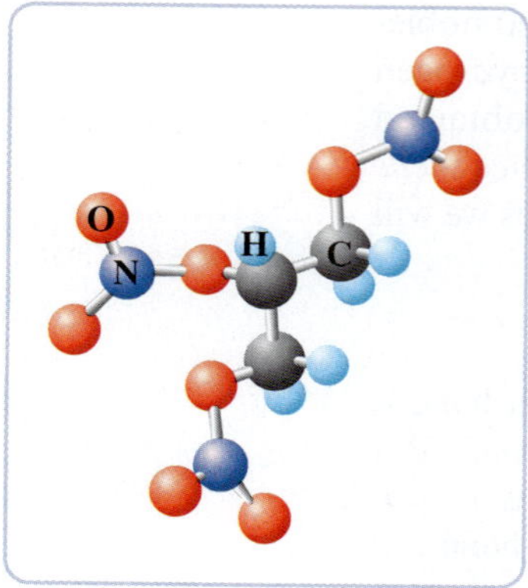

Figure 9.12 ▲

Molecular model of nitroglycerin In this ball-and-stick model, carbon atoms are gray, hydrogen atoms are light blue, oxygen atoms are red, and nitrogen atoms are dark blue.

With just a little jostling, nitroglycerin, $C_3H_5(ONO_2)_3$, can rearrange its atoms to give stable products:

$$4C_3H_5(ONO_2)_3(l) \longrightarrow 6N_2(g) + 12CO_2(g) + 10H_2O(g) + O_2(g)$$

The stability of the products results from their strong bonds, which are much stronger than those in nitroglycerin. Nitrogen, for example, has a strong nitrogen–nitrogen triple bond, and carbon dioxide has two strong carbon–oxygen double bonds. The explosive force of the reaction results from both the rapid reaction and from the large volume increase on forming gaseous products.

In 1867, the Swedish chemist Alfred Nobel discovered that nitroglycerin behaved better when absorbed on diatomaceous earth, a crumbly rock, giving an explosive mixture that Nobel called *dynamite*. Part of the wealth he made from his explosive factories was left in trust to establish the Nobel Prizes. Although explosives are often associated in people's minds with war, they do have many peacetime uses, including road building, mining, and demolition (Figure 9.13).

© David Hoffman Photo Library/Alamy

Figure 9.13 ▲

Demolition of a building Explosives are placed at predetermined positions so that when they are detonated, the building collapses on itself.

■ See Problems 9.113 and 9.114.

9.5 Polar Covalent Bonds; Electronegativity

A covalent bond involves the sharing of at least one pair of electrons between two atoms. When the atoms are alike, as in the case of the H—H bond of H_2, the bonding electrons are shared equally. That is, the electrons spend the same amount of time in the vicinity of each atom. But when the two atoms are of different elements, the bonding electrons need not be shared equally. A **polar covalent bond** (or simply *polar bond*) is *a covalent bond in which the bonding electrons spend more time near one atom than the other.* For example, in the case of the HCl molecule (Figure 9.14), the bonding electrons spend more time near the chlorine atom than the hydrogen atom.

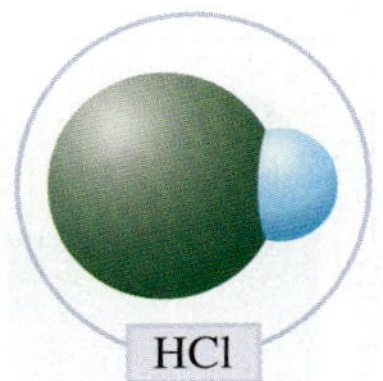

Figure 9.14 ▲

The HCl molecule A molecular model.

You can consider the polar covalent bond as intermediate between a *nonpolar* covalent bond, as in H_2, and an ionic bond, as in NaCl. From this point of view, an ionic bond is simply an extreme example of a polar covalent bond. To illustrate, we can represent the bonding in H_2, HCl, and NaCl with electron-dot formulas as follows:

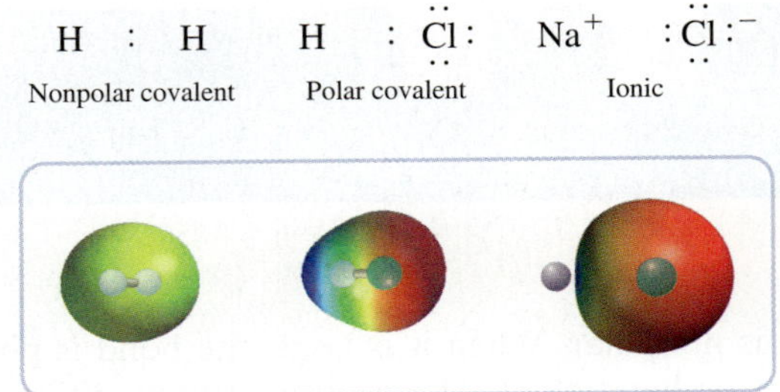

The bonding pairs of electrons are equally shared in H_2, unequally shared in HCl, and essentially not shared in NaCl. Thus, it is possible to arrange different bonds to form a gradual transition from nonpolar covalent to ionic.

Note that a polar bond results when the bonding pair is drawn more toward one atom than the other. The concept of electronegativity is useful in judging whether a bond will be polar or not. **Electronegativity** is *a measure of the ability of an atom in a molecule to draw bonding electrons to itself.* Several electronegativity scales have been proposed. In 1934 Robert S. Mulliken suggested on theoretical grounds that the electronegativity (X) of an atom be given as half its ionization energy ($I.E.$) plus electron affinity ($E.A.$). ▶

Robert S. Mulliken received the Nobel Prize in chemistry in 1966 for his work on molecular orbital theory (discussed in Chapter 10).

$$X = \frac{I.E. + E.A.}{2}$$

An atom such as fluorine that tends to pick up electrons easily (large $E.A.$) and hold on to them strongly (large $I.E.$) has a large electronegativity. On the other hand, an atom such as lithium or cesium that loses electrons readily (small $I.E.$) and has little tendency to gain electrons (small $E.A.$) has a small electronegativity. Until recently, only a few electron affinities had been measured. For this reason, Mulliken's scale has had limited utility. A more widely used scale was derived earlier by Linus Pauling from bond enthalpies, which are discussed later in this chapter. Pauling's electronegativity values are given in Figure 9.15. Because electronegativities depend somewhat on the bonds formed, these values are actually average ones. ▶

Linus Pauling received the Nobel Prize in chemistry in 1954 for his work on the nature of the chemical bond. In 1962 he received the Nobel Peace Prize.

Fluorine, the most electronegative element, was assigned a value of 4.0 on Pauling's scale. Lithium, at the left end of the same period, has a value of 1.0. Cesium, in the same column but below lithium, has a value of 0.7. *In general, electronegativity increases from left to right and decreases from top to bottom in the periodic table.* Metals are the least electronegative elements (they are *electropositive*) and nonmetals the most electronegative.

The absolute value of the difference in electronegativity of two bonded atoms gives a rough measure of the polarity to be expected in a bond. When this difference

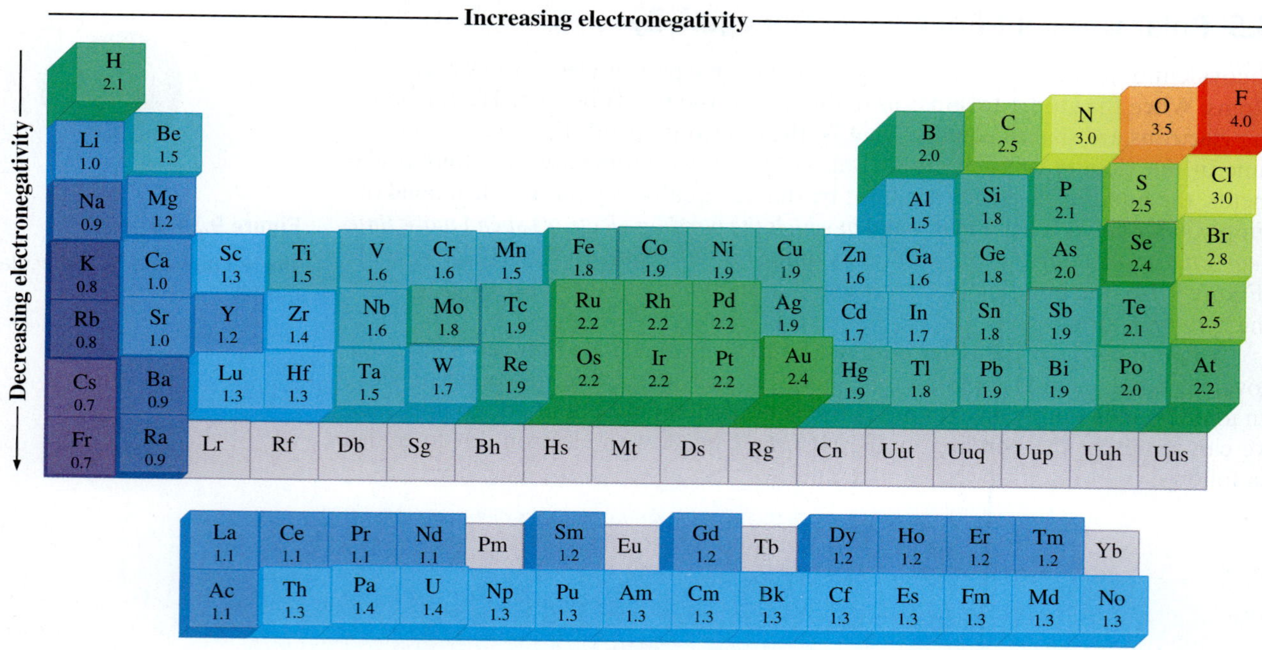

Figure 9.15 ▲

Electronegativities of the elements
The values given are those of Pauling (those in gray are not known).

The electronegativity difference in metal–metal bonding would also be small. This bonding frequently involves delocalized metallic bonding, *briefly described in Section 9.7.*

is small, the bond is nonpolar. When it is large, the bond is polar (or, if the difference is very large, perhaps ionic). The electronegativity differences for the bonds H—H, H—Cl, and Na—Cl are 0.0, 0.9, and 2.1, respectively, following the expected order. Differences in electronegativity explain why ionic bonds usually form between a metal atom and a nonmetal atom; the electronegativity difference would be largest between these elements. On the other hand, covalent bonds form primarily between two nonmetals because the electronegativity differences are small. ◀

Example 9.5 Using Electronegativities to Obtain Relative Bond Polarities

Gaining Mastery Toolbox

Critical Concept 9.5
The absolute value of the difference in electronegativity of two bonded atoms is a rough measure of the polarity of the bond. (A useful rule to remember is that electronegativity increases left to right and decreases top to bottom in the periodic table.)

Solution Essentials:
- Electronegativity

Use electronegativity values (Figure 9.15) to arrange the following bonds in order of increasing polarity: P—H, H—O, C—Cl.

Problem Strategy Order the bonds by the increasing positive value of the difference of electronegativities of the atoms forming the bond. The bonds should then be roughly in order of increasing polarity.

Solution The absolute values of the electronegativity differences are P—H, 0.0; H—O, 1.4; C—Cl, 0.5. Hence, the order is **P—H, C—Cl, H—O.**

Answer Check Make sure you have the correct electronegativities and differences.

Exercise 9.8 Using electronegativities, decide which of the following bonds is most polar: C—O, C—S, H—Br.

■ See Problems 9.57 and 9.58.

You can use an electronegativity scale to predict the direction in which the electrons shift during bond formation; the electrons are pulled toward the more electronegative atom. For example, consider the H—Cl bond. Because the Cl atom ($X = 3.0$) is more electronegative than the H atom ($X = 2.1$), the bonding electrons in H—Cl are pulled toward Cl. Because the bonding electrons spend most of their time around the Cl atom, that end of the bond acquires a partial negative charge (indicated $\delta-$). The H-atom end of the bond has a partial positive charge ($\delta+$). You can show this as follows:

$$\overset{\delta+}{\text{H}}—\overset{\delta-}{\text{Cl}}$$

The HCl molecule is said to be a *polar molecule*. We will say more about polar molecules when we look at molecular structures in Chapter 10.

9.6 Writing Lewis Electron-Dot Formulas

The Lewis electron-dot formula of a molecule is similar to the structural formula in that it shows how atoms are bonded. Bonding electron pairs are indicated either by two dots or by a dash. In addition to the bonding electrons, however, an electron-dot formula shows the positions of lone pairs of electrons, whereas the structural formula does not. Thus, the electron-dot formula is a simple two-dimensional representation of the positions of electrons in a molecule. In the next chapter we will see how to predict the three-dimensional shape of a molecule from the two-dimensional electron-dot formula. In this section we will discuss the steps for writing the electron-dot formula for a molecule made from atoms of the main-group elements. ▶

Lewis formulas do not directly convey information about molecular shape. For example, the Lewis formula of methane, CH_4, is written as the flat (two-dimensional) formula

```
   H
   ..
H:C:H
   ..
   H
```

The actual methane molecule, however, is not flat; it has a three-dimensional structure, as explained in Chapter 10.

Before you can write the Lewis formula of a molecule (or a polyatomic ion), you must know the *skeleton structure* of the molecule. The skeleton structure tells you which atoms are bonded to one another (without regard to whether the bonds are single or not).

Normally the skeleton structure must be found by experiment. For simple molecules, you can often predict the skeleton structure. For instance, many small molecules or polyatomic ions consist of a central atom around which are bonded atoms of greater electronegativity, such as F, Cl, and O. Note the structural formula (and molecular model) of phosphorus oxychloride, $POCl_3$ (Figure 9.16). The phosphorus atom is surrounded by more electronegative atoms. (In some cases, H atoms surround a more electronegative atom as in H_2O and NH_3, but H cannot be a central atom because it normally forms only one bond.) Similarly, *oxoacids* are substances in which O atoms (and possibly other electronegative atoms) are bonded to a central atom, with one or more H atoms usually bonded to O atoms. An example is chlorosulfonic acid, HSO_3Cl (Figure 9.17).

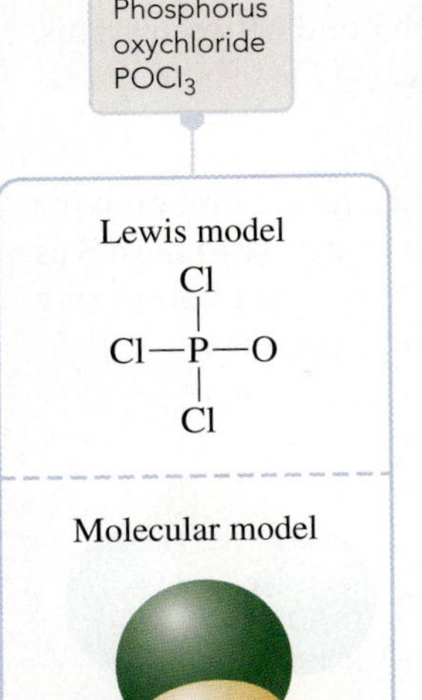

Figure 9.16 ◀ **Lewis formula and molecular model of phosphorus oxychloride molecule** The central atom, P, is surrounded by O and Cl atoms.

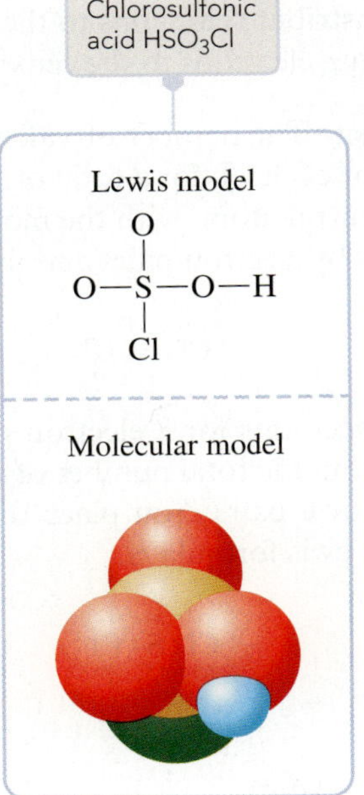

Figure 9.17 ◀ **Lewis formula and molecular model of chlorosulfonic acid molecule** This oxoacid has the central atom, S, surrounded by O atoms and a Cl atom.

Another useful idea for predicting skeleton structures is that molecules or polyatomic ions with symmetrical formulas often have symmetrical structures. For example, disulfur dichloride, S_2Cl_2, is symmetrical, with the more electronegative Cl atoms around the S atoms: Cl—S—S—Cl.

Once you know the skeleton structure of a molecule, you can find the Lewis formula by applying the following four steps. These steps allow you to write electron-dot formulas even when the central atom does not follow the octet rule.

Step 1: *Calculate the total number of valence electrons* for the molecule by summing the number of valence electrons (= group number) for each atom. If you are writing the Lewis formula of a polyatomic anion, you *add* the number of negative charges to this total. (For CO_3^{2-} you add 2 because the −2 charge indicates that there are two more electrons than are provided by the neutral atoms.) For a polyatomic cation, you *subtract* the number of positive charges from the total. (For NH_4^+ you subtract 1.)

Step 2: *Write the skeleton structure of the molecule or ion,* connecting every bonded pair of atoms by a pair of dots (or a dash).

Step 3: *Distribute electrons to the atoms surrounding the central atom (or atoms)* to satisfy the octet rule for these surrounding atoms.

Step 4: *Distribute the remaining electrons as pairs to the central atom (or atoms),* after subtracting the number of electrons already distributed from the total found in Step 1. If there are fewer than eight electrons on the central atom, this suggests that a multiple bond is present. (Two electrons fewer than an octet suggests a double bond; four fewer suggests a triple bond or two double bonds.) To obtain a multiple bond, move one or two electron pairs (depending on whether the bond is to be double or triple) from a surrounding atom to the bond connecting the central atom. Atoms that often form multiple bonds are C, N, O, and S.

The next several examples illustrate how to write the Lewis electron-dot formula for a small molecule, given the molecular formula.

Example 9.6 Writing Lewis Formulas (Single Bonds Only)

Gaining Mastery Toolbox

Critical Concept 9.6
To write the Lewis electron-dot formula for a molecule, first obtain the total number of valence electrons for the molecule; second, write the skeleton structure, allowing two electrons for each bond between atoms; third, distribute electrons to the outer atoms in order to satisfy the octet rule for each atom; and finally distribute the remaining electrons to the central atom.

Solution Essentials:

- Skeleton structure of a molecule
- Octet rule
- Number of valence electrons in a main-group atom

Sulfur dichloride, SCl_2, is a red, fuming liquid used in the manufacture of insecticides. Write the Lewis formula for the molecule.

Problem Strategy You follow the four steps: (1) Calculate the total number of valence electrons. (2) Write the skeleton structure with two electrons to each bond between atoms. (3) Distribute electrons to the outer atoms to satisfy the octet rule. (4) Distribute the remaining electrons to the central atom.

Solution The number of valence electrons from an atom equals the group number: 6 for S, 7 for each Cl, for a total of 20 electrons. You expect the skeleton structure to have S as the central atom, with the more electronegative Cl atoms bonded to it. After connecting atoms by electron pairs and distributing electrons to the outer atoms, you have

$$:\ddot{\underset{..}{Cl}}:S:\ddot{\underset{..}{Cl}}: \quad \text{or} \quad :\ddot{\underset{..}{Cl}}—S—\ddot{\underset{..}{Cl}}:$$

This accounts for 8 electron pairs, or 16 electrons. Subtracting this from the total number of electrons (20) gives 4 electrons, or 2 electron pairs. You place these on the central atom (S). The final Lewis formula is

SCl_2

$$:\ddot{\underset{..}{Cl}}:\ddot{\underset{..}{S}}:\ddot{\underset{..}{Cl}}: \quad \text{or} \quad :\ddot{\underset{..}{Cl}}—\ddot{\underset{..}{S}}—\ddot{\underset{..}{Cl}}:$$

(continued)

(continued)

Answer Check In general, atoms surrounding the central atom have four electron pairs (an octet) about them. Note that our answer follows this "octet" rule. Even the central atom often follows this octet rule, as it does here. (We will see some exceptions to this octet rule for the central atom, however.)

Exercise 9.9 Dichlorodifluoromethane, CCl_2F_2, is a gas used as a refrigerant and aerosol propellant. Write the Lewis formula for CCl_2F_2.

■ See Problems 9.61 and 9.62.

Example 9.7 Writing Lewis Formulas (Including Multiple Bonds)

Gaining Mastery Toolbox

Critical Concept 9.7
If after following the rules for writing an electron-dot formula, you find that the central atom does not have enough electrons for an octet, transfer electron pairs from outer atoms to form multiple bonds in order to give an octet to the central atom.

Solution Essentials:
- Skeleton structure of a molecule
- Lewis representation of multiple bonds
- Octet rule
- Number of valence electrons in a main-group atom

Carbonyl chloride, or phosgene, $COCl_2$, is a highly toxic gas used as a starting material for the preparation of polyurethane plastics. What is the electron-dot formula of $COCl_2$?

Problem Strategy You follow the four steps as in the previous example, but after distributing the remaining electrons to the central atom, you find that it does not have enough electrons to give an octet. To rectify this, you transfer an electron pair from an outer atom to form a double bond (part of Step 4). Do you move the electron pair from the Cl or the O atom? It helps to remember that C, N, O, and S atoms often form double bonds. In this case, you move the electron pair from the O atom.

Solution The total number of valence electrons is $4 + 6 + (2 \times 7) = 24$. You expect the more electropositive atom, C, to be central, with the O and Cl atoms bonded to it. After distributing electron pairs to these surrounding atoms, you have

```
 ..       ..              ..       ..
:Cl : C : O :     or     :Cl—C—O:
 ..   ..  ..              ..  |  ..
     :Cl:                    :Cl:
      ..                      ..
```

$COCl_2$

This accounts for all 24 valence electrons, leaving only 6 electrons on C. This is two fewer than an octet, so you move a pair of electrons on the O atom to give a carbon–oxygen double bond. The electron-dot formula of $COCl_2$ is

```
 ..       ..              ..    ..
:Cl : C :: O      or     :Cl—C=O
 ..   ..  ..              ..  |  ..
     :Cl:                    :Cl:
      ..                      ..
```

Answer Check Note that the atoms surrounding the central atom have four electron pairs about them; that is, they follow the octet rule. The central atom also follows the octet rule, although exceptions occur.

Exercise 9.10 Write the electron-dot formula of carbon dioxide, CO_2.

■ See Problems 9.63 and 9.64.

Example 9.8 Writing Lewis Formulas (Ionic Species)

Gaining Mastery Toolbox

Critical Concept 9.8
In obtaining the Lewis formula for an ion, note that the total number of valence electrons equals that for the neutral molecule less the charge on the ion (that is, subtract if a positive ion and add if a negative ion).

Solution Essentials:
- Skeleton structure of a molecule
- Lewis representation of multiple bonds
- Octet rule
- Number of valence electrons in a main-group atom

Obtain the electron-dot formula (Lewis formula) of the BF_4^- ion.

Problem Strategy After calculating the total number of valence electrons for the neutral atoms, you add or subtract electrons to account for the ion charge. If the ion is negative, add electrons to give the negative charge. If the ion is positive, subtract electrons to give the positive charge. Once you have the total number of valence electrons in the ion, proceed with the rest of the steps to write the electron-dot formula.

Solution The total valence electrons provided by the boron and four fluorine atoms is $3 + (4 \times 7) = 31$. Because the anion has a charge of -1, it has one more electron than is provided by the neutral atoms. Thus, the total number of valence electrons is 32 (or 16 electron pairs). You assume that the skeleton structure has boron as the central atom, with the more electronegative F atoms bonded to it. After connecting the B and F atoms by bonds and placing electron pairs around the F atoms to satisfy the octet rule, you obtain

:F:B:F: (with :F: above and below B) or :F—B—F: (with F bonded above and below B)

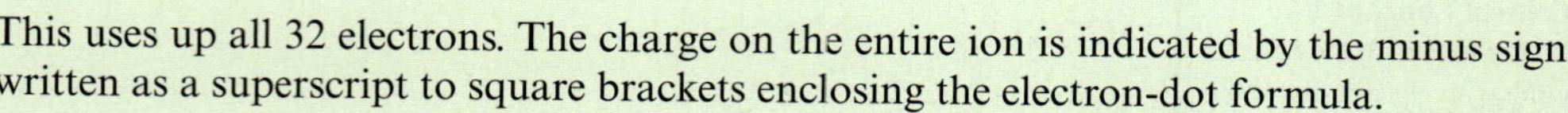

This uses up all 32 electrons. The charge on the entire ion is indicated by the minus sign written as a superscript to square brackets enclosing the electron-dot formula.

$[\,$:F:B:F: (with :F: above and below B)$\,]^-$ or $[\,$:F—B—F: (with F bonded above and below B)$\,]^-$

BF_4^-

Answer Check Note that the atoms surrounding the central atom have four electron pairs about them; that is, they follow the octet rule. The central atom also follows the octet rule, although exceptions occur.

Exercise 9.11 Write the electron-dot formula of **a** the hydronium ion, H_3O^+; **b** the chlorite ion, ClO_2^-.

■ See Problems 9.65 and 9.66.

CONCEPT CHECK 9.2

Each of the following may seem, at first glance, to be plausible electron-dot formulas for the molecule N_2F_2. Most, however, are incorrect for some reason. What concepts or rules apply to each, either to cast it aside or to keep it as the correct formula?

a :F:N:N:F: **b** :F:N::N:F: **c** :F::N:N:F:

d :F:N:N:F: **e** :F:N::F:N: **f** :F:N N:F:

9.7 Delocalized Bonding: Resonance

We have assumed up to now that the bonding electrons are localized in the region between two atoms. In some cases, however, this assumption does not fit the experimental data. Suppose, for example, that you try to write an electron-dot formula for ozone, O_3. You find that you can write two formulas:

O=O—O (A) and O—O=O (B)

In formula A, the oxygen–oxygen bond on the left is a double bond and the oxygen–oxygen bond on the right is a single bond. In formula B, the situation is just the opposite. Experiment shows, however, that both bonds in O_3 are identical. Therefore, neither formula can be correct.

The lengths of the two oxygen–oxygen bonds (that is, the distances between the atomic nuclei) are both 128 pm.

According to theory, one of the bonding pairs in ozone is spread over the region of all three atoms rather than associated with a particular oxygen–oxygen bond. This is called **delocalized bonding,** *a type of bonding in which a bonding pair of electrons is spread over a number of atoms rather than localized between two.* We might symbolically describe the delocalized bonding in ozone as follows:

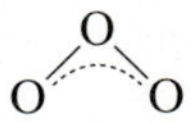

O
O O

(For clarity, only the bonding pairs are given.) The broken line indicates a bonding pair of electrons that spans three nuclei rather than only two. In effect, the oxygen–oxygen bond is neither a single bond nor a double bond but an intermediate type.

A single electron-dot formula cannot properly describe delocalized bonding. Instead, a resonance description is often used. According to the **resonance description,** *you describe the electron structure of a molecule having delocalized bonding by writing all possible electron-dot formulas.* These are called the *resonance formulas* of the molecule. The actual electron distribution of the molecule is a composite of these resonance formulas.

The electron structure of ozone can be described in terms of the two resonance formulas presented at the start of this section. By convention, we usually write all of the resonance formulas and connect them by double-headed arrows. For ozone we would write

$$\underset{\textbf{A}}{:\ddot{O}=\ddot{O}-\ddot{O}:} \longleftrightarrow \underset{\textbf{B}}{:\ddot{O}-\ddot{O}=\ddot{O}:}$$

Unfortunately, this notation can be misinterpreted. It does not mean that the ozone molecule flips back and forth between two forms. There is only one ozone molecule. The double-headed arrow means that you should form a mental picture of the molecule by fusing the various resonance formulas. The left oxygen–oxygen bond is double in formula A and the right one is double in formula B, so you must picture an electron pair that actually encompasses both bonds.

Attempting to write electron-dot formulas leads you to recognize that delocalized bonding exists in many molecules. Whenever you can write several plausible electron-dot formulas—which often differ merely in the allocation of single and double bonds to the same kinds of atoms (as in ozone)—you can expect delocalized bonding.

Example 9.9 Writing Resonance Formulas

Gaining Mastery Toolbox

Critical Concept 9.9
Having written one electron-dot formula containing single and double bonds, note whether it is possible to write other electron-dot formulas that differ in the placement of these single and double bonds.

Solution Essentials:

- Resonance
- Skeleton structure of a molecule
- Lewis representation of multiple bonds
- Octet rule
- Number of valence electrons in a main-group atom

Describe the electron structure of the carbonate ion, CO_3^{2-}, in terms of electron-dot formulas.

Problem Strategy Write at least one electron-dot formula for a molecule or ion. If the formula has both single and multiple bonds, note whether it is possible to write other electron-dot formulas differing only in the placement of single and double bonds.

Solution One possible electron-dot formula for the carbonate ion is

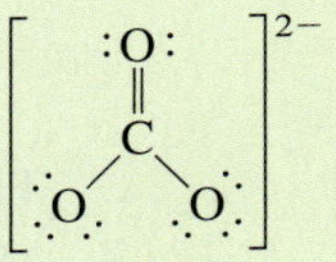

$$\left[\begin{array}{c} :\ddot{O}: \\ \| \\ C \\ :\ddot{O}:\diagup \quad \diagdown :\ddot{O}: \end{array}\right]^{2-}$$

(continued)

(continued)

Because you expect all carbon–oxygen bonds to be equivalent, you must describe the electron structure in resonance terms.

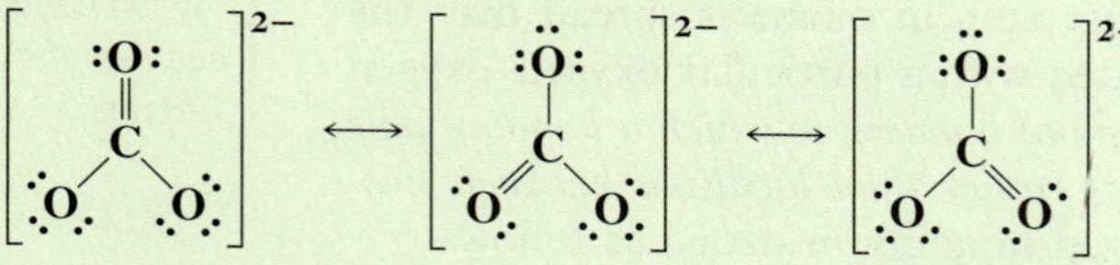

You expect one electron pair to be delocalized over the region of all three carbon–oxygen bonds.

CO_3^{2-}

Answer Check Note that the skeleton structure is the same for all resonance formulas. This must be so, because they represent the same molecule. Also, note that the atoms about the central atom follow the octet rule. (The central atom in this case also follows the octet rule.)

Exercise 9.12 Describe the bonding in NO_3^- using resonance formulas.

■ See Problems 9.67, 9.68, 9.69, and 9.70.

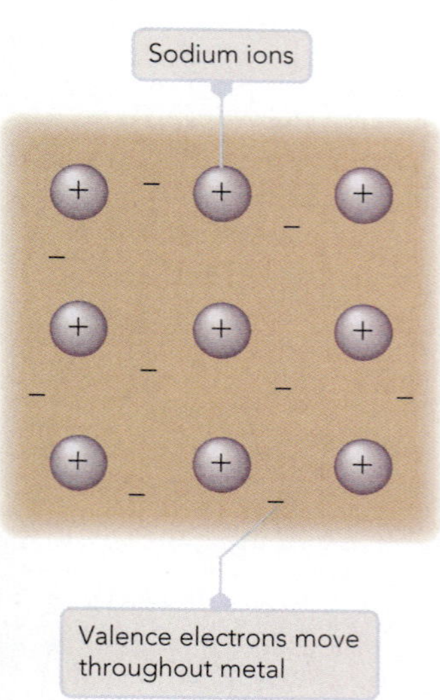

Figure 9.18 ▲

Delocalized bonding in sodium metal The metal consists of positive sodium ions in a "sea" of valence electrons. Valence (bonding) electrons are free to move throughout the metal crystal (beige area).

Metals are extreme examples of delocalized bonding. A sodium metal crystal, for example, can be regarded as an array of Na^+ ions surrounded by a "sea" of electrons (Figure 9.18). The valence, or bonding, electrons are delocalized over the entire metal crystal. The freedom of these electrons to move throughout the crystal is responsible for the electrical conductivity of a metal.

9.8 Exceptions to the Octet Rule

Many molecules composed of atoms of the main-group elements have electronic structures that satisfy the octet rule, but a number of them do not. A few molecules, such as NO, have an odd number of electrons and so cannot satisfy the octet rule. Other exceptions to the octet rule fall into two groups—one a group of molecules with an atom having fewer than eight valence electrons around it and the other a group of molecules with an atom having more than eight valence electrons around it.

The exceptions in which the central atom has more than eight valence electrons around it are fairly numerous. Phosphorus pentafluoride is a simple example. This is a colorless gas of PF_5 molecules. Each molecule consists of a phosphorus atom surrounded by the more electronegative fluorine atoms (Figure 9.19). Following the steps outlined in Section 9.6, you arrive at the following electron-dot formula, in which the phosphorus atom has ten valence electrons around it:

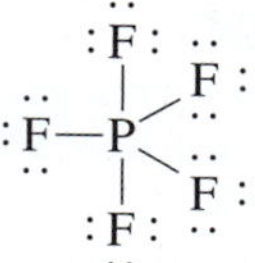

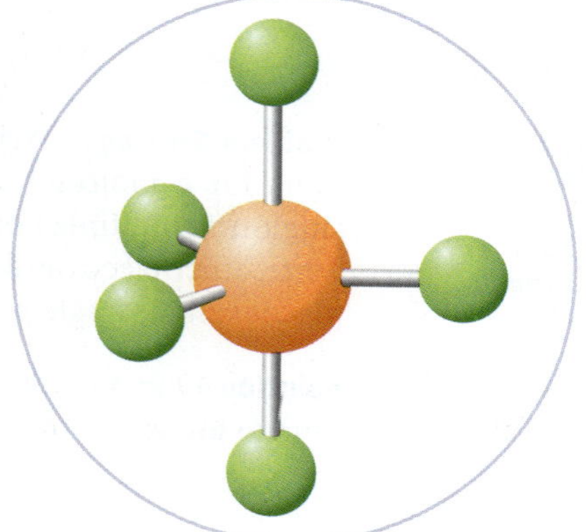

Figure 9.19 ▲

Phosphorus pentafluoride, PF_5 A ball-and-stick molecular model.

The octet rule stems from the fact that the main-group elements in most cases employ only an *ns* and three *np* valence-shell orbitals in bonding, and these orbitals hold eight electrons. Elements of the second period are restricted to these orbitals, but from the third period on, the elements also have unfilled *nd* orbitals, which may be used in bonding. For example, the valence-shell configuration of phosphorus is $3s^2 3p^3$. Using just these $3s$ and $3p$ orbitals, the phosphorus atom can accept only three additional electrons, forming three covalent bonds (as in PF_3). However, more bonds can be formed if the empty $3d$ orbitals of the atom are used. If each of the five electrons of the phosphorus atom is paired with unpaired electrons of fluorine atoms, PF_5 can be formed. In this way, phosphorus forms both the trifluoride and the pentafluoride. By contrast, nitrogen (which has no available d orbitals in its valence shell) forms only the trifluoride, NF_3.

Example 9.10 Writing Lewis Formulas (Exceptions to the Octet Rule)

Gaining Mastery Toolbox

Critical Concept 9.10
In the final step in writing an electron-dot formula (in which you distribute the remaining electrons to the central atom), you may discover that this atom now has more than an octet of electrons.

Solution Essentials:
- Skeleton structure of a molecule
- Lewis representation of multiple bonds
- Octet rule
- Number of valence electrons in a main-group atom

Xenon, a noble gas, forms a number of compounds. One of these is xenon tetrafluoride, XeF_4, a white, crystalline solid first prepared in 1962. What is the electron-dot formula of the XeF_4 molecule?

Problem Strategy You follow the four steps outlined in Section 9.6. In Step 4, when you distribute the remaining electrons to the central atom, you find that it has more than an octet.

Solution There are 8 valence electrons from the Xe atom and 7 from each F atom, for a total of 36 valence electrons. For the skeleton structure, you draw the Xe atom surrounded by the electronegative F atoms. After placing electron pairs on the F atoms to satisfy the octet rule for them, you have:

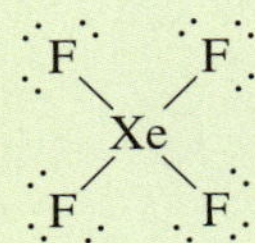

This accounts for 16 pairs, or 32 electrons. A total of 36 electrons is available, so you put an additional 36 − 32 = 4 electrons (2 pairs) on the Xe atom. The Lewis formula is

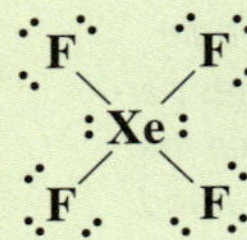

Answer Check Note that the atoms surrounding the central atom follow the octet rule, although the central atom itself does not. The central atom is from a period greater than 2, so it has a valence shell that can accommodate more than eight electrons.

Exercise 9.13 Sulfur tetrafluoride, SF_4, is a colorless gas. Write the electron-dot formula of the SF_4 molecule.

■ See Problems 9.71 and 9.72.

The other group of exceptions to the octet rule consists mostly of molecules containing Group IIA or IIIA atoms. Consider boron trifluoride, BF_3. The molecule consists of boron surrounded by the much more electronegative fluorine atoms. The total number of valence electrons is $3 + (3 \times 7) = 24$. If you connect boron and fluorine atoms by electron pairs and fill out the fluorine atoms with octets of electrons, you obtain

```
      ..
     :F:
      |
 ..   |
:F—B
 ..   |
     :F:
      ..
```

If you now follow with Step 4 of the procedure outlined in Section 9.6, you note that the boron atom has only six electrons on it. You write the following resonance formulas, each having a double bond:

```
                      ..              ..
    :F:              :F:             :F:
     ‖                |               |
 ..  ‖         ..     |        ..     |
:F—B   ⟷   F=B    ⟷  :F—B
 ..  |         ..     |        ..     ‖
    :F:              :F:             :F:
     ..               ..
```

In fact, there is evidence to suggest that the first formula written—the one with all single bonds and in which boron has only six electrons around it—describes the chemistry of boron trifluoride very well. For example, boron trifluoride reacts with molecules having a lone pair, such as with ammonia, NH_3, to give the compound BF_3NH_3. The reaction is easy to describe in terms of the formula in which boron has only six electrons around it.

```
  :F:       H                :F: H
   |        |                 |   |
:F—B  +  :N—H  ⟶        :F—B : N—H
   |        |                 |   |
  :F:       H                :F: H
```

coordinate covalent bond

In this reaction, a coordinate covalent bond forms between the boron and nitrogen atoms, and the boron achieves an octet of electrons.

The chemistry of BF_3 thus appears to support an electron structure with boron having only six electrons around it. No doubt, resonance involving all four of the Lewis formulas that we have drawn best describes the actual electron structure of boron trifluoride, but the relative importance of the different resonance formulas is not settled.

The B–F bond length (130 pm) is shorter than expected for a single bond (152 pm), which indicates partial double-bond character. See the discussion in Section 9.10 on bond length and bond order.

Other examples of molecules with Group IIIA atoms (such as Al) or Group IIA atoms (such as Be) display electron structures similar to that of boron trifluoride. For example, the BeF_2 molecule (found in the vapor over the heated solid BeF_2) has the Lewis formula

```
:F:Be:F:
```

Aluminum chloride, $AlCl_3$, offers an interesting study in bonding. At room temperature, the substance is a white, crystalline solid and an ionic compound, as might be expected for a binary compound of a metal and a nonmetal. However, the substance has a relatively low melting point (192°C) for an ionic compound. Apparently this is due to the fact that instead of melting to a liquid of ions, as happens with most ionic solids, the compound forms Al_2Cl_6 molecules (Figure 9.20), with Lewis formula

```
:Cl:    :Cl:    :Cl:
    \   /   \   /
     Al       Al
    /   \   /   \
:Cl:    :Cl:    :Cl:
```

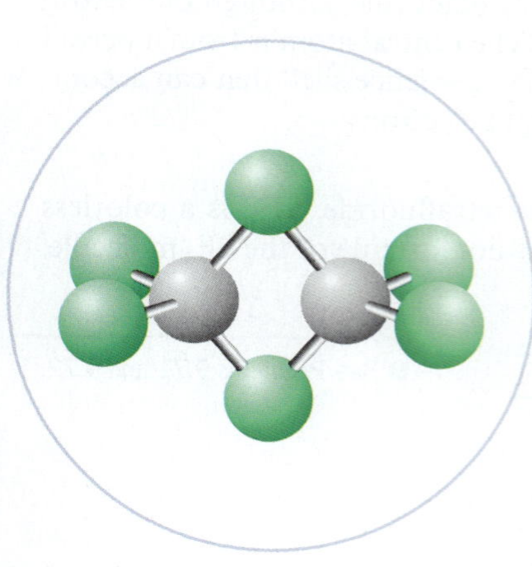

Figure 9.20 ▲

The Al_2Cl_6 molecule A ball-and-stick molecular model.

Each atom has an octet of electrons around it. Note that two of the Cl atoms are in *bridge* positions, with each Cl atom having two covalent bonds. When this liquid is heated, it vaporizes as Al_2Cl_6 molecules. As the vapor is further heated, these molecules break up into $AlCl_3$ molecules. These molecules have an electron structure similar to that of BF_3.

Exercise 9.14 Beryllium chloride, $BeCl_2$, is a solid substance consisting of long (essentially infinite) chains of atoms with Cl atoms in bridge positions.

```
 /:Cl:\     /:Cl:\     /:Cl:\
       Be         Be         Be
 \:Cl:/     \:Cl:/     \:Cl:/
```

However, if the solid is heated, it forms a vapor of $BeCl_2$ molecules. Write the electron-dot formula of the $BeCl_2$ molecule.

■ See Problems 9.73 and 9.74.

9.9 Formal Charge and Lewis Formulas

Earlier we looked at writing the Lewis formula for carbonyl chloride, $COCl_2$ (Example 9.7). In the solution to that example, we wrote the answer as follows:

$$:\ddot{\underset{..}{Cl}}-\underset{\underset{:\underset{..}{Cl}:}{|}}{C}=\ddot{\underset{..}{O}}$$

You may recall that we had to decide whether to draw the double bond between C and O or between C and Cl. We decided in favor of a carbon-oxygen double bond by applying the simple idea that multiple bonds most likely involve C, N, O, and S atoms. In this section, we will describe how you can use the concept of formal charge to determine the best Lewis formula. Formal charge is also a help in writing the skeleton structure of a molecule.

The **formal charge** of an atom in a Lewis formula is *the hypothetical charge you obtain by assuming that bonding electrons are equally shared between bonded atoms and that the electrons of each lone pair belong completely to one atom.* ▶ When you use the Lewis formula that most clearly describes the bonding, the formal charges give you an approximate distribution of electrons in the molecule. Let us see how you would obtain the formal charges, given several possible Lewis formulas of a molecule, and how you would use these formal charges to determine the best Lewis formula.

The rules for computing oxidation number are similar, except that both bonding electrons are assigned to the more electronegative atom.

You begin by writing the possible Lewis formulas. For $COCl_2$, you can write

$$:\ddot{\underset{..}{Cl}}-\underset{\underset{:\underset{..}{Cl}:}{|}}{C}=\ddot{\underset{..}{O}} \qquad \ddot{\underset{..}{Cl}}=\underset{\underset{:\underset{..}{Cl}:}{|}}{C}-\ddot{\underset{..}{O}}: \qquad :\ddot{\underset{..}{Cl}}-\underset{\underset{:Cl:}{\|}}{C}-\ddot{\underset{..}{O}}:$$

For each formula, you apply the following *rules for formal charge* to assign the valence electrons to individual atoms:

1. Half of the electrons of a bond are assigned to each atom in the bond (counting each dash as two electrons).
2. Both electrons of a lone pair are assigned to the atom to which the lone pair belongs.

You calculate the formal charge on an atom by taking the number of valence electrons on the free atom (equal to the group number) and subtracting the number of electrons you assigned the atom by the rules of formal charge; that is,

$$\text{Formal charge} = \text{valence electrons on free atom} - \tfrac{1}{2}(\text{number of electrons in a bond}) - (\text{number of lone-pair electrons})$$

As a check on your work, you should note that the sum of the formal charges equals the charge on the molecular species (zero for a neutral molecule).

Consider a Cl atom in the first Lewis formula we wrote for $COCl_2$. You begin by counting the electrons assigned to the Cl atom by the rules of formal charge. You have 1 electron from the single bond and 6 from the lone-pair electrons, for a total of 7. The formal charge of Cl equals the number of valence electrons (7) less the number of assigned electrons (7). This gives a formal charge of 0 for the Cl atom. Similar calculations of formal charge give 0 for each of the other atoms in the Lewis formula.

Now look at the second Lewis formula we wrote for $COCl_2$ and assign the electrons by the rules for formal charge. For the Cl atom having the double bond, you count 2 electrons from the double bond and 4 from the lone pairs, for a total of 6. The formal charge of the Cl atom is $7 - 6 = +1$. For the O atom, you count 1 electron from the single bond and 6 from the lone-pair electrons, for a total of 7. The number of valence electrons on the free O atom is 6, so the formal charge of the O atom is $6 - 7 = -1$. The formal charge of each of the other atoms is 0.

We indicate the formal charges in a Lewis formula by inserting circled numbers near the atoms (writing + for +1 and − for −1). For the formulas given earlier for $COCl_2$, the formal charges are shown as follows:

```
 ..     ..           ..     ..              ..     ..
:Cl—C=O         ⊕ Cl=C—O:⊖          :Cl—C—O:⊖
 ..  |  ..             |  ..              ..  ||  ..
   :Cl:              :Cl:                   :Cl:
    ..                ..                     ⊕
```

Now we can discuss three rules that are useful in deciding which of several resonance formulas best approximates the electron distribution of a molecule or ion.

RULE A Whenever you can write several Lewis formulas for a molecule, choose the one having the lowest magnitudes of formal charges.

RULE B When two proposed Lewis formulas for a molecule have the same magnitudes of formal charges, choose the one having the negative formal charge on the more electronegative atom.

RULE C When possible, choose Lewis formulas that do not have like charges on adjacent atoms.

The second and third formulas for $COCl_2$ have formal charges on the Cl and O atoms. The first formula, however, has zero formal charges for all atoms. So by Rule A, this formula most closely approximates the actual electron distribution. You would write it as the best description by a single Lewis formula. (To be more precise, you could consider a resonance description involving all three Lewis formulas. However, the Lewis formulas would not participate equally in the resonance description; the first one would predominate.)

You can also use formal charges to help you choose the most likely skeleton structure from several possibilities. Consider thionyl chloride, $SOCl_2$. Which of the following Lewis formulas is most plausible? Note that these Lewis formulas have quite different atomic arrangements, so they should not be considered as simply different resonance formulas of the same molecule.

```
                                                ⊖
   ..              ..                          ..
  :Cl:            :Cl:                         :S:
 ..  |  ..       ..  |  ..                ..   |   ..
:Cl—S—O:        :Cl—O—S:                  :Cl—Cl—O:
 ..  ..  ..      ..  ..  ..               ..   ..  ..
     ⊕  ⊖            ⊕  ⊖                     (+2)  ⊖
```

You can verify these formal charges using the method we have outlined here. The last formula has a +2 formal charge on the Cl atom, which is larger in magnitude than the formal charges on any atoms in the other two formulas. Therefore, using Rule A, you would not consider the last formula a plausible structure. To choose between the first and second structures, you apply Rule B. According to this rule, you would choose the first structure over the second, because the first one associates a negative formal charge with the more electronegative atom (O), whereas the second structure associates a positive charge with the more electronegative atom. Note that this application of formal charge gives the same result that you would obtain by assuming that the least electronegative atom is the central atom.

Now that you have one Lewis formula with the appropriate skeleton structure for $SOCl_2$, you should explore the possibility of other Lewis formulas with this skeleton structure. You can write a Lewis formula that has a sulfur–oxygen double bond.

$SOCl_2$

```
   ..
  :Cl:
 ..  |  ..
:Cl—S=O
 ..  ..  ..
```

This formula has formal charges of zero for each atom; so according to Rule A, it should be closer to the actual electron distribution than the previous formula. (You could write a resonance description involving both formulas.) Note that the sulfur atom has 10 electrons around it. As long as the atom has available *d* orbitals for bonding (as any element in the third and higher periods will have), the atom need not obey the octet rule. The next example further illustrates the application of formal charges in choosing the appropriate Lewis formula.

Example 9.11 Using Formal Charges to Determine the Best Lewis Formula

Gaining Mastery Toolbox

Critical Concept 9.11

Given several resonance formulas for a molecule, decide the most important contributor by applying one or more of the following rules: (a) choose the formula with the lowest magnitudes of formal charges, (b) choose the formula with a negative formal charge on the more electronegative atom, and (c) when possible, choose the formula not having like charges on adjacent atoms.

Solution Essentials:

- Formal charge
- Resonance
- Skeleton structure of a molecule
- Lewis representation of multiple bonds
- Octet rule
- Number of valence electrons in a main-group atom

H_2SO_4

Write the Lewis formula that best describes the charge distribution in the sulfuric acid molecule, H_2SO_4, according to the rules of formal charge.

Problem Strategy Draw the skeleton structure of the molecule or ion, and then follow the steps for writing electron-dot formulas. Note that you can move electron pairs into bonding regions to obtain additional formulas (having multiple bonds). The formal charge of an atom in a formula equals the group number of the atom minus the number of electrons assigned to the atom by the rules of formal charge (half of the electrons of connecting bonds plus all of the lone-pair electrons on the atom). Apply Rules A, B, and C given before this example to decide which Lewis formula is best.

Solution Assume a skeleton structure in which the S atom is surrounded by the more electronegative O atoms; the H atoms are then attached to two of the O atoms. The following is a possible Lewis formula having single bonds:

```
         ..
        :O:
   ..    |   ..
H—O—S—O—H
   ..    |   ..
        :O:
         ..
```

You calculate the formal charge of each O atom bonded to an H atom as follows: You assign the O atom 2 electrons from the two single bonds and 4 electrons from the two lone pairs, for a total of 6 electrons. Because the number of valence electrons on the O atom is 6, the formal charge is 6 − 6 = 0. For each of the other two O atoms, you assign 1 electron from the one single bond and 6 electrons from the three lone pairs, for a total of 7. The formal charge is 6 − 7 = −1. For the S atom, you assign 4 electrons from the four single bonds, for a total of 4. Because the number of valence electrons is 6, the formal charge is 6 − 4 = +2. The formal charge of each of the H atoms is 1 − 1 = 0. The Lewis formula with formal charges is

```
         ..
        :O:⊖
   ..    |(+2) ..
H—O—S—O—H
   ..    |   ..
        :O:⊖
         ..
```

You can also write a Lewis formula that has zero formal charges for atoms, if you form sulfur–oxygen double bonds:

```
        :O:
   ..    ‖   ..
H—O—S—O—H
   ..    ‖   ..
        :O:
```

To the top O atom, you assign 2 electrons from the double bond and 4 electrons from the two lone pairs, for a total of 6 electrons. The formal charge is 6 − 6 = 0. To the S atom, you assign a total of 6 electrons from bonds. The formal charge is 6 − 6 = 0. The formal

(continued)

(*continued*)

charges of the other atoms are also 0. Thus, this Lewis formula should be a better representation of the electron distribution in the H_2SO_4 molecule.

Answer Check The sum of the formal charges in any formula should equal the charge on the species, which is zero for a neutral molecule.

Exercise 9.15 Write the Lewis formula that best describes the phosphoric acid molecule, H_3PO_4.

■ See Problems 9.77 and 9.78.

CONCEPT CHECK 9.3

Which of the models shown below most accurately represents the hydrogen cyanide molecule, HCN? Write the electron-dot formula that most closely agrees with this model. State any concept or rule you used in arriving at your answer.

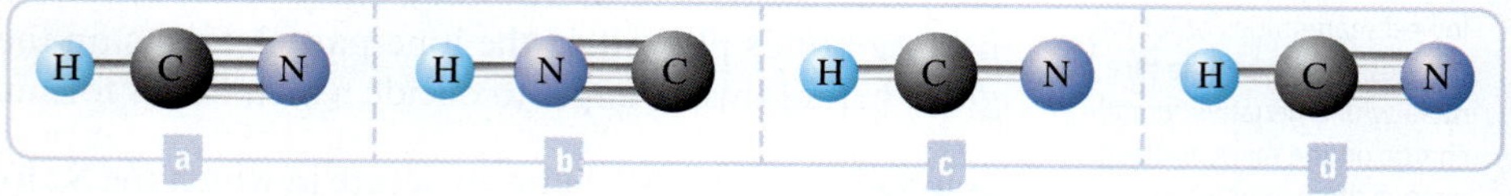

9.10 Bond Length and Bond Order

Bond length (or **bond distance**) is *the distance between the nuclei in a bond.* Bond lengths are determined experimentally using x-ray diffraction or the analysis of molecular spectra. Knowing the bond length in a particular molecule can sometimes provide a clue to the type of bonding present.

Bond lengths for a given bonding situation can often be predicted within a few picometers from a set of covalent radii. The **covalent radius** of an atom is *the value for that atom in a set of covalent radii assigned to atoms in such a way that the sum of the covalent radii of atoms A and B predicts the approximate A—B bond length.* A number of different sets of covalent radii have been derived for various bonding situations. Those given in Table 9.4 were obtained by a statistical analysis of known *single bond* lengths. To illustrate how we might use this table, let's suppose we would like to have an approximate C—Cl single bond length. Looking at Table 9.4, we see that the C and Cl covalent radii are 76 pm and 102 pm, respectively. The sum of these two radii (76 + 102 pm = 178 pm) gives us an approximate value for the C—Cl single bond length. Here are actual C—Cl bond lengths in several compounds:

chloromethane, CH_3Cl, 178.4 pm;

tetrachloromethane, CCl_4, 176.6 pm;

hexachloroethane, CCl_3CCl_3, 174 pm.

The value predicted from Table 9.4 compares favorably with these known bond lengths.

Covalent radii are one type of atomic radii (recall the discussion in Section 8.6), and as such, the values for any given set of radii should follow the two major periodic trends for atomic radii:

1. Within a period, the covalent radius tends to decrease with increasing atomic number.
2. Within a group, the covalent radius tends to increase with period number.

Note that the radii in Table 9.4 do follow these trends rather well.

Table 9.4 Single-Bond Covalent Radii

Atomic Number	Symbol	Name	Covalent Radius (pm)
1	H	Hydrogen	31
2	He	Helium	28
3	Li	Lithium	128
4	Be	Beryllium	96
5	B	Boron	84
6	C	Carbon	76
7	N	Nitrogen	71
8	O	Oxygen	66
9	F	Fluorine	57
10	Ne	Neon	58
11	Na	Sodium	166
12	Mg	Magnesium	141
13	Al	Aluminum	121
14	Si	Silicon	111
15	P	Phosphorus	107
16	S	Sulfur	105
17	Cl	Chlorine	102
18	Ar	Argon	106
19	K	Potassium	203
20	Ca	Calcium	176
21	Sc	Scandium	170
22	Ti	Titanium	160
23	V	Vanadium	153
24	Cr	Chromium	139
25	Mn	Manganese	139
26	Fe	Iron	132
27	Co	Cobalt	126
28	Ni	Nickel	124
29	Cu	Copper	132
30	Zn	Zinc	122
31	Ga	Gallium	122
32	Ge	Germanium	120
33	As	Arsenic	119
34	Se	Selenium	120
35	Br	Bromine	120
36	Kr	Krypton	116
37	Rb	Rubidium	220
38	Sr	Strontium	195
39	Y	Yttrium	190
40	Zr	Zirconium	175
41	Nb	Niobium	164
42	Mo	Molybdenum	154
43	Tc	Technetium	147
44	Ru	Ruthenium	146
45	Rh	Rhodium	142
46	Pd	Palladium	139
47	Ag	Silver	145
48	Cd	Cadmium	144
49	In	Indium	142
50	Sn	Tin	139
51	Sb	Antimony	139
52	Te	Tellurium	138
53	I	Iodine	139
54	Xe	Xenon	140
55	Cs	Cesium	244
56	Ba	Barium	215
57	La	Lanthanum	207
58	Ce	Cerium	204
59	Pr	Praseodymium	203
60	Nd	Neodymium	201
61	Pm	Promethium	199
62	Sm	Samarium	198
63	Eu	Europium	198
64	Gd	Gadolinium	196
65	Tb	Terbium	194
66	Dy	Dysprosium	192
67	Ho	Holmium	192
68	Er	Erbium	189
69	Tm	Thulium	190
70	Yb	Ytterbium	187
71	Lu	Lutetium	187
72	Hf	Hafnium	175
73	Ta	Tantalum	170
74	W	Tungsten	162
75	Re	Rhenium	151
76	Os	Osmium	144
77	Ir	Iridium	141
78	Pt	Platinum	136
79	Au	Gold	136
80	Hg	Mercury	132
81	Tl	Thallium	145
82	Pb	Lead	146
83	Bi	Bismuth	148
84	Po	Polonium	140
85	At	Astatine	150
86	Rn	Radon	150

Exercise 9.16 Estimate the O—H bond length in H_2O from the covalent radii listed in Table 9.4.

■ See Problems 9.79, 9.80, 9.81, and 9.82.

Although a double bond is stronger than a single bond, it is not necessarily less reactive. Ethylene, $CH_2{=}CH_2$, for example, is more reactive than ethane, $CH_3{-}CH_3$, where carbon atoms are linked through a single bond.

The **bond order,** defined in terms of the Lewis formula, is *the number of pairs of electrons in a bond.* For example, in C : C the bond order is 1 (single bond); in C : : C the bond order is 2 (double bond). Bond length depends on bond order. As the bond order increases, the bond strength increases and the nuclei are pulled inward, decreasing the bond length. Look at carbon–carbon bonds. The average C—C bond length is 154 pm, whereas C═C is 134 pm long and C≡C is 120 pm long. ◀

Example 9.12 Relating Bond Order and Bond Length

Gaining Mastery Toolbox

Critical Concept 9.12
As the bond order between two atoms increases, the bond length decreases.

Solution Essentials:
- Bond order
- Bond length

Consider the molecules N_2H_4, N_2, and N_2F_2. Which molecule has the shortest nitrogen–nitrogen bond? Which has the longest nitrogen–nitrogen bond?

Problem Strategy First, write the electron-dot formulas for the molecules. Then note that the length of a bond between atoms decreases with bond order.

Solution First write the Lewis formulas:

```
   ..  ..                        ..  ..  ..  ..
H—N—N—H        :N≡N:          :F—N═N—F:
  |   |                        ..          ..
  H   H
  N2H4          N2               N2F2
```

The nitrogen–nitrogen bond should be shortest in N_2, where it is a triple bond, and longest in N_2H_4, where it is a single bond. (Experimental values for the nitrogen–nitrogen bond lengths are 109 pm for N_2, 122 pm for N_2F_2, and 147 pm for N_2H_4.)

Answer Check Check that you have written the electron-dot formulas correctly.

Exercise 9.17 Formic acid, isolated in 1670, is the irritant in ant bites. The structure of formic acid is

```
H—C═O
  |
  O
  |
  H
```

One of the carbon–oxygen bonds has a length of 136 pm; the other is 123 pm long. What is the length of the C═O bond in formic acid?

■ See Problems 9.83 and 9.84.

9.11 Bond Enthalpy

In Section 9.4, when we described the formation of a covalent bond, we introduced the concept of bond dissociation energy, the energy required to break a particular bond in a molecule (see Figure 9.11). This bond dissociation energy is one measure of the strength of a bond. In this section, we want to look at a related idea, the *bond enthalpy,* which we can use as a measure of the average strength of a bond in its compounds. Bond enthalpies are often obtained from enthalpies of reaction, ΔH, so the term "bond enthalpies" is appropriate, but you will often see the term *bond energies* used for the same values. As a practical matter, the approximations made in arriving at average bond values tend to obscure the small differences between enthalpy changes and energy changes for bonds (which are small). So, for that reason, the terms "bond enthalpy" and "bond energy" are often used interchangeably. We will, however, continue to use the term bond enthalpy.

Let's explore the possibility of assigning a value for the enthalpy change involved in breaking a particular type of bond, whatever the compound. Consider the experimentally determined enthalpy changes for the breaking, or dissociation, of a C—H bond in methane, CH_4, and in ethane, C_2H_6, in the gas phase:

$$\mathrm{H{-}\overset{\displaystyle H}{\underset{\displaystyle H}{\overset{|}{\underset{|}{C}}}}{-}H}(g) \longrightarrow \mathrm{H{-}\overset{\displaystyle H}{\underset{\displaystyle H}{\overset{|}{\underset{|}{C}}}}}(g) + \mathrm{H}(g) \qquad \Delta H = 435\ \mathrm{kJ}$$

$$\mathrm{H{-}\overset{\displaystyle H}{\underset{\displaystyle H}{\overset{|}{\underset{|}{C}}}}{-}\overset{\displaystyle H}{\underset{\displaystyle H}{\overset{|}{\underset{|}{C}}}}{-}H}(g) \longrightarrow \mathrm{H{-}\overset{\displaystyle H}{\underset{\displaystyle H}{\overset{|}{\underset{|}{C}}}}{-}\overset{\displaystyle H}{\underset{\displaystyle H}{\overset{|}{\underset{|}{C}}}}}(g) + \mathrm{H}(g) \qquad \Delta H = 410\ \mathrm{kJ}$$

Note that the ΔH values are approximately the same in the two cases. This suggests that the enthalpy change for the dissociation of a C—H bond may be about the same in other molecules. Comparisons of this sort lead to the conclusion that we can obtain approximate values of the enthalpy changes for given bond types.

We define the A—B **bond enthalpy** (denoted *BE*) as *the average enthalpy change for the breaking of an A—B bond in a molecule in the gas phase.* For example, to calculate a value for the C—H bond enthalpy, or *BE*(C—H), we might look at the experimentally determined enthalpy change for the breaking of all the C—H bonds in methane:

$$CH_4(g) \longrightarrow C(g) + 4H(g);\ \Delta H = 1662\ \mathrm{kJ}$$

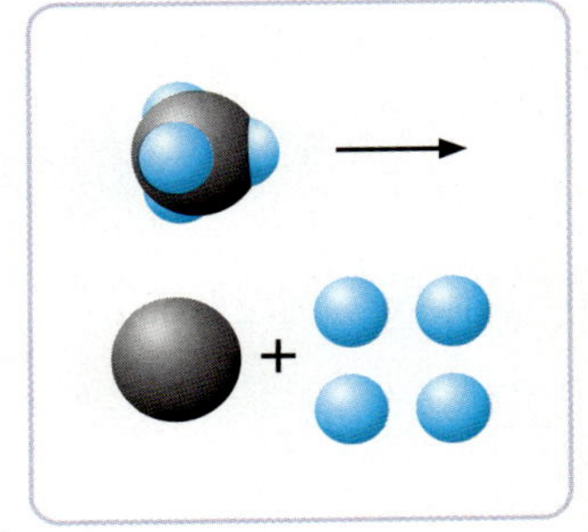

Because four C—H bonds are broken, we obtain an average value for breaking one C—H bond by dividing the enthalpy change for the reaction by 4:

$$BE(\mathrm{C{-}H}) = \tfrac{1}{4} \times 1662\ \mathrm{kJ} = 416\ \mathrm{kJ}$$

Similar calculations with other molecules, such as ethane, would yield approximately the same values for the C—H bond enthalpy.

Table 9.5 lists values of some bond enthalpies, which are based on an approximate fit to known thermodynamic data. Note that the value given in this table for the C—H bond is 413 kJ/mol, which is in fair agreement with the value we just obtained (416 kJ/mol). Because it takes energy to break a bond, bond enthalpies are always positive numbers. When a bond is formed, the enthalpy change is equal to the negative of the bond enthalpy (heat is released).

Table 9.5 Bond Enthalpies (in kJ/mol)*

Single Bonds

	H	C	N	O	S	F	Cl	Br	I
H	436								
C	413	348							
N	391	393	163						
O	463	358	201	146					
S	339	259	—	—	266				
F	567	485	272	190	327	159			
Cl	431	328	200	203	253	253	242		
Br	366	276	243	—	218	237	218	193	
I	299	240	—	234	—	—	208	175	151

Multiple Bonds

C═C	614	C═N	615	C═O	804 (in CO_2)
C≡C	839	C≡N	891	C≡O	1076
N═N	418	N═O	607	S═O	323
N≡N	945	O═O	498	S═S	418

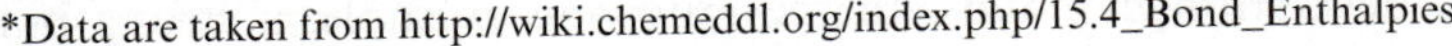

*Data are taken from http://wiki.chemeddl.org/index.php/15.4_Bond_Enthalpies.

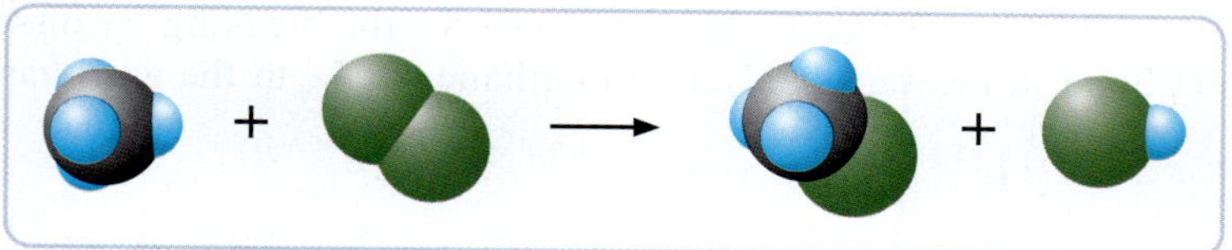

Figure 9.21 ▶

Reaction of methane with chlorine Molecular models illustrate the reaction $CH_4 + Cl_2 \longrightarrow CH_3Cl + HCl$.

Bond enthalpy is a measure of the strength of a bond: the larger the bond enthalpy, the stronger the chemical bond. Note from Table 9.5 that the bonds C—C, C═C, and C≡C have bond enthalpies of 348, 614, and 839 kJ/mol, respectively. These numbers indicate that the triple bond is stronger than the double bond, which in turn is stronger than the single bond.

You can use this table of bond enthalpies to estimate heats of reaction, or enthalpy changes, ΔH, for gaseous reactions. To illustrate this, let's find ΔH for the following reaction (Figure 9.21):

$$CH_4(g) + Cl_2(g) \longrightarrow CH_3Cl(g) + HCl(g)$$

You can *imagine* that the reaction takes place in steps involving the breaking and forming of bonds. Starting with the reactants, you suppose that one C—H bond and the Cl—Cl bond break.

$$\begin{array}{c}\text{H}\\|\\\text{H—C—H}\\|\\\text{H}\end{array} + \text{Cl—Cl} \longrightarrow \begin{array}{c}\text{H}\\|\\\text{H—C}\\|\\\text{H}\end{array} + \text{H} + \text{Cl} + \text{Cl}$$

The enthalpy change is BE(C—H) + BE(Cl—Cl). Now you reassemble the fragments to give the products.

$$\begin{array}{c}\text{H}\\|\\\text{H—C}\\|\\\text{H}\end{array} + \text{H} + \text{Cl} + \text{Cl} \longrightarrow \begin{array}{c}\text{H}\\|\\\text{H—C—Cl}\\|\\\text{H}\end{array} + \text{H—Cl}$$

In this case, C—Cl and H—Cl bonds are formed, and the enthalpy change equals the negative of the bond enthalpies: $-BE$(C—Cl) $-$ BE(H—Cl). Substituting bond-enthalpy values from Table 9.5, you get the enthalpy of reaction.

$$\begin{aligned}\Delta H &\simeq BE(\text{C—H}) + BE(\text{Cl—Cl}) - BE(\text{C—Cl}) - BE(\text{H—Cl})\\ &= (413 + 242 - 328 - 431)\text{ kJ}\\ &= -104\text{ kJ}\end{aligned}$$

The negative sign means that heat is released by the reaction. Because the bond-enthalpy concept is only approximate, this value is only approximate (the experimental value is -101 kJ).

In general, the enthalpy of reaction is (approximately) equal to the sum of the bond enthalpies for bonds broken minus the sum of the bond enthalpies for bonds formed.

Rather than calculating a heat of reaction from bond enthalpies, you usually obtain the heat of reaction from thermochemical data, because these are generally known more accurately. If the thermochemical data are not known, however, bond enthalpies can give you a reasonable estimate of the heat of reaction.

Bond enthalpies are perhaps of greatest value when you try to explain heats of reaction or to understand the relative stabilities of compounds. In general, a reaction is exothermic (gives off heat) if weak bonds are replaced by strong bonds (see Figure 9.22). In the reaction we just discussed, two bonds were broken and replaced by two new, stronger bonds. In the following example, a strong bond

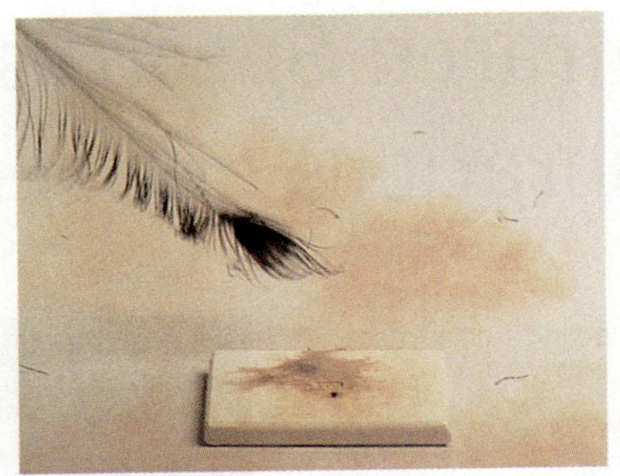

Figure 9.22 ▲

Explosion of nitrogen triiodide–ammonia complex The red-brown complex of nitrogen triiodide and ammonia is so sensitive to explosion that it can be detonated with a feather. Nitrogen–iodine single bonds are replaced by very stable nitrogen–nitrogen triple bonds (N_2) and iodine–iodine single bonds (I_2).

(C═C) is replaced by two weaker ones (two C—C bonds). Although the single bond is weaker than the double bond, as you would expect, two C—C bonds *together* create an energetically more stable situation than one C═C bond.

Example 9.13 Estimating ΔH from Bond Enthalpies

Gaining Mastery Toolbox

Critical Concept 9.13
The enthalpy of a reaction equals approximately the sum of the bond enthalpies for bonds broken minus the sum of bond enthalpies formed.

Solution Essentials:
- Bond enthalpy
- Enthalpy of reaction

Polyethylene is formed by linking many ethylene molecules into long chains. Estimate the enthalpy change per mole of ethylene for this reaction (shown below), using bond enthalpies.

$$+ H_2C{=}CH_2 + H_2C{=}CH_2 + H_2C{=}CH_2 + H_2C{=}CH_2 + \longrightarrow$$

$$-CH_2-CH_2-CH_2-CH_2-CH_2-CH_2-CH_2-CH_2-$$

Problem Strategy Break bonds in the reactants, and then form new bonds to give the products. The approximate enthalpy change equals the sum of the bond enthalpies for the bonds broken minus the sum of the bond enthalpies for the bonds formed.

Solution Imagine the reaction to involve the breaking of the carbon–carbon double bonds and the formation of carbon–carbon single bonds. For a very long chain, the net result is that for every C═C bond broken, two C—C bonds are formed.

$$\Delta H \simeq 614 - (2 \times 348) = \mathbf{-82\ kJ}$$

Answer Check Note whether the sign of ΔH agrees with what you would expect. Perhaps weak bonds have been replaced by strong bonds, in which case ΔH should be negative. Here, every double bond is replaced by two single bonds, so ΔH should be negative, as it is.

Exercise 9.18 Use bond enthalpies to estimate the enthalpy change for the combustion of ethylene, C_2H_4, according to the equation

$$C_2H_4(g) + 3O_2(g) \longrightarrow 2CO_2(g) + 2H_2O(g)$$

■ See Problems 9.85 and 9.86.

Instrumental Methods

Infrared Spectroscopy and Vibrations of Chemical Bonds

A chemical bond acts like a stiff spring connecting nuclei. As a result, the nuclei in a molecule vibrate, rather than maintaining fixed positions relative to each other. Nuclear vibration is depicted in Figure 9.23, which shows a spring model of HCl.

This vibration of molecules is revealed in their absorption of infrared radiation. (An instrument for observing the absorption of infrared radiation is shown in Figure 9.24.) The frequency of radiation absorbed equals the frequencies of nuclear vibrations. For example, the H—Cl bond vibrates at a frequency of 8.652×10^{13} vibrations per second. If radiation of this frequency falls on the molecule, it absorbs the radiation, which is in the infrared region, and begins vibrating more strongly.

The infrared absorption spectrum of a molecule of even moderate size can have a rather complicated appearance. Figure 9.25 shows the infrared (IR) spectrum of ethyl butyrate, a compound present in pineapple flavor. The complicated appearance of the IR spectrum is actually an advantage. Two different compounds are unlikely to have exactly the same IR spectrum. Therefore, the IR spectrum can act as a compound's "fingerprint."

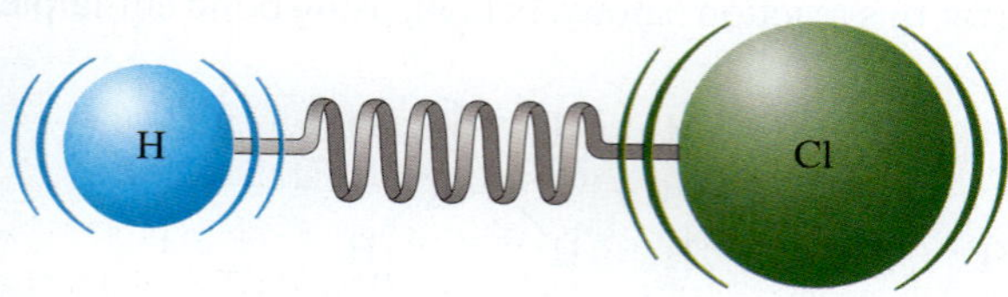

Figure 9.23 ▲

Vibration of the HCl molecule The vibrating molecule is represented here by a spring model. The atoms in the molecule vibrate; that is, they move constantly back and forth.

Courtesy of Nicolet

Figure 9.24 ▲

A Fourier transform infrared (FTIR) spectrometer A Nicolet 560 E.S.P.FT-IR spectrometer.

The IR spectrum of a compound can also yield structural information. Suppose you would like to obtain the structural formula of ethyl butyrate. The molecular formula, determined from combustion analysis, is $C_6H_{12}O_2$. Important information about this structure can be obtained from Figure 9.25.

You first need to be able to read such a spectrum. Instead of plotting an IR spectrum in frequency units (since the frequencies are very large), you usually would give the frequencies in *wavenumbers*, which are proportional to frequency. To get the wavenumber, you divide the frequency by the speed of light expressed in centimeters per second. For example, HCl absorbs at $(8.652 \times 10^{13}\ s^{-1})/(2.998 \times 10^{10}\ cm/s) = 2886\ cm^{-1}$ (wavenumbers).

A Checklist for Review

OWL and Go Chemistry Sign in at **www.cengage.com/owl** to:

- View tutorials and simulations, develop problem-solving skills, and complete online homework assigned by your professor.
- For quick review and exam prep, download Go Chemistry mini lecture modules from OWL or purchase them at **www.cengagebrain.com.**

Summary of Facts and Concepts

An *ionic bond* is a strong attractive force holding ions together. An ionic bond can form between two atoms by the transfer of electrons from the valence shell of one atom to the valence shell of the other. Many similar ions attract one another to form a crystalline solid, in which positive ions are surrounded by negative ions and negative ions are surrounded by positive ions. As a result, ionic solids are typically high-melting solids. Monatomic cations of the main-group elements have charges equal to the group number (or in some cases, the group number minus two). Monatomic anions of the main-group elements have charges equal to the group number minus eight.

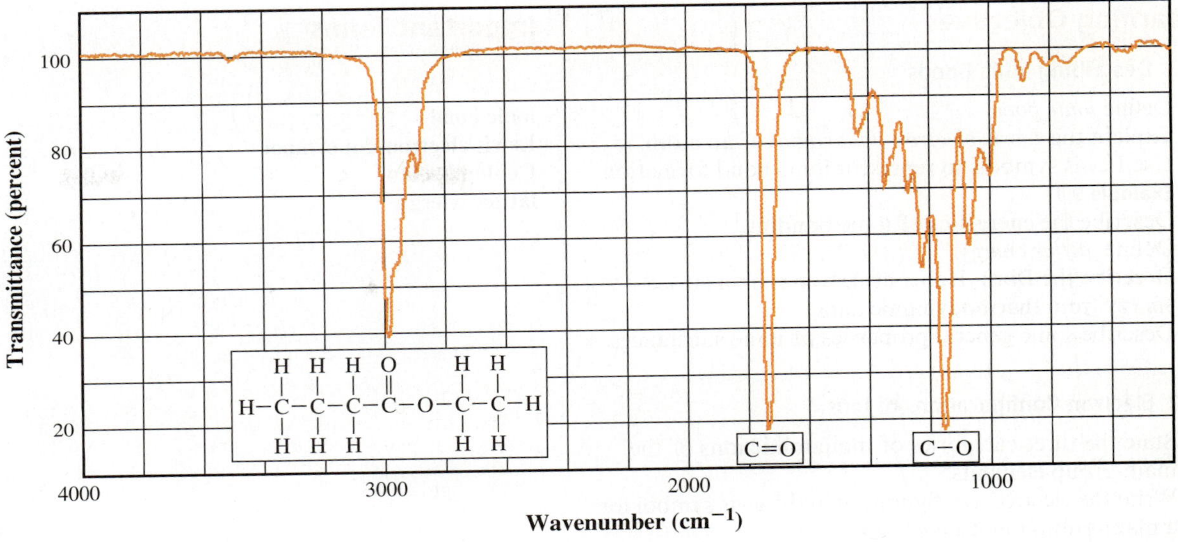

Figure 9.25

Infrared spectrum of ethyl butyrate Note the peaks corresponding to vibrations of C═O and C—O bonds. The molecular structure is shown at the bottom left. (From NIST Mass Spec Data Center, S.E. Stein, director, "IR and Mass Spectra" in *NIST Chemistry WebBook, NIST Standard Reference Database Number 69,* Eds. W.G. Mallard and P.J. Linstrom, February 2000, National Institute of Standards and Technology, Gaithersburg, MD, 20899 [http://webbook.nist.gov]. © 1991, 1994, 1996, 1997, 1998, 1999, 2000. Copyright by the U.S. Secretary of Commerce on behalf of the United States of America. All rights reserved.)

Wavenumber, or sometimes wavelength, is plotted along the horizontal axis.

Percent transmittance—that is, the percent of radiation that passes through a sample—is plotted on the vertical axis. When a molecule absorbs radiation of a given frequency or wavenumber, this is seen in the spectrum as an inverted spike (peak) at that wavenumber. Certain structural features of molecules appear as absorption peaks in definite regions of the infrared spectrum. For example, the absorption peak at 1730 cm^{-1} is characteristic of the C═O bond. With some knowledge of where various bonds absorb, one can identify other peaks, including that of C—O at 1180 cm^{-1}. (Generally, the IR peak for an A—B bond occurs at lower wavenumber than for an A═B bond.) The IR spectrum does not reveal the complete structure, but it provides important clues. Data from other instruments, such as the mass spectrometer (page 100), give additional clues.

■ See Problems 9.115 and 9.116.

A *covalent bond* is a strong attractive force that holds two atoms together by their sharing of electrons. These bonding electrons are attracted simultaneously to both atomic nuclei, and they spend part of the time near one atom and part of the time near the other. If an electron pair is not equally shared, the bond is *polar.* This polarity results from the difference in *electronegativities* of the atoms—that is, from the unequal abilities of the atoms to draw bonding electrons to themselves.

Lewis electron-dot formulas are simple representations of the valence-shell electrons of atoms in molecules and ions. You can apply simple rules to draw these formulas. In molecules with *delocalized bonding,* it is not possible to describe accurately the electron distribution with a single Lewis formula. For these molecules, you must use *resonance.* Although the atoms in Lewis formulas often satisfy the octet rule, *exceptions to the octet rule* are not uncommon. You can obtain the Lewis formulas for these exceptions by following the rules for writing Lewis formulas. The concept of *formal charge* will often help you decide which of several Lewis formulas gives the best description of a molecule or ion.

Bond lengths can be estimated from the *covalent radii* of atoms. Bond length depends on *bond order;* as the bond order increases, the bond length decreases. The A—B *bond enthalpy* is the average enthalpy change when an A—B bond is broken. You can use bond enthalpies to estimate ΔH for gas-phase reactions.

Learning Objectives	Important Terms
9.1 Describing Ionic Bonds	
▪ Define *ionic bond.* ▪ Explain the *Lewis electron-dot symbol* of an atom. ▪ Use Lewis symbols to represent ionic bond formation. **Example 9.1** ▪ Describe the energetics of ionic bonding. ▪ Define *lattice energy.* ▪ Describe the Born–Haber cycle to obtain a lattice energy from thermodynamic data. ▪ Describe some general properties of ionic substances.	**ionic bond** **Lewis electron-dot symbol** **Coulomb's law** **lattice energy**
9.2 Electron Configurations of Ions	
▪ State the three categories of monatomic ions of the main-group elements. ▪ Write the electron configuration and Lewis symbol for a main-group ion. **Example 9.2** ▪ Note the polyatomic ions given earlier in Table 2.5. ▪ Note the formation of 2+ and 3+ transition-metal ions. ▪ Write electron configurations of transition-metal ions. **Example 9.3**	
9.3 Ionic Radii	
▪ Define *ionic radius.* ▪ Define *isoelectronic ions.* ▪ Use periodic trends to obtain relative ionic radii. **Example 9.4**	**ionic radius** **isoelectronic**
9.4 Describing Covalent Bonds	
▪ Describe the formation of a covalent bond between two atoms. ▪ Define *Lewis electron-dot formula.* ▪ Define *bonding pair* and *lone (nonbonding) pair* of electrons. ▪ Define *coordinate covalent bond.* ▪ State the octet rule. ▪ Define *single bond, double bond,* and *triple bond.*	**covalent bond** **Lewis electron-dot formula** **bonding pair** **lone (nonbonding) pair** **coordinate covalent bond** **octet rule** **single bond** **double bond** **triple bond**
9.5 Polar Covalent Bonds; Electronegativity	
▪ Define *polar covalent bond.* ▪ Define *electronegativity.* ▪ State the general periodic trends in electronegativities. ▪ Use electronegativity to obtain relative bond polarity. **Example 9.5**	**polar covalent bond** **electronegativity**
9.6 Writing Lewis Electron-Dot Formulas	
▪ Write Lewis formulas with single bonds only. **Example 9.6** ▪ Write Lewis formulas having multiple bonds. **Example 9.7** ▪ Write Lewis formulas for ionic species. **Example 9.8**	
9.7 Delocalized Bonding; Resonance	
▪ Define *delocalized bonding.* ▪ Define *resonance description.* ▪ Write resonance formulas. **Example 9.9**	**delocalized bonding** **resonance description**

9.8 Exceptions to the Octet Rule	
■ Write Lewis formulas (exceptions to the octet rule). Example 9.10 ■ Note exceptions to the octet rule in Group IIA and Group IIIA elements.	
9.9 Formal Charge and Lewis Formulas	
■ Define *formal charge.* ■ State the rules for obtaining formal charge. ■ State two rules useful in writing Lewis formulas. ■ Use formal charges to determine the best Lewis formula. Example 9.11	**formal charge**
9.10 Bond Length and Bond Order	
■ Define *bond length* (*bond distance*). ■ Define *covalent radii.* ■ Define *bond order*. ■ Explain how bond order and bond length are related. Example 9.12	**bond length (bond distance)** **covalent radii** **bond order**
9.11 Bond Enthalpy	
■ Define *bond enthalpy.* ■ Estimate ΔH from bond enthalpies. Example 9.13	**bond enthalpy**

Questions and Problems

OWL Interactive versions of these problems may be assigned in OWL.

Self-Assessment and Review Questions

Key: These questions test your understanding of the ideas you worked with in the chapter. These problems vary in difficulty and often can be used for the basis of discussion.

9.1 Describe the formation of a sodium chloride crystal from atoms.

9.2 Why does sodium chloride normally exist as a crystal rather than as a molecule composed of one cation and one anion?

9.3 Explain what energy terms are involved in the formation of an ionic solid from atoms. In what way should these terms change (become larger or smaller) to give the lowest energy possible for the solid?

9.4 Define lattice energy for potassium bromide.

9.5 Why do most monatomic cations of the main-group elements have a charge equal to the group number? Why do most monatomic anions of these elements have a charge equal to the group number minus eight?

9.6 The 2+ ions of transition elements are common. Explain why this might be expected.

9.7 Explain how ionic radii are obtained from known distances between nuclei in crystals.

9.8 Describe the trends shown by the radii of the monatomic ions for the main-group elements both across a period and down a column.

9.9 Describe the formation of a covalent bond in H_2 from atoms. What does it mean to say that the bonding electrons are shared by the two atoms?

9.10 Draw a potential-energy diagram for a molecule such as Cl_2. Indicate the bond length (194 pm) and the bond dissociation energy (240 kJ/mol).

9.11 Give an example of a molecule that has a coordinate covalent bond.

9.12 The octet rule correctly predicts the Lewis formula of many molecules involving main-group elements. Explain why this is so.

9.13 Describe the general trends in electronegativities of the elements in the periodic table both across a period and down a column.

9.14 What is the qualitative relationship between bond polarity and electronegativity difference?

9.15 What is a resonance description of a molecule? Why is this concept required if we wish to retain Lewis formulas as a description of the electron structure of molecules?

9.16 Describe the kinds of exceptions to the octet rule that we encounter in compounds of the main-group elements. Give examples.

9.17 What is the relationship between bond order and bond length? Use an example to illustrate it.

9.18 Define bond enthalpy. Explain how one can use bond enthalpies to estimate the heat of reaction.

9.19 Which of the following contains *both* ionic and covalent bonds in the same compound?

a $BaCO_3$ b $MgCl_2$
c BaO d H_2S
e SO_4^{2-}

9.20 The radii of the species S, S^+, and S^- *decrease* in the following order:

a $S^+ > S > S^-$
b $S^+ > S > S^-$
c $S > S^- > S^+$
d $S > S^+ > S^-$
e $S^- > S > S^+$

9.21 Which of the following is the ground-state electron configuration of a C^{3-} ion?

a $1s^22s^22p^4$
b $[He]2s^22p^6$
c $[He]2s^1$
d $1s^22s^22p^5$
e $1s^22s^22p^63s^1$

9.22 The element X below could be

$\cdot\ddot{\underset{\cdot\cdot}{X}}\cdot$

a sulfur
b iodine
c aluminum
d phosphorus
e silicon

Concept Explorations

Key: Concept explorations are comprehensive problems that provide a framework that will enable you to explore and learn many of the critical concepts and ideas in each chapter. If you master the concepts associated with these explorations, you will have a better understanding of many important chemistry ideas and will be more successful in solving all types of chemistry problems. These problems are well suited for group work and for use as in-class activities.

9.23 Forming Ionic Compounds

a Consider a metal atom, which we will give the symbol M. Metal M can readily form the M^+ cation. If the sphere on the left below represents the metal atom M, which of the other three spheres would probably represent the M^+ cation? Explain your choice.

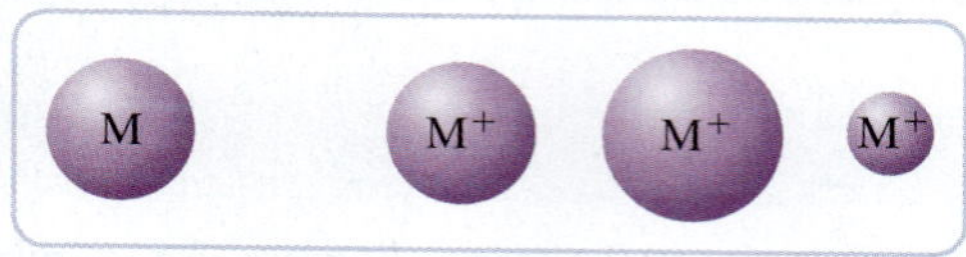

b Consider a nonmetal atom, which we will give the symbol X. The element X is a gas at room temperature and can easily form the X^- anion. If atom X is represented by the sphere on the left below, which of the other three spheres would probably represent the X^- anion. Explain your choice.

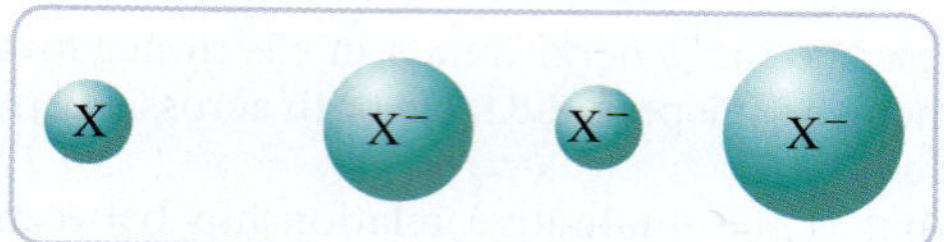

c On the basis of the information given previously, write a balanced chemical equation for the reaction that occurs when metal M reacts with nonmetal X. Include state symbols for the reactants and products.

d Is the product of the reaction of M and X a molecule or an ionic compound? Provide justification for your answer.

e Which of the representations below is the most appropriate to describe the compound of M and X, according to your answer for part d? Explain how you arrived at your choice.

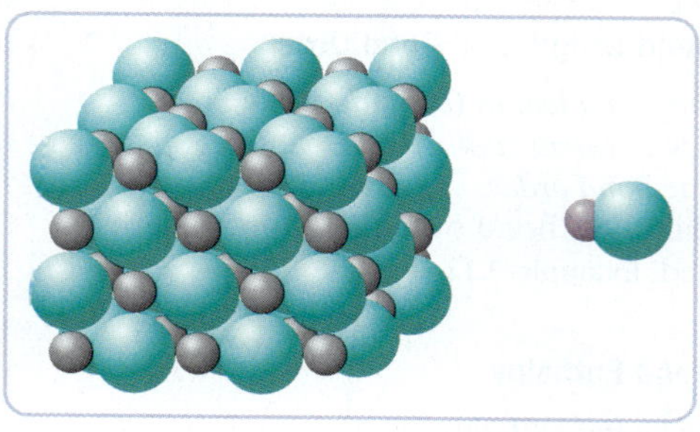

9.24 Bond Enthalpy

When atoms of the hypothetical element X are placed together, they rapidly undergo reaction to form the X_2 molecule:

$$X(g) + X(g) \longrightarrow X_2(g)$$

a Would you predict that this reaction is exothermic or endothermic? Explain.

b Is the bond enthalpy of X_2 a positive or a negative quantity? Why?

c Suppose ΔH for the reaction is -500 kJ/mol. Estimate the bond enthalpy of the X_2 molecule.

d Another hypothetical molecular compound, $Y_2(g)$, has a bond enthalpy of 750 kJ/mol, and the molecular compound $XY(g)$ has a bond enthalpy of 1500 kJ/mol. Using bond enthalpy information, calculate ΔH for the following reaction.

$$X_2(g) + Y_2(g) \longrightarrow 2XY(g)$$

e Given the following information, as well as the information previously presented, predict whether or not the hypothetical ionic compound AX is likely to form. In this compound, A forms the A^+ cation, and X forms the X^- anion. Be sure to justify your answer.

Reaction: $A(g) + \frac{1}{2}X_2(g) \longrightarrow AX(s)$

The first ionization energy of $A(g)$ is 400 kJ/mol.

The electron affinity of $X(g)$ is 525 kJ/mol.

The lattice energy of $AX(s)$ is 100 kJ/mol.

f If you predicted that no ionic compound would form from the reaction in part e, what minimum amount of $AX(s)$ lattice energy might lead to compound formation?

Conceptual Problems

Key: These problems are designed to check your understanding of the concepts associated with some of the main topics presented in each chapter. A strong conceptual understanding of chemistry is the foundation for both applying chemical knowledge and solving chemical problems. These problems vary in level of difficulty and often can be used as a basis for group discussion.

9.25 You land on a distant planet in another universe and find that the $n = 1$ level can hold a maximum of 4 electrons, the $n = 2$ level can hold a maximum of 5 electrons, and the $n = 3$ level can hold a maximum of 3 electrons. Like our universe, protons have a charge of $+1$, electrons have a charge of -1, and opposite charges attract. Also, a filled shell results in greater stability of an atom, so the atom tends to gain or lose electrons to give a filled shell. Predict the formula of a compound that results from the reaction of a neutral metal atom X, which has 7 electrons, and a neutral nonmetal atom Y, which has 3 electrons.

9.26 Which of the following represent configurations of thallium ions in compounds? Explain your decision in each case.

a. Tl^{2+} [Xe]$4f^{14}5d^{10}6p^1$
b. Tl^{3+} [Xe]$4f^{14}5d^{10}$
c. Tl^{4+} [Xe]$4f^{14}5d^9$
d. Tl^{+} [Xe]$4f^{14}5d^{10}6s^2$

9.27 Below on the left side are models of two atoms, one from a metal, the other from a nonmetal. On the right side are corresponding monatomic ions of those atoms. Decide which of these ions is the cation and which is the anion. Label atoms as "metal" or "nonmetal," as appropriate.

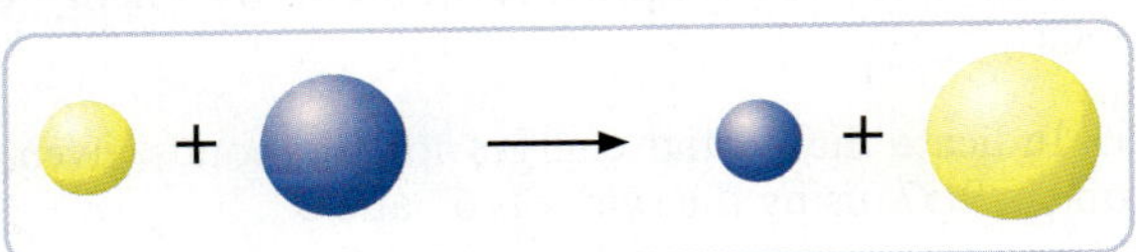

9.28 Predict a possible monatomic ion for Element 117, Uus. Given the spherical models below, which should be labeled "atom" and which should be labeled "ion"?

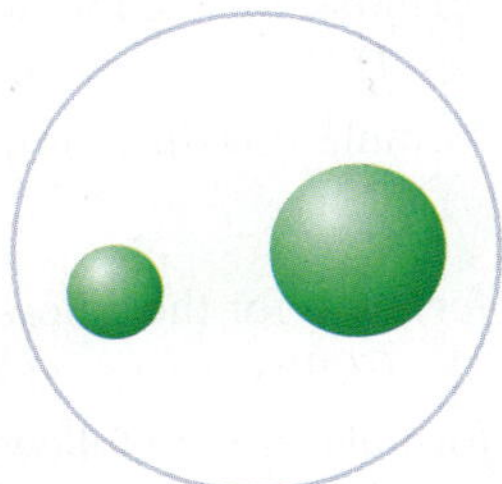

9.29 Examine each of the following electron-dot formulas and decide whether the formula is correct, or whether you could write a formula that better approximates the electron structure of the molecule. State which concepts or rules you use in each case to arrive at your conclusion.

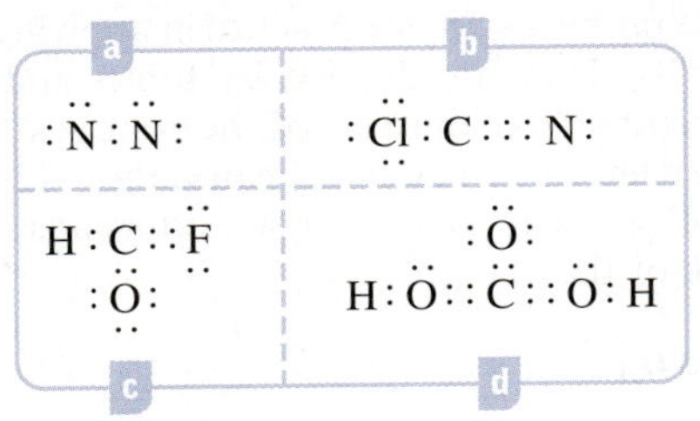

9.30 For each of the following molecular models, write an appropriate Lewis formula.

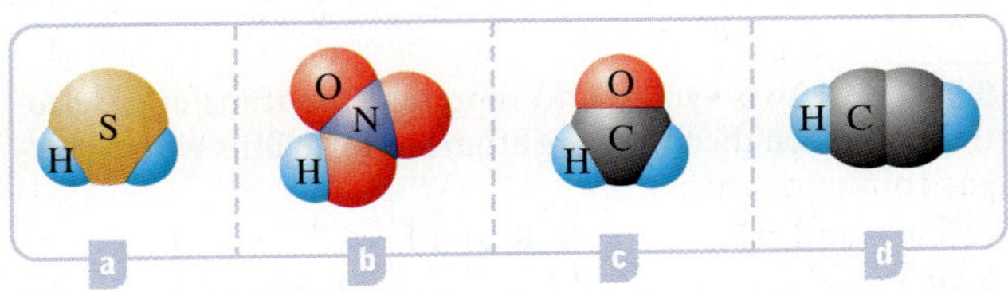

9.31 For each of the following molecular formulas, draw the most reasonable skeleton structure.

a. CH_2Cl_2 b. HNO_2 c. NOF d. N_2O_4

What rule or concept did you use to obtain each structure?

9.32 Below are three resonance formulas for N_2O (nitrous oxide). Rank these in terms of how closely you think each one represents the true electron structure of the molecule. State the rules and concepts you use to do this ranking.

a. $:\ddot{N}{-}N{\equiv}O:$ b. $:N{\equiv}N{-}\ddot{O}:$ c. $:\ddot{N}{=}N{=}\ddot{O}$

9.33 Sodium, Na, reacts with element X to form an ionic compound with the formula Na_3X.

a. What is the formula of the compound you expect to form when calcium, Ca, reacts with element X?
b. Would you expect this compound to be ionic or molecular?

9.34 The enthalpy change for each of the following reactions was calculated using bond enthalpies. The bond enthalpies of X—O, Y—O, and Z—O are all equal.

$$X{-}X + O{=}O \longrightarrow X{-}O{-}O{-}X;\ \Delta H = -275\ \text{kJ}$$
$$Y{-}Y + O{=}O \longrightarrow Y{-}O{-}O{-}Y;\ \Delta H = +275\ \text{kJ}$$
$$Z{-}Z + O{=}O \longrightarrow Z{-}O{-}O{-}Z;\ \Delta H = -100\ \text{kJ}$$

a. Rank the bonds X—X, Y—Y, and Z—Z from strongest to weakest.
b. Compare the enthalpies required to completely dissociate each of the products to atoms.
c. If O_2 molecules were O—O instead of O=O, how would this change ΔH for each reaction?

Practice Problems

Key: These problems are for practice in applying problem-solving skills. They are divided by topic, and some are keyed to exercises (see the ends of the exercises). The problems are arranged in matching pairs; the odd-numbered problem of each pair is listed first, and its answer is given in the back of the book.

Ionic Bonding

9.35 Write Lewis symbols for the following:
a P b P^{3-} c Ga d Ga^{3+}

9.36 Write Lewis symbols for the following:
a O b O^{2-} c Ca d Ca^{2+}

9.37 Use Lewis symbols to represent the transfer of electrons between the following atoms to form ions with noble-gas configurations:
a Ca and Br b K and I

9.38 Use Lewis symbols to represent the electron transfer between the following atoms to give ions with noble-gas configurations:
a Ba and I b Sr and O

9.39 For each of the following, write the electron configuration and Lewis symbol:
a As b As^{3+} c Se d Se^{2-}

9.40 For each of the following, write the electron configuration and Lewis symbol:
a Ge b Ge^{2+} c K^{+} d I^{-}

9.41 Write the electron configurations of Bi and Bi^{3+}.

9.42 Write the electron configurations of Sn and Sn^{2+}.

9.43 Give the electron configurations of Ni^{2+} and Ni^{3+}.

9.44 Give the electron configurations of Cu^{+} and Cu^{2+}.

Ionic Radii

9.45 Arrange the members of each of the following pairs in order of increasing radius and explain the order:
a Sr, Sr^{2+} b Br, Br^{-}

9.46 Arrange the members of each of the following pairs in order of increasing radius and explain the order:
a Te, Te^{2-} b Al, Al^{3+}

9.47 Without looking at Table 9.3, arrange the following in order of increasing ionic radius: Se^{2-}, Te^{2-}, S^{2-}. Explain how you arrived at this order. (You may use a periodic table.)

9.48 Which has the larger radius, N^{3-} or P^{3-}? Explain. (You may use a periodic table.)

9.49 Arrange the following in order of increasing ionic radius: F^{-}, Na^{+}, and N^{3-}. Explain this order. (You may use a periodic table.)

9.50 Arrange the following in order of increasing ionic radius: I^{-}, Cs^{+}, and Te^{2+}. Explain this order. (You may use a periodic table.)

Covalent Bonding

9.51 Use Lewis symbols to show the reaction of atoms to form hydrogen selenide, H_2Se. Indicate bonding pairs and lone pairs in the electron-dot formula of this compound.

9.52 Use Lewis symbols to show the reaction of atoms to form arsine, AsH_3. Indicate which electron pairs in the Lewis formula of AsH_3 are bonding and which are lone pairs.

9.53 Assuming that the atoms form the normal number of covalent bonds, give the molecular formula of the simplest compound of arsenic and bromine atoms.

9.54 Assuming that the atoms form the normal number of covalent bonds, give the molecular formula of the simplest compound of germanium and fluorine atoms.

Polar Covalent Bonds; Electronegativity

9.55 Using a periodic table (not Figure 9.15), arrange the following in order of increasing electronegativity:
a P, O, N b Na, Al, Mg c C, Al, Si

9.56 With the aid of a periodic table (not Figure 9.15), arrange the following in order of increasing electronegativity:
a Sr, Cs, Ba b Ca, Ge, Ga c P, As, S

9.57 Arrange the following bonds in order of increasing polarity using electronegativities of atoms: P—O, C—Cl, As—Br.

9.58 Decide which of the following bonds is least polar on the basis of electronegativities of atoms: H—S, Si—Cl, N—Cl.

9.59 Indicate the partial charges for the bonds given in Problem 9.57, using the symbols δ^{+} and δ^{-}.

9.60 Indicate the partial charges for the bonds given in Problem 9.58, using the symbols δ^{+} and δ^{-}.

Writing Lewis Formulas

9.61 Write Lewis formulas for the following molecules:
a Br_2 b H_2S c NF_3

9.62 Write Lewis formulas for the following molecules:
a BrF b PBr_3 c NOF

9.63 Write Lewis formulas for the following molecules:
a P_2 b $COBr_2$ c HNO_2

9.64 Write Lewis formulas for the following molecules:
a CO b BrCN c N_2F_2

9.65 Write Lewis formulas for the following ions:
a ClO^{-} b $SnCl_3^{-}$ c S_2^{2-}

9.66 Write Lewis formulas for the following ions:
a IBr_2^{+} b ClF_2^{+} c CN^{-}

Resonance

9.67 Write resonance descriptions for the following:
a HNO_3 b SO_3

9.68 Write resonance descriptions for the following:
a $ClNO_2$ b NO_2^-

9.69 Use resonance to describe the electron structure of nitromethane, CH_3NO_2. The skeleton structure is

```
    H  O
    |  |
 H—C—N—O
    |
    H
```

9.70 Give the resonance description of the formate ion. The skeleton structure is

```
 [     O   ]⁻
 [     |   ]
 [ H—C—O  ]
```

Exceptions to the Octet Rule

9.71 Write Lewis formulas for the following:
a XeF_2 b SeF_4 c TeF_6 d XeF_5^+

9.72 Write Lewis formulas for the following:
a I_3^- b ClF_3 c IF_4^- d BrF_5

9.73 Write Lewis formulas for the following:
a BCl_3 b $TlCl_2^+$ c $BeBr_2$

9.74 Write Lewis formulas for the following:
a BeF_2 b BeF_3^- c $AlBr_3$

Formal Charge and Lewis Formulas

9.75 Write a Lewis formula for each of the following, assuming that the octet rule holds for the atoms. Then obtain the formal charges of the atoms.
a O_3 b CO c HNO_3

9.76 Write a Lewis formula for each of the following, assuming that the octet rule holds for the atoms. Then obtain the formal charges of the atoms.
a ClNO b $POCl_3$ c N_2O (NNO)

9.77 For each of the following, use formal charges to choose the Lewis formula that gives the best description of the electron distribution:
a SOF_2
b H_2SO_3
c $HClO_2$

9.78 For each of the following, use formal charges to choose the Lewis formula that gives the best description of the electron distribution:
a ClO_2F
b SO_2
c ClO_3^-

Bond Length, Bond Order, and Bond Enthalpy

9.79 Use covalent radii (Table 9.4) to estimate the length of the P—F bond in phosphorus trifluoride, PF_3.

9.80 What do you expect for the B—Cl bond length in boron trichloride, BCl_3, on the basis of covalent radii (Table 9.4)?

9.81 Calculate the bond length for each of the following single bonds, using covalent radii (Table 9.4):
a C—H
b S—Cl
c Br—Cl
d Si—O

9.82 Calculate the C—H and C—Cl bond lengths in chloroform, $CHCl_3$, using values for the covalent radii from Table 9.4. How do these values compare with the experimental values: C—H, 107 pm; C—Cl, 177 pm?

9.83 One of the following compounds has a carbon–nitrogen bond length of 116 pm; the other has a carbon–nitrogen bond length of 147 pm. Match a bond length with each compound.

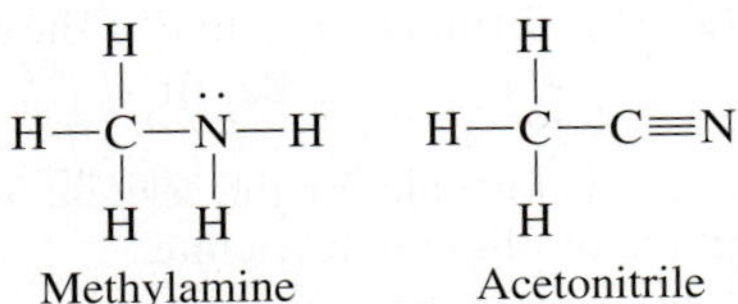

9.84 Which of the following two compounds has the shorter carbon–oxygen bond?

```
    H                 :O:
    |      ..          ‖
 H—C—O—H         H—C—H
    |      ..
    H
 Methanol         Formaldehyde
```

9.85 Use bond enthalpies (Table 9.5) to estimate ΔH for the following gas-phase reaction.

```
 H  H                   H  H
 |  |                   |  |
 C=C + H—Br  ⟶  H—C—C—Br
 |  |                   |  |
 H  H                   H  H
```

This is called an "addition" reaction, because a compound (HBr) is added across the double bond.

9.86 A commercial process for preparing ethanol (ethyl alcohol), C_2H_5OH, consists of passing ethylene gas, C_2H_4, and steam over an acid catalyst (to speed up the reaction). The gas-phase reaction is

```
 H  H                     H  H
 |  |                     |  |
 C=C + H—O—H  ⟶  H—C—C—O—H
 |  |                     |  |
 H  H                     H  H
```

Estimate ΔH for this reaction, using bond enthalpies (Table 9.5).

General Problems

Key: These problems provide more practice but are not divided by topic or keyed to exercises. Each section ends with essay questions, each of which is color coded to refer to the *A Chemist Looks at Frontiers* (purple), *A Chemist Looks at Daily Life* (orange), and *Instrumental Methods* (brown) chapter essay on which it is based. Odd-numbered problems and the even-numbered problems that follow are similar; answers to all odd-numbered problems except the essay questions are given in the back of the book.

9.87 For each of the following pairs of elements, state whether the binary compound formed is likely to be ionic or covalent. Give the formula and name of the compound.
a Sr, O b C, Br c Ga, F d N, Br

9.88 For each of the following pairs of elements, state whether the binary compound formed is likely to be ionic or covalent. Give the formula and name of the compound.
a K, Se b Al, F c Ba, Br d Si, Cl

9.89 Give the Lewis formula for the selenite ion, SeO_3^{2-}. Write the formula of aluminum selenite.

9.90 Give the Lewis formula for the arsenate ion, AsO_4^{3-}. Write the formula of lead(II) arsenate.

9.91 Iodic acid, HIO_3, is a colorless, crystalline compound. What is the electron-dot formula of iodic acid?

9.92 Selenic acid, H_2SeO_4, is a crystalline substance and a strong acid. Write the electron-dot formula of selenic acid.

9.93 Sodium amide, known commercially as sodamide, is used in preparing indigo, the dye used to color blue jeans. It is an ionic compound with the formula $NaNH_2$. What is the electron-dot formula of the amide anion, NH_2^-?

9.94 Lithium aluminum hydride, $LiAlH_4$, is an important reducing agent (an element or compound that generally has a strong tendency to give up electrons in its chemical reactions). Write the electron-dot formula of the AlH_4^- ion.

9.95 Nitronium perchlorate, NO_2ClO_4, is a reactive salt of the nitronium ion NO_2^+. Write the electron-dot formula of NO_2^+.

9.96 Solid phosphorus pentabromide, PBr_5, has been shown to have the ionic structure $[PBr_4^+][Br^-]$. Write the electron-dot formula of the PBr_4^+ cation.

9.97 Write electron-dot formulas for the following:
a $SeOCl_2$ b CSe_2 c $GaCl_4^-$ d C_2^{2-}

9.98 Write electron-dot formulas for the following:
a NO^+ b IF_2^+ c Si_2H_6 d $POBr_3$

9.99 Write Lewis formulas for the following:
a $SbCl_3$ b ICN c ICl_3 d IF_5

9.100 Write Lewis formulas for the following:
a $AlCl_4^-$ b AlF_6^{3-} c BrF_3 d IF_6^+

9.101 Give resonance descriptions for the following:
a SeO_2 b N_2O_4

9.102 Give resonance descriptions for the following:
a $C_2O_4^{2-}$ b CH_3NO_2

9.103 The compound S_2N_2 has a cyclic structure with alternating sulfur and nitrogen atoms. Draw all resonance formulas in which the atoms obey the octet rule. Of these, select those in which the formal charges of all atoms are closest to zero.

9.104 Acetic acid has the structure $CH_3CO(OH)$, in which the OH group is bonded to a C atom. The two carbon–oxygen bonds have different lengths. When an acetic acid molecule loses the H from the OH group to form the acetate ion, the two carbon–oxygen bonds become equal in length. Explain.

9.105 The atoms in N_2O_5 are connected as follows:

```
O           O
 \         /
  N—O—N
 /         \
O           O
```

No attempt has been made here to indicate whether a bond is single or double or whether there is resonance. Obtain the Lewis formula (or formulas). The N—O bond lengths are 118 pm and 136 pm. Indicate the lengths of the bonds in the compound.

9.106 Methyl nitrite has the structure

```
    H
    |
H—C—O—N—O
    |
    H
```

No attempt has been made here to indicate whether a bond is single or double or whether there is resonance. Obtain the Lewis formula (or formulas). The N—O bond lengths are 122 pm and 137 pm. Indicate the lengths of the N—O bonds in the compound.

9.107 Use bond enthalpies to estimate ΔH for the reaction

$$H_2(g) + O_2(g) \longrightarrow H_2O_2(g)$$

9.108 Use bond enthalpies to estimate ΔH for the reaction

$$2F_2(g) + N_2(g) \longrightarrow N_2F_4(g)$$

9.109 Use bond enthalpies to estimate ΔH for the reaction

$$N_2F_2(g) + F_2(g) \longrightarrow N_2F_4(g)$$

9.110 Use bond enthalpies to estimate ΔH for the reaction

$$HCN(g) + 2H_2(g) \longrightarrow CH_3NH_2(g)$$

■ **9.111** Compare the properties of an ionic material such as sodium chloride with a room-temperature ionic liquid. Explain this difference.

■ **9.112** What advantages does using an ionic liquid as a solvent have over using an organic solvent?

■ **9.113** Explain the decomposition of nitroglycerin in terms of relative bond enthalpies.

■ **9.114** How did the Swedish chemist Alfred Nobel manage to tame the decomposition of nitroglycerin?

■ **9.115** What property of a chemical bond gives rise to the infrared spectrum of a compound?

■ **9.116** What kind of information can be obtained about a compound from its infrared spectrum?

Strategy Problems

Key: As noted earlier, all of the practice and general problems are matched pairs. This section is a selection of problems that are not in matched-pair format. These challenging problems require that you employ many of the concepts and strategies that were developed in the chapter. In some cases, you will have to integrate several concepts and operational skills in order to solve the problem successfully.

9.117 Which of the following Lewis symbols is (or are) *not* correct?

a $[\dot{\text{Al}}\cdot]^{3+}$ b $[:\ddot{\text{F}}:]^{-}$ c $[:\ddot{\text{O}}:]^{2-}$ d $\cdot\text{Mg}$ e $\cdot\dot{\text{Si}}\cdot$

9.118 Calculate the lattice energy of potassium fluoride, KF, using the Born-Haber cycle. Use thermodynamic data from Appendix C to obtain the enthalpy changes for each step. (Note: You will obtain a slightly different answer if you use values given in Chapter 8 for the ionization energy and electron affinity, which are energy values at 0 K rather than the enthalpy changes at 298 K.)

9.119 Draw a figure similar to Figure 9.2 but for the energetics of ionic bonding in KF. Use the ionization energy and electron affinity from values given in Tables 8.3 and 8.4 in Chapter 8. Calculate the sublimation energy from thermodynamic data (Appendix C). For the lattice energy, see Problem 9.118.

9.120 Consider the following ions: Al^{3+}, Mg^{2+}, Na^{+}, S^{2-}, Cl^{-}. Write down all possible formulas of ionic compounds between each of the oppositely charged ions. Of these compounds, note the ones with a metal-to-nonmetal ratio of one-to-one. Which compound do you expect has the highest melting point? Why?

9.121 Which of the following are isoelectronic with the potassium ion, K^{+}? What are the electron configurations of these species?

a Ar b Cl^{+} c Kr d Cl^{-} e Ca^{+}

9.122 An ion M^{2+} has the configuration $[Ar]3d^2$, and an atom has the configuration $[Ar]4s^2$. Identify the ion and the atom.

9.123 Give the symbol of an atomic ion for each of the following electron configurations:

a $1s^22s^22p^63s^23p^6$
b $1s^22s^22p^63s^23p^64s^1$
c $1s^22s^22p^63s^23p^63d^8$
d $1s^22s^22p^63s^23p^63d^{10}4s^2$
e $1s^22s^22p^63s^23p^63d^{10}4s^24p^6$

9.124 Using the ionic radii given in Table 9.3, estimate the energy to form a mole of $Na^{+}F^{-}$ ion pairs from the corresponding atomic ions.

9.125 Calculate the difference in electronegativities between the atoms in SrF_2 and between the atoms in SnF_2. Which substance would you expect to be more ionic in character? One of these compounds melts at 213°C, the other one at about 1400°C. What is the melting point of tin(II) fluoride?

9.126 Which of the following molecules possesses a double bond? If none do, so state. Which species has an atom with a nonzero formal charge?

a I_2 b SF_4 c $COCl_2$ d C_2H_4 e CN^{-}

9.127 Which of the following molecules contains *only* double bonds? If none do, so state.

a NCCN b CO_2 c C_2H_4 d O_3 e N_2

9.128 Draw resonance formulas for the azide ion, N_3^{-}, and for the nitronium ion, NO_2^{+}. Decide which resonance formula is the best description of each ion.

9.129 Two fourth-period atoms, one of a transition metal, M, and the other of a main-group nonmetal, X, form a compound with the formula M_2X_3. What is the electron configuration of atom X if M is Fe? What is the configuration of X if M is Co?

9.130 In Section 9.8, several resonance formulas of BF_3 were discussed. Which of these would be favored based on the concept of formal charge? Explain how you arrived at this answer. Does this answer agree with the chemistry described in the text?

9.131 Draw resonance formulas of the phosphoric acid molecule, $(HO)_3PO$. Obtain formal charges for the atoms in these resonance formulas. From this result, which resonance formula would you expect to most closely approximate the actual electron distribution?

9.132 Consider all A—B bonds that can be formed between any two of the four elements germanium (Ge) to bromine (Br) in Period 4. Which bond should be the most polar? What is the covalent single bond length for this bond? What is the likely molecular formula of a binary compound of these two elements? Draw an electron-dot formula for the molecule.

9.133 Consider hypothetical elements X and Y. Suppose the enthalpy of formation of the compound XY is −336 kJ/mol, the bond enthalpy for X_2 is 414 kJ/mol, and the bond enthalpy for Y_2 is 159 kJ/mol. Estimate the XY bond enthalpy in units of kJ/mol.

9.134 Nitrous oxide, N_2O, has a linear structure NNO. Write resonance formulas for this molecule and from them estimate the NN bond length in the molecule. Use the data in Example 9.12.

9.135 Write the electron-dot formula for the nitrous acid molecule, H—O—N—O (bond lines are drawn only to show how the atoms are bonded, not the multiplicity of the bonds). The experimental values for the nitrogen-oxygen bonds in this molecule are 120 pm and 146 pm. Assign these values to the N—O bonds in the nitrous acid molecule. Explain how you arrived at your answer.

9.136 Using bond enthalpies, estimate the heat obtained in burning 10.0 g of methane gas, CH_4, in oxygen gas, O_2, to obtain carbon dioxide gas and water vapor.

Cumulative-Skills Problems

Key: The problems under this heading combine skills introduced in previous chapters with those given in the current one.

9.137 Phosphorous acid, H_3PO_3, has the structure $(HO)_2PHO$, in which one H atom is bonded to the P atom, and two H atoms are bonded to O atoms. For each bond to an H atom, decide whether it is polar or nonpolar. Assume that only polar-bonded H atoms are acidic. Write the balanced equation for the complete neutralization of phosphorous acid with sodium hydroxide. A 200.0-mL sample of H_3PO_3 requires 22.50 mL of 0.1250 *M* NaOH for complete neutralization. What is the molarity of the H_3PO_3 solution?

9.138 Hypophosphorous acid, H_3PO_2, has the structure $(HO)PH_2O$, in which two H atoms are bonded to the P atom, and one H atom is bonded to an O atom. For each bond to an H atom, decide whether it is polar or nonpolar. Assume that only polar-bonded H atoms are acidic. Write the balanced equation for the complete neutralization of hypophosphorous acid with sodium hydroxide. A 200.0-mL sample of H_3PO_2 requires 22.50 mL of 0.1250 *M* NaOH for complete neutralization. What is the molarity of the H_3PO_2 solution?

9.139 An ionic compound has the following composition (by mass): Mg, 10.9%; Cl, 31.8%; O, 57.3%. What are the formula and name of the compound? Write the Lewis formulas for the ions.

9.140 An ionic compound has the following composition (by mass): Ca, 30.3%; N, 21.2%; O, 48.5%. What are the formula and name of the compound? Write the Lewis formulas for the ions.

9.141 A gaseous compound has the following composition by mass: C, 25.0%; H, 2.1%; F, 39.6%; O, 33.3%. Its molecular mass is 48.0 amu. Write the Lewis formula for the molecule.

9.142 A liquid compound used in dry cleaning contains 14.5% C and 85.5% Cl by mass and has a molecular mass of 166 amu. Write the Lewis formula for the molecule.

9.143 A compound of tin and chlorine is a colorless liquid. The vapor has a density of 7.49 g/L at 151°C and 1.00 atm. What is the molecular mass of the compound? Why do you think the compound is molecular and not ionic? Write the Lewis formula for the molecule.

9.144 A compound of arsenic and fluorine is a gas. A sample weighing 0.100 g occupies 14.2 mL at 23°C and 765 mmHg. What is the molecular mass of the compound? Write the Lewis formula for the molecule.

9.145 Calculate the enthalpy of reaction for

$$HCN(g) \longrightarrow H(g) + C(g) + N(g)$$

from enthalpies of formation (see Appendix C). Given that the C—H bond enthalpy is 411 kJ/mol, obtain a value for the C≡N bond enthalpy. Compare your result with the value given in Table 9.5.

9.146 Assume the values of the C—H and C—C bond enthalpies given in Table 9.5. Then, using data given in Appendix C, calculate the C═O bond enthalpy in acetaldehyde,

```
     H  O
     |  ‖
  H—C—C—H
     |
     H
```

Compare your result with the value given in Table 9.5.

9.147 According to Pauling, the A—B bond enthalpy is equal to the average of the A—A and B—B bond enthalpies plus an enthalpy contribution from the polar character of the bond:

$$BE(\text{A—B}) = \tfrac{1}{2}[BE(\text{A—A}) + BE(\text{B—B})] + k(X_A - X_B)^2]$$

Here X_A and X_B are the electronegativities of atoms A and B, and k is a constant equal to 98.6 kJ. Assume that the electronegativity of H is 2.1. Use the formula to calculate the electronegativity of oxygen.

9.148 Because known compounds with N—I bonds tend to be unstable, there are no thermodynamic data available with which to calculate the N—I bond enthalpy. However, we can estimate a value from Pauling's formula relating electronegativities and bond enthalpies (see Problem 9.147). Using Pauling's electronegativities and the bond enthalpies given in Table 9.5, calculate the N—I bond enthalpy.

9.149 Using Mulliken's formula, calculate a value for the electronegativity of chlorine. Use values of the ionization energy from Figure 8.18 and values of the electron affinity from Table 8.4. Divide this result (in kJ/mol) by 230 to get a value comparable to Pauling's scale.

9.150 Using Mulliken's formula, calculate a value for the electronegativity of oxygen. Convert the result to a value on Pauling's scale. See Problem 9.149.

10

Molecular Geometry and Chemical Bonding Theory

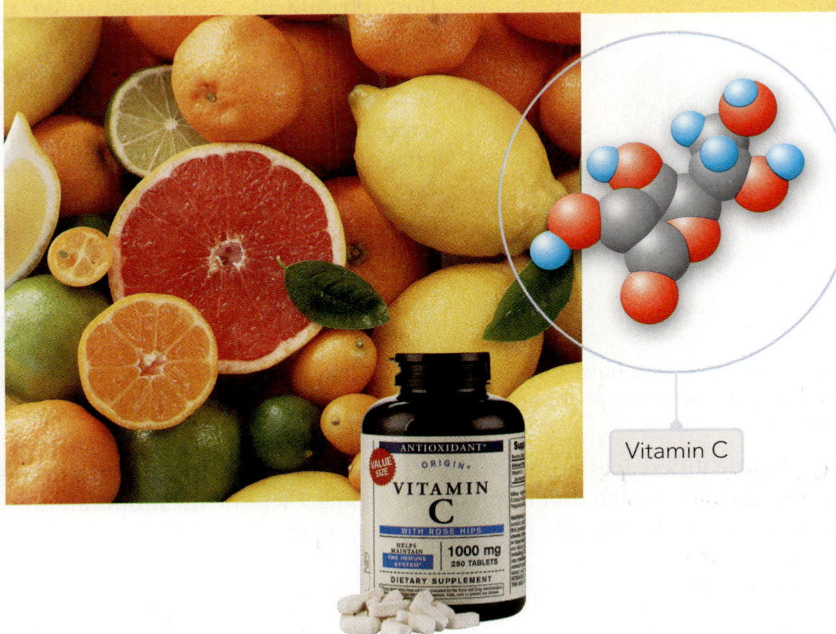

Humans are one of the few animals unable to produce vitamin C (ascorbic acid) naturally and must obtain it through the diet (such as citrus fruits) or from supplements. The space-filling model shows the three-dimensional shape of the ascorbic acid molecule.

Contents and Concepts

Molecular Geometry and Directional Bonding

We can predict the *molecular geometry* of a molecule, its general shape as determined by the relative positions of atomic nuclei, with a simple model: the valence-shell electron-pair repulsion model. After exploring molecular geometry, we explain chemical bonding by means of *valence bond theory*, which gives us insight into why bonds form and why they have definite directions in space, giving particular molecular geometries.

Molecular Orbital Theory

Although valence bond theory satisfactorily describes most molecules, it has trouble explaining the bonding in a molecule such as oxygen, O_2, which has even numbers of electrons but is paramagnetic. *Molecular orbital theory* is an alternative theory that views the electronic structure of molecules much the way we think of atoms: in terms of orbitals that are successively occupied by electrons.

OWL Sign in to OWL at **www.cengage.com/owl** to view tutorials and simulations, develop problem-solving skills, and complete online homework assigned by your professor.

go Chemistry Download mini lecture videos for key concept review and exam prep from OWL or purchase them from **www.cengagebrain.com**.

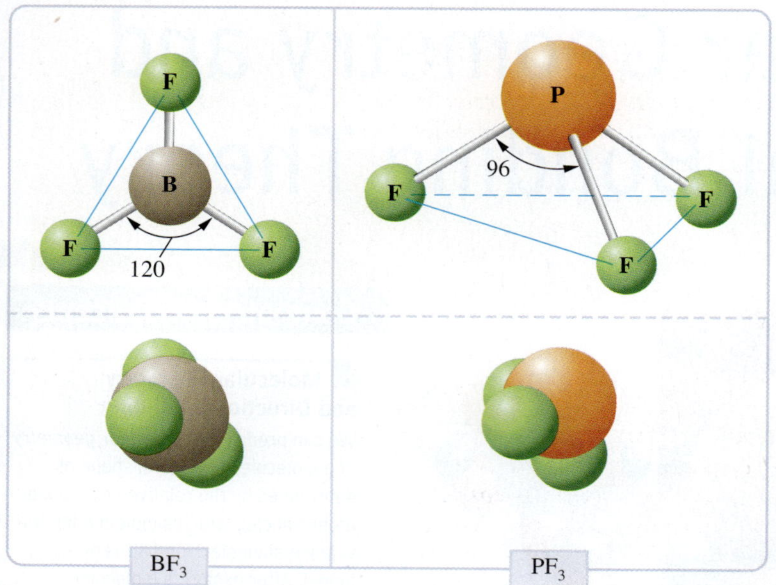

Figure 10.1 ▲

Molecular models of BF_3 and PF_3 Although both molecules have the general formula AX_3, boron trifluoride is planar (flat), whereas phosphorus trifluoride is pyramidal (pyramid shaped).

We know from various experiments that molecules have definite structures; that is, the atoms of a molecule have definite positions relative to one another in three-dimensional space. Consider two molecules: boron trifluoride, BF_3, and phosphorus trifluoride, PF_3. Both have the same general formula, AX_3, but their molecular structures are very different (Figure 10.1). Boron trifluoride is a planar, or flat, molecule. The angle between any two B—F bonds is 120°. The geometry or general shape of the BF_3 molecule is said to be *trigonal planar* (trigonal because the figure described by imaginary lines connecting the F atoms has three sides). Phosphorus trifluoride is nonplanar, and the angle between any two P—F bonds is 96°. If you imagine lines connecting the fluorine atoms, these lines and those of the P—F bonds describe a pyramid with the phosphorus atom at its apex. The geometry of the PF_3 molecule is said to be *trigonal pyramidal.* How do you explain such different molecular geometries?

Subtle differences in structure are also possible. Consider the following molecular structures, which differ only in the arrangement of the atoms about the C═C bond; such molecules are referred to as *isomers.*

H, H / C═C / Cl, Cl	H, Cl / C═C / Cl, H
cis-1,2-Dichloroethene	*trans*-1,2-Dichloroethene

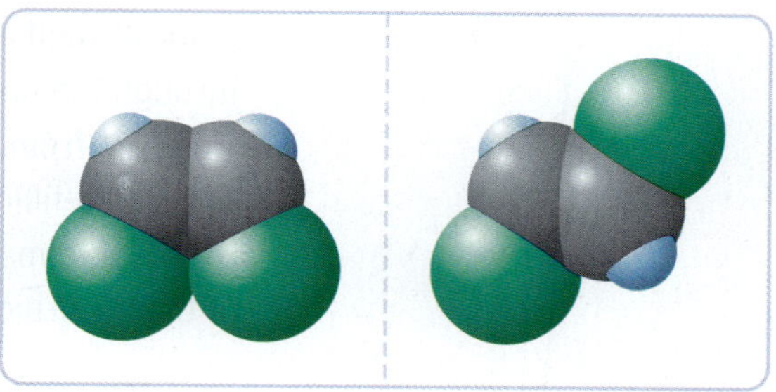

In the *cis* compound, both H atoms are on the same side of the C═C bond; in the *trans* compound, they are on opposite sides. These structural formulas represent entirely different compounds, as their boiling points clearly demonstrate. *Cis*-1,2-dichloroethene boils at 60°C; *trans*-1,2-dichloroethene boils at 48°C. The differences between such *cis* and *trans* compounds can be quite important. The central molecular event in the detection of light by the human eye involves the transformation of a *cis* compound to its corresponding *trans* compound after the absorption of a photon of light. How do you explain the existence of *cis* and *trans* compounds?

In this chapter we discuss how to explain the geometries of molecules in terms of their electronic structures. We also explore two theories of chemical bonding: valence bond theory and molecular orbital theory.

Molecular Geometry and Directional Bonding

Molecular geometry is *the general shape of a molecule, as determined by the relative positions of the atomic nuclei.* There is a simple model that allows you to predict molecular geometries, or shapes, from Lewis formulas. This valence-shell electron-pair model usually predicts the correct general shape of a molecule. It does not explain chemical bonding, however. For this you must look at a theory, such as valence bond theory, that is based on quantum mechanics. Valence bond theory provides further insight into why bonds form and, at the same time, reveals that bonds have definite directions in space, resulting in particular molecular geometries.

10.1 The Valence-Shell Electron-Pair Repulsion (VSEPR) Model

The **valence-shell electron-pair repulsion (VSEPR) model** *predicts the shapes of molecules and ions by assuming that the valence-shell electron pairs are arranged about each atom so that electron pairs are kept as far away from one another as possible, thus minimizing electron-pair repulsions.* ▶ The possible arrangements assumed by different numbers of electron pairs about an atom are shown in Figure 10.2. (If you tie various numbers of similar balloons together, they assume these same arrangements; see Figure 10.3.)

The acronym VSEPR is pronounced "vesper."

For example, if there are only two electron pairs in the valence shell of an atom, these pairs tend to be at opposite sides of the nucleus, so that repulsion is minimized. This gives a *linear* arrangement of electron pairs; that is, the electron pairs mainly occupy regions of space at an angle of 180° to one another (Figure 10.2).

If three electron pairs are in the valence shell of an atom, they tend to be arranged in a plane directed toward the corners of a triangle of equal sides (equilateral triangle). This arrangement is *trigonal planar,* in which the regions of space occupied by electron pairs are directed at 120° angles to one another (Figure 10.2).

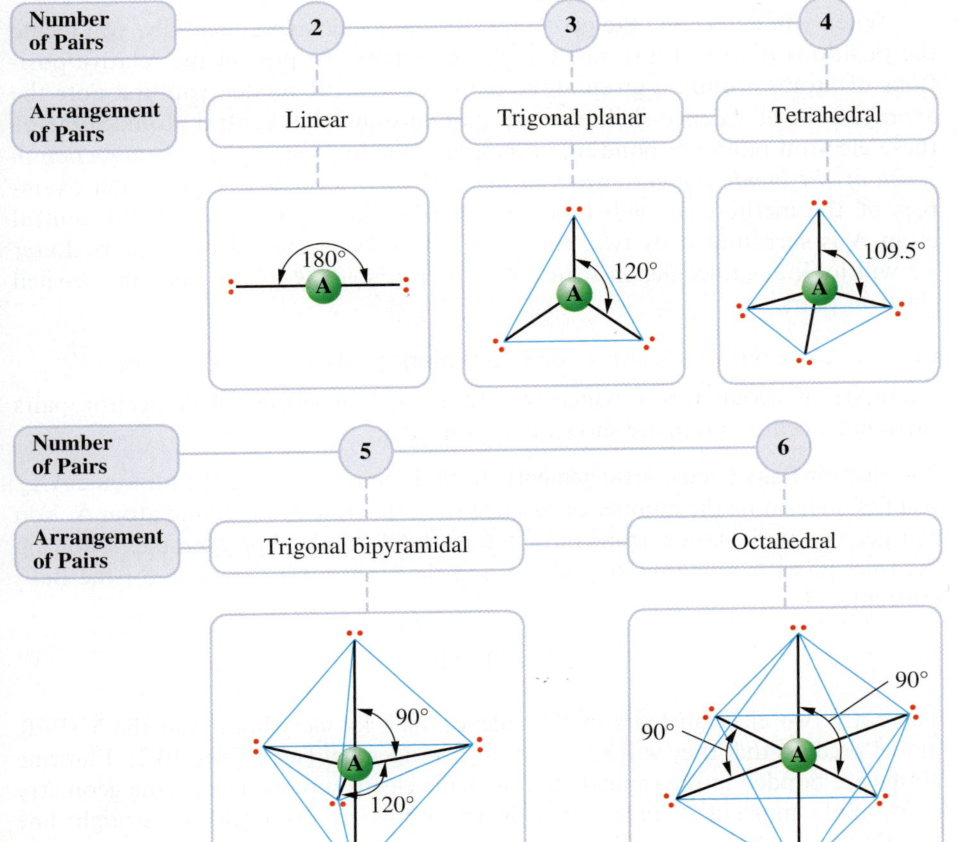

Figure 10.2 ◀ **Arrangement of electron pairs about an atom** Black lines give the directions of electron pairs about atom A. The blue lines merely help depict the geometric arrangement of electron pairs.

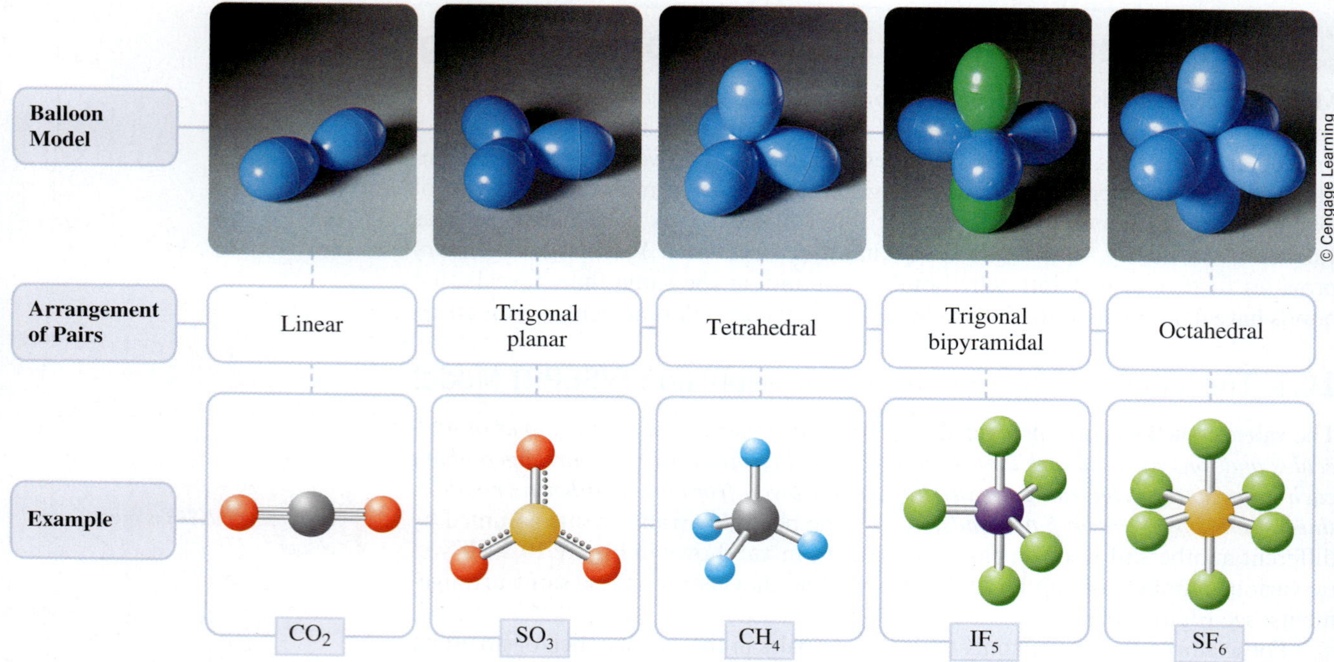

Figure 10.3 ▲

Balloon models of electron-pair arrangements When you tie similar balloons together, they assume the arrangements shown. These arrangements are the same as those assumed by valence electron pairs about an atom, as in the examples.

Four electron pairs in the valence shell of an atom tend to have a *tetrahedral* arrangement. That is, if you imagine the atom at the center of a regular tetrahedron, each region of space in which an electron pair mainly lies extends toward a corner, or vertex (Figure 10.2). (A regular tetrahedron is a geometrical shape with four faces, each an equilateral triangle. Thus, it has the form of a triangular pyramid.) The regions of space mainly occupied by electron pairs are directed at approximately 109.5° angles to one another.

When you determine the geometry of a molecule experimentally, you locate the positions of the atoms, not the electron pairs. To predict the relative positions of atoms around a given atom using the VSEPR model, you first note the arrangement of valence-shell electron pairs around that central atom. Some of these electron pairs are bonding pairs and some are lone pairs. *The direction in space of the bonding pairs gives you the molecular geometry.* To consider examples of the method, we will first look at molecules AX_n in which the central atom A is surrounded by two, three, or four valence-shell electron pairs. Later we will look at molecules in which A is surrounded by five or six valence-shell electron pairs.

Central Atom with Two, Three, or Four Valence-Shell Electron Pairs

Examples of geometries in which two, three, or four valence-shell electron pairs surround a central atom are shown in Figure 10.4.

Beryllium fluoride, BeF_2, is normally a solid. The BeF_2 molecule exists in the vapor phase at high temperature.

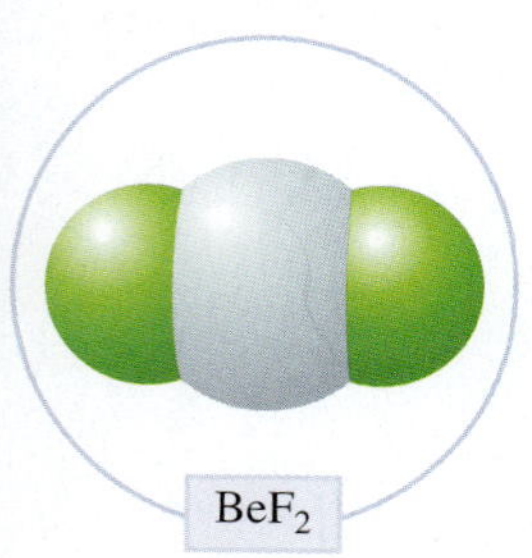

Two Electron Pairs (Linear Arrangement) To find the geometry of the molecule AX_n, you first determine the number of valence-shell electron pairs around atom A. You can get this information from the electron-dot formula. For example, following the rules given in Section 9.6, you can find the electron-dot formula for the BeF_2 molecule. ◀

$$:\ddot{\underset{..}{F}}:Be:\ddot{\underset{..}{F}}:$$

There are two electron pairs in the valence shell for beryllium, and the VSEPR model predicts that they will have a linear arrangement (see Figure 10.2). Fluorine atoms are bonded in the same direction as the electron pairs. Hence, the geometry of the BeF_2 molecule is *linear*—that is, the atoms are arranged in a straight line (see Figure 10.4).

Electron Pairs Total	Bonding	Lone	Arrangement of Pairs	Molecular Geometry	Example	
2	2	0	Linear	Linear AX_2	BeF_2	F—Be—F
3	3	0	Trigonal planar	Trigonal planar AX_3	BF_3	
	2	1		Bent (or angular) AX_2 (Lone pair)	SO_2	
4	4	0	Tetrahedral	Tetrahedral AX_4	CH_4	
	3	1		Trigonal pyramidal AX_3	NH_3	
	2	2		Bent (or angular) AX_2	H_2O	

Figure 10.4 ▲

Molecular geometries Central atom with two, three, or four valence-shell electron pairs. The notation in the column at the right provides a convenient way to represent the shape of a molecule drawn on a flat surface. A straight line represents a bond or electron pair in the plane of the page. A dashed line represents a bond or electron pair extending behind the page. A wedge represents a bond or electron pair extending above (in front of) the page.

The VSEPR model can also be applied to molecules with multiple bonds. In this case, each multiple bond is treated as though it were a single electron pair. (All pairs of a multiple bond are required to be in approximately the same region.) To predict the geometry of carbon dioxide, for example, you first write the electron-dot formula.

$$:\ddot{O}::C::\ddot{O}:$$

You have two double bonds, or two electron groups, about the carbon atom, and these are treated as though there were two pairs on carbon. Thus, according to the VSEPR model, the bonds are arranged linearly and the geometry of carbon dioxide is linear.

Three Electron Pairs (Trigonal Planar Arrangement) Let us now predict the geometry of the boron trifluoride molecule, BF_3, introduced in the chapter opening. The electron-dot formula is

$$\begin{array}{c} :\ddot{F}: \\ :\ddot{F}:\ddot{B}:\ddot{F}: \end{array}$$

The three electron pairs on boron have a trigonal planar arrangement, and the molecular geometry, which is determined by the directions of the bonding pairs, is also *trigonal planar* (Figures 10.1 and 10.4).

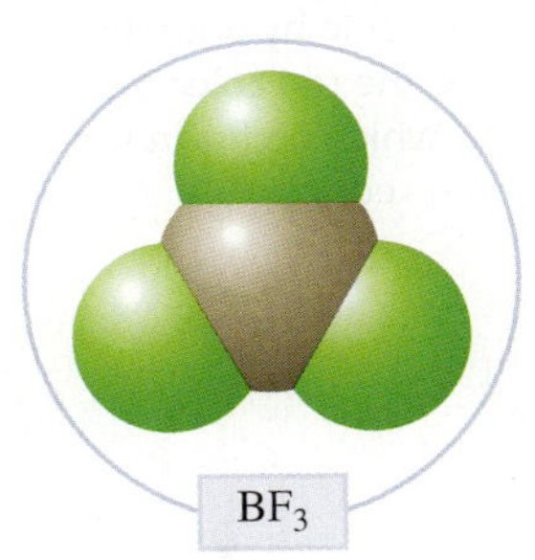

Sulfur dioxide, SO_2, provides an example of a molecule with three electron groups about the S atom; one group is a lone pair. This molecule requires a resonance description, so it gives us a chance to see how the VSEPR model can be applied in such cases. The resonance description can be written

$$\ddot{\mathrm{S}}(=\mathrm{O})(-\mathrm{O}) \longleftrightarrow \ddot{\mathrm{S}}(-\mathrm{O})(=\mathrm{O}) \longleftrightarrow \ddot{\mathrm{S}}(-\mathrm{O})(-\mathrm{O})$$

Whichever formula you consider, sulfur has three groups of electrons about it. These groups have a trigonal planar arrangement, so sulfur dioxide is a *bent,* or *angular,* molecule (Figure 10.4). In general, whenever you consider a resonance description, you can use any of the resonance formulas to predict the molecular geometry with the VSEPR model.

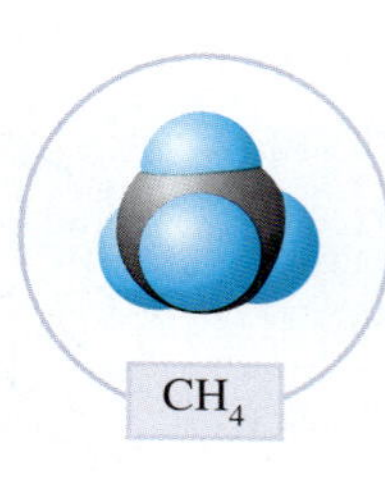

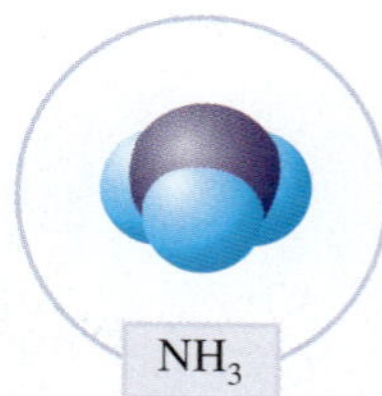

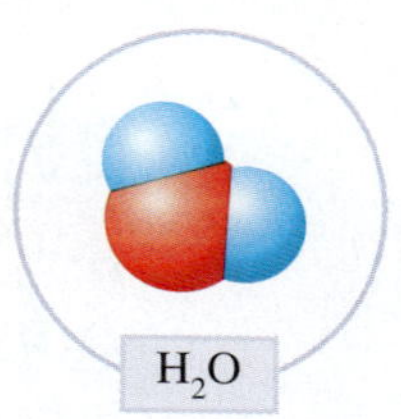

Note the difference between the *arrangement of electron pairs* and the *molecular geometry,* or the arrangement of nuclei. What you usually "see" by means of x-ray diffraction and similar methods are the nuclear positions—that is, the molecular geometry. Unseen, but nevertheless important, are the lone pairs, which occupy definite positions about an atom according to the VSEPR model.

Four Electron Pairs (Tetrahedral Arrangement) The common and most important case of four electron pairs about the central atom (the octet rule) leads to three different molecular geometries, depending on the number of bonds formed. Note the following examples.

```
     H            H           H
     ..           ..
   H:C:H         :N:H        :O:H
     ..           ..          ..
     H            H
```

	CH_4	NH_3	H_2O
Molecular geometry:	tetrahedral	trigonal pyramidal	bent

In each of these examples, the electron pairs are arranged tetrahedrally, and two or more atoms are bonded in these tetrahedral directions to give the different geometries (Figure 10.4).

When all electron pairs are bonding, as in methane, CH_4, the molecular geometry is *tetrahedral* (Figure 10.5). When three of the pairs are bonding and one pair is nonbonding, as in ammonia, NH_3, the molecular geometry is *trigonal pyramidal.* Note that the nitrogen atom is at the apex of the pyramid, and the three hydrogen atoms extend downward to form the triangular base of the pyramid.

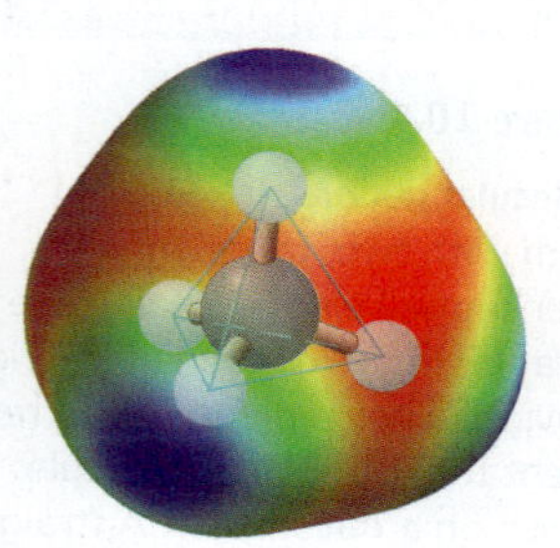

Figure 10.5 ▲

The methane molecule A ball-and-stick model shows the tetrahedral arrangement of bonds. This model is within an electrostatic-potential map, showing the electron density about the molecule by gradations of color. Thus, the red color means high density, which occurs in the C—H bonding regions.

Now you can answer the question raised in the chapter opening: Why is the BF_3 molecule trigonal planar and the PF_3 molecule trigonal pyramidal? The difference occurs because in BF_3 there are three pairs of electrons around boron, and in PF_3 there are four pairs of electrons around phosphorus. As in NH_3, there are three bonding pairs and one lone pair in PF_3, which give rise to the trigonal pyramidal geometry.

Steps in the Prediction of Geometry by the VSEPR Model Let us summarize the steps to follow in order to predict the geometry of an AX_n molecule or ion by the VSEPR method. (All X atoms of AX_n need not be identical.)

Step 1: Write the electron-dot formula from the molecular formula.

Step 2: Determine from the electron-dot formula the number of electron pairs around the central atom, including bonding and nonbonding pairs. (Count a multiple bond as one pair. If resonance occurs, use one resonance formula to determine this number.)

Step 3: Determine the arrangement of these electron pairs about the central atom (see Figure 10.2).

Step 4: Obtain the molecular geometry from the directions of the bonding pairs for this arrangement (see Figure 10.4).

CONCEPT CHECK 10.1

An atom in a molecule is surrounded by four pairs of electrons: one lone pair and three bonding pairs. Describe how the four electron pairs are arranged about the atom. How are any three of these pairs arranged in space? What is the geometry about this central atom, taking into account just the bonded atoms?

OWL Interactive Example 10.1 Predicting Molecular Geometries (Two, Three, or Four Electron Pairs)

Gaining Mastery Toolbox

Critical Concept 10.1
The arrangement of electron groups about a central atom depends on the number of electron groups: two electron groups have a linear arrangement, three electron groups a trigonal planar arrangement, and four electron groups a tetrahedral arrangement.

Solution Essentials:
- Molecular geometries for a given arrangement of electrons
- Arrangement of electron groups about an atom
- Electron-dot formulas

Predict the geometry of the following molecules or ions, using the VSEPR method: a $BeCl_2$; b NO_2^-; c $SiCl_4$.

Problem Strategy After drawing the Lewis formula for a molecule, determine the number of electron groups about the central atom. This number determines the arrangement of the electron groups (Figure 10.2). The geometry is determined by the arrangement of bonds to the central atom (Figure 10.4).

Solution

a Following the steps outlined in Sections 9.6 and 9.8 for writing Lewis formulas, you distribute the valence electrons to the skeleton structure of $BeCl_2$ as follows:

$$:\ddot{\underset{..}{Cl}}:Be:\ddot{\underset{..}{Cl}}:$$

The two pairs on Be have a linear arrangement, indicating a **linear** molecular geometry for $BeCl_2$. (See Figure 10.4; 2 electron pairs, 0 lone pairs.)

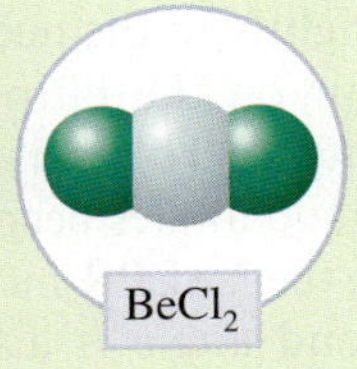
$BeCl_2$

b The nitrite ion, NO_2^-, has the following resonance description:

$$\left[:\ddot{\underset{..}{O}}-\ddot{N}=\ddot{O}:\right]^- \longleftrightarrow \left[:\ddot{O}=\ddot{N}-\ddot{\underset{..}{O}}:\right]^-$$

The N atom has three electron groups (one single bond, one double bond, and one lone pair) about it. Therefore the molecular geometry of the NO_2^- ion is **bent.** (See Figure 10.4; 2 bonding pairs, 1 lone pair.)

NO_2^-

c The electron-dot formula of $SiCl_4$ is

$$\begin{array}{c} :\ddot{Cl}: \\ :\ddot{\underset{..}{Cl}}:\underset{..}{\overset{..}{Si}}:\ddot{\underset{..}{Cl}}: \\ :\underset{..}{Cl}: \end{array}$$

The molecular geometry is **tetrahedral.** (See Figure 10.4; 4 bonding pairs, 0 lone pairs.)

$SiCl_4$

Answer Check Be careful not to confuse the name for the arrangement of valence electron pairs with the name for the molecular geometry. The geometry refers only to the arrangement of bonds. In this example, the nitrite ion, NO_2^-, has a trigonal planar arrangement of valence electron pairs and a bent geometry (arrangement of the bonds).

Exercise 10.1 Use the VSEPR method to predict the geometry of the following ion and molecules: a ClO_3^-; b OF_2; c SiF_4.

■ See Problems 10.33, 10.34, 10.35, and 10.36.

Bond Angles and the Effect of Lone Pairs

The VSEPR model allows you to predict the approximate angles between bonds in molecules. For example, it tells you that CH_4 should have a tetrahedral geometry and that the H—C—H bond angles should be 109.5° (see Figure 10.2). Because all of the

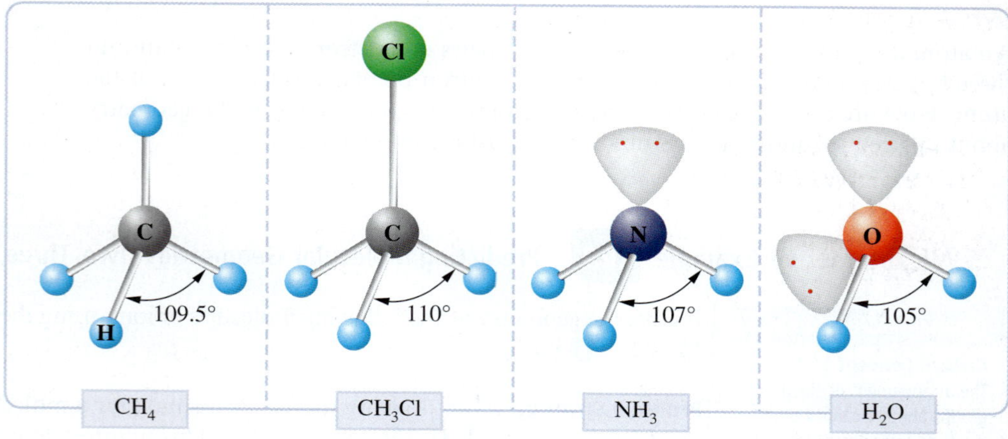

Figure 10.6 ▶

H—A—H bond angles in some molecules Experimentally determined bond angles are shown for CH_4, CH_3Cl, NH_3, and H_2O, represented here by models.

valence-shell electron pairs about the carbon atom are bonding and all of the bonds are alike, you expect the CH_4 molecule to have an exact tetrahedral geometry. However, if one or more of the electron pairs is nonbonding or if there are dissimilar bonds, all four valence-shell electron pairs will not be alike. Then you expect the bond angles to deviate from the 109.5° predicted by an exact tetrahedral geometry. This is indeed what you see. Experimentally determined H—A—H bond angles (A is the central atom) in CH_4, CH_3Cl, NH_3, and H_2O are shown in Figure 10.6.

The increase or decrease of bond angles from the ideal values is often predictable. *A lone pair tends to require more space than a corresponding bonding pair.* You can explain this as follows: A lone pair of electrons is attracted to only one atomic core, whereas a bonding pair is attracted to two. As a result, the lone pair is spatially more diffuse, while the bonding pair is drawn more tightly to the nuclei. Consider the trigonal pyramidal molecule NH_3. The lone pair on the nitrogen atom requires more space than the bonding pairs. Therefore, the N—H bonds are effectively pushed away from the lone pair, and the H—N—H bond angles become smaller than the tetrahedral value of 109.5°. How much smaller, the VSEPR model cannot tell you. The experimental value of an H—N—H bond angle is a few degrees smaller (107°). But the trigonal pyramidal molecule PF_3, mentioned in the chapter opening, has an F—P—F bond angle of 96°, which is significantly smaller than the exact tetrahedral value.

Multiple bonds require more space than single bonds because of the greater number of electrons. You therefore expect the C=O bond in the formaldehyde molecule, CH_2O, to require more space than the C—H bonds. You predict that the H—C—H bond angle will be smaller than the 120° seen in an exact trigonal planar geometry, such as the F—B—F angle in BF_3. Similarly, you expect the H—C—H bond angles in the ethylene molecule, CH_2CH_2, to be smaller than the trigonal planar value. Experimental values are shown in Figure 10.7.

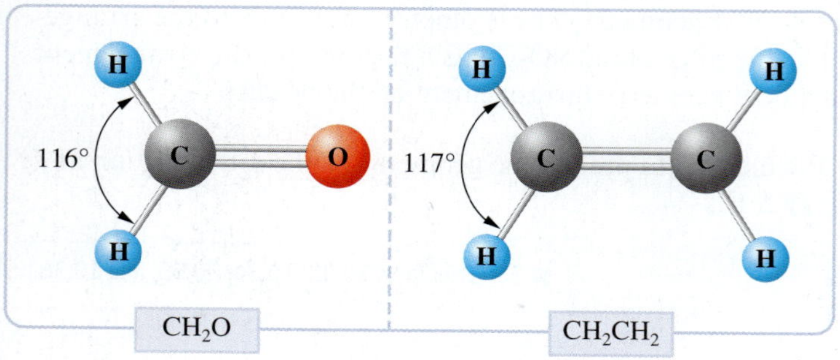

Figure 10.7 ▲

H—C—H bond angles in molecules with a carbon double bond Bond angles are shown for formaldehyde, CH_2O, and ethylene, CH_2CH_2, represented here by models.

Central Atom with Five or Six Valence-Shell Electron Pairs

Five electron pairs tend to have a *trigonal bipyramidal arrangement.* The electron pairs tend to be directed to the corners of a trigonal bipyramid, a figure formed by placing the face of one tetrahedron onto the face of another tetrahedron (see Figure 10.2).

Six electron pairs tend to have an *octahedral arrangement.* The electron pairs tend to be directed to the corners of a regular octahedron, a figure that has eight triangular faces and six vertexes, or corners (see Figure 10.2).

Five Electron Pairs (Trigonal Bipyramidal Arrangement) Large atoms like phosphorus can accommodate more than eight valence electrons. The phosphorus atom in phosphorus pentachloride, PCl_5, has five electron pairs in its valence shell. With five electron pairs around phosphorus, all bonding, PCl_5 should have a *trigonal bipyramidal geometry* (Figure 10.8). Note, however, that the vertexes of the trigonal bipyramid are not all equivalent (that is, the angles between the electron pairs are not all the same). Thus, the directions to which the electron pairs point are

Figure 10.8 ▼

Molecular geometries
Central atom with five or six valence-shell electron pairs.

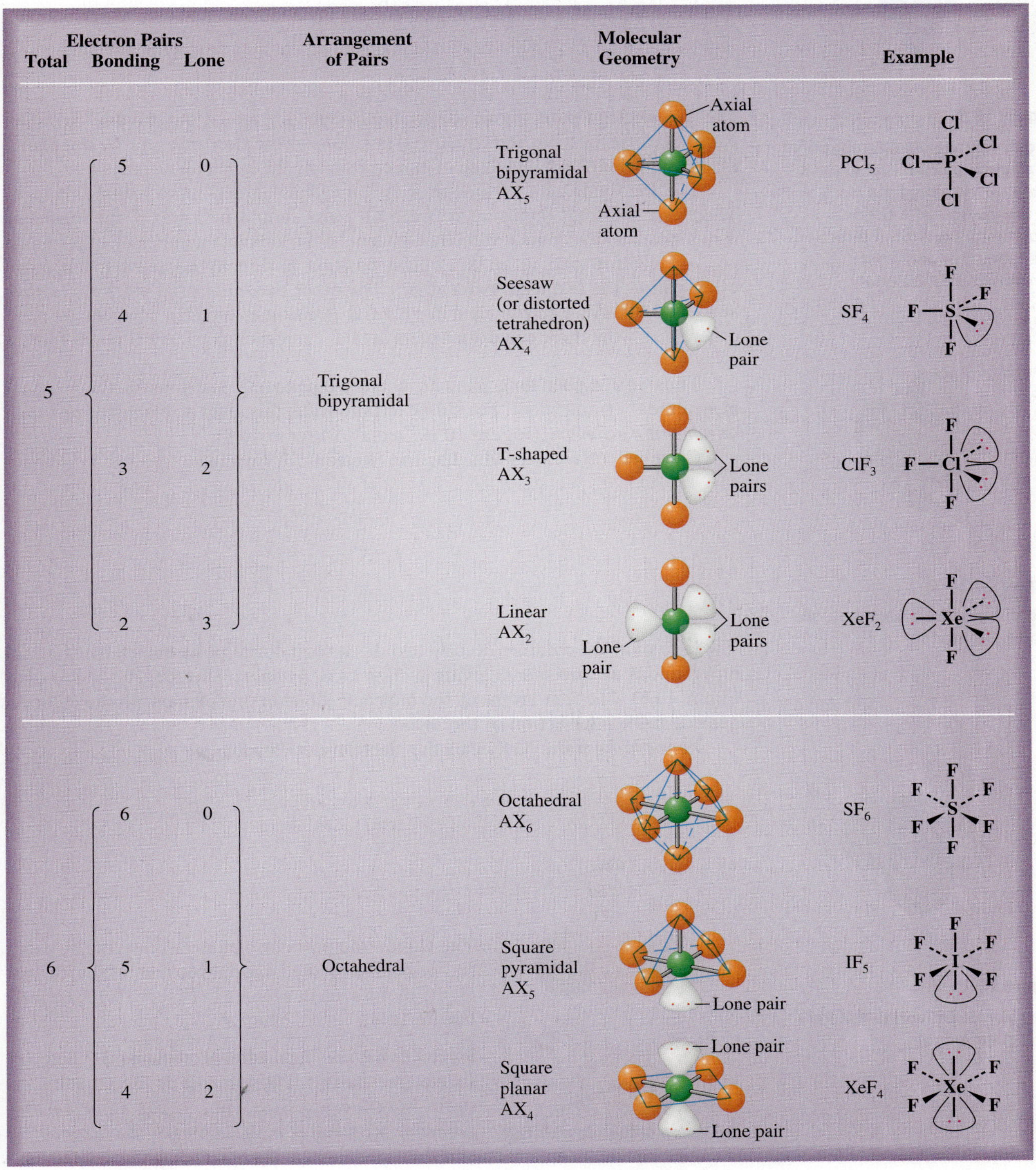

Electron Pairs Total	Electron Pairs Bonding	Electron Pairs Lone	Arrangement of Pairs	Molecular Geometry	Example
5	5	0	Trigonal bipyramidal	Trigonal bipyramidal AX_5 (Axial atom, Axial atom)	PCl_5
	4	1		Seesaw (or distorted tetrahedron) AX_4 (Lone pair)	SF_4
	3	2		T-shaped AX_3 (Lone pairs)	ClF_3
	2	3		Linear AX_2 (Lone pair, Lone pairs)	XeF_2
6	6	0	Octahedral	Octahedral AX_6	SF_6
	5	1		Square pyramidal AX_5 (Lone pair)	IF_5
	4	2		Square planar AX_4 (Lone pair, Lone pair)	XeF_4

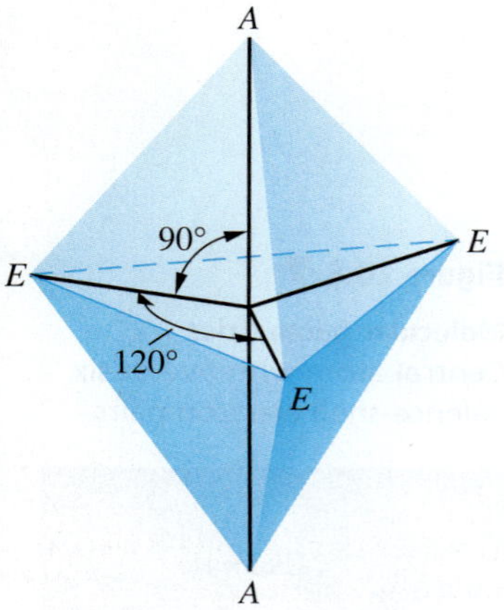

Figure 10.9 ▲

Trigonal bipyramidal arrangement of electron pairs Electron pairs are directed along the black lines to the vertices of a trigonal bipyramid. Equatorial directions are labeled *E*, and axial directions are labeled *A*.

not equivalent. Two of the directions, called *axial directions,* form an axis through the central atom (see Figure 10.9). They are 180° apart. The other three directions are called *equatorial directions.* These point toward the corners of the equilateral triangle that lies on a plane through the central atom, perpendicular (at 90°) to the axial directions. The equatorial directions are 120° from each other.

Molecular geometries other than trigonal bipyramidal are possible when one or more of the five electron pairs are lone pairs. Consider the sulfur tetrafluoride molecule, SF_4. The Lewis electron-dot formula is

The five electron pairs about sulfur should have a trigonal bipyramidal arrangement. Because the axial and equatorial positions of the electron pairs are not equivalent, you must decide in which of these positions the lone pair appears.

The lone pair acts as though it is somewhat "larger" than a bonding pair. Therefore, the total repulsion between all pairs should be lower if the lone pair is in a position that puts it directly adjacent to the smallest number of other pairs.

An electron pair in an equatorial position is directly adjacent to only *two* other pairs—the two axial pairs at 90°. The other two equatorial pairs are farther away at 120°. An electron pair in an axial position is directly adjacent to *three* other pairs—the three equatorial pairs at 90°. The other axial pair is much farther away at 180°.

Thus you expect lone pairs to occupy equatorial positions in the trigonal bipyramidal arrangement. For sulfur tetrafluoride, this gives a *seesaw* (or *distorted tetrahedral*) *geometry* (Figure 10.10; see also Figure 10.8).

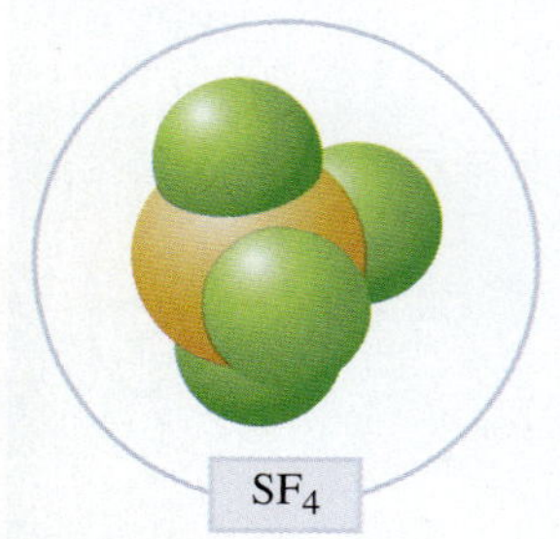

Figure 10.10 ▲

The sulfur tetrafluoride molecule Molecular model.

Chlorine trifluoride, ClF_3, has the electron-dot formula

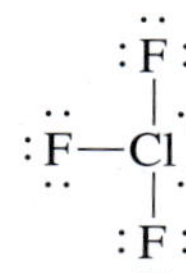

The lone pairs on chlorine occupy two of the equatorial positions of the trigonal bipyramidal arrangement, giving a *T-shaped geometry* (Figure 10.11; see also Figure 10.8). The four atoms of the molecule all lie in one plane, with the chlorine nucleus at the intersection of the "T."

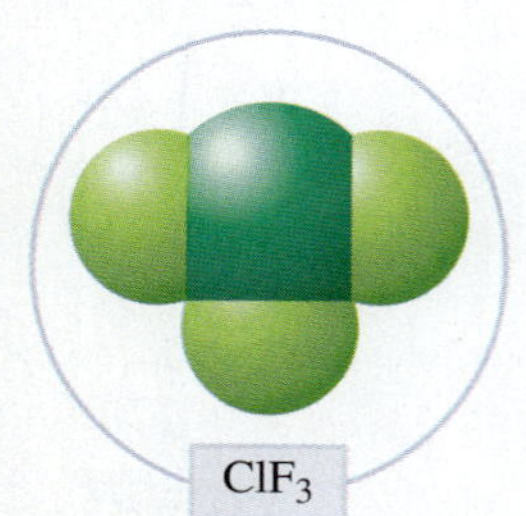

Figure 10.11 ▲

The chlorine trifluoride molecule Molecular model.

Xenon difluoride, XeF_2, has the electron-dot formula

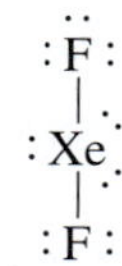

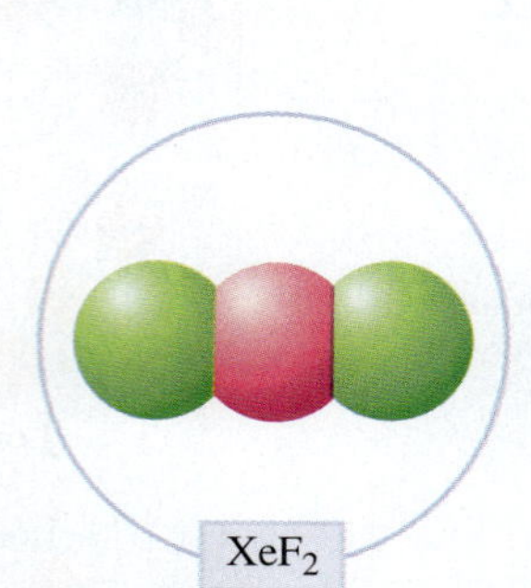

Figure 10.12 ▲

The xenon difluoride molecule Molecular model.

The three lone pairs on xenon occupy the equatorial positions of the trigonal bipyramidal arrangement, giving a *linear geometry* (Figure 10.12; see also Figure 10.8).

Six Electron Pairs (Octahedral Arrangement) There are six electron pairs (all bonding pairs) about sulfur in sulfur hexafluoride, SF_6. Thus, it has an *octahedral geometry,* with sulfur at the center of the octahedron and fluorine atoms at the vertexes (Figure 10.8).

Iodine pentafluoride, IF_5, has the electron-dot formula

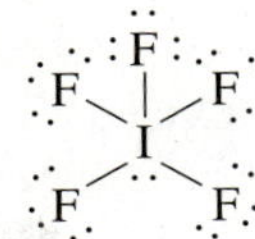

The lone pair on iodine occupies one of the six equivalent positions in the octahedral arrangement, giving a *square pyramidal geometry* (Figure 10.13; see also Figure 10.8). The name derives from the shape formed by drawing lines between atoms.

Xenon tetrafluoride, XeF_4, has the electron-dot formula

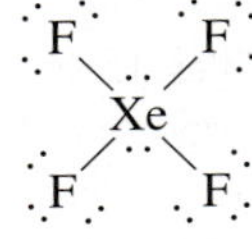

The two lone pairs on xenon occupy opposing positions in the octahedral arrangement to minimize their repulsion. The result is a *square planar geometry* (Figure 10.14; see also Figure 10.8).

Figure 10.13 ▲
The iodine pentafluoride molecule Molecular model.

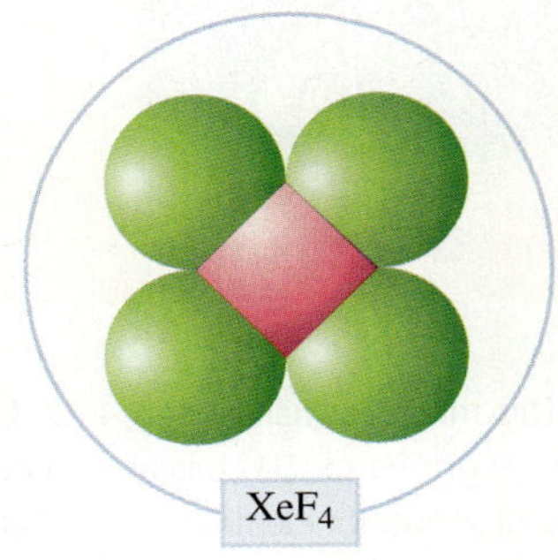

Figure 10.14 ▲
The xenon tetrafluoride molecule Molecular model.

OWL Interactive Example 10.2 Predicting Molecular Geometries (Five or Six Electron Pairs)

Gaining Mastery Toolbox

Critical Concept 10.2
The arrangement of electron groups about a central atom depends on the number of electron groups: five electron groups have a trigonal bipyramidal arrangement and six electron groups an octahedral arrangement.

Solution Essentials:
- Molecular geometries for a given arrangement of electrons
- Arrangement of electron groups about an atom
- Electron-dot formulas

What do you expect for the geometry of tellurium tetrachloride, $TeCl_4$?

Problem Strategy After drawing the Lewis formula for a molecule, determine the number of electron groups about the central atom. This number determines the arrangement of the electron groups (Figure 10.2). The geometry is determined by the arrangement of bonds to the central atom (Figure 10.8).

Solution First you distribute the valence electrons to the Cl atoms to satisfy the octet rule. Then you allocate the remaining valence electrons to the central atom, Te (following the steps outlined in Section 9.6). The electron-dot formula is

:Cl:
:Cl — Te: — :Cl:
:Cl:

There are five electron pairs in the valence shell of Te in $TeCl_4$. Of these, four are bonding pairs and one is a lone pair. The arrangement of electron pairs is trigonal bipyramidal. You expect the lone pair to occupy an equatorial position, so $TeCl_4$ has a **seesaw** molecular geometry. (See Figure 10.8; 4 bonding pairs, 1 lone pair.)

Answer Check Again, be careful to distinguish the name for the arrangement of valence electron pairs from the name for the molecular geometry.

Exercise 10.2 According to the VSEPR model, what molecular geometry would you predict for iodine trichloride, ICl_3?

■ See Problems 10.39, 10.40, 10.41, and 10.42.

Applying the VSEPR Model to Larger Molecules

In the preceding discussion, we looked at the geometry of molecules having a single central atom with other atoms or groups attached to that central atom. We can also apply what we have learned there to more complicated molecules. What we need to

A **CHEMIST** Looks at...

Left-Handed and Right-Handed Molecules

This morning, half asleep, I (D. E.) tried to put my left glove on my right hand. Silly mistake, of course. A person's two hands are similar but not identical. If I look at my hands, say, with both palms toward me, the thumbs point in opposite directions. But, if I now hold the palm of my right hand toward a mirror and compare that image with my actual left hand, I see that the thumbs point in the same direction. A person's left hand looks like the mirror image of his or her right hand, and vice versa. You can mentally superimpose one hand onto the mirror image of the other hand. Yet the two hands are themselves nonsuperimposable. Many molecules have this "handedness" property too, and this possibility gives rise to a subtle kind of isomerism.

The presence of handedness in molecules depends on the fact that the atoms in the molecules occupy specific places in three-dimensional space. In most organic molecules, this property depends on the tetrahedral bonding of the carbon atom. Consider a carbon atom bonded to four different kinds of groups, say, —H, —OH, —CH_3, and —COOH. (The resulting molecule is called lactic acid.) Two isomers are possible in which one isomer is the mirror image of the other, but the two isomers are themselves not identical (they are nonsuperimposable; see Figure 10.15). However, all four groups bonded to this carbon atom must be different to have this isomerism. If you replace the —OH group of lactic acid by, say, an H atom, the molecule and its mirror image become identical. There are no isomers; there is just one kind of molecule.

The two lactic acid isomers shown in Figure 10.15 are labeled D-lactic acid and L-lactic acid (D is for *dextro*, meaning right; L is for *levo*, meaning left). They might be expected to have quite similar properties, and they do. They have identical melting points, boiling points, and solubilities, for instance. And if we try to prepare lactic acid in the laboratory, we almost always get an equal molar mixture of the D and L isomers. But the biological origins of the molecules are different. Only D-lactic acid occurs in sour milk, whereas only L-lactic acid forms in muscle tissue after exercising.

Biological molecules frequently have the handedness character. The chemical substance responsible for the flavor of spearmint is L-carvone. The substance responsible for the flavor of caraway seeds, which you see in some rye bread, is D-carvone (see Figure 10.16). Here are two substances whose molecules differ only in being mirror images of one another. Yet, they have strikingly different flavors: one is minty; the other is pungent aromatic.

do is to imagine a large molecule broken into pieces to which we can apply the VSEPR model. As an example, let's consider the lactic acid molecule.

Lactic acid has the following electron-dot formula:

```
      H    H   :O:
      |    |    ||     ..
   H—C————C————C————O—H
      |    |           ..
      H   :O:
           |
           H
```

We have divided this formula into several "central atoms," each surrounded by a number of bonding or lone pairs of electrons. Note that the carbon atom at the far left is surrounded by bonds to three hydrogen atoms plus a bond to another carbon atom. These four bonds would be expected to have a tetrahedral arrangement about this carbon atom. Similarly, the next carbon atom is surrounded by four bonds and would also be expected to have a tetrahedral arrangement of bonds about it. The carbon atom on the right, though, has a single bond to the carbon atom on its left and a single bond to the oxygen atom on its right plus a double bond to the oxygen atom above it. We would expect these three groups of electrons to have a trigonal

Figure 10.15 ▶

Isomers of lactic acid Note that the two molecules labeled D-lactic acid and L-lactic acid are mirror images and cannot be superimposed on one another. See Figure 10.4 for an explanation of the notation used to represent the tetrahedral geometry.

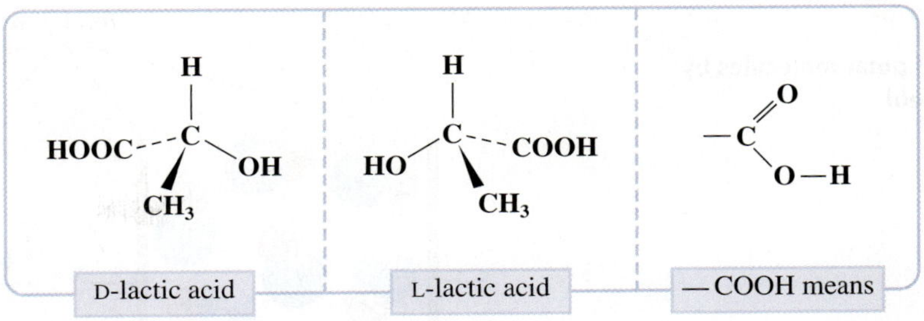

D-lactic acid

L-lactic acid

—COOH means

D-Carvone

caraway seeds

spearmint

L-Carvone

© Cengage Learning

Figure 10.16 ▲

Spearmint and caraway
Spearmint leaves contain L-carvone; caraway seeds contain D-carvone.

■ See Problems 10.83 and 10.84.

planar arrangement about that carbon atom. Then, there are two —O—H groups, with each oxygen atom having a tetrahedral arrangement of electron pairs and a bent or angular geometry of bonds. Here is a molecular model of lactic acid.

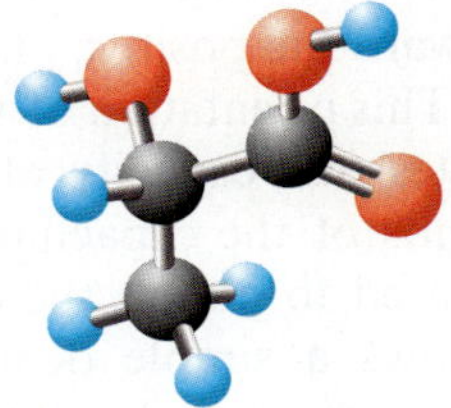

You might find it instructive to use a molecular model kit to build this structure. The essay on the preceding page describes a subtle type of isomerism that arises in lactic acid.

10.2 Dipole Moment and Molecular Geometry

The VSEPR model provides a simple procedure for predicting the geometry of a molecule. However, predictions must be verified by experiment. Information about the geometry of a molecule can sometimes be obtained from an experimental quantity called the dipole moment, which is related to the polarity of the bonds in a molecule.

Figure 10.17 ▶

Alignment of polar molecules by an electric field

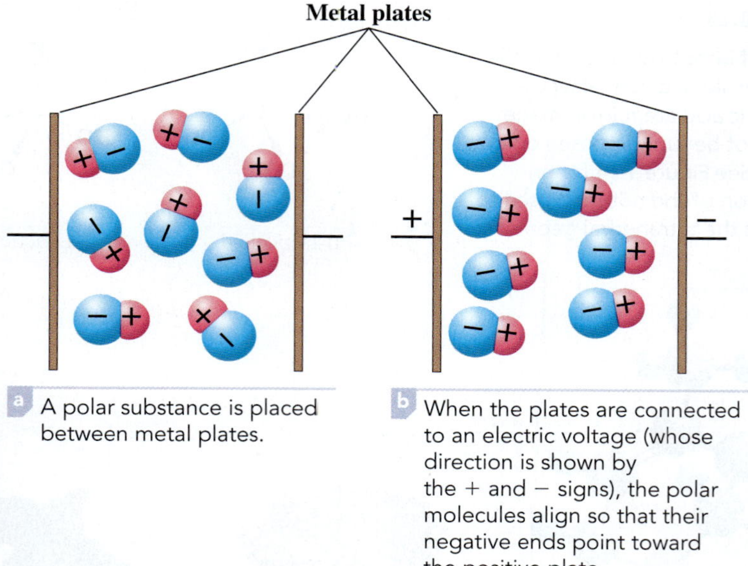

a A polar substance is placed between metal plates.

b When the plates are connected to an electric voltage (whose direction is shown by the + and − signs), the polar molecules align so that their negative ends point toward the positive plate.

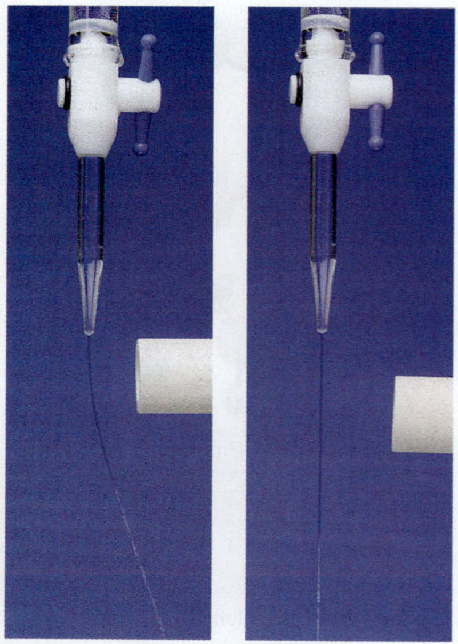

left Water is a polar liquid; it is attracted to the electrically charged rod. The capacitance of the polar liquid (due to the alignment of polar molecules) stabilizes a charge separation induced in the column of liquid surface near the rod, resulting in a net attraction.

right Carbon tetrachloride, CCl_4, is a nonpolar liquid; it is not attracted to the glass rod.

Figure 10.18 ▲

Attraction of a polar liquid to an electrified rod

The **dipole moment** is *a quantitative measure of the degree of charge separation in a molecule and is therefore an indicator of the polarity of the molecule.* In general, a positive charge q and a negative charge $-q$, separated by a distance d has a dipole moment, μ, equal to

$$\mu = q \times d$$

Consider the bond in the H—Cl molecule. Because the Cl atom is more electronegative than the H atom, the electrons of the bond are pulled toward it, giving this atom a partial negative charge, which we indicate as δ^-. The H atom then has an equal but positive charge, δ^+:

$$\overset{\delta^+}{\text{H}}\text{—}\overset{\delta^-}{\text{Cl}}$$

So, the dipole moment in HCl is δd. Dipole moments are usually measured in units of *debyes* (D); the dipole moment of HCl is 1.08 D. In SI units, dipole moments are measured in coulomb-meters (C·m), and 1 D = 3.34×10^{-30} C·m.

Measurements of dipole moments are based on the fact that polar molecules (molecules having a dipole moment) can be oriented by an electric field. Figure 10.17 shows an electric field generated by charged plates. Note that the polar molecules tend to align themselves so that the negative ends of the molecules point toward the positive plate, and the positive ends point toward the negative plate. This orientation of the molecules affects the *capacitance* of the charged plates (the capacity of the plates to hold a charge). Consequently, measurements of the capacitance of plates separated by different substances can be used to obtain the dipole moments of those substances. Figure 10.18 shows a simple demonstration that distinguishes between a polar liquid (one whose molecules have a dipole moment) and a nonpolar liquid. The polar liquid, but not the nonpolar one, is attracted to a charged rod.

You can sometimes relate the presence or absence of a dipole moment in a molecule to its molecular geometry. For example, consider the carbon dioxide molecule. Each carbon–oxygen bond has a polarity in which the more electronegative oxygen atom has a partial negative charge.

$$\overset{\delta^-}{\text{O}}\text{=}\overset{2\delta^+}{\text{C}}\text{=}\overset{\delta^-}{\text{O}}$$

We will denote the dipole-moment contribution from each bond (the bond dipole) by an arrow with a positive sign at one end (⟷). The dipole-moment arrow points from the positive partial charge toward the negative partial charge. Thus, you can rewrite the formula for carbon dioxide as

$$\overset{\leftrightarrow}{O{=}}\overset{\leftrightarrow}{C{=}}O$$

Each bond dipole, like a force, is a *vector* quantity; that is, it has both magnitude and direction. Like forces, two bond dipoles of equal magnitude but opposite direction cancel each other. (Think of two groups of people in a tug of war. As long as each group pulls on the rope with the same force but in the opposite direction, there is no movement—the net force is zero.) Because the two carbon–oxygen bond dipoles in CO_2 are equal but point in opposite directions, they give a net dipole moment of zero for the molecule.

For comparison, consider the water molecule. The bond dipoles point from the hydrogen atoms toward the more electronegative oxygen.

H—O—H (bent, with bond dipole arrows pointing from each H toward O)

Here, the two bond dipoles do not point directly toward or away from each other. As a result, they add together to give a nonzero dipole moment for the molecule.

H—O—H (bent, with net dipole arrow pointing toward O)

The dipole moment of H_2O has been observed to be 1.94 D.

The fact that the water molecule has a dipole moment is excellent experimental evidence for a bent geometry. If the H_2O molecule were linear, the dipole moment would be zero.

The analysis we have just made for two different geometries of AX_2 molecules can be extended to other AX_n molecules (in which all X atoms are identical). Table 10.1 summarizes the relationship between molecular geometry and dipole moment. Those geometries in which A—X bonds are directed symmetrically about the central atom (for example, linear, trigonal planar, and tetrahedral) give molecules of zero dipole moment; that is, the molecules are *nonpolar.* Those geometries in which the X atoms tend to be on one side of the molecule (for example, bent and trigonal pyramidal) can have nonzero dipole moments; that is, they can give *polar* molecules.

Table 10.1 Relationship Between Molecular Geometry and Dipole Moment

Formula	Molecular Geometry	Dipole Moment*
AX	Linear	Can be nonzero
AX_2	Linear	Zero
	Bent	Can be nonzero
AX_3	Trigonal planar	Zero
	Trigonal pyramidal	Can be nonzero
	T-shaped	Can be nonzero
AX_4	Tetrahedral	Zero
	Square planar	Zero
	Seesaw	Can be nonzero
AX_5	Trigonal bipyramidal	Zero
	Square pyramidal	Can be nonzero
AX_6	Octahedral	Zero

*All X atoms are assumed to be identical.

Example 10.3 Relating Dipole Moment and Molecular Geometry

Gaining Mastery Toolbox

Critical Concept 10.3
The dipole moment is a vector quantity; it is possible for similar bond dipoles that are arranged symmetrically to cancel one another in order to give an overall zero dipole moment.

Solution Essential:
- Relationship between molecular geometry and dipole moment

Each of the following molecules has a nonzero dipole moment. Select the molecular geometry that is consistent with this information. Explain your reasoning.

a SO_2 linear, bent

b PH_3 trigonal planar, trigonal pyramidal

Problem Strategy Because the dipole moment is a *vector* quantity, it can happen that nonzero bond dipoles cancel one another to give a molecule with zero dipole moment. Look for bonds that are symmetrically arranged to oppose one another (think of a tug of war between bonds). Table 10.1 summarizes the relationship between geometry and dipole moment.

Solution a In the linear geometry, the S—O bond contributions to the dipole moment would cancel, giving a zero dipole moment. That would not happen in the **bent** geometry; hence, this must be the geometry for the SO_2 molecule. b In the trigonal planar geometry, the bond contributions to the dipole moment would cancel, giving a zero dipole moment. That would not occur in the **trigonal pyramidal** geometry; hence, this is a possible molecular geometry for PH_3.

Answer Check Make sure that you have the correct geometry for the molecule, and note how the bonds either oppose one another, to give zero dipole moment, or else do not oppose and so give a dipole moment.

Exercise 10.3 Bromine trifluoride, BrF_3, has a nonzero dipole moment. Indicate which of the following geometries are consistent with this information: a trigonal planar; b trigonal pyramidal; c T-shaped.

■ See Problems 10.43 and 10.44.

Exercise 10.4 Which of the following would be expected to have a dipole moment of zero on the basis of symmetry? Explain. a $SOCl_2$; b SiF_4; c OF_2.

■ See Problems 10.45 and 10.46.

In Example 10.3, the fact that certain molecular geometries necessarily imply a zero dipole moment allowed us to eliminate these geometries when considering a molecule with a nonzero dipole moment. The reverse argument does not follow, however. For example, suppose a molecule of the type AX_3 is found to have no measurable dipole moment. It may be that the geometry is trigonal planar. Another possibility, however, is that the geometry is trigonal pyramidal or T-shaped, but the lone pairs on the central atom offset the polarity of the bonds.

You can see this effect of lone pairs on the dipole moment of nitrogen trifluoride, NF_3. Judging by the electronegativity difference, you would expect each N—F bond to be quite polar. Yet the dipole moment of nitrogen trifluoride is only 0.2 D. (By contrast, ammonia has a dipole moment of 1.47 D.) The explanation for the small dipole moment in NF_3 appears in Figure 10.19. The lone pair on nitrogen has a dipole-moment contribution that is directed outward from the nucleus, because the electrons are offset from the nuclear center. This dipole-moment contribution thus opposes the N—F bond moments. ◀

Can you explain why NH_3 has such a large dipole moment compared with NF_3?

Figure 10.19 ▶

Explanation of the small dipole moment of NF_3

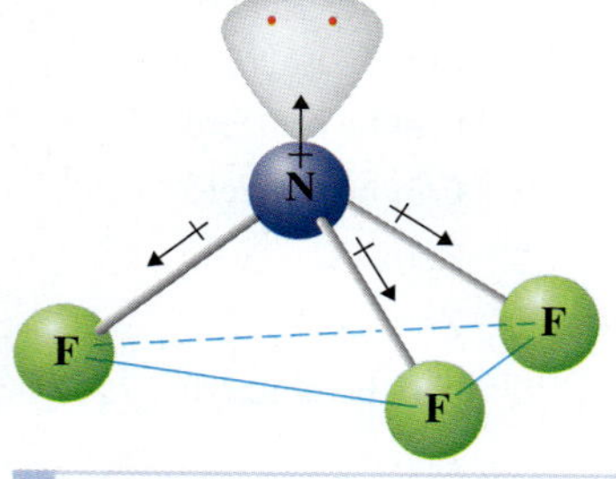

a The lone-pair and bond contributions to the dipole moment of NF_3 tend to offset one another.

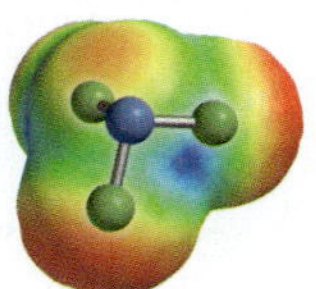

b The electrostatic-potential model of NF_3 shows the electron density, which increases as the color changes from blue to yellow to red.

Effect of Polarity on Molecular Properties

Whether a molecule has a dipole moment (that is, whether it is polar or nonpolar) can affect its properties. We will explore this issue in detail in the next chapter (Section 11.5). Briefly, though, we can see why *cis*- and *trans*-1,2-dichloroethene, mentioned in the chapter opening, have different boiling points. The C—Cl bonds are quite polar, because of the relatively large electronegativity difference between C and Cl atoms (0.9); the H—C bonds have moderate polarity (electronegativity difference of 0.4). The bond dipoles (light arrows) add together in *cis*-1,2-dichloroethene to give a polar molecule (the total dipole moment is indicated by the heavy arrow). (Note that the electronegative Cl atoms are on the same side in the *cis* molecule.) In the *trans* molecule, the bond dipoles subtract to give a nonpolar molecule.

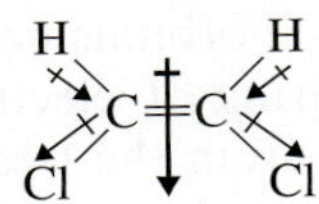

cis-1,2-Dichloroethene

trans-1,2-Dichloroethene

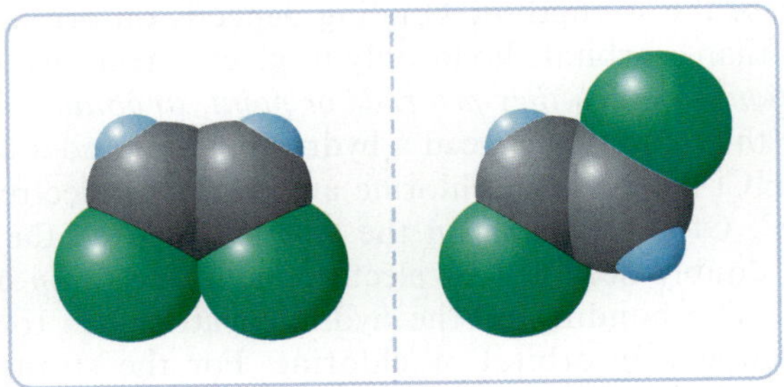

In a liquid, polar molecules tend to orient so that positive ends of molecules are attracted to negative ends. This results in an attractive force between molecules (an *intermolecular* force), which increases the energy required for the liquid to boil, and therefore increases the boiling point. Recall that the *trans* compound boils at 48°C, whereas the *cis* compound boils at a higher temperature, 60°C.

CONCEPT CHECK 10.2

Two molecules, each with the general formula AX_3, have different dipole moments. Molecule Y has a dipole moment of zero, whereas molecule Z has a nonzero dipole moment. From this information, what can you say about the geometries of Y and Z?

10.3 Valence Bond Theory

The VSEPR model is usually a satisfactory method for predicting molecular geometries. To understand bonding and electronic structure, however, you must look to quantum mechanics. We will consider two theories stemming from quantum mechanics: valence bond theory and molecular orbital theory. Both use the methods of quantum mechanics but make different simplifying assumptions. In this section, we will look in a qualitative way at the basic ideas involved in **valence bond theory,** *an approximate theory to explain the electron pair or covalent bond by quantum mechanics.*

Basic Theory

According to valence bond theory, a bond forms between two atoms when the following conditions are met:

1. An orbital on one atom comes to occupy a portion of the same region of space as an orbital on the other atom. The two orbitals are said to *overlap*.
2. The total number of electrons in both orbitals is no more than two.

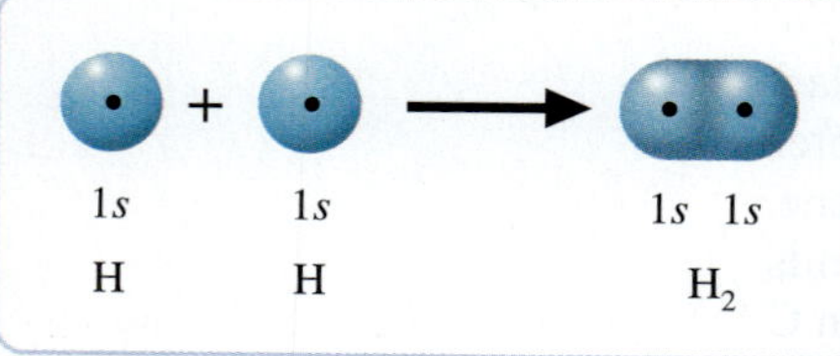

Figure 10.20 ▲

Formation of H_2 The H—H bond forms when the 1*s* orbitals, one from each atom, overlap.

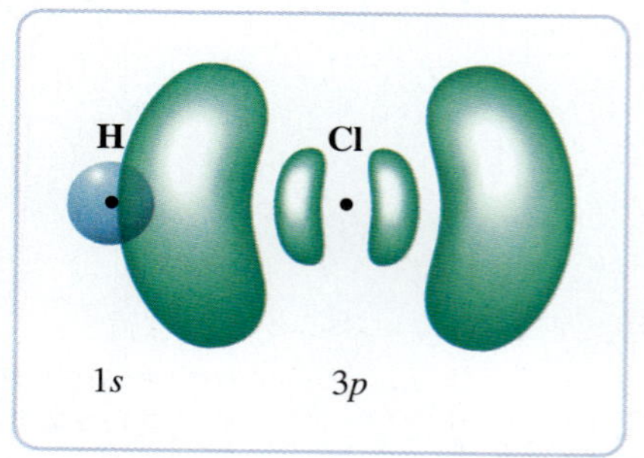

Figure 10.21 ▲

Bonding in HCl The bond forms by the overlap of a hydrogen 1*s* orbital (blue) along the axis of a chlorine 3*p* orbital (green).

As the orbital of one atom overlaps the orbital of another, the electrons in the orbitals begin to move about both atoms. Because the electrons are attracted to both nuclei at once, they pull the atoms together. Strength of bonding depends on the amount of overlap; the greater the overlap, the greater the bond strength. The two orbitals cannot contain more than two electrons because a given region of space can hold only two electrons with opposite spin.

For example, consider the formation of the H_2 molecule from atoms. Each atom has the electron configuration $1s^1$. As the H atoms approach each other, their 1*s* orbitals begin to overlap and a covalent bond forms (Figure 10.20). Valence bond theory also explains why two He atoms (each with electron configuration $1s^2$) do not bond. Suppose two He atoms approach one another and their 1*s* orbitals begin to overlap. Each orbital is doubly occupied, so the sharing of electrons between atoms would place the four valence electrons from the two atoms in the same region. This, of course, could not happen. As the orbitals begin to overlap, each electron pair strongly repels the other. The atoms come together, then fly apart.

Because the strength of bonding depends on orbital overlap, orbitals other than *s* orbitals bond only in given directions. *Orbitals bond in the directions in which they protrude or point, to obtain maximum overlap.* Consider the bonding between a hydrogen atom and a chlorine atom to give the HCl molecule. A chlorine atom has the electron configuration $[Ne]3s^23p^5$. Of the orbitals in the valence shell of the chlorine atom, three are doubly occupied by electrons and one (a 3*p* orbital) is singly occupied. The bonding of the hydrogen atom has to occur with the singly occupied 3*p* orbital of chlorine. For the strongest bonding to occur, the maximum overlap of orbitals is required. The 1*s* orbital of hydrogen must overlap along the axis of the singly occupied 3*p* orbital of chlorine (see Figure 10.21).

Hybrid Orbitals

From what has been said, you might expect the number of bonds formed by a given atom to equal the number of unpaired electrons in its valence shell. Chlorine, whose orbital diagram is

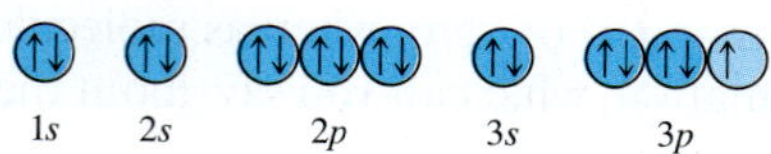

has one unpaired electron and forms one bond. Oxygen, whose orbital diagram is

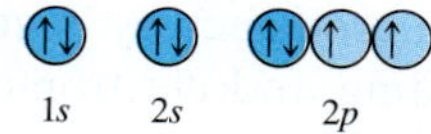

has two unpaired electrons and forms two bonds, as in H_2O.

However, consider the carbon atom, whose orbital diagram is

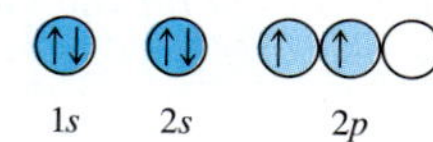

You might expect this atom to bond to two hydrogen atoms to form the CH_2 molecule. Although this molecule is known to be present momentarily during some reactions, it is very reactive and cannot be isolated. But methane, CH_4, in which the carbon atom bonds to four hydrogen atoms, is well known. In fact, a carbon atom usually forms four bonds.

You might explain this as follows: Four unpaired electrons are formed when an electron from the 2s orbital of the carbon atom is *promoted* (excited) to the vacant 2p orbital.

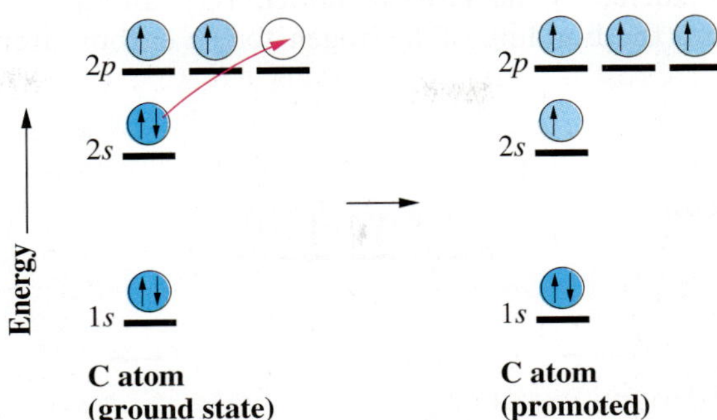

It would require energy to promote the carbon atom this way, but more than enough energy would be obtained from the formation of two additional covalent bonds. One bond would form from the overlap of the carbon 2s orbital with a hydrogen 1s orbital. Each of the other three bonds would form from a carbon 2p orbital and a hydrogen 1s orbital.

Experiment shows, however, that the four C—H bonds in methane are identical. ▶ This implies that the carbon orbitals involved in bonding are also equivalent. For this reason, valence bond theory assumes that the four valence orbitals of the carbon atom combine during the bonding process to form four new, but equivalent, hybrid orbitals. **Hybrid orbitals** are *orbitals used to describe bonding that are obtained by taking combinations of atomic orbitals of the isolated atoms.* In this case, a set of hybrid orbitals is constructed from one s orbital and three p orbitals, so they are called sp^3 hybrid orbitals. Calculations from theory show that each sp^3 hybrid orbital has a large lobe pointing in one direction and a small lobe pointing in the opposite direction. The four sp^3 hybrid orbitals point in tetrahedral directions. Figure 10.22a shows the shape of a single sp^3 orbital; Figure 10.22b shows a stylized set of four orbitals pointing in tetrahedral directions.

Nuclear magnetic resonance (page 302) and infrared spectroscopy (page 370) both show that CH_4 has four equivalent C—H bonds.

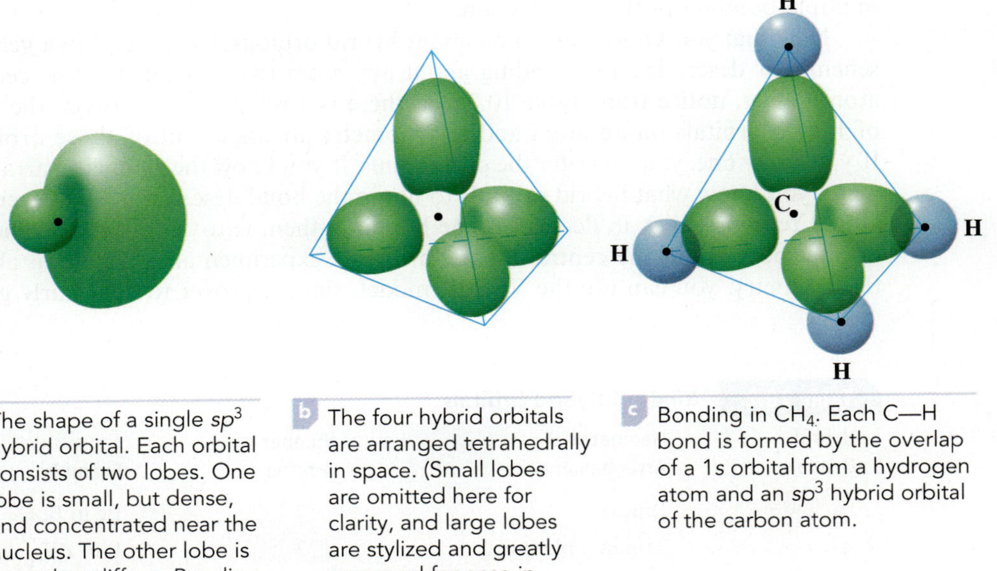

Figure 10.22 ◀ **Spatial arrangement of sp^3 hybrid orbitals**

a The shape of a single sp^3 hybrid orbital. Each orbital consists of two lobes. One lobe is small, but dense, and concentrated near the nucleus. The other lobe is large, but diffuse. Bonding occurs with the large lobe, since it extends farther from the nucleus.

b The four hybrid orbitals are arranged tetrahedrally in space. (Small lobes are omitted here for clarity, and large lobes are stylized and greatly narrowed for ease in depicting the directional bonding)

c Bonding in CH_4. Each C—H bond is formed by the overlap of a 1s orbital from a hydrogen atom and an sp^3 hybrid orbital of the carbon atom.

The C—H bonds in methane, CH_4, are described by valence bond theory as the overlapping of each sp^3 hybrid orbital of the carbon atom with 1*s* orbitals of hydrogen atoms (see Figure 10.22*C*). Thus, the bonds are arranged tetrahedrally, which is predicted by the VSEPR model. You can represent the hybridization of carbon and the bonding of hydrogen to the carbon atom in methane as follows:

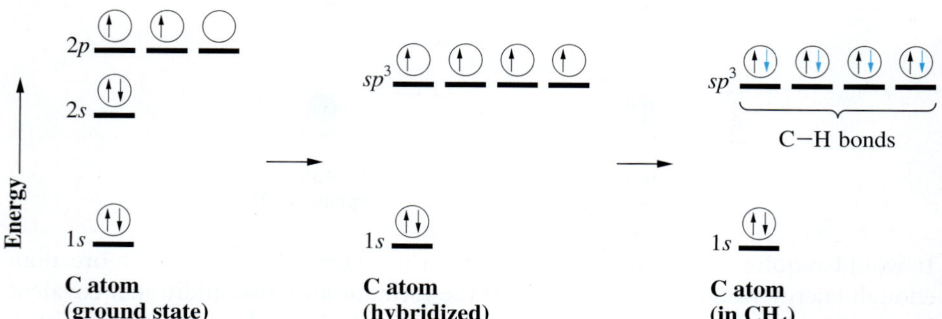

Here the blue arrows represent electrons originally belonging to hydrogen atoms.

Hybrid orbitals can be formed from various numbers of atomic orbitals. *The number of hybrid orbitals formed always equals the number of atomic orbitals used.* For example, if you combine an *s* orbital and two *p* orbitals to get a set of equivalent orbitals, you get three hybrid orbitals (called sp^2 hybrid orbitals). A set of hybrid orbitals always has definite directional characteristics. Here, all three sp^2 hybrid orbitals lie in a plane and are directed at 120° angles to one another; that is, they have a trigonal planar arrangement. Some possible hybrid orbitals and their geometric arrangements are listed in Table 10.2 and shown in Figure 10.23. Note that the geometric arrangements of hybrid orbitals are the same as those for electron pairs in the VSEPR model.

Only two of the three *p* orbitals are used to form sp^2 hybrid orbitals. The unhybridized *p* orbital is perpendicular to the plane of the sp^2 hybrid orbitals. Similarly, only one of the three *p* orbitals is used to form *sp* hybrid orbitals. The two unhybridized *p* orbitals are perpendicular to the axis of the *sp* hybrid orbitals and perpendicular to each other. We will use these facts when we discuss multiple bonding in the next section.

Now that you know something about hybrid orbitals, let us develop a general scheme for describing the bonding about any atom (we will call this the central atom). First, notice from Table 10.2 that there is a relationship between the type of hybrid orbitals on an atom and the geometric arrangement of those orbitals. If you know one, you can infer the other. Thus, if you know the geometric arrangement, you know what hybrid orbitals to use in the bond description of the central atom. Your first task in describing the bonding, then, is to obtain the geometric arrangement about the central atom. In lieu of experimental information about the geometry, you can use the VSEPR model, since it proves to be a fairly good

Table 10.2 Kinds of Hybrid Orbitals

Hybrid Orbitals	Geometric Arrangement	Number of Orbitals	Example
sp	Linear	2	Be in BeF_2
sp^2	Trigonal planar	3	B in BF_3
sp^3	Tetrahedral	4	C in CH_4
sp^3d	Trigonal bipyramidal	5	P in PCl_5
sp^3d^2	Octahedral	6	S in SF_6

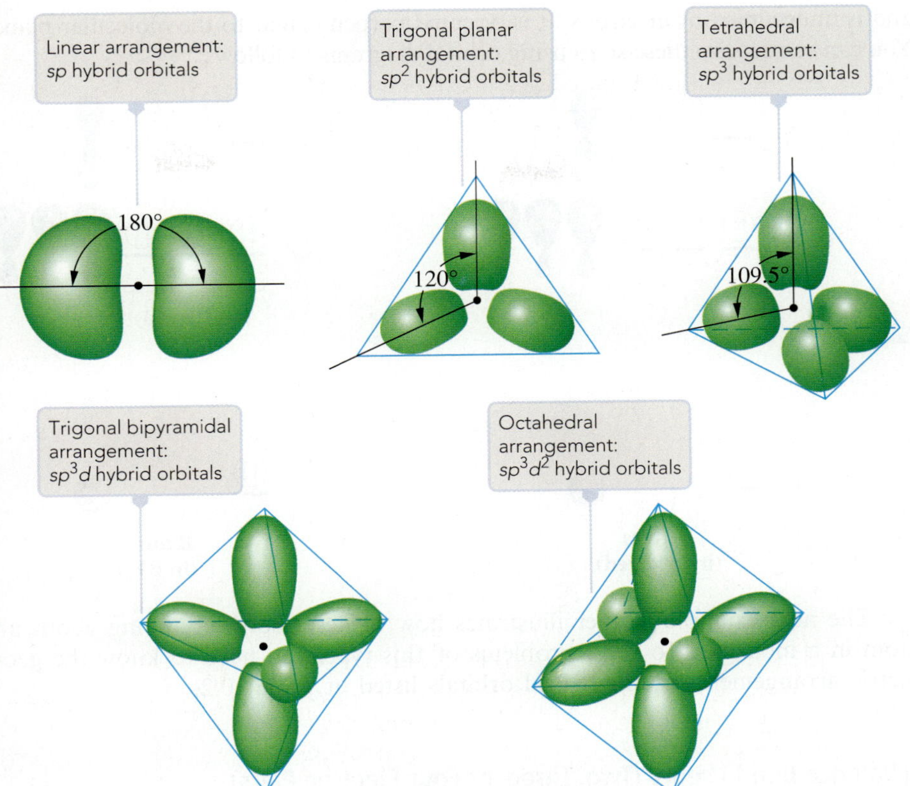

Figure 10.23

Diagrams of hybrid orbitals showing their spatial arrangements Each lobe shown is one hybrid orbital (small lobes are omitted for clarity).

predictor of molecular geometry. To obtain the bonding description about any atom in a molecule, you proceed as follows:

Step 1: Write the Lewis electron-dot formula of the molecule.

Step 2: From the Lewis formula, use the VSEPR model to obtain the arrangement of electron pairs about this atom.

Step 3: From the geometric arrangement of the electron pairs, deduce the type of hybrid orbitals on this atom required for the bonding description (see Table 10.2).

Step 4: Assign valence electrons to the hybrid orbitals of this atom one at a time, pairing them only when necessary.

Step 5: Form bonds to this atom by overlapping singly occupied orbitals of other atoms with the singly occupied hybrid orbitals of this atom.

As an application of this scheme, let us look at the BF_3 molecule and obtain the bond description of the boron atom. Following Step 1, you write the Lewis formula of BF_3:

```
 ..
:F:
 |  ..
 B—F:
 |  ..
:F:
 ..
```

Now you apply the VSEPR model to the boron atom (Step 2). There are three electron pairs about the boron atom, so they are expected to have a planar trigonal arrangement. Looking at Table 10.2 (Step 3), you note that three sp^2 hybrid orbitals have a trigonal planar arrangement. In Step 4, you assign the valence electrons of the boron atom to the hybrid orbitals. Finally, in Step 5, you imagine three fluorine atoms approaching the boron atom. The singly occupied $2p$ orbital on a fluorine atom overlaps one of the sp^2 hybrid orbitals on boron, forming a covalent bond. Three such B—F bonds form. Note that one of the $2p$ orbitals of boron remains unhybridized and is unoccupied by electrons. It is oriented perpendicular to the molecular plane. You can summarize these steps using orbital diagrams as follows:

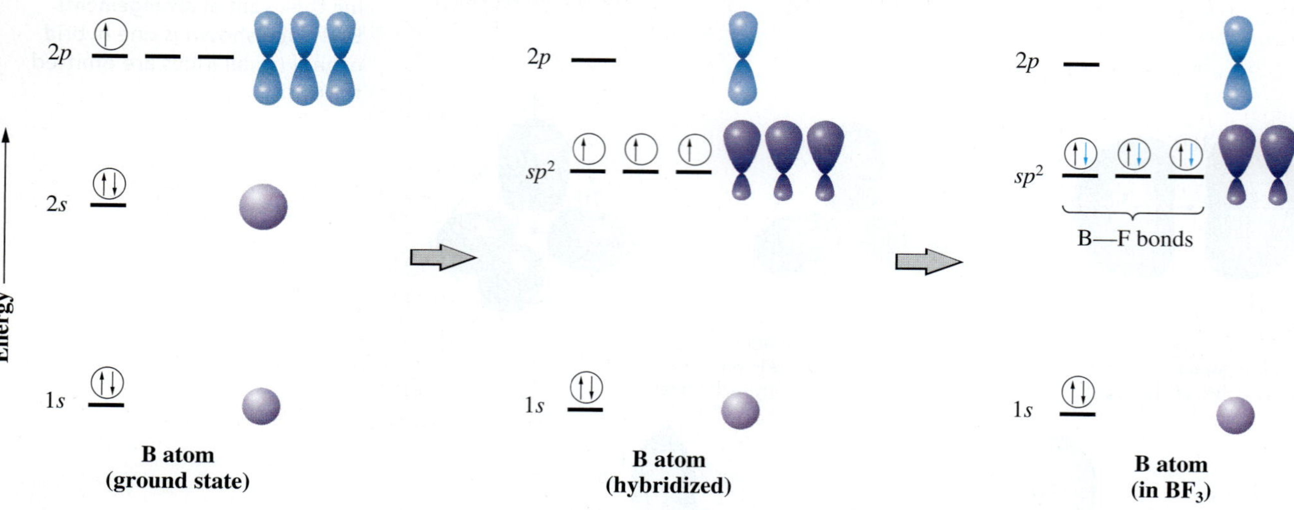

The next example further illustrates how to describe the bonding about an atom in a molecule. To solve problems of this type, you need to know the geometric arrangements of the hybrid orbitals listed in Table 10.2.

Example 10.4 Applying Valence Bond Theory (Two, Three, or Four Electron Pairs)

Gaining Mastery Toolbox

Critical Concept 10.4
Using the VSEPR model, determine the geometry about the central atom and then the hybrid orbitals of the atom. Now form bonds by overlapping these hybrid orbitals with orbitals from bonding atoms; use nonbonded hybrid orbitals for the lone pairs.

Solution Essentials:
- Bond formation by orbital overlap
- Kinds of hybrid orbitals
- VSEPR model
- Electron-dot formulas

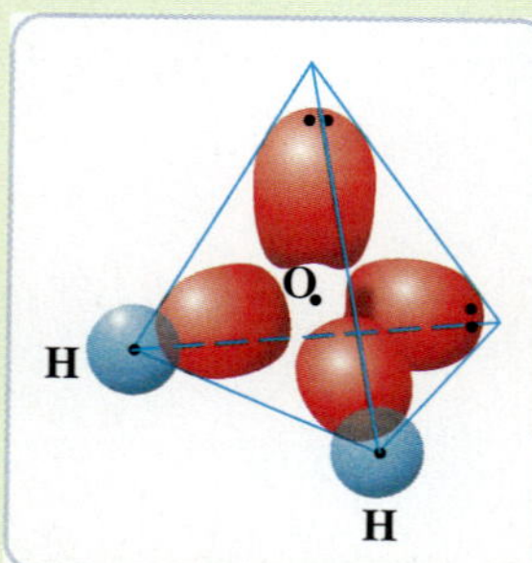

Figure 10.24 ▲

Bonding in H_2O Orbitals on oxygen are sp^3 hybridized; bonding is tetrahedral.

Describe the bonding in H_2O according to valence bond theory. Assume that the molecular geometry is the same as given by the VSEPR model.

Problem Strategy From the Lewis formula for a molecule, determine its geometry about the central atom using the VSEPR model. From this geometry, determine the hybrid orbitals on this atom, assigning its valence electrons to these orbitals one at a time. Finally, form bonds by overlapping these hybrid orbitals with orbitals of the other atoms.

Solution The Lewis formula for H_2O is

$$\mathrm{H-\ddot{O}:}\\ \quad\ \ |\\ \quad\ \ \mathrm{H}$$

Note that there are four pairs of electrons about the oxygen atom. According to the VSEPR model, these are directed tetrahedrally, and from Table 10.2 you see that you should use sp^3 hybrid orbitals. Each O—H bond is formed by the overlap of a 1*s* orbital of a hydrogen atom with one of the singly occupied sp^3 hybrid orbitals of the oxygen atom. You can represent the bonding to the oxygen atom in H_2O as follows (see also Figure 10.24):

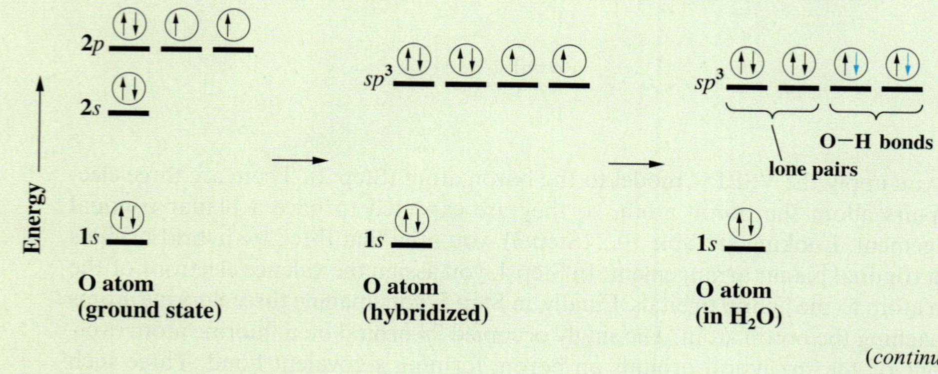

(continued)

(continued)

Answer Check Check that you have the correct Lewis formula for the molecule and that you have the correct geometry about the central atom and the correct hybrid orbitals (Table 10.2).

Exercise 10.5 Using hybrid orbitals, describe the bonding in NH_3 according to valence bond theory.

■ See Problems 10.47, 10.48, 10.49, and 10.50.

The actual angle between O—H bonds in H_2O has been experimentally determined to be 104.5°. Because this is close to the tetrahedral angle (109.5°), the description of bonding given in Example 10.4 is essentially correct. To be more precise, you should note that the hybrid orbitals for bonds and lone pairs are not exactly equivalent. The lone pairs are somewhat larger than bonding pairs. Because they take up more space, the lone pairs push the bonding pairs closer together than they are in the exact tetrahedral case.

As the next example illustrates, hybrid orbitals are also useful for describing bonding when the central atom of a molecule is surrounded by more than eight valence electrons.

Example 10.5 Applying Valence Bond Theory (Five or Six Electron Pairs)

Gaining Mastery Toolbox

Critical Concept 10.5
Using the VSEPR model, determine the geometry about the central atom and then the hybrid orbitals of the atom. Now form bonds by overlapping these hybrid orbitals with orbitals from bonding atoms; use nonbonded hybrid orbitals for the lone pairs.

Solution Essentials:
- Bond formation by orbital overlap
- Kinds of hybrid orbitals
- VSEPR model
- Electron-dot formula

Describe the bonding in XeF_4 using hybrid orbitals.

Problem Strategy From the Lewis formula for a molecule, determine its geometry about the central atom using the VSEPR model. From this geometry, determine the hybrid orbitals on this atom, assigning its valence electrons to these orbitals one at a time. Finally, form bonds by overlapping these hybrid orbitals with orbitals of the other atoms.

Solution The Lewis formula of XeF_4 is

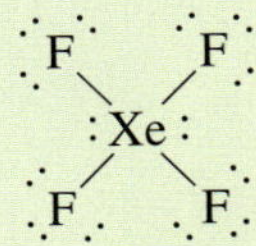

The xenon atom has four single bonds and two lone pairs. It will require six orbitals to describe the bonding. This suggests that you use sp^3d^2 hybrid orbitals on xenon (according to Table 10.2). Each fluorine atom (valence-shell configuration $2s^22p^5$) has one singly occupied orbital, so you assume that this orbital is used in bonding. Each Xe—F bond is formed by the overlap of a xenon sp^3d^2 hybrid orbital with a singly occupied fluorine $2p$ orbital. You can summarize this as follows:

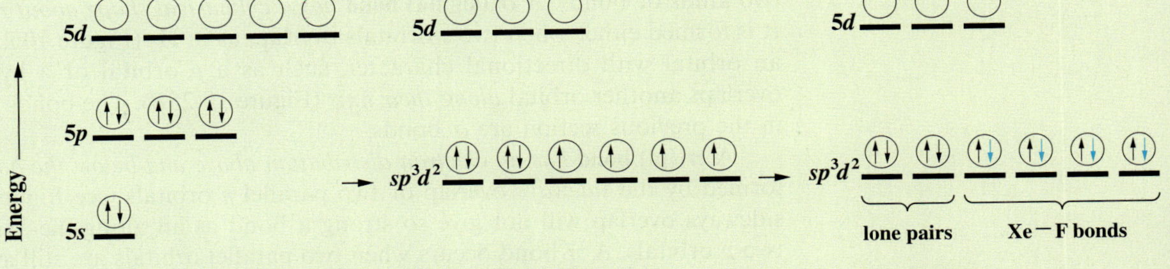

(continued)

(continued)

Answer Check Check that you have the correct Lewis formula for the molecule and that you have the correct geometry about the central atom and the correct hybrid orbitals (Table 10.2).

Exercise 10.6 Describe the bonding in PCl_5 using hybrid orbitals.

■ See Problems 10.53, 10.54, 10.55, and 10.56.

10.4 Description of Multiple Bonding

In the previous section, we described bonding as the overlap of *one* orbital from each of the bonding atoms. Now we consider the possibility that *more than one* orbital from each bonding atom might overlap, resulting in a multiple bond.

As an example, consider the ethylene molecule.

H H
C=C
H H

One hybrid orbital is needed for each bond (whether a single or a multiple bond) and for each lone pair. Because each carbon atom is bonded to three other atoms and there are no lone pairs, three hybrid orbitals are needed. This suggests the use of sp^2 hybrid orbitals on each carbon atom (there are three sp^2 hybrid orbitals; see Table 10.2). Thus, during bonding, the 2*s* orbital and two of the 2*p* orbitals of each carbon atom form three hybrid orbitals having trigonal planar orientation. A third 2*p* orbital on each carbon atom remains unhybridized and is perpendicular to the plane of the three sp^2 hybrid orbitals.

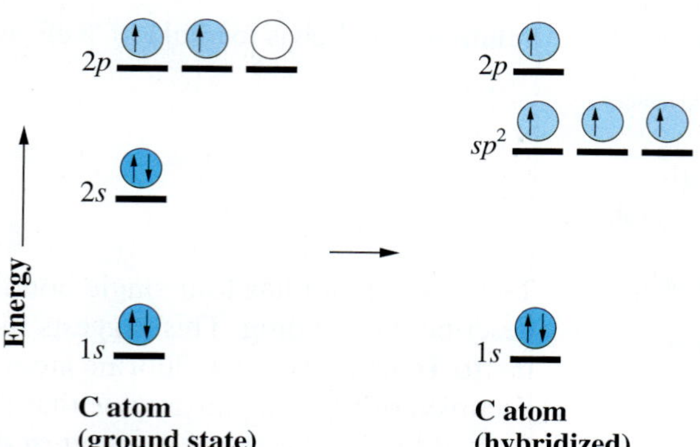

To describe the multiple bonding in ethylene, we must distinguish between two kinds of bonds. A **σ (sigma) bond** *has a cylindrical shape about the bond axis.* It is formed either when two *s* orbitals overlap, as in H_2 (Figure 10.25*A*), or when an orbital with directional character, such as a *p* orbital or a hybrid orbital, overlaps another orbital *along their axis* (Figure 10.25*B*). The bonds we discussed in the previous section are σ bonds.

A **π (pi) bond** *has an electron distribution above and below the bond axis.* It is formed by the *sideways* overlap of two parallel *p* orbitals (see Figure 10.25*C*). A sideways overlap will not give so strong a bond as an along-the-axis overlap of two *p* orbitals. A π bond occurs when two parallel orbitals are still available after strong σ bonds have formed.

Now imagine that the separate atoms of ethylene move into their normal molecular positions. Each sp^2 hybrid carbon orbital overlaps a 1*s* orbital of a

Figure 10.25 ◀
Sigma and pi bonds

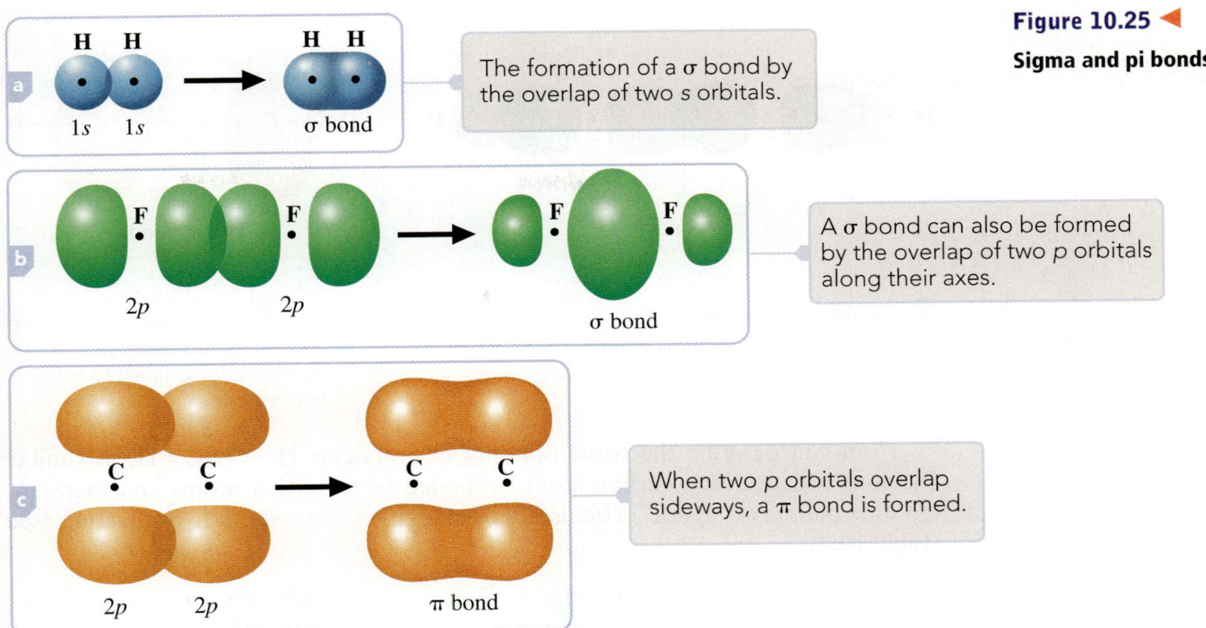

hydrogen atom or an sp^2 hybrid orbital of another carbon atom to form a σ bond (Figure 10.26*A*). Together, the σ bonds give the molecular framework of ethylene.

As you see from the orbital diagram for the hybridized C atom, a single 2*p* orbital still remains on each carbon atom. These orbitals are perpendicular to the plane of the hybrid orbitals; that is, they are perpendicular to the —CH_2 plane. Note that the two —CH_2 planes can rotate about the carbon–carbon axis without affecting the overlap of the hybrid orbitals. As these planes rotate, the 2*p* orbitals also rotate. When the —CH_2 planes rotate so that the 2*p* orbitals become parallel, the orbitals overlap to give a π bond (Figure 10.26*B*).

You therefore describe the carbon–carbon double bond as one σ bond and one π bond. Note that when the two 2*p* orbitals are parallel, the two —CH_2 ends of the molecule lie in the same plane. Thus, the formation of a π bond "locks" the two ends into a flat, rigid molecule.

Figure 10.26 ▼
Bonding in ethylene

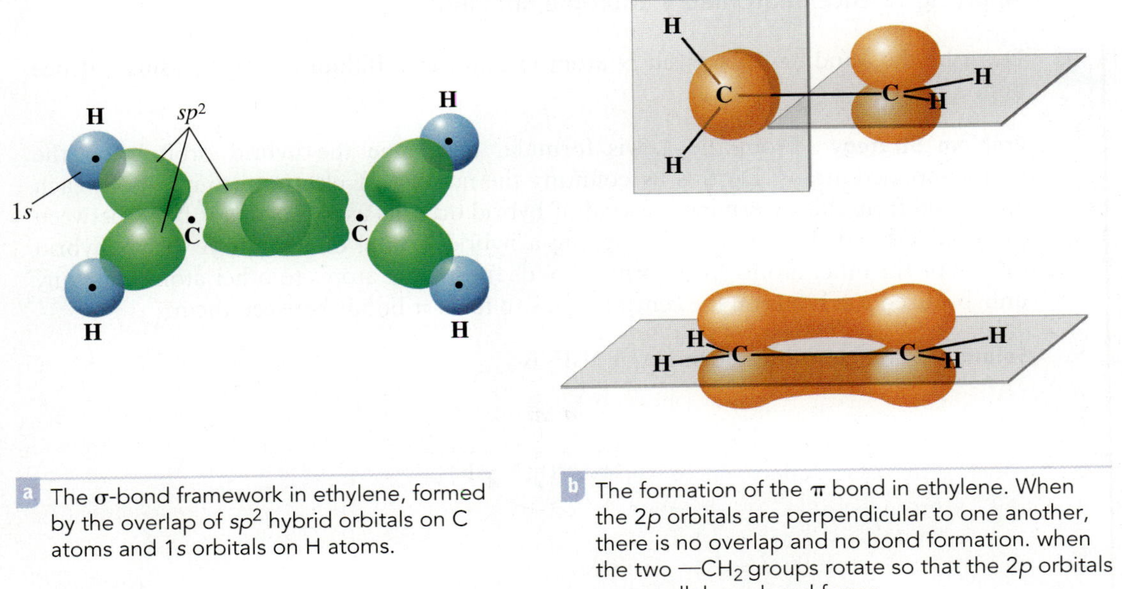

a The σ-bond framework in ethylene, formed by the overlap of sp^2 hybrid orbitals on C atoms and 1*s* orbitals on H atoms.

b The formation of the π bond in ethylene. When the 2*p* orbitals are perpendicular to one another, there is no overlap and no bond formation. when the two —CH_2 groups rotate so that the 2*p* orbitals are parallel, a π bond forms.

Figure 10.27 ▶
Bonding in acetylene

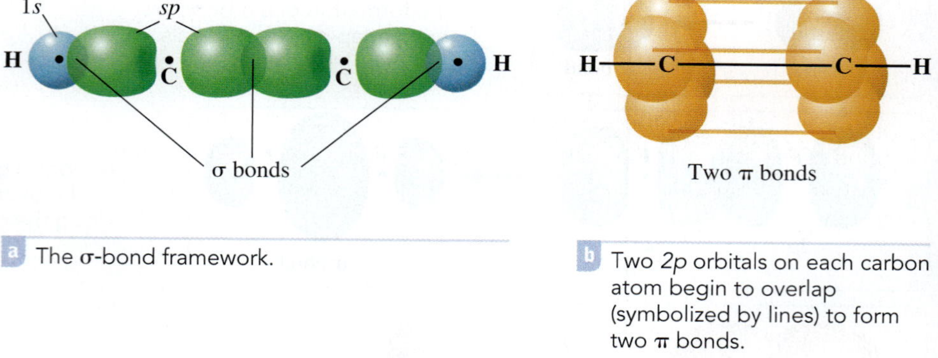

a The σ-bond framework.

b Two 2*p* orbitals on each carbon atom begin to overlap (symbolized by lines) to form two π bonds.

You can describe the triple bonding in acetylene, H—C≡C—H, in similar fashion. Because each carbon atom is bonded to two other atoms and there are no lone pairs, two hybrid orbitals are needed. This suggests *sp* hybridization (see Table 10.2).

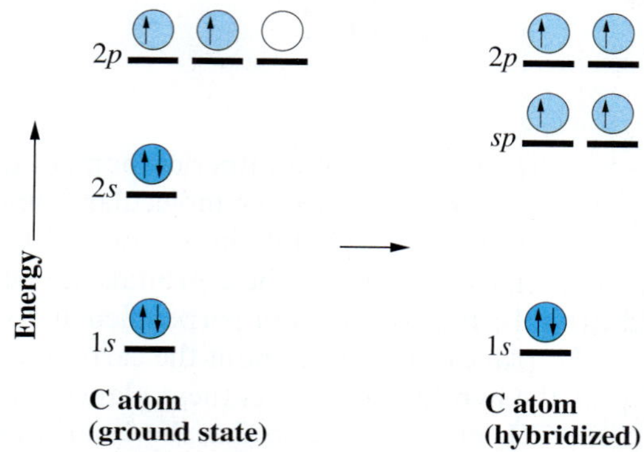

These *sp* hybrid orbitals have a linear arrangement, so the H—C—C—H geometry is linear. Bonds formed by the overlap of these hybrid orbitals are σ bonds. The two 2*p* orbitals not used to construct hybrid orbitals are perpendicular to the bond axis and to each other. They are used to form two π bonds. Thus, the carbon–carbon triple bond consists of one σ bond and two π bonds (see Figure 10.27).

Example 10.6 Applying Valence Bond Theory (Multiple Bonding)

Gaining Mastery Toolbox

Critical Concept 10.6
A double bond is described as a π bond plus a σ bond. To describe the bonding, first determine the number of hybrid orbitals on the central atom (equal to the number of electron groups about the atom). Use these orbitals to form sigma bonds and lone pairs. Finally, use unhybridized *p* orbitals to form π bonds.

Solution Essentials:
- σ- and π-bond formation by orbital overlap
- Kinds of hybrid orbitals
- VSEPR model
- Electron-dot formulas

Describe the bonding on a given N atom in dinitrogen difluoride, N_2F_2, using valence bond theory.

Problem Strategy From the Lewis formula, determine the hybrid orbitals on the double-bonded atoms. Do this by counting the number of electron groups about each atom, and from this determine the kind of hybrid orbitals used. Form a σ bond between the double-bonded atoms by overlapping a hybrid orbital of one atom with a hybrid orbital of the other atom. Also, form σ bonds from these atoms to other atoms. Use any unhybridized *p* orbitals on the central atoms to form π bonds between them.

Solution The electron-dot formula of N_2F_2 is

:F̤—N̈=N̈—F̤: (π and σ bonds indicated)

(continued)

(*continued*)

Note that a double bond is described as a π bond plus a σ bond. Hybrid orbitals are needed to describe each σ bond and each lone pair (a total of three hybrid orbitals for each N atom). This suggests sp^2 hybridization (see Table 10.2). According to this description, one of the sp^2 hybrid orbitals is used to form the N—F bond, another to form the σ bond of N═N, and the third to hold the lone pair on the N atom. The 2*p* orbitals on each N atom overlap to form the π bond of N═N. Hybridization and bonding of the N atoms are shown as follows:

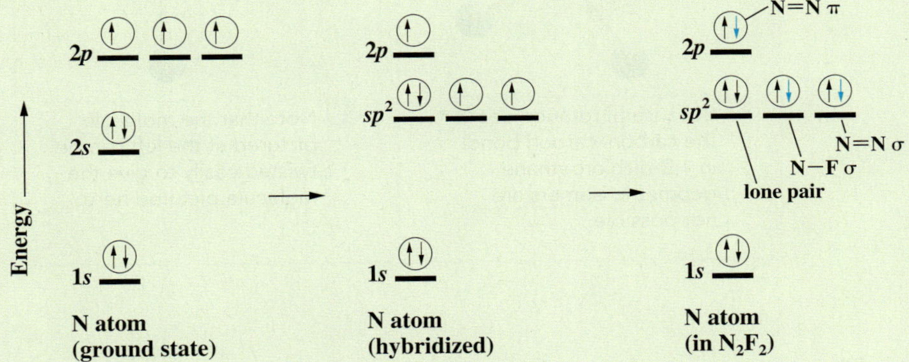

Answer Check Check that the number of hybrid orbitals on an atom of a multiple bond equals the number of sigma bonds plus lone pairs on that atom.

Exercise 10.7 Describe the bonding on the carbon atom in carbon dioxide, CO_2, using valence bond theory.

■ See Problems 10.57 and 10.58.

The π-bond description of the double bond agrees well with experiment. The *geometric,* or *cis–trans, isomers* of the compound 1,2-dichloroethene (described in the chapter opening) illustrate this. *Isomers* are compounds of the same molecular formula but with different arrangements of the atoms. (The numbers in the name 1,2-dichloroethene refer to the positions of the chlorine atoms; one chlorine atom is attached to carbon atom 1 and the other to carbon atom 2.) The structures of these isomers of 1,2-dichloroethene are

Cl(H)C═C(Cl)H	Cl(H)C═C(H)Cl
cis-1,2-Dichloroethene	*trans*-1,2-Dichloroethene

To transform one isomer into the other, one end of the molecule must be rotated as the other remains fixed. For this to happen, the π bond must be broken. Breaking the π bond requires considerable energy, so the *cis* and *trans* compounds are not easily interconverted. Contrast this with 1,2-dichloroethane (Figure 10.28), in which the two ends of the molecule can rotate without breaking any bonds. Here isomers corresponding to different spatial orientations of the two chlorine atoms cannot be prepared, because the two ends rotate freely with respect to one another. There is only one compound.

As we noted in the chapter opening, *cis* and *trans* isomers have different properties. The *cis* isomer of 1,2-dichloroethene boils at 60°C, the *trans* compound at 48°C.

Figure 10.28 ▶ **Lack of geometric isomers in 1,2-dichloroethane**

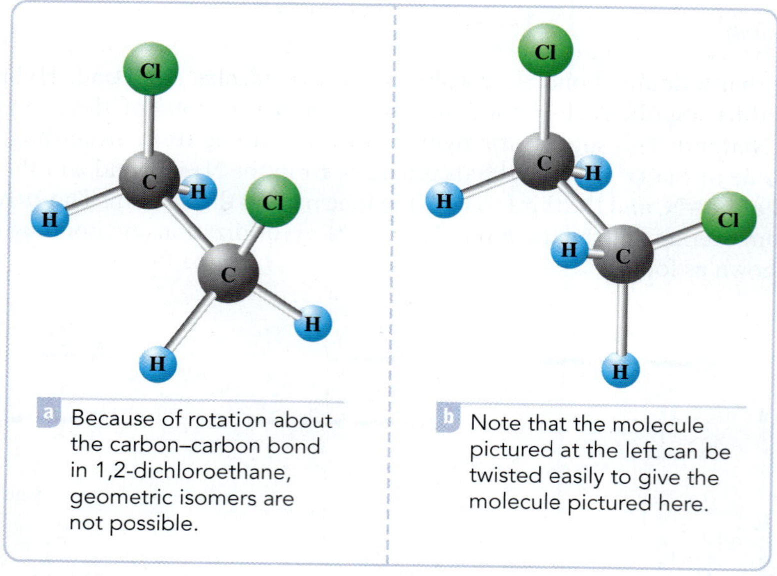

a Because of rotation about the carbon–carbon bond in 1,2-dichloroethane, geometric isomers are not possible.

b Note that the molecule pictured at the left can be twisted easily to give the molecule pictured here.

These isomers can also be differentiated on the basis of dipole moment. The *trans* compound has no dipole moment because it is symmetrical (the polar C—Cl bonds point in opposite directions and so cancel). However, the *cis* compound has a dipole moment of 1.85 D.

It is possible to convert one isomer to another if sufficient energy is supplied—say by chemical reaction. The role of the conversion of a *cis* isomer to its *trans* isomer in human vision is discussed in the essay at the end of Section 10.7.

Exercise 10.8 Dinitrogen difluoride (see Example 10.6) exists as *cis* and *trans* isomers. Write structural formulas for these isomers and explain (in terms of the valence bond theory of the double bond) why they exist.

■ See Problems 10.59 and 10.60.

CONCEPT CHECK 10.3

An atom in a molecule has one single bond and one triple bond to other atoms. What hybrid orbitals do you expect for this atom? Describe how you arrive at your answer.

Molecular Orbital Theory

In the preceding sections, we looked at a simple version of valence bond theory. Although simple valence bond theory satisfactorily describes most of the molecules we encounter, it does not apply to all molecules. For example, according to this theory any molecule with an even number of electrons should be diamagnetic (not attracted to a magnet), because we assume the electrons to be paired and to have opposite spins. In fact, a few molecules with an even number of electrons are paramagnetic (attracted to a magnet), indicating that some of the electrons are not paired. The best-known example of such a paramagnetic molecule is O_2. Because of the paramagnetism of O_2, liquid oxygen sticks to a magnet when poured over it

(Figure 10.29). Although valence bond theory can be extended to explain the electronic structure of O_2, molecular orbital theory (an alternative bonding theory) provides a straightforward explanation of the paramagnetism of O_2.

Molecular orbital theory is *a theory of the electronic structure of molecules in terms of molecular orbitals, which may spread over several atoms or the entire molecule.* This theory views the electronic structure of molecules to be much like the electronic structure of atoms. Each molecular orbital has a definite energy. To obtain the ground state of a molecule, electrons are put into orbitals of lowest energy, consistent with the Pauli exclusion principle, just as in atoms.

© Yoav Levy/Phototake

Figure 10.29 ▲

Paramagnetism of oxygen, O_2 Liquid oxygen is poured between the poles of a strong magnet. Oxygen adheres to the poles, showing that it is paramagnetic.

10.5 Principles of Molecular Orbital Theory

You can think of a molecular orbital as being formed from a combination of atomic orbitals. As atoms approach each other and their atomic orbitals overlap, molecular orbitals are formed.

Bonding and Antibonding Orbitals

Consider the H_2 molecule. As the atoms approach to form the molecule, their 1*s* orbitals overlap. One molecular orbital is obtained by adding the two 1*s* orbitals (see Figure 10.30). Note that where the atomic orbitals overlap, their values sum to give a larger result. This means that in this molecular orbital, electrons are often found in the region between the two nuclei where the electrons can hold the nuclei together. *Molecular orbitals that are concentrated in regions between nuclei* are called **bonding orbitals.** The bonding orbital in H_2, which we have just described, is denoted σ_{1s}. The σ (sigma) means that the molecular orbital has a cylindrical shape about the bond axis. The subscript 1*s* tells us that the molecular orbital was obtained from 1*s* atomic orbitals.

Another molecular orbital is obtained by subtracting the 1*s* orbital on one atom from the 1*s* orbital on the other (Figure 10.30). When the orbitals are subtracted, the resulting values in the region of the overlap are close to zero. This means that in this molecular orbital, the electrons spend little time between the nuclei. *Molecular orbitals having zero values in the region between two nuclei and therefore concentrated in other regions* are called **antibonding orbitals.** The antibonding orbital in H_2, which we have just described, is denoted σ^*_{1s}. The asterisk, which you read as "star," tells us that the molecular orbital is antibonding.

Figure 10.31 shows the energies of the molecular orbitals σ_{1s} and σ^*_{1s} relative to the atomic orbitals. Energies of the separate atomic orbitals are represented

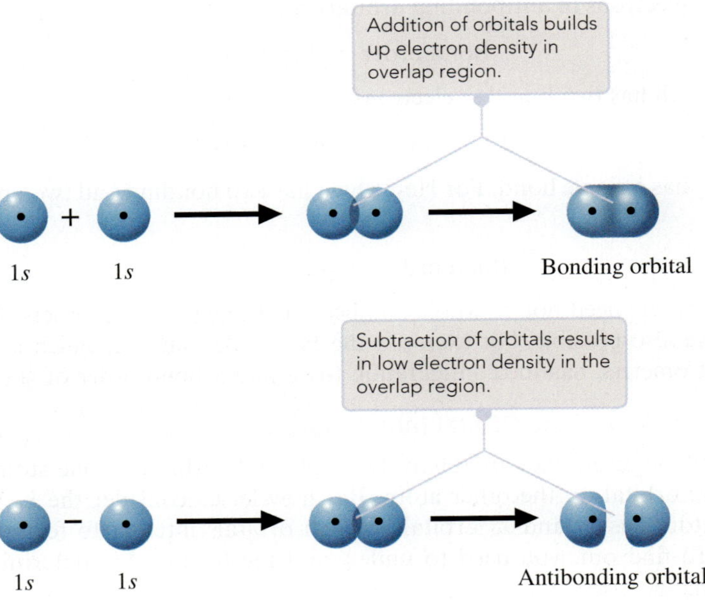

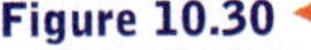

Figure 10.30 ◄

Formation of bonding and antibonding orbitals from 1*s* orbitals of hydrogen atoms When the two 1*s* orbitals overlap, they can either add to give a bonding molecular orbital or subtract to give an antibonding molecular orbital.

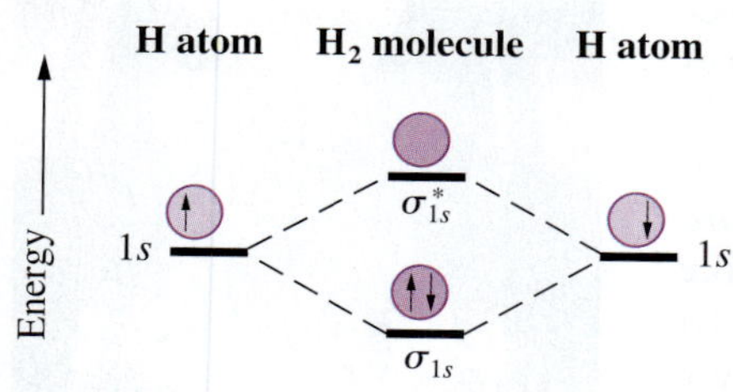

Figure 10.31 ▲

Relative energies of the 1s orbital of the H atom and the σ_{1s} and σ^*_{1s} molecular orbitals of H_2 Arrows denote occupation of the σ_{1s} orbital by electrons in the ground state of H_2.

by heavy black lines at the far left and right. The energies of the molecular orbitals are shown by heavy black lines in the center. (These are connected by dashed lines to show which atomic orbitals were used to obtain the molecular orbitals.) Note that the energy of a bonding orbital is less than that of the separate atomic orbitals, whereas the energy of an antibonding orbital is higher.

You obtain the electron configuration for the ground state of H_2 by placing the two electrons (one from each atom) into the lower-energy orbital (see Figure 10.31). The orbital diagram is

and the electron configuration is $(\sigma_{1s})^2$. Because the energy of the two electrons is lower than their energies in the isolated atoms, the H_2 molecule is stable.

Configurations involving the σ^*_{1s} orbital describe excited states of the molecule. As an example of an excited state, you can write the orbital diagram

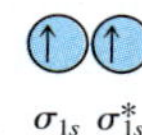

The corresponding electron configuration is $(\sigma_{1s})^1(\sigma^*_{1s})^1$.

You obtain a similar set of orbitals when you consider the approach of two helium atoms. To obtain the ground state of He_2, you note that there are four electrons, two from each atom, that would fill the molecular orbitals. Two electrons go into the σ_{1s} orbital and two go into the σ^*_{1s} orbital. The orbital diagram is

and the configuration is $(\sigma_{1s})^2(\sigma^*_{1s})^2$. The energy decrease from the bonding electrons is offset by the energy increase from the antibonding electrons. Hence, He_2 is not a stable molecule. Molecular orbital theory therefore explains why the element helium exists as a monatomic gas, whereas hydrogen is diatomic.

Bond Order

The term *bond order* refers to the number of bonds that exist between two atoms. In molecular orbital theory, the bond order of a diatomic molecule is defined as one-half the difference between the number of electrons in bonding orbitals, n_b, and the number of electrons in antibonding orbitals, n_a. ◀

For a Lewis formula, or electron-dot formula, the bond order equals the number of electron pairs shared between two atoms (see Section 9.10).

$$\text{Bond order} = \tfrac{1}{2}(n_b - n_a)$$

For H_2, which has two bonding electrons,

$$\text{Bond order} = \tfrac{1}{2}(2 - 0) = 1$$

That is, H_2 has a single bond. For He_2, which has two bonding and two antibonding electrons,

$$\text{Bond order} = \tfrac{1}{2}(2 - 2) = 0$$

Bond orders need not be whole numbers; half-integral bond orders of $\frac{1}{2}$, $\frac{3}{2}$, and so forth are also possible. For example, the H_2^+ molecular ion, which is formed in mass spectrometers, has the configuration $(\sigma_{1s})^1$ and a bond order of $\frac{1}{2}(1 - 0) = \frac{1}{2}$.

Factors That Determine Orbital Interaction

The H_2 and He_2 molecules are relatively simple. A 1*s* orbital on one atom interacts with the 1*s* orbital on the other atom. But now let us consider the Li_2 molecule. Each Li atom has 1*s* and 2*s* orbitals. Which orbitals interact to form molecular orbitals? To find out, you need to understand the factors that determine orbital interaction.

The strength of the interaction between two atomic orbitals to form molecular orbitals is determined by two factors: (1) the energy difference between the interacting orbitals and (2) the magnitude of their overlap. *For the interaction to be strong, the energies of the two orbitals must be approximately equal and the overlap must be large.*

From this last statement, you see that when two Li atoms approach one another to form Li_2, only like orbitals on the two atoms interact appreciably. The 2*s* orbital of one lithium atom interacts with the 2*s* orbital of the other atom, but the 2*s* orbital from one atom does not interact with the 1*s* orbital of the other atom, because their energies are quite different. Also, because the 2*s* orbitals are outer orbitals, they are able to overlap and interact strongly when the atoms approach. As in H_2, these atomic orbitals interact to give a bonding orbital (denoted σ_{2s}) and an antibonding orbital (denoted σ^*_{2s}). However, even though the 1*s* orbitals of the two atoms have the same energy, they do not overlap appreciably and so interact weakly. (The difference in energy between σ_{1s} and σ^*_{1s} is very small.) Figure 10.32 gives the relative energies of the orbitals.

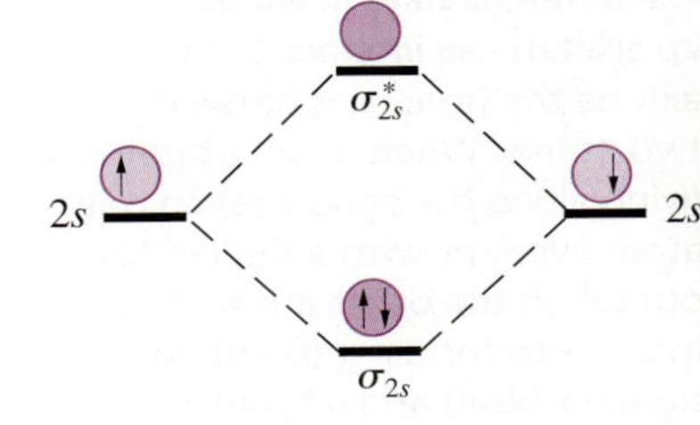
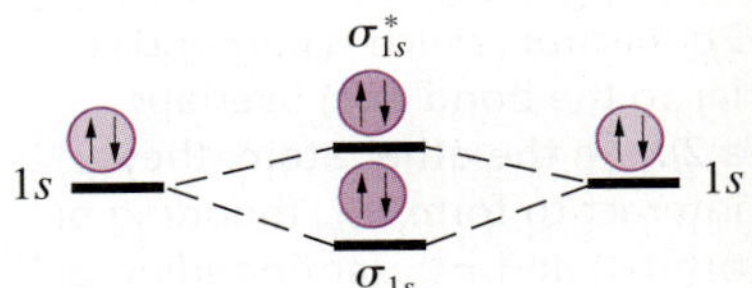

Figure 10.32

The energies of the molecular orbitals of Li_2 Arrows indicate the occupation of these orbitals by electrons in the ground state.

You obtain the ground-state configuration of Li_2 by putting six electrons (three from each atom) into the molecular orbitals of lowest energy. The configuration of the diatomic molecule Li_2 is

$$Li_2 \quad (\sigma_{1s})^2(\sigma^*_{1s})^2(\sigma_{2s})^2$$

The $(\sigma_{1s})^2(\sigma^*_{1s})^2$ part of the configuration is often abbreviated KK (which denotes the K shells, or inner shells, of the two atoms). These electrons do not have a significant effect on bonding.

However, small shifts in the energy of the core electrons due to bonding can be measured by x-ray photoelectron spectroscopy (page 310).

$$Li_2 \quad KK(\sigma_{2s})^2$$

In calculating bond order, we can ignore KK (it includes two bonding and two antibonding electrons). Thus, the bond order is 1, as in H_2.

In Be_2, the energy diagram is similar to that in Figure 10.32. You have eight electrons to distribute (four from each Be atom). The ground-state configuration of Be_2 is

$$Be_2 \quad KK(\sigma_{2s})^2(\sigma^*_{2s})^2$$

Note that the configuration has two bonding and two antibonding electrons outside the K shells. Thus, the bond order is $\frac{1}{2}(2 - 2) = 0$. No bond is formed, so the Be_2 molecule, like the He_2 molecule, is unstable.

10.6 Electron Configurations of Diatomic Molecules of the Second-Period Elements

The previous section looked at the electron configurations of some simple molecules: H_2, He_2, Li_2, and Be_2. These are **homonuclear diatomic molecules**—that is, *molecules composed of two like nuclei.* (**Heteronuclear diatomic molecules** are *molecules composed of two different nuclei*—for example, CO and NO.) To find the electron configurations of other homonuclear diatomic molecules, we need to have additional molecular orbitals.

We have already looked at the formation of molecular orbitals from *s* atomic orbitals. Now we need to consider the formation of molecular orbitals from *p* atomic orbitals. There are two different ways in which 2*p* atomic orbitals can interact. One set of 2*p* orbitals can overlap along their axes to give one bonding and one antibonding σ orbital (σ_{2p} and σ^*_{2p}). The other two sets of 2*p* orbitals then overlap sideways to give two bonding and two antibonding π orbitals (π_{2p} and π^*_{2p}). (See Figure 10.33.)

Figure 10.34 shows the relative energies of the molecular orbitals obtained from 2*s* and 2*p* atomic orbitals. This order of molecular orbitals reproduces the known electron configurations of homonuclear diatomic molecules composed of elements in the second row of the periodic table. The order of filling is

This order gives the correct number of electrons in subshells, even though in O_2 and F_2 the energy of σ_{2p} is below π_{2p}.

$$\sigma_{2s} \; \sigma^*_{2s} \; \pi_{2p} \; \sigma_{2p} \; \pi^*_{2p} \; \sigma^*_{2p}$$

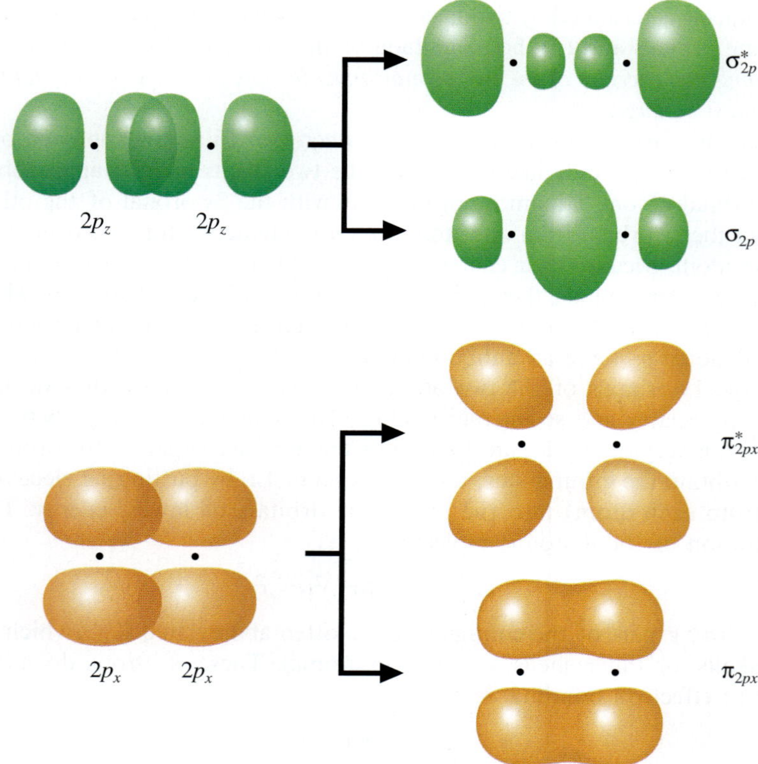

Figure 10.33 **The different ways in which 2*p* orbitals can interact** Let the *z* axis be the bond axis between two atoms. When a $2p_z$ orbital (lying along the bond axis) on one atom overlaps with a similar $2p_z$ orbital on the other atom, they interact to form σ_{2p} (bonding sigma orbital) and σ^*_{2p} (antibonding sigma orbital). See at the top of the figure. When a $2p_x$ orbital (which is perpendicular to the bond axis) overlaps a $2p_x$ on the other atom, they interact to form π_{2px} (bonding pi orbital) and π^*_{2px} (antibonding pi orbital). These are shown at the bottom of the figure. Similar pi orbitals (π_{2py} and π^*_{2py}) are formed by overlap of the $2p_y$ orbitals on the two atoms. (The $2p_y$ orbital is perpendicular to the bond axis and perpendicular to the $2p_x$ orbital.)

Note that there are two orbitals in the π subshell and two orbitals in the π^* subshell. Because each orbital can hold two electrons, a π or π^* subshell can hold four electrons. The next example shows how you can use the order of filling to obtain the orbital diagram, magnetic character, electron configuration, and bond order of a homonuclear diatomic molecule.

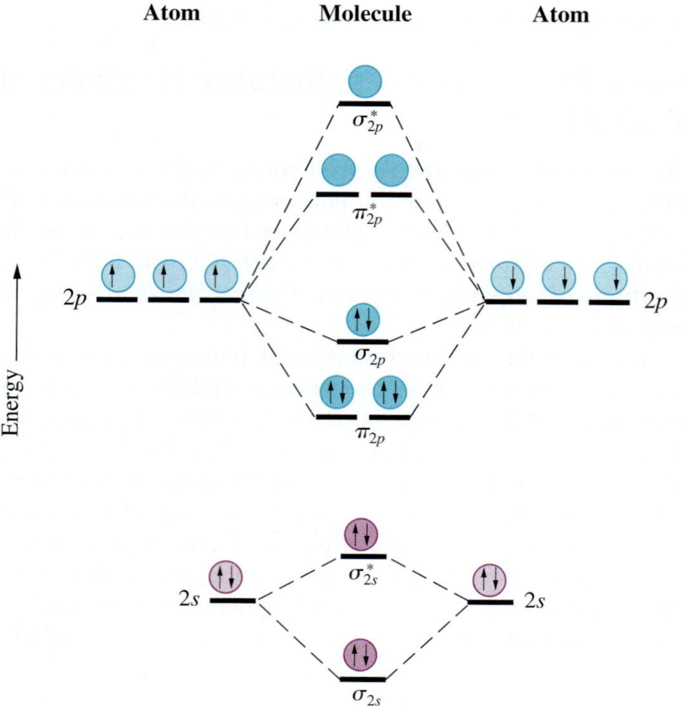

Figure 10.34 **Relative energies of molecular orbitals of homonuclear diatomic molecules (excluding K shells)** The arrows show the occupation of molecular orbitals by the valence electrons of N_2.

Example 10.7 Describing Molecular Orbital Configurations (Homonuclear Diatomic Molecules)

Gaining Mastery Toolbox

Critical Concept 10.7
Place the electrons of a homonuclear diatomic molecule into the molecular orbitals. As in an atom, observe the order of filling of the orbitals, the Pauli exclusion principle, and Hund's rule.

Solution Essentials:
- Order of filling of molecular orbitals
- Hund's rule
- Pauli exclusion principle

Give the orbital diagram of the O_2 molecule. Is the molecular substance diamagnetic or paramagnetic? What is the electron configuration? What is the bond order of O_2?

Problem Strategy Note the order of filling of molecular orbitals for the second-row homonuclear diatomic molecules (given just before this example). Now assign the valence electrons to these orbitals in this order. Pay attention to both the Pauli exclusion principle (no more than two electrons to an orbital, and then only with opposed spins) and Hund's rule (fill all orbitals of a subshell with one electron each—and with the same spins—before pairing electrons in any orbital). Decide whether the substance is diamagnetic (all electrons paired) or paramagnetic (unpaired electrons). Then calculate the bond order equal to $\frac{1}{2}$(number of bonding electrons − number of antibonding electrons).

Solution There are 12 valence electrons in O_2 (six from each atom), which occupy the molecular orbitals as shown in the following orbital diagram:

KK (↑↓) (↑↓) (↑↓)(↑↓) (↑↓) (↑)(↑) ( )

σ_{2s} σ^*_{2s} π_{2p} σ_{2p} π^*_{2p} σ^*_{2p}

Note that the two electrons in the π^*_{2p} subshell must go into different orbitals with their spins in the same direction (Hund's rule). Because there are two unpaired electrons, the molecular substance is **paramagnetic.** The electron configuration is

$$\mathbf{KK(\sigma_{2s})^2(\sigma^*_{2s})^2(\pi_{2p})^4(\sigma_{2p})^2(\pi^*_{2p})^2}$$

There are eight bonding electrons and four antibonding electrons. Therefore,

$$\text{Bond order} = \tfrac{1}{2}(8 - 4) = \mathbf{2}$$

Answer Check Check that the number of electrons assigned to orbitals equals the number of valence electrons of the two atoms. Make sure that the Pauli exclusion principle and Hund's rule have been properly applied.

Exercise 10.9 The C_2 molecule exists in the vapor phase over carbon at high temperature. Describe the molecular orbital structure of this molecule; that is, give the orbital diagram and electron configuration. Would you expect the molecular substance to be diamagnetic or paramagnetic? What is the bond order for C_2?

■ See Problems 10.61 and 10.62.

Table 10.3 compares experimentally determined bond lengths, bond dissociation energies, and magnetic character of the second-period homonuclear diatomic molecules with the bond order calculated from molecular orbital theory, as in the preceding example. Note that as the bond order increases, bond length tends to decrease and bond dissociation energy tends to increase. You should be able to verify that

Table 10.3 Theoretical Bond Orders and Experimental Data for the Second-Period Homonuclear Diatomic Molecules

Molecule	Bond Order	Bond Length (pm)	Bond Dissociation Energy (kJ/mol)	Magnetic Character
Li_2	1	267	110	Diamagnetic
Be_2	0	*	*	*
B_2	1	159	290	Paramagnetic
C_2	2	124	602	Diamagnetic
N_2	3	110	942	Diamagnetic
O_2	2	121	494	Paramagnetic
F_2	1	142	155	Diamagnetic
Ne_2	0	*	*	*

The symbol * means that no stable molecule has been observed.

the experimentally determined magnetic character of the molecule, given in the last column of the table, is correctly predicted by molecular orbital theory.

When the atoms in a heteronuclear diatomic molecule are close to one another in a row of the periodic table, the molecular orbitals have the same relative order of energies as those for homonuclear diatomic molecules. In this case you can obtain the electron configurations in the same way, as the next example illustrates.

Example 10.8 Describing Molecular Orbital Configurations (Heteronuclear Diatomic Molecules)

Gaining Mastery Toolbox

Critical Concept 10.8
If the two different atoms of a heteronuclear diatomic molecule are close to one another in the periodic table, you can use the same order of the filling of orbitals that you used for homonuclear diatomic molecules.

Solution Essentials:
- Order of filling of molecular orbitals
- Hund's rule
- Pauli exclusion principle

Write the molecular orbital diagram for nitrogen monoxide (nitric oxide), NO. What is the bond order of NO?

Problem Strategy If the two different atoms of the molecule are close to one another in the periodic table, then you can assume that the order of filling is the same as that for homonuclear diatomic molecules. After that, the procedure is similar to that for a homonuclear molecule (see Example 10.7).

Solution You assume that the order of filling of orbitals is the same as for homonuclear diatomic molecules.

There are 11 valence electrons in NO. Thus, the orbital diagram is

KK (↑↓) (↑↓) (↑↓)(↑↓) (↑↓) (↑)( ) ( )

σ_{2s} σ^*_{2s} π_{2p} σ_{2p} π^*_{2p} σ^*_{2p}

Because there are eight bonding and three antibonding electrons,

$$\text{Bond order} = \tfrac{1}{2}(8 - 3) = \tfrac{5}{2}$$

Answer Check Check that the number of electrons assigned to orbitals equals the number of valence electrons of the two atoms. Make sure that the Pauli exclusion principle and Hund's rule have been properly applied.

Exercise 10.10 Give the orbital diagram and electron configuration for the carbon monoxide molecule, CO. What is the bond order of CO? Is the molecule diamagnetic or paramagnetic?

See Problems 10.63 and 10.64.

When the two atoms in a heteronuclear diatomic molecule differ appreciably, you can no longer use the scheme appropriate for homonuclear diatomic molecules. The HF molecule is an example. Figure 10.35 shows the relative energies of the molecular orbitals that form. Sigma bonding and antibonding orbitals are formed by combining the 1*s* orbital on the H atom with the 2*p* orbital on the F atom that lies along the bond axis. The 2*s* orbital and the other 2*p* orbitals on fluorine remain as *nonbonding* orbitals (neither bonding nor antibonding). Note that the energy of the 2*p* subshell in fluorine is lower than the energy of the 1*s* orbital in hydrogen, because the fluorine electrons are more tightly held. As a result, the bonding orbital is made up of a greater percentage of the 2*p* fluorine orbital than of the 1*s* hydrogen orbital. This means that electrons in the bonding molecular orbital spend more time in the vicinity of the fluorine atom; that is, the H—F bond is polar, with the F atom having a small negative charge.

10.7 Molecular Orbitals and Delocalized Bonding

One of the advantages of using molecular orbital theory is the simple way in which it describes molecules with delocalized bonding. Whereas valence bond theory requires two or more resonance formulas, molecular orbital theory describes the bonding in terms of a single electron configuration.

Consider the ozone molecule as an example. In the previous chapter, we described this molecule in terms of resonance as follows:

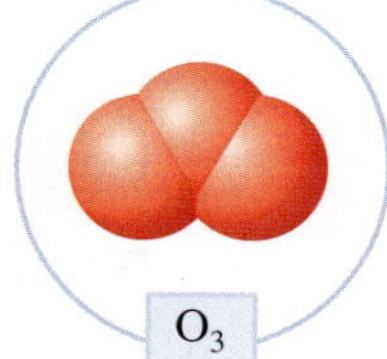
O_3

O=O–O ⟷ O–O=O

Some of these electrons are localized either on one atom or between an oxygen–oxygen bond (these are printed in black). The other electrons are delocalized (these are printed in blue). They are the electron pairs that change position in going from one resonance formula to the other.

Each O atom has three localized electron pairs about it, which suggests that each atom uses sp^2 hybrid orbitals (see Table 10.2). The overlap of one hybrid orbital on the center O atom with a hybrid orbital on the end O atom at the left forms the left O—O bond (Figure 10.36). The overlap of another hybrid orbital on the center O atom with a hybrid orbital on the end O atom at the right forms the other O—O bond. This leaves one hybrid orbital on the center O atom and two on each of the end O atoms for lone pairs.

After the hybrid orbitals are formed, one unhybridized p orbital remains on each O atom. These three p orbitals are perpendicular to the plane of the molecule and parallel to one another. They can overlap sideways to give three π molecular orbitals. (Three interacting atomic orbitals produce three molecular orbitals.) All three will span the entire molecule (Figure 10.37). One of these orbitals turns out to be antibonding; one is nonbonding (that is, it has an energy equal to that of the isolated atoms); and the third is bonding.

The bonding and nonbonding π orbitals are both doubly occupied. This agrees with the resonance description, in which one delocalized pair is bonding and the other is a lone (nonbonding) pair on the end oxygen atoms.

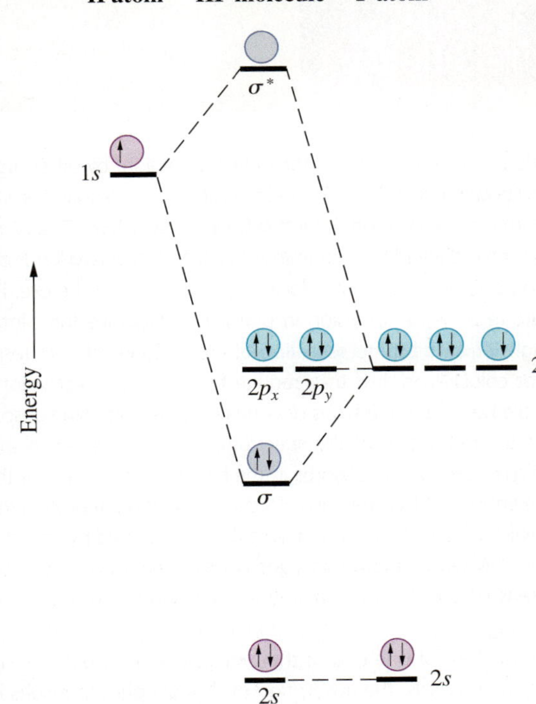

Figure 10.35 ▲

Relative energies of molecular orbitals in the valence shell of HF Arrows show the occupation of orbitals by electrons.

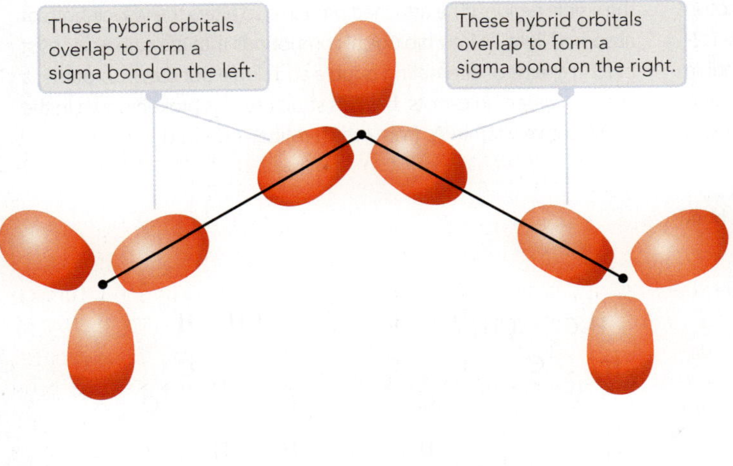

Figure 10.36 ◄

Hybrid orbitals on oxygen atoms of ozone, O_3 Each oxygen atom has sp^2 hybrid orbitals. One hybrid orbital on the center atom and one hybrid orbital on the left atom overlap to form a single bond at the left. Similar orbitals at the right form another single bond. Each atom has a π orbital perpendicular to the plane of these hybrid orbitals.

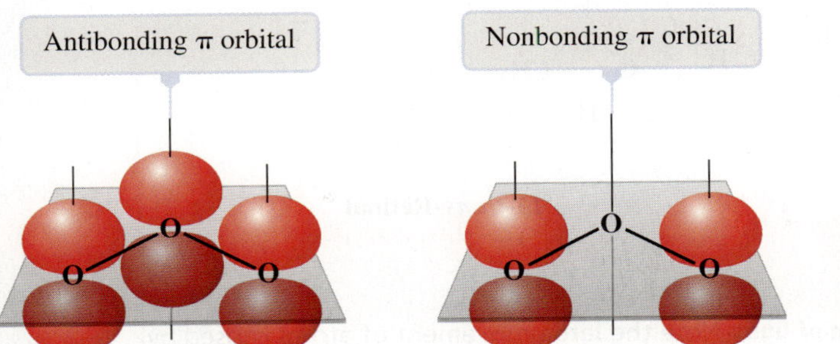

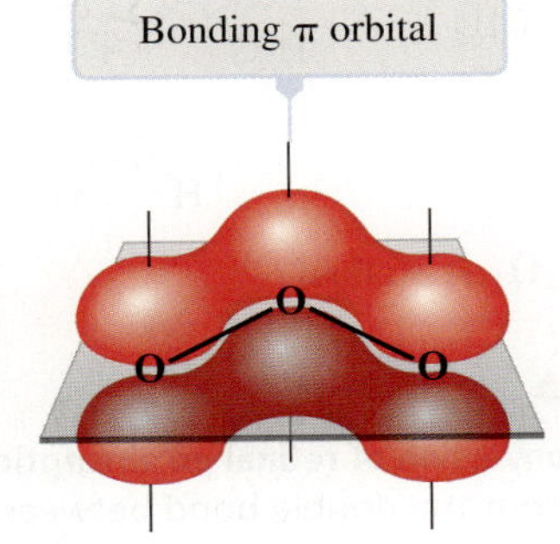

Figure 10.37 ▲

The electron distributions of the π molecular orbitals of O_3 Note that one molecular orbital is nonbonding. It describes a lone pair of electrons that can be on either end-oxygen atom.

A **CHEMIST** Looks at...

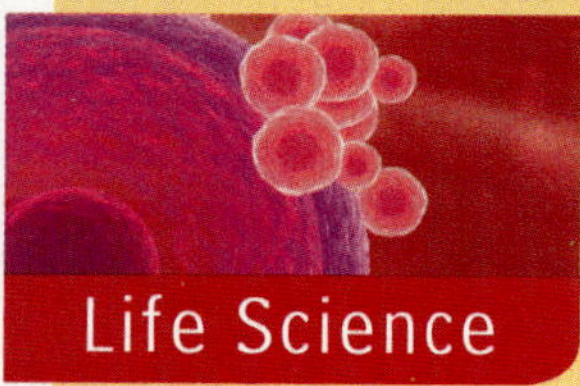

Human Vision

Human vision, as well as the detection of light by other organisms, is known to involve a compound called 11-*cis*-retinal. The structural formula is shown on the left side of Figure 10.38. This compound, which is derived from vitamin A, is contained in two kinds of cells—rod cells and cone cells—located in the retina of the eye. Rod cells are responsible for vision in dim light. They give the sensation of light or dark but no sensation of color. Cone cells are responsible for color vision, but they require bright light. Color vision is possible because three types of cone cells exist: one type absorbs light in the red region of the spectrum, another absorbs in the green region, and a third absorbs in the blue region. In each of these different cells, 11-*cis*-retinal is attached to a different protein molecule, which affects the region of light that is absorbed by retinal.

When a photon of light is absorbed by 11-*cis*-retinal, it is transformed to 11-*trans*-retinal (see Figure 10.38). This small change at one carbon–carbon double bond results in a large movement of one end of the molecule with respect to the other. In other words, the absorption of a single photon results in a significant change in molecular geometry. This change in shape of retinal also affects the shape of the protein molecule to which it is attached, and *this* change results in a sequence of events, not completely understood, in which an electrical signal is generated.

To understand the *cis–trans* conversion when a photon is absorbed, consider the similar, but simpler, molecule *cis*-1,2-dichloroethene (see structural formula on page 382). The double bond consists of σ bond and a π bond. The π bond results from two electrons in a π-type molecular orbital. A corresponding antibonding orbital, π^*, has higher energy than the π orbital and is unoccupied in the ground state. When a photon of light in the ultraviolet region (near 180 nm) is absorbed by *cis*-1,2-dichloroethene, an electron in the π orbital undergoes a transition to the π^* orbital. Whereas there were previously two electrons in the bonding π orbital (contributing one bond), there is now one electron in the bonding orbital and one in the antibonding orbital (with zero net bonding). As a result of the absorption of a photon, the double bond becomes a single bond and rotation about the bond is now possible. The *cis* isomer tends to rotate to the *trans* isomer, which is more stable. Then energy is lost as heat, and the electron in the π^* orbital undergoes a transition back to the π orbital; the single bond becomes a double bond again. The net result is the conversion of the *cis* isomer to the *trans* isomer.

Retinal differs from dichloroethene in having six double bonds alternating with single bonds instead of only one double bond. The alternating double and single bonds give a rigid structure, and when the middle double bond undergoes the *cis*-to-*trans* change, a large movement of atoms occurs. Another result of a long chain of alternating double and single bonds is a change in the wavelength of light absorbed in the π-to-π^* transition. Dichloroethene absorbs in the ultraviolet region, but retinal attached to its protein absorbs in the visible region. (The attached protein also alters the wavelength of absorbed light.) Many biological compounds that are colored consist of long chains of alternating double and single bonds. β-Carotene, a yellow pigment in carrots, has this structure. It is broken down in the body to give vitamin A, from which retinal is derived.

Rotation about this bond

light

11-*cis*-Retinal **11-*trans*-Retinal**

Figure 10.38 ▲

The *cis–trans* conversion of retinal on absorption of light Note the large movement of atoms caused by the rotation about the double bond between carbon atoms 11 and 12.

■ See Problems 10.85 and 10.86.

A CHEMIST Looks at...

Stratospheric Ozone (An Absorber of Ultraviolet Rays)

Ozone (also known as trioxygen, O_3) is a reactive form of oxygen having a bent molecular geometry, as you would predict from its resonance formulas:

$$\ddot{O}=\ddot{O}-\ddot{O}: \longleftrightarrow :\ddot{O}-\ddot{O}=\ddot{O}$$

Ozone forms in the atmosphere whenever O_2 is irradiated by ultraviolet light or subjected to electrical discharges. On bright sunny days, you might notice the fresh-air smell of dilute ozone (trioxygen, O_3). Around electrical equipment, however, ozone often occurs in higher concentrations and has a disagreeable odor.

Although ozone occurs at ground level and is an ingredient of *photochemical smog,* it is an essential component of the *stratosphere,* which occurs at about 10 to 15 km above ground level. Ozone in the stratosphere absorbs ultraviolet radiation between 200 and 300 nm, which is vitally important to life on earth. Radiation from the sun contains ultraviolet rays that are harmful to the DNA of biological organisms. Oxygen, O_2, absorbs the most energetic of these ultraviolet rays in the earth's upper atmosphere, but only ozone in the stratosphere absorbs the remaining ultraviolet radiation that is destructive to life on earth (Figure 10.39).

In 1974, Mario J. Molina and F. Sherwood Rowland expressed concern that chlorofluorocarbons (CFCs), such as $CClF_3$ and CCl_2F_2, would be a source of chlorine atoms that could catalyze the decomposition of ozone in the stratosphere, so that the ozone would be destroyed faster than it could be produced. The chlorofluorocarbons are relatively inert compounds that were used as refrigerants, spray-can propellants, and blowing agents (substances used to produce plastic foams). As a result of their inertness, however, CFCs concentrate in the atmosphere, where they steadily rise into the stratosphere. Once they are in the stratosphere, ultraviolet light decomposes them to form chlorine atoms, which react with ozone to form ClO and O_2. The ClO molecules react with oxygen atoms in the stratosphere to regenerate Cl atoms.

$$\begin{aligned} Cl(g) + O_3(g) &\longrightarrow ClO(g) + O_2(g) \\ ClO(g) + O(g) &\longrightarrow Cl(g) + O_2(g) \\ \hline O_3(g) + O(g) &\longrightarrow 2O_2(g) \end{aligned}$$

The net result is the decomposition of ozone to dioxygen. Chlorine atoms are consumed in the first step but regenerated in the second. Thus, chlorine atoms are not used up, so they function as a catalyst.

In 1985, British researchers reported that the ozone over the Antarctic had been declining precipitously each spring for several years, although it returned the following winter. Then in August 1987, an expedition was assembled to begin flights through the "ozone hole" with instruments (Figure 10.40). Researchers found that wherever ozone is depleted in the stratosphere, chlorine monoxide (ClO) appears, as expected if chlorine atoms are catalyzing the depletion (see the previous reactions). The chlorofluorocarbons are the suspected source of these chlorine atoms, and by international treaty they are being phased out of production. Refrigerators and

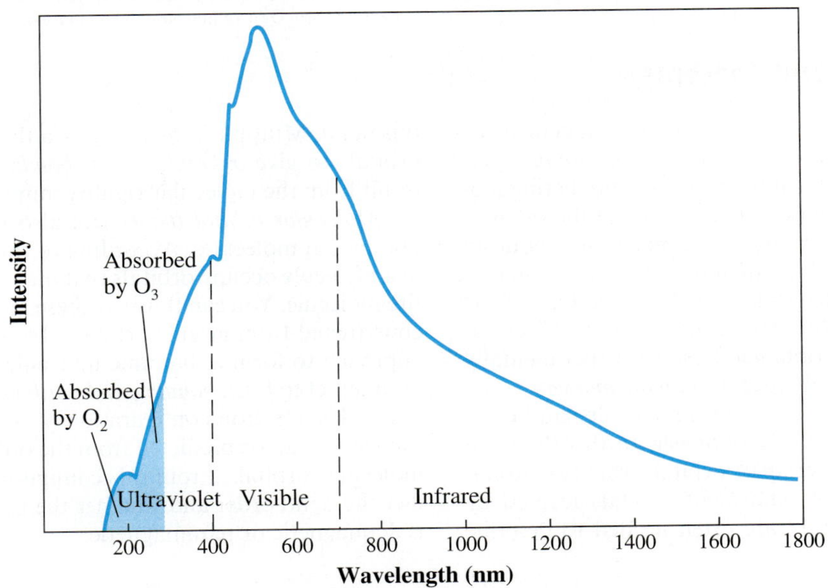

Figure 10.39 ▲

The solar spectrum Radiation below 200 nm is absorbed in the upper atmosphere by O_2. Ultraviolet radiation between 200 and 300 nm is absorbed by O_3 in the stratosphere.

(continued)

air conditioners, for example, now use hydrofluorocarbons (HFCs), such as R410A, which is a mixture of CH_2F_2 and CF_3CHF_2.

In October 1995, Sherwood Rowland (University of California, Irvine), Mario Molina (Massachusetts Institute of Technology), and Paul Crutzen (Max Planck Institute, Mainz, Germany) were awarded the Nobel Prize in chemistry for their work on stratospheric ozone depletion. Crutzen studied the effect of nitrogen oxides in catalyzing the decomposition of ozone.

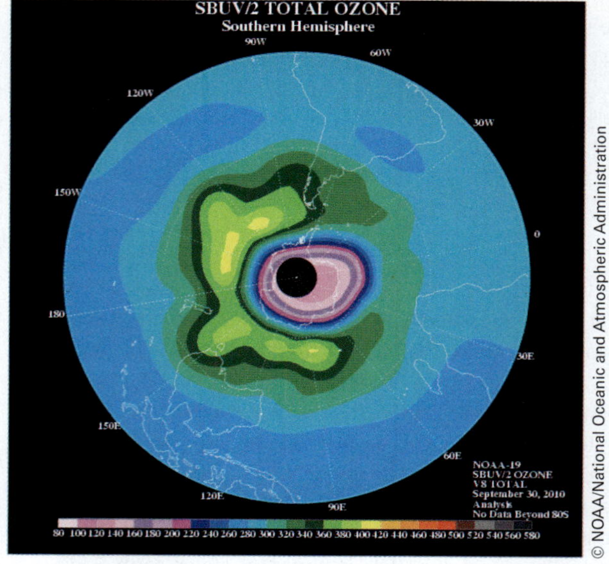

Figure 10.40 ▶

Total stratospheric ozone over the Southern Hemisphere This computer image shows the total quantity of ozone over the Southern Hemisphere in Dobson units (equivalent to the thickness, in units of 10^{-2} mm, of total ozone, assuming it to be compressed to 1 atm at 0°C). The ozone depletion is shown near the South Pole (pink color; black color indicates no data). The data are from the SBUV/2 instrument on board the National Oceanic and Atmospheric Administration's polar orbiting satellite.

■ See Problems 10.87 and 10.88.

A Checklist for Review

OWL and Go Chemistry Sign in at **www.cengage.com/owl** to:

- View tutorials and simulations, develop problem-solving skills, and complete online homework assigned by your professor.
- For quick review and exam prep, download Go Chemistry mini lecture modules from OWL or purchase them at **www.cengagebrain.com.**

Summary of Facts and Concepts

Molecular geometry refers to the spatial arrangement of atoms in a molecule. The *valence-shell electron-pair repulsion (VSEPR) model* is a simple model for predicting molecular geometries. It is based on the idea that the valence-shell electron pairs are arranged symmetrically about an atom to minimize electron-pair repulsion. The geometry about an atom is then determined by the directions of the bonding pairs. Information about the geometry of a molecule can sometimes be obtained from the experimentally determined presence or absence of a *dipole moment.*

The bonding and geometry in a molecule can be described in terms of *valence bond theory.* In this theory, a bond is formed by the overlap of orbitals from two atoms. *Hybrid orbitals,* a set of equivalent orbitals formed by combining atomic orbitals, are often needed to describe this bond. Multiple bonds occur via the overlap of atomic orbitals to give *σ bonds* and *π bonds. Cis–trans* isomers result from the molecular rigidity imposed by a π bond.

Molecular orbital theory can also be used to explain bonding in molecules. According to this theory, electrons in a molecule occupy orbitals that may spread over the entire molecule. You can think of these molecular orbitals as constructed from atomic orbitals. Thus, when two atoms approach to form a diatomic molecule, the atomic orbitals interact to form *bonding* and *antibonding* molecular orbitals. The electron configuration of a diatomic molecule such as O_2 can be predicted from the order of filling of the molecular orbitals. From this configuration, you can predict the bond order and whether the molecular substance is diamagnetic or paramagnetic.

Learning Objectives	Important Terms
10.1 The Valence-Shell Electron-Pair Repulsion (VSEPR) Model	
■ Define *molecular geometry.*	**molecular geometry**
■ Define *valence-shell electron-pair repulsion model.*	**valence-shell electron-pair repulsion (VSEPR) model**

▪ Note the difference between the arrangement of electron pairs about a central atom and molecular geometry. ▪ Note the four steps in the prediction of geometry by the VSEPR model. ▪ Predict the molecular geometry (two, three, or four electron pairs). **Example 10.1** ▪ Note that a lone pair tends to require more space than a corresponding bonding pair and that a multiple bond requires more space than a single bond. ▪ Predict the molecular geometry (five or six electron pairs). **Example 10.2**	
10.2 Dipole Moment and Molecular Geometry	
▪ Define *dipole moment.* ▪ Explain the relationship between dipole moment and molecular geometry. **Example 10.3** ▪ Note that the polarity of a molecule can affect certain properties, such as a boiling point.	**dipole moment**
10.3 Valence Bond Theory	
▪ Define *valence bond theory.* ▪ State the two conditions needed for bond formation, according to valence bond theory. ▪ Define *hybrid orbitals.* ▪ State the five steps in describing bonding, following the valence bond theory. ▪ Apply valence bond theory (two, three, or four electron pairs). **Example 10.4** ▪ Apply valence bond theory (five or six electron pairs). **Example 10.5**	**valence bond theory** **hybrid orbitals**
10.4 Description of Multiple Bonding	
▪ Define *σ (sigma) bond.* ▪ Define *π (pi) bond.* ▪ Apply valence bond theory (multiple bonding). **Example 10.6** ▪ Explain geometric, or *cis–trans,* isomers in terms of the π-bond description of a double bond.	**σ (sigma) bond** **π (pi) bond**
10.5 Principles of Molecular Orbital Theory	
▪ Define *molecular orbital theory.* ▪ Define *bonding orbitals* and *antibonding orbitals.* ▪ Define *bond order.* ▪ State the two factors that determine the strength of interaction between two atomic orbitals. ▪ Describe the electron configurations of H_2, He_2, Li_2, and Be_2.	**molecular orbital theory** **bonding orbitals** **antibonding orbitals**
10.6 Electron Configurations of Diatomic Molecules of the Second-Period Elements	
▪ Define *homonuclear diatomic molecules* and *heteronuclear diatomic molecules.* ▪ Describe molecular orbital configurations (homonuclear diatomic molecules). **Example 10.7** ▪ Describe molecular orbital configurations (heteronuclear diatomic molecules). **Example 10.8**	**homonuclear diatomic molecules** **heteronuclear diatomic molecules**
10.7 Molecular Orbitals and Delocalized Bonding	
▪ Describe the delocalized bonding in molecules such as O_3.	

Questions and Problems

OWL Interactive versions of these problems may be assigned in OWL.

Self-Assessment and Review Questions

Key: These questions test your understanding of the ideas you worked with in the chapter. These problems vary in difficulty and often can be used for the basis of discussion.

10.1 Describe the main features of the VSEPR model.

10.2 According to the VSEPR model, what are the arrangements of two, three, four, five, and six valence-shell electron pairs about an atom?

10.3 Why is a lone pair expected to occupy an equatorial position instead of an axial position in the trigonal bipyramidal arrangement?

10.4 Why is it possible for a molecule to have polar bonds, yet have a dipole moment of zero?

10.5 Explain why nitrogen trifluoride has a small dipole moment even though it has polar bonds in a trigonal pyramidal arrangement.

10.6 Explain in terms of valence bond theory why the orbitals used on an atom give rise to a particular geometry about that atom.

10.7 What is the angle between two sp^3 hybrid orbitals?

10.8 What is the difference between a sigma bond and a pi bond?

10.9 Describe the bonding in ethylene, C_2H_4, in terms of valence bond theory.

10.10 How does the valence bond description of a carbon–carbon double bond account for *cis–trans* isomers?

10.11 What are the differences between a bonding and an antibonding molecular orbital of a diatomic molecule?

10.12 What factors determine the strength of interaction between two atomic orbitals to form a molecular orbital?

10.13 Describe the formation of bonding and antibonding molecular orbitals resulting from the interaction of two 2*s* orbitals.

10.14 Describe the formation of molecular orbitals resulting from the interaction of two 2*p* orbitals.

10.15 How does molecular orbital theory describe the bonding in the HF molecule?

10.16 Describe the bonding in O_3, using molecular orbital theory. Compare this with the resonance description.

10.17 According to the VSEPR model, the H—P—H bond angle in PH_3 is

a 120° b 109.5° c 90°
d a little less than 120° e a little less than 109.5°

10.18 Which of the following molecular geometries does the XeF_5^+ cation exhibit?

a tetradral
b T-shaped
c octahedral
d square pyramidal
e trigonal bipyramidal

10.19 Which of the following would be a polar molecule?

a CO_2 b H_2S c CH_4 d SF_6 e BeF_2

10.20 What is the bond order of NO?

a $\frac{1}{2}$ b 1 c $\frac{3}{2}$ d 3 e $\frac{5}{2}$

Concept Explorations

Key: Concept explorations are comprehensive problems that provide a framework that will enable you to explore and learn many of the critical concepts and ideas in each chapter. If you master the concepts associated with these explorations, you will have a better understanding of many important chemistry ideas and will be more successful in solving all types of chemistry problems. These problems are well suited for group work and for use as in-class activities.

10.21 Best Lewis Formula and Molecular Geometry

A student writes the Lewis electron-dot formula for the carbonate anion, CO_3^{2-}, as

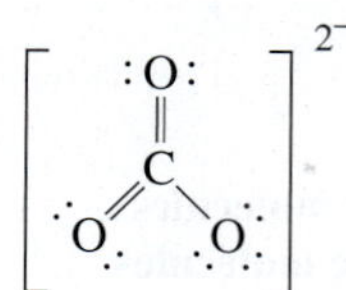

a Does this Lewis formula obey the octet rule? Explain. What are the formal charges on the atoms? Try describing the bonding for this formula in valence bond terms. Do you have any difficulty doing this?

b Does this Lewis formula give a reasonable description of the electron structure, or is there a better one? If there is a better Lewis formula, write it down and explain why it is better.

c The same student writes the following resonance description for CO_2:

$$\ddot{\underset{..}{O}}-\ddot{C}=\ddot{O} \longleftrightarrow :\ddot{\underset{..}{O}}-C=\ddot{\underset{..}{O}}$$

Is there something wrong with this description? (What would you predict as the geometries of these formulas?)

d Is one or the other formula a better description? Could a value for the dipole moment help you decide?

e Can you write a Lewis formula that gives an even better description of CO_2? Explain your answer.

10.22 Molecular Geometry and Bonding

a Shown here are several potential Lewis electron-dot formulas for the SeO_2 molecule.

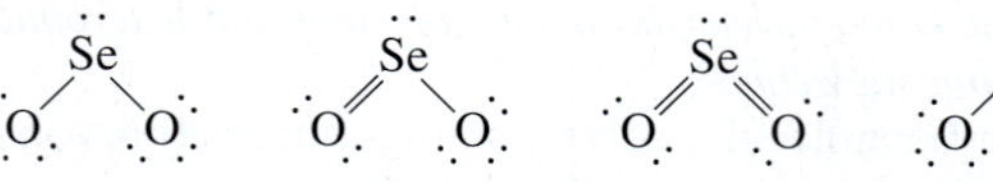

Are there any electron-dot formulas that you expect to give the best description? How would you describe the bonding in terms of electron-dot formulas?

b Describe the bonding in SeO_2 in valence bond terms. If there is delocalized bonding, note this in your description and explain how molecular orbital theory can be useful here.

c Determine the arrangement of electron pairs about Se in the SeO_2 molecule. What would you expect for the molecular geometry of the SeO_2 molecule?

d Would you expect the O—Se—O angle in the SeO_2 molecule to be greater than, equal to, or less than 120°. Explain your answer.

e Draw an electron-dot formula for the H_2Se molecule and determine its molecular geometry.

f Compare the H—Se—H bond angle to the O—Se—O bond angle. Is it larger or smaller? Explain.

g Determine whether the H_2Se and SeO_2 molecules have dipole moments. Describe how you arrived at your answer.

Conceptual Problems

Key: These problems are designed to check your understanding of the concepts associated with some of the main topics presented in each chapter. A strong conceptual understanding of chemistry is the foundation for both applying chemical knowledge and solving chemical problems. These problems vary in level of difficulty and often can be used as a basis for group discussion.

10.23 Match the following molecular substances a–d with one of the molecular models (i) to (iv) that correctly depicts the geometry of the corresponding molecule.

a SeO_2 b $BeCl_2$ c PBr_3 d BCl_3

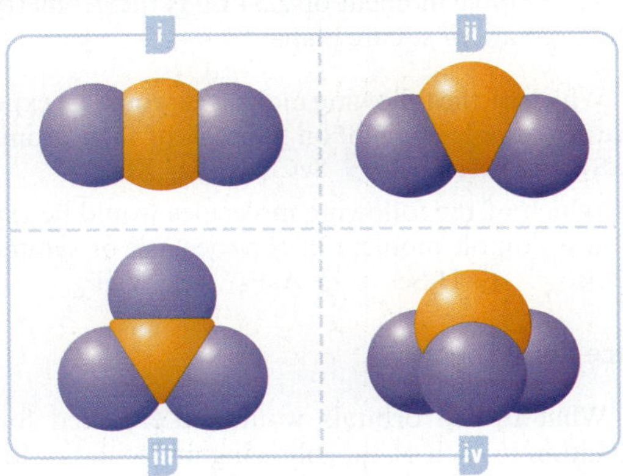

10.24 Which of the following molecular models correctly depicts the geometry of ClCN?

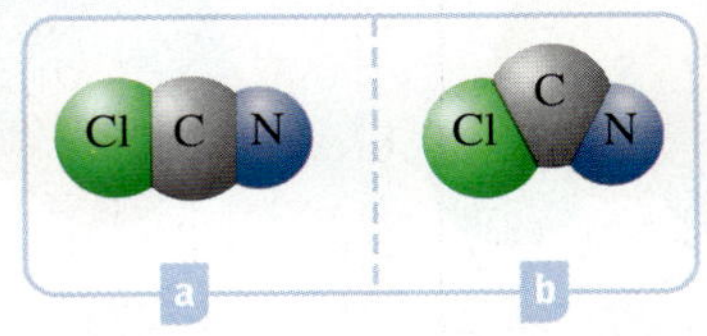

10.25 Suppose that an ethane molecule, CH_3CH_3, is broken into two CH_3 molecules in such a way that one $:CH_3$ molecule retains the electron pair that was originally the one making up the C—C bond. The other CH_3 molecule has two fewer electrons. Imagine that momentarily this CH_3 molecule has the geometry it had in the ethane molecule. Describe the electron repulsions present in this molecule, and indicate how they would be expected to rearrange its geometry.

10.26 Suppose that a BF_3 molecule approaches the lone pair on the N atom of an $:NH_3$ molecule, and that a bond forms between the B atom and the N atom. Consider the arrangement of electron pairs about the B atom at the moment of this bond formation, and describe the repulsions among the electron pairs and how they might be expected to change the geometry about the B atom.

10.27 Indicate what hybrid orbital depicted below is expected for the central atom of each of the following:

a BeF_2 b SiF_4 c SeF_4 d RnF_4

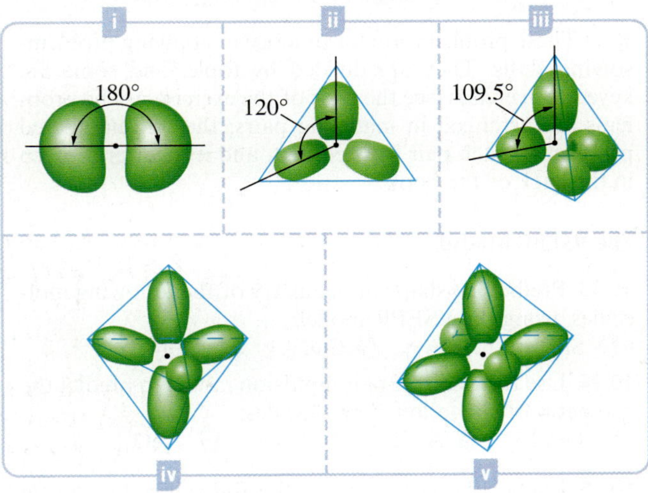

10.28 An atom in a molecule has two bonds to other atoms and one lone pair. What kind of hybrid orbitals do you expect for this atom? Describe how you arrived at your answer.

10.29 Two compounds have the same molecular formula, $C_2H_2Br_2$. One has a dipole moment; the other does not. Both compounds react with bromine, Br_2, to produce the same compound. This reaction is a generally accepted test for double bonds, and each bromine atom of Br_2 adds to a different atom of the double bond. What is the identity of the original compounds? Describe the argument you use.

10.30 A neutral molecule is identified as a tetrafluoride, XF_4, where X is an unknown atom. If the molecule has a dipole moment of 0.63 D, can you give some possibilities for the identity of X?

10.31 Acetic acid, the sour constituent of vinegar, has the following structure:

$$\underset{a}{CH_3}\underset{b}{\overset{\overset{\displaystyle O}{\|}}{C}}—\underset{c}{OH}$$

Indicate what geometry i–iv is expected to be found about each of the atoms labeled a, b, and c.

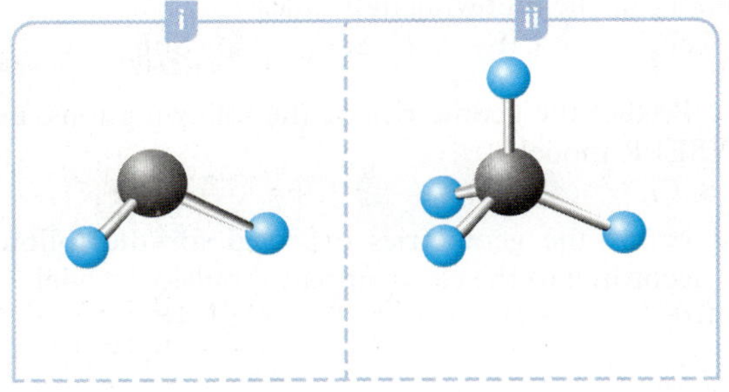

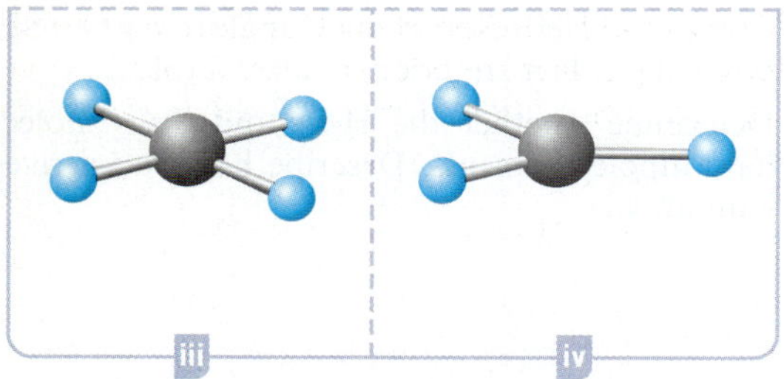

10.32 What are the bond angles predicted by the VSEPR model about the carbon atom in the formate ion, HCO_2^-? Considering that the bonds to this atom are not identical, would you expect the experimental values to agree precisely with the VSEPR values? How might they differ?

Practice Problems

Key: These problems are for practice in applying problem-solving skills. They are divided by topic, and some are keyed to exercises (see the ends of the exercises). The problems are arranged in matching pairs; the odd-numbered problem of each pair is listed first, and its answer is given in the back of the book.

The VSEPR Model

10.33 Predict the shape or geometry of the following molecules, using the VSEPR model.
a SiF_4 b SF_2 c COF_2 d PCl_3

10.34 Use the electron-pair repulsion model to predict the geometry of the following molecules:
a $GeCl_2$ b $AsCl_3$ c SO_3 d XeO_4

10.35 Predict the geometry of the following ions, using the electron-pair repulsion model.
a ClO_3^- b PO_4^{3-} c SCN^- d H_3O^+

10.36 Use the VSEPR model to predict the geometry of the following ions:
a NO_2^- b SO_3^{2-} c N_3^- d BH_4^-

10.37 For each of the following molecules, state the bond angle (or bond angles, as appropriate) that you would expect to see on the central atom based on the simple VSEPR model. Would you expect the actual bond angles to be greater or less than this?
a CCl_4 b SCl_2 c $COCl_2$ d AsH_3

10.38 For each of the following molecules, state the bond angle (or bond angles, as appropriate) that you would expect to see on the central atom based on the VSEPR model. Would you expect the actual bond angles to be greater or less than this?
a OF_2 b NCl_3 c CF_2CF_2 d GeF_4

10.39 What geometry is expected for the following molecules, according to the VSEPR model?
a PF_5 b BrF_3 c BrF_5 d SCl_4

10.40 From the electron-pair repulsion model, predict the geometry of the following molecules:
a TeF_6 b ClF_5 c SeF_4 d SbF_5

10.41 Predict the geometries of the following ions, using the VSEPR model.
a $SnCl_5^-$ b PF_6^- c ClF_2^- d IF_4^-

10.42 Name the geometries expected for the following ions, according to the electron-pair repulsion model.
a BrF_6^+ b IF_2^- c ICl_4 d IF_4^+

Dipole Moment and Molecular Geometry

10.43 a The molecule AsF_3 has a dipole moment of 2.59 D. Which of the following geometries are possible: trigonal planar, trigonal pyramidal, or T-shaped? b The molecule H_2S has a dipole moment of 0.97 D. Is the geometry linear or bent?

10.44 a The molecule BrF_3 has a dipole moment of 1.19 D. Which of the following geometries are possible: trigonal planar, trigonal pyramidal, or T-shaped? b The molecule $TeCl_4$ has a dipole moment of 2.54 D. Is the geometry tetrahedral, seesaw, or square planar?

10.45 Which of the following molecules would be expected to have zero dipole moment on the basis of their geometry?
a CS_2 b TeF_2 c $SeCl_4$ d XeF_4

10.46 Which of the following molecules would be expected to have a dipole moment of zero because of symmetry?
a $BeBr_2$ b H_2Se c AsF_3 d SeF_6

Valence Bond Theory

10.47 What hybrid orbitals would be expected for the central atom in each of the following molecules or ions?

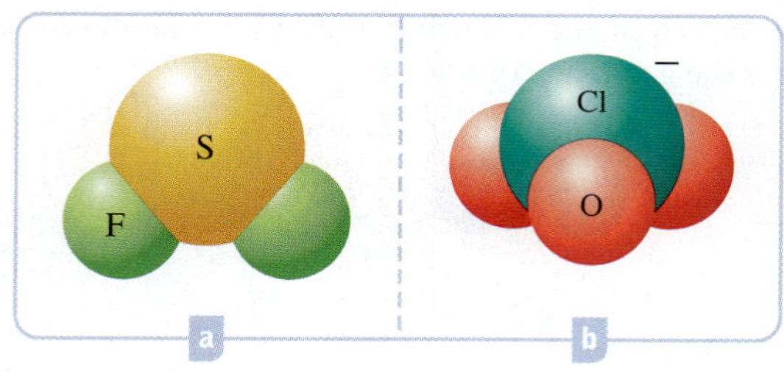

10.48 What hybrid orbitals would be expected for the central atom in each of the following molecules or ions?

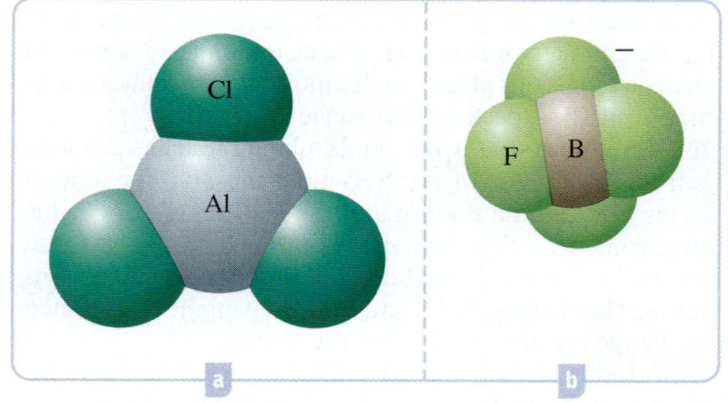

10.49 What hybrid orbitals would be expected for the central atom in each of the following molecules or ions? a $SeCl_2$ b NO_2^- c CO_2 d COF_2

10.50 What hybrid orbitals would be expected for the central atom in each of the following molecules or ions? a PCl_3 b $SiCl_4$ c BeF_2 d SO_2

10.51 a Mercury(II) chloride dissolves in water to give poorly conducting solutions, indicating that the compound is largely nonionized in solution—it dissolves as $HgCl_2$ molecules. Describe the bonding of the $HgCl_2$ molecule, using valence bond theory. b Phosphorus trichloride, PCl_3, is a colorless liquid with a highly irritating vapor. Describe the bonding in the PCl_3 molecule, using valence bond theory. Use hybrid orbitals.

10.52 a Nitrogen trifluoride, NF_3, is a relatively unreactive, colorless gas. How would you describe the bonding in the NF_3 molecule in terms of valence bond theory? Use hybrid orbitals. b Silicon tetrafluoride, SiF_4, is a colorless gas formed when hydrofluoric acid attacks silica (SiO_2) or glass. Describe the bonding in the SiF_4 molecule, using valence bond theory.

10.53 What hybrid orbitals would be expected for the central atom in each of the following? a XeF_2 b BrF_5 c PCl_5 d ClF_4^+

10.54 What hybrid orbitals would be expected for the central atom in each of the following? a SF_4 b $AsCl_5$ c XeF_4 d IF_4^-

10.55 Phosphorus pentachloride is normally a white solid. It exists in this state as the ionic compound $[PCl_4^+][PCl_6^-]$. Describe the electron structure of the PCl_6^- ion in terms of valence bond theory.

10.56 Iodine, I_2, dissolves in an aqueous solution of iodide ion, I^-, to give the triiodide ion, I_3^-. This ion consists of one iodine atom bonded to two others. Describe the bonding of I_3^- in terms of valence bond theory.

10.57 a Formaldehyde, H_2CO, is a colorless, pungent gas used to make plastics. Give the valence bond description of the formaldehyde molecule. (Both hydrogen atoms are attached to the carbon atom.) b Nitrogen, N_2, makes up about 80% of the earth's atmosphere. Give the valence bond description of this molecule.

10.58 a The molecule HN═NH exists as a transient species in certain reactions. Give the valence bond description of this species. b Hydrogen cyanide, HCN, is a very poisonous gas or liquid with the odor of bitter almonds. Give the valence bond description of HCN. (Carbon is the central atom.)

10.59 The hyponitrite ion, $^-$O—N═N—O$^-$, exists in solid compounds as the *trans* isomer. Using valence bond theory, explain why *cis–trans* isomers might be expected for this ion. Draw structural formulas of the *cis–trans* isomers.

10.60 Fumaric acid, $C_4H_4O_4$, occurs in the metabolism of glucose in the cells of plants and animals. It is used commercially in beverages. The structural formula of fumaric acid is

```
H     COOH      (—COOH is an abbreviation
 \   /                  O
  C                     ‖
  ‖             for —C—O—H)
  C
 /   \
HOOC   H
```

Maleic acid is the *cis* isomer of fumaric acid. Using valence bond theory, explain why these isomers are possible.

Molecular Orbital Theory

10.61 Describe the electronic structure of each of the following, using molecular orbital theory. Calculate the bond order of each and decide whether it should be stable. For each, state whether the substance is diamagnetic or paramagnetic.
a B_2 b B_2^+ c O_2^-

10.62 Use molecular orbital theory to describe the bonding in the following. For each one, find the bond order and decide whether it is stable. Is the substance diamagnetic or paramagnetic?
a C_2^+ b Be_2^+ c Ne_2

10.63 Assume that the cyanide ion, CN^-, has molecular orbitals similar to those of a homonuclear diatomic molecule. Write the configuration and bond order of CN^-. Is a substance of the ion diamagnetic or paramagnetic?

10.64 Write the molecular orbital configuration of the diatomic molecule BN. What is the bond order of BN? Is the substance diamagnetic or paramagnetic? Use the order of energies that was given for homonuclear diatomic molecules.

General Problems

Key: These problems provide more practice but are not divided by topic or keyed to exercises. Each section ends with essay questions, each of which is color coded to refer to the *A Chemist Looks at Daily Life* (orange), *A Chemist Looks at Life Science* (pink), and *A Chemist Looks at Environment* (green) chapter essay on which it is based. Odd-numbered problems and the even-numbered problems that follow are similar; answers to all odd-numbered problems except the essay questions are given in the back of the book.

10.65 Predict the molecular geometry of the following:
a $SnCl_2$ b $COBr_2$ c ICl_2^- d PCl_6^-

10.66 Predict the molecular geometry of the following:
a NH_4^+ b HOF c ClF_4^+ d PF_5

10.67 Which of the following molecules or ions are linear?
a NH_4^+ b HOF c ClF_4^+ d PF_5

10.68 Which of the following molecules or ions are trigonal planar?
a AsH_3 b TeF_5^- c BBr_3 d ClF_4^-

10.69 Describe the hybrid orbitals used by each carbon atom in the following molecules:

(a)
```
H     H  H
 \    |  |
  C=C—C—OH
 /       |
H        H
```
(b) N≡C—C≡N

10.70 Describe the hybrid orbitals used by each nitrogen atom in the following molecules:

(a)
```
  H   H
  |   |
H—N—N—H
```
(b) N≡C—C≡N

10.71 Explain how the dipole moment could be used to distinguish between the *cis* and *trans* isomers of 1,2-dibromoethene:

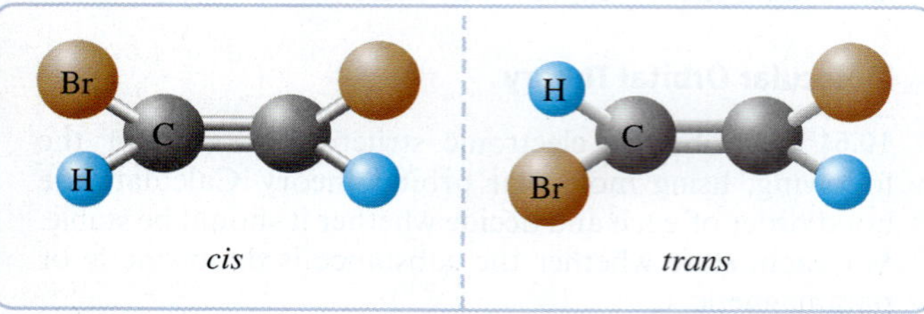

10.72 Two compounds have the formula $Pt(NH_3)_2Cl_2$. (Compound B is *cisplatin,* mentioned in the opening to Chapter 1.) They have square planar structures. One is expected to have a dipole moment; the other is not. Which one would have a dipole moment?

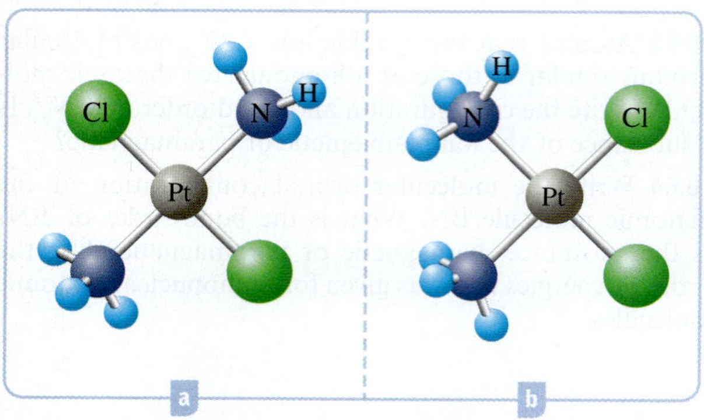

10.73 Explain in terms of bonding theory why all four hydrogen atoms of allene, $H_2C{=}C{=}CH_2$, cannot lie in the same plane.

10.74 Explain in terms of bonding theory why all atoms of $H_2C{=}C{=}C{=}CH_2$ must lie in the same plane.

10.75 What is the molecular orbital configuration of HeH^+? Do you expect the ion to be stable?

10.76 What is the molecular orbital configuration of He_2^+? Do you expect the ion to be stable?

10.77 Calcium carbide, CaC_2, consists of Ca^{2+} and C_2^{2-} (acetylide) ions. Write the molecular orbital configuration and bond order of the acetylide ion, C_2^{2-}.

10.78 Sodium peroxide, Na_2O_2, consists of Na^+ and O_2^{2-} (peroxide) ions. Write the molecular orbital configuration and bond order of the peroxide ion, O_2^{2-}.

10.79 The oxygen–oxygen bond in O_2^+ is 112 pm and in O_2 is 121 pm. Explain why the bond length in O_2^+ is shorter than in O_2. Would you expect the bond length in O_2^- to be longer or shorter than that in O_2? Why?

10.80 The nitrogen–nitrogen bond distance in N_2 is 109 pm. On the basis of bond orders, would you expect the bond distance in N_2^+ to be less than or greater than 109 pm? Answer the same question for N_2^-.

10.81 Using molecular orbital theory, determine the electronic structure of the first excited electronic state of N_2. What differences are expected in the properties of the excited state of N_2 compared with the same properties of the ground state?

10.82 The ionization energy of O_2 is smaller than the ionization energy of atomic O; the opposite is true for the ionization energies of N_2 and atomic N. Explain this behavior in terms of the molecular orbital energy diagrams of O_2 and N_2.

■ **10.83** Describe the property of "handedness." What is it about the bonding of atoms in molecules that can give rise to handedness in molecules?

■ **10.84** Biological molecules frequently display handedness. Describe the difference between the molecule responsible for the flavor of spearmint and the molecule responsible for the flavor of caraway seeds. In which way are the molecules similar and in which way do they differ?

■ **10.85** Color vision results from the absorption of light by the cone cells of the retina. Briefly describe how different types of cone cells give rise to color vision. What is the chemical substance primarily responsible for human vision? How can its color absorption be changed?

■ **10.86** Describe the *cis–trans* conversion that occurs in the ultraviolet absorption of *cis*-1,2-dichloroethene.

■ **10.87** What is the biological importance of stratospheric ozone? Explain.

■ **10.88** In 1995, Sherwood Rowland and Mario Molina received the Nobel Prize in chemistry. Explain in some detail the work they did for which they were awarded this prize.

Strategy Problems

Key: As noted earlier, all of the practice and general problems are matched pairs. This section is a selection of problems that are not in matched-pair format. These challenging problems require that you employ many of the concepts and strategies that were developed in the chapter. In some cases, you will have to integrate several concepts and operational skills in order to solve the problem successfully.

10.89 Each of the following compounds has a nitrogen–nitrogen bond: N_2, N_2H_4, N_2F_2. Match each compound with one of the following bond lengths: 110 pm, 122 pm, 145 pm. Describe the geometry about one of the N atoms in each compound. What hybrid orbitals are needed to describe the bonding in valence bond theory?

10.90 The bond length in C_2 is 131 pm. Compare this with the bond lengths in C_2H_2 (120 pm), C_2H_4 (134 pm), and C_2H_6 (153 pm). What bond order would you predict for C_2 from its bond length? Does this agree with the molecular orbital configuration you would predict for C_2?

10.91 Calcium carbide, CaC_2, has an ionic structure with ions Ca^{2+} and C_2^{2-}. Give the valence bond description of the bonding in the C_2^{2-} ion. Now write the MO configuration of this ion. What is the bond order according to this configuration? Compare the valence bond and molecular orbital descriptions.

10.92 Write Lewis formulas for the BF molecule (two with a single B—F bond, two with a double B—F bond, and one with a triple B—F bond) in which the octet rule is satisfied for at least one of the atoms. Obtain the formal charges of the atoms. Based on the information you have, is there one formula that you think best describes the molecule? Explain. For this best Lewis formula, give the valence bond description of the bonding.

10.93 Boron trifluoride, BF_3, reacts with ammonia, NH_3, to form an addition compound, BF_3NH_3. Describe the geometries about the B and the N atoms in this compound. Describe the hybridization on these two atoms. Now describe the bonding between the B and N atoms, using valence bond theory. Compare the geometries and hybridization of these atoms in this addition compound with that in the reactant molecules BF_3 and NH_3.

10.94 Acetylsalicylic acid (aspirin) has the structure shown by the following molecular model:

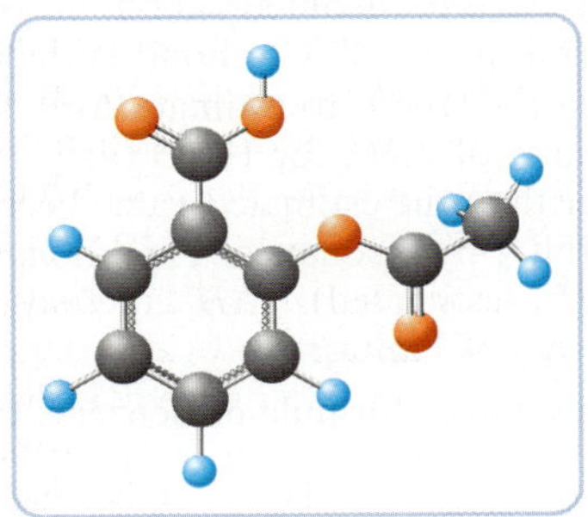

Describe the geometry and bonding about the carbon atom shown at the top of the model, in which the carbon atom (in black) is bonded to two oxygen atoms (red) and to another carbon atom. One of the oxygen atoms is bonded to a hydrogen atom (blue).

10.95 Allene (1,2-propadiene), a gas, has the following structure:

$$\begin{array}{ccc} H & & H \\ & C{=}C{=}C & \\ H & & H \end{array}$$

What is the hybridization at each carbon atom? What is the geometry about each carbon atom? Using valence-bond theory, describe the bonding about the center carbon atom. 1,3-dichloroallene is a derivative of allene with the following structure:

$$\begin{array}{ccc} Cl & & H \\ & C{=}C{=}C & \\ H & & Cl \end{array}$$

Is this a polar molecule? (If you have difficulty visualizing the structure of this molecule, you might build a molecular model of it.)

10.96 When allene (described in the previous problem) is heated with sodium metal, it rearranges to give the isomeric molecule methyl acetylene, which has the following structure:

$$\begin{array}{c} H \\ H{-}C{-}C{\equiv}C{-}H \\ H \end{array}$$

What is the hybridization at each carbon atom? What is the geometry about each carbon atom? Using valence-bond theory, describe the bonding about the center carbon atom.

10.97 The triiodide ion, I_3^-, and the azide ion, N_3^-, have similar skeleton formulas:

$$I{-}I{-}I \qquad N{-}N{-}N$$

Otherwise, how similar are they? To answer, draw electron-dot formulas for each ion, including resonance formulas, if there are any. What is the geometry about the center atom in each molecule? Describe the hybridization about this atom in each molecule.

10.98 Hydrogen azide, HN_3, is a covalent molecule in which the hydrogen atom is bonded to a nitrogen atom at one end of the azide ion (see previous problem). Draw an electron-dot formula of this molecule. Describe the hybridization about the nitrogen atom to which the hydrogen atom is bonded. Use valence-bond theory to describe the bonding about this nitrogen atom. What do you expect for the approximate H—N—N angle?

10.99 There are a number of known pentahalides of phosphorus. One of them is the trifluorodibromophosphorus compound, PF_3Br_2, an unstable liquid. Draw the Lewis electron-dot formula of PF_3Br_2. What is the hybridization of the phosphorus atom expected in this molecule? What is the geometry about this atom? Describe the placement of fluorine and bromine atoms about the phosphorus atom in this geometry. Explain how you come to this conclusion.

10.100 A molecule XF_6 (having no lone pairs) has a dipole moment of zero. (X denotes an unidentified element.) When two atoms of fluorine have been taken away, you get the molecule XF_4, which has dipole moment of 0.632 D.

$$XF_6 \longrightarrow XF_4 + 2F$$

Describe the molecular geometry and bonding for each molecule. What are some possibilities for X?

10.101 Describe the molecular orbital configurations of C_2^+, C_2, and C_2^{2-}. What are the bond orders of these species? Arrange the three species by increasing bond length. Arrange the species by increasing bond enthalpy. Explain these arrangements of bond length and bond enthalpy.

10.102 If the HCl molecule were 100% ionic, the molecule would consist of a positive charge e and a negative charge $-e$ separated by a distance d equal to the bond length. The experimental value of the bond length is 127 pm. Calculate the dipole moment for such a completely ionic molecule. The actual dipole moment, μ, of HCl is 1.08 D. Linus Pauling, Nobel laureate, used the ratio of the actual dipole moment to that of the 100% ionic molecule as an estimate of the fraction of ionic character in the HCl molecule. What is the fraction of ionic character in this molecule according to this estimate?

10.103 Three different compounds have the same molecular formula, $C_2H_2F_2$; call them A, B, and C. Compound A has a dipole moment of 2.42 D, compound B has a dipole moment of 1.38 D, and compound C has a dipole moment of 0 D. All three compounds react with hydrogen, H_2:

$$C_2H_2F_2 + H_2 \longrightarrow C_2H_4F_2$$

Compound A gives product D (molecular formula $C_2H_4F_2$), compound B gives product E (same molecular formula, $C_2H_4F_2$, but different from product D), and compound C gives product D. Identify the structural formulas of compounds A, B, and C.

10.104 Obtain a Lewis formula of H_2N_2. This molecule exhibits isomerism (it has two isomers). Sketch the electron-dot formulas of these isomers. Describe the bonding in terms of valence bond theory, and explain this isomerism in terms of this bonding description.

10.105 Dioxygen, O_2, has sometimes been represented by the electron-dot formula

$$\ddot{\underset{..}{O}}{=}\ddot{\underset{..}{O}}$$

Note, however, that this formula implies that all electrons are paired. Give the molecular orbital description of the state of the O_2 molecule of lowest energy *in which all electrons are paired in this way.* Compare this configuration with that of the ground state of O_2. In what way are they similar, and in what way are they different?

10.106 Solid sulfur normally consists of crystals of S_8 molecules, but when heated strongly, the solid vaporizes to give S_2 molecules (among other molecular species). Describe the bonding in S_2 in molecular orbital terms, assuming the orbitals are analogous to those of the preceding period. What would you expect to happen to the sulfur–sulfur bond length if two electrons were added to give the S_2^{2-} ion? What would you expect to happen to the bond length if, instead, two electrons were taken away to give S_2^{2+}?

10.107 Formaldehyde has the formula H_2CO. Describe the bonding between the C and O atoms in VB terms. Formaldehyde absorbs UV light at about 270 nm. The absorption is described as a π to π^* transition. Explain in molecular orbital terms what is involved.

10.108 Consider the bonding in nitrate ion, NO_3^-. First draw resonance formulas of this ion. Now describe the bonding of this ion in terms of molecular orbitals. (Refer to the delocalized bonding of the ozone molecule described in the text.) Suppose each atom uses sp^2 hybrid orbitals. How many π molecular orbitals can you form from the $2p$ orbitals that remain on these atoms? How many of these π orbitals will be occupied?

Cumulative-Skills Problems

Key: The problems under this heading combine skills introduced in previous chapters with those given in the current one.

10.109 A molecular compound is composed of 60.4% Xe, 22.1% O, and 17.5% F, by mass. If the molecular weight is 217.3 amu, what is the molecular formula? What is the Lewis formula? Predict the molecular geometry using the VSEPR model. Describe the bonding, using valence bond theory.

10.110 A molecular compound is composed of 58.8% Xe, 7.2% O, and 34.0% F, by mass. If the molecular weight is 223 amu, what is the molecular formula? What is the Lewis formula? Predict the molecular geometry using the VSEPR model. Describe the bonding, using valence bond theory.

10.111 A compound of chlorine and fluorine, ClF_n, reacts at about 75°C with uranium metal to produce uranium hexafluoride, UF_6, and chlorine monofluoride, $ClF(g)$. A quantity of uranium produced 3.53 g UF_6 and 343 mL ClF at 75°C and 2.50 atm. What is the formula (n) of the compound? Describe the bonding in the molecule, using valence bond theory.

10.112 Excess fluorine, $F_2(g)$, reacts at 150°C with bromine, $Br_2(g)$, to give a compound BrF_n. If 423 mL $Br_2(g)$ at 150°C and 748 mmHg produced 4.20 g BrF_n, what is n? Describe the bonding in the molecule, using valence bond theory.

10.113 Draw resonance formulas of the nitric acid molecule, HNO_3. What is the geometry about the N atom? What is the hybridization on N? Use bond enthalpies and one Lewis formula for HNO_3 to estimate ΔH°_f for $HNO_3(g)$. The actual value of ΔH°_f for $HNO_3(g)$ is −135 kJ/mol, which is lower than the estimated value because of stabilization of HNO_3 by resonance. The *resonance energy* is defined as ΔH°_f (estimated) $-\Delta H^\circ_f$ (actual). What is the resonance energy of HNO_3?

10.114 One resonance formula of benzene, C_6H_6, is

```
          H
          |
          C
        /   \\
   H—C       C—H
     ||      |
   H—C       C—H
        \   //
          C
          |
          H
```

What is the other resonance formula? What is the geometry about a carbon atom? What hybridization would be used in valence bond theory to describe the bonding? The ΔH°_f for $C_6H_6(g)$ is –83 kJ/mol; ΔH°_f for C(g) is 715 kJ/mol. Obtain the resonance energy of benzene. (See Problem 10.113.)

STUDENT SOLUTIONS MANUAL

CHAPTER 1

Chemistry and Measurement

■ SOLUTIONS TO EXERCISES

Note on significant figures: If the final answer to a solution needs to be rounded off, it is given first with one nonsignificant figure, and the last significant figure is underlined. The final answer is then rounded to the correct number of significant figures. In multistep problems, intermediate answers are given with at least one nonsignificant figure; however, only the final answer has been rounded off.

1.1. From the law of conservation of mass,

Mass of wood + mass of air = mass of ash + mass of gases

Substituting, you obtain

1.85 grams + 9.45 grams = 0.28 grams + mass of gases

or,

Mass of gases = (1.85 + 9.45 − 0.28) grams = 11.02 grams

Thus, the mass of gases in the vessel at the end of the experiment is 11.02 grams.

1.2. Physical properties: soft, silvery-colored metal; melts at 64°C.
Chemical properties: reacts vigorously with water, with oxygen, and with chlorine.

1.3. a. The factor 9.1 has the fewest significant figures, so the answer should be reported to two significant figures.

$$\frac{5.61 \times 7.891}{9.1} = 4.\underline{8}6 = 4.9$$

b. The number with the least number of decimal places is 8.91. Therefore, round the answer to two decimal places.

$8.91 - 6.435 = 2.4\underline{7}5 = 2.48$

c. The number with the least number of decimal places is 6.81. Therefore, round the answer to two decimal places.

$6.81 - 6.730 = 0.0\underline{8}0 = 0.08$

d. You first do the subtraction within parentheses. In this step, the number with the least number of decimal places is 6.81, so the result of the subtraction has two decimal places. The least significant figure for this step is underlined.

$38.91 \times (6.81 - 6.730) = 38.91 \times 0.0\underline{8}0$

Next, perform the multiplication. In this step, the factor $0.0\underline{8}0$ has the fewest significant figures, so round the answer to one significant figure.

$38.91 \times 0.0\underline{8}0 = \underline{3}.11 = 3$

1.4. a. 1.84×10^{-9} m = 1.84 nm

b. 5.67×10^{-12} s = 5.67 ps

c. 7.85×10^{-3} g = 7.85 mg

d. 9.7×10^{3} m = 9.7 km

e. 0.000732 s = 0.732 ms, or 732 μs

f. 0.000000000154 m = 0.154 nm, or 154 pm

1.5. a. Substituting, we find that

$$t_C = \frac{5°C}{9°F} \times (t_F - 32°F) = \frac{5°C}{9°F} \times (102.5°F - 32°F) = 39.\underline{1}67°C$$

$= 39.2°C$

b. Substituting, we find that

$$T_K = \left(t_c \times \frac{1\ K}{1°C}\right) + 273.15\ K = \left(-78°C \times \frac{1\ K}{1°C}\right) + 273.15\ K = 19\underline{5}.15\ K$$

$= 195$ K

1.6. Recall that density equals mass divided by volume. You substitute 159 g for the mass and 20.2 g/cm^3 for the volume.

$$d = \frac{m}{V} = \frac{159\ g}{20.2\ cm^3} = 7.8\underline{7}1\ g/cm^3 = 7.87\ g/cm^3$$

The density of the metal equals that of iron.

1.7. Rearrange the formula defining the density to obtain the volume.

$$V = \frac{m}{d}$$

Substitute 30.3 g for the mass and 0.789 g/cm^3 for the density.

$$V = \frac{30.3\ g}{0.789\ g/cm^3} = 38.\underline{4}0\ cm^3 = 38.4\ cm^3$$

1.8. Since one pm = 10^{-12} m, and the prefix milli- means 10^{-3}, you can write

$$121\ pm \times \frac{10^{-12}\ m}{1\ pm} \times \frac{1\ mm}{10^{-3}\ m} = 1.21 \times 10^{-7}\ mm$$

1.9. $$67.6\ Å^3 \times \left(\frac{10^{-10}\ m}{1\ Å}\right)^3 \times \left(\frac{1\ dm}{10^{-1}\ m}\right)^3 = 6.76 \times 10^{-26}\ dm^3$$

1.10. From the definitions, you obtain the following conversion factors:

$$1 = \frac{36\ in}{1\ yd} \qquad 1 = \frac{2.54\ cm}{1\ in} \qquad 1 = \frac{10^{-2}\ m}{1\ cm}$$

The conversion factor for yards to meters is as follows:

$$1.000 \text{ yd} \times \frac{36 \text{ in}}{1 \text{ yd}} \times \frac{2.54 \text{ cm}}{1 \text{ in}} \times \frac{10^{-2} \text{ m}}{1 \text{ cm}} = 0.9144 \text{ m (exact)}$$

Finally,

$$3.5\underline{4} \text{ yd} \times \frac{0.9144 \text{ m}}{1 \text{ yd}} = 3.2\underline{3}7 \text{ m} = 3.24 \text{ m}$$

ANSWERS TO CONCEPT CHECKS

1.1. Box A contains a collection of identical units; therefore, it must represent an element. Box B contains a compound because a compound is the chemical combination of two or more elements (two elements in this case). Box C contains a mixture because it is made up of two different substances.

1.2. a. For a person who weighs less than 100 pounds, two significant figures are typically used, although one significant figure is possible (for example, 60 pounds). For a person who weighs 100 pounds or more, three significant figures are typically used to report the weight (given to the whole pound), although people often round to the nearest unit of 10, which may result in reporting the weight with two significant figures (for example, 170 pounds).

b. Assuming a weight of 165 pounds, rounded to two significant figures this would be reported as 1.7×10^2 pounds.

c. For example, 165 lb weighed on a scale that can measure in 100-lb increments would be 200 lb. Using the conversion factor 1 lb = 0.4536 kg, 165 lb is equivalent to 74.8 kg. Thus, on a scale that can measure in 50-kg increments, 165 lb would be 50 kg.

1.3. a. If your leg is approximately 32 inches long, this would be equivalent to 0.81 m, 8.1 dm, or 81 cm.

b. One story is approximately 10 feet, so three stories is 30 feet. This would be equivalent to approximately 9 m.

c. Normal body temperature is 98.6°F, or 37.0°C. Thus, if your body temperature were 39°C (102°F), you would feel as if you had a moderate fever.

d. Room temperature is approximately 72°F, or 22°C. Thus, if you were sitting in a room at 23°C (73°F), you would be comfortable in a short-sleeve shirt.

1.4. Gold is a very unreactive substance, so comparing physical properties is probably your best option. However, color is a physical property you cannot rely on in this case to get your answer.

One experiment you could perform is to determine the densities of the metal and the chunk of gold. You could measure the mass of the nugget on a balance and the volume of the nugget by water displacement. Using this information, you could calculate the density of the nugget. Repeat the experiment and calculations for the sample of gold. If the nugget is gold, the two densities should be equal and be 19.3 g/cm^3.

Also, you could determine the melting points of the metal and the chunk of pure gold. The two melting points should be the same (1338 K) if the metal is gold.

■ ANSWERS TO SELF-ASSESSMENT AND REVIEW QUESTIONS

1.1. One area of technology that chemistry has changed is the characteristics of materials. The liquid-crystal displays (LCDs) in devices such as watches, cell phones, computer monitors, and televisions are materials made of molecules designed by chemists. Electronics and communications have been transformed by the development of optical fibers to replace copper wires. In biology, chemistry has changed the way scientists view life. Biochemists have found that all forms of life share many of the same molecules and molecular processes.

1.2. An experiment is an observation of natural phenomena carried out in a controlled manner so that the results can be duplicated and rational conclusions obtained. A theory is a tested explanation of basic natural phenomena. They are related in that a theory is based on the results of many experiments and is fruitful in suggesting other, new experiments. Also, an experiment can disprove a theory but can never prove it absolutely. A hypothesis is a tentative explanation of some regularity of nature.

1.3. Rosenberg conducted controlled experiments and noted a basic relationship that could be stated as a hypothesis—that is, that certain platinum compounds inhibit cell division. This led him to do new experiments on the anticancer activity of these compounds.

1.4. Matter is the general term for the material things around us. It is whatever occupies space and can be perceived by our senses. Mass is the quantity of matter in a material. The difference between mass and weight is that mass remains the same wherever it is measured, but weight is proportional to the mass of the object divided by the square of the distance between the center of mass of the object and that of the earth.

1.5. The law of conservation of mass states that the total mass remains constant during a chemical change (chemical reaction). To demonstrate this law, place a sample of wood in a sealed vessel with air, and weigh it. Heat the vessel to burn the wood, and weigh the vessel after the experiment. The weight before the experiment and that after it should be the same.

1.6. Mercury metal, which is a liquid, reacts with oxygen gas to form solid mercury(II) oxide. The color changes from that of metallic mercury (silvery) to a color that varies from red to yellow depending on the particle size of the oxide.

1.7. Gases are easily compressible and fluid. Liquids are relatively incompressible and fluid. Solids are relatively incompressible and rigid.

1.8. An example of a substance is the element sodium. Among its physical properties: It is a solid, and it melts at 98°C. Among its chemical properties: It reacts vigorously with water, and it burns in chlorine gas to form sodium chloride.

1.9. An example of an element: sodium; of a compound: sodium chloride, or table salt; of a heterogeneous mixture: salt and sugar; of a homogeneous mixture: sodium chloride dissolved in water to form a solution.

1.10. A glass of bubbling carbonated beverage with ice cubes contains three phases: gas, liquid, and solid.

1.11. A compound may be decomposed by chemical reactions into elements. An element cannot be decomposed by any chemical reaction. Thus, a compound cannot also be an element in any case.

1.12. The precision refers to the closeness of the set of values obtained from identical measurements of a quantity. The number of digits reported for the value of a measured or calculated quantity (significant figures) indicates the precision of the value.

1.13. Multiplication and division rule: In performing the calculation 100.0 x 0.0634 ÷ 25.31, the calculator display shows 0.2504938. We would report the answer as 0.250 because the factor 0.0634 has the least number of significant figures (three).

Addition and subtraction rule: In performing the calculation 184.2 + 2.324, the calculator display shows 186.524. Because the quantity 184.2 has the least number of decimal places (one), the answer is reported as 186.5.

1.14. An exact number is a number that arises when you count items or sometimes when you define a unit. For example, a foot is defined to be 12 inches. A measured number is the result of a comparison of a physical quantity with a fixed standard of measurement. For example, a steel rod measures 9.12 centimeters, or 9.12 times the standard centimeter unit of measurement.

1.15. For a given unit, the SI system uses prefixes to obtain units of different sizes. Units for all other possible quantities are obtained by deriving them from any of the seven base units. You do this by using the base units in equations that define other physical quantities.

1.16. An absolute temperature scale is a scale in which the lowest temperature that can be attained theoretically is zero. Degrees Celsius and kelvins have units of equal size and are related by the formula

$$t_C = (T_K - 273.15\text{ K}) \times \frac{1°\text{C}}{1\text{ K}}$$

1.17. The density of an object is its mass per unit volume. Because the density is characteristic of a substance, it can be helpful in identifying it. Density can also be useful in determining whether a substance is pure. It also provides a useful relationship between mass and volume.

1.18. Units should be carried along because (1) the units for the answers will come out in the calculations, and (2), if you make an error in arranging factors in the calculation, this will become apparent because the final units will be nonsense.

1.19. The answer is c, three significant figures.

1.20. The answer is a, 4.43×10^2 mm.

1.21. The answer is e, 75 mL.

1.22. The answer is c, 0.23 mg.

■ ANSWERS TO CONCEPTUAL PROBLEMS

1.25. a. Two phases: liquid and solid.

b. Three phases: liquid water, solid quartz, and solid seashells.

1.27. a. You need to establish two points on the thermometer with known (defined) temperatures—for example, the freezing point (0°C) and boiling point (100°C) of water. You could first immerse the thermometer in an ice-water bath and mark the level at this point as 0°C. Then, immerse the thermometer in boiling water, and mark the level at this point as 100°C. As long as the two points are far enough apart to obtain readings of the desired accuracy, the thermometer can be used in experiments.

b. You could make 19 evenly spaced marks on the thermometer between the two original points, each representing a difference of 5°C. You may divide the space between the two original points into fewer spaces as long as you can read the thermometer to obtain the desired accuracy.

1.29. a. To answer this question, you need to develop an equation that converts between °F and °YS. To do so, you need to recognize that one degree on the Your Scale does not correspond to one degree on the Fahrenheit scale and that −100°F corresponds to 0° on Your Scale (different "zero" points). As stated in the problem, in the desired range of 100 Your Scale degrees, there are 120 Fahrenheit degrees. Therefore, the relationship can be expressed as 120°F = 100°YS, since it covers the same temperature range. Now you need to "scale" the two systems so that they correctly convert from one scale to the other. You could set up an equation with the known data points and then employ the information from the relationship above.

For example, to construct the conversion between °YS and °F, you could perform the following steps:

Step 1: $°F = °YS$

Not a true statement, but one you would like to make true.

Step 2: $°F = °YS \times \dfrac{120°F}{100°YS}$

This equation takes into account the difference in the size between the temperature unit on the two scales but will not give you the correct answer because it doesn't take into account the different zero points.

Step 3: By subtracting 100°F from your equation from Step 2, you now have the complete equation that converts between °F and °YS.

$$°F = (°YS \times \frac{120°F}{100°YS}) - 100°F$$

b. Using the relationship from part a, 66°YS is equivalent to

$$(66°YS \times \frac{120°F}{100°YS}) - 100°F = -2\underline{0}.8°F = -21°F$$

1.31. The empty boxes are identical, so they do not contribute to any mass or density difference. Since the edge of the cube and the diameter of the sphere are identical, they will occupy the same volume in each of the boxes; therefore, each box will contain the same number of cubes or spheres. If you view the spheres as cubes that have been rounded by removing wood, you can conclude that the box containing the cubes must have a greater mass of wood; hence, it must have a greater density.

1.33. a. A paper clip has a mass of about 1 g.

b. Answers will vary depending on your particular sample. Keeping in mind that the SI unit for mass is kg, the approximate weights for the items presented in the problem are as follows: a grain of sand, 1×10^{-5} kg; a paper clip, 1×10^{-3} kg; a nickel, 5×10^{-3} kg; a 5.0-gallon bucket of water, 2.0×10^{1} kg; a brick, 3 kg; a car, 1×10^{3} kg.

1.35. a. The number of significant figures in this answer follows the rules for multiplication and division. Here, the measurement with the fewest significant figures is the reported volume 0.310 m^3, which has three. Therefore, the answer will have three significant figures. Since Volume = $L \times W \times H$, you can rearrange and solve for one of the measurements, say the length.

$$L = \frac{V}{W \times H} = \frac{0.310\ \text{m}^3}{(0.7120\ \text{m})\ (0.52145\ \text{m})} = 0.83\underline{4}96\ \text{m} = 0.835\ \text{m}$$

b. The number of significant figures in this answer follows the rules for addition and subtraction. The measurement with the least number of decimal places is the result 1.509 m, which has three. Therefore, the answer will have three decimal places. Since the result is the sum of the three measurements, the third length is obtained by subtracting the other two measurements from the total.

$$\text{Length} = 1.509\ \text{m} - 0.7120\ \text{m} - 0.52145\ \text{m} = 0.27\underline{5}55\ \text{m} = 0.276\ \text{m}$$

■ SOLUTIONS TO PRACTICE PROBLEMS

Note on significant figures: If the final answer to a solution needs to be rounded off, it is given first with one nonsignificant figure, and the last significant figure is underlined. The final answer is then rounded to the correct number of significant figures. In multistep problems, intermediate answers are given with at least one nonsignificant figure; however, only the final answer has been rounded off.

1.37 . By the law of conservation of mass:

Mass of sodium carbonate + mass of acetic acid solution = mass of contents of reaction vessel + mass of carbon dioxide

Plugging in gives

$$15.9\ \text{g} + 20.0\ \text{g} = 29.3\ \text{g} + \text{mass of carbon dioxide}$$

$$\text{Mass of carbon dioxide} = 15.9\ \text{g} + 20.0\ \text{g} - 29.3\ \text{g} = 6.6\ \text{g}$$

1.39. By the law of conservation of mass:

Mass of zinc + mass of sulfur = mass of zinc sulfide

Rearranging and plugging in give

$$\text{Mass of zinc sulfide} = 65.4\ \text{g} + 32.1\ \text{g} = 97.5\ \text{g}$$

For the second part, let x = mass of zinc sulfide that could be produced. By the law of conservation of mass:

$$20.0\ \text{g} + \text{mass of sulfur} = x$$

Write a proportion that relates the mass of zinc reacted to the mass of zinc sulfide formed, which should be the same for both cases.

$$\frac{\text{mass zinc}}{\text{mass zinc sulfide}} = \frac{65.4\text{ g}}{97.5\text{ g}} = \frac{20.0\text{ g}}{x}$$

Solving gives $x = 29.\underline{8}1\text{ g} = 29.8\text{ g}$

1.41 . a. Solid b. Liquid c. Gas d. Solid

1.43. a. Physical change
b. Physical change
c. Chemical change
d. Physical change

1.45. Physical change: Liquid mercury is cooled to solid mercury.

Chemical changes: (1) Solid mercury oxide forms liquid mercury metal and gaseous oxygen; (2) glowing wood and oxygen form burning wood (form ash and gaseous products).

1.47. a. Physical property
b. Chemical property
c. Physical property
d. Physical property
e. Chemical property

1.49. Physical properties: (1) Iodine is solid; (2) the solid has lustrous blue-black crystals; (3) the crystals vaporize readily to a violet-colored gas.

Chemical properties: (1) Iodine combines with many metals, such as with aluminum to give aluminum iodide.

1.51. a. Physical process
b. Chemical reaction
c. Physical process
d. Chemical reaction
e. Physical process

1.53. a. Solution
b. Substance
c. Substance
d. Heterogeneous mixture

1.55. a. A pure substance with two phases present, liquid and gas.

b. A mixture with two phases present, solid and liquid.

c. A pure substance with two phases present, solid and liquid.

d. A mixture with two phases present, solid and solid.

1.57. a. four

b. three

c. four

d. five

e. three

f. four

1.59. $4\underline{0},000 \text{ km} = 4.0 \times 10^4 \text{ km}$

1.61. a. $\dfrac{8.71 \times 0.0301}{0.031} = 8.\underline{4}57 = 8.5$

b. $0.71 + 89.3 = 90.\underline{0}1 = 90.0$

c. $934 \times 0.00435 + 107 = 4.0\underline{6}29 + 107 = 11\underline{1}.06 = 111$

d. $(847.89 - 847.73) \times 14673 = 0.1\underline{6} \times 14673 = 2\underline{3}47 = 2.3 \times 10^3$

1.63. The volume of the first sphere is

$$V_1 = (4/3)\pi r^3 = (4/3)\pi \times (5.15 \text{ cm})^3 = 57\underline{2}.15 \text{ cm}^3$$

The volume of the second sphere is

$$V_2 = (4/3)\pi r^3 = (4/3)\pi \times (5.00 \text{ cm})^3 = 52\underline{3}.60 \text{ cm}^3$$

The difference in volume is

$$V_1 - V_2 = 57\underline{2}.15 \text{ cm}^3 - 52\underline{3}.60 \text{ cm}^3 = 4\underline{8}.55 \text{ cm}^3 = 49 \text{ cm}^3$$

1.65. a. $5.89 \times 10^{-12} \text{ s} = 5.89 \text{ ps}$

b. $0.2010 \text{ m} = 2.01 \text{ dm}$

c. $2.560 \times 10^{-9} \text{ g} = 2.560 \text{ ng}$

d. $6.05 \times 10^3 \text{ m} = 6.05 \text{ km}$

1.67. a. $6.15 \text{ ps} = 6.15 \times 10^{-12} \text{ s}$

b. $3.781\ \mu\text{m} = 3.781 \times 10^{-6} \text{ m}$

c. $1.546\ \text{Å} = 1.546 \times 10^{-10} \text{ m}$

d. $9.7 \text{ mg} = 9.7 \times 10^{-3} \text{ g}$

1.69. a. $t_C = \frac{5°C}{9°F} \times (t_F - 32°F) = \frac{5°C}{9°F} \times (68°F - 32°F) = 2\underline{0}.0°C = 20.°C$

b. $t_C = \frac{5°C}{9°F} \times (t_F - 32°F) = \frac{5°C}{9°F} \times (-23°F - 32°F) = -3\underline{0}.56°C = -31°C$

c. $t_F = (t_C \times \frac{9°F}{5°C}) + 32°F = (26°C \times \frac{9°F}{5°C}) + 32°F = 7\underline{8}.8°F = 79°F$

d. $t_F = (t_C \times \frac{9°F}{5°C}) + 32°F = (-81°C \times \frac{9°F}{5°C}) + 32°F = -11\underline{3}.8°F = -114°F$

1.71. $t_F = (t_C \times \frac{9°F}{5°C}) + 32°F = (-20.0°C \times \frac{9°F}{5°C}) + 32°F = -4.\underline{0}°F = -4.0°F$

1.73. $d = \frac{m}{V} = \frac{12.4\text{ g}}{1.64\text{ cm}^3} = 7.5\underline{6}0\text{ g/cm}^3 = 7.56\text{ g/cm}^3$

1.75. First, determine the density of the liquid.

$$d = \frac{m}{V} = \frac{6.71\text{ g}}{8.5\text{ mL}} = 0.7\underline{8}94 = 0.79\text{ g/mL}$$

The density is closest to ethanol (0.789 g/cm^3).

1.77. The mass of platinum is obtained as follows.

Mass = $d \times V = 21.4\text{ g/cm}^3 \times 5.9\text{ cm}^3 = 1\underline{2}6\text{ g} = 1.3 \times 10^2\text{ g}$

1.79. The volume of ethanol is obtained as follows. Recall that 1 mL = 1 cm^3.

$$\text{Volume} = \frac{m}{d} = \frac{19.8\text{ g}}{0.789\text{ g/cm}^3} = 25.\underline{0}9\text{ cm}^3 = 25.1\text{ cm}^3 = 25.1\text{ mL}$$

1.81. Since 1 kg = 10^3 g, and 1 mg = 10^{-3} g, you can write

$$0.480\text{ kg x } \frac{10^3\text{ g}}{1\text{ kg}} \times \frac{1\text{ mg}}{10^{-3}\text{ g}} = 4.80 \times 10^5\text{ mg}$$

1.83. Since 1 nm = 10^{-9} m, and 1 cm = 10^{-2} m, you can write

$$555\text{ nm} \times \frac{10^{-9}\text{ m}}{1\text{ nm}} \times \frac{1\text{ cm}}{10^{-2}\text{ m}} = 5.55 \times 10^{-5}\text{ cm}$$

1.85. Since 1 km = 10^3 m, you can write

$$3.73 \times 10^8\text{ km}^3 \times \left(\frac{10^3\text{ m}}{1\text{ km}}\right)^3 = 3.73 \times 10^{17}\text{ m}^3$$

Now, 1 dm = 10^{-1} m. Also, note that 1 dm^3 = 1 L. Therefore, you can write

$$3.73 \times 10^{17}\text{ m}^3 \times \left(\frac{1\text{ dm}}{10^{-1}\text{ m}}\right)^3 = 3.73 \times 10^{20}\text{ dm}^3 = 3.73 \times 10^{20}\text{ L}$$

1.87. $3.58 \text{ short ton} \times \frac{2000 \text{ lb}}{1 \text{ short ton}} \times \frac{16 \text{ oz}}{1 \text{ lb}} \times \frac{1 \text{ g}}{0.03527 \text{ oz}} = 3.2\underline{4}8 \times 10^6 \text{ g} = 3.25 \times 10^6 \text{ g}$

1.89. $2425 \text{ fathoms} \times \frac{6 \text{ ft}}{1 \text{ fathom}} \times \frac{12 \text{ in.}}{1 \text{ ft}} \times \frac{2.54 \times 10^{-2} \text{ m}}{1 \text{ in.}} = 443\underline{4}.8 \text{ m} = 4.435 \times 10^3 \text{ m}$

1.91. $(20.0 \text{ in.}) \times (20.0 \text{ in.}) \times (10.0 \text{ in.}) \times \left(\frac{2.54 \text{ cm}}{1 \text{ in.}}\right)^3 \times \frac{1 \text{ L}}{1000 \text{ cm}^3} = 65.\underline{5}4 \text{ L} = 65.5 \text{ L}$

SOLUTIONS TO GENERAL PROBLEMS

1.93. From the law of conservation of mass,

Mass of sodium + mass of water = mass of hydrogen + mass of solution

Substituting, you obtain

$$19.70 \text{ g} + 126.22 \text{ g} = \text{mass of hydrogen} + 145.06 \text{ g}$$

or,

$$\text{Mass of hydrogen} = 19.70 \text{ g} + 126.22 \text{ g} - 145.06 \text{ g} = 0.8\underline{6} \text{ g}$$

Thus, the mass of hydrogen produced was 0.86 g.

1.95. From the law of conservation of mass,

Mass of aluminum + mass of iron(III) oxide = mass of iron +

mass of aluminum oxide + mass of unreacted iron(III) oxide

$$5.40 \text{ g} + 18.50 \text{ g} = 11.17 \text{ g} + 10.20 \text{ g} + \text{mass of iron(III) oxide unreacted}$$

Mass of iron(III) oxide unreacted = 5.40 g + 18.50 g − 11.17 g − 10.20 g = 2.53 g

Thus, the mass of unreacted iron(III) oxide is 2.53 g.

1.97. $53.10 \text{ g} + 5.348 \text{ g} + 56.1 \text{ g} = 114.\underline{5}4 \text{ g} = 114.5 \text{ g total}$

1.99. a. Chemical b. Physical c. Physical d. Chemical

1.101. Compounds always contain the same proportions of the elements by mass. Thus, if we let X be the proportion of iron in a sample, we can calculate the proportion of iron in each sample as follows.

Sample A: $X = \frac{\text{mass of iron}}{\text{mass of sample}} = \frac{1.094 \text{ g}}{1.518 \text{ g}} = 0.720\underline{6}8 = 0.7207$

Sample B: $X = \frac{\text{mass of iron}}{\text{mass of sample}} = \frac{1.449 \text{ g}}{2.056 \text{ g}} = 0.704\underline{7}6 = 0.7048$

Sample C: $X = \frac{\text{mass of iron}}{\text{mass of sample}} = \frac{1.335 \text{ g}}{1.873 \text{ g}} = 0.712\underline{7}6 = 0.7128$

Since each sample has a different proportion of iron by mass, the material is not a compound.

1.103. $V = (\text{edge})^3 = (39.3\ \text{cm})^3 = 6.0\underline{6}9 \times 10^4\ \text{cm}^3 = 6.07 \times 10^4\ \text{cm}^3$

1.105. $V = LWH = 47.8\ \text{in.} \times 12.5\ \text{in.} \times 19.5\ \text{in.} \times \dfrac{1\ \text{gal}}{231\ \text{in}^3} = 50.\underline{4}3\ \text{gal} = 50.4\ \text{gal}$

1.107. The volume of the first sphere is given by

$$V_1 = (4/3)\pi r^3 = (4/3)\pi \times (5.61\ \text{cm})^3 = 73\underline{9}.5\ \text{cm}^3$$

The volume of the second sphere is given by

$$V_2 = (4/3)\pi r^3 = (4/3)\pi \times (5.85\ \text{cm})^3 = 83\underline{8}.6\ \text{cm}^3$$

The difference in volume between the two spheres is given by

$$V = V_2 - V_1 = 83\underline{8}.6\ \text{cm}^3 - 73\underline{9}.5\ \text{cm}^3 = 9.\underline{9}1 \times 10^1 = 9.9 \times 10^1\ \text{cm}^3$$

1.109. a. $\dfrac{56.1 - 51.1}{6.58} = 7.59 \times 10^{-1} = 7.6 \times 10^{-1}$

b. $\dfrac{56.1 + 51.1}{6.58} = 1.6\underline{2}9 \times 10^1 = 1.63 \times 10^1$

c. $(9.1 + 8.6) \times 26.91 = 4.7\underline{6}3 \times 10^2 = 4.76 \times 10^2$

d. $0.0065 \times 3.21 + 0.0911 = 1.1\underline{1}9 \times 10^{-1} = 1.12 \times 10^{-1}$

1.111. a. 9.12 cg

b. 66 pm

c. 7.1 μm

d. 56 nm

1.113. a. 1.07×10^{-12} s

b. 5.8×10^{-6} m

c. 3.19×10^{-7} m

d. 1.53×10^{-2} s

1.115. $t_F = (t_C \times \dfrac{9°\text{F}}{5°\text{C}}) + 32°\text{F} = (3410°\text{C} \times \dfrac{9°\text{F}}{5°\text{C}}) + 32°\text{F} = 617\underline{0}°\text{F} = 6170°\text{F}$

1.117. $t_F = (t_C \times \dfrac{9°\text{F}}{5°\text{C}}) + 32°\text{F} = (825°\text{C} \times \dfrac{9°\text{F}}{5°\text{C}}) + 32°\text{F} = 15\underline{1}7°\text{F} = 1.52 \times 10^3°\text{F}$

1.119. The temperature in kelvins is

$$T_K = (t_C \times \frac{1\ \text{K}}{1°\text{C}}) + 273.15\ \text{K} = (29.8°\text{C} \times \frac{1\ \text{K}}{1°\text{C}}) + 273.15\ \text{K} = 302.\underline{9}5\ \text{K}$$
$$= 303.0\ \text{K}$$

The temperature in degrees Fahrenheit is

$$t_F = (t_C \times \frac{9°F}{5°C}) + 32°F = (29.8°C \times \frac{9°F}{5°C}) + 32°F = 85.\underline{6}4°F = 85.6°F$$

1.121. The temperature in degrees Celsius is

$$t_C = \frac{5°C}{9°F} \times (t_F - 32°F) = \frac{5°C}{9°F} \times (1666°F - 32°F) = 907.\underline{7}7°C = 907.8°C$$

The temperature in kelvins is

$$T_K = (t_C \times \frac{1\ K}{1°C}) + 273.15\ K = (907.\underline{7}7°C \times \frac{1\ K}{1°C}) + 273.15\ K = 1180.\underline{9}2\ K$$
$$= 1180.9\ K$$

1.123. $\text{Density} = \frac{2.70\ g}{1\ cm^3} \times \frac{1\ kg}{10^3\ g} \times \left(\frac{1\ cm}{10^{-2}\ m}\right)^3 = 2.70 \times 10^3\ kg/m^3$

1.125. The volume of the quartz is 65.7 mL − 51.2 mL = 14.5 mL. Then, the density is

$$\text{Density} = \frac{\text{mass}}{\text{volume}} = \frac{38.4\ g}{14.5\ mL} = 2.6\underline{4}8\ g/mL = 2.65\ g/mL = 2.65\ g/cm^3$$

1.127. First, determine the density of the liquid sample.

$$\text{Density} = \frac{\text{mass}}{\text{volume}} = \frac{22.3\ g}{15.0\ mL} = 1.4\underline{8}6\ g/mL = 1.49\ g/mL = 1.49\ g/cm^3$$

This density is closest to that of chloroform (1.489 g/cm^3), so the unknown liquid is chloroform.

1.129. First, determine the volume of the cube of platinum.

$$V = (\text{edge})^3 = (4.40\ cm)^3 = 85.\underline{1}8\ cm^3$$

Now, use the density to determine the mass of the platinum.

$$\text{Mass} = d \times V = 21.4\ g/cm^3 \times 85.\underline{1}8\ cm^3 = 18\underline{2}2.9\ g = 1.82 \times 10^3\ g$$

1.131. $\text{Volume} = \frac{\text{mass}}{\text{density}} = \frac{35.00\ g}{1.053\ g/mL} = 33.2\underline{3}8\ mL = 33.24\ mL$

1.133. a. $8.45\ kg \times \frac{10^3\ g}{1\ kg} \times \frac{1\ \mu g}{10^{-6}\ g} = 8.45 \times 10^9\ \mu g$

b. $318\ \mu s \times \frac{10^{-6}\ s}{1\ \mu s} \times \frac{1\ ms}{10^{-3}\ s} = 3.18 \times 10^{-1}\ ms$

c. $93\ km \times \frac{10^3\ m}{1\ km} \times \frac{1\ nm}{10^{-9}\ m} = 9.3 \times 10^{13}\ nm$

d. $37.1\ mm \times \frac{10^{-3}\ m}{1\ mm} \times \frac{1\ cm}{10^{-2}\ m} = 3.71\ cm$

1.135. a. $5.91 \text{ kg} \times \frac{10^3 \text{ g}}{1 \text{ kg}} \times \frac{1 \text{ mg}}{10^{-3} \text{ g}} = 5.91 \times 10^6 \text{ mg}$

b. $753 \text{ mg} \times \frac{10^{-3} \text{ g}}{1 \text{ mg}} \times \frac{1 \text{ μg}}{10^{-6} \text{ g}} = 7.53 \times 10^5 \text{ μg}$

c. $90.1 \text{ MHz} \times \frac{10^6 \text{ Hz}}{1 \text{ MHz}} \times \frac{1 \text{ kHz}}{10^3 \text{ Hz}} = 9.01 \times 10^4 \text{ kHz}$

d. $498 \text{ mJ} \times \frac{10^{-3} \text{ J}}{1 \text{ mJ}} \times \frac{1 \text{ kJ}}{10^3 \text{ J}} = 4.98 \times 10^{-4} \text{ kJ}$

1.137. $\text{Volume} = 12{,}230 \text{ km}^3 \times \left(\frac{10^3 \text{ m}}{1 \text{ km}}\right)^3 \times \left(\frac{1 \text{ dm}}{10^{-1} \text{ m}}\right)^3 \times \frac{1 \text{ L}}{1 \text{ dm}^3} = 1.2230 \times 10^{16} \text{ L}$

1.139. First, calculate the volume of the room in cubic feet.

$$\text{Volume} = LWH = 10.0 \text{ ft} \times 11.0 \text{ ft} \times 9.0 \text{ ft} = 9\underline{9}0 \text{ ft}^3$$

Next, convert the volume to liters.

$$V = 9\underline{9}0 \text{ ft}^3 \text{ x } \left(\frac{12 \text{ in}}{1 \text{ ft}}\right)^3 \times \left(\frac{2.54 \text{ cm}}{1 \text{ in}}\right)^3 \times \frac{1 \text{ L}}{10^3 \text{ cm}^3} = 2.\underline{8}0 \times 10^4 \text{ L} = 2.8 \times 10^4 \text{ L}$$

1.141. $\text{Mass} = 275 \text{ carats} \times \frac{200 \text{ mg}}{1 \text{ carat}} \times \frac{10^{-3} \text{ g}}{1 \text{ mg}} = 55.\underline{0}0 \text{ g} = 55.0 \text{ g}$

1.143. The adhesive is not permanent, is easily removable, and does no harm to the object.

1.145. Chromatography depends on how fast a substance moves in a stream of gas or liquid, past a stationary phase to which the substance is slightly attracted.

■ SOLUTIONS TO STRATEGY PROBLEMS

1.147. $5 \text{ x } 10^{-2} \text{ mg} = 0.05 \text{ mg}$. So $4.7 \text{ mg} - 0.05 \text{ mg} = 4.\underline{65} \text{ mg} = 4.7 \text{ mg}$

1.149. $V = V_A + V_B = \frac{m_A}{d_A} + 50.0 \text{ mL} = \frac{175 \text{ g}}{3.00 \text{ g/mL}} + 50.0 \text{ mL} = 58.\underline{33} \text{ mL} + 50.0 \text{ mL}$

$= 108.\underline{33} = 108.3 \text{ mL}$

1.151. Here it is necessary to convert the data into comparable units, mL for example. Converting 0.100 qt to mL gives (0.100 qt = 0.106 L) 106 mL. Next, using the density relationship, and the fact that 1 mL = 1 cm^3, 50.0 g of Pb corresponds to 4.42 mL Pb. The last quantity is first converted to gram mass units, and then to mL using the density relationship. In this case 0.0250 lb = 11.3 g Pb = 1.00 cm^3 = 1.00 mL Pb. Thus, ranking from smallest volume to greatest:

0.0250 lb Pb (or 1.00 mL) < 50.0 g Pb (or 4.42 mL) < 50.0 mL Pb < 0.100 qt Pb (or 106 mL)

1.153. $2 \text{ converters} \times \dfrac{5.0 \times 10^3 \text{ beads}}{\text{converter}} \times \dfrac{1.0 \times 10^6 \text{ cm}^2}{\text{bead}} \times \left(\dfrac{1 \text{ m}}{100 \text{ cm}}\right)^2 \times \left(\dfrac{1 \text{ km}}{10^3 \text{ m}}\right)^2 = 1.\underline{0}0 = 1.0 \text{ km}^2$

1.155 $55.0 \text{ cm}^3 \times \dfrac{1025 \text{ kg}}{\text{m}^3} \times \left(\dfrac{10^{-2} \text{ m}}{1 \text{ cm}}\right)^3 \times \dfrac{1000 \text{ g}}{1 \text{ kg}} = 56.\underline{3}8 = 56.4$ g sample of ocean water originally

Assuming a density of 1.0 g/mL for the evaporated water, and recognizing the equivalency between cm^3 and mL units, when 5.0 mL of water evaporates, the sample loses 5.0 g in mass and 5.0 cm^3 in volume. The density of the partially evaporated ocean water sample is obtained by dividing the remaining mass (56.4 g – 5.0 g = 51.4 g) by the remaining volume (55.0 cm^3 - 5.0 cm^3 = 50.0 cm^3).

$$d = \frac{mass}{volume} = \frac{51.4 \text{ g}}{50.0 \text{ cm}^3} = 1.03 \text{ g/cm}^3 \left(= \frac{1030 \text{ kg}}{\text{m}^3}\right)$$

As one would expect, the remaining salt solution is now slightly more dense than the original ocean water sample.

1.157. a. Since there is the same number of atoms of the gas in each container, the mass is the same in each container.

b. Since $d = m/V$, for the same mass, when the volume is smaller, the density is greater. Since the volume is less in container A, the density is greater.

c. If the volume of container A was doubled, the density would decrease and become equal to the density in container B

1.159. $\dfrac{300 \text{ million people}}{1} \times \dfrac{1 \times 10^{10} \text{ miles}}{1 \text{ person}} \times \dfrac{1.609 \text{ km}}{1 \text{ mi}} \times \dfrac{1000 \text{ m}}{1 \text{ km}} \times \dfrac{1 \text{ lightyear}}{9.46 \times 10^{15} \text{ m}}$

$= \underline{5}.102 \times 10^5 = 5 \times 10^5$ lightyears

1.161. a. Since $d = m/V$, an increased mass for the same volume means a higher density for the solution.

b. There would be less water for the same mass of salt. Since the salt is more dense than water, the density would be higher than in part a.

c. Since there would be a greater volume of water for the same mass of salt, the density would be lower than in part a.

■ SOLUTIONS TO CUMULATIVE-SKILLS PROBLEMS

1.163. The mass of hydrochloric acid is obtained from the density and the volume.

Mass = density x volume = 1.096 g/mL × 50.0 mL = 54.$\underline{8}$0 g

Next, from the law of conservation of mass,

Mass of marble + mass of acid = mass of solution + mass of carbon dioxide gas

Plugging in gives

10.0 g + 54.$\underline{8}$0 g = 60.4 g + mass of carbon dioxide gas

Mass of carbon dioxide gas = 10.0 g + 54.$\underline{8}$0 g − 60.4 g = 4.$\underline{4}$0 g

Finally, use the density to convert the mass of carbon dioxide gas to volume.

$$\text{Volume} = \frac{\text{mass}}{\text{density}} = \frac{4.\underline{4}0\text{ g}}{1.798\text{ g/ L}} = 2.\underline{4}47\text{ L} = 2.4\text{ L}$$

1.165. First, calculate the volume of the steel sphere.

$$V = (4/3)\pi r^3 = (4/3)\pi \times (1.58\text{ in})^3 \times \left(\frac{2.54\text{ cm}}{1\text{ in}}\right)^3 = 27\underline{0}.7\text{ cm}^3$$

Next, determine the mass of the sphere using the density.

Mass = density x volume = 7.88 g/cm^3 × 27$\underline{0}$.7 cm^3 = 21$\underline{3}$3 g = 2.13 × 10^3 g

1.167. The area of the ice is 840,000 mi^2 − 132,000 mi^2 = 70$\underline{8}$,000 mi^2. Now, determine the volume of this ice.

Volume = area × thickness

$$= 70\underline{8},000\text{ mi}^2 \times \underline{5}000\text{ ft} \times \left(\frac{5280\text{ ft}}{1\text{ mi}}\right)^2 \times \left(\frac{12\text{ in}}{1\text{ ft}}\right)^3 \times \left(\frac{2.54\text{ cm}}{1\text{ in}}\right)^3$$

$$= \underline{2}.794 \times 10^{21}\text{ cm}^3$$

Now use the density to determine the mass of the ice.

Mass = density x volume = 0.917 g/cm^3 × $\underline{2}$.794 × 10^{21} cm^3 = $\underline{2}$.56 × 10^{21} g = 3 × 10^{21} g

1.169. Let x = mass of ethanol and y = mass of water. Then, use the total mass to write $x + y$ = 49.6 g, or y = 49.6 g − x. Thus, the mass of water is 49.6 g − x. Next,

Total volume = volume of ethanol + volume of water

Since the volume is equal to the mass divided by density, you can write

$$\text{Total volume} = \frac{\text{mass of ethanol}}{\text{density of ethanol}} + \frac{\text{mass of water}}{\text{density of water}}$$

Substitute in the known and unknown values to get an equation for x.

$$54.2\text{ cm}^3 = \frac{x}{0.789\text{ g/cm}^3} + \frac{49.6\text{ g} - x}{0.998\text{ g/cm}^3}$$

Multiply both sides of this equation by (0.789)(0.998). Also, multiply both sides by g/cm^3 to simplify the units. This gives the following equation to solve for x.

$$(0.789)(0.998)(54.2)\text{ g} = (0.998)\,x + (0.789)(49.6\text{ g} - x)$$

$$42.\underline{6}78\text{ g} = 0.998\,x + 39.\underline{1}34\text{ g} - 0.789\,x$$

$$0.209\,x = 3.\underline{5}44\text{ g}$$

$$x = \text{mass of ethanol} = 1\underline{6}.95\text{ g}$$

The percentage of ethanol (by mass) in the solution can now be calculated.

$$\text{Percent (mass)} = \frac{\text{mass of ethanol}}{\text{mass of solution}} \times 100\% = \frac{1\underline{6}.95\text{ g}}{49.6\text{ g}} \times 100\% = 3\underline{4}.1\% = 34\%$$

To determine the proof, you must first find the percentage by volume of ethanol in the solution. The volume of ethanol is obtained using the mass and the density.

$$\text{Volume} = \frac{\text{mass of ethanol}}{\text{density of ethanol}} = \frac{1\underline{6}.95\text{ g}}{0.789\text{ g/ cm}^3} = 2\underline{1}.48\text{ cm}^3$$

The percentage of ethanol (by volume) in the solution can now be calculated.

$$\text{Percent (volume)} = \frac{\text{volume of ethanol}}{\text{volume of solution}} \times 100\% = \frac{2\underline{1}.48\text{ cm}^3}{54.2\text{ cm}^3} \times 100\% = 3\underline{9}.63\%$$

The proof can now be calculated.

$$\text{Proof} = 2 \times \text{Percent (volume)} = 2 \times 3\underline{9}.63 = 7\underline{9}.27 = 79\text{ proof}$$

1.171. The volume of the mineral can be obtained from the mass difference between the water displaced and the air displaced, and the densities of water and air.

$$\text{Mass difference} = 18.49\text{ g} - 16.21\text{ g} = 2.28\text{ g}$$

$$\text{Volume of mineral} = \frac{\text{mass difference}}{\text{density of water - density of air}}$$

$$= \frac{2.28\text{ g}}{0.9982\text{ g/cm}^3 - 1.205 \times 10^{-3}\text{ g/cm}^3} = 2.2\underline{8}6\text{ cm}^3$$

The mass of the mineral is equal to its mass in air plus the weight of the displaced air. The weight of the displaced air is obtained from the volume of the mineral and the density of air.

Mass of displaced air = density x volume

$$= 1.205\text{ g/L x } 2.2\underline{8}6\text{ cm}^3 \times \frac{1\text{ L}}{10^3\text{ cm}^3} = 2.7\underline{5}5 \times 10^{-3}\text{ g}$$

$$\text{Mass of mineral} = 18.49\text{ g} + 2.7\underline{5}5 \times 10^{-3}\text{ g} = 18.4\underline{9}27\text{ g}$$

The density of the mineral can now be calculated.

$$\text{Density} = \frac{\text{mass}}{\text{volume}} = \frac{18.4\underline{9}27\text{ g}}{2.2\underline{8}6\text{ cm}^3} = 8.0\underline{8}9\text{ g/cm}^3 = 8.09\text{ g/cm}^3$$

1.173. The volume of the object can be obtained from the mass of the ethanol displaced and the density of ethanol.

Mass of ethanol displaced = 15.8 g − 10.5 g = 5.3 g

$$\text{Volume of object} = \frac{\text{mass of ethanol}}{\text{density of ethanol}} = \frac{5.3\ \text{g}}{0.789\ \text{g/ cm}^3} = 6.\underline{7}17\ \text{cm}^3$$

The density of the object can now be calculated.

$$\text{Density} = \frac{\text{mass}}{\text{volume}} = \frac{15.8\ \text{g}}{6.\underline{7}17\ \text{cm}^3} = 2.\underline{3}52\ \text{g/cm}^3 = 2.4\ \text{g/cm}^3$$

CHAPTER 2

Atoms, Molecules, and Ions

SOLUTIONS TO EXERCISES

Note on significant figures: If the final answer to a solution needs to be rounded off, it is given first with one nonsignificant figure, and the last significant figure is underlined. The final answer is then rounded to the correct number of significant figures. In multistep problems, intermediate answers are given with at least one nonsignificant figure; however, only the final answer has been rounded off.

2.1. The element with atomic number 17 (the number of protons in the nucleus) is chlorine, symbol Cl. The mass number is 17 + 18 = 35. The symbol is ${}^{35}_{17}Cl$.

2.2. Multiply each isotopic mass by its fractional abundance; then sum:

$$34.96885 \text{ amu} \times 0.75771 = 26.49\underline{6}247$$

$$36.96590 \text{ amu} \times 0.24229 = \underline{8.956467}$$

$$35.45\underline{2}714 = 35.453 \text{ amu}$$

The atomic mass of chlorine is 35.453 amu.

2.3. a. Se: Group VIA, Period 4; nonmetal

b. Cs: Group IA, Period 6; metal

c. Fe: Group VIIIB, Period 4; metal

d. Cu: Group IB, Period 4; metal

e. Br: Group VIIA, Period 4; nonmetal

2.4. Take as many cations as there are units of charge on the anion and as many anions as there are units of charge on the cation. Two K^+ ions have a total charge of 2+, and one CrO_4^{2-} ion has a charge of 2−, giving a net charge of zero. The simplest ratio of K^+ to CrO_4^{2-} is 2:1, and the formula is K_2CrO_4.

2.5. a. CaO: Calcium, a Group IIA metal, is expected to form only a 2+ ion (Ca^{2+}, the calcium ion). Oxygen (Group VIA) is expected to form an anion of charge equal to the group number minus 8 (O^{2-}, the oxide ion). The name of the compound is calcium oxide.

b. $PbCrO_4$: Lead has more than one monatomic ion. You can find the charge on the Pb ion if you know the formula of the anion. From Table 2.5, the CrO_4 refers to the anion CrO_4^{2-} (the chromate ion). Therefore, the Pb cation must be Pb^{2+} to give electrical neutrality. The name of Pb^{2+} is lead(II) ion, so the name of the compound is lead(II) chromate.

2.6. Thallium(III) nitrate contains the thallium(III) ion, Tl^{3+}, and the nitrate ion, NO_3^-. The formula is $Tl(NO_3)_3$.

2.7. a. Dichlorine hexoxide

b. Phosphorus trichloride

c. Phosphorus pentachloride

2.8. a. CS_2 b. SO_3

2.9. a. Boron trifluoride b. Hydrogen selenide

2.10. When you remove one H^+ ion from $HBrO_4$, you obtain the BrO_4^- ion. You name the ion from the acid by replacing *-ic* with *-ate*. The anion is called the perbromate ion.

2.11. Sodium carbonate decahydrate

2.12. Sodium thiosulfate is composed of sodium ions (Na^+) and thiosulfate ions ($S_2O_3^{2-}$), so the formula of the anhydrous compound is $Na_2S_2O_3$. Since the material is a pentahydrate, the formula of the compound is $Na_2S_2O_3{\bullet}5H_2O$.

2.13. Balance O first in parts (a) and (b) because it occurs in only one product. Balance S first in part (c) because it appears in only one product. Balance H first in part (d) because it appears in just one reactant as well as in the product.

a. Write a 2 in front of $POCl_3$ for O; this requires a 2 in front of PCl_3 for final balance:

$$O_2 + 2PCl_3 \rightarrow 2POCl_3$$

b. Write a 6 in front of N_2O to balance O; this requires a 6 in front of N_2 for final balance:

$$P_4 + 6N_2O \rightarrow P_4O_6 + 6N_2$$

c. Write $2As_2S_3$ and $6SO_2$ to achieve an even number of oxygens on the right to balance what will always be an even number of oxygens on the left. The $2As_2S_3$ then requires $2As_2O_3$. Finally, to balance (6 + 12) O's on the right, write $9O_2$.

$$2As_2S_3 + 9O_2 \rightarrow 2As_2O_3 + 6SO_2$$

d. Write a 4 in front of H_3PO_4; this requires a 3 in front of $Ca(H_2PO_4)_2$ for twelve H's.

$$Ca_3(PO_4)_2 + 4H_3PO_4 \rightarrow 3Ca(H_2PO_4)_2$$

ANSWERS TO CONCEPT CHECKS

2.1. CO_2 is a compound that is a combination of 1 carbon atom and 2 oxygen atoms. Therefore, the chemical model must contain a chemical combination of 3 atoms stuck together with 2 of the atoms being the same (oxygen). Since each "ball" represents an individual atom, the three models on the left can be eliminated since they don't contain the correct number of atoms. Keeping in mind that balls of the same color represent the same element, only the model on the far right contains two elements with the correct ratio of atoms, 1:2; therefore, it must be CO_2.

2.2. If 7999 out of 8000 alpha particles deflected back at the alpha-particle source, this would imply that the atom was a solid, impenetrable mass. Keep in mind that this is in direct contrast to what was observed in the actual experiments, where the majority of the alpha particles passed through without being deflected.

2.3. Elements are listed together in groups because they have similar chemical and/or physical properties.

2.4. Statement (a) is the best statement regarding molecular compounds. Although you may have wanted to classify Br_2 as a molecular compound, it is an element and not a compound. Regarding statement (b), quite a few molecular compounds exit that don't contain carbon. Water and the nitrogen oxides associated with smog are prime examples. Statement (c) is false; ionic compounds consist of anions and cations. Statement (d) is very close to the right selection but it is too restrictive. Some molecular compounds containing both metal and nonmetal atoms are known to exist, e.g., cisplatin, $Ni(CO)_4$, etc. Because numerous molecular compounds are either solids or liquids at room temperature, statement (e) is false.

2.5. a. This compound is an ether because it has a functional group of an oxygen atom between two carbon atoms (–O–).

b. This compound is an alcohol because it has an –OH functional group.

c. This compound is a carboxylic acid because it has the –COOH functional group.

d. This compound is a hydrocarbon because it contains only carbon and hydrogen atoms.

2.6. The SO_4^{2-}, NO_2^-, and I_3^- are considered to be polyatomic ions. Statement (a) is true based on the prefix *poly*. By definition, any ion must have a negative or positive charge; thus, statement (b) is true. Bring that the triiodide ion has only iodine atoms bonded together, and no other elements present, statement (c) is false. There are numerous examples to show that statement (d) is true, e.g., chromate, dichromate, permanganate to name a few. Oxoanions are polyatomic ions containing a central characteristic element surrounded by a number of oxygen atoms, e.g., sulfate and nitrite given in this concept check's. Statement (e) is true.

2.7. A bottle containing a compound with the formula Al_2Q_3 would have an anion, Q, with a charge of 2–. The total positive charge in the compound due to the Al^{3+} is 6+ (2 x 3+), so the total negative charge must be 6–; therefore, each Q ion must have a charge of 2–. Thus, Q would probably be an element from Group VIA on the periodic table.

■ ANSWERS TO SELF-ASSESSMENT AND REVIEW QUESTIONS

2.1. Atomic theory is an explanation of the structure of matter in terms of different combinations of very small particles called atoms. Since compounds are composed of atoms of two or more elements, there is no limit to the number of ways in which the elements can be combined. Each compound has its own unique properties. A chemical reaction consists of the rearrangement of the atoms present in the reacting substances to give new chemical combinations present in the substances formed by the reaction.

2.2. Divide each amount of chlorine, 1.270 g and 1.904 g, by the lower amount, 1.270 g. This gives 1.000 and 1.499, respectively. Convert these to whole numbers by multiplying by 2, giving 2.000 and 2.998. The ratio of these amounts of chlorine is essentially 2:3. This is consistent with the law of multiple proportions because, for a fixed mass of iron (1 gram), the masses of chlorine in the other two compounds are in a ratio of small whole numbers.

2.3. A cathode-ray tube consists of a negative electrode, or cathode, and a positive electrode, or anode, in an evacuated tube. Cathode rays travel from the cathode to the anode when a high voltage is turned on. Some of the rays pass through the hole in the anode to form a beam, which is then bent toward positively charged electric plates in the tube. This implies that a cathode ray consists of a beam of negatively charged particles (or electrons) and that electrons are constituents of all matter.

2.4. Millikan performed a series of experiments in which he obtained the charge on the electron by observing how a charged drop of oil falls in the presence and in the absence of an electric field. An atomizer introduces a fine mist of oil drops into the top chamber (Figure 2.6). Several drops happen to fall through a small hole into the lower chamber, where the experimenter follows the motion of one drop with a microscope. Some of these drops have picked up one or more electrons as a result of friction in the atomizer and have become negatively charged. A negatively charged drop will be attracted upward when the experimenter turns on a current to the electric plates. The drop's upward speed (obtained by timing its rise) is related to its mass-to-charge ratio, from which you can calculate the charge on the electron.

2.5. The nuclear model of the atom is based on experiments of Geiger, Marsden, and Rutherford. Rutherford stated that most of the mass of an atom is concentrated in a positively charged center called the nucleus around which negatively charged electrons move. The nucleus, although it contains most of the mass, occupies only a very small portion of the space of the atom. Most of the alpha particles passed through the metal atoms of the foil undeflected by the lightweight electrons. When an alpha particle does happen to hit a metal-atom nucleus, it is scattered at a wide angle because it is deflected by the massive, positively charged nucleus (Figure 2.8).

2.6. The atomic nucleus consists of two kinds of particles, protons and neutrons. The mass of each is about the same, on the order of 1.67×10^{-27} kg, and about 1800 times that of the electron. An electron has a much smaller mass, on the order of 9.11×10^{-31} kg. The neutron is electrically neutral, but the proton is positively charged. An electron is negatively charged. The charges on the proton and the electron are equal in magnitude.

2.7. Protons (hydrogen nuclei) were discovered as products of experiments involving the collision of alpha particles with nitrogen atoms that resulted in a proton being knocked out of the nitrogen nucleus. Neutrons were discovered as the radiation product of collisions of alpha particles with beryllium atoms. The resulting radiation was discovered to consist of particles having a mass approximately equal to that of a proton and having no charge (neutral).

2.8. Oxygen consists of three different isotopes, each having 8 protons but a different number of neutrons.

2.9. The percentages of the different isotopes in most naturally occurring elements have remained essentially constant over time and in most cases are independent of the origin of the element. Thus, what Dalton actually calculated were average atomic masses (relative masses). He could not weigh individual atoms, but he could find the average mass of one atom relative to the average mass of another.

2.10. A mass spectrometer measures the mass-to-charge ratio of positively charged atoms (and molecules). It produces a mass spectrum, which shows the relative numbers of atoms (fractional abundances) of various masses (isotopic masses). The mass spectrum gives us all the information needed to calculate the atomic weight.

2.11. The atomic mass of an element is the average atomic mass for the naturally occurring element expressed in atomic mass units. The atomic mass would be different elsewhere in the universe if the percentages of isotopes in the element were different from those on earth. Recent research has shown that isotopic abundances actually do differ slightly depending on the location found on earth.

2.12. The element in Group IVA and Period 5 is tin (atomic number 50).

2.13. A metal is a substance or mixture that has characteristic luster, or shine, and is generally a good conductor of heat and electricity.

2.14. The formula for ethane is C_2H_6.

2.15. A molecular formula gives the exact number of different atoms of an element in a molecule. A structural formula is a chemical formula that shows how the atoms are bonded to one another in a molecule.

2.16. Organic molecules contain carbon combined with other elements such as hydrogen, oxygen, and nitrogen. An inorganic molecule is composed of elements other than carbon. Some inorganic molecules that contain carbon are carbon monoxide (CO), carbon dioxide (CO_2), carbonates, and cyanides.

2.17. An ionic binary compound: NaCl; a molecular binary compound: H_2O.

2.18. a. The elements are represented by B, F, and I.

b. The compounds are represented by A, E, and G.

c. The mixtures are represented by C, D, and H.

d. The ionic solid is represented by A.

e. The gas made up of an element and a compound is represented by C.

f. The mixtures of elements are represented by D and H.

g. The solid element is represented by F.

h. The solids are represented by A and F.

i. The liquids are represented by E, H, and I.

2.19. In the Stock system, CuCl is called copper(I) chloride, and $CuCl_2$ is called copper(II) chloride. One of the advantages of the Stock system is that more than two different ions of the same metal can be named with this system. In the former (older) system, a new suffix other than *-ic* and *-ous* must be established and/or memorized.

2.20. A balanced chemical equation has the numbers of atoms of each element equal on both sides of the arrow. The coefficients are the smallest possible whole numbers.

2.21. The answer is a: 50 p, 69 n, and 48 e^-.

2.22. The answer is d: 65%.

2.23. The answer is c: magnesium hydroxide, $Mg(OH)_2$.

2.24. The answer is b: Li.

ANSWERS TO CONCEPTUAL PROBLEMS

2.27. If atoms were balls of positive charge with the electrons evenly distributed throughout, there would be no massive, positive nucleus to deflect the beam of alpha particles when it is shot at the gold foil.

2.29. You could group elements by similar physical properties such as density, mass, color, conductivity, etc., or by chemical properties, such as reaction with air, reaction with water, etc.

2.31. a. In each case, the total positive charge and the total negative charge in the compounds must cancel. Therefore, the compounds with the cations X^+, X^{2+}, and X^{5+}, combined with the SO_4^{2-} anion, are X_2SO_4, XSO_4, and $X_2(SO_4)_5$, respectively.

b. You recognize the fact that whenever a cation can have multiple oxidation states (1+, 2+, and 5+ in this case), the name of the compound must indicate the charge. Therefore, the names of the compounds in part (a) would be exy(I) sulfate, exy(II) sulfate, and exy(V) sulfate, respectively.

2.33. A potassium-39 atom in this case would contain 19 protons and 20 neutrons. If the charge of the proton were twice that of an electron, it would take twice as many electrons as protons, or 38 electrons, to maintain a charge of zero.

2.35. a. $2Li + Cl_2 \rightarrow 2LiCl$

b. $16Na + S_8 \rightarrow 8Na_2S$

c. $2Al + 3I_2 \rightarrow 2AlI_3$

d. $3Ba + N_2 \rightarrow Ba_3N_2$

e. $12V + 5P_4 \rightarrow 4V_3P_5$

SOLUTIONS TO PRACTICE PROBLEMS

Note on significant figures: If the final answer to a solution needs to be rounded off, it is given first with one nonsignificant figure, and the last significant figure is underlined. The final answer is then rounded to the correct number of significant figures. In multistep problems, intermediate answers are given with at least one nonsignificant figure; however, only the final answer has been rounded off.

2.37. a. Argon b. Zinc c. Silver d. Magnesium

2.39. a. K b. S c. Fe d. Mn

2.41. The mass of the electron is found by multiplying the two values:

$$1.602 \times 10^{-19}\ C \times \frac{5.64 \times 10^{-12}\ kg}{1\ C} = 9.0\underline{3}5 \times 10^{-31}\ kg = 9.04 \times 10^{-31}\ kg$$

2.43. The isotope of atom A is the atom with 18 protons, atom C; the atom that has the same mass number as atom A (37) is atom D.

2.45. Each isotope of chlorine (atomic number 17) has 17 protons. Each neutral atom will also have 17 electrons. The number of neutrons for Cl-35 is 35 – 17 = 18 neutrons. The number of neutrons for Cl-37 is 37 – 17 = 20 neutrons.

2.47. The element with 14 protons in its nucleus is silicon (Si). The mass number = 14 + 14 = 28. The notation for the nucleus is $^{28}_{14}\text{Si}$.

2.49. Since the atomic ratio of nitrogen to hydrogen is 1:3, divide the mass of N by one-third of the mass of hydrogen to find the relative mass of N.

$$\frac{\text{Atomic mass of N}}{\text{Atomic mass of H}} = \frac{7.933\text{ g}}{1/3 \text{ x } 1.712\text{ g}} = \frac{13.9\underline{0}1\text{ g N}}{1\text{ g H}} = \frac{13.90}{1}$$

2.51. Multiply each isotopic mass by its fractional abundance, and then sum:

X-63: $62.930 \times 0.6909 = 43.4\underline{7}83$

X-65: $64.928 \times 0.3091 = 20.0692$

$63.5\underline{4}75 = 63.55$ amu

The element is copper, atomic mass 63.546 amu.

2.53. Multiply each isotopic mass by its fractional abundance, and then sum:

$38.964 \times 0.9326 = 36.3\underline{3}78$

$39.964 \times 1.00 \times 10^{-4} = 0.00399\underline{6}4$

$40.962 \times 0.0673 = 2.75674$

$= 39.0\underline{9}853 = 39.10$ amu

The atomic mass of this element is 39.10 amu. The element is potassium (K).

2.55. According to the picture, there are 20 atoms, 5 of which are brown and 15 of which are green. Using the isotopic masses in the problem, the atomic mass of element X is

$$\frac{5}{20}(23.02\text{ amu}) + \frac{15}{20}(25.147\text{ amu}) = 5.75\underline{5} + 18.86\underline{0}2 = 24.61\underline{5}2 = 24.615\text{ amu}$$

2.57.
a. C: Group IVA, Period 2; nonmetal
b. Po: Group VIA, Period 6; metal
c. Cr: Group VIB, Period 4; metal
d. Mg: Group IIA, Period 3; metal
e. B: Group IIIA, Period 2; metalloid

2.59. a. Tellurium b. Aluminum

2.61. Examples are:
a. O (oxygen)
b. F (fluorine)
c. Fe (iron)
d. Ce (cerium)

2.63. They are different in that the solid sulfur consists of S_8 molecules, whereas the hot vapor consists of S_2 molecules. The S_8 molecules are four times as heavy as the S_2 molecules. Hot sulfur is a mixture of S_8 and S_2 molecules, but at high enough temperatures only S_2 molecules are formed. Both hot sulfur and solid sulfur consist of molecules with only sulfur atoms.

2.65. The number of nitrogen atoms in the 1.50-g sample of N_2O is

$$2.05 \times 10^{22}\ N_2O \text{ molecules} \times \frac{2\ N \text{ atoms}}{1\ N_2O \text{ molecule}} = 4.10 \times 10^{22}\ N \text{ atoms}$$

The number of nitrogen atoms in 44.0 g of N_2O is

$$44.0\ g\ N_2O \times \frac{4.10 \times 10^{22}\ N \text{ atoms}}{1.50\ g\ N_2O} = 1.2\underline{0}3 \times 10^{24}\ N \text{ atoms} = 1.20 \times 10^{24}\ N \text{ atoms}$$

2.67. $$3.3 \times 10^{21}\ H \text{ atoms} \times \frac{1\ NH_3 \text{ molecule}}{3\ H \text{ atoms}} = 1.1 \times 10^{21}\ NH_3 \text{ molecules}$$

2.69. a. N_2H_4

b. H_2O_2

c. C_3H_8O

d. PCl_3

2.71. a. PCl_5

b. NO_2

c. $C_3H_6O_2$

2.73. $$\frac{1\ Fe \text{ atom}}{1\ Fe(NO_3)_2 \text{ unit}} \times \frac{1\ Fe(NO_3)_2 \text{ unit}}{2\ NO_3^- \text{ ions}} \times \frac{1\ NO_3^- \text{ ion}}{3\ O \text{ atoms}} = \frac{1\ Fe \text{ atom}}{6\ O \text{ atoms}} = \frac{1}{6}$$

Thus, the ratio of iron atoms to oxygen atoms is one Fe atom to six O atoms.

2.75. a. $Fe(CN)_3$

b. K_2CO_3

c. Li_3N

d. Ca_3P_2

2.77. a. Na_2SO_4: sodium sulfate (Group IA forms only 1+ cations.)

b. Na_3N: sodium nitride (Group IIA forms only 1+ cations.)

c. CuCl: copper(I) chloride (Group IB forms 1+ and 2+ cations.)

d. Cr_2O_3: chromium(III) oxide (Group VIB forms numerous oxidation states.)

2.79. a. Lead(II) permanganate: $Pb(MnO_4)_2$ (Permanganate is in Table 2.6.)

b. Barium hydrogen carbonate: $Ba(HCO_3)_2$ (The HCO_3^- ion is in Table 2.6.)

c. Cesium sulfide: Cs_2S (Group 1A ions form 1+ cations.)

d. Iron(III) acetate: $Fe(C_2H_3O_2)_3$ (The acetate ion = 1– [from Table 2.6]; for the sum of charges to be zero, three must be used.)

2.81. a. Molecular

b. Ionic

c. Molecular

d. Ionic

2.83. a. Dinitrogen monoxide

b. Tetraphosphorus dec(a)oxide

c. Arsenic trichloride

d. Dichlorine hept(a)oxide

2.85. a. NBr_3

b. XeF_6

c. CO

d. Cl_2O_5

2.87. a. Selenium trioxide

b. Disulfur dichloride

c. Carbon monoxide

2.89. a. Sulfurous acid: H_2SO_3

b. Hyponitrous acid: $H_2N_2O_2$

c. Disulfurous acid: $H_2S_2O_5$

d. Arsenic acid: H_3AsO_4

2.91. $Na_2SO_4 \bullet 10H_2O$ is sodium sulfate decahydrate.

2.93. Iron(II) sulfate heptahydrate is $FeSO_4 \bullet 7H_2O$.

2.95. $$1\ PbCO_3 \times \frac{3\ \text{O atoms}}{1\ PbCO_3\ \text{unit}} + 2\ KNO_3 \times \frac{3\ \text{O atoms}}{1\ KNO_3\ \text{unit}} = 9\ \text{O atoms}$$

2.97. a. Balance: $Sn + NaOH \rightarrow Na_2SnO_2 + H_2$

If Na is balanced first by writing a 2 in front of NaOH, the entire equation is balanced.

$$Sn + 2NaOH \rightarrow Na_2SnO_2 + H_2$$

b. Balance: $Al + Fe_3O_4 \rightarrow Al_2O_3 + Fe$

First balance O (it appears once on each side) by writing a 3 in front of Fe_3O_4 and a 4 in front of Al_2O_3:

$$Al + 3Fe_3O_4 \rightarrow 4Al_2O_3 + Fe$$

Now balance Al against the 8 Al's on the right and Fe against the 9 Fe's on the left:

$$8Al + 3Fe_3O_4 \rightarrow 4Al_2O_3 + 9Fe$$

c. Balance: $CH_3OH + O_2 \rightarrow CO_2 + H_2O$

First balance H (it appears once on each side) by writing a 2 in front of H_2O:

$$CH_3OH + O_2 \rightarrow CO_2 + 2H_2O$$

To avoid fractional coefficients for O, multiply the equation by 2:

$$2CH_3OH + 2O_2 \rightarrow 2CO_2 + 4H_2O$$

Finally, balance O by changing $2O_2$ to "$3O_2$"; this balances the entire equation:

$$2CH_3OH + 3O_2 \rightarrow 2CO_2 + 4H_2O$$

d. Balance: $P_4O_{10} + H_2O \rightarrow H_3PO_4$

First balance P (it appears once on each side) by writing a 4 in front of H_3PO_4:

$$P_4O_{10} + H_2O \rightarrow 4H_3PO_4$$

Finally, balance H by writing a 6 in front of H_2O; this balances the entire equation:

$$P_4O_{10} + 6H_2O \rightarrow 4H_3PO_4$$

e. Balance: $PCl_5 + H_2O \rightarrow H_3PO_4 + HCl$

First balance Cl (it appears once on each side) by writing a 5 in front of HCl:

$$PCl_5 + H_2O \rightarrow H_3PO_4 + 5HCl$$

Finally, balance H by writing a 4 in front of H_2O; this balances the entire equation:

$$PCl_5 + 4H_2O \rightarrow H_3PO_4 + 5HCl$$

2.99. Balance: $Ca_3(PO_4)_2(s) + H_2SO_4(aq) \rightarrow CaSO_4(s) + H_3PO_4(aq)$

Balance Ca first with a 3 in front of $CaSO_4$:

$$Ca_3(PO_4)_2(s) + H_2SO_4(aq) \rightarrow 3CaSO_4(s) + H_3PO_4(aq)$$

Next, balance the P with a 2 in front of H_3PO_4:

$$Ca_3(PO_4)_2(s) + H_2SO_4(aq) \rightarrow 3CaSO_4(s) + 2H_3PO_4(aq)$$

Finally, balance the S with a 3 in front of H_2SO_4; this balances the equation:

$$Ca_3(PO_4)_2(s) + 3H_2SO_4(aq) \rightarrow 3CaSO_4(s) + 2H_3PO_4(aq)$$

2.101. Balance: $NH_4Cl(aq) + Ba(OH)_2(aq) \rightarrow NH_3(g) + BaCl_2(aq) + H_2O(l)$

Balance O first with a 2 in front of H_2O:

$$NH_4Cl + Ba(OH)_2 \rightarrow NH_3 + BaCl_2 + 2H_2O$$

Balance H with a 2 in front of NH_4Cl and a 2 in front of NH_3; this balances the equation:

$$2NH_4Cl(aq) + Ba(OH)_2(aq) \xrightarrow{\Delta} 2NH_3(g) + BaCl_2(aq) + 2H_2O(l)$$

■ SOLUTIONS TO GENERAL PROBLEMS

2.103. Calculate the ratio of oxygen for 1 g (fixed amount) of nitrogen in both compounds:

$$\text{A:} \quad \frac{2.755 \text{ g O}}{1.206 \text{ g N}} = \frac{2.28\underline{4}4 \text{ g O}}{1 \text{ g N}} \qquad \text{B:} \quad \frac{4.714 \text{ g O}}{1.651 \text{ g N}} = \frac{2.85\underline{5}2 \text{ g O}}{1 \text{ g N}}$$

Next, find the ratio of oxygen per gram of nitrogen for the two compounds.

$$\frac{\text{g O in B/1 g N}}{\text{g O in A/1 g N}} = \frac{2.85\underline{5}2 \text{ g O}}{2.28\underline{4}4 \text{ g O}} = \frac{1.24\underline{9}8 \text{ g O}}{1 \text{g O}}$$

B contains 1.25 times as many O atoms as A does (there are five O's in B for every four O's in A).

2.105. The smallest difference is between -1.12×10^{-18} C and 9.60×10^{-19} C and is equal to -1.6×10^{-19} C. If this charge is equivalent to one electron, the number of excess electrons on a drop may be found by dividing the negative charge by the charge of one electron.

$$\text{Drop 1:} \quad \frac{-3.20 \times 10^{-19} \text{ C}}{-1.6 \times 10^{-19} \text{ C}} = 2.\underline{0} \cong 2 \text{ electrons}$$

$$\text{Drop 2:} \quad \frac{-6.40 \times 10^{-19} \text{ C}}{-1.6 \times 10^{-19} \text{ C}} = 4.\underline{0} \cong 4 \text{ electrons}$$

$$\text{Drop 3:} \quad \frac{-9.60 \times 10^{-19} \text{ C}}{-1.6 \times 10^{-19} \text{ C}} = 6.\underline{0} \cong 6 \text{ electrons}$$

$$\text{Drop 4:} \quad \frac{-1.12 \times 10^{-18} \text{ C}}{-1.6 \times 10^{-19} \text{ C}} = 7.\underline{0} \cong 7 \text{ electrons}$$

2.107. For the Eu atom to be neutral, the number of electrons must equal the number of protons, so a neutral europium atom has 63 electrons. The 3+ charge on the Eu^{3+} indicates there are three more protons than electrons, so the number of electrons is 63 – 3 = 60.

2.109. The number of protons = mass number – number of neutrons = 81 – 46 = 35. The element with Z = 35 is bromine (Br).

The ionic charge = number of protons – number of electrons = 35 – 36 = –1.

Symbol: ${}^{81}_{35}Br^{-}$.

2.111. The sum of the fractional abundances must equal 1. Let y equal the fractional abundance of ${}^{63}Cu$. Then the fractional abundance of ${}^{65}Cu$ equals $(1 - y)$. We write one equation in one unknown:

$$\text{Atomic mass} = 63.546 = 62.9298y + 64.9278(1 - y)$$

$$63.546 = 64.9278 - 1.9980y$$

$$y = \frac{64.9278 - 63.546}{1.9980} = 0.691\underline{5}9$$

The fractional abundance of $^{63}Cu = 0.691\underline{5}9 = 0.6916$.

The fractional abundance of $^{65}Cu = 1 - 0.69159 = 0.308\underline{4}1 = 0.3084$.

2.113. a. Bromine, Br

b. Hydrogen, H

c. Niobium, Nb

d. Fluorine, F

2.115. a. Chromium(III) ion

b. Lead(IV) ion

c. Titanium(II) ion

d. Copper(II) ion

2.117. All possible ionic compounds: Na_2SO_4, NaCl, $CoSO_4$, and $CoCl_2$.

2.119. a. Tin(II) phosphate

b. Ammonium nitrite

c. Magnesium hydroxide

d. Nickel(II) sulfite

2.121. a. Hg_2S [Mercury(I) exists as the polyatomic Hg_2^{2+} ion (Table 2.6).]

b. $Co_2(SO_3)_3$

c. $(NH_4)_2Cr_2O_7$

d. AlF_3

2.123. a. Arsenic tribromide

b. Hydrogen telluride (dihydrogen telluride)

c. Diphosphorus pent(a)oxide

d. Silicon dioxide

2.125. a. Balance the C and H first:

$$C_2H_6 + O_2 \rightarrow 2CO_2 + 3H_2O$$

Avoid a fractional coefficient for O on the left by doubling all coefficients except O_2's, and then balance the O's:

$$2C_2H_6 + 7O_2 \rightarrow 4CO_2 + 6H_2O$$

b. Balance the P first:

$$P_4O_6 + H_2O \rightarrow 4H_3PO_3$$

Then balance the O (or H), which also gives the H (or O) balance:

$$P_4O_6 + 6H_2O \rightarrow 4H_3PO_3$$

c. Balancing the O first is the simplest approach. (Starting with K and Cl and then proceeding to O will cause the initial coefficient for $KClO_3$ to be changed in balancing O last.)

$$4KClO_3 \rightarrow KCl + 3KClO_4$$

d. Balance the N first:

$$(NH_4)_2SO_4 + NaOH \rightarrow 2NH_3 + H_2O + Na_2SO_4$$

Then balance the Na, followed by O; this also balances the H:

$$(NH_4)_2SO_4 + 2NaOH \rightarrow 2NH_3 + 2H_2O + Na_2SO_4$$

e. Balance the N first:

$$2NBr_3 + NaOH \rightarrow N_2 + NaBr + HOBr$$

Note that NaOH and HOBr each have one O and that NaOH and NaBr each have one Na; thus the coefficients of all three are equal; from $2NBr_3$, this coefficient must be 6Br/2 = 3:

$$2NBr_3 + 3NaOH \rightarrow N_2 + 3NaBr + 3HOBr$$

2.127. Let: x = number of protons. Then $1.21x$ is the number of neutrons. Since the mass number is 62, you get

$$62 = x + 1.21x = 2.21x$$

Thus, $x = 2\underline{8}.054$, or 28. The element is nickel (Ni). Since the ion has a +2 charge, there are 26 electrons.

2.129. The average atomic mass would be

Natural carbon: 12.011 × 1/2	=	$6.00\underline{5}500$
Carbon-13: 13.00335 × 1/2	=	$\underline{6.501675}$
Average	=	$12.50\underline{7}175$

The average atomic mass of the sample is 12.507 amu.

2.131. The island of stability is a region of the periodic table where a relatively stable super-heavy nuclide is at the peak of stability and is surrounded by foothills consisting of less stable nuclides. It is centered around the most stable nuclide, which is predicted to have an atomic number of 114 and a mass number of 298.

■ SOLUTIONS TO STRATEGY PROBLEMS

2.133. SO_3, sulfur trioxide; NO_2, nitrogen dioxide; PO_4^{3-}, phosphate ion;

N_2, nitrogen; $Mg(OH)_2$, magnesium hydroxide

2.135. The name of the product is aluminum oxide. The reaction is

$$4Al(s) + 3O_2(g) \rightarrow 2Al_2O_3(s)$$

2.137. (0.7721)(37.24 amu) + (1 – 0.7721)(x) = 37.45 amu

$$x = \frac{37.45 - (0.7721)(37.24)}{(1 - 0.7721)} = 38.\underline{1}61 \text{ amu} = 38.2 \text{ amu}$$

2.139. 6.5×10^{20} formula units $CaCl_2 \times \dfrac{3 \text{ ions}}{1 \text{ formula unit}} = 1.95 \times 10^{21}$ ions

2.141. SO_3, sulfur trioxide

HNO_2, nitrous acid

Mg_3N_2, magnesium nitride

$HI(aq)$, hydroiodic acid

$Cu_3(PO_4)_2$, copper(II) phosphate

$CuSO_4{\bullet}5H_2O$, copper(II) sulfate pentahydrate

2.143. a. An aqueous solution of lead(II) chloride is mixed with an aqueous solution of sodium sulfide to form an aqueous solution of sodium chloride and a lead(II) sulfide precipitate.

b. When gaseous sulfur trioxide is passed into liquid water an aqueous solution of sulfuric acid is formed.

c. Graphite is combusted in an oxygen atmosphere to form gaseous carbon dioxide.

d. Gaseous hydrogen iodide forms when hydrogen gas and gaseous iodine are mixed.

2.145. Each ${}^{1}H_2{}^{16}O$ molecule contains 8 neutrons, 10 protons, and 10 electrons.

neutrons in 6.0×10^{23} molecules $= 8 \times (6.0 \times 10^{23}) = 4.8 \times 10^{24}$ neutrons

protons in 6.0×10^{23} molecules $= 10 \times (6.0 \times 10^{23}) = 6.0 \times 10^{24}$ protons

Because the water molecules are neutrally charged there must also be 6.0×10^{24} electrons.

2.147. We are told that a ${}^{238}U$ nucleus decays by emitting a ${}^{4}He$ atom while the remaining subatomic particles remain intact. In equation form, we can write this as follows:

$${}^{238}_{92}U \rightarrow {}^{4}_{2}He + {}^{y}_{z}X$$

Such nuclear decay processes must follow conservation laws. In this regard, the total mass number of the reactants must equal that of the products. It follows that if $238 = 4 + y$, then $y = 234$. Likewise, the total atomic number of the reactants must equal that of the products. Again, if $92 = 2 + z$, then $z = 90$. The symbol for the other nuclide being produced in this decay process is ${}^{234}_{90}Th$. Thorium (Th) is produced.

■ SOLUTIONS TO CUMULATIVE-SKILLS PROBLEMS

2.149. The spheres occupy a diameter of 2×1.86 Å $= 3.72$ Å. The line of sodium atoms would stretch a length of

$$\text{Length} = \frac{3.72 \text{ Å}}{1 \text{ Na atom}} \times 2.619 \times 10^{22} \text{ Na atoms} = 9.7\underline{4}2 \times 10^{22} \text{ Å}$$

Now, convert this to miles.

$$9.7\underline{4}2 \times 10^{22} \text{ Å} \times \frac{10^{-10} \text{ m}}{1 \text{ Å}} \times \frac{1 \text{ mile}}{1.609 \times 10^{3} \text{ m}} = 6.0\underline{5}5 \times 10^{9} \text{ miles} = 6.06 \times 10^{9} \text{ miles}$$

2.151. $NiSO_4 \bullet 7H_2O(s) \rightarrow NiSO_4 \bullet 6H_2O(s) + H_2O(g)$

[8.753 g] = [8.192 g + (8.753 − 8.192 = 0.561 g)]

The 8.192 g of $NiSO_4 \bullet 6H_2O$ must contain 6 × 0.561 = 3.366 g H_2O.

Mass of anhydrous $NiSO_4$ = 8.192 g $NiSO_4 \bullet 6H_2O$ − 3.366 g $6H_2O$ = 4.826 g $NiSO_4$

2.153. $$\text{Mass of O} = 0.6015\ \text{L} \times \frac{1.330\ \text{g O}}{1\ \text{L}} = 0.799\underline{9}95\ \text{g} = 0.8000\ \text{g oxygen}$$

$$15.9994\ \text{amu O} \times \frac{3.177\ \text{g X}}{0.799\underline{9}95\ \text{g O}} = 63.5\underline{3}8\ \text{amu X} = 63.54\ \text{amu}$$

The atomic mass of X is 63.54 amu; X is copper.

CHAPTER 3

Calculations with Chemical Formulas and Equations

SOLUTIONS TO EXERCISES

Note on significant figures: If the final answer to a solution needs to be rounded off, it is given first with one nonsignificant figure, and the last significant figure is underlined. The final answer is then rounded to the correct number of significant figures. In multistep problems, intermediate answers are given with at least one nonsignificant figure; however, only the final answer has been rounded off.

3.1. a. NO_2

1 × AM of N	=		14.0067 amu
2 × AM of O	=	2 × 15.9994 =	31.9988 amu
MM of NO_2	=		46.0055 = 46.0 amu (3 s.f.)

b. $C_6H_{12}O_6$

6 × AM of C	=	6 × 12.011 =	72.066 amu
12 × AM of H	=	12 × 1.0079 =	12.0948 amu
6 × AM of O	=	6 × 15.9994 =	95.9964 amu
MM of $C_6H_{12}O_6$	=		180.1572 amu = 180. amu (3 s.f.)

c. NaOH

1 × AM of Na	=	22.98977 amu
1 × AM of O	=	15.9994 amu
1 × AM of H	=	1.0079 amu
MM of NaOH	=	39.9971 amu = 40.0 amu (3 s.f.)

d. $Mg(OH)_2$

1 × AM of Mg	=		24.305 amu
2 × AM of O	=	2 × 15.9994 =	31.9988 amu
2 × AM of H	=	2 × 1.0079 =	2.0158 amu
MM of $Mg(OH)_2$	=		58.3196 amu = 58.3 amu (3 s.f.)

3.2. a. The molecular model represents a molecule made up of one S and three O. The chemical formula is SO_3. Calculating the formula mass by using the same approach as in Example 3.1 in the text yields 80.1 amu.

b. The molecular model represents one S, four O, and two H. The chemical formula is then H_2SO_4. The formula mass is 98.1 amu.

3.3. a. The atomic mass of Ca = 40.08 amu; thus, the molar mass = 40.08 g/mol, and 1 mol Ca = 6.022×10^{23} Ca atoms.

$$\text{Mass of one Ca} = \frac{40.08 \text{ g}}{1 \text{ mol Ca}} \times \frac{1 \text{ mol}}{6.022 \times 10^{23} \text{ atoms}} = 6.65\underline{5}\underline{6} \times 10^{-23}$$

$$= 6.656 \times 10^{-23} \text{ g/atom}$$

b. The molecular mass of C_2H_5OH, or C_2H_6O, = $(2 \times 12.01) + (6 \times 1.008) + 16.00 = 46.0\underline{6}\underline{8}$. Its molar mass = 46.07 g/mol, and 1 mol = 6.022×10^{23} molecules of C_2H_6O.

$$\text{Mass of one } C_2H_6O = \frac{46.07 \text{ g}}{1 \text{ mol } C_2H_6O} \times \frac{1 \text{ mol}}{6.022 \times 10^{23} \text{ molecules}}$$

$$= 7.65\underline{0}\underline{3} \times 10^{-23} = 7.650 \times 10^{-23} \text{ g/molecule}$$

3.4. The molar mass of H_2O_2 is 34.02 g/mol. Therefore,

$$0.909 \text{ mol } H_2O_2 \times \frac{34.02 \text{ g } H_2O_2}{1 \text{ mol } H_2O_2} = 30.\underline{9}2 = 30.9 \text{ g } H_2O_2$$

3.5. The molar mass of HNO_3 is 63.01 g/mol. Therefore,

$$28.5 \text{ g } HNO_3 \times \frac{1 \text{ mol } HNO_3}{63.01 \text{ g } HNO_3} = 0.45\underline{2}3 = 0.452 \text{ mol } HNO_3$$

3.6. Convert the mass of HCN from milligrams to grams. Then convert grams of HCN to moles of HCN. Finally, convert moles of HCN to the number of HCN molecules.

$$56 \text{ mg HCN} \times \frac{1 \text{ g}}{1000 \text{ mg}} \times \frac{1 \text{ mol HCN}}{27.02 \text{ g HCN}} \times \frac{6.022 \times 10^{23} \text{ HCN molecules}}{1 \text{ mol HCN}}$$

$$= 1.\underline{2}48 \times 10^{21} = 1.2 \times 10^{21} \text{ HCN molecules}$$

3.7. The molecular mass of NH_4NO_3 = 80.05; thus, its molar mass = 80.05 g/mol. Hence

$$\text{Percent N} = \frac{28.02 \text{ g}}{80.05 \text{ g}} \times 100\% = 35.\underline{0}0 = 35.0\%$$

$$\text{Percent H} = \frac{4.032 \text{ g}}{80.05 \text{ g}} \times 100\% = 5.0\underline{3}6 = 5.04\%$$

$$\text{Percent O} = \frac{48.00 \text{ g}}{80.05 \text{ g}} \times 100\% = 59.\underline{9}6 = 60.0\%$$

3.8. From the previous exercise, NH_4NO_3 is 35.0% N (fraction N = 0.350), so the mass of N in 48.5 g of NH_4NO_3 is

$$48.5 \text{ g } NH_4NO_3 \times (0.350 \text{ g N}/1 \text{ g } NH_4NO_3) = 16.\underline{9}75 = 17.0 \text{ g N}$$

3.9. First, convert the mass of CO_2 to moles of CO_2. Next, convert this to moles of C (1 mol CO_2 is equivalent to 1 mol C). Finally, convert to mass of carbon, changing milligrams to grams first:

$$5.80 \times 10^{-3} \text{ g } CO_2 \times \frac{1 \text{ mol } CO_2}{44.01 \text{ g}} \times \frac{1 \text{ mol C}}{1 \text{ mol } CO_2} \times \frac{12.01 \text{ g C}}{1 \text{ mol C}} = 1\ 5\underline{8}3 \times 10^{-3} \text{g C}$$

Do the same series of calculations for water, noting that 1 mol H_2O contains 2 mol H.

$$1.58 \times 10^{-3} \text{ g } H_2O \times \frac{1 \text{ mol } H_2O}{18.02 \text{ g}} \times \frac{2 \text{ mol H}}{1 \text{ mol } H_2O} \times \frac{1.008 \text{ g H}}{1 \text{ mol H}} = 1.7\underline{6}7 \times 10^{-4} \text{g H}$$

The mass percentages of C and H can be calculated using the masses from the previous calculations:

$$\text{Percent C} = \frac{1.583 \text{ mg}}{3.87 \text{ mg}} \times 100\% = 40.\underline{9}0 = 40.9\% \text{ C}$$

$$\text{Percent H} = \frac{0.1767 \text{ mg}}{3.87 \text{ mg}} \times 100\% = 4.5\underline{6}58 = 4.57\% \text{ H}$$

The mass percentage of O can be determined by subtracting the sum of the above percentages from 100%:

$$\text{Percent O} = 100.000\% - (40.90 + 4.5658) = 54.\underline{5}342 = 54.5\% \text{ O}$$

3.10. Convert the masses to moles that are proportional to the subscripts in the empirical formula:

$$33.4 \text{ g S} \times \frac{1 \text{ mol S}}{32.07 \text{ g S}} = 1.0\underline{4}14 \text{ mol S}$$

$$(83.5 - 33.4) \text{ g O} \times \frac{1 \text{ mol O}}{16.00 \text{ g O}} = 3.1\underline{3}12 \text{ mol O}$$

Next, obtain the smallest integers from the moles by dividing each by the smallest number of moles:

$$\text{For O: } \frac{3.1312 \text{ mol O}}{1.0414 \text{ mol S}} = 3.01 \qquad \text{For S: } \frac{1.0414 \text{ mol S}}{1.0414 \text{ mol S}} = 1.00$$

The empirical formula is SO_3.

3.11. For a 100.0-g sample of benzoic acid, 68.8 g are C, 5.0 g are H, and 26.2 g are O. Using the molar masses, convert these masses to moles:

$$68.8 \text{ g C} \times \frac{1 \text{ mol C}}{12.01 \text{ g C}} = 5.7\underline{2}9 \text{ mol C}$$

$$5.0 \text{ g H} \times \frac{1 \text{ mol H}}{1.008 \text{ g H}} = 4.\underline{9}6 \text{ mol H}$$

$$26.2 \text{ g O} \times \frac{1 \text{ mol O}}{16.00 \text{ g O}} = 1.6\underline{3}8 \text{ mol O}$$

These numbers are in the same ratio as the subscripts in the empirical formula. They must be changed to integers. First, divide each one by the smallest number of moles:

$$\text{For C: } \frac{5.729}{1.638} = 3.497 \quad \text{For H: } \frac{4.96}{1.638} = 3.03 \quad \text{For O: } \frac{1.638}{1.638} = 1.000$$

Rounding off, we obtain $C_{3.5}H_{3.0}O_{1.0}$. Multiplying the numbers by 2 gives whole numbers, for an empirical formula of $C_7H_6O_2$.

3.12. For a 100.0-g sample of acetaldehyde, 54.5 g are C, 9.2 g are H, and 36.3 g are O. Using the molar masses, convert these masses to moles:

$$54.5 \text{ g C} \times \frac{1 \text{ mol C}}{12.01 \text{ g C}} = 4.5\underline{3}7 \text{ mol C}$$

$$9.2 \text{ g H} \times \frac{1 \text{ mol H}}{1.008 \text{ g H}} = 9.\underline{1}2 \text{ mol H}$$

$$36.3 \text{ g O} \times \frac{1 \text{ mol O}}{16.00 \text{ g O}} = 2.2\underline{6}8 \text{ mol O}$$

These numbers are in the same ratio as the subscripts in the empirical formula. They must be changed to integers. First, divide each one by the smallest number of moles:

For C: $\frac{4.537}{2.268} = 2.000$ For H: $\frac{9.12}{2.268} = 4.02$ For O: $\frac{2.268}{2.268} = 1.000$

Rounding off, we obtain C_2H_4O, the empirical formula, which is also the molecular formula.

3.13.

H_2	+	Cl_2	→	2HCl
1 molec. (mol) H_2	+	1 molec. (mol) Cl_2	→	2 molec. (mol) HCl (molec., mole interp.)
2.016 g H_2	+	70.9 g Cl_2	→	2 × 36.5 g HCl (mass interp.)

3.14. Equation: $Na + H_2O \rightarrow 1/2H_2 + NaOH$, or $2Na + 2H_2O \rightarrow H_2 + 2NaOH$.From this equation, one mole of Na corresponds to one-half mole of H_2, or two moles of Na corresponds to one mole of H_2. Therefore,

$$7.81 \text{ g } H_2 \times \frac{1 \text{ mol } H_2}{2.016 \text{ g } H_2} \times \frac{2 \text{ mol Na}}{1 \text{ mol } H_2} \times \frac{22.99 \text{ g Na}}{1 \text{ mol Na}} = 17\underline{8}.1 = 178 \text{ g Na}$$

3.15. Balanced equation: $2ZnS + 3O_2 \rightarrow 2ZnO + 2SO_2$

Convert grams of ZnS to moles of ZnS. Then determine the relationship between ZnS and O_2 (2ZnS is equivalent to $3O_2$). Finally, convert to mass of O_2.

$$5.00 \times 10^3 \text{g ZnS x } \frac{1 \text{ mol ZnS}}{97.45 \text{ g ZnS}} \times \frac{3 \text{ mol } O_2}{2 \text{ mol ZnS}} \times \frac{32.00 \text{ g } O_2}{1 \text{ mol } O_2} \times \frac{1 \text{ kg}}{1000 \text{ g}}$$

$$= 2.4\underline{6}3 = 2.46 \text{ kg } O_2$$

3.16. Balanced equation: $2HgO \rightarrow 2Hg + O_2$

Convert the mass of O_2 to moles of O_2. Using the fact that one mole of O_2 is equivalent to two moles of Hg, determine the number of moles of Hg, and convert to mass of Hg.

$$6.47 \text{ g } O_2 \times \frac{1 \text{ mol } O_2}{32.00 \text{ g } O_2} \times \frac{2 \text{ mol Hg}}{1 \text{ mol } O_2} \times \frac{200.59 \text{ g Hg}}{1 \text{ mol Hg}} = 81.\underline{1}1 = 81.1 \text{ g Hg}$$

3.17. First, determine the limiting reactant by calculating the moles of $AlCl_3$ that would be obtained if Al and HCl were totally consumed:

$$0.15 \text{ mol Al} \times \frac{2 \text{ mol AlCl}_3}{2 \text{ mol Al}} = 0.1\underline{5}0 \text{ mol AlCl}_3$$

$$0.35 \text{ mol HCl} \times \frac{2 \text{ mol AlCl}_3}{6 \text{ mol HCl}} = 0.1\underline{1}66 \text{ mol AlCl}_3$$

Because the HCl produces the smaller amount of $AlCl_3$, the reaction will stop when HCl is totally consumed but before the Al is consumed. The limiting reactant is therefore HCl. The amount of $AlCl_3$ produced must be 0.1$\underline{1}$66, or 0.12 mol.

3.18. First, determine the limiting reactant by calculating the moles of ZnS produced by totally consuming Zn and S_8:

$$7.36 \text{ g Zn} \times \frac{1 \text{ mol Zn}}{65.38 \text{ g Zn}} \times \frac{8 \text{ mol ZnS}}{8 \text{ mol Zn}} = 0.11\underline{2}57 \text{ mol ZnS}$$

$$6.45 \text{ g S}_8 \times \frac{1 \text{ mol S}_8}{256.52 \text{ g S}_8} \times \frac{8 \text{ mol ZnS}}{1 \text{ mol S}_8} = 0.20\underline{1}2 \text{ mol ZnS}$$

The reaction will stop when Zn is totally consumed; S_8 is in excess, and not all of it is converted to ZnS. The limiting reactant is therefore Zn. Now convert the moles of ZnS obtained from the Zn to grams of ZnS:

$$0.11\underline{2}57 \text{ mol ZnS} \times \frac{97.45 \text{ g ZnS}}{1 \text{ mol ZnS}} = 10.\underline{9}7 = 11.0 \text{ g ZnS}$$

3.19. First, write the balanced equation:

$$CH_3OH + CO \rightarrow HC_2H_3O_2$$

Convert grams of each reactant to moles of acetic acid:

$$15.0 \text{ g CH}_3\text{OH} \times \frac{1 \text{ mol CH}_3\text{OH}}{32.04 \text{ g CH}_3\text{OH}} \times \frac{1 \text{ mol HC}_2\text{H}_3\text{O}_2}{1 \text{ mol CH}_3\text{OH}} = 0.46\underline{8}1 \text{ mol HC}_2\text{H}_3\text{O}_2$$

$$10.0 \text{ g CO} \times \frac{1 \text{ mol CO}}{28.01 \text{ g CO}} \times \frac{1 \text{ mol HC}_2\text{H}_3\text{O}_2}{1 \text{ mol CO}} = 0.35\underline{7}0 \text{ mol HC}_2\text{H}_3\text{O}_2$$

Thus, CO is the limiting reactant, and 0.3570 mol $HC_2H_3O_2$ is obtained. The theoretical mass of product is

$$0.35\underline{7}0 \text{ mol HC}_2\text{H}_3\text{O}_2 \times \frac{60.05 \text{ g HC}_2\text{H}_3\text{O}_2}{1 \text{ mol HC}_2\text{H}_3\text{O}_2} = 21.\underline{4}4 = 21.4 \text{ g HC}_2\text{H}_3\text{O}_2$$

The percentage yield is

$$\frac{19.1 \text{ g actual yield}}{21.44 \text{ g theoretical yield}} \times 100\% = 89.\underline{0}9 = 89.1\%$$

■ ANSWERS TO CONCEPT CHECKS

3.1. a. Each tricycle has one seat, so you have a total of 1.5 mol of seats.

b. Each tricycle has three tires, so you have 1.5 mol × 3 = 4.5 mol of tires.

c. Each $Mg(OH)_2$ has two OH^- ions, so there are 1.5 mol × 2 = 3.0 mol OH^- ions.

3.2. a. When conducting this type of experiment, you are assuming that all of the carbon and hydrogen show up in the CO_2 and H_2O, respectively. In this experiment, where all of the carbon and hydrogen do not show up, when you analyze the CO_2 for carbon and H_2O for hydrogen, you find that the masses in the products are less than those in the carbon and hydrogen you started with.

b. Since you collected less carbon and hydrogen than were present in the original sample, the calculated mass percentage will be less than the expected (real) value. For example, say you have a 10.0-g sample that contains 7.5 g of carbon. You run the experiment on the 10.0-g sample and collect only 5.0 g of carbon. The calculated percent carbon based on your experimental results would be 50% instead of the correct amount of 75%.

3.3. a. $C_2H_8O_2$ is not an empirical formula because each of the subscripts can be divided by 2 to obtain a possible empirical formula of CH_4O. (The empirical formula is not the smallest integer ratio of subscripts.)

b. $C_{1.5}H_4$ is not a correct empirical formula because one of the subscripts is not an integer. Multiply each of the subscripts by 2 to obtain the possible empirical formula C_3H_8. (Since the subscript of carbon is the decimal number 1.5, the empirical formula is not the smallest integer ratio of subscripts.)

c. Yes, the empirical formula and the molecular formula can be the same, as is the case in this problem, where the formula is written with the smallest integer subscripts.

3.4. a. Correct. Coefficients in balanced equations can represent amounts in atoms and molecules.

b. Incorrect. The coefficients in a balanced chemical equation do not represent amounts in grams. One gram of carbon and one gram of oxygen represent different molar amounts.

c. Incorrect. The coefficients in a balanced chemical equation do not represent amounts in grams. Furthermore, the data do not support the law of mass conservation.

d. Correct. You might initially think this is an incorrect representation; however, 12 g of C, 32 g of O_2, and 44 g of CO_2 all represent one mole of the substance, so the relationship of the chemical equation is obeyed.

e. Correct. The coefficients in balanced equations can represent amounts in moles.

f. Incorrect. The amount of O_2 present is not enough to react completely with one mole of carbon. Only one-half of the carbon would react, and one-half mole of CO_2 would form.

g. Incorrect. In this representation, oxygen is being shown as individual atoms of O, not as molecules of O_2, so the drawings are not correctly depicting the chemical reaction.

h. Correct. The molecular models correctly depict a balanced chemical reaction since the same number of atoms of each element appears on both sides of the equation.

3.5. a. The balanced chemical equation stipulates that for 7 mol of ReO_3 to form, the molar ratio of Re to Re_2O_7 must be 1 to 3. Statement a is false.

b. Rhenium is in excess by 2 moles. The rhenium(VII) oxide is the limiting reagent. Statement b is false.

c. Statement c implies that 3 moles of rhenium(VII) oxide actually reacts with at least enough Re or even an excess of Re. According to the balanced equation, 7 mol of ReO_3 will then form. Statement c is true and is the best reason that 7 mol of ReO_3 are produced during the reaction.

d. This is equivalent to statement b and it also is false.

e. The excess Re undergoes no further reaction since it is limited by how much rhenium(VII) is present. Statement c is false.

3.6. a. $X_2(g) + 2Y(g) \rightarrow 2XY(g)$

b. Since the product consists of a combination of X and Y in a 1:1 ratio, it must consist of two atoms hooked together. If you count the total number of X atoms (split apart the X_2 molecules) and Y atoms present prior to the reaction, you find that there are four X atoms and three Y atoms. From these starting quantities, you are limited to three XY molecules and left with an unreacted X. Option #1 represents this situation and is therefore the correct answer.

c. Since $Y(g)$ was completely used up during the course of the reaction, it is the limiting reactant.

ANSWERS TO SELF-ASSESSMENT AND REVIEW QUESTIONS

3.1. The molecular mass is the sum of the atomic masses of all the atoms in a molecule of the substance whereas the formula mass is the sum of the atomic masses of all the atoms in one formula unit of the compound, whether the compound is molecular or not. A given substance could have both a molecular mass and a formula mass if it existed as discrete molecules.

3.2. To obtain the formula mass of a substance, sum up the atomic masses of all atoms in the formula of the compound.

3.3. A mole of N_2 contains Avogadro's number (6.02×10^{23}) of N_2 molecules and $2 \times 6.02 \times 10^{23}$ N atoms. One mole of $Fe_2(SO_4)_3$ contains three moles of SO_4^{2-} ions, and it contains twelve moles of O atoms.

3.4. A sample of the compound of known mass is burned, and CO_2 and H_2O are obtained as products. Next, you relate the masses of CO_2 and H_2O to the masses of carbon and hydrogen. Then you calculate the mass percentages of C and H. You find the mass percentage of O by subtracting the mass percentages of C and H from 100.

3.5. The empirical formula is obtained from the percentage composition by assuming for the purposes of the calculation a sample of 100 g of the substance. Then the mass of each element in the sample equals the numerical value of the percentage. Convert the masses of the elements to moles of the elements using the atomic mass of each element. Divide the moles of each by the smallest number to obtain the smallest ratio of each atom. If necessary, find a whole-number factor to multiply these results by to obtain integers for the subscripts in the empirical formula.

3.6. The empirical formula is the formula of a substance written with the smallest integer (whole-number) subscripts. Each of the subscripts in the formula $C_6H_{12}O_2$ can be divided by 2, so the empirical formula of the compound is C_3H_6O.

3.7. The number of empirical formula units in a compound, n, equals the molecular mass divided by the empirical formula mass.

$$n = \frac{34.0 \text{ amu}}{17.0 \text{ amu}} = 2.00$$

The molecular formula of hydrogen peroxide is therefore $(HO)_2$, or H_2O_2.

3.8. The coefficients in a chemical equation can be interpreted directly in terms of molecules or moles. For the mass interpretation, you will need the molar masses of CH_4, O_2, CO_2, and H_2O, which are 16.0, 32.0, 44.0, and 18.0 g/mol, respectively. A summary of the three interpretations is given below the balanced equation:

CH_4	+	$2O_2$	→	CO_2	+	$2H_2O$
1 molecule	+	2 molecules	→	1 molecule	+	2 molecules
1 mole	+	2 moles	→	1 mole	+	2 moles
16.0 g	+	2 × 32.0 g	→	44.0 g	+	2 × 18.0 g

3.9. A chemical equation yields the mole ratio of a reactant to a second reactant or product. Once the mass of a reactant is converted to moles, this can be multiplied by the appropriate mole ratio to give the moles of a second reactant or product. Multiplying this number of moles by the appropriate molar mass gives mass. Thus, the masses of two different substances are related by a chemical equation.

3.10. The limiting reactant is the reactant that is entirely consumed when the reaction is complete. Because the reaction stops when the limiting reactant is used up, the moles of product are always determined by the starting number of moles of the limiting reactant.

3.11. Two examples are given in the book. The first involves making cheese sandwiches. Each sandwich requires two slices of bread and one slice of cheese. The limiting reactant is the cheese because some bread is left unused. The second example is assembling automobiles. Each auto requires one steering wheel, four tires, and other components. The limiting reactant is the tires, since they will run out first.

3.12. Since the theoretical yield represents the maximum amount of product that can be obtained by a reaction from given amounts of reactants under any conditions, in an actual experiment you can never obtain more than this amount.

3.13. The answer is a, 3.27 g of NH_3.

3.14. The answer is b, 1 g of formaldehyde.

3.15. The answer is d, 4.85×10^{24} atoms.

3.16. The answer is a, $C_3H_4O_3$.

ANSWERS TO CONCEPTUAL PROBLEMS

3.19. a. $3H_2(g) + N_2(g) \rightarrow 2NH_3(g)$

b. Since there is no H_2 present in the container, it was entirely consumed during the reaction, which makes it the limiting reactant.

c. According to the chemical reaction, three molecules of H_2 are required for every molecule of N_2. Since there are two molecules of unreacted N_2, you would need six additional molecules of H_2 to complete the reaction.

3.21. a. This answer is unreasonable because 1.0×10^{-3} g is too small a mass for 0.33 mol of an element. For example, 0.33 mol of hydrogen (as H_2), the lightest element, would have a mass of 0.67 g.

b. This answer is unreasonable because 1.80×10^{-10} g is too large for one water molecule. (The mass of one water molecule is 2.99×10^{-23} g.)

c. This answer is reasonable because 3.01×10^{23} is one-half of Avogadro's number.

d. This answer is unreasonable because the units for molar mass should be g/mol, so this quantity is 1000 times too large.

3.23. a. The limiting reactant would be the charcoal because the air would supply as much oxygen as is needed.

b. The limiting reactant would be the magnesium because the beaker would contain much more water than is needed for the reaction (approximately 18 mL of water is 1 mole).

c. The limiting reactant would be the H_2 because the air could supply as much nitrogen as is needed.

3.25. a. The problem is that Avogadro's number was inadvertently used for the molar mass of calcium, which should be 40.08 g/mol. The correct calculation is

$$27.0 \text{ g Ca} \times \frac{1 \text{ mol Ca}}{40.08 \text{ g Ca}} = 0.67\underline{3}6 = 0.674 \text{ mol Ca}$$

b. The problem here is an incorrect mole ratio. There are 2 mol K^+ ions per 1 mol K_2SO_4. The correct calculation is

$$2.5 \text{ mol } K_2SO_4 \times \frac{2 \text{ mol } K^+ \text{ ions}}{1 \text{ mol } K_2SO_4} \times \frac{6.022 \text{ x } 10^{23} \text{ } K^+ \text{ ions}}{1 \text{ mol } K^+ \text{ ions}}$$

$$= 3.\underline{0}1 \times 10^{24} = 3.0 \times 10^{24} \text{ } K^+ \text{ ions}$$

c. The problem here is an incorrect mole ratio. The result should be

$$0.50 \text{ mol Na} \times \frac{2 \text{ mol } H_2O}{2 \text{ mol Na}} = 0.50 \text{ mol } H_2O$$

■ SOLUTIONS TO PRACTICE PROBLEMS

Note on significant figures: If the final answer to a solution needs to be rounded off, it is given first with one nonsignificant figure, and the last significant figure is underlined. The final answer is then rounded to the correct number of significant figures. In multistep problems, intermediate answers are given with at least one nonsignificant figure; however, only the final answer has been rounded off.

3.27. a. FM of CH_3OH = AM of C + 4(AM of H) + AM of O

= 12.01 amu + (4 × 1.008 amu) + 16.00 amu
= $32.0\underline{4}2$ = 32.0 amu (3 s.f.)

b. FM of NO_3 = AM of N + 3(AM of O)

= 14.01 amu + (3 × 16.00 amu) = 62.01 = 62.0 amu (3 s.f.)

c. FM of K_2CO_3 = 2(AM of K) + AM of C + 3(AM of O)

= (2 × 39.10 amu) + 12.01 amu + (3 × 16.00 amu)
= $138.2\underline{1}0$ = 138 amu (3 s.f.)

d. FM of $Ni_3(PO_4)_2$ = 3(AM of Ni) + 2(AM of P) + 8(AM of O)

= (3 × 58.70 amu) + (2 × 30.97 amu) + (8 × 16.00 amu)
= $366.0\underline{4}0$ = 366 amu (3 s.f.)

3.29. a. SO_2

1 × AM of S			=	32.07 amu
2 × AM of O	=	2 × 16.00	=	32.00 amu
MM of SO_2	=			$64.\underline{0}7$ amu = 64.1 amu (3 s.f.)

b. PCl_3

1 × AM of P			=	30.97 amu
3 × AM of Cl	=	3 × 35.45	=	106.35 amu
MM of PCl_3			=	137.32 amu = 137 amu (3 s.f.)

3.31. First, find the formula mass of NH_4NO_3 by adding the respective atomic masses. Then convert it to the molar mass:

FM of NH_4NO_3 = 2(AM of N) + 4(AM of H) + 3(AM of O)

= (2 × 14.01 amu) + (4 × 1.008 amu) + (3 × 16.00 amu)

= $80.0\underline{5}2$ amu

The molar mass of NH_4NO_3 is 80.05 g/mol.

3.33. a. The atomic mass of Na equals 22.99 amu; thus, the molar mass equals 22.99 g/mol. Because 1 mol of Na atoms equals 6.022×10^{23} Na atoms, we calculate

$$\text{Mass of one Na atom} = \frac{22.99 \text{ g/mol}}{6.022 \times 10^{23} \text{ atom/mol}} = 3.81\underline{7}7 \times 10^{-23}$$

$= 3.818 \times 10^{-23}$ g/atom

b. The atomic mass of N equals 14.01 amu; thus, the molar mass equals 14.01 g/mol. Because 1 mol of N atoms equals 6.022×10^{23} N atoms, we calculate

$$\text{Mass of one N atom} = \frac{14.01 \text{ g/mol}}{6.022 \times 10^{23} \text{ atom/mol}} = 2.32\underline{6}4 \times 10^{-23}$$

$= 2.326 \times 10^{-23}$ g/atom

c. The formula mass of CH_3Cl = [12.01 + (3 × 1.008) + 35.45] = 50.48 amu; thus, the molar mass equals 50.48 g/mol. Because 1 mol of CH_3Cl molecules equals 6.022×10^{23} CH_3Cl molecules, we calculate

$$\text{Mass of one CH}_3\text{Cl molecule} = \frac{50.48 \text{ g/mol}}{6.022 \times 10^{23} \text{ molecules/mol}} = 8.38\underline{2}6 \times 10^{-23}$$

$= 8.383 \times 10^{-23}$ g/molecule

d. The formula mass of $Hg(NO_3)_2$ = 200.59 + (2 × 14.01) + (6 × 16.00)] = 324.61 amu; thus, the molar mass equals 324.61 g/mol. Because 1 formula mass of $Hg(NO_3)_2$ equals 6.022×10^{23} $Hg(NO_3)_2$ formula units, we calculate

$$\text{Mass of one Hg(NO}_3)_2 = \frac{324.61 \text{ g/mol}}{6.022 \times 10^{23} \text{ units/mol}} = 5.39\underline{0}4 \times 10^{-22}$$

$= 5.390 \times 10^{-22}$ g/unit

3.35. First, find the formula mass (in amu) using the periodic table (inside front cover):

FM of $(CH_3CH_2)_2O$ = (4 × 12.01 amu) + (10 × 1.008 amu) + 16.00 amu = 74.12 amu

$$\text{Mass of one (CH}_3\text{CH}_2)_2\text{O molecule} = \frac{74.12 \text{ g/mol}}{6.022 \times 10^{23} \text{ molecules/mol}}$$

$= 1.23\underline{0}8 \times 10^{-22} = 1.231 \times 10^{-22}$ g/molecule

3.37. From the table of atomic masses, we obtain the following molar masses for parts a through d: Na = 22.99 g/mol; S = 32.07 g/mol; C = 12.01 g/mol; H = 1.008 g/mol; Cl = 35.45 g/mol; and N = 14.01 g/mol.

a. $0.15 \text{ mol Na} \times \frac{22.99 \text{ g}}{1 \text{ mol Na}} = 3.\underline{4}48 = 3.4 \text{ g Na}$

b. $0.594 \text{ mol S} \times \frac{32.07 \text{ g S}}{1 \text{ mol S}} = 19.\underline{0}4 = 19.0 \text{ g S}$

c. Using molar mass = 84.93 g/mol for CH_2Cl_2, we obtain

$$2.78 \text{ mol CH}_2\text{Cl}_2 \times \frac{84.93 \text{ g CH}_2\text{Cl}_2}{1 \text{ mol CH}_2\text{Cl}_2} = 23\underline{6}.1 = 236 \text{ g CH}_2\text{Cl}_2$$

d. Using molar mass = 68.14 g/mol for $(NH_4)_2S$, we obtain

$$38 \text{ mol (NH}_4)_2\text{S} \times \frac{68.14 \text{ g (NH}_4)_2\text{S}}{1 \text{ mol (NH}_4)_2\text{S}} = 2.\underline{5}8 \times 10^3 = 2.6 \times 10^3 \text{ g (NH}_4)_2\text{S}$$

3.39. First, find the molar mass of H_3BO_3: (3 × 1.008 amu) + 10.81 amu + (3 × 16.00 amu) = 61.83. Therefore, the molar mass of H_3BO_3 = 61.83 g/mol. The mass of H_3BO_3 is calculated as follows:

$$0.543 \text{ mol H}_3\text{BO}_3 \times \frac{61.83 \text{ g H}_3\text{BO}_3}{1 \text{ mol H}_3\text{BO}_3} = 33.\underline{5}7 = 33.6 \text{ g H}_3\text{BO}_3$$

3.41. From the table of atomic masses, we obtain the following rounded molar masses for parts a through d: C = 12.01 g/mol; Cl = 35.45 g/mol; H = 1.008 g/mol; Al = 26.98 g/mol; and O = 16.00 g/mol.

a. $2.86 \text{ g C} \times \dfrac{1 \text{ mol C}}{12.01 \text{ g C}} = 0.23\underline{8}1 = 0.238 \text{ mol C}$

b. $7.05 \text{ g Cl}_2 \times \dfrac{1 \text{ mol Cl}_2}{70.90 \text{ g Cl}_2} = 0.099\underline{4}3 = 0.0994 \text{ mol Cl}_2$

c. The molar mass of C_4H_{10} = (4 × 12.01) + (10 × 1.008) = 58.12 g C_4H_{10}/mol C_4H_{10}. The mass of C_4H_{10} is calculated as follows:

$$76 \text{ g C}_4\text{H}_{10} \times \frac{1 \text{ mol C}_4\text{H}_{10}}{58.12 \text{ g C}_4\text{H}_{10}} = 1.\underline{3}07 = 1.3 \text{ mol C}_4\text{H}_{10}$$

d. The molar mass of $Al_2(CO_3)_3$ = (2 × 26.98) + (3 × 12.01) + (9 × 16.00 g) =233.99 g/mol $Al_2(CO_3)_3$. The mass of $Al_2(CO_3)_3$ is calculated as follows:

$$26.2 \text{ g Al}_2(\text{CO}_3)_3 \times \frac{1 \text{ mol Al}_2(\text{CO}_3)_3}{233.99 \text{ g Al}_2(\text{CO}_3)_3} = 0.11\underline{1}9 = 0.112 \text{ mol Al}_2(\text{CO}_3)_3$$

3.43. Calculate the formula mass of calcium sulfate: 40.08 amu + 32.07 amu + (4 × 16.00 amu) = 136.15 amu. Therefore, the molar mass of $CaSO_4$ is 136.15 g/mol. Use this to convert the mass of $CaSO_4$ to moles:

$$0.791 \text{ g CaSO}_4 \times \frac{1 \text{ mol CaSO}_4}{136.15 \text{ g CaSO}_4} = 5.8\underline{0}9 \times 10^{-3} = 5.81 \times 10^{-3} \text{ mol CaSO}_4$$

Calculate the molecular mass of water: (2 × 1.008 amu) + 16.00 amu = 18.02 amu. Therefore, the molar mass of H_2O equals 18.02 g/mol. Use this to convert the rest of the sample to moles of water:

$$0.209 \text{ g H}_2\text{O} \times \frac{1 \text{ mol H}_2\text{O}}{18.02 \text{ g H}_2\text{O}} = 1.1\underline{5}9 \times 10^{-2} = 1.16 \times 10^{-2} \text{ mol H}_2\text{O}$$

Because 0.01159 mol is about twice 0.005811 mol, both numbers of moles are consistent with the formula, $CaSO_4 \bullet 2H_2O$.

3.45. The following rounded atomic masses are used: Li = 6.94 g/mol; Br = 79.90 g/mol; N = 14.01 g/mol; H = 1.008 g/mol; Pb = 207.2 g/mol; Cr = 52.00 g/mol; O = 16.00 g/mol; and S = 32.07 g/mol. Also, Avogadro's number is 6.022×10^{23} atoms, so

a. $\text{No. Li atoms} = 8.21 \text{ g Li} \times \dfrac{6.022 \times 10^{23} \text{ atoms}}{6.941 \text{ g Li}} = 7.1\underline{2}2 \times 10^{23} = 7.12 \times 10^{23} \text{ atoms}$

b. $\text{No. Br atoms} = 32.0 \text{ g Br}_2 \times \dfrac{2 \text{ x } 6.022 \times 10^{23} \text{ atoms}}{(2 \text{ x } 79.90) \text{ g Br}_2} = 2.4\underline{1}2 \times 10^{23} = 2.41 \times 10^{23} \text{ atoms}$

c. $\text{No. NH}_3 \text{ molecules} = 45 \text{ g NH}_3 \times \dfrac{6.022 \times 10^{23} \text{ molecules}}{17.03 \text{ g NH}_3} = 1.\underline{5}9 \times 10^{24}$

$= 1.6 \times 10^{24}$ molecules

d. $$\text{No. PbCrO}_4 \text{ units} = 201 \text{ g PbCrO}_4 \times \frac{6.022 \times 10^{23} \text{ units}}{323.2 \text{ g PbCrO}_4} = 3.7\underline{4}5 \times 10^{23}$$

$= 3.75 \times 10^{23}$ units

e. $$\text{No. SO}_4^{2-} \text{ ions} = 14.3 \text{ g Cr}_2(\text{SO}_4)_3 \times \frac{3 \times 6.022 \times 10^{23} \text{ ions}}{392.21 \text{ g Cr}_2(\text{SO}_4)_3} = 6.5\underline{8}7 \times 10^{22}$$

$= 6.59 \times 10^{22}$ ions

3.47. Calculate the molecular mass of CCl_4: 12.01 amu + (4 × 35.45 amu) = 153.81 amu. Use this and Avogadro's number to express it as 153.81 g/N_A in order to calculate the number of molecules:

$$7.58 \text{ mg CCl}_4 \times \frac{1 \text{ g}}{1000 \text{ mg}} \times \frac{6.022 \times 10^{23} \text{ molecules}}{153.81 \text{ g CCl}_4} = 2.9\underline{6}8 \times 10^{19}$$

$= 2.97 \times 10^{19}$ molecules

3.49. $$\text{Mass percentage carbon} = \frac{\text{mass of C in sample}}{\text{mass of sample}} \times 100\%$$

$$\text{Percent carbon} = \frac{1.584 \text{ g}}{1.836 \text{ g}} \times 100\% = 86.2\underline{7}4 = 86.27\%$$

3.51. $$\text{Mass percentage phosphorus} = \frac{\text{mass of P in sample}}{\text{mass of sample}} \times 100\%$$

$$\text{Percent P} = \frac{1.72 \text{ mg}}{8.53 \text{ mg}} \times 100\% = 20.\underline{1}6 = 20.2\%$$

3.53. Start with the definition for percentage nitrogen, and rearrange this equation to find the mass of N in the fertilizer.

$$\text{Mass percentage nitrogen} = \frac{\text{mass of N in fertilizer}}{\text{mass of fertilizer}} \times 100\%$$

$$\text{Mass N} = \frac{\text{mass \% N}}{100\%} \times \text{mass of fertilizer} = \frac{14.0\%}{100\%} \times 4.15 \text{ kg} = 0.58\underline{1}0 = 0.581 \text{ kg N}$$

3.55. Convert moles to mass using the molar masses from the respective atomic masses. Then calculate the mass percentages from the respective masses.

$$0.0898 \text{ mol Al} \times \frac{26.98 \text{ g Al}}{1 \text{ mol Al}} = 2.4\underline{2}2 \text{ g Al}$$

$$0.0381 \text{ mol Mg} \times \frac{24.31 \text{ g Mg}}{1 \text{ mol Mg}} = 0.92\underline{6}2 \text{ g Mg}$$

$$\text{Percent Al} = \frac{\text{mass of Al}}{\text{mass of alloy}} = \frac{2.422 \text{ g Al}}{3.349 \text{ g alloy}} \times 100\% = 72.\underline{3}4 = 72.3\% \text{ Al}$$

$$\text{Percent Mg} = \frac{\text{mass of Mg}}{\text{mass of alloy}} = \frac{0.9262 \text{ g Mg}}{3.349 \text{ g alloy}} \times 100\% = 27.\underline{6}55 = 27.7\% \text{ Mg}$$

3.57. In each part, the numerator consists of the mass of the element in one mole of the compound; the denominator is the mass of one mole of the compound. Use the atomic weights of C = 12.01 g/mol; O = 16.00 g/mol; Na = 22.99 g/mol; H = 1.008 g/mol; P = 30.97 g/mol; Co = 58.93 g/mol; and N = 14.01 g/mol.

a. $$\text{Percent C} = \frac{\text{mass of C}}{\text{mass of CO}} = \frac{12.01 \text{ g C}}{28.01 \text{ g CO}} \times 100\% = 42.878 = 42.9\%$$

$$\text{Percent O} = 100.000\% - 42.878\%\text{C} = 57.122 = 57.1\%$$

b. $$\text{Percent C} = \frac{\text{mass of C}}{\text{mass of CO}_2} = \frac{12.01 \text{ g C}}{44.01 \text{ g CO}_2} \times 100\% = 27.289 = 27.3\%$$

$$\text{Percent O} = 100.000\% - 27.2\underline{89}\% \text{ C} = 72.711 = 72.7\%$$

c. $$\text{Percent Na} = \frac{\text{mass of Na}}{\text{mass of NaH}_2\text{PO}_4} = \frac{22.99 \text{ g Na}}{119.98 \text{ g NaH}_2\text{PO}_4} \times 100\% = 19.161 = 19.2\%$$

$$\text{Percent H} = \frac{\text{mass of H}}{\text{mass of NaH}_2\text{PO}_4} = \frac{2.016 \text{ g H}}{119.98 \text{ g NaH}_2\text{PO}_4} \times 100\% = 1.6802 = 1.68\%$$

$$\text{Percent P} = \frac{\text{mass of P}}{\text{mass of NaH}_2\text{PO}_4} = \frac{30.97 \text{ g P}}{119.98 \text{ g NaH}_2\text{PO}_4} \times 100\% = 25.812 = 25.8\%$$

$$\text{Percent O} = 100.000\% - (19.161 + 1.6802 + 25.812)\% = 53.346 = 53.3\%$$

d. $$\text{Percent Co} = \frac{\text{mass of Co}}{\text{mass of Co(NO}_3)_2} = \frac{58.93 \text{ g Co}}{182.95 \text{ g Co(NO}_3)_2} \times 100\% = 32.211 = 32.2\%$$

$$\text{Percent N} = \frac{\text{mass of N}}{\text{mass of Co(NO}_3)_2} = \frac{2 \text{ x } 14.01 \text{ g N}}{182.95 \text{ g Co(NO}_3)_2} \times 100\% = 15.316 = 15.3\%$$

$$\text{Percent O} = 100.000\% - (32.211 + 15.316)\% = 52.473 = 52.5\%$$

3.59. The molecular model of toluene contains seven carbon atoms and eight hydrogen atoms, so the molecular formula of toluene is C_7H_8. The molar mass of toluene is 92.134 g/mol. The mass percentages are

$$\text{Percent C} = \frac{\text{mass of C}}{\text{mass of C}_7\text{H}_8} = \frac{7 \text{ x } 12.01 \text{ g}}{92.134 \text{ g}} \times 100\% = 91.247 = 91.2\%$$

$$\text{Percent H} = 100\% - 91.247 = 8.753 = 8.75\%$$

3.61. Find the moles of C in each amount in one-step operations. Calculate the moles of each compound using the molar mass; then multiply by the number of moles of C per mole of compound:

$$\text{Mol C (glucose)} = 6.01 \text{ g} \times \frac{1 \text{ mol}}{180.2 \text{ g}} \times \frac{6 \text{ mol C}}{1 \text{ mol glucose}} = 0.20\underline{0} \text{ mol}$$

$$\text{Mol C (ethanol)} = 5.85 \text{ g} \times \frac{1 \text{ mol}}{46.07 \text{ g}} \times \frac{2 \text{ mol C}}{1 \text{ mol ethanol}} = 0.25\underline{4} \text{ mol (more C)}$$

3.63. First, calculate the mass of C in the glycol by multiplying the mass of CO_2 by the molar mass of C and the reciprocal of the molar mass of CO_2. Next, calculate the mass of H in the glycol by multiplying the mass of H_2O by the molar mass of 2H and the reciprocal of the molar mass of H_2O. Then use the masses to calculate the mass percentages. Calculate O by difference.

$$9.06 \text{ mg } CO_2 \times \frac{1 \text{ mol } CO_2}{44.01 \text{ g } CO_2} \times \frac{12.01 \text{ g C}}{1 \text{ mol C}} = 2.4\underline{7}2 \text{ mg C}$$

$$5.58 \text{ mg } H_2O \times \frac{1 \text{ mol } H_2O}{18.02 \text{ g } H_2O} \times \frac{2 \text{ H}}{1 \text{ } H_2O} \times \frac{1.008 \text{ g H}}{1 \text{ mol H}} = 0.62\underline{4}3 \text{ mg H}$$

Mass O = 6.38 mg − (2.472 + 0.6243) = $3.2\underline{8}4$ mg O

Percent C = (2.472 mg C/6.38 mg glycol) × 100% = $38.\underline{7}4$ = 38.7%

Percent H = (0.6243 mg H/6.38 mg glycol) × 100% = $9.7\underline{8}5$ = 9.79%

Percent O = (3.284 mg O/6.38 mg glycol) × 100% = $51.\underline{4}7$ = 51.5%

3.65. Start by calculating the moles of Os and O; then divide each by the smaller number of moles to obtain integers for the empirical formula.

$$\text{Mol Os} = 2.16 \text{ g Os} \times \frac{1 \text{ mol Os}}{190.2 \text{ g Os}} = 0.011\underline{3}6 \text{ mol (smaller number)}$$

$$\text{Mol O} = (2.89 - 2.16) \text{ g O} \times \frac{1 \text{ mol O}}{16.00 \text{ g O}} = 0.04\underline{5}6 \text{ mol}$$

Integer for Os = 0.01136 ÷ 0.01136 = $1.0\underline{0}0$

Integer for O = 0.0456 ÷ 0.01136 = $4.\underline{0}1$

Within experimental error, the empirical formula is OsO_4.

3.67. Assume a sample of 100.0 g of potassium manganate. By multiplying this by the percentage composition, we obtain 39.6 g of K, 27.9 g of Mn, and 32.5 g of O. Convert each of these masses to moles by dividing by molar mass.

$$\text{Mol K} = 39.6 \text{ g K} \times \frac{1 \text{ mol K}}{39.10 \text{ g K}} = 1.0\underline{1}3 \text{ mol}$$

$$\text{Mol Mn} = 27.9 \text{ g Mn} \times \frac{1 \text{ mol Mn}}{54.94 \text{ g Mn}} = 0.50\underline{7}8 \text{ mol (smallest number)}$$

$$\text{Mol O} = 32.5 \text{ g O} \times \frac{1 \text{ mol O}}{16.00 \text{ g O}} = 2.0\underline{3}1 \text{ mol}$$

Now, divide each number of moles by the smallest number to obtain the smallest set of integers for the empirical formula.

Integer for K = 1.013 ÷ 0.5078 = 1.998, or 2

Integer for Mn = 0.5078 ÷ 0.5078 = 1.000, or 1

Integer for O = 2.031 ÷ 0.5078 = 3.999, or 4

The empirical formula is thus K_2MnO_4.

3.69. Assume a sample of 100.0 g of acrylic acid. By multiplying this by the percentage composition, we obtain 50.0 g C, 5.6 g H, and 44.4 g O. Convert each of these masses to moles by dividing by the molar mass.

$$\text{Mol C} = 50.0 \text{ g C} \times \frac{1 \text{ mol C}}{12.01 \text{ g C}} = 4.1\underline{6}3 \text{ mol}$$

$$\text{Mol H} = 5.6 \text{ g H} \times \frac{1 \text{ mol H}}{1.008 \text{ g H}} = 5.\underline{5}6 \text{ mol}$$

$$\text{Mol O} = 44.0 \text{ g O} \times \frac{1 \text{ mol O}}{16.00 \text{ g O}} = 2.7\underline{7}5 \text{ mol (smallest number)}$$

Now, divide each number of moles by the smallest number to obtain the smallest number of moles and the tentative integers for the empirical formula.

Tentative integer for C = 4.163 ÷ 2.775 = 1.50, or 1.5

Tentative integer for H = 5.56 ÷ 2.775 = 2.00, or 2

Tentative integer for O = 2.775 ÷ 2.775 = 1.00, or 1

Because 1.5 is not a whole number, multiply each tentative integer by 2 to obtain the final integer for the empirical formula:

C: 2 × 1.5 = 3

H: 2 × 2 = 4

O: 2 × 1 = 2

The empirical formula is thus $C_3H_4O_2$.

3.71. a. Assume for the calculation that you have 100.0 g; of this quantity, 92.25 g is C and 7.75 g is H. Now, convert these masses to moles:

$$92.25 \text{ g C} \times \frac{1 \text{ mol C}}{12.01 \text{ g C}} = 7.68\underline{1}09 \text{ mol C}$$

$$7.75 \text{ g H} \times \frac{1 \text{ mol H}}{1.008 \text{ g H}} = 7.6\underline{8}8 \text{ mol H}$$

Usually, you divide all the mole numbers by the smaller one, but in this case the mole numbers are equal, so the ratio of the number of C atoms to the number of H atoms is 1:1. Thus, the empirical formula for both compounds is CH.

b. Obtain n, the number of empirical formula units in the molecule, by dividing the molecular mass of 52.03 amu and 78.05 amu by the empirical formula mass of 13.018 amu:

$$\text{For 52.03: } n = \frac{52.03 \text{ amu}}{13.018 \text{ amu}} = 3.99\underline{6}8, \text{ or } 4$$

$$\text{For 78.05: } n = \frac{78.05 \text{ amu}}{13.018 \text{ amu}} = 5.99\underline{5}5, \text{ or } 6$$

The molecular formulas are as follows: for 52.03, $(CH)_4$ or C_4H_4; for 78.05, $(CH)_6$ or C_6H_6.

3.73. The formula mass corresponding to the empirical formula C_2H_6N may be found by adding the respective atomic masses.

$$\text{Formula mass} = (2 \times 12.01 \text{ amu}) + (6 \times 1.008 \text{ amu}) + 14.01 \text{ amu} = 44.08 \text{ amu}$$

Dividing the molecular mass by the formula mass gives the number of times the C_2H_6N unit occurs in the molecule. Because the molecular mass is an average of 88.5 ([90 + 87] ÷ 2), this quotient is

$$88.5 \text{ amu} \div 44.1 \text{ amu} = 2.\underline{0}06, \text{ or } 2$$

Therefore, the molecular formula is $(C_2H_6N)_2$, or $C_4H_{12}N_2$.

3.75. Assume a sample of 100.0 g of oxalic acid. By multiplying this by the percentage composition, we obtain 26.7 g C, 2.2 g H, and 71.1 g O. Convert each of these masses to moles by dividing by the molar mass.

$$\text{Mol C} = 26.7 \text{ g C} \times \frac{1 \text{ mol C}}{12.01 \text{ g C}} = 2.2\underline{2}3 \text{ mol}$$

$$\text{Mol H} = 2.2 \text{ g H} \times \frac{1 \text{ mol H}}{1.008 \text{ g H}} = 2.\underline{1}8 \text{ mol (smallest number)}$$

$$\text{Mol O} = 71.1 \text{ g O} \times \frac{1 \text{ mol O}}{16.00 \text{ g O}} = 4.4\underline{4}3 \text{ mol}$$

Now, divide each number of moles by the smallest number to obtain the smallest set of integers for the empirical formula.

$$\text{Integer for C} = 2.223 \div 2.18 = 1.02, \text{ or } 1$$

$$\text{Integer for H} = 2.18 \div 2.18 = 1.00, \text{ or } 1$$

$$\text{Integer for O} = 4.443 \div 2.18 = 2.0\underline{3}8, \text{ or } 2$$

The empirical formula is thus CHO_2. The formula mass corresponding to this formula may be found by adding the respective atomic masses:

$$\text{Formula mass} = 12.01 \text{ amu} + 1.008 \text{ amu} + (2 \times 16.00 \text{ amu}) = 45.02 \text{ amu}$$

Dividing the molecular mass by the formula mass gives the number of times the CHO_2 unit occurs in the molecule. Because the molecular mass is 90 amu, this quotient is

$$90 \text{ amu} \div 45.02 \text{ amu} = 2.\underline{0}0, \text{ or } 2$$

The molecular formula is thus $(CHO_2)_2$, or $C_2H_2O_4$.

3.77.

C_2H_4	+	$3O_2$	→	$2CO_2$	+	$2H_2O$
1 molecule C_2H_4	+	3 molecules O_2	→	2 molecules CO_2	+	2 molecules H_2O
1 mole C_2H_4	+	3 moles O_2	→	2 moles CO_2	+	2 moles H_2O
28.052 g C_2H_4	+	3 × 32.00 g O_2	→	2 × 44.01 g CO_2	+	2 × 18.016 g H_2O

3.79. By inspecting the balanced equation, obtain a conversion factor of eight mol CO_2 to two mol C_4H_{10}. Multiply the given amount of 0.41 mol of C_4H_{10} by the conversion factor to obtain the moles of CO_2.

$$0.41 \text{ mol } C_4H_{10} \times \frac{8 \text{ mol } CO_2}{2 \text{ mol } C_4H_{10}} = 1.\underline{6}4 = 1.6 \text{ mol } CO_2$$

3.81. By inspecting the balanced equation, obtain a conversion factor of three mol O_2 to two mol Fe_2O_3. Multiply the given amount of 3.91 mol Fe_2O_3 by the conversion factor to obtain moles of O_2.

$$3.91 \text{ mol } Fe_2O_3 \times \frac{3 \text{ mol } O_2}{2 \text{ mol } Fe_2O_3} = 5.8\underline{6}5 = 5.87 \text{ mol } O_2$$

3.83. $3NO_2 + H_2O \rightarrow 2HNO_3 + NO$

Three moles of NO_2 are equivalent to two moles of HNO_3 (from equation).

One mole of NO_2 is equivalent to 46.01 g NO_2 (from molecular mass of NO_2).

One mole of HNO_3 is equivalent to 63.02 g HNO_3 (from molecular mass of HNO_3).

$$7.50 \text{ g } HNO_3 \times \frac{1 \text{ mol } HNO_3}{63.02 \text{ g } HNO_3} \times \frac{3 \text{ mol } NO_2}{2 \text{ mol } HNO_3} \times \frac{46.01 \text{ g } NO_2}{1 \text{ mol } NO_2} = 8.2\underline{1}3 = 8.21 \text{ g } NO_2$$

3.85. $WO_3 + 3H_2 \rightarrow W + 3H_2O$

One mole of W is equivalent to three moles of H_2 (from equation).

One mole of H_2 is equivalent to 2.016 g H_2 (from molecular mass of H_2).

One mole of W is equivalent to 183.8 g W (from atomic mass of W).

4.81 kg of H_2 is equivalent to 4.81×10^3 g of H_2.

$$4.81 \times 10^3 \text{ g } H_2 \times \frac{1 \text{ mol } H_2}{2.016 \text{ g } H_2} \times \frac{1 \text{ mol W}}{3 \text{ mol } H_2} \times \frac{183.85 \text{ g W}}{1 \text{ mol W}} = 1.4\underline{6}2 \times 10^5 = 1.46 \times 10^5 \text{ g W}$$

3.87. Write the equation, and set up the calculation below the equation (after calculating the two molecular masses):

$CS_2 + 3Cl_2 \rightarrow CCl_4 + S_2Cl_2$

$$62.7 \text{ g } Cl_2 \times \frac{1 \text{ mol } Cl_2}{70.90 \text{ g } Cl_2} \times \frac{1 \text{ mol } CS_2}{3 \text{ mol } Cl_2} \times \frac{76.15 \text{ g } CS_2}{1 \text{ mol } CS_2} \; 22.\underline{4}48 = 22.4 \text{ g } CS_2$$

3.89. Write the equation, and set up the calculation below the equation (after calculating the two molecular masses):

$$2N_2O_5 \rightarrow 4NO_2 + O_2$$

$$1.381 \text{ g } O_2 \times \frac{1 \text{ mol } O_2}{32.00 \text{ g } O_2} \times \frac{4 \text{ mol } NO_2}{1 \text{ mol } O_2} \times \frac{46.01 \text{ g } NO_2}{1 \text{ mol } NO_2} = 7.94\underline{2}48 = 7.942 \text{ g } NO_2$$

3.91. First determine whether KO_2 or H_2O is the limiting reactant by calculating the moles of O_2 that each would form if it were the limiting reactant. Identify the limiting reactant by the smaller number of moles of O_2 formed.

$$0.15 \text{ mol } H_2O \times \frac{3 \text{ mol } O_2}{2 \text{ mol } H_2O} = 0.2\underline{2}5 \text{ mol } O_2$$

$$0.25 \text{ mol } KO_2 \times \frac{3 \text{ mol } O_2}{4 \text{ mol } KO_2} = 0.1\underline{8}7 \text{ mol } O_2$$ (KO_2 is the limiting reactant.)

Moles of O_2 produced = 0.19 mol

3.93. First determine whether CO or H_2 is the limiting reactant by calculating the moles of CH_3OH that each would form if it were the limiting reactant. Identify the limiting reactant by the smaller number of moles of CH_3OH formed. Use the molar mass of CH_3OH to calculate the mass of CH_3OH formed. Then calculate the mass of the unconsumed reactant.

$$CO + 2H_2 \rightarrow CH_3OH$$

$$10.2 \text{ g } H_2 \times \frac{1 \text{ mol } H_2}{2.016 \text{ g } H_2} \times \frac{1 \text{ mol } CH_3OH}{2 \text{ mol } H_2} = 2.5\underline{2}9 \text{ mol } CH_3OH$$

$$35.4 \text{ g CO} \times \frac{1 \text{ mol CO}}{28.01 \text{ g CO}} \times \frac{1 \text{ mol } CH_3OH}{1 \text{ mol CO}} = 1.2\underline{6}3 \text{ mol } CH_3OH$$

CO is the limiting reactant.

$$\text{Mass } CH_3OH \text{ formed} = 1.2\underline{6}3 \text{ mol } CH_3OH \times \frac{32.042 \text{ g } CH_3OH}{1 \text{ mol } CH_3OH} = 40.\underline{4}7 = 40.5 \text{ g } CH_3OH$$

Hydrogen is left unconsumed at the end of the reaction. The mass of H_2 that reacts can be calculated from the moles of product obtained:

$$1.2\underline{6}3 \text{ mol } CH_3OH \times \frac{2 \text{ mol } H_2}{1 \text{ mol } CH_3OH} \times \frac{2.016 \text{ g } H_2}{1 \text{ mol } H_2} = 5.0\underline{9}2 \text{ g } H_2$$

Unreacted H_2 = 10.2 g total H_2 − 5.092 g reacted H_2 = $5.\underline{1}08$ = 5.1 g H_2

3.95. First, determine which of the three reactants is the limiting reactant by calculating the moles of $TiCl_4$ that each would form if it were the limiting reactant. Identify the limiting reactant by the smallest number of moles of $TiCl_4$ formed. Use the molar mass of $TiCl_4$ to calculate the mass of $TiCl_4$ formed.

$$3TiO_2 + 4C + 6Cl_2 \rightarrow 3TiCl_4 + 2CO_2 + 2CO$$

$$4.15 \text{ g } TiO_2 \times \frac{1 \text{ mol } TiO_2}{79.88 \text{ g } TiO_2} \times \frac{3 \text{ mol } TiCl_4}{3 \text{ mol } TiO_2} = 0.051\underline{9}5 \text{ mol } TiCl_4$$

$$5.67 \text{ g C} \times \frac{1 \text{ mol C}}{12.01 \text{ g C}} \times \frac{3 \text{ mol TiCl}_4}{4 \text{ mol C}} = 0.35\underline{4}07 \text{ mol TiCl}_4$$

$$6.78 \text{ g Cl}_2 \times \frac{1 \text{ mol Cl}_2}{70.90 \text{ g Cl}_2} \times \frac{3 \text{ mol TiCl}_4}{6 \text{ mol Cl}_2} = 0.047\underline{8}1 \text{ mol TiCl}_4$$

Cl_2 is the limiting reactant.

$$\text{Mass TiCl}_4 \text{ formed} = 0.047\underline{8}1 \text{ mol TiCl}_4 \times \frac{189.68 \text{ g TiCl}_4}{1 \text{ mol TiCl}_4} = 9.0\underline{6}8 = 9.07 \text{ g TiCl}_4$$

3.97. First, determine which of the two reactants is the limiting reactant by calculating the moles of aspirin that each would form if it were the limiting reactant. Identify the limiting reactant by the smallest number of moles of aspirin formed. Use the molar mass of aspirin to calculate the theoretical yield in grams of aspirin. Then calculate the percentage yield.

$$C_7H_6O_3 + C_4H_6O_3 \rightarrow C_9H_8O_4 + C_2H_4O_2$$

$$2.00 \text{ g } C_7H_6O_3 \times \frac{1 \text{ mol } C_7H_6O_3}{138.12 \text{ g } C_7H_6O_3} \times \frac{1 \text{ mol } C_9H_8O_4}{1 \text{ mol } C_7H_6O_3} = 0.014\underline{4}8 \text{ mol } C_9H_8O_4$$

$$4.00 \text{ g } C_4H_6O_3 \times \frac{1 \text{ mol } C_4H_6O_3}{102.09 \text{ g } C_4H_6O_3} \times \frac{1 \text{ mol } C_9H_8O_4}{1 \text{ mol } C_4H_6O_3} = 0.039\underline{1}8 \text{ mol } C_9H_8O_4$$

Thus, $C_7H_6O_3$ is the limiting reactant. The theoretical yield of $C_9H_8O_4$ is

$$0.014\underline{4}8 \text{ mol } C_9H_8O_4 \times \frac{180.15 \text{ g } C_9H_8O_4}{1 \text{ mol } C_9H_8O_4} = 2.6\underline{0}9 \text{ g } C_9H_8O_4$$

The percentage yield is

$$\text{Percentage yield} = \frac{\text{actual yield}}{\text{theoretical yield}} \times 100\% = \frac{1.86 \text{ g}}{2.609 \text{ g}} \times 100\% = 71.\underline{2}9 = 71.3\%$$

SOLUTIONS TO GENERAL PROBLEMS

3.99. For 1 mol of caffeine, there are eight mol of C, ten mol of H, four mol of N, and two mol of O. Convert these amounts to masses by multiplying them by their respective molar masses:

8 mol C × 12.01 g C/1 mol C	=	96.08 g C
10 mol H × 1.008 g H/1 mol H	=	10.08 g H
4 mol N × 14.01 g N/1 mol N	=	56.04 g N
2 mol O × 16.00 g O/1 mol O	=	32.00 g O
1 mol of caffeine (total)	=	194.20 g (molar mass)

Each mass percentage is calculated by dividing the mass of the element by the molar mass of caffeine and multiplying by 100 percent: Mass percentage = (mass element ÷ mass caffeine) × 100%.

Mass percentage C = (96.08 g ÷ 194.20 g) × 100% = 49.5% (3 s.f.)

Mass percentage H = (10.08 g ÷ 194.20 g) × 100% = 5.19% (3 s.f.)

Mass percentage N = (56.04 g ÷ 194.20 g) × 100% = 28.9% (3 s.f.)

Mass percentage O = (32.00 g ÷ 194.20 g) × 100% = 16.5% (3 s.f.)

3.101. Assume a sample of 100.0 g of *p*-dichlorobenzene. By multiplying this by the percentage composition, we obtain 49.1 g C, 2.7 g of H, and 48.2 g of Cl. Convert each mass to moles by dividing by the molar mass:

$$49.1 \text{ g C} \times \frac{1 \text{ mol C}}{12.01 \text{ g C}} = 4.0\underline{88} \text{ mol C}$$

$$2.7 \text{ g H} \times \frac{1 \text{ mol H}}{1.008 \text{ g H}} = 2.\underline{6}8 \text{ mol H}$$

$$48.2 \text{ g Cl} \times \frac{1 \text{ mol Cl}}{35.45 \text{ g Cl}} = 1.3\underline{6}0 \text{ mol Cl}$$

Divide each number of moles by the smallest number to obtain the smallest set of integers for the empirical formula.

Integer for C = 4.088 mol ÷ 1.360 mol = 3.00, or 3

Integer for H = 2.68 mol ÷ 1.360 mol = 1.97, or 2

Integer for Cl = 1.360 mol ÷ 1.360 mol = 1.00, or 1

The empirical formula is thus C_3H_2Cl. Find the formula mass by adding the atomic masses:

Formula mass = (3 × 12.01 amu) + (2 × 1.008 amu) + 35.45 amu = 73.4$\underline{9}$6 = 73.50 amu

Divide the molecular mass by the formula mass to find the number of times the C_3H_2Cl unit occurs in the molecule. Because the molecular mass is 147 amu, this quotient is

147 amu ÷ 73.50 amu = 2.00, or 2

The molecular formula is $(C_3H_2Cl)_2$, or $C_6H_4Cl_2$.

3.103. Find the percentage composition of C and S from the analysis:

$$0.01665 \text{ g } CO_2 \times \frac{1 \text{ mol } CO_2}{44.01 \text{ g } CO_2} \times \frac{1 \text{ mol C}}{1 \text{ mol } CO_2} \times \frac{12.01 \text{ g C}}{1 \text{ mol C}} = 0.00454\underline{4} \text{ g C}$$

Percent C = (0.004544 g C ÷ 0.00796 g comp.) × 100% = 57.$\underline{0}$9%

$$0.01196 \text{ g } BaSO_4 \times \frac{1 \text{ mol } BaSO_4}{233.39 \text{ g } BaSO_4} \times \frac{1 \text{ mol S}}{1 \text{ mol } BaSO_4} \times \frac{32.07 \text{ g S}}{1 \text{ mol S}} = 0.0016\underline{4}3 \text{ g S}$$

Percent S = (0.001643 g S ÷ 0.00431 g comp.) × 100% = 38.$\underline{1}$2%

Percent H = 100.00% − (57.09 + 38.12)% = 4.$\underline{7}$9%

We now obtain the empirical formula by calculating moles from the grams corresponding to each mass percentage of element:

$$57.09 \text{ g C} \times \frac{1 \text{ mol C}}{12.01 \text{ g C}} = 4.75\underline{4} \text{ mol C}$$

$$38.12 \text{ g S} \times \frac{1 \text{ mol S}}{32.07 \text{ g S}} = 1.18\underline{9} \text{ mol S}$$

$$4.79 \text{ g H} \times \frac{1 \text{ mol H}}{1.008 \text{ g H}} = 4.7\underline{5}2 \text{ mol H}$$

Dividing the moles of the elements by the smallest number (1.189), we obtain for C: 3.997, or 4; for S: 1.000, or 1; and for H: 3.996, or 4. Thus, the empirical formula is C_4H_4S (formula mass = 84). Because the formula mass was given as 84 amu, the molecular formula is also C_4H_4S.

3.105. For g $CaCO_3$, use this equation: $CaCO_3 + H_2C_2O_4 \rightarrow CaC_2O_4 + H_2O + CO_2$.

$$0.469 \text{ g } CaC_2O_4 \times \frac{1 \text{ mol } CaC_2O_4}{128.10 \text{ g } CaC_2O_4} \times \frac{1 \text{ mol } CaCO_3}{1 \text{ mol } CaC_2O_4} \times \frac{100.09 \text{ g } CaCO_3}{1 \text{ mol } CaCO_3}$$

$$= 0.36\underline{6}4 \text{ g } CaCO_3$$

$$\text{Mass percentage } CaCO_3 = \frac{\text{mass } CaCO_3}{\text{mass limestone}} \times 100\% = \frac{0.36\underline{6}4 \text{ g}}{0.438 \text{ g}} \times 100\%$$

$$= 83.\underline{6}6 = 83.7\%$$

3.107. Calculate the theoretical yield using this equation: $2C_2H_4 + O_2 \rightarrow 2C_2H_4O$.

$$10.6 \text{ g } C_2H_4 \times \frac{1 \text{ mol } C_2H_4}{28.05 \text{ g } C_2H_4} \times \frac{2 \text{ mol } C_2H_4O}{2 \text{ mol } C_2H_4} \times \frac{44.05 \text{ g } C_2H_4O}{1 \text{ mol } C_2H_4O} = 16.\underline{6}5 \text{ g } C_2H_4O$$

$$\text{Percentage yield} = \frac{\text{actual yield}}{\text{theoretical yield}} \times 100\% = \frac{9.91 \text{ g}}{16.65 \text{ g}} \times 100\% = 59.\underline{5}3 = 59.5\%$$

3.109. To find Zn, use these equations:

$2C + O_2 \rightarrow 2CO$ and $ZnO + CO \rightarrow Zn + CO_2$

Two moles of C produces 2 mol CO; because 1 mol ZnO reacts with 1 mol CO, 2 mol ZnO will react with 2 mol CO. Thus, 2 mol C is equivalent to 2 mol ZnO, or 1 mol C is equivalent to 1 mol ZnO.

Using this to calculate mass of C from mass of ZnO, we have

$$75.0 \text{ g ZnO} \times \frac{1 \text{ mol ZnO}}{81.39 \text{ g ZnO}} \times \frac{1 \text{ mol C}}{1 \text{ mol ZnO}} \times \frac{12.01 \text{ g C}}{1 \text{ mol C}} = 11.\underline{0}7 \text{ g C}$$

Thus, all of the ZnO is used up in reacting with just 11.07 g of C, making ZnO the limiting reactant. Use the mass of ZnO to calculate the mass of Zn formed:

$$75.0 \text{ g ZnO} \times \frac{1 \text{ mol ZnO}}{81.39 \text{ g ZnO}} \times \frac{1 \text{ mol Zn}}{1 \text{ mol ZnO}} \times \frac{65.39 \text{ g Zn}}{1 \text{ mol Zn}} = 60.\underline{2}56 = 60.3 \text{ g Zn}$$

3.111. For $CaO + 3C \rightarrow CaC_2 + CO$, find the limiting reactant in terms of moles of CaC_2 obtainable:

$$\text{Mol } CaC_2 = 2.60 \times 10^3 \text{ g C} \times \frac{1 \text{ mol C}}{12.01 \text{ g C}} \times \frac{1 \text{ mol } CaC_2}{3 \text{ mol C}} = 72.\underline{1}6 \text{ mol}$$

$$\text{Mol } CaC_2 = 2.60 \times 10^3 \text{ g CaO} \times \frac{1 \text{ mol CaO}}{56.08 \text{ g CaO}} \times \frac{1 \text{ mol } CaC_2}{1 \text{ mol CaO}} = 46.\underline{3}62 \text{ mol}$$

Because CaO is the limiting reactant, calculate the mass of CaC_2 from it:

$$\text{Mass } CaC_2 = 46.\underline{3}62 \text{ mol } CaC_2 \times \frac{64.10 \text{ g } CaC_2}{1 \text{ mol } CaC_2} = 2.9\underline{7}1 \times 10^3 = 2.97 \times 10^3 \text{ g } CaC_2$$

3.113. From the equation $2Na + H_2O \rightarrow 2NaOH + H_2$, convert mass of H_2 to mass of Na, and then use the mass to calculate the percentage:

$$0.108 \text{ g } H_2 \times \frac{1 \text{ mol } H_2}{2.016 \text{ g } H_2} \times \frac{2 \text{ mol Na}}{1 \text{ mol } H_2} \times \frac{22.99 \text{ g Na}}{1 \text{ mol Na}} = 2.4\underline{6}3 \text{ g Na}$$

$$\text{Percent Na} = \frac{\text{mass Na}}{\text{mass amalgam}} \times 100\% = \frac{2.463 \text{ g}}{15.23 \text{ g}} \times 100\% = 16.\underline{1}7 = 16.2\%$$

3.115. The mass spectrometer measures the masses of positive ions produced from a very small sample and displays the data as a mass spectrum. This mass spectrum can be used to identify a substance or to obtain the molecular formula of a compound. The mass spectrum contains a wealth of information about molecular structure.

SOLUTIONS TO STRATEGY PROBLEMS

3.117. a. $$48 \text{ } O_2 \text{ molecules} \times \frac{2 \text{ } H_2 \text{ molec}}{1 \text{ } O_2 \text{ molec}} = 96 \text{ } H_2 \text{ molecules}$$

b. $$5 \text{ mol } H_2O \times \frac{1 \text{ mol } O_2}{2 \text{ mol } H_2O} = 2.5 \text{ mol } O_2$$

$$5 \text{ mol } H_2O \times \frac{2 \text{ mol } H_2}{2 \text{ mol } H_2O} = 5.0 \text{ mol } H_2$$

c. $$37.5 \text{ g } O_2 \times \frac{4.0 \text{ g } H_2}{32 \text{ g } O_2} = 4.\underline{6}8 = 4.7 \text{ g } H_2$$

d. The mass of water formed is 4.0 g + 32 g = $3\underline{6}.0$ g

$$30.0 \text{ g } H_2O \times \frac{32 \text{ g } O_2}{36 \text{ g } H_2O} = 2\underline{6}.6 = 27 \text{ g } O_2$$

$$30.0 \text{ g } H_2O \times \frac{4.0 \text{ g } H_2}{36 \text{ g } H_2O} = 3.\underline{3}3 = 3.3 \text{ g } H_2$$

3.119. a. First, determine the limiting reactant.

$$25 \text{ g } H_2SO_4 \times \frac{1 \text{ mol } H_2SO_4}{98.086 \text{ g}} \times \frac{1 \text{ mol } K_2SO_4}{1 \text{ mol } H_2SO_4} \times \frac{174.27 \text{ g}}{1 \text{ mol } K_2SO_4} = 4\underline{4}.4 = 44 \text{ g}$$

$$7.7 \text{ g KOH} \times \frac{1 \text{ mol KOH}}{56.108 \text{ g}} \times \frac{1 \text{ mol } K_2SO_4}{2 \text{ mol KOH}} \times \frac{174.27 \text{ g}}{1 \text{ mol } K_2SO_4} = 1\underline{1}.95 = 12 \text{ g}$$

KOH is the limiting reactant, and 12 g K_2SO_4 is produced.

b. KOH is the limiting reactant. The mass of H_2SO_4 that reacts is

$$7.7 \text{ g KOH} \times \frac{1 \text{ mol KOH}}{56.108 \text{ g}} \times \frac{1 \text{ mol } H_2SO_4}{2 \text{ mol KOH}} \times \frac{98.086 \text{ g}}{1 \text{ mol } H_2SO_4} = 6.\underline{7}3 \text{ g reacted}$$

The mass remaining after reaction is 25 g − 6.73 g = 1$\underline{8}$.27 = 18 g.

c. The theoretical yield was determined in part a to be 1$\underline{1}$.95 g. For an actual yield of 67.1%, the amount of K_2SO_4 formed is

$$1\underline{1}.95 \text{ g} \times 0.671 = 8.\underline{0}2 = 8.0 \text{ g}$$

3.121. The balance chemical equation is $4NH_3 + 5O_2 \rightarrow 4NO + 6H_2O$. The theoretical yield of NO is

$$8.5 \text{ g } NH_3 \times \frac{1 \text{ mol } NH_3}{17.034 \text{ g } NH_3} \times \frac{4 \text{ mol NO}}{4 \text{ mol } NH_3} \times \frac{30.01 \text{ g NO}}{1 \text{ mol NO}} = 1\underline{4}.97 \text{ g NO}$$

The percentage yield is

$$\text{Percentage Yield} = \frac{12.0 \text{ g}}{14.97 \text{ g}} \times 100\% = 8\underline{0}.1\% = 80.\%$$

3.123. The fraction of O in Na_2SO_3 is: $X_1 = \dfrac{48.00 \text{ g O}}{126.03 \text{ g } Na_2SO_3} = 0.380\underline{8}6$

The fraction of O in $MgSO_4$ is: $X_2 = \dfrac{64.00 \text{ g O}}{120.38 \text{ g } MgSO_4} = 0.531\underline{6}5$

The total mass of oxygen present is

$$\text{Mass O} = (0.38086 \times 12.1 \text{ g}) + (0.53165 \times 14.6 \text{ g}) = 4.6\underline{0}8 \text{ g} + 7.7\underline{6}2 \text{ g} = 12.3\underline{7}0 = 12.37 \text{ g}$$

3.125. $175 \text{ tablets} \times \dfrac{68.4 \times 10^{-3} \text{ g}}{1 \text{ tablet}} \times \dfrac{1 \text{ mol } CaCO_3}{100.09 \text{ g } CaCO_3} = 0.11\underline{9}6 = 0.120 \text{ mol } CaCO_3$

3.127. First, using the successive reactions, calculate the mass of sulfur required to produce 87.0 g of H_2SO_4:

$$87.0 \text{ g } H_2SO_4 \times \frac{1 \text{ mol } H_2SO_4}{98.08 \text{ g } H_2SO_4} \times \frac{1 \text{ mol } SO_3}{1 \text{ mol } H_2SO_4} \times \frac{2 \text{ mol } SO_2}{2 \text{ mol } SO_3} \times \frac{1 \text{ mol } S_8}{8 \text{ mol } SO_2}$$

$$\times \frac{256.52 \text{ g } S_8}{1 \text{ mol } S_8} = 28.\underline{4}4 \text{ g } S_8 \text{ (sulfur)}$$

Finally, calculate the mass of the starting material required using the fact that every 100 g of the material contains only 65.0 g of sulfur (presumably as S_8):

$$28.\underline{44}\text{ g sulfur} \times \frac{100.0\text{ g sulfur source}}{65.0\text{ g sulfur}} = 43.\underline{7}6\text{ g} = 43.8\text{ g of starting material required.}$$

3.129. The balanced combustion reaction for this problem is

$$CH_4 + 2\ O_2 \rightarrow CO_2 + 2\ H_2O$$

Using the volumes of the reactants and the density relationship, the amount of CO_2 that can be produced by either reactant is calculated as follows:

$$20.\underline{0}\text{ L } CH_4 \times \frac{1\text{ m}^3}{1000\text{ L}} \times \frac{1.82\text{ kg}}{1\text{ m}^3} \times \frac{1\text{ kmol } CH_4}{16.04\text{ kg } CH_4} \times \frac{1\text{ kmol } CO_2}{1\text{ kmol } CH_4} \times \frac{44.01\text{ kg } CO_2}{1\text{ kmol } CO_2}$$

$$= 0.099\underline{8}7 = 0.0999\text{ kg } CO_2$$

$$30.\underline{0}\text{ L } O_2 \times \frac{1\text{ m}^3}{1000\text{ L}} \times \frac{1.31\text{ kg}}{1\text{ m}^3} \times \frac{1\text{ kmol } O_2}{32.00\text{ kg } O_2} \times \frac{1\text{ kmol } CO_2}{2\text{ kmol } O_2} \times \frac{44.01\text{ kg } CO_2}{1\text{ kmol } CO_2}$$

$$= 0.027\underline{0}2 = 0.0270\text{ kg } CO_2$$

Thus, the limiting reactant is O_2 which yields 2.70×10^{-2} kg of CO_2.

3.131. This problem is best approached by first combining the two steps given in the synthesis of NaN_3. In the process of doing this, the intermediate $NaNH_2$(s) and one NH_3(g) cancel giving the following result:

$$2Na(s) + NH_3(g) + N_2O(g) \rightarrow NaN_3(s) + H_2(g) + NaOH(s)$$

The problem clearly becomes a limiting reagent problem where we have the starting masses of each reactant. Determining which provides the least amount of NaN_3 product will yield the desired result:

$$1.\underline{0}\text{ kg Na} \times \frac{1\text{ kmol Na}}{22.99\text{ kg Na}} \times \frac{1\text{ kmol } NaN_3}{2\text{ kmol Na}} \times \frac{65.02\text{ kg } NaN_3}{1\text{ kmol } NaN_3} = 2.\underline{8}3 = 2.8\text{ kg } NaN_3$$

$$6.\underline{0}\text{ kg } NH_3 \times \frac{1\text{ kmol } NH_3}{17.03\text{ kg } NH_3} \times \frac{1\text{ kmol } NaN_3}{1\text{ kmol } NH_3} \times \frac{65.02\text{ kg } NaN_3}{1\text{ kmol } NaN_3} = 2\underline{2}.9 = 23\text{ kg } NaN_3$$

$$1.\underline{0}\text{ kg } N_2O \times \frac{1\text{ kmol } N_2O}{44.02\text{ kg } N_2O} \times \frac{1\text{ kmol } NaN_3}{1\text{ kmol } N_2O} \times \frac{65.02\text{ kg } NaN_3}{1\text{ kmol } NaN_3} = 1.\underline{4}8 = 1.5\text{ kg } NaN_3$$

Thus, the limiting reactant is N_2O which yields 1.5 kg of NaN_3.

SOLUTIONS TO CUMULATIVE-SKILLS PROBLEMS

3.133. Let y equal the mass of CuO in the mixture. Then 0.500 g − y equals the mass of Cu_2O in the mixture. Multiplying the appropriate conversion factors for Cu times the mass of each oxide will give one equation in one unknown for the mass of 0.425 g Cu:

$$0.425 = y\left[\frac{63.55 \text{ g Cu}}{79.55 \text{ g CuO}}\right] + (0.500 - y)\left[\frac{127.10 \text{ g Cu}}{143.10 \text{ g Cu}_2\text{O}}\right]$$

Simplifying the equation by dividing the conversion factors and combining terms gives

$$0.425 = 0.798\underline{8}7\, y + 0.8881\underline{9}0\,(0.500 - y)$$

$$0.08932\, y = 0.01\underline{9}095$$

$$y = 0.2\underline{1}39 = 0.21 \text{ g} = \text{mass of CuO}$$

3.135. If one heme molecule contains one iron atom, then the number of moles of heme in 35.2 mg of heme must be the same as the number of moles of iron in 3.19 mg of iron. Start by calculating the moles of Fe (equals moles of heme):

$$3.19 \times 10^{-3} \text{ g Fe} \times \frac{1 \text{ mol Fe}}{55.85 \text{ g Fe}} = 5.7\underline{1}2 \times 10^{-5} \text{ mol Fe or heme}$$

$$\text{Molar mass of heme} = \frac{35.2 \times 10^{-3} \text{ g}}{5.712 \times 10^{-5} \text{ mol}} = 61\underline{6}.2 = 616 \text{ g/mol}$$

The molecular mass of heme is 616 amu.

3.137. Use the data to find the molar mass of the metal and anion. Start with X_2.

$$\text{Mass X}_2 \text{ in MX} = 4.52 \text{ g} - 3.41 \text{ g} = 1.11 \text{ g}$$

$$\text{Molar mass X}_2 = \frac{1.11 \text{ g}}{0.0158 \text{ mol}} = 70.\underline{2}5 \text{ g/mol}$$

$$\text{Molar mass X} = \frac{70.25 \text{ g/mol X}_2}{2 \text{ mol X/mol X}_2} = 35.\underline{1}4 = 35.1 \text{ g/mol}$$

Thus X is Cl, chlorine.

$$\text{Moles of M in 4.52 g MX} = 0.0158 \times 2 = 0.0316 \text{ mol}$$

$$\text{Molar mass of M} = \frac{3.41 \text{ g}}{0.0316 \text{ mol}} = 10\underline{7}.9 = 108 \text{ g/mol}$$

Thus M is Ag, silver.

3.139. After finding the volume of the alloy, convert it to mass Fe using density and percent Fe. Then use Avogadro's number and the atomic mass for the number of atoms.

$$\text{Volume} = 10.0\text{ cm} \times 20.0\text{ cm} \times 15.0\text{ cm} = 3.00 \times 10^3\text{ cm}^3$$

$$\text{Mass Fe} = 3.00 \times 10^3\text{ cm}^3 \times \frac{8.17\text{ g alloy}}{1\text{ cm}^3} \times \frac{54.7\text{ g Fe}}{100.0\text{ g alloy}} = 1.3\underline{4}07 \times 10^4\text{ g}$$

$$\text{No. of Fe atoms} = 1.3\underline{4}07 \times 10^{-4}\text{ g Fe} \times \frac{1\text{ mol Fe}}{55.85\text{ g Fe}} \times \frac{6.022 \times 10^{23}\text{ Fe atoms}}{1\text{ mol Fe}}$$

$$= 1.4\underline{4}56 \times 10^{26} = 1.45 \times 10^{26}\text{ Fe atoms}$$

CHAPTER 4

Chemical Reactions

SOLUTIONS TO EXERCISES

Note on significant figures: If the final answer to a solution needs to be rounded off, it is given first with one nonsignificant figure, and the last significant figure is underlined. The final answer is then rounded to the correct number of significant figures. In multistep problems, intermediate answers are given with at least one nonsignificant figure; however, only the final answer has been rounded off.

4.1. a. According to Table 4.1, all compounds that contain sodium, Na^+, are soluble. Thus, NaBr is soluble in water.

b. According to Table 4.1, most compounds that contain hydroxides, OH^-, are insoluble in water. However, $Ba(OH)_2$ is listed as one of the exceptions to this rule, so it is soluble in water.

c. Calcium carbonate is $CaCO_3$. According to Table 4.1, most compounds that contain carbonate, CO_3^{2-}, are insoluble. $CaCO_3$ is not one of the exceptions, so it is insoluble in water.

4.2. a. The problem states that HNO_3 is a strong electrolyte, but $Mg(OH)_2$ is a solid, so retain its formula. On the product side, $Mg(NO_3)_2$ is a soluble ionic compound, but water is a nonelectrolyte, so retain its formula. The resulting complete ionic equation is

$$2H^+(aq) + 2NO_3^-(aq) + Mg(OH)_2(s) \rightarrow 2H_2O(l) + Mg^{2+}(aq) + 2NO_3^-(aq)$$

The corresponding net ionic equation is

$$2H^+(aq) + Mg(OH)_2(s) \rightarrow 2H_2O(l) + Mg^{2+}(aq)$$

b. Both reactants are soluble ionic compounds, and on the product side, $NaNO_3$ is also a soluble ionic compound. $PbSO_4$ is a solid, so retain its formula. The resulting complete ionic equation is

$$Pb^{2+}(aq) + 2NO_3^-(aq) + 2Na^+(aq) + SO_4^{2-}(aq) \rightarrow PbSO_4(s) + 2Na^+(aq) + 2NO_3^-(aq)$$

The corresponding net ionic equation is

$$Pb^{2+}(aq) + SO_4^{2-}(aq) \rightarrow PbSO_4(s)$$

4.3. The formulas of the compounds are NaI and $Pb(C_2H_3O_2)_2$. Exchanging anions, you get sodium acetate, $NaC_2H_3O_2$, and lead(II) iodide, PbI_2. The equation for the exchange reaction is

$$NaI + Pb(C_2H_3O_2)_2 \rightarrow NaC_2H_3O_2 + PbI_2$$

From Table 4.1, you see that NaI is soluble, $Pb(C_2H_3O_2)_2$ is soluble, $NaC_2H_3O_2$ is soluble, and PbI_2 is insoluble. Thus, lead(II) iodide precipitates. The balanced molecular equation with phase labels is

$$Pb(C_2H_3O_2)_2(aq) + 2NaI(aq) \rightarrow PbI_2(s) + 2NaC_2H_3O_2(aq)$$

To get the net ionic equation, you write the soluble ionic compounds as ions and cancel the spectator ions, ($C_2H_3O_2^-$ and Na^+). The final result is

$$Pb^{2+}(aq) + 2I^-(aq) \rightarrow PbI_2(s)$$

4.4. a. H_3PO_4 is not listed as a strong acid in Table 4.3, so it is a weak acid.

b. Hypochlorous acid, HClO, is not one of the strong acids listed in Table 4.3, so we assume that HClO is a weak acid.

c. As noted in Table 4.3, $HClO_4$ is a strong acid.

d. As noted in Table 4.3, $Sr(OH)_2$ is a strong base.

4.5. The salt consists of the cation from the base (Li^+) and the anion from the acid (CN^-); its formula is LiCN. You will need to add H_2O as a product to complete and balance the molecular equation:

$$HCN(aq) + LiOH(aq) \rightarrow LiCN(aq) + H_2O(l)$$

Note that LiOH (a strong base) and LiCN (a soluble ionic substance) are strong electrolytes; HCN is a weak electrolyte (it is not one of the strong acids in Table 4.3). After eliminating the spectator ions (Li^+ and CN^-), the net ionic equation is

$$HCN(aq) + OH^-(aq) \rightarrow H_2O(l) + CN^-(aq)$$

4.6. The first step in the neutralization is described by the following molecular equation:

$$H_2SO_4(aq) + KOH(aq) \rightarrow KHSO_4(aq) + H_2O(l)$$

The corresponding net ionic equation is

$$H^+(aq) + OH^-(aq) \rightarrow H_2O(l)$$

The reaction of the acid salt $KHSO_4$ is given by the following molecular equation:

$$KHSO_4(aq) + KOH(aq) \rightarrow K_2SO_4(aq) + H_2O(l)$$

The corresponding net ionic equation is

$$HSO_4^-(aq) + OH^-(aq) \rightarrow H_2O(l) + SO_4^{2-}(aq)$$

4.7. First, write the molecular equation for the exchange reaction, noting that the products of the reaction would be soluble $Ca(NO_3)_2$ and H_2CO_3. The carbonic acid decomposes to water and carbon dioxide gas. The molecular equation for the process is

$$CaCO_3(s) + 2HNO_3(aq) \rightarrow Ca(NO_3)_2(aq) + H_2O(l) + CO_2(g)$$

The corresponding net ionic equation is

$$CaCO_3(s) + 2H^+(aq) \rightarrow Ca^{2+}(aq) + H_2O(l) + CO_2(g)$$

4.8. a. For potassium dichromate, $K_2Cr_2O_7$,

$$2 \times (\text{oxidation number of K}) + 2 \times (\text{oxidation number of Cr}) + 7 \times (\text{oxidation number of O}) = 0$$

For oxygen, the oxidation number is −2 (rule 3), and for potassium ion, the oxidation number is +1 (rule 2)

$$[2 \times (+1)] + 2 \times (\text{oxidation number of Cr}) + [7 \times (-2)] = 0$$

Therefore,

$$2 \times \text{oxidation number of Cr} = -\,[2 \times (+1)] - [7 \times (-2)] = +12$$

or, oxidation number of Cr = +6.

b. For the permanganate ion, MnO_4^-,

$$(\text{Oxidation number of Mn}) + 4 \times (\text{oxidation number of O}) = -1$$

For oxygen, the oxidation number is −2 (rule 3).

$$(\text{oxidation number of Mn}) + [4 \times (-2)] = -1$$

Therefore,

$$\text{Oxidation number of Mn} = -1 - [4 \times (-2)] = +7$$

4.9. Identify the oxidation states of the elements.

$$\overset{0}{Ca} + \overset{0}{Cl_2} \rightarrow \overset{+2\ -1}{CaCl_2}$$

Break the reaction into two half-reactions, making sure that both mass and charge are balanced.

$$Ca \rightarrow Ca^{2+} + 2e^-$$

$$Cl_2 + 2e^- \rightarrow 2Cl^-$$

Since each half-reaction has two electrons, it is not necessary to multiply the reactions by any factors to cancel them out. Adding the two half-reactions together and canceling out the electrons, you get

$$Ca(s) + Cl_2(g) \rightarrow CaCl_2(s)$$

4.10. Convert mass of NaCl (molar mass, 58.44 g) to moles of NaCl. Then divide moles of solute by liters of solution. Note that 25.0 mL = 0.0250 L.

$$0.0678\text{ g NaCl} \times \frac{1\text{ mol NaCl}}{58.44\text{ g NaCl}} = 1.1\underline{6}0 \times 10^{-3}\text{ mol NaCl}$$

$$\text{Molarity} = \frac{1.160 \times 10^{-3}\text{ mol NaCl}}{0.0250\text{ L soln}} = 0.046\underline{4}1 = 0.0464\ M$$

4.11. Convert grams of NaCl (molar mass, 58.44 g) to moles NaCl and then to volume of NaCl solution.

$$0.0958\text{ g NaCl} \times \frac{1\text{ mol NaCl}}{58.44\text{ g NaCl}} \times \frac{1\text{ L solution}}{0.163\text{ mol NaCl}} \times \frac{1000\text{ mL}}{1\text{ L}} = 10.\underline{0}6 = 10.1\text{ mL NaCl}$$

4.12. One (1) liter of solution is equivalent to 0.15 mol NaCl. The amount of NaCl in 50.0 mL of solution is

$$50.0\text{ mL} \times \frac{1\text{ L}}{1000\text{ mL}} \times \frac{0.15\text{ mol NaCl}}{1\text{ L soln}} = 0.007\underline{5}0\text{ mol NaCl}$$

Convert to grams using the molar mass of NaCl (58.44 g/mol).

$$0.00750\text{ mol NaCl} \times \frac{58.4\text{ g NaCl}}{1\text{ mol NaCl}} = 0.4\underline{3}8 = 0.44\text{ g NaCl}$$

4.13. Use the rearranged version of the dilution formula from the text to calculate the initial volume of 1.5 *M* sulfuric acid required:

$$V_i = \frac{M_f V_f}{M_i} = \frac{0.18\,M \times 100.0\text{ mL}}{1.5\,M} = 1\underline{2}.0 = 12\text{ mL}$$

4.14. There are two different reactions taking place in forming the CaC_2O_4 (molar mass 128.10 g/mol) precipitate. These are

$$CaCO_3(s) + 2HCl(aq) \rightarrow CaCl_2(aq) + CO_2(g) + H_2O(l)$$

$$CaCl_2(aq) + Na_2C_2O_4(aq) \rightarrow CaC_2O_4(s) + 2NaCl(aq)$$

The overall stoichiometry of the reactions is one mol $CaCO_3$/one mol CaC_2O_4. Also note that each $CaCO_3$ contains one Ca atom, so this gives an overall conversion factor of one mol Ca/one mol CaC_2O_4.

The mass of Ca can now be calculated.

$$0.1402\text{ g }CaC_2O_4 \times \frac{1\text{ mol }CaC_2O_4}{128.10\text{ g }CaC_2O_4} \times \frac{1\text{ mol Ca}}{1\text{ mol }CaC_2O_4} \times \frac{40.08\text{ g Ca}}{1\text{ mol Ca}} = 0.0438\underline{6}6\text{ g Ca}$$

Now, calculate the percentage of calcium in the 128.3 mg (0.1283 g) limestone:

$$\frac{0.0438\underline{6}6\text{ g Ca}}{0.1283\text{ g limestone}} \times 100\% = 34.1\underline{9}0 = 34.19\%$$

4.15. Convert the volume of Na_3PO_4 to moles using the molarity of Na_3PO_4. Note that 45.7 mL = 0.0457 L.

$$0.0457\text{ L }Na_3PO_4 \times \frac{0.265\text{ mol }Na_3PO_4}{1\text{ L}} = 0.012\underline{1}1\text{ mol }Na_3PO_4$$

Finally, calculate the amount of $NiSO_4$ required to react with this amount of Na_3PO_4:

$$0.12\underline{1}1\text{ mol }Na_3PO_4 \times \frac{3\text{ mol }NiSO_4}{2\text{ mol }Na_3PO_4} \times \frac{1\text{ L }NiSO_4}{0.375\text{ mol }NiSO_4} = 0.048\underline{4}4\text{ L (48.4 mL)}$$

4.16. Convert the volume of NaOH solution (0.0391 L) to moles NaOH (from the molarity of NaOH). Then convert moles NaOH to moles $HC_2H_3O_2$ (from the chemical equation). Finally, convert moles of $HC_2H_3O_2$ (molar mass 60.05 g/mol) to grams $HC_2H_3O_2$.

$$0.0391\text{ L NaOH} \times \frac{0.108\text{ mol NaOH}}{1\text{ L}} \times \frac{1\text{ mol }HC_2H_3O_2}{1\text{ mol NaOH}} \times \frac{60.05\text{ g }HC_2H_3O_2}{1\text{ mol }HC_2H_3O_2} = 0.25\underline{3}59\text{ g}$$

The mass percentage of acetic acid in the vinegar can now be calculated.

$$\text{Percentage mass} = \frac{0.25\underline{3}59 \text{ g } HC_2H_3O_2}{5.00 \text{ g vinegar}} \times 100\% = 5.0\underline{7}1 = 5.07\%$$

■ ANSWERS TO CONCEPT CHECKS

4.1. Of the six compounds, only NH_4Cl, $MgBr_2$, and HCl are strong electrolytes. H_2O is nearly a nonelectrolyte, $Ca_3(PO_4)_2$ would be considered a weak electrolyte because of its low solubility in water, and the molecular compound methanol, CH_3OH, dissolves molecularly in water. Statements c and e are valid statements. Statement a is invalid in the context of insoluble ionic compounds. Statement b is invalid in the context of the highly ionizable behavior of molecular strong acids. Statement d is invalid because water by itself is not a strong conductor of electricity, i.e., it's nearly a nonconductor.

4.2. The left beaker contains two types of individual atoms (ions) and no solid; therefore, it must represent the soluble, LiI. Because LiI is a soluble ionic compound, it is an electrolyte. The beaker on the right represents a molecular compound that is soluble but not dissociated in solution. Therefore, it must be the CH_3OH. Because the CH_3OH is not dissociated in solution, and no ions are present, it is a nonelectrolyte.

4.3. a. In order to solve this part of the problem, keep in mind that this is an exchange (metathesis) reaction. Since you are given the products in the picture, you need to work backward to determine the reactants. Starting with the solid $SrSO_4(s)$, you know that the SO_4^{2-} anion started the reaction with a different cation (not Sr^{2+}). Since Na^+ is the only option, you can conclude that one of the reactants must be Na_2SO_4. Based on solubility rules, you know that Na_2SO_4 is soluble, so you represent it as $Na_2SO_4(aq)$. The remaining cation and anion indicate that the other reactant is the soluble $Sr(C_2H_3O_2)_2$. Observing the soluble and insoluble species in the picture, you can conclude that the molecular equation is

$$Na_2SO_4(aq) + Sr(C_2H_3O_2)_2(aq) \rightarrow SrSO_4(s) + 2NaC_2H_3O_2(aq)$$

b. Writing the strong electrolytes in the form of ions and the solid with its molecular formula, the complete ionic equation for the reaction is

$$2Na^+(aq) + SO_4^{2-}(aq) + Sr^{2+}(aq) + 2C_2H_3O_2^-(aq)$$
$$\rightarrow SrSO_4(s) + 2Na^+(aq) + 2C_2H_3O_2^-(aq)$$

c. After canceling the spectator ions, the net ionic equation for the reaction is

$$Sr^{2+}(aq) + SO_4^{2-}(aq) \rightarrow SrSO_4(s)$$

4.4.

$$HClO_4(aq) + H_2O(l) \rightarrow H_3O^+(aq) + ClO_4^-(aq)$$

$$HBr(aq) + H_2O(l) \rightarrow H_3O^+(aq) + Br^-(aq)$$

a. HBr is a strong acid and completely ionizes in aqueous solution. Statement a is false.

b. $HClO_4$ is a strong acid and completely ionizes in aqueous solution. Statement b is true.

c. Since $HClO_4$ is a strong acid and completely ionizes, very few of the unionized molecules are expected to be found. On the other hand, there should be as many ClO_4^- ions present as the number of $HClO_4$ molecules originally placed in solution. Statement c is true

d. Since HBr is a strong acid, very few unionized HBr molecules will be present in solution. Statement d is false.

e. With statements b and c being true, statement e is false.

4.5. e. MOH must be a base since OH^- is being produced in solution. It must be a strong base because the reaction indicates that MOH is completely soluble (strong electrolyte). In order to maintain charge balance in the formula, the element M must be a 1+ cation, probably a metal from Group IA of the periodic table. Examples of bases that fall into this category include NaOH and KOH.

f. This must be an acid since H^+ is being produced in solution. It is a weak acid because the double arrow is used, indicating only a partial ionization in solution. From the chemical reaction, A^- represents an anion with a 1− charge. Acetic acid, $HC_2H_3O_2$, is a weak acid of this type.

g. This must be an acid since H^+ is being produced in solution. $H_2A(aq)$ is a weak acid because the equation indicates only partial ionization in solution. A^{2-} represents an anion with a 2− charge. Carbonic acid, H_2CO_3, is a weak acid of this type.

h. Examples of M include Na^+, K^+, and Li^+. Examples of A for reaction b include F^-, $C_2H_3O_2^-$, and CN^-. Examples of A for reaction c. include S^{2-}, CO_3^{2-}, and $C_4H_4O_6^{2-}$.

4.6. a. In order to answer this question, you need to compare the number of atoms of X per unit of volume. In order to compare volumes, use the lines on the sides of the beakers. Beaker A has concentration of five atoms per two volume units, 5/2 or 2.5/1. Beaker B has a concentration of ten atoms per one volume unit, 10/1. Beaker C has a concentration of ten atoms per two volume units, 10/2 or 5/1. Beaker D has a concentration of five atoms per volume unit, 5/1. Comparing the concentrations reveals that the ranking from lowest to highest concentration is Beaker A < Beaker C = Beaker D < Beaker B.

b. To make the concentrations of X equal in each beaker, they all have to be made to match the beaker with the lowest concentration. This is Beaker A, which has five atoms of X in one-half a beaker of solution. To make the concentrations equal, do the following: double the volume of Beakers C and D, and quadruple the volume of Beaker B by adding the solvent. Overall, Beakers A and B will contain a full beaker of solution, and Beakers C and D will contain a half-beaker of solution.

4.7. a. Since flask C required three times the amount of titrant (NaOH) as of acid A, you have learned that acid C has three times as many acidic protons as acid A. Since flask B required two times the amount of titrant as of acid A, you have also learned that acid B has two times as many acidic protons as acid A.

b. If you assume that acid A contains a monoprotic acid, then you know the number of moles of A in the flask. After performing the titration, you know that the moles of NaOH must equal the moles of acid in flask A. You take the number of moles of NaOH and divide it by the volume of NaOH added during the titration to determine the concentration of the NaOH solution.

ANSWERS TO SELF-ASSESSMENT AND REVIEW QUESTIONS

4.1. Some electrolyte solutions are strongly conducting because they are almost completely ionized, and others are weakly conducting because they are weakly ionized. The former solutions will have many more ions to conduct electricity than will the latter solutions if both are present at the same concentrations.

4.2. A strong electrolyte is an electrolyte that exists in solution almost entirely as ions. An example is NaCl. When NaCl dissolves in water, it dissolves almost completely to give Na^+ and Cl^- ions. A weak electrolyte is an electrolyte that dissolves in water to give a relatively small percentage of ions. An example is NH_3. When NH_3 dissolves in water, it reacts very little with the water, so the level of NH_3 is relatively high, and the level of the NH_4^+ and OH^- ions is relatively low.

4.3. Soluble means the ability of a substance to dissolve in water. A compound is insoluble if it does not dissolve appreciably in water. An example of a soluble ionic compound is sodium chloride, NaCl, and an example of an insoluble ionic compound is calcium carbonate, $CaCO_3$.

4.4. The advantage of using a molecular equation to represent an ionic equation is that it states explicitly what chemical species have been added and what chemical species are obtained as products. It also makes stoichiometric calculations easy to perform. The disadvantages are (1) the molecular equation does not represent the fact that the reaction actually involves ions, and (2) the molecular equation does not indicate which species exist as ions and which exist as molecular solids or molecular gases.

4.5. A spectator ion is an ion that does not take part in the reaction. In the following ionic reaction, the Na^+ and Cl^- are spectator ions:

$$Na^+(aq) + OH^-(aq) + H^+(aq) + Cl^-(aq) \rightarrow Na^+(aq) + Cl^-(aq) + H_2O(l)$$

4.6. A net ionic equation is an ionic equation from which spectator ions have been canceled. The value of such an equation is that it shows the reaction that actually occurs at the ionic level. An example is the ionic equation representing the reaction of calcium chloride ($CaCl_2$) with potassium carbonate(K_2CO_3).

$$CaCl_2(aq) + K_2CO_3(aq) \rightarrow CaCO_3(s) + 2\ KCl(aq):$$

$$Ca^{2+}(aq) + CO_3^{2-}(aq) \rightarrow CaCO_3(s) \qquad \text{(net ionic equation)}$$

4.7. The three major types of chemical reactions are precipitation reactions, acid–base reactions, and oxidation–reduction reactions. Oxidation–reduction reactions can be further classified as combination reactions, decomposition reactions, displacement reactions, and combustion reactions. Brief descriptions and examples of each are given below.

A precipitation reaction is a reaction that involves the formation of an insoluble solid compound. An example is $2KCl\ (aq) + Pb(NO_3)_2(aq) \rightarrow 2KNO_3(aq) + PbI_2(s)$.

An acid–base reaction, or neutralization reaction, results in an ionic compound and possibly water. An example is $HCl(aq) + NaOH(aq) \rightarrow NaCl(aq) + H_2O(l)$.

A combination reaction is a reaction in which two substances combine to form a third substance. An example is $2Na(s) + Cl_2(g) \rightarrow 2NaCl(s)$.

A decomposition reaction is a reaction in which a single compound reacts to give two or more substances. An example is $2HgO(s) \xrightarrow{\Delta} 2Hg(l) + O_2(g)$.

A displacement reaction, or single replacement reaction, is a reaction in which an element reacts with a compound displacing an element from it. An example is $Cu(s) + 2AgNO_3(aq) \rightarrow 2Ag(s) + Cu(NO_3)_2(aq)$.

A combustion reaction is a reaction of a substance with oxygen, usually with rapid release of heat to produce a flame. The products include one or more oxides. An example is $CH_4(g) + 2O_2(g) \rightarrow CO_2(g) + 2H_2O(l)$.

4.8. To prepare crystalline AgCl and $NaNO_3$, first make solutions of $AgNO_3$ and NaCl by weighing equivalent molar amounts of both solid compounds. Then mix the two solutions together, forming a precipitate of silver chloride and a solution of soluble sodium nitrate. Filter off the silver chloride, and wash it with water to remove the sodium nitrate solution. Then allow it to dry to obtain pure crystalline silver chloride. Finally, take the filtrate containing the sodium nitrate and evaporate it, leaving pure crystalline sodium nitrate.

4.9. An example of a neutralization reaction is

HBr	+	KOH	$\rightarrow$	KBr	+	$H_2O(l)$
acid		base		salt		

4.10. An example of a polyprotic acid is carbonic acid, H_2CO_3. The successive neutralization is given by the following molecular equations:

$$H_2CO_3(aq) + NaOH(aq) \rightarrow NaHCO_3(aq) + H_2O(l)$$

$$NaHCO_3(aq) + NaOH(aq) \rightarrow Na_2CO_3(aq) + H_2O(l)$$

4.11. Since an oxidation–reduction reaction is an electron transfer reaction, one substance must lose the electrons and be oxidized while another substance must gain electrons and be reduced.

4.12. A displacement reaction is an oxidation–reduction reaction in which a free element reacts with a compound, displacing an element from it.

$$Cu(s) + 2AgNO_3(aq) \rightarrow 2Ag(s) + Cu(NO_3)_2(aq)$$

Ag^+ is the oxidizing agent, and Cu is the reducing agent.

4.13. The number of moles present does not change when the solution is diluted.

4.14. The reaction is

$$HCl + NaOH \rightarrow NaCl + H_2O$$

After titration, the volume of hydrochloric acid is converted to moles of HCl using the molarity. Since the stoichiometry of the reaction is 1 mol HCl to 1 mol NaOH, these quantities are equal.

Moles HCl = moles NaOH = molarity × volume

You could then multiply by the molar mass of NaOH to obtain the amount in the mixture.

4.15. The answer is e, $HF(aq) + OH^-(aq) \rightarrow F^-(aq) + H_2O(l)$.

4.16. The answer is c, magnesium hydroxide(*s*).

4.17. The answer is d, $MgCl_2$.

4.18. The answer is d, HNO_2.

ANSWERS TO CONCEPTUAL PROBLEMS

4.21. a. Any soluble salt that will form a precipitate when reacted with Ag^+ ions in solution will work, for example: $CaCl_2$, Na_2S, $(NH_4)_2CO_3$.

b. No, no precipitate will form.

c. You would underestimate the amount of silver present in the solution.

4.23. a. $3Ca(C_2H_3O_2)_2(aq) + 2(NH_4)_3PO_4(aq) \rightarrow Ca_3(PO_4)_2(s) + 6NH_4C_2H_3O_2(aq)$

b. $3Ca^{2+}(aq) + 6C_2H_3O_2^-(aq) + 6NH_4^+(aq) + 2PO_4^{3-}(aq)$

$\rightarrow Ca_3(PO_4)_2(s) + 6NH_4^+(aq) + 6C_2H_3O_2^-(aq)$

c. $3Ca^{2+}(aq) + 2PO_4^{3-}(aq) \rightarrow Ca_3(PO_4)_2(s)$

4.25. Probably not, since the ionic compound that is a nonelectrolyte is not soluble.

4.27. a. Since both solutions are made with compounds that contain chloride ions, the total chloride ion concentration is highest.

b. First, determine the concentrations of the compounds after mixing them together. Use the dilution relationship, $M_1V_1 = M_2V_2$. Since equal volumes of equal molar solutions are mixed, the resulting concentrations are 0.50 *M* KBr and 0.50 *M* K_3PO_4.There is one Br^- ion per KBr, so the concentration of Br^- is 0.50 *M*. There is also one PO_4^{3-} ion per K_3PO_4, so its concentration is also 0.50 *M*. Potassium ion can be determined as follows:

$$0.50\ M\ \text{KBr} \times \frac{1\ \text{mol K}^+}{1\ \text{mol KBr}} + 0.50\ M\ \text{K}_3\text{PO}_4 \times \frac{3\ \text{mol K}^+}{1\ \text{mol K}_3\text{PO}_4} = 2.0\ \text{M K}^+$$

SOLUTIONS TO PRACTICE PROBLEMS

Note on significant figures: If the final answer to a solution needs to be rounded off, it is given first with one nonsignificant figure, and the last significant figure is underlined. The final answer is then rounded to the correct number of significant figures. In multistep problems, intermediate answers are given with at least one nonsignificant figure; however, only the final answer has been rounded off.

4.29. a. Insoluble

b. Soluble

c. Soluble

d. Soluble

4.31. a. Insoluble

b. Soluble; the ions present would be Li^+ and SO_4^{2-}.

$$Li_2SO_4(s) \rightarrow 2Li^+(aq) + SO_4^{2-}(aq)$$

c. Insoluble

d. Soluble; the ions present would be Na^+ and CO_3^{2-}.

$$Na_2CO_3(s) \rightarrow 2Na^+(aq) + CO_3^{2-}(aq)$$

4.33. a. $H^+(aq) + OH^-(aq) \rightarrow H_2O(l)$

b. $Ag^+(aq) + Br^-(aq) \rightarrow AgBr(s)$

c. $S^{2-}(aq) + 2H^+(aq) \rightarrow H_2S(g)$

d. $OH^-(aq) + NH_4^+(aq) \rightarrow NH_3(g) + H_2O(l)$

4.35. Molecular equation: $Pb(NO_3)_2(aq) + Na_2SO_4(aq) \rightarrow PbSO_4(s) + 2NaNO_3(aq)$

Net ionic equation: $Pb^{2+}(aq) + SO_4^{2-}(aq) \rightarrow PbSO_4(s)$

4.37. a. $FeSO_4(aq) + NaCl(aq) \rightarrow NR$

b. $Na_2CO_3(aq) + MgBr_2(aq) \rightarrow MgCO_3(s) + 2NaBr(aq)$

$CO_3^{2-}(aq) + Mg^{2+}(aq) \rightarrow MgCO_3(s)$

c. $MgSO_4(aq) + 2NaOH(aq) \rightarrow Mg(OH)_2(s) + Na_2SO_4(aq)$

$Mg^{2+}(aq) + 2OH^-(aq) \rightarrow Mg(OH)_2(s)$

d. $NiCl_2(aq) + NaBr(aq) \rightarrow NR$

4.39. a. $Ba(NO_3)_2(aq) + Li_2SO_4(aq) \rightarrow BaSO_4(s) + 2LiNO_3(aq)$

$Ba^{2+}(aq) + SO_4^{2-}(aq) \rightarrow BaSO_4(s)$

b. $Ca(NO_3)_2(aq) + NaBr(aq) \rightarrow NR$

c. $Al_2(SO_4)_3(aq) + 6NaOH(aq) \rightarrow 2Al(OH)_3(s) + 3Na_2SO_4(aq)$

$Al^{3+}(aq) + 3OH^-(aq) \rightarrow Al(OH)_3(s)$

d. $3CaBr_2(aq) + 2Na_3PO_4(aq) \rightarrow Ca_3(PO_4)_2(s) + 6NaBr(aq)$

$3Ca^{2+}(aq) + 2PO_4^{3-}(aq) \rightarrow Ca_3(PO_4)_2(s)$

4.41. a. Weak acid b. Strong base

c. Strong acid d. Weak acid

4.43. a. $NaOH(aq) + HNO_3(aq) \rightarrow H_2O(l) + NaNO_3(aq)$

$H^+(aq) + OH^-(aq) \rightarrow H_2O(l)$

b. $2HCl(aq) + Ba(OH)_2(aq) \rightarrow 2H_2O(l) + BaCl_2(aq)$

$H^+(aq) + OH^-(aq) \rightarrow H_2O(l)$

c. $2HC_2H_3O_2(aq) + Ca(OH)_2(aq) \rightarrow 2H_2O(l) + Ca(C_2H_3O_2)_2(aq)$

$HC_2H_3O_2(aq) + OH^-(aq) \rightarrow H_2O(l) + C_2H_3O_2^-(aq)$

d. $NH_3(aq) + HNO_3(aq) \rightarrow NH_4NO_3(aq)$

$NH_3(aq) + H^+(aq) \rightarrow NH_4^+(aq)$

4.45. a. $2HBr(aq) + Ca(OH)_2(aq) \rightarrow 2H_2O(l) + CaBr_2(aq)$

$H^+(aq) + OH^-(aq) \rightarrow H_2O(l)$

b. $3HNO_3(aq) + Al(OH)_3(s) \rightarrow 3H_2O(l) + Al(NO_3)_3(aq)$

$3H^+(aq) + Al(OH)_3(s) \rightarrow 3H_2O(l) + Al^{3+}(aq)$

c. $2HCN(aq) + Ca(OH)_2(aq) \rightarrow 2H_2O(l) + Ca(CN)_2(aq)$

$HCN(aq) + OH^-(aq) \rightarrow H_2O(l) + CN^-(aq)$

d. $HCN(aq) + LiOH(aq) \rightarrow H_2O(l) + LiCN(aq)$

$HCN(aq) + OH^-(aq) \rightarrow H_2O(l) + CN^-(aq)$

4.47. a. $2KOH(aq) + H_3PO_4(aq) \rightarrow K_2HPO_4(aq) + 2H_2O(l)$

$2OH^-(aq) + H_3PO_4(aq) \rightarrow HPO_4^{2-}(aq) + 2H_2O(l)$

b. $3H_2SO_4(aq) + 2Al(OH)_3(s) \rightarrow 6H_2O(l) + Al_2(SO_4)_3(aq)$

$3H^+(aq) + Al(OH)_3(s) \rightarrow 3H_2O(l) + Al^{3+}(aq)$

c. $2HC_2H_3O_2(aq) + Ca(OH)_2(aq) \rightarrow 2H_2O(l) + Ca(C_2H_3O_2)_2(aq)$

$HC_2H_3O_2(aq) + OH^-(aq) \rightarrow H_2O(l) + C_2H_3O_2^-(aq)$

d. $H_2SO_3(aq) + NaOH(aq) \rightarrow H_2O(l) + NaHSO_3(aq)$

$H_2SO_3(aq) + OH^-(aq) \rightarrow HSO_3^-(aq) + H_2O(l)$

4.49. Molecular equations: $2H_2SO_3(aq) + Ca(OH)_2(aq) \rightarrow 2H_2O(l) + Ca(HSO_3)_2(aq)$

$Ca(HSO_3)_2(aq) + Ca(OH)_2(aq) \rightarrow 2H_2O(l) + 2CaSO_3(s)$

Ionic equations: $H_2SO_3(aq) + OH^-(aq) \rightarrow H_2O(l) + HSO_3^-(aq)$

$Ca^{2+}(aq) + HSO_3^-(aq) + OH^-(aq) \rightarrow CaSO_3(s) + H_2O(l)$

4.51. a. Molecular equation: $CaS(s) + 2HBr(aq) \rightarrow CaBr_2(aq) + H_2S(g)$

Ionic equation: $CaS(s) + 2H^+(aq) \rightarrow Ca^{2+}(aq) + H_2S(g)$

b. Molecular equation: $MgCO_3(s) + 2HNO_3(aq) \rightarrow CO_2(g) + H_2O(l) + Mg(NO_3)_2(aq)$

Ionic equation: $MgCO_3(s) + 2H^+(aq) \rightarrow CO_2(g) + H_2O(l) + Mg^{2+}(aq)$

c. Molecular equation: $K_2SO_3(aq) + H_2SO_4(aq) \rightarrow K_2SO_4(aq) + SO_2(g) + H_2O(l)$

Ionic equation: $SO_3^{2-}(aq) + 2H^+(aq) \rightarrow SO_2(g) + H_2O(l)$

4.53. Molecular equation: $FeS(s) + 2HCl(aq) \rightarrow H_2S(g) + FeCl_2(aq)$

Ionic equation: $FeS(s) + 2H^+(aq) \rightarrow H_2S(g) + Fe^{2+}(aq)$

4.55. a. Because all three O's = a total of −6, both Ga's = +6; thus, the oxidation number of Ga = +3.

b. Because both O's = a total of −4, the oxidation number of Nb = +4.

c. Because the four O's = a total of −8 and K = +1, the oxidation number of Br = +7.

d. Because the four O's = a total of −8 and the 2 K's = +2, the oxidation number of Mn = +6.

4.57. a. Because the charge of −1 = [x_N + 2 (from 2 H's)], x_N must equal −3.

b. Because the charge of −1 = [x_I − 6 (from 3 O's)], x_I must equal +5.

c. Because this is a neutral element, the oxidation number is 0.

d. Because the charge of 0 = [x_{Cl} − 8 (4 O's) + 1 (1 H's)], x_{Cl} must equal +7.

4.59. a. From the list of common polyatomic anions in Table 2.6, the formula of the ClO_2 anion must be ClO_2^-. Thus, the formula of Mn is Mn^{2+} (see also Tables 2.5 and 4.5). Since the oxidation state of O is −2 and the net ionic charge is −1, the oxidation state of chlorine is determined by $x_{Cl} - 4 = -1$, so x_{Cl} must equal +3.

b. From the list of common polyatomic anions in Table 2.6, the formula of the CrO_4 anion must be CrO_4^{2-}. Thus, the formula of Fe is Fe^{3+}. Since the oxidation state of O is −2 and the net ionic charge is −2, the oxidation state of Cr is determined by $x_{Cr} - 8 = -2$, so x_{Cr} must equal +6.

c. From the list of common polyatomic anions in Table 2.6, the formula of the Cr_2O_7 anion must be $Cr_2O_7^{2-}$. Thus, the formula of Hg is Hg^{2+}. Since the oxidation state of O is −2 and the net ionic charge is −2, the oxidation state of Cr is determined by $2x_{Cr} - 14 = -2$, so x_{Cr} must equal +6.

d. From the list of common polyatomic anions in Table 2.6, the formula of the PO_4 anion must be PO_4^{3-}. Thus, the formula of Co is Co^{2+}. Since the oxidation state of O is −2 and the net ionic charge is −3, the oxidation state of P is determined by $x_P - 8 = -3$, so x_P must equal +5.

4.61. a. Phosphorus changes from an oxidation number of zero in P_4 to +5 in P_4O_{10}, losing electrons and acting as a reducing agent. Oxygen changes from an oxidation number of zero in O_2 to −2 in P_4O_{10}, gaining electrons and acting as an oxidizing agent.

b. Cobalt changes from an oxidation number of zero in Co(*s*) to +2 in $CoCl_2$, losing electrons and acting as a reducing agent. Chlorine changes from an oxidation number of zero in Cl_2 to −1 in $CoCl_2$, gaining electrons and acting as an oxidizing agent.

4.63. a. Al changes from oxidation number zero to +3; Al is the reducing agent.

F changes from oxidation number zero to −1; F_2 is the oxidizing agent.

b. Hg changes from oxidation state +2 to 0; Hg^{2+} is the oxidizing agent.

N changes from oxidation state +3 to +5; NO_2^- is the reducing agent.

4.65. a. First, identify the species being oxidized and reduced, and assign the appropriate oxidation states. Since $CuCl_2$ and $AlCl_3$ are both soluble ionic compounds, Cl^- is a spectator ion and can be removed from the equation. The resulting net ionic equation is

$$\overset{+2}{Cu^{2+}(aq)} + \overset{0}{Al(s)} \rightarrow \overset{+3}{Al^{3+}(aq)} + \overset{0}{Cu(s)}$$

Next, write the half-reactions in an unbalanced form.

$Al \rightarrow Al^{3+}$ (oxidation)

$Cu^{2+} \rightarrow Cu$ (reduction)

Next, balance the charge in each equation by adding electrons to the more positive side to create balanced half-reactions.

$Al \rightarrow Al^{3+} + 3e^-$ (oxidation half-reaction)

$Cu^{2+} + 2e^- \rightarrow Cu$ (reduction half-reaction)

Multiply each half-reaction by a factor that will cancel out the electrons.

$2 \times (Al \rightarrow Al^{3+} + 3e^-)$

$\underline{3 \times (Cu^{2+} + 2e^- \rightarrow Cu)}$

$3Cu^{2+} + 2Al + \cancel{6e^-} \rightarrow 2Al^{3+} + 3Cu + \cancel{6e^-}$

Therefore, the balanced oxidation-reduction reaction is

$3Cu^{2+} + 2Al \rightarrow 2Al^{3+} + 3Cu$

Finally, add six Cl^- ions to each side, and add phase labels. The resulting balanced equation is

$3CuCl_2(aq) + 2Al(s) \rightarrow 2AlCl_3(aq) + 3Cu(s)$

b. First, identify the species being oxidized and reduced, and assign the appropriate oxidation states.

$$\overset{+3}{Cr^{3+}(aq)} + \overset{0}{Zn(s)} \rightarrow \overset{0}{Cr(s)} + \overset{+2}{Zn^{2+}(aq)}$$

Next, write the half-reactions in an unbalanced form.

$Zn \rightarrow Zn^{2+}$ (oxidation)

$Cr^{3+} \rightarrow Cr$ (reduction)

Next, balance the charge in each equation by adding electrons to the more positive side to create balanced half-reactions.

$Zn \rightarrow Zn^{2+} + 2e^-$ (oxidation half-reaction)

$Cr^{3+} + 3e^- \rightarrow Cr$ (reduction half-reaction)

Multiply each half-reaction by a factor that will cancel out the electrons.

$3 \times (Zn \rightarrow Zn^{2+} + 2e^-)$

$\underline{2 \times (Cr^{3+} + 3e^- \rightarrow Cr)}$

$2Cr^{3+} + 3Zn + \cancel{6e^-} \rightarrow 2Cr + 3Zn^{2+} + \cancel{6e^-}$

Therefore, the balanced oxidation-reduction reaction, including phase labels, is

$$2Cr^{3+}(aq) + 3Zn(s) \rightarrow 2Cr(s) + 3Zn^{2+}(aq)$$

4.67. $\text{Molarity} = \dfrac{\text{moles solute}}{\text{liters solution}} = \dfrac{0.0512 \text{ mol}}{0.0250 \text{ L}} = 2.0\underline{4}8 = 2.05\ M$

4.69. Find the number of moles of solute ($KMnO_4$) using the molar mass of 158.03 g $KMnO_4$ per 1 mol $KMnO_4$:

$$0.798 \text{ g } KMnO_4 \times \frac{1 \text{ mol } KMnO_4}{158.03 \text{ g } KMnO_4} = 5.0\underline{4}97 \times 10^{-3} \text{ mol } KMnO_4$$

$$\text{Molarity} = \frac{\text{moles solute}}{\text{liters of solution}} = \frac{5.0497 \times 10^{-3} \text{ mol}}{0.0500 \text{ L}} = 0.10\underline{0}99 = 0.101\ M$$

4.71. $0.150 \text{ mol } CuSO_4 \times \dfrac{1 \text{ L solution}}{0.120 \text{ mol } CuSO_4} = 1.2\underline{5}0 = 1.25 \text{ L solution}$

4.73. $0.0353 \text{ g KOH} \times \dfrac{1 \text{ mol KOH}}{56.10 \text{ g KOH}} \times \dfrac{1 \text{ L solution}}{0.0176 \text{ mol KOH}} = 0.035\underline{7}51 \text{ L} = 35.8 \text{ mL}$

4.75. From the molarity, one L of heme solution is equivalent to 0.0019 mol of heme solute. Before starting the calculation, note that 150 mL of solution is equivalent to 150×10^{-3} L of solution:

$$150 \times 10^{-3} \text{ L soln} \times \frac{0.0019 \text{ mol heme}}{1 \text{ L solution}} = 2.\underline{8}50 \times 10^{-4} = 2.9 \times 10^{-4} \text{ mol heme}$$

4.77. Multiply the volume of solution by molarity to convert it to moles; then convert to mass of solute by multiplying by the molar mass:

$$100.0 \times 10^{-3} \text{ L soln} \times \frac{0.025 \text{ mol } Na_2Cr_2O_7}{1 \text{ L solution}} \times \frac{262.0 \text{ g } Na_2Cr_2O_7}{1 \text{ mol } Na_2Cr_2O_7} = 0.6\underline{5}50$$

$$= 0.66 \text{ g } Na_2Cr_2O_7$$

4.79. Use the rearranged version of the dilution formula to calculate the initial volume of 15.8 M HNO_3 required:

$$V_i = \frac{M_f V_f}{M_i} = \frac{0.12\ M \times 1000 \text{ mL}}{15.8\ M} = 7.\underline{5}9 = 7.6 \text{ mL}$$

4.81. The initial concentration of KCl (molar mass, 74.55 g/mol) is

$$4.00 \text{ g KCl} \times \frac{1 \text{ mol KCl}}{74.55 \text{ g KCl}} \times \frac{1}{0.0100 \text{ L}} = 5.3\underline{6}6\ M$$

Using the dilution factor, $M_1V_1 = M_2V_2$, with $V_2 = 10.0\text{ mL} + 60.0\text{ mL} = 70.0\text{ mL}$, after the solutions are mixed, the concentration of KCl is

$$M_2 = \frac{M_1V_1}{V_2} = \frac{5.366\ M \times 10.0\text{ mL}}{70.0\text{ mL}} = 0.76\underline{6}5\ M\ \text{KCl}$$

For $CaCl_2$, the concentration is

$$M_2 = \frac{M_1V_1}{V_2} = \frac{0.500\ M \times 60.0\text{ mL}}{70.0\text{ mL}} = 0.42\underline{8}57\ M\ CaCl_2$$

Therefore, the concentrations of the ions are 0.767 M K^+ and 0.429 M Ca^{2+}. For Cl^-, it is 0. 76$\underline{6}$5 M + 2 V 0.42$\underline{8}$57 M = 1.62$\underline{3}$6 M = 1.624 M

4.83. Use the appropriate conversion factors to convert the mass of $BaSO_4$ to the mass of Ba^{2+} ions:

$$0.513\text{ g }BaSO_4 \times \frac{1\text{ mol }BaSO_4}{233.40\text{ g }BaSO_4} \times \frac{1\text{ mol }Ba^{2+}}{1\text{ mol }BaSO_4} \times \frac{137.33\text{ g }Ba^{2+}}{1\text{ mol }Ba^{2+}}$$

$$= 0.30\underline{1}84\text{ g }Ba^{2+}$$

Then calculate the percentage of barium in the 458 mg (0.458 g) compound:

$$\frac{0.30184\text{ g }Ba^{2+}}{0.458\text{ g}} \times 100\% = 65.\underline{9}039 = 65.9\%\ Ba^{2+}$$

4.85. a. The mass of chloride ion in the AgCl from the copper chloride compound is

$$86.00\text{ mg AgCl} \times \frac{35.45\text{ mg }Cl^-}{143.32\text{ mg AgCl}} = 21.2\underline{7}1\text{ mg }Cl^-$$

The percentage of chlorine in the 59.40 mg sample is

$$\frac{21.2\underline{7}1\text{ mg }Cl^-}{59.40\text{ mg sample}} \times 100\% = 35.8\underline{0}9 = 35.81\%\ Cl^-$$

b. Of the various approaches, it is as easy to calculate the theoretical percentage of Cl^- in both CuCl and $CuCl_2$ as it is to use another approach:

$$\text{CuCl: } \frac{35.45\text{ mg }Cl^-}{99.00\text{ mg CuCl}} \times 100\% = 35.8\underline{0}8\%$$

$$CuCl_2\text{: } \frac{70.90\text{ mg }Cl^-}{134.45\text{ mg }CuCl_2} \times 100\% = 52.7\underline{3}3\%$$

The compound must be CuCl.

4.87. First, calculate the moles of chlorine in the compound:

$$0.3048\text{ g AgCl} \times \frac{1\text{ mol AgCl}}{143.32\text{ g AgCl}} \times \frac{1\text{ mol }Cl^-}{1\text{ mol AgCl}} = 0.00212\underline{6}7\text{ mol }Cl^-$$

Then, calculate the g Fe^{x+} from the g Cl^-:

$$g\ Fe^{x+} = 0.1348\ g\ comp - \left(0.0021267\ mol\ Cl^- \times \frac{35.45\ g\ Cl^-}{1\ mol\ Cl^-}\right) = 0.0594\underline{0}\underline{8}\ g\ Fe^{x+}$$

Now, calculate the moles of Fe^{x+} using the molar mass:

$$0.0594\underline{0}\underline{8}\ g\ Fe^{x+} \times \frac{1\ mol\ Fe^{x+}}{55.85\ g\ Fe^{x+}} = 0.00106\underline{3}\underline{7}\ mol\ Fe^{x+}$$

Finally, divide the mole numbers by the smallest mole number:

$$\text{For Cl: } \frac{0.002127\ mol\ Cl^-}{0.0010637\ mol} = 2.00;\ \text{for } Fe^{x+}\text{: } \frac{0.0010637\ mol\ Fe^{x+}}{0.0010637\ mol} = 1.00$$

Thus, the formula is $FeCl_2$.

4.89. Using molarity, convert the volume of Na_2CO_3 to moles of Na_2CO_3; then use the equation to convert to moles of HNO_3 and finally to volume:

$$2HNO_3 + Na_2CO_3 \rightarrow 2NaNO_3 + H_2O + CO_2$$

$$44.8 \times 10^{-3}\ L\ Na_2CO_3 \times \frac{0.150\ mol\ Na_2CO_3}{1\ L\ soln} \times \frac{2\ mol\ HNO_3}{1\ mol\ Na_2CO_3} \times \frac{1\ L\ HNO_3}{0.250\ mol\ HNO_3}$$

$$= 0.053\underline{7}\underline{6}\ L = 0.0538\ L = 53.8\ mL$$

4.91. The reaction is $H_2SO_4 + 2NaHCO_3 \rightarrow Na_2SO_4 + 2H_2O + CO_2$.

$$8.20\ g\ NaHCO_3 \times \frac{1\ mol\ NaHCO_3}{84.01\ g\ NaHCO_3} \times \frac{1\ mol\ H_2SO_4}{2\ mol\ NaHCO_3} \times \frac{1\ L\ soln}{0.150\ mol\ H_2SO_4}$$

$$= 0.32\underline{5}3\ L\ (325\ mL)\ soln$$

4.93. First, find the mass of H_2O_2 required to react with $KMnO_4$.

$$5H_2O_2 + 2KMnO_4 + 3H_2SO_4 \rightarrow 5O_2 + 2MnSO_4 + K_2SO_4 + 8H_2O$$

$$51.7 \times 10^{-3}\ L\ soln \times \frac{0.145\ mol\ KMnO_4}{1\ L\ soln} \times \frac{5\ mol\ H_2O_2}{2\ mol\ KMnO_4} \times \frac{34.02\ g\ H_2O_2}{1\ mol\ H_2O_2}$$

$$= 0.63\underline{7}5\ g\ H_2O_2$$

Percent H_2O_2 = (mass H_2O_2 ÷ mass sample) × 100% = (0.6375 g ÷ 20.0 g) × 100% = 3.1$\underline{8}$7 = 3.19%

SOLUTIONS TO GENERAL PROBLEMS

4.95. For the reaction of magnesium metal and hydrobromic acid, the equations are as follows.

Molecular equation: $Mg(s) + 2HBr(aq) \rightarrow H_2(g) + MgBr_2(aq)$

Ionic equation: $Mg(s) + 2H^+(aq) \rightarrow H_2(g) + Mg^{2+}(aq)$

4.97. For the reaction of nickel(II) sulfate and sodium hydroxide, the equations are as follows.

Molecular equation: $NiSO_4(aq) + 2NaOH(aq) \rightarrow Ni(OH)_2(s) + Na_2SO_4(aq)$

Ionic equation: $Ni^{2+}(aq) + 2OH^-(aq) \rightarrow Ni(OH)_2(s)$

4.99. a. Molecular equation: $LiOH(aq) + HCN(aq) \rightarrow LiCN(aq) + H_2O(l)$

Ionic equation: $OH^-(aq) + HCN(aq) \rightarrow CN^-(aq) + H_2O(l)$

b. Molecular equation: $Li_2CO_3(aq) + 2HNO_3(aq) \rightarrow 2LiNO_3(aq) + CO_2(g) + H_2O(l)$

Ionic equation: $CO_3^{2-}(aq) + 2H^+(aq) \rightarrow CO_2(g) + H_2O(l)$

c. Molecular equation: $LiCl(aq) + AgNO_3(aq) \rightarrow LiNO_3(aq) + AgCl(s)$

Ionic equation: $Cl^-(aq) + Ag^+(aq) \rightarrow AgCl(s)$

d. Molecular equation: $MgSO_4(aq) + NaCl(aq) \rightarrow$ NR

(Na_2SO_4 and $MgCl_2$ are soluble.)

4.101. a. Molecular equation: $Sr(OH)_2(aq) + 2HC_2H_3O_2(aq) \rightarrow Sr(C_2H_3O_2)_2(aq) + 2H_2O(l)$

Ionic equation: $HC_2H_3O_2(aq) + OH^-(aq) \rightarrow C_2H_3O_2^-(aq) + H_2O(l)$

b. Molecular equation: $NH_4I(aq) + CsCl(aq) \rightarrow$ NR

(NH_4Cl and CsI are soluble.)

c. Molecular equation: $NaNO_3(aq) + CsCl(aq) \rightarrow$ NR

(NaCl and $CsNO_3$ are soluble.)

d. Molecular equation: $NH_4I(aq) + AgNO_3(aq) \rightarrow NH_4NO_3(aq) + AgI(s)$

Ionic equation: $I^-(aq) + Ag^+(aq) \rightarrow AgI(s)$

4.103. For each preparation, the compound to be prepared is given first, followed by the compound from which it is to be prepared. Then the method of preparation is given, followed by the molecular equation for the preparation reaction. Steps such as evaporation are not given in the molecular equation.

a. To prepare $CuCl_2$ from $CuSO_4$, add a solution of $BaCl_2$ to a solution of the $CuSO_4$, precipitating $BaSO_4$. The $BaSO_4$ can be filtered off, leaving aqueous $CuCl_2$, which can be obtained in solid form by evaporation. Molecular equation:

$$CuSO_4(aq) + BaCl_2(aq) \rightarrow BaSO_4(s) + CuCl_2(aq)$$

b. To prepare $Ca(C_2H_3O_2)_2$ from $CaCO_3$, add a solution of acetic acid, $HC_2H_3O_2$, to the solid $CaCO_3$, forming CO_2, H_2O, and aqueous $Ca(C_2H_3O_2)_2$. The aqueous $Ca(C_2H_3O_2)_2$ can be converted to the solid form by evaporation, which also removes the CO_2 and H_2O products. Molecular equation:

$$CaCO_3(s) + 2HC_2H_3O_2(aq) \rightarrow Ca(C_2H_3O_2)_2(aq) + CO_2(g) + H_2O(l)$$

c. To prepare $NaNO_3$ from Na_2SO_3, add a solution of nitric acid, HNO_3, to the solid Na_2SO_3, forming SO_2, H_2O, and aqueous $NaNO_3$. The aqueous $NaNO_3$ can be converted to the solid by evaporation, which also removes the SO_2 and H_2O products. Molecular equation:

$$Na_2SO_3(s) + 2HNO_3(aq) \rightarrow 2NaNO_3(aq) + SO_2(g) + H_2O(l)$$

d. To prepare $MgCl_2$ from $Mg(OH)_2$, add a solution of hydrochloric acid (HCl) to the solid $Mg(OH)_2$, forming H_2O and aqueous $MgCl_2$. The aqueous $MgCl_2$ can be converted to the solid form by evaporation. Molecular equation:

$$Mg(OH)_2(s) + 2HCl(aq) \rightarrow MgCl_2(aq) + 2H_2O(l)$$

4.105. a. Decomposition

b. Decomposition

c. Combination

d. Displacement

4.107. a. $Pb(NO_3)_2 + H_2SO_4$ $[\rightarrow PbSO_4(s) + HNO_3(aq)]$

$Pb(NO_3)_2 + MgSO_4$ $[\rightarrow PbSO_4(s) + Mg(NO_3)_2(aq)]$

$Pb(NO_3)_2 + Ba(OH)_2$ $[\rightarrow Pb(OH)_2(s) + Ba(NO_3)_2(aq)]$

b. $Ba(OH)_2 + MgSO_4$ $[\rightarrow BaSO_4(s) + Mg(OH)_2(s)]$

c. $Ba(OH)_2 + H_2SO_4$ $[\rightarrow BaSO_4(s) + H_2O(l)]$

4.109. Divide the mass of $CaCl_2$ by its molar mass and volume to find molarity:

$$5.00 \text{ g } CaCl_2 \times \frac{1 \text{ mol } CaCl_2}{110.98 \text{ g } CaCl_2} \times \frac{1}{1.000 \text{ L soln}} = 0.045\underline{0}53 = 0.0451\ M\ CaCl_2$$

The $CaCl_2$ dissolves to form Ca^{2+} and $2Cl^-$ ions. Therefore, the molarities of the ions are 0.0451 M Ca^{2+} and $2 \times 0.045\underline{0}53$, or 0.0901, M Cl^- ions.

4.111. Divide the mass of $K_2Cr_2O_7$ by its molar mass and volume to find molarity. Then calculate the volume needed to prepare 1.00 L of a 0.100 M solution.

$$89.3 \text{ g } K_2Cr_2O_7 \times \frac{1 \text{ mol } K_2Cr_2O_7}{294.20 \text{ g } K_2Cr_2O_7} = 0.30\underline{3}5 \text{ mol } K_2Cr_2O_7$$

$$\text{Molarity} = \frac{0.3035 \text{ mol } K_2Cr_2O_7}{1.00 \text{ L}} = 0.30\underline{3}5\ M$$

$$V_i = \frac{M_f \text{ x } V_f}{M_i} = \frac{0.100\ M \times 1.00 \text{ L}}{0.3035\ M} = 0.32\underline{9}4 \text{ L (329 mL)}$$

4.113. Assume a volume of 1.000 L (1000 cm^3) for the 6.00% NaBr solution, and convert to moles and then to molarity.

$$1000 \text{ cm}^3 \times \frac{1.046 \text{ g soln}}{1 \text{ cm}^3} \times \frac{6.00 \text{ g NaBr}}{100 \text{ g soln}} \times \frac{1 \text{ mol NaBr}}{102.89 \text{ g NaBr}} = 0.60\underline{9}9 \text{ mol}$$

$$\text{Molarity NaBr} = \frac{0.6099 \text{ mol NaBr}}{1.000 \text{ L}} = 0.60\underline{9}9 = 0.610\ M$$

4.115. First, calculate the moles of $BaCl_2$:

$$1.128\text{ g }BaSO_4 \times \frac{1\text{ mol }BaSO_4}{233.40\text{ g }BaSO_4} \times \frac{1\text{ mol }BaCl_2}{1\text{ mol }BaSO_4} = 0.00483\underline{2}9\text{ mol }BaCl_2$$

Then calculate the molarity from the moles and volume (0.0500 L):

$$\text{Molarity} = \frac{0.0048329\text{ mol }BaCl_2}{0.0500\text{ L}} = 0.096\underline{6}58 = 0.0967\ M$$

4.117. First, calculate the grams of thallium(I) sulfate:

$$0.2122\text{ g TlI} \times \frac{1\text{ mol TlI}}{331.28\text{ g TlI}} \times \frac{1\text{ mol }Tl_2SO_4}{2\text{ mol TlI}} \times \frac{504.83\text{ g }Tl_2SO_4}{1\text{ mol }Tl_2SO_4} = 0.161\underline{6}8\text{ g }Tl_2SO_4$$

Then calculate the percent Tl_2SO_4 in the rat poison:

$$\text{Percent }Ti_2SO_4 = \frac{0.16168\text{ g}}{0.7590\text{ g}} \times 100\% = 21.3\underline{0}1 = 21.30\%$$

4.119. The mass of copper(II) ion and the mass of sulfate ion in the 98.77-mg sample are

$$0.09877\text{ g} \times 0.3250 = 0.0321\underline{0}0\text{ g }Cu^{2+}\text{ ion}$$

$$0.11666\text{ g }BaSO_4 \times \frac{96.07\text{ g }SO_4^{2-}}{233.40\text{ g }BaSO_4} = 0.0480\underline{1}9 = 0.04802\text{ g }SO_4^{2-}$$

Mass of water left = $0.09877\text{ g} - (0.03210\text{ g }Cu^{2+} + 0.04802\text{ g }SO_4^{2-}) = 0.01865\text{ g }H_2O$

Moles of water left = $0.01865\text{ g} \div 18.02\text{ g/mol} = 1.035 \times 10^{-3}\text{ mol}$

Moles of Cu^{2+} = $0.03210\text{ g} \div 63.55\text{ g/mol} = 5.05\underline{1}1 \times 10^{-4}\text{ mol}$

Ratio of water to Cu^{2+}, or $CuSO_4$ = $1.035 \times 10^{-3} \div 5.0511 \times 10^{-4} = 2.05$, or 2

The formula is thus $CuSO_4 \bullet 2H_2O$.

4.121. For these calculations, the relative numbers of moles of gold and chlorine must be determined. These can be found from the masses of the two elements in the sample:

Total mass = mass of Au + mass of Cl = 328 mg

The mass of chlorine in the precipitated AgCl is equal to the mass of chlorine in the compound of gold and chlorine. The mass of Cl in the 0.464 g of AgCl is

$$0.464\text{ g AgCl} \times \frac{1\text{ mol AgCl}}{143.32\text{ g AgCl}} \times \frac{1\text{ mol Cl}}{1\text{ mol AgCl}} \times \frac{35.45\text{ g Cl}}{1\text{ mol Cl}}$$

$$= 0.11\underline{4}76\text{ g Cl (114.76 mg Cl)}$$

$$\text{Mass percentage Cl} = \frac{\text{mass Cl}}{\text{mass comp}} \times 100\% = \frac{114.76\text{ mg}}{328\text{ mg}} \times 100\% = 35.0\%\text{ Cl}$$

To find the empirical formula, convert each mass to moles:

Mass percentage Au = 328 mg − 114.76 mg Cl = 213.23 mg Au

$$0.11476 \text{ g Cl} \times \frac{1 \text{ mol Cl}}{35.45 \text{ g Cl}} = 0.003238 \text{ mol Cl}$$

$$0.21323 \text{ g Au} \times \frac{1 \text{ mol Au}}{196.97 \text{ g Au}} = 0.001082 \text{ mol Au}$$

Divide both numbers of moles by the smaller number (0.001082) to find the integers.

Integer for Cl: 0.003238 mol ÷ 0.001082 mol = 2.99, or 3

Integer for Au: 0.001082 mol ÷ 0.001082 mol = 1.00, or 1

The empirical formula is $AuCl_3$.

4.123. From the equations $NH_3 + HCl \rightarrow NH_4Cl$ and $NaOH + HCl \rightarrow NaCl + H_2O$, we write

Mol NH_3 = mol $HCl(NH_3)$

Mol NaOH = mol HCl(NaOH)

We can calculate the mol NaOH and the sum [mol $HCl(NH_3)$ + mol HCl(NaOH)] from the titration data. Because the sum equals mol NH_3 plus mol NaOH, we can calculate the unknown mol NH_3 from the difference: Mol NH_3 = sum − mol NaOH.

$$\text{Mol HCl (NaOH)} + \text{mol HCl (NH}_3\text{)} = 0.0463 \text{ L} \times \frac{0.213 \text{ mol HCl}}{1.000 \text{ L}} = 0.009862 \text{ mol HCl}$$

$$\text{Mol NaOH} = 0.0443 \text{ L} \times \frac{0.128 \text{ mol NaOH}}{1.000 \text{ L}} = 0.005670 \text{ mol NaOH}$$

Mol $HCl(NH_3)$ = 0.009862 mol − 0.005670 mol = 0.004192 mol

Mol NH_3 = mol $HCl(NH_3)$ = 0.004192 mol NH_3

Because all the N in the $(NH_4)_2SO_4$ was liberated as and titrated as NH_3, the amount of N in the fertilizer is equal to the amount of N in the NH_3. Thus, the moles of NH_3 can be used to calculate the mass percentage of N in the fertilizer:

$$0.004192 \text{ mol NH}_3 \times \frac{1 \text{ mol N}}{1 \text{ mol NH}_3} \times \frac{14.01 \text{ g N}}{1 \text{ mol N}} = 0.05873 \text{ g N}$$

$$\text{Mass percentage N} = \frac{\text{mass N}}{\text{mass fert.}} \times 100\% = \frac{0.05873 \text{ g N}}{0.608 \text{ g}} \times 100\% = 9.659 = 9.66\%$$

SOLUTIONS TO STRATEGY PROBLEMS

4.125. The final volume to which the solution must be diluted is

$$V_2 = \frac{M_1V_1}{M_2} = \frac{(3.47\ M)\ (100\ \text{mL})}{1.78\ M} = 19\underline{4}.94\ \text{mL}$$

The amount of water that must be added is 19$\underline{4}$.94 mL − 100 mL = 9$\underline{4}$.94 mL. Since the density of water is 1.00 g/mL, this is equivalent to 9$\underline{4}$.94 g of water, so the amount of ice is

$$9\underline{4}.94\ \text{g water} \times \frac{3\ \text{g ice}}{1\ \text{g water}} = 28\underline{4}.83 = 285\ \text{g ice}$$

(~3 × 10^2 g is more appropriate considering the precision of some the data given in the problem)

4.127. $0.248\ M \times 38.2 \times 10^{-3}\ \text{L} \times \frac{2\ \text{mol Al}^{3+}}{1\ \text{mol Al}_2(\text{SO}_4)_3} \times \frac{26.98\ \text{g}}{1\ \text{mol Al}^{3+}} = 0.51\underline{1}19 = 0.511\ \text{g}$

4.129. $3.33 \times 10^{23}\ \text{ions} \times \frac{1\ \text{mol}}{6.022 \times 10^{23}\ \text{ions}} \times \frac{1\ \text{mol Zn(C}_2\text{H}_3\text{O}_2)_2}{2\ \text{mol C}_2\text{H}_3\text{O}_2^-} \times \frac{183.478\ \text{g}}{1\ \text{mol}} = 50.\underline{7}2$

$= 50.7\ \text{g Zn(C}_2\text{H}_3\text{O}_2)_2$

4.131. The reactions are

$6 \times (\text{K} \rightarrow \text{K}^+ + \text{e}^-)$

$\underline{1 \times (\text{N}_2 + 6\text{e}^- \rightarrow 2\text{N}^{3-})}$

$6\text{K}(s) + \text{N}_2(g) + \cancel{6\text{e}^-} \rightarrow 2\text{K}_3\text{N}(s) + \cancel{6\text{e}^-}$

Thus six electrons are canceled.

4.133. Since the solution is homogeneous, the concentration of the solution in the beaker is the same as the original solution, 0.196 *M*. The concentration of the solution in the large flask is also the same.

4.135. Because portions of solutions of identical concentration are being added together, the resulting mixture will have the same concentration as the original solutions, i.e., 0.50 *M* NaOH. It's instructive to show this using the molarity concept. The total volume of solution prepared is 0.100 L. From the 25-mL sample of 0.50 *M* NaOH, 0.0125 mol NaOH is provided; the 75-mL portion of 0.50 *M* NaOH provides another 0.0375 mol NaOH. Thus, a total of 0.0500 mol of NaOH is present in the 0.100 L mixture. The molarity is thus calculated:

$$[\text{NaOH}] = \frac{\text{total mol NaOH}}{\text{total volume of solution}} = \frac{0.050\ \text{mol}}{0.100\ \text{L}} = 0.500M\ \text{(as expected)}$$

4.137. This is a limiting reagent problem. Using the balanced chemical equation we first calculate the amounts of product possible from the given amounts of both reagents:

$$3.75\ \text{g Zn} \times \frac{1\ \text{mol Zn}}{65.38\ \text{g Zn}} \times \frac{2\ \text{mol NO}_2}{1\ \text{mol Zn}} \times \frac{46.01\ \text{g NO}_2}{1\ \text{mol NO}_2} = 5.2\underline{7}8\ \text{g NO}_2$$

$$0.175\text{ L} \times \frac{0.500\text{ mol HNO}_3}{1\text{ L}} \times \frac{2\text{ mol NO}_2}{4\text{ mol HNO}_3} \times \frac{46.01\text{ g NO}_2}{1\text{ mol NO}_2} = 2.0\underline{1}3\text{ g NO}_2$$

Thus, the nitric acid is the limiting reagent yielding 2.01 g of NO_2.

4.139. The key to solving this problem is a sound foundation in chemical nomenclature. The 0.250-L solution you are to prepare is to have 0.500 mol of NO_3^- ions per liter of solution. The source of the nitrate ions is a solution of 1.00 *M* $Ca(NO_3)_2$. Since there are two nitrate ions for every formula unit of calcium nitrate, the source solution must have 2.00 mol of NO_3^- ions present per liter of solution. With this information you can use the dilution equation:

$$M_1V_1 = M_2V_2$$

$$\frac{0.500\text{ mol NO}_3^-}{1\text{ L}} \times 0.250\text{ L} = \frac{2.00\text{ mol NO}_3^-}{1\text{ L}} \times V_2$$

Solving for V_2 yields 0.0625 L, or 62.5 mL. To make the desired solution, you would take 62.5-mL of the source solution and dilute it with water to a total volume 250 mL with mixing.

■ SOLUTIONS TO CUMULATIVE-SKILLS PROBLEMS

4.141. For this reaction, the formulas are listed first, followed by the molecular and net ionic equations, the names of the products, and the molecular equation for another reaction giving the same precipitate.

Lead(II) nitrate is $Pb(NO_3)_2$, and cesium sulfate is Cs_2SO_4.

Molecular equation: $Pb(NO_3)_2(aq) + Cs_2SO_4(aq) \rightarrow PbSO_4(s) + 2CsNO_3(aq)$

Net ionic equation: $Pb^{2+}(aq) + SO_4^{2-}(aq) \rightarrow PbSO_4(s)$

$PbSO_4$ is lead(II) sulfate, and $CsNO_3$ is cesium nitrate.

Molecular equation: $Pb(NO_3)_2(aq) + Na_2SO_4(aq) \rightarrow PbSO_4(s) + 2NaNO_3(aq)$

4.143. Net ionic equation: $2Br^-(aq) + Cl_2(g) \rightarrow 2Cl^-(aq) + Br_2(l)$

Molecular equation: $CaBr_2(aq) + Cl_2(g) \rightarrow CaCl_2(aq) + Br_2(l)$

Mass Br_2 = 40.0 g + 14.2 g − 22.2 g = 32.0 g

Combining the three known masses gives the unknown mass of Br_2. Now, use a ratio of the known masses of $CaBr_2$ to Br_2 to convert pounds of Br_2 to grams of $CaBr_2$:

$$10.0\text{ lb Br}_2 \times \frac{40.0\text{ g CaBr}_2}{32.0\text{ g Br}_2} \times \frac{453.6\text{ g}}{1\text{ lb}} = 56\underline{7}0 = 5.67 \times 10^3\text{ g CaBr}_2$$

4.145. Molecular equation: $Hg(NO_3)_2 + H_2S(g) \rightarrow HgS(s) + 2HNO_3(aq)$

Net ionic equation: $Hg^{2+} + H_2S(g) \rightarrow HgS(s) + 2H^+(aq)$

The acid formed is nitric acid, a strong acid. The other product is mercury(II) sulfide.

Mass HNO_3 = (81.15 g + 8.52 g) − 58.16 g = 31.5$\underline{1}$ g

Mass of solution = 550.$\underline{0}$ g H_2O + 31.5$\underline{1}$ g HNO_3 = 581.$\underline{51}$ = 581.5 g

4.147. Let the number of Fe^{3+} ions equal y; then the number of Fe^{2+} ions equals $(7 - y)$. Since Fe_7S_8 is neutral, the number of positive charges must equal the number of negative charges. If the signs are omitted, then

Total charge on both Fe^{2+} and Fe^{3+} = total charge on all eight sulfide ions

$$3y + 2(7 - y) = 8 \times 2$$

$$y + 14 = 16$$

$$y = 2$$

Thus, the ratio of Fe^{2+} to Fe^{3+} is 5/2.

4.149. Use the density, formula mass,, and percentage to convert to molarity. Then combine the 0.200 mol with mol/L to obtain the volume in liters.

$$\frac{0.807 \text{ g soln}}{1 \text{ mL}} \times \frac{0.940 \text{ g ethanol}}{1.00 \text{ g soln}} \times \frac{1 \text{ mol ethanol}}{46.07 \text{ g ethanol}} \times \frac{1000 \text{ mL}}{1 \text{ L}} = \frac{16.\underline{4}65 \text{ mol ethanol}}{1 \text{ L ethanol}}$$

$$\text{L ethanol} = 0.200 \text{ mol ethanol} \times \frac{1 \text{ L ethanol}}{16.465 \text{ mol ethanol}} = 0.012\underline{1}46 = 0.0121 \text{ L}$$

4.151. Convert the 2.183 g of AgI to mol AgI, which is chemically equivalent to mol KI. Use that to calculate the molarity of the KI.

$$2.183 \text{ g AgI} \times \frac{1 \text{ mol AgI}}{234.77 \text{ g AgI}} = 9.29\underline{8}4 \times 10^{-3} \text{ mol AgI (equivalent to mol KI)}$$

$$\text{Molarity} = \frac{9.29\underline{8}4 \times 10^{-3} \text{ mol KI}}{0.0100 \text{ L}} = 0.92\underline{9}8 = 0.930\ M$$

4.153. Convert the 6.026 g of $BaSO_4$ to moles of $BaSO_4$; then, from the equation, deduce that 3 mol of $BaSO_4$ is equivalent to 1 mol of $M_2(SO_4)_3$ and is equivalent to 2 mol of M. Use that with 1.200 g of the metal M to calculate the atomic mass of M.

$$6.026 \text{ g BaSO}_4 \times \frac{1 \text{ mol BaSO}_4}{233.40 \text{ g BaSO}_4} \times \frac{2 \text{ mol M}}{3 \text{ mol BaSO}_4} = 0.0172\underline{1}2 \text{ mol M}$$

$$\text{Atomic mass of M in g/mol} = \frac{1.200 \text{ g M}}{0.0172\underline{1}2 \text{ mol M}} = 69.7\underline{1}9 \text{ g/mol (gallium)}$$

4.155. Use the density, formula mass, percentage, and volume to convert to moles of H_3PO_4. Then, from the equation $P_4O_{10} + 6H_2O \rightarrow 4H_3PO_4$, deduce that 4 mol H_3PO_4 is equivalent to 1 mol of P_4O_{10}, and use that to convert to moles of P_4O_{10}.

$$15\underline{0}0 \text{ mL} \times \frac{1.025 \text{ g soln}}{1 \text{ mL}} \times \frac{0.0500 \text{ g H}_3\text{PO}_4}{1 \text{ g soln}} \times \frac{1 \text{ mol H}_3\text{PO}_4}{98.00 \text{ g H}_3\text{PO}_4} = 0.78\underline{4}4 \text{ mol H}_3\text{PO}_4$$

$$0.78\underline{4}4 \text{ mol H}_3\text{PO}_4 \times \frac{1 \text{ mol P}_4\text{O}_{10}}{4 \text{ mol H}_3\text{PO}_4} = 0.19\underline{6}1 \text{ mol P}_4\text{O}_{10}$$

$$\text{Mass P}_4\text{O}_{10} = 0.19\underline{6}1 \text{ mol P}_4\text{O}_{10} \times \frac{283.92 \text{ g P}_4\text{O}_{10}}{1 \text{ mol P}_4\text{O}_{10}} = 55.\underline{6}77 = 55.7 \text{ g P}_4\text{O}_{10}$$

4.157. Convert the 0.1068 g of hydrogen to moles of H_2; then deduce from the equation that 3 moles of H_2 is equivalent to 2 mol of Al. Use the moles of Al to calculate mass of Al and the percentage Al.

$$0.1068 \text{ g } H_2 \times \frac{1 \text{ mol } H_2}{2.016 \text{ g } H_2} \times \frac{2 \text{ mol Al}}{3 \text{ mol } H_2} = 0.035\underline{3}17 \text{ mol Al}$$

$$\text{Percentage Al} = \frac{0.035\underline{3}17 \text{ mol Al} \times \frac{26.98 \text{ g Al}}{1 \text{ mol Al}}}{1.118 \text{ g alloy}} \times 100\% = 85.2\underline{2}9 = 85.23\%$$

4.159. Use the formula mass of $Al_2(SO_4)_3$ to convert to moles of $Al_2(SO_4)_3$. Then deduce from the equation that 1 mol of $Al_2(SO_4)_3$ is equivalent to 3 mol of H_2SO_4, and calculate the moles of H_2SO_4 needed. Combine density, percentage, and formula mass to obtain molarity of H_2SO_4. Then combine molarity and moles to obtain volume.

$$37.4 \text{ g } Al_2(SO_4)_3 \times \frac{1 \text{ mol } Al_2(SO_4)_3}{342.17 \text{ g } Al_2(SO_4)_3} \times \frac{3 \text{ mol } H_2SO_4}{1 \text{ mol } Al_2(SO_4)_3} = 0.32\underline{7}9 \text{ mol } H_2SO_4$$

$$\frac{1.104 \text{ g soln}}{1 \text{ mL}} \times \frac{15.0 \text{ g } H_2SO_4}{100 \text{ g soln}} \times \frac{1 \text{ mol } H_2SO_4}{98.09 \text{ g } H_2SO_4} \times \frac{1000 \text{ mL}}{1 \text{ L}} = 1.6\underline{8}8 \text{ mol } H_2SO_4\text{/L}$$

$$0.3279 \text{ mol } H_2SO_4 \times \frac{1 \text{ L } H_2SO_4}{1.688 \text{ mol } H_2SO_4} = 0.19\underline{4}2 = 0.194 \text{ L (194 mL)}$$

4.161. The equations for the neutralization are $2HCl + Mg(OH)_2 \rightarrow MgCl_2 + 2H_2O$ and $3HCl + Al(OH)_3 \rightarrow AlCl_3 + 3H_2O$. Calculate the moles of HCl, and set it equal to the total moles of hydroxide ion, OH^-.

$$0.0485 \text{ L HCl} \times \frac{0.187 \text{ mol HCl}}{1 \text{ L HCl}} = 0.0090\underline{6}95 \text{ mol HCl}$$

$$0.0090695 \text{ mol HCl} = 2 \text{ [mol } Mg(OH)_2] + 3 \text{ [mol } Al(OH)_3]$$

Rearrange the last equation, and solve for the moles of $Al(OH)_3$.

$$\text{(Eq 1) mol } Al(OH)_3 = 0.0030231 \text{ mol HCl} - 2/3 \text{ [mol } Mg(OH)_2]$$

The total mass of chloride salts is equal to the sum of the masses of $MgCl_2$ (molar mass = 95.21 g/mol) and $AlCl_3$ (molar mass = 133.33 g/mol).

$$[95.21 \text{ g/mol} \times \text{mol } MgCl_2] + [133.33 \text{ g/mol} \times \text{mol } AlCl_3] = 0.4200 \text{ g}$$

Since the moles of $Mg(OH)_2$ equals the moles of $MgCl_2$, and the moles of $Al(OH)_3$ equals the moles of $AlCl_3$, you can substitute these quantities into the last equation and get

$$[95.21 \text{ g/mol} \times \text{mol } Mg(OH)_2] + [133.33 \text{ g/mol} \times \text{mol } Al(OH)_3] = 0.4200 \text{ g}$$

Substitute Equation 1 into this equation for the moles of $Al(OH)_3$:

$$[95.21 \text{ g/mol} \times \text{mol } Mg(OH)_2] + [133.33 \text{ g/mol} \times (0.0030231 \text{ mol} - 2/3 \text{ mol } Mg(OH)_2)]$$
$$= 0.4200 \text{ g}$$

Rearrange the equation, and solve for the moles of $Mg(OH)_2$.

$$6.323 \text{ mol } Mg(OH)_2 + 0.403\underline{0}70 = 0.4200$$

$$\text{mol } Mg(OH)_2 = 0.016\underline{9}3 \div 6.323 = 0.0026\underline{7}28 \text{ mol}$$

Calculate the mass of $Mg(OH)_2$ in the antacid tablet (molar mass = 58.33 g/mol).

$$0.0026728 \text{ mol } Mg(OH)_2 \times \frac{58.33 \text{ g } Mg(OH)_2}{1 \text{ mol } Mg(OH)_2} = 0.15\underline{5}90 \text{ g } Mg(OH)_2$$

Use Eq 1 to find the moles and mass of $Al(OH)_3$ (molar mass 78.00 g/mol) in a similar fashion.

$$\text{mol } Al(OH)_3 = 0.0030231 - 2/3[0.0026\underline{7}28 \text{ mol } Mg(OH)_2] = 0.0012\underline{4}12 \text{ mol}$$

$$0.0012\underline{4}12 \text{ mol } Al(OH)_3 \times \frac{78.00 \text{ g } Al(OH)_3}{1 \text{ mol } Al(OH)_3} = 0.096\underline{6}81 \text{ g } Al(OH)_3$$

The mass percentage $Mg(OH)_2$ in the antacid is

$$\text{Mass percentage of } Mg(OH)_2 = [0.15590 \text{ g} \div (0.15590 + 0.096\underline{6}81) \text{ g}] \times 100\%$$

$$= 61.\underline{7}2 = 61.7\%$$

CHAPTER 7

Quantum Theory of the Atom

SOLUTIONS TO EXERCISES

Note on significant figures: If the final answer to a solution needs to be rounded off, it is given first with one nonsignificant figure, and the last significant figure is underlined. The final answer is then rounded to the correct number of significant figures. In multistep problems, intermediate answers are given with at least one nonsignificant figure; however, only the final answer has been rounded off.

7.1. Rearrange the equation $c = \nu\lambda$, which relates wavelength to frequency and the speed of light (3.00 × 10^8 m/s):

$$\lambda = \frac{c}{\nu} = \frac{3.00 \times 10^8 \text{ m/s}}{3.91 \times 10^{14} \text{ /s}} = 7.6\underline{7}2 \times 10^{-7} = 7.67 \times 10^{-7} \text{ m, or 767 nm}$$

7.2. Rearrange the equation $c = \nu\lambda$, which relates frequency to wavelength and the speed of light (3.00 × 10^8 m/s). Recognize that 456 nm = 4.56 × 10^{-7} m.

$$\nu = \frac{c}{\lambda} = \frac{3.00 \times 10^8 \text{ m/s}}{4.56 \times 10^{-7} \text{ m}} = 6.5\underline{7}8 \times 10^{14} = 6.58 \times 10^{14} \text{ /s}$$

7.3. First, use the wavelengths to calculate the frequencies from $c = \nu\lambda$. Then calculate the energies using $E = h\nu$.

$$\nu = \frac{c}{\lambda} = \frac{3.00 \times 10^8 \text{ m/s}}{1.0 \times 10^{-6} \text{ m}} = 3.\underline{0}0 \times 10^{14} \text{ /s}$$

$$\nu = \frac{c}{\lambda} = \frac{3.00 \times 10^8 \text{ m/s}}{1.0 \times 10^{-8} \text{ m}} = 3.\underline{0}0 \times 10^{16} \text{ /s}$$

$$\nu = \frac{c}{\lambda} = \frac{3.00 \text{ x } 10^8 \text{ m/s}}{1.0 \text{ x } 10^{-10} \text{ m}} = 3.\underline{0}0 \times 10^{18} \text{ /s}$$

$$E = h\nu = 6.63 \times 10^{-34} \text{ J•s} \times 3.00 \times 10^{14} \text{ /s} = 1.\underline{9}89 \times 10^{-19} = 2.0 \times 10^{-19} \text{ J (IR)}$$

$$E = h\nu = 6.63 \times 10^{-34} \text{ J•s} \times 3.00 \times 10^{16} \text{ /s} = 1.\underline{9}89 \times 10^{-17} = 2.0 \times 10^{-17} \text{ J (UV)}$$

$$E = h\nu = 6.63 \times 10^{-34} \text{ J•s} \times 3.00 \times 10^{18} \text{ /s} = 1.\underline{9}89 \times 10^{-15} = 2.0 \times 10^{-15} \text{ J (x ray)}$$

The x-ray photon (shortest wavelength) has the greatest amount of energy; the infrared photon (longest wavelength) has the least amount of energy.

7.4. From the formula for the energy levels, $E = -R_H/n^2$, obtain the expressions for both E_i and E_f. Then calculate the energy change for the transition from $n = 3$ to $n = 1$ by subtracting the lower value from the upper value. Set this equal to $h\nu$. The result is

$$\left[\frac{-R_H}{9}\right] - \left[\frac{-R_H}{1}\right] = \frac{8R_H}{9} = h\nu$$

The frequency of the emitted radiation is

$$\nu = \frac{8R_H}{9} = \frac{8}{9} \times \frac{2.179 \times 10^{-18}\text{ J}}{6.63 \times 10^{-34}\text{ J}\cdot\text{s}} = 2.9\underline{2}1 \times 10^{15} = 2.92 \times 10^{15}\text{ /s}$$

Since $\lambda = c/\nu$,

$$\lambda = \frac{3.00 \times 10^8\text{ m/s}}{2.92 \times 10^{15}\text{ /s}} = 1.0\underline{2}7 \times 10^{-7} = 1.03 \times 10^{-7}\text{ m, or 103 nm}$$

7.5. Calculate the frequency from $c = \nu\lambda$, recognizing that 589 nm is 5.89×10^{-7} m.

$$\nu = \frac{c}{\lambda} = \frac{3.00 \times 10^8\text{ m/s}}{5.89 \times 10^{-7}\text{ m}} = 5.0\underline{9}3 \times 10^{14} = 5.09 \times 10^{14}\text{ /s}$$

Finally, calculate the energy difference.

$$E = h\nu = 6.63 \times 10^{-34}\text{ J}\bullet\text{s} \times 5.093 \times 10^{14}\text{ /s} = 3.3\underline{7}66 \times 10^{-19} = 3.38 \times 10^{-19}\text{ J}$$

7.6. To calculate wavelength, use the mass of an electron ($m = 9.11 \times 10^{-31}$ kg), and Planck's constant ($h = 6.63 \times 10^{-34}$ J•s, or 6.63×10^{-34} kg•m^2/s).

$$\lambda = \frac{h}{mv} = \frac{6.63 \times 10^{-34}\text{ kg}\bullet\text{m}^2/\text{s}}{9.11 \times 10^{-31}\text{ kg} \times 2.19 \times 10^6\text{ m/s}} = 3.3\underline{2}3 \times 10^{-10}\text{ m (332 pm)}$$

7.7.
a. The value of n must be a positive whole number greater than zero. Here, it is zero. Also, if n is zero, there are no allowed values for l and m_l.

b. The values for l can range only from zero to $(n - 1)$. Here, l has a value greater than n.

c. The values for m_l range from $-l$ to $+l$. Here, m_l has a value greater than l.

d. The value for m_s is either + ½ or −½. Here, it is zero.

■ ANSWERS TO CONCEPT CHECKS

7.1. The frequency and wavelength are inversely related. Therefore, if the frequency is doubled, the wavelength is halved. Red light has a wavelength around 700 nm, so doubling its frequency halves its wavelength to about 350 nm, which is in the ultraviolet range just beyond the visible spectrum.

7.2. Since the transitions are between adjacent levels, the energy-level diagram must look something like the following diagram, with the red transition between two close levels and the blue transition between two more widely spaced levels. (The three levels could be spaced so the red and blue transitions are interchanged, with the blue transition above the red one.)

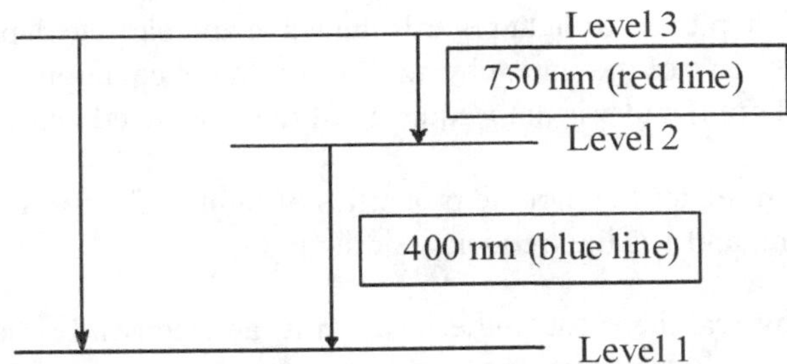

The transition from the top level to the lowest level would correspond to a transition that is greater in energy change than either of the other two transitions. Thus, the three transitions, from lowest to highest energy change, are in this order: red, blue, and the transition from the highest to lowest level. The last transition would have the highest frequency and therefore the shortest wavelength. It would lie just beyond the blue portion of the visible spectrum in the ultraviolet region.

7.3. The de Broglie relation says the wavelength of a particle is inversely proportional to both mass and speed. So, to maintain the wavelength constant while the mass increases would mean the speed would have to decrease. In going from a particle with the mass of an electron to one with that of a proton, the speed would have to decrease by a factor of about 2000 in order to maintain the same wavelength. The proton would have to have a speed approximately 2000 times slower than an electron of the same wavelength.

■ ANSWERS TO SELF-ASSESSMENT AND REVIEW QUESTIONS

7.1. Light is a wave, which is a form of electromagnetic radiation. In terms of waves, light can be described as a continuously repeating change, or oscillation, in electric and magnetic fields that can travel through space. Two characteristics of light are wavelength (often given in nanometers, nm) and frequency.

7.2. The relationship among the different characteristics of light waves is $c = \nu\lambda$, where ν is the frequency, λ is the wavelength, and c is the speed of light.

7.3. Starting with the shortest wavelengths, the electromagnetic spectrum consists of gamma rays, x rays, far ultraviolet (UV), near UV, visible light, near infrared (IR), far IR, microwaves, radar, and TV/FM radio waves (longest wavelengths).

7.4. The term *quantized* means the possible values of the energies of an atom are limited to only certain values. Planck was trying to explain the intensity of light of various frequencies emitted by a hot solid at different temperatures. The formula he arrived at was $E = nh\nu$, where E is energy, n is a whole number ($n = 1, 2, 3, \ldots$), h is Planck's constant, and ν is frequency.

7.5. *Photoelectric effect* is the term applied to the ejection of electrons from the surface of a metal or from other materials when light shines on it. Electrons are ejected only when the frequency (or energy) of light is larger than a certain minimum, or threshold, value that is constant for each metal. If a photon has a frequency equal to or greater than this minimum value, then it will eject one electron from the metal surface.

7.6. The wave–particle picture of light regards the wave and particle depictions of light as complementary views of the same physical entity. In the equation $E = h\nu$, E is the energy of a light particle (photon) and ν is the frequency of the associated wave.

7.7. The equation that relates the particle properties of light is $E = h\nu$. The symbol E is energy, h is Planck's constant, and ν is the frequency of the light.

7.8. According to physical theory at Rutherford's time, an electrically charged particle revolving around a center would continuously lose energy as electromagnetic radiation. As an electron in an atom lost energy, it would spiral into the nucleus (in about 10^{-10} s). Thus, the stability of the atom could not be explained.

7.9. According to Bohr, an electron in an atom can have only specific energy values. An electron in an atom can change energy only by going from one energy level (of allowed energy) to another energy level (of allowed energy). An electron in a higher energy level can go to a lower energy level by emitting a photon of an energy equal to the difference in energy. However, when an electron is in its lowest energy level, no further changes in energy can occur. Thus, the electron does not continuously radiate energy as was thought at Rutherford's time. These features solve the difficulty alluded to in Question 7.8.

7.10. Emission of a photon occurs when an electron in a higher energy level undergoes a transition to a lower energy level. The energy lost is emitted as a photon.

7.11. Absorption, the reverse of emission, occurs when a photon of a certain required energy is absorbed by a certain electron in an atom. The energy of the photon must be equal to the energy necessary to excite the electron of the atom from a lower energy level, usually the lowest, to a higher energy level.

7.12. The quantum corral, shown in Figure 7.21, shows evidence of electron waves. A practical example of diffraction is the operation of the scanning tunneling microscope.

7.13. The square of a wave function equals the probability of finding an electron within a region of space.

7.14. The uncertainty principle says we can no longer think of the electron as having a precise orbit in an atom similar to the orbit of the planets around the sun. This principle says it is impossible to know with absolute precision both the position and the speed of a particle such as an electron.

7.15. Quantum mechanics vastly changes Bohr's original picture of the hydrogen atom in that we can no longer think of the electron as having a precise orbit around the nucleus in this atom. Recall that Bohr's theory depended on the hydrogen electron having specific energy values and thus specific positions and speeds around the nucleus. But quantum mechanics and the uncertainty principle say it is impossible to know with absolute precision both the speed and the position of an electron. So Bohr's energy levels are only the most probable paths of the electrons.

7.16. a. The principal quantum number can have an integer value between one and infinity.

b. The angular momentum quantum number can have any integer value between zero and $(n - 1)$.

c. The magnetic quantum number can have any integer value between $-l$ and $+l$.

d. The spin quantum number can be either $+½$ or $-½$.

7.17. The notation is 4*f*. This subshell contains seven orbitals.

7.18. An *s* orbital has a spherical shape. A *p* orbital has two lobes positioned along a straight line through the nucleus at the center of the line (a dumbbell shape).

7.19. The answer is d, an electron in the $n = 1$ level is higher in energy than an electron in the $n = 4$ level.

7.20. The answer is a, I only.

7.21. The answer is c, transition from the $n = 3$ to the $n = 1$ level.

7.22. The answer is a, 8.66×10^{-6} m.

■ ANSWERS TO CONCEPTUAL PROBLEMS

7.25. Wavelength and frequency are inversely related. Moreover, ultraviolet light is at higher frequency than yellow light. Doubling the frequency of a beam of light would give that beam a higher frequency than yellow light, whereas doubling the wavelength would give that beam a lower frequency than yellow light. Consequently, the beam with frequency doubled must be the one in the ultraviolet region. Here is another way to look at the problem. Energy is directly related to the frequency and inversely related to the wavelength. Thus, the beam whose frequency is doubled will increase in energy, whereas the beam whose wavelength is doubled will decrease in energy. Since yellow light is in the visible region of the spectrum, which is lower in energy than the ultraviolet region (Figure 7.5), the beam whose frequency is doubled will be higher in energy and thus in the UV region of the spectrum.

7.27. That one color of light does not result in an ejection of electrons implies that that color has too little energy per photon. Of the two colors, red and green, red light has less energy per photon. Thus, you expect the experiment with red light to result in no ejection of photons, whereas the experiment with green light must be the one that ejects electrons. (Two red photons have more than enough energy to eject an electron, but this energy needs to be concentrated in only one photon to be effective.) In the photoelectric effect, one photon of light ejects at most one electron. Therefore, in the experiment with green light, one electron is ejected.

7.29. Energy is inversely proportional to the wavelength of the radiation. The transition from the highest energy level to the lowest energy level would involve the greatest energy change and thus the shortest wavelength, *x* nm.

7.31. A proton is approximately 2000 times the weight of an electron. Also, from the de Broglie relation, $l = h/mv$, you see that the wavelength is inversely proportional to both the mass and the speed of the particle. Considering the protons in parts b and c, since the mass is the same in both parts, the proton with the smaller speed, part b, will have a longer wavelength. Now, comparing the electron in part a with the proton in part b, since both have the same speed, the electron in part a with the smaller mass will have the longer wavelength. Therefore, the electron in part a will have the longest wavelength.

7.33. a. The frequency is directly proportional to the energy difference between the two ransition levels ($\Delta E = h\nu$). The lowest frequency corresponds with the smallest energy difference. Thus the transition from $n = 3$ to $n = 2$ will emit the lowest-frequency light.

b. The highest frequency corresponds with the largest energy difference. Thus the $n = 3$ to $n = 5$ transition will require (absorb) the highest-frequency light.

c. Since the frequency is proportional to the energy difference between two transition levels, the energy difference is the same for absorption and transmission, and the color light is the same for both. Thus, green light is absorbed.

SOLUTIONS TO PRACTICE PROBLEMS

Note on significant figures: If the final answer to a solution needs to be rounded off, it is given first with one nonsignificant figure, and the last significant figure is underlined. The final answer is then rounded to the correct number of significant figures. In multistep problems, intermediate answers are given with at least one nonsignificant figure; however, only the final answer has been rounded off. Starting with Problem 7.35, the value 2.998 x 10^8 m/s will be used for the speed of light.

7.35. Solve $c = \lambda\nu$ for λ:

$$\lambda = \frac{c}{\nu} = \frac{2.998 \times 10^8 \text{ m/s}}{1.365 \times 10^6 \text{ /s}} = 219.\underline{6}3 = 219.6 \text{ m}$$

7.37. Solve $c = \lambda\nu$ for ν. Recognize that 478 nm = 478×10^{-9} m, or 4.78×10^{-7} m.

$$\nu = \frac{c}{\lambda} = \frac{2.998 \times 10^8 \text{ m/s}}{4.78 \times 10^{-7} \text{ m}} = 6.2\underline{7}1 \times 10^{14} = 6.27 \times 10^{14} \text{ /s}$$

7.39. Radio waves travel at the speed of light, so divide the distance by c:

$$56 \times 10^9 \text{ m} \times \frac{1 \text{ s}}{2.998 \times 10^8 \text{ m}} = 1\underline{8}6.7 = 1.9 \times 10^2 \text{ s}$$

7.41. To do the calculation, divide 1 meter by the number of wavelengths in 1 meter to find the wavelength of this transition. Then use the speed of light (with nine digits for significant figures) to calculate the frequency:

$$\lambda = \frac{1 \text{ m}}{1{,}650{,}763.73} = 6.0578021\underline{0}6 \times 10^{-7} \text{ m} = 6.05780211 \times 10^{-7} \text{ m}$$

$$\nu = \frac{c}{\lambda} = \frac{2.99792458 \times 10^8 \text{ m/s}}{6.057802106 \times 10^{-7} \text{ m}} = 4.9488651\underline{6}2 \times 10^{14} = 4.94886516 \times 10^{14} \text{ /s}$$

7.43. Solve for E, using $E = h\nu$, and use four significant figures for h:

$$E = h\nu = (6.626 \times 10^{-34} \text{ J}\bullet\text{s}) \times (1.365 \times 10^6 \text{ /s}) = 9.04\underline{4}4 \times 10^{-28} = 9.044 \times 10^{-28} \text{ J}$$

7.45. Recognize that 535 nm = 535 × 10^{-9} m = 5.35 × 10^{-7} m. Then calculate ν and E.

$$\nu = \frac{c}{\lambda} = \frac{2.998 \times 10^8 \text{ m/s}}{5.35 \times 10^{-7} \text{ m}} = 5.6\underline{0}37 \times 10^{14} \text{ /s}$$

$$E = h\nu = (6.626 \times 10^{-34} \text{ J•s}) \times (5.6037 \times 10^{14}\text{/s}) = 3.7\underline{1}3 \times 10^{-19} = 3.71 \times 10^{-19} \text{ J}$$

7.47. First, calculate the wavelength of this transition from the frequency using the speed of light:

$$\lambda = \frac{c}{\nu} = \frac{2.998 \times 10^8 \text{ m/s}}{3.84 \times 10^{14} \text{ /s}} = 7.8\underline{0}72 \times 10^{-7} = 7.81 \times 10^{-7} \text{ m (781 nm)}$$

Using Figure 7.5, note that 781 nm is just on the edge of the red end of the spectrum and is barely visible to the eye.

7.49. Solve the equation $E = -R_H/n^2$ for both E_5 and E_3; equate to $h\nu$, and solve for ν.

$$\Delta E = h\nu = = E_5 - E_3 = \frac{-R_H}{5^2} - \frac{-R_H}{3^2} = \frac{-R_H}{25} - \frac{-R_H}{9} = \frac{16R_H}{225}$$

If you reversed the order and obtained a negative energy change, such a change actually corresponds to the relaxing atom. However, the frequency of the emitted radiation is related to the absolute energy released in the form of a photon, which must be a positive quantity:

$$\nu = \frac{16R_H}{225h} = \frac{16}{225} \times \frac{2.179 \times 10^{-18} \text{ J}}{6.626 \times 10^{-34} \text{ J} \bullet \text{s}} = 2.3\underline{3}8 \times 10^{14} = 2.34 \times 10^{14} \text{ /s}$$

7.51. Solve the equation $E = -R_H/n^2$ for both E_2 and E_1; solve for ν, and convert to λ.

$$\Delta E = h\nu = E_2 - E_1 = \frac{-R_H}{2^2} - \frac{-R_H}{1^2} = \frac{-R_H}{4} - \frac{-R_H}{1} = \frac{3R_H}{4}$$

If you reversed the order and obtained a negative energy change, such a change actually corresponds to the relaxing atom. However, the frequency of the emitted radiation is related to the absolute energy released in the form of a photon, which must be a positive quantity

$$\nu = \frac{3R_H}{4h} = \frac{3}{4} \times \frac{2.179 \times 10^{-18} \text{ J}}{6.626 \times 10^{-34} \text{ J} \bullet \text{s}} = 2.4\underline{6}6 \times 10^{15} \text{ /s}$$

The wavelength can now be calculated.

$$\lambda = \frac{c}{\nu} = \frac{2.998 \times 10^8 \text{ m/s}}{2.466 \times 10^{15} \text{ /s}} = 1.2\underline{1}55 \times 10^{-7} = 1.22 \times 10^{-7} \text{ m (near UV)}$$

7.53. This is the highest energy transition from the $n = 6$ level, so the electron must undergo a transition to the $n = 1$ level. Solve the Balmer equation using Bohr's approach:

$$\Delta E = h\nu = E_6 - E_1 = \frac{-R_H}{6^2} - \frac{-R_H}{1^2} = \frac{-R_H}{36} - \frac{-R_H}{1} = \frac{35R_H}{36}$$

If you reversed the order and obtained a negative energy change, such a change actually corresponds to the relaxing atom. However, the frequency of the emitted radiation is related to the absolute energy released in the form of a photon, which must be a positive quantity:

$$\nu = \frac{35R_H}{36h} = \frac{35}{36} \times \frac{2.179 \times 10^{-18}\ \text{J}}{6.626 \times 10^{-34}\ \text{J}\bullet\text{s}} = 3.1\underline{9}7 \times 10^{15}\ \text{/s}$$

The wavelength can now be calculated.

$$\lambda = \frac{c}{\nu} = \frac{2.998 \times 10^{8}\ \text{m/s}}{3.197 \times 10^{15}\ \text{/s}} = 9.3\underline{7}69 \times 10^{-8} = 9.38 \times 10^{-8}\ \text{m (93.8 nm)}$$

7.55. Noting that 422.7 nm = 4.227×10^{-7} m, convert the 422.7 nm to frequency. Then convert the frequency to energy using $E = h\nu$.

$$\nu = \frac{c}{\lambda} = \frac{2.998 \times 10^{8}\ \text{m/s}}{4.227 \times 10^{-7}\ \text{m}} = 7.09\underline{2}5 \times 10^{14}\ \text{/s}$$

$$E = h\nu = (6.626 \times 10^{-34}\ \text{J}\bullet\text{s}) \times 7.09\underline{2}5 \times 10^{14}\ \text{/s}) = 4.69\underline{9}4 \times 10^{-19} = 4.699 \times 10^{-19}\ \text{J}$$

7.57. The mass of a neutron = 1.67493×10^{-27} kg. Its speed or velocity, v, of 4.15 km/s equals 4.15×10^{3} m/s. Substitute these parameters into the de Broglie relation, and solve for λ:

$$\lambda = \frac{h}{mv} = \frac{6.626 \times 10^{-34}\ \text{kg}\bullet\text{m}^2\text{/s}}{1.67493 \times 10^{-27}\ \text{kg} \times 4.15 \times 10^{3}\ \text{m/s}} = 9.5\underline{3}2 \times 10^{-11} = 9.53 \times 10^{-11}\ \text{m}$$

A wavelength of 9.53×10^{-11} m (95.3 pm) would be in the x-ray region of the spectrum.

7.59. The mass of an electron equals 9.109×10^{-31} kg. The wavelength, λ, given as 10.0 pm, is equivalent to 1.00×10^{-11} m. Substitute these parameters into the de Broglie relation, and solve for the speed, v:

$$v = \frac{h}{m\lambda} = \frac{6.626 \times 10^{-34}\ \text{kg}\bullet\text{m}^2\text{/s}}{9.109 \times 10^{-31}\ \text{kg} \times 1.00 \times 10^{-11}\ \text{m}} = 7.2\underline{7}4 \times 10^{7} = 7.27 \times 10^{7}\ \text{m/s}$$

7.61. Substitute the 1.45×10^{-1} kg mass of the baseball and the 30.0 m/s velocity, v, into the de Broglie relation, and solve for wavelength (recall that 1 pm = 10^{-12} m).

$$\lambda = \frac{h}{mv} = \frac{6.626 \times 10^{-34}\ \text{kg}\bullet\text{m}^2\text{/s}}{1.45 \times 10^{-1}\ \text{kg} \times 30.0\ \text{m/s}} = 1.5\underline{2}3 \times 10^{-34} = 1.52 \times 10^{-34}\ \text{m} = 1.52 \times 10^{-22}\ \text{pm}$$

Because this is much smaller than 100 pm, the wavelength is much smaller than the diameter of one atom.

7.63. The possible values of l range from zero to $(n - 1)$, so l may be 0, 1, 2, or 3. The possible values of m_l range from $-l$ to $+l$. If $l = 3$, m_l may be −3, −2, −1, 0, +1, +2, or +3.

7.65. For the M shell, $n = 3$; there are three subshells in this shell ($l = 0$, 1, and 2). A d subshell has $l = 2$; the number of orbitals in this subshell is 2(2) + 1 = 5 (m_l = −2, −1, 0, 1, and 2).

7.67. a. $6d$
b. $5g$
c. $4f$
d. $6p$

7.69. a. Not permissible; m_s can be only +½ or −½.
b. Not permissible; l can be only as large as $(n - 1)$.
c. Not permissible; m_l cannot exceed +2 in magnitude.
d. Not permissible; n cannot be zero.
e. Not permissible; m_s can be only +½ or −½.

■ SOLUTIONS TO GENERAL PROBLEMS

7.71. Use $c = \nu\lambda$ to calculate frequency; then use $E = h\nu$ to calculate energy.

$$\nu = \frac{c}{\lambda} = \frac{2.998 \times 10^8 \text{ m/s}}{4.61 \times 10^{-7} \text{ m}} = 6.5\underline{0}3 \times 10^{14} = 6.50 \times 10^{14} \text{ /s}$$

$$E = h\nu = (6.626 \times 10^{-34} \text{ J}\bullet\text{s}) \times (6.503 \times 10^{14}\text{/s}) = 4.3\underline{0}9 \times 10^{-19} = 4.31 \times 10^{-19} \text{ J}$$

7.73. Calculate the frequency corresponding to 4.10×10^{-19} J. Then convert that to wavelength.

$$\nu = \frac{E}{h} = \frac{4.10 \times 10^{-19} \text{ J}}{6.626 \times 10^{-34} \text{ J}\bullet\text{s}} = 6.1\underline{8}7 \times 10^{14} \text{ /s}$$

$$\lambda = \frac{c}{\nu} = \frac{2.998 \times 10^8 \text{ m/s}}{6.187 \times 10^{14} \text{ /s}} = 4.8\underline{4}5 \times 10^{-7} = 4.85 \times 10^{-7} \text{ m} = 485 \text{ nm (blue-green)}$$

7.75. Solve for frequency using $E = h\nu$.

$$\nu = \frac{E}{h} = \frac{4.34 \times 10^{-19} \text{ J}}{6.626 \times 10^{-34} \text{ J}\bullet\text{s}} = 6.5\underline{4}9 \times 10^{14} = 6.55 \times 10^{14} \text{ /s}$$

7.77. First calculate E_p, the energy of the 345-nm photon, noting that it is equivalent to 3.45×10^{-7} m.

$$E_p = \frac{hc}{\lambda} = \frac{(6.626 \times 10^{-34} \text{ J}\bullet\text{s})(2.998 \times 10^8 \text{ m/s})}{3.45 \times 10^{-7} \text{ m}} = 5.7\underline{5}78 \times 10^{-19} \text{ J}$$

Now, subtract the work function of Ca = 4.34×10^{-19} J (Problem 7.75) from E_p:

$$5.7578 \times 10^{-19} \text{ J} - 4.34 \times 10^{-19} \text{ J} = 1.4\underline{1}78 \times 10^{-19} \text{ J}$$

Note that for this situation, $E = \frac{1}{2}mv^2$. Recall that the mass of the electron is 9.1094×10^{-31} kg. Now calculate speed, v:

$$v = \sqrt{\frac{2E}{m}} = \sqrt{\frac{2 \times 1.4178 \times 10^{-19} \text{ J}}{9.1095 \times 10^{-31} \text{ kg}}} = 5.5\underline{7}9 \times 10^5 = 5.58 \times 10^5 \text{ m/s}$$

7.79. This is a transition from the $n = 5$ level to the $n = 2$ level. Solve the Balmer equation using Bohr's approach.

$$E = h\nu = = E_5 - E_2 = \frac{-R_H}{5^2} - \frac{-R_H}{2^2} = \frac{-R_H}{25} - \frac{-R_H}{4} = \frac{21R_H}{100}$$

The wavelength can now be calculated.

$$\nu = \frac{21R_H}{100h} = \frac{21}{100} \times \frac{2.179 \times 10^{-18}\ \text{J}}{6.626 \times 10^{-34}\ \text{J}\bullet\text{s}} = 6.9\underline{0}59 \times 10^{14}\ \text{/s}$$

$$\lambda = \frac{c}{\nu} = \frac{2.998 \times 10^8\ \text{m/s}}{6.9059 \times 10^{14}\ \text{/s}} = 4.3\underline{4}1 \times 10^{-7} = 4.34 \times 10^{-7}\ \text{m (434 nm)}$$

7.81. Use 397 nm = 3.97×10^{-7} m, and convert to frequency and then to energy.

$$\nu = \frac{c}{\lambda} = \frac{2.998 \times 10^8\ \text{m/s}}{3.97 \times 10^{-7}\ \text{m}} = 7.5\underline{5}1 \times 10^{14}\ \text{/s}$$

$$E = h\nu = (6.626 \times 10^{-34}\ \text{J}\bullet\text{s}) \times (7.551 \times 10^{14}) = 5.0\underline{0}37 \times 10^{-19}\ \text{J}$$

Substitute this energy into the Balmer formula, recalling that the Balmer series is an emission spectrum, so ΔE is negative:

$$E = E_2 - E_i = \frac{-R_H}{2^2} - \frac{-R_H}{n_i^2} = \frac{-R_H}{4} - \frac{-R_H}{n_i^2} = 5.0037 \times 10^{-19}\ \text{J}$$

$$\frac{1}{4} - \frac{1}{n_i^2} = \frac{\Delta E}{-R_H} = \frac{-5.0037 \times 10^{-19}\ \text{J}}{-2.179 \times 10^{-18}\ \text{J}} = 0.22\underline{9}63$$

$$\frac{1}{n_i^2} = \frac{1}{4} - 0.22\underline{9}63 = 0.02\underline{0}36$$

$$n_i = \left[\frac{1}{0.02036}\right]^{\frac{1}{2}} = 7.\underline{0}07 = 7.0\ (= n)$$

7.83. Employ the Balmer formula using $Z = 2$ for the He^+ ion.

$$E = h\nu = E_3 - E_2 = \frac{-(2^2)R_H}{3^2} - \frac{-(2^2)R_H}{2^2} = \frac{-4R_H}{9} - \frac{-4R_H}{4} = \frac{20R_H}{36}$$

The frequency of the radiation is

$$\nu = \frac{20R_H}{36h} = \frac{20}{36} \times \frac{2.179 \times 10^{-18}\ \text{J}}{6.626 \times 10^{-34}\ \text{J}\bullet\text{s}} = 1.82\underline{6}9 \times 10^{15}\ \text{/s}$$

$$\lambda = \frac{c}{\nu} = \frac{2.998 \times 10^8\ \text{m/s}}{1.8269 \times 10^{15}\ \text{/s}} = 1.64\underline{0}9 \times 10^{-7} = 1.641 \times 10^{-7}\ \text{m (164.1 nm; near UV)}$$

7.85. First, use the wavelength of 10.0 pm (1.00×10^{-11} m) and the mass of 9.1094×10^{-31} kg to calculate the velocity, v. Then use the kinetic energy equation to calculate kinetic energy from velocity.

$$v = \frac{h}{m\lambda} = \frac{6.626 \times 10^{-34}\ \text{kg} \bullet \text{m}^2/\text{s}}{9.1095 \times 10^{-31}\ \text{kg} \times 1.00 \times 10^{-11}\ \text{m}} = 7.2\underline{7}37 \times 10^{7}\ \text{m/s}$$

$$E = ½mv^2 = ½ \times (9.1094 \times 10^{-31}\ \text{kg}) \times (7.273 \times 10^{7}\ \text{m/s})^2 = 2.4\underline{0}9 \times 10^{-15}\ \text{J}$$

$$E_{\text{eV}} = 2.409 \times 10^{-15}\ \text{J} \times \frac{1\ \text{eV}}{1.602 \times 10^{-19}\ \text{J}} = 1.5\underline{0}3 \times 10^{4} = 1.50 \times 10^{4}\ \text{eV}$$

7.87. a. Five b. Seven c. Three d. One

7.89. The possible subshells for the $n = 6$ shell are 6*s*, 6*p*, 6*d*, 6*f*, 6*g*, and 6*h*.

7.91. Gamma rays are a form of electromagnetic radiation similar to x rays, but the photons of gamma rays have higher energy. High-energy radiation kills bacteria and molds in foods by breaking up the DNA molecules within their cells.

7.93. When a photon of appropriate wavelength is used to flash the laser crystal, it stimulates a transition, and a photon of exactly the same wavelength as the original photon is emitted. In place of just one photon, there are now two photons. The net effect is to increase the intensity (or amplitude) of the light at this wavelength.

7.95. Tunneling depends on the probability interpretation of quantum mechanics. The probability of an electron in an atom being at a location far from atom A and near atom B is very small but not zero. This means that an electron that normally belongs to atom A can find itself near atom B without any extra energy having been supplied. The electron is said to have tunneled from one atom to another.

■ SOLUTIONS TO STRATEGY PROBLEMS

7.97. $$\lambda = \frac{c}{\nu} = \frac{3.00 \times 10^{8}\ \text{m/s}}{7.76 \times 10^{9}\ /\text{s}} = 0.038\underline{6}5\ \text{m} = 0.0387\ \text{m}$$

7.99. The energy required to heat the water is

$$E = s \times m \times \Delta t = (4.18\ \text{J/mol}\bullet°\text{C})(1.00\ \text{g})(1.00°\text{C}) = 4.18\ \text{J}$$

The microwave radiation has a wavelength of 12.2 cm (0.122 m), so the frequency and energy are

$$\nu = \frac{c}{\lambda} = \frac{3.00 \times 10^{8}\ \text{m/s}}{0.122\ \text{m}} = 2.4\underline{5}9 \times 10^{9}\ /\text{s}$$

$$E = h\nu = (6.63 \times 10^{-34}\ \text{J}\bullet\text{s})(2.459 \times 10^{9}\ /\text{s}) = 1.6\underline{3}0 \times 10^{-24}\ \text{J/photon}$$

Therefore, the number of photons needed to heat the water is

$$n = \frac{4.18\ \text{J}}{1.630 \times 10^{-24}\ \text{J/photon}} = 2.5\underline{6}3 \times 10^{24} = 2.56 \times 10^{24}\ \text{photons}$$

7.101. The number of photons in each pulse is determined by dividing the energy of each pulse by the energy of each photon:

$$\#\text{photons} = \frac{\text{Energy /pulse}}{\text{Energy/photon}} = \frac{E_{pulse}}{E_{photon}}$$

The energy delivered by each pulse, E_{pulse}, is given in the problem statement:

$$E_{pulse} = 1.6\underline{0} \times 10^{-16}\ \frac{\text{J}}{\text{pulse}}$$

The energy of each 498 nm photon, E_{photon}, is

$$E_{photon} = h\nu = \frac{hc}{\lambda} = \frac{(6.626 \times 10^{-34}\ \frac{\text{J·s}}{\text{photon}})(2.998 \times 10^{8}\ \text{m/s})}{49\underline{8} \times 10^{-9}\ \text{m}} = 3.9\underline{8}9 \times 10^{-19}\ \frac{\text{J}}{\text{photon}}$$

Therefore, the number of photons in each pulse is:

$$\#\text{photons} = \frac{E_{pulse}}{E_{photon}} = \frac{1.60 \times 10^{-16}\ \text{J/pulse}}{3.9\underline{8}9 \times 10^{-19}\ \text{J/photon}} = 40\underline{1}.1 = 401\ \text{photons /pulse}$$

7.103. The longest wavelength of light that can eject an electron from the calcium will be associated with a photon that has just enough energy to match the work function. Setting that energy equal to the energy of the photon, solve for the frequency and then find the wavelength

$$\nu = \frac{E}{h} = \frac{4.60 \times 10^{-19}\ \text{J}}{6.63 \times 10^{-34}\ \text{Js}} = 6.9\underline{3}8 \times 10^{14}\ /\text{s}$$

$$\lambda = \frac{c}{\nu} = \frac{3.00 \times 10^{8}\ \text{m/s}}{6.938 \times 10^{14}\ /\text{s}} = 4.3\underline{2}3 \times 10^{-7} = 4.32 \times 10^{-7}\ \text{m} = 432\ \text{nm}$$

7.105. The sum of the kinetic energy of the emitted electron, E_{kin}, and the work function of the metal, φ_{metal}, equals the energy of the photon, $E_{h\nu}$.

$$E_{h\nu} = E_{kin} + \varphi_{metal}$$

φ_{metal} is calculated by subtracting the kinetic energy of the electron, $E_{kin} = ½mv^2$, from the energy of the photon, $E_{kin} = hc / \lambda$.

$$\varphi_{metal} = E_{h\nu} - E_{kin} = (hc/\lambda) - \frac{1}{2}mv^2$$

$$\varphi_{metal} = \frac{(6.626 \times 10^{-34}\ \text{J}\bullet\text{s})(2.998 \times 10^{8}\ \text{m/s})}{3.8\underline{9} \times 10^{-7}\ \text{m}} - \frac{1}{2}(9.1094 \times 10^{-31}\ \text{kg})(3.3\underline{4} \times 10^{5}\ \text{m/s})^2$$

$$\varphi_{metal} = 5.1\underline{0}7 \times 10^{-19}\ \text{J} - 5.0\underline{8}1 \times 10^{-20}\ \text{J} = 4.5\underline{9}9 \times 10^{-19}\ \text{J} = 4.60 \times 10^{-19}\ \text{J}$$

A final conversion to eV units is made.

$$\varphi_{metal} = 4.5\underline{9}9 \times 10^{-19}\text{ J}\left(\frac{1\text{ eV}}{1.602 \times 10^{-19}\text{ J}}\right) = 2.8\underline{7}0 = 2.87\text{ eV}$$

The *CRC Handbook of Chemistry and Physics* provides the closest match value of 2.87 eV for the work function of calcium, Ca. The metal is most likely Ca.

7.107. Light of wavelength 1.03×10^{-7} m (103 nm) is in the ultraviolet region of the spectrum. The frequency and energy of this radiation are

$$\nu = \frac{c}{\lambda} = \frac{3.00 \times 10^{8}\text{ m/s}}{1.03 \times 10^{-7}\text{ m}} = 2.9\underline{1}2 \times 10^{15}\text{ /s}$$

$$E = h\nu = (6.63 \times 10^{-34}\text{ J}\bullet\text{s})(2.912 \times 10^{15}\text{ /s}) = 1.9\underline{3}1 \times 10^{-18}\text{ J}$$

Now use the appropriate formula to determine the quantum number of the excited (n_i) level. Remember, ΔE is negative for an emission spectrum.

$$\Delta E = -R_{\text{H}}\left(\frac{1}{n_f^2} - \frac{1}{n_i^2}\right) = -R_{\text{H}}\left(\frac{1}{1^2} - \frac{1}{n_i^2}\right) = -R_{\text{H}}\left(1 - \frac{1}{n_i^2}\right)$$

$$1 - \frac{1}{n_i^2} = \frac{-\Delta E}{-R_{\text{H}}} = \frac{1.931 \times 10^{-18}\text{ J}}{2.179 \times 10^{-18}\text{ J}} = 0.88\underline{6}2$$

$$n_i = \sqrt{\frac{1}{1 - 0.8862}} = 2.9\underline{6}4 = 2.96$$

Thus, the quantum number of the excited level is 3.

7.109. The shortest wavelength photon corresponds to that having the highest possible energy. This would occur if an electron relaxed from an infinitely high quantum state, for example n = ∞, to the ground state at n = 1. Employing the Balmer formula using $Z = 3$ for the Li^{2+} ion.

$$\Delta E = h\nu = E_1 - E_\infty = \frac{-(3^2)R_{\text{H}}}{1^2} - \frac{-(3^2)R_{\text{H}}}{\infty^2}$$

With an infinitely large denominator, the second term in this equation has a infinitesimally small value, essentially 0. Thus,

$$\Delta E = h\nu = E_1 - E_\infty = E_1 - 0 = \frac{-(3^2)R_{\text{H}}}{1^2}$$

$$\Delta E = \frac{hc}{\lambda} = \frac{-(3^2)(2.179 \times 10^{-18}\text{ J})}{1^2} = -1.9611 \times 10^{-17}\text{ J}$$

The negative result here relates to the energy change for the relaxing Li^{2+} ion. The emitted photon carries this energy away and the wavelength of that photon is

$$\lambda = \frac{hc}{-\Delta E} = \frac{(6.626 \times 10^{-34}\ \text{J}\bullet\text{s})(2.998 \times 10^{8}\ \text{m/s})}{-1.9611 \times 10^{-17}\ \text{J}} = 1.0129 \times 10^{-8}\ \text{m} = 10.13\ \text{nm}$$

A photon of this wavelength would be found on the fringe of the x ray-UV region of the electromagnetic spectrum.

7.111. First, calculate the mass of one oxygen molecule.

$$m = \frac{32.00\ \text{g O}_2}{1\ \text{mol}} \times \frac{1\ \text{mol}}{6.02 \times 10^{23}\ \text{molec}} \times \frac{1\ \text{kg}}{1000\ \text{g}} = 5.314 \times 10^{-26}\ \text{kg}$$

Now the de Broglie wavelength can be calculated.

$$\lambda = \frac{h}{mv} = \frac{6.63 \times 10^{-34}\ \text{J}\bullet\text{s}}{(5.314 \times 10^{-26}\ \text{kg})(479\ \text{m/s})} = 2.604 \times 10^{-11} = 2.60 \times 10^{-11}\ \text{m (26.0 pm)}$$

This wavelength can be compared to the length of the molecule (242 pm) as follows.

$$\frac{26.04\ \text{pm}}{242\ \text{pm}} \times 100\% = 10.76\% = 10.8\%$$

Thus, the de Broglie wavelength is 10.8% of the length of the molecule.

7.113. From the given wavelength, λ, the de Broglie relation can be used to determine the velocity of the electron, $v = h/m\lambda$. In turn, the electron's velocity and mass can be used to find the kinetic energy of the electron, $E_{kin} = ½mv^2$.

$$v(\text{m/s}) = \frac{h}{m\lambda} = \frac{6.626 \text{ x } 10^{-34}\ \text{J}\bullet\text{s}}{(9.1094 \times 10^{-31}\text{kg})(0.1476 \times 10^{-9}\ \text{m})} = 4.9281 \times 10^{6}\ \text{m/s}$$

$$E_{kin} = \frac{1}{2}mv^2 = \frac{1}{2}(9.1094 \times 10^{-31}\text{kg})(4.9281 \times 10^{6}\ \text{m/s})^2 = 1.1061 \times 10^{-17} = 1.106 \times 10^{-17}\ \text{J}$$

7.115. For the $n = 5$ level, the sublevels are 5*s*, 5*p*, 5*d*, 5*f*, and 5*g*. The numbers of orbitals for the sublevels are 1, 3, 5, 7, and 9, respectively. The total number of orbitals for the $n = 5$ level is therefore 1 + 3 + 5 + 7 + 9 = 25. Thus, the degeneracy of the $n = 5$ level in the H atom is 25.

■ SOLUTIONS TO CUMULATIVE-SKILLS PROBLEMS

7.117. First, use Avogadro's number to calculate the energy for one Cl_2 molecule.

$$\frac{239\ \text{kJ}}{1\ \text{mol}} \times \frac{1000\ \text{J}}{1\ \text{kJ}} \times \frac{1\ \text{mol}}{6.022 \times 10^{23}\ \text{molecules}} = 3.9687 \times 10^{-19}\ \text{J/molecule}$$

Then convert energy to frequency and finally to wavelength.

$$\nu = \frac{E}{h} = \frac{3.9687 \times 10^{-19}\ \text{J}}{6.626 \times 10^{-34}\ \text{J}\bullet\text{s}} = 5.9897 \times 10^{14}\ \text{/s}$$

$$\lambda = \frac{c}{\nu} = \frac{2.998 \times 10^{8}\ \text{m/s}}{5.9897 \times 10^{14}\ \text{/s}} = 5.0052 \times 10^{-7}\ \text{m (501 nm; visible region)}$$

7.119. First, calculate the energy needed to heat the 0.250 L of water from 20.0°C to 100.0°C.

$$0.250\text{ L} \times \frac{1000\text{ g}}{1\text{ L}} \times \frac{4.184\text{ J}}{(\text{g} \bullet {}^\circ\text{C})} \times (100.0\ {}^\circ\text{C} - 20.0{}^\circ\text{C}) = 8.3\underline{6}8 \times 10^4\text{ J}$$

Then calculate the frequency, the energy of one photon, and the number of photons.

$$\nu = \frac{c}{\lambda} = \frac{2.998 \times 10^8\text{ m/s}}{0.125\text{ m}} = 2.3\underline{9}8 \times 10^9\text{ /s}$$

$$E\text{ of one photon} = h\nu = (6.626 \times 10^{-34}\text{ J}\bullet\text{s}) \times (2.398 \times 10^9\text{/s}) = 1.5\underline{8}9 \times 10^{-24}\text{ J}$$

$$n = 8.368 \times 10^4\text{ J} \times \frac{1\text{ photon}}{1.589 \times 10^{-24}\text{ J}} = 5.2\underline{6}56 \times 10^{28} = 5.27 \times 10^{28}\text{ photons}$$

7.121. First, write the following equality for the energy to remove one electron, E_{removal}:

$$E_{\text{removal}} = E_{425\text{ nm}} - E_k\text{ of ejected photon}$$

Use $E = h\nu$ to calculate the energy of the photon. Then recall that E_k, the kinetic energy, is $\frac{1}{2}mv^2$. Use this to calculate E_k.

$$E_{425\text{ nm}} = \frac{hc}{\lambda} = \frac{(6.626 \times 10^{-34}\text{ J} \bullet \text{s})(2.998 \times 10^8\text{ m/s})}{4.25 \times 10^{-7}\text{ m}} = 4.6\underline{7}40 \times 10^{-19}\text{ J}$$

$$E_k = \tfrac{1}{2}mv^2 = \tfrac{1}{2} \times (9.1094 \times 10^{-31}\text{ kg}) \times (4.88 \times 10^5\text{ m/s})^2 = 1.0\underline{8}46 \times 10^{-19}\text{ J}$$

Subtract to find E_{removal}, and convert it to kJ/mol:

$$E_{\text{removal}} = 4.6740 \times 10^{-19}\text{ J} - (1.0847 \times 10^{-19}\text{ J}) = 3.5\underline{8}93 \times 10^{-19} = 3.59 \times 10^{-19}\text{ J/electron}$$

$$E_{\text{removal}} = \frac{3.5893 \times 10^{-19}\text{ J}}{1\text{ e}^-} \times \frac{6.022 \times 10^{23}\text{ e}^-}{1\text{ mol}} \times \frac{1\text{ kJ}}{1000\text{ J}} = 2.1\underline{6}1 \times 10^2$$

$$E_{\text{removal}} = 2.16 \times 10^2\text{ kJ/mol}$$

7.123. First, calculate the energy, E_{kin}, in joules, using the product of voltage and charge:

$$E_{kin} = (4.00 \times 10^3\text{ V}) \times (1.602 \times 10^{-19}\text{ C}) = 6.4\underline{0}8 \times 10^{-16}\text{ J}$$

Now, use the kinetic energy equation, $E_k = \frac{1}{2}mv^2$, and solve for velocity:

$$v = \sqrt{\frac{2E_{kin}}{m}} = \sqrt{\frac{2 \times 6.408 \times 10^{-16}\text{ J}}{9.1095 \times 10^{-31}\text{ kg}}} = 3.7\underline{5}08 \times 10^7\text{ m/s}$$

$$\lambda = \frac{h}{mv} = \frac{6.626 \times 10^{-34}\text{ J} \bullet \text{s}}{(9.1095 \times 10^{-31}\text{ kg}) \times (3.7508 \times 10^7\text{ m/s})} = 1.9\underline{3}9 \times 10^{-11}$$

$$= 1.94 \times 10^{-11}\text{ m (19.4 pm)}$$

CHAPTER 8

Electron Configurations and Periodicity

SOLUTIONS TO EXERCISES

Note on significant figures: If the final answer to a solution needs to be rounded off, it is given first with one nonsignificant figure, and the last significant figure is underlined. The final answer is then rounded to the correct number of significant figures. In multistep problems, intermediate answers are given with at least one nonsignificant figure; however, only the final answer has been rounded off.

8.1. a. Possible orbital diagram.

b. Possible orbital diagram.

c. Impossible orbital diagram; there are two electrons in a $2p$ orbital with the same spin.

d. Possible electron configuration.

e. Impossible electron configuration; only two electrons are allowed in an s subshell.

f. Impossible electron configuration; only six electrons are allowed in a p subshell.

8.2. Look at the periodic table. Start with hydrogen and go through the periods, writing down the subshells being filled, stopping with manganese ($Z = 25$). You obtain the following order:

Order:	$1s$	$2s2p$	$3s3p$	$4s3d4p$
Period:	first	second	third	fourth

Now fill the subshells with electrons, remembering that you have a total of twenty-five electrons to distribute. You obtain

$1s^22s^22p^63s^23p^64s^23d^5$, or $1s^22s^22p^63s^23p^63d^54s^2$

8.3. Arsenic is a main-group element in Period 4, Group VA, of the periodic table. The five outer electrons should occupy the $4s$ and $4p$ subshells; the five valence electrons have the configuration $4s^24p^3$.

8.4. Because the sum of the $6s^2$ and $6p^2$ electrons gives four outer (valence) electrons, lead should be in Group IVA, which it is. Looking at the table, you find lead in Period 6. From its position, it would be classified as a main-group element.

8.5. The electron configuration of phosphorus is $1s^22s^22p^63s^23p^3$. The orbital diagram is

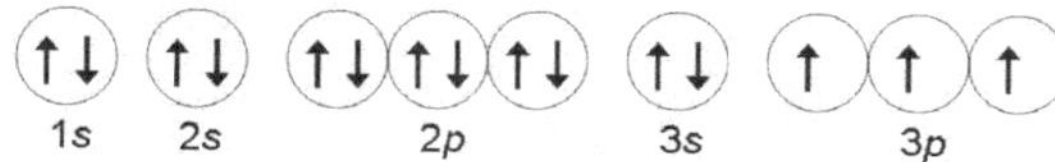

8.6. The radius tends to decrease across a row of the periodic table from left to right, and it tends to increase from the top of a column to the bottom. Therefore, in order of increasing radius,

Be < Mg < Na

8.7. It is more likely that answer a, 1000 kJ/mol, is the ionization energy for iodine because ionization energies tend to decrease with atomic number in a group (I is below Cl in Group VIIA).

8.8. Fluorine should have a more negative electron affinity because (1) carbon has only two electrons in the *p* subshell, (2) the 1− fluoride ion has a stable noble-gas configuration, and (3) the electron can approach the fluorine nucleus more closely than the carbon nucleus. This follows the general trend, which is toward more negative electron affinities from left to right in any period.

■ ANSWERS TO CONCEPT CHECKS

8.1. The second-period elements are those in which the 2*s* and 2*p* orbitals fill. Each orbital can hold only one electron, so all four orbitals will be filled after four electrons. Therefore, the second period will have four elements.

8.2. The *s* orbital fills in the first two elements of the period (Groups IA and IIA); then the *p* orbital starts to fill (Group IIIA). Thus, the first element is in Group IIA (Mg), and the next element is in Group IIIA (Al).

8.3. From the information given, the element must be in Group IIA. These elements have positive electron affinities and also have large third ionization energies.

8.4. A metalloid is an element near the staircase line in the periodic table (the green elements in the periodic table on the inside front cover of the text). The formula R_2O_5 suggests a Group VA element. There are two metalloids in Group VA, arsenic and antimony. That this is an acidic oxide indicates this metalloid has considerable nonmetal character. So, of the two metalloids, the one nearer the top of the column, arsenic, seems more likely. This agrees with the text, which notes that arsenic(V) oxide is acidic, whereas antimony(V) oxide is amphoteric.

■ ANSWERS TO SELF-ASSESSMENT AND REVIEW QUESTIONS

8.1. In the original Stern-Gerlach experiment, a beam of silver atoms is directed into the field of a specially designed magnet. (The same can be done with hydrogen atoms.) The beam of atoms is split into two by the magnetic field; half are bent toward one magnetic pole face and the other half toward the other magnetic pole face. This effect shows that the atoms themselves act as magnets with a positive or a negative component, as indicated by the positive or negative spin quantum numbers.

8.2. In effect, the electron acts as though it were a sphere of spinning charge (Figure 8.3). Like any circulating electric charge, it creates a magnetic field with a spin axis that has more than one possible direction relative to a magnetic field. Electron spin is subject to a quantum restriction to one of two directions corresponding to the m_s quantum numbers +1/2 and −1/2.

8.3. The Pauli exclusion principle limits the configurations of an atom by excluding configurations in which two or more electrons have the same four quantum numbers. For example, each electron in the same orbital must have different m_s values. This also implies that only two electrons occupy one orbital.

8.4. According to the principles discussed in Section 7.5, the number of orbitals in the *g* subshell ($l = 4$) is given by $2l + 1$ and is thus equal to 9 Because each orbital can hold a maximum of two electrons, the *g* subshell can hold a maximum of eighteen electrons.

8.5. The orbitals, in order of increasing energy up to and including the 3*p* orbitals (but not including the 3*d* orbitals), are as follows: 1*s*, 2*s*, 2*p*, 3*s*, and 3*p* (Figure 8.7).

8.6. The noble-gas core is an inner-shell configuration corresponding to one of the noble gases. The pseudo-noble-gas core is an inner-shell configuration corresponding to one of the noble gases together with $(n-1)d^{10}$ electrons. Like the noble-gas core electrons, the d^{10} electrons are not involved in chemical reactions. The valence electron is an electron (of an atom) located outside the noble-gas core or pseudo-noble-gas core. It is an electron primarily involved in chemical reactions.

8.7. The orbital diagram for the $1s^22s^22p^4$ ground state of oxygen is

Another possible oxygen orbital diagram, but not a ground state, is

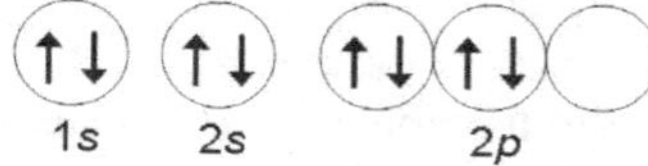

8.8. A diamagnetic substance is a substance that is not attracted by a magnetic field or is very slightly repelled by such a field. This property generally indicates the substance has only paired electrons. A paramagnetic substance is a substance that is weakly attracted by a magnetic field. This property generally indicates the substance has one or more unpaired electrons. Ground-state oxygen has two unpaired 2*p* electrons and is therefore paramagnetic.

8.9. In Groups IA and IIA, the outer *s* subshell is being filled: s^1 for Group IA and s^2 for Group IIA. In Groups IIIA to VIIIA, the outer *p* subshell is being filled: p^1 for IIIA, p^2 for IVA, p^3 for VA, p^4 for VIA, p^5 for VIIA, and p^6 for VIIIA. In the transition elements, the $(n-1)d$ subshell is being filled from d^1 to d^{10} electrons. In the lanthanides and actinides, the *f* subshell is being filled from f^1 to f^{14} electrons.

8.10. Mendeleev arranged the elements in order of increasing atomic weight, an arrangement that was later changed to atomic numbers. His periodic table was divided into rows (periods) and columns (groups). In his first attempt, he left spaces for what he believed to be undiscovered elements. In his row 5, under aluminum and above indium in Group III, he left a blank space. This Group III element he called eka-aluminum, and he predicted its properties from those of aluminum and indium. Later, the French chemist de Boisbaudran discovered this element and named it gallium.

8.11. In a plot of atomic radii versus atomic number (Figure 8.16), the major trends that emerge are the following: (1) Within each period (horizontal row), the atomic radius tends to decrease with increasing atomic number or nuclear charge. The largest atom in a period is thus the Group 1A atom, and the smallest atom in a period is thus the noble-gas atom. (2) Within each group (vertical column), the atomic radius tends to increase with the period number.

In a plot of ionization energy versus atomic number (Figure 8.18), the major trends are that (1) the ionization energy within a period increases with atomic number, and (2) the ionization energy within a group tends to decrease going down the group.

8.12. The alkaline earth element with the smallest radius is beryllium (Be).

8.13. Group VIIA (halogens) is the main group with the largest (and most negative) electron affinities. Configurations with filled subshells (ground states of the noble-gas elements) would form unstable negative ions when adding one electron per atom. Group VIIIA (noble gases) is the group of elements having only unstable negative ions.

8.14. The Na^{+} and Mg^{2+} ions are stable because they have the same electronic configuration as the noble gas neon. If Na^{2+} and Mg^{3+} ions were to exist, they would be very unstable because they would not have the same electronic configuration as a noble-gas structure and because of the energy needed to remove an electron from an inner shell.

8.15. The elements tend to increase in metallic character from right to left in any period. They also tend to increase in metallic character down any column (group) of elements.

8.16. A basic oxide is an oxide that reacts with acids. An example is calcium oxide, CaO. An acidic oxide is an oxide that reacts with bases. An example is carbon dioxide, CO_2.

8.17. Rubidium is the alkali metal atom with a $5s^1$ configuration.

8.18. Atomic number equals 117 (protons in last known element plus those needed to reach Group VIIA).

8.19. The following elements are in Groups IIIA to VIA:

Group IIIA	Group IVA	Group VA	Group VIA
B: metalloid	C: nonmetal	N: nonmetal	O: nonmetal
Al: metal	Si: metalloid	P: nonmetal	S: nonmetal
Ga: metal	Ge: metalloid	As: metalloid	Se: nonmetal
In: metal	Sn: metal	Sb: metalloid	Te: metalloid
Tl: metal	Pb: metal	Bi: metal	Po: metal

Yes, each column displays the expected increasing metallic character.

8.20. The oxides of the following elements are listed as acidic, basic, or amphoteric:

Group IIIA	Group IVA	Group VA	Group VIA
B: acidic	C: acidic	N: acidic	O: amphoteric (H_2O)
Al: amphoteric	Si: acidic	P: acidic	S: acidic
Ga: amphoteric	Ge: acidic	As: acidic	Se: acidic
In: basic	Sn: amphoteric	Sb: amphoteric	Te: amphoteric
Tl: basic	Pb: amphoteric	Bi: basic	Po: amphoteric

8.21. $2K(s) + 2H_2O(l) \rightarrow 2KOH(aq) + H_2(g)$

8.22. Barium should be a soft, reactive metal. Barium should form the basic oxide, BaO. Barium metal, for example, would be expected to react with water according to the equation

$$Ba(s) + 2H_2O(l) \rightarrow Ba(OH)_2(aq) + H_2(g)$$

8.23. The two oxides of carbon are carbon monoxide, CO, and carbon dioxide, CO_2.

8.24.
a. White phosphorus
b. Sulfur
c. Bromine
d. Sodium

8.25. The answer is c, 3.

8.26. The answer is b, $[Ne]3s^23p^3$

8.27. The answer is a. For Be, you would expect the first two ionization potentials to be successively getting larger with the third ionization energy much larger than the first two. The first two electrons are taken from the valence shell; the third electron is much harder to ionize because it is removed from the core.

8.28. The answer is b, II only.

ANSWERS TO CONCEPTUAL PROBLEMS

8.31. This statement of the Pauli principle implies there can be two electrons with the same spin in a given orbital. Because an electron can have either one of two spins, any orbital can hold a maximum of four electrons. The first six elements of the periodic table would have the following electron configurations:

(1) $1s^1$

(2) $1s^2$

(3) $1s^3$

(4) $1s^4$

(5) $1s^4 2s^1$

(6) $1s^4 2s^2$

8.33. The elements are in Group IIA (only *s* electrons) and IIIB (*d* electrons). They are also in Period 5. Therefore, the elements are strontium (Sr) and yttrium (Y).

8.35. Keeping in mind that a filled orbital is usually a stable configuration for an atom, an element in this universe with five electrons would probably lose the two electrons in the second orbital and form a cation with a charge of positive two. The other possible option is for the atom to gain seven additional electrons to fill the second orbital. However, this is unlikely given that the nuclear charge would be relatively small, and electron-electron repulsions in such an atom would be large.

8.37. The elements that form oxides of the form RO_2 are in Groups IVA and VIA. However, the metalloid oxide in Group VIA is amphoteric. Therefore, the elements are in Group IVA. The metalloid is germanium (Ge), and the metal is tin (Sn). GeO_2 is the acidic oxide, and SnO_2 is the amphoteric oxide.

8.39. a. Only heavier elements of Group IIA form slightly stable negative ions whereas the lighter elements do not form negative ions; thus, elements in Group IIA have small electron affinities as a rule. The only other main group of elements that would be in question here would be the noble gases in Group VIIIA. Noble gases do not form negative ions.

b. The large difference between the second and third ionization energies means this would be a Group IIA element.

c. Luster and conductivity are properties of metals. Group IIA is the only main group that contains just metallic elements.

Considering the above, element "E" must be a Group IIA alkaline earth metal.

SOLUTIONS TO PRACTICE PROBLEMS

Note on significant figures: If the final answer to a solution needs to be rounded off, it is given first with one nonsignificant figure, and the last significant figure is underlined. The final answer is then rounded to the correct number of significant figures. In multistep problems, intermediate answers are given with at least one nonsignificant figure; however, only the final answer has been rounded off.

8.41. a. Not allowed; the paired electrons in the 2*p* orbital should have opposite spins.

b. Allowed; electron configuration is $1s^22s^22p^4$.

c. Not allowed; the electrons in the 1*s* orbital must have opposite spins.

b. Not allowed; the 2*s* orbital can hold at most two electrons, with opposite spins.

8.43. a. Impossible state; the 2*p* orbitals can hold no more than six electrons.

b. Possible state.

c. Impossible state; the 3*s* orbital can hold no more than two electrons.

d. Possible state; however, the 3*p* and 4*s* orbitals should be filled before the 3*d* orbital.

8.45. The six possible orbital diagrams for $1s^22p^1$ are

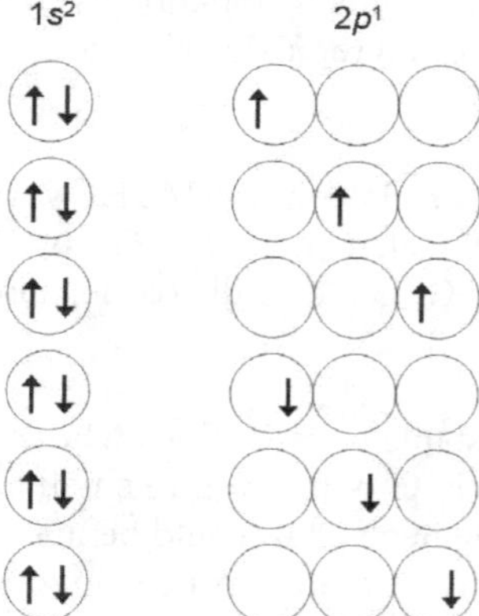

8.47. Iodine ($Z = 53$): $1s^22s^22p^63s^23p^63d^{10}4s^24p^64d^{10}5s^25p^5$

8.49. Manganese ($Z = 25$): $1s^22s^22p^63s^23p^63d^54s^2$

8.51. Bromine ($Z = 35$): $4s^24p^5$

8.53. Zirconium ($Z = 40$): $4d^25s^2$

8.55. The highest value of *n* is six, so thallium (Tl) is in the sixth period. The 5*d* subshell is filled, and there is a 6*p* electron, so Tl belongs in an A group. There are three valence electrons, so Tl is in Group IIIA. It is a main-group element.

8.57. Cobalt ($Z = 27$): [Ar]

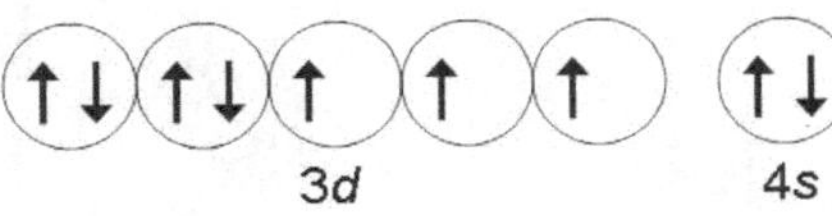

8.59. Potassium ($Z = 19$): [Ar]

All the subshells are filled in the argon core; however, the 4*s* electron is unpaired, causing the ground state of the potassium atom to be a paramagnetic substance.

8.61. Atomic radius increases going down a column (group), from S to Se, and increases going from right to left in a row, from Se to As. Thus, the order by increasing atomic radius is S, Se, As.

8.63. Ionization energy increases going left to right in a row. Thus, the order by increasing ionization energy is Na, Al, Cl, Ar.

8.65. a. In general, the electron affinity becomes larger going from left to right within a period. Thus, Br has a larger electron affinity than As. When Br gains an electron it also results in a stable noble gas electron configuration.

b. In general, a nonmetal has a larger electron affinity than a metal. Thus, F has a larger electron affinity than Li. When F gains an electron it also results in a stable noble gas electron configuration.

8.67. Chlorine forms the ClO_3^- ion, so bromine should form the BrO_3^- ion, and potassium forms the K^+ ion, so lithium should be Li^+. Thus, the expected formula of lithium bromate is $LiBrO_3$.

■ SOLUTIONS TO GENERAL PROBLEMS

8.69. Strontium: $1s^22s^22p^63s^23p^63d^{10}4s^24p^65s^2$

8.71. Polonium: $6s^26p^4$

8.73. The orbital diagram for arsenic is

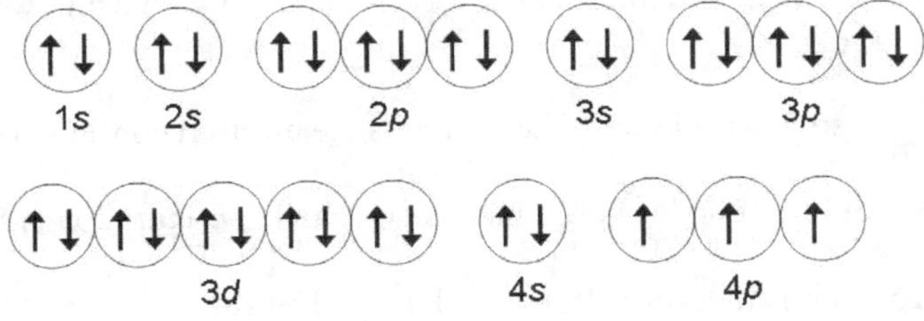

8.75. For eka-lead: [Rn] $5f^{14}6d^{10}7s^27p^2$. It is a metal; the oxide is eka-PbO or eka-PbO_2.

8.77. The ionization energy of Fr is ~370 kJ/mol (slightly less than that of Cs).

8.79. Niobium: [Kr]

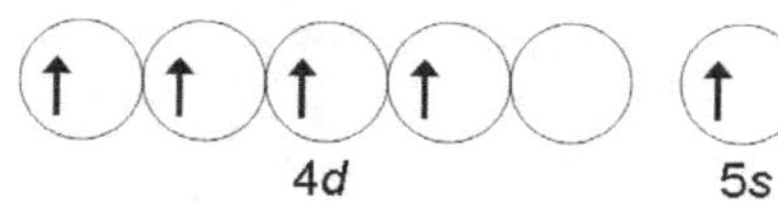

8.81. a. Cl_2

b. Na

c. Sb

d. Ar

8.83. Element with $Z = 23$: $1s^22s^22p^63s^23p^63d^34s^2$. The element is in Group VB (three of the five valence electrons are d electrons) and in Period 4 (largest n is 4). It is a d-block transition element.

8.85. Nuclear magnetic resonance depends upon the property of nuclear spin. Modern NMR uses the frequency range 300 MHz to 900 MHz.

8.87. When an electron in the cathode ray hits a metal atom in the target, it can (if it has sufficient energy) knock an electron from an inner shell of the atom. This produces a metal ion with an electron missing from an inner orbital. This electron configuration is unstable, so an electron from an orbital of higher energy drops into the half-filled orbital, and a photon is emitted. The photon corresponds to electromagnetic radiation in the x-ray region.

8.89. When you place a diamagnetic material in an external magnetic field, its electrons move so as to induce, or generate, a smaller magnetic field that is opposite in direction to the external field. This results in a repulsive force between the diamagnetic material and the external field. This repulsive force can be made upward to balance the downward force of gravity so the diamagnetic material can be levitated by a magnetic field.

■ SOLUTIONS TO STRATEGY PROBLEMS

8.91. Diagram (a) violates Hund's rule in the $2p$ sublevel by not having all the spins in the same direction. Diagram (b) violates Pauli's principle in the $2s$ sublevel with both electrons having the same spin in the same orbital. Diagram (c) violates Pauli's principle in the $2p$ sublevel with one orbital having two electrons with the same spin. Diagram (d) does not violate either condition. Diagram (e) violates Pauli's principle in the $3p$ sublevel with one orbital having two electrons with the same spin.

8.93. The element is in Group IA, so it is rubidium. The ground-state electron configuration is $[Kr]5s^1$.

8.95. The element is in Group VIA, so it is tellurium. The valence-shell configuration is $5s^25p^4$.

8.97. The element is in Period 5, so it is tellurium. Its symbol is Te. The value of b must be 4.

8.99. If the electron configuration $1s^22s^22p^53s^23p^2$ is for a neutral atom (M), it must be an excited state atom (M^*) because there are only 5 electrons in the $2p$ sublevel. Allowing one of the $3p$ electrons to transition (by way of relaxation) to the $2p$ sublevel results in the following change:

$$M^* (1s^22s^22p^53s^23p^2) \rightarrow M (1s^22s^22p^63s^23p^1) + hv$$

The resulting electron configuration $1s^22s^22p^63s^23p^1$ is the electron configuration for an atom in period 3 and group IIIA. The element is aluminum, Al. The second electron configuration given in the problem statement, i.e., $1s^22s^22p^63s^23p^5$, has no partly filled sublevels below the last sublevel so it is a ground state electron configuration associated with an atom in period 3 and Group VIIA. The element is chlorine, Cl. Since Al forms the 3+ ion and chlorine forms the 1- ion, the formula of the compound of these elements is $AlCl_3$.

8.101. The alkaline earths in Group IIA are moderately reactive metals. The element in this group that has a referenced melting point near 850 °C is calcium, Ca. An element that is a green gas under normal conditions is chlorine, Cl_2. When Cl_2 reacts with Ca, calcium chloride, $CaCl_2$ forms as a white solid.

$$Ca(s) + Cl_2(g) \rightarrow CaCl_2(s)$$

Calcium chloride dissolves in water to form $Ca^{2+}(aq)$ and $Cl^-(aq)$ ions.

$$CaCl_2(s) + xH_2O \rightarrow Ca^{2+}(aq) + Cl^-(aq)$$

When carbon dioxide gas, $CO_2(g)$, is bubbled into water, it forms a solution of carbonic acid, a polyprotic acid which dissociates stepwise to form some carbonate ions.

$$CO_2(g) + H_2O(1) \rightarrow H_2CO_3(aq)$$

$$H_2CO_3(aq) + H_2O(l) \rightarrow HCO_3^-(aq) + H_3O^+(aq)$$

$$HCO_3^-(aq) + H_2O(l) \rightarrow CO_3^{2-}(aq) + H_3O^+(aq)$$

The presence of $Ca^{2+}(aq)$ produces a reaction with the carbonate ions yielding a white precipitate of calcium carbonate, $CaCO_3$.

$$Ca^{2+}(aq) + CO_3^{2-}(aq) \rightarrow CaCO_3(s)$$

In summary, the white solid is $CaCl_2(s)$ and the white precipitate is $CaCO_3(s)$.

8.103. Element 120 is expected to be associated with the alkaline earth metal family in Group IIA. Applying the Aufbau principle, element 120 should have a ground state electron configuration of $[Uuo]8s^2$. Here, [Uuo] represents the shorthand e^- configuration of element 118. Note the similarity in the valence shell configuration with the other alkaline earth metals (ns^2).

8.105. The main-group atom in Period 5 with the smallest radius is xenon. The trend in atomic radius is to decrease as you go from left to right across a period. This is because electrons are being added to the same valence shell, but the charge on the nucleus is increasing. The net effect is to pull the valence electrons closer to the nucleus. The orbital diagram for xenon is

(↑↓) (↑↓)(↑↓)(↑↓)
5s 5p

8.107. The element in Period 6 that forms the ion R^{2+} and forms the oxides RO and RO_2 is lead (Group IVA). The element that immediately follows lead is bismuth ($Z = 83$) and is in Group VA. The orbital diagram for the ground-state valence-shell electrons is

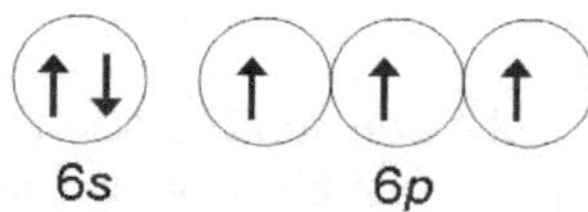

8.109. Applying Hess's law, we add the thermochemical equation associated with the dissociation process given in the problem statement to the thermochemical equation associated with Cl's electron affinity as follows:

$Cl_2\ (g)$	$\rightarrow$	$2Cl(g)$	$\Delta H = +240$ kJ/mol
$2Cl(g) + 2e^-$	$\rightarrow$	$2Cl^-(g)$	2 x ($\Delta H = -349$ kJ/mol)
$Cl_2(g) + 2e^-$	$\rightarrow$	$2Cl^-(g)$	$\Delta H = -458$ kJ/mol

The net result in line 3 above shows that 2 electrons are being added to a Cl_2 molecule to produce two Cl^- ions. The negative enthalpy change of −458 kJ/mol shows the process to be exothermic.

■ SOLUTIONS TO CUMULATIVE-SKILLS PROBLEMS

8.111. The equation is

$$Ba(s) + 2H_2O(l) \rightarrow Ba(OH)_2(aq) + H_2(g)$$

Using the equation, calculate the moles of H_2; then use the ideal gas law to convert to volume.

$$\text{mol } H_2 = 2.50 \text{ g Ba} \times \frac{1 \text{ mol Ba}}{137.33 \text{ g Ba}} \times \frac{1 \text{ mol } H_2}{1 \text{ mol Ba}} = 0.018\underline{2}04 \text{ mol}$$

$$V = \frac{nRT}{P} = \frac{(0.018204 \text{ mol})(0.082057 \text{ L} \bullet \text{atm/K} \bullet \text{mol})(294.2 \text{ K})}{(748/760) \text{ atm}} = 0.44\underline{6}51 \text{ L (447 mL)}$$

8.113. Radium is in Group IIA; hence, the radium cation is Ra^{2+}, and its oxide is RaO. Use the atomic weights to calculate the percentage of Ra in RaO.

$$\text{Percent Ra} = \frac{226 \text{ amu Ra}}{226 \text{ amu Ra } + \text{ } 16.00 \text{ amu O}} \times 100\% = 93.\underline{3}8 = 93.4\% \text{ Ra}$$

8.115. Convert 5.00 mg (0.00500 g) Na to moles of Na; then convert to energy using the first ionization energy of 496 kJ/mol Na.

$$\text{mol Na} = 0.00500 \text{ g Na} \times \frac{1 \text{ mol Na}}{22.99 \text{ g Na}} = 2.1\underline{7}4 \times 10^{-4} \text{ mol Na}$$

$$2.1\underline{7}4 \times 10^{-4} \text{ mol Na} \times \frac{496 \text{ kJ}}{1 \text{ mol Na}} = 0.10\underline{7}8 = 0.108 \text{ kJ} = 108 \text{ J}$$

8.117. Use the Bohr formula, where $n_f = \infty$ and $n_i = 1$.

$$\Delta E = -R_H \left[\frac{1}{\infty^2} - \frac{1}{1^2} \right] = -R_H[-1] = R_H = \frac{2.179 \times 10^{-18}\ \text{J}}{1\ \text{H atom}}$$

$$I.E. = \frac{2.179 \times 10^{-18}\ \text{J}}{1\ \text{H atom}} \times \frac{6.022 \times 10^{23}\ \text{H atoms}}{1\ \text{mol H}} = \frac{1.31\underline{2}19 \times 10^{6}\ \text{J}}{1\ \text{mol H}}$$

$$I.E. = 1.312 \times 10^3\ \text{kJ/mol H}$$

8.119. Add the three equations after reversing the equation for the lattice energy and its ΔH:

$Na(g)$	$\rightarrow$	$Na^+(g) + e^-$	$\Delta H = +496$ kJ/mol
$Cl(g) + e^-$	$\rightarrow$	$Cl^-(g)$	$\Delta H = -349$ kJ/mol
$Na^+(g) + Cl^-(g)$	$\rightarrow$	$NaCl(s)$	$-1(\Delta H = 786$ kJ/mol)
$Na(g) + Cl(g)$	$\rightarrow$	$NaCl(s)$	$\Delta H = -639$ kJ/mol

CHAPTER 9

Ionic and Covalent Bonding

SOLUTIONS TO EXERCISES

9.1. The Lewis symbol for oxygen is $:\ddot{\underset{..}{O}}\cdot$ and the Lewis symbol for magnesium is $\cdot$ Mg $\cdot$. The magnesium atom loses two electrons, and the oxygen atom accepts two electrons. You can represent this electron transfer as follows:

$$\cdot Mg \cdot + \cdot\ddot{\underset{\cdot}{O}}: \longrightarrow Mg^{2+} + \left[:\ddot{\underset{..}{O}}:\right]^{2-}$$

9.2. The electron configuration of the Ca atom is $[Ar]4s^2$. By losing two electrons, the atom assumes a 2+ charge and the argon configuration, [Ar]. The Lewis symbol is Ca^{2+}. The S atom has the configuration $[Ne]3s^23p^4$. By gaining two electrons, the atom assumes a 2− charge and the argon configuration $[Ne]3s^23p^6$ and is the same as [Ar]. The Lewis symbol is

$$\left[:\ddot{\underset{..}{S}}:\right]^{2-}$$

9.3. The electron configuration of lead (Pb) is $[Xe]4f^{14}5d^{10}6s^26p^2$. The electron configuration of Pb^{2+} is $[Xe]4f^{14}5d^{10}6s^2$.

9.4. The electron configuration of manganese ($Z = 25$) is $[Ar]3d^54s^2$. To find the ion configuration, remove first the 4*s* electrons, then the 3*d* electrons. In this case, only two electrons need to be removed. The electron configuration of Mn^{2+} is $[Ar]3d^5$.

9.5. S^{2-} has a larger radius than S. The anion has more electrons than the atom. The electron-electron repulsion is greater; hence, the valence orbitals expand. The anion radius is larger than the atomic radius.

9.6. The ionic radii increase down any column because of the addition of electron shells. All of these ions are from the Group IIA family; therefore, $Mg^{2+} < Ca^{2+} < Sr^{2+}$.

9.7. Cl^-, Ca^{2+}, and P^{3-} are isoelectronic with an electron configuration equivalent to [Ar]. In an isoelectronic sequence, the ionic radius decreases with increasing nuclear charge. Therefore, in order of increasing ionic radius, we have Ca^{2+}, Cl^-, and P^{3-}.

9.8. The absolute values of the electronegativity differences are C–O, 1.0; C–S, 0.0; and H–Br, 0.7. Therefore, C–O is the most polar bond.

9.9. First, calculate the total number of valence electrons. C has four, Cl has seven, and F has seven. The total number is $4 + (2 \times 7) + (2 \times 7) = 32$. The expected skeleton consists of a carbon atom surrounded by Cl and F atoms. Distribute the electron pairs to the surrounding atoms to satisfy the octet rule. All 32 electrons (16 pairs) are accounted for.

```
        ..
      : Cl :
  ..    ..    ..
: F  :  C  :  F :
  ..    ..    ..
      : Cl :
        ..
```

9.10. The total number of electrons in CO_2 is $4 + (2 \times 6) = 16$. Because carbon is more electropositive than oxygen, it is expected to be the central atom. Distribute the electrons to the surrounding atoms to satisfy the octet rule.

```
  ..       ..
: O : C  : O :
  ..       ..
```

All sixteen electrons have been used, but note that there are only four electrons on carbon. This is four electrons short of a complete octet, which suggests the existence of double bonds. Move a pair of electrons from each oxygen to the carbon-oxygen bonds.

```
..        ..                  ..         ..
O :: C :: O        or         O ══ C ══ O
..        ..                  ..         ..
```

9.11. a. There are $(3 \times 1) + 6 = 9$ valence electrons in H_3O. The H_3O^+ ion has one less electron than is provided by the neutral atoms because the charge on the ion is 1+. Hence, there are eight valence electrons in H_3O^+. The electron-dot formula is

```
[      H      ]+
       ..
[  H : O : H  ]
       ..
```

b. Cl has seven valence electrons, and O has six valence electrons. The total number of valence electrons from the neutral atoms is $7 + (2 \times 6) = 19$. The charge on the ClO_2^- is 1−, which provides one more electron than the neutral atoms. This makes a total of 20 valence electrons. The electron-dot formula for ClO_2^- is

```
[   ..   ..   ..   ]-
[ : O : Cl : O :   ]
[   ..   ..   ..   ]
```

9.12. The resonance formulas for NO_3^- are

```
[     : O :     ]-         [     ..        ]-         [      ..       ]-
[      ::       ]          [    : O :      ]          [    : O :      ]
[   ..     ..   ]  <--->   [   ..  ..  ..  ]  <--->   [   ..   ..     ]
[ : O : N : O : ]          [ : O :: N : O :]          [ : O : N :: O :]
[   ..     ..   ]          [   ..      ..  ]          [   ..      ..  ]
```

9.13. The number of valence electrons in SF_4 is $6 + (4 \times 7) = 34$. The skeleton structure is a sulfur atom surrounded by fluorine atoms. After the electron pairs are placed on the F atoms to satisfy the octet rule, two electrons remain.

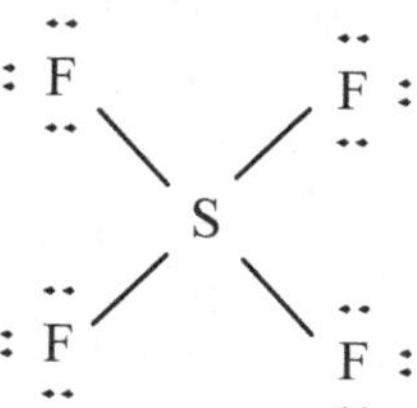

These additional two electrons are put on the sulfur atom because it had *d* orbitals and, therefore, can expand its octet.

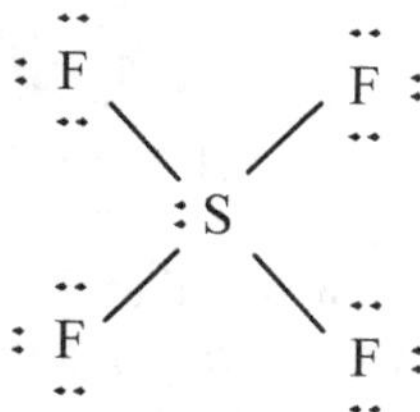

9.14. Be has two valence electrons, and Cl has seven valence electrons. The total number of valence electrons is $2 + (2 \times 7) = 16$ in the $BeCl_2$ molecule. Be, a Group IIA element, can have fewer than eight electrons around it. The electron-dot formula of $BeCl_2$ is

```
  ..        ..
: Cl : Be : Cl :
  ..        ..
```

9.15. The total number of electrons in H_3PO_4 is $3 + 5 + 24 = 32$. Assume a skeleton structure in which the phosphorus atom is surrounded by the more electronegative four oxygen atoms. The hydrogen atoms are then attached to the oxygen atoms. Distribute the electron pairs to the surrounding atoms to satisfy the octet rule. If you assume all single bonds (structure on the left), the formal charge on the phosphorus is 1+ and the formal charge on the top oxygen is 1−. Using the principle of forming a double bond with a pair of electrons on the atom with the negative formal charge, you obtain the structure on the right. The formal charge on all oxygens in this structure is zero; the formal charge on phosphorus is $5 - 5 = 0$. This is the better structure.

```
           ..
          : O :                          : O :
            |                              ||
     ..           ..               ..           ..
H — O — P — O — H            H — O — P — O — H
     ..     |     ..               ..     |     ..
          : O :                          : O :
            |                              |
            H                              H
```

9.16. The bond length can be predicted by adding the covalent radii of the two atoms. For O–H, we have 66 pm + 31 pm = 97 pm.

9.17. As the bond order increases, the bond length decreases. Since the C=O is a double bond, we would expect it to be the shorter one, 123 pm.

9.18.

$$\mathrm{H_2C{=}CH_2} + 3\,O_2 \longrightarrow 2\,\mathrm{O{=}C{=}O} + 2\,\mathrm{H{-}O{-}H}$$

One C=C bond, four C–H bonds, and three O=O bonds are broken. There are four C=O bonds and four O–H bonds formed.

$\Delta H = \{[614 + (4 \times 413) + (3 \times 498)] - [(4 \times 804) + (4 \times 463)]\}$ kJ = −1308 kJ

ANSWERS TO CONCEPT CHECKS

9.1. a. The 2+ ions are common in transition elements, but it is the outer *s* electrons that are lost to form these ions in compounds. Iron, whose configuration is $[Ar]3d^64s^2$, would be expected to lose two 4*s* electrons to give the configuration $[Ar]3d^6$ for the Fe^{2+} ion in compounds. The configuration given in the problem is for an excited state; you would not expect to see it in compounds.

b. Nitrogen, whose ground-state atomic configuration is $[He]2s^22p^3$, would be expected to form an anion with a noble-gas configuration by gaining three electrons. This would give the anion N^{3-} with the configuration $[He]2s^22p^6$. You would not expect to see the anion N^{2-} in compounds.

c. The zinc atom has the ground-state configuration $[Ar]3d^{10}4s^2$. The element is often considered to be a transition element. In any case, you would expect the atom to form compounds by losing its 4*s* electrons to give Zn^{2+} with the pseudo-noble-gas configuration $[Ar]3d^{10}$. This is the ion configuration given in the problem.

d. The configuration of the ground-state sodium atom is $[He]2s^22p^63s^1$. You would expect the atom to lose one electron to give the Na^+ ion with the noble-gas configuration $[He]2s^22p^6$. You would not expect to see compounds with the Na^{2+} ion.

e. The ground state of the calcium atom is $[Ne]3s^23p^64s^2$. You would expect the atom to lose its two outer electrons to give Ca^{2+} with the noble-gas configuration $[Ne]3s^23p^6$, which is the configuration given in the problem.

9.2. a. There are two basic points to consider in assessing the validity of each of the formulas given in the problem. One is whether the formula has the correct skeleton structure. You expect the F atoms to be bonded to the central N atoms because the F atoms are more electronegative. The second point is the number of dots in the formula. This should equal the total number of electrons in the valence shell of the atoms (five for each nitrogen atom and seven for each fluorine atom), which is $(2 \times 5) + (2 \times 7) = 24$, or twelve pairs. The number showing in the formula here is thirteen, which is incorrect.

b. This formula has the correct skeleton structure and the correct number of dots. All of the atoms have octets, so the formula would appear to be correct. As a final check, however, you might try drawing the formula beginning with the skeleton structure. In drawing an electron-dot formula, after connecting atoms by single bonds (a single electron pair), you would place electron pairs around the outer atoms (the F atoms in this formula) to give octets. After doing that, you would have used up nine electron pairs (three for the single bonds and three for each F atom to fill out its octet). This leaves three pairs, which you might distribute as follows:

```
 ..   ..   ..   ..
: F : N : N : F :
 ..   ..        ..
```

One of the nitrogen atoms (the one on the right) does not have an octet. The lack of an octet on this atom suggests trying for a double bond. This suggests that you move one of the lone pairs on the left N atom into one of the adjacent bonding regions. Moving this lone pair into the N–N region would give a symmetrical result, whereas moving the lone pair into the F–N region would not. The text notes that the atoms often showing multiple bonds are C, N, O, and S. The formula given here with a nitrogen-nitrogen double bond appears quite reasonable.

c. This formula is similar to the previous one, b, but the double bond is between the F and N atoms. In this case, however, there are 10 electrons around the left F and only 6 electrons around the right N. Both of these violate the octet rule for Period 2 elements. This formula is incorrect.

d. This formula is similar to the one drawn earlier in describing how one would get to the formula in b. The left N atom does not have an octet, which suggests you move a lone pair on the other N atom into the N–N bond region to give a double bond.

e. This formula does not have the correct skeleton structure.

f. This formula has the correct skeleton structure and the correct total number of electron dots, but neither N atom has an octet. In fact, there is no bond between the two N atoms.

9.3. a. This model and corresponding Lewis structure, H:C:::N:, has the expected skeleton structure. (H must be an exterior atom, but either C or N might be in the center; however, you expect the more electropositive, or less electronegative, atom, C, to be in this position). This formula also has the correct number of electron dots ($1 + 4 + 5 = 10$, or 5 pairs). Finally, the formal charge of each atom is zero. Therefore, this model should be an accurate representation of the HCN molecule.

b. This structure has the more electronegative atom, N, in the central position; you don't expect this to be the correct structure. You can also look at this from the point of view of formal charges. This formula has a 1+ charge on the N atom and a 1− charge on the C atom. You would not expect the more electronegative atom to have the positive formal charge. Moreover, the previous formula, a, has zero charges for each atom, which would be preferred.

c. If you draw the Lewis structure, you will see that each atom has an octet, but the formula has too many electron pairs (seven instead of five). You can remove two pairs and still retain octets if you move two lone pairs into the bonding region, to give a triple bond.

d. If you draw the Lewis structure, you will see that each atom has an octet, but the formula has too many electron pairs (six instead of five). You would arrive at a similar conclusion using formal charges. The formal charges are 0 for H, 1− for C, and 1− for N. Since these don't add to 0 in relation to the neutral molecule, the formula must be incorrect.

ANSWERS TO SELF-ASSESSMENT AND REVIEW QUESTIONS

9.1. As an Na atom approaches a Cl atom, the outer electron of the Na atom is transferred to the Cl atom. The result is an Na^+ and a Cl^- ion. Positively charged ions attract negatively charged ions, so, finally, the NaCl crystal consists of Na^+ ions surrounded by six Cl^- ions surrounded by six Na^+ ions.

9.2. Ions tend to attract as many ions of opposite charge about them as possible. The result is that ions tend to form crystalline solids rather than molecular substances.

9.3. The energy terms involved in the formation of an ionic solid from atoms are the ionization energy of the metal atom, the electron affinity of the nonmetal atom, and the energy of the attraction of the ions forming the ionic solid. The energy of the solid will be low if the ionization energy of the metal is low, the electron affinity of the nonmetal is high, and the energy of the attraction of the ions is large.

9.4. The lattice energy for potassium bromide is the change in energy that occurs when KBr(*s*) is separated into isolated $K^+(g)$ and $Br^-(g)$ ions in the gas phase.

$$KBr(s) + \text{lattice energy} \rightarrow K^+(g) + Br^-(g)$$

9.5. A monatomic cation with a charge equal to the group number corresponds to the loss of all valence electrons. This loss of electrons would give a noble-gas configuration, which is especially stable. A monatomic anion with a charge equal to the group number minus eight would have a noble-gas configuration.

9.6. Most of the transition elements have configurations in which the outer *s* subshell is doubly occupied. These electrons will be lost first, and we might expect each to be lost with almost equal ease, resulting in +2 ions.

9.7. If we assume the ions are spheres that are just touching, the distances between centers of the spheres will be related to the radii of the spheres. For example, in LiI, we assume that the −1 ions are large spheres that are touching. The distance between centers of the I^- ions equals two times the radius of the I^- ion.

9.8. In going across a period, the cations decrease in radius. When we reach the anions, there is an abrupt increase in radius, and then the radii again decrease. Ionic radii increase going down any column of the periodic table.

9.9. As the H atoms approach one another, their 1*s* orbitals begin to overlap. Each electron can then occupy the space around both atoms; that is, the two electrons are shared by the atoms.

9.10.

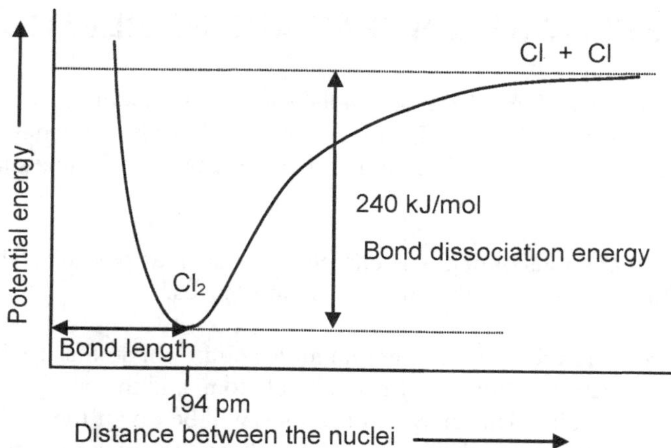

9.11. An example is thionyl chloride, $SOCl_2$:

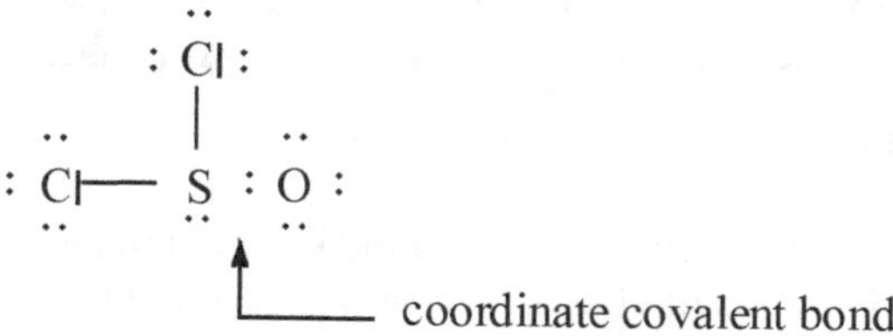

Note that the O atom has eight electrons around it; that is, it has two more electrons than the neutral atom. These two electrons must have come from the S atom. Thus, this bond is a coordinate covalent bond.

9.12. In many atoms of the main-group elements, bonding uses an *s* orbital and the three *p* orbitals of the valence shell. These four orbitals are filled with eight electrons, thus accounting for the octet rule.

9.13. Electronegativity increases from left to right (with the exception of the noble gases) and decreases from top to bottom in the periodic table.

9.14. The absolute difference in the electronegativities of the two atoms in a bond gives a rough measure of the polarity of the bond.

9.15. Resonance is used to describe the electron structure of a molecule in which bonding electrons are delocalized. In a resonance description, the molecule is described in terms of two or more Lewis formulas. If we want to retain Lewis formulas, resonance is required because each Lewis formula assumes that a bonding pair of electrons occupies the region between two atoms. We must imagine that the actual electron structure of the molecule is a composite of all resonance formulas.

9.16. Molecules having an odd number of electrons do not obey the octet rule. An example is nitrogen monoxide, NO. The other exceptions fall into two groups. In one group are molecules with an atom having fewer than eight valence electrons around it. An example is borane, BH_3. In the other group are molecules with an atom having more than eight valence electrons around it. An example is sulfur hexafluoride, SF_6.

9.17. As the bond order increases, the bond length decreases. For example, the average carbon-carbon single-bond length is 154 pm, whereas the carbon-carbon double-bond length is 134 pm, and the carbon-carbon triple-bond length is 120 pm.

9.18. Bond enthalpy is the average enthalpy change for the breaking of a bond in a molecule. The enthalpy of a reaction for gaseous reactions can be determined by summing the bond enthalpies of all the bonds that are broken and subtracting the sum of the bond enthalpies of all the bonds that are formed.

9.19. The answer is a, $BaCO_3$.

9.20. The answer is e, $S^- > S > S^+$.

9.21. The answer is d, $1s^2 2s^2 2p^5$.

9.22. The answer is a, sulfur.

ANSWERS TO CONCEPTUAL PROBLEMS

9.25. Because the compound that forms is a combination of a metal and a nonmetal, we would expect it to be ionic. If we assume that metal atoms tend to lose electrons to obtain filled shells, then the metal atom X would lose three electrons from the $n = 2$ level, forming the X^{3+} cation. We can expect the nonmetal atom Y to gain electrons to obtain a filled shell, so it requires an additional electron to fill the $n = 1$ level, forming the Y^- anion. To produce an ionic compound with an overall charge of zero, the compound formed from these two elements would be XY_3.

9.27. The smaller atom on the left (yellow) becomes a larger ion on the right, while the larger atom on the left (blue) becomes the smaller ion on the right. Since cations are smaller than their parent atom, and anions are larger than their parent atom, the cation on the right is the smaller ion (blue), and the anion is the larger ion (yellow). Finally, since metals tend to form cations, and nonmetals form anions, the metal on the left is the larger atom (blue), and the nonmetal is the smaller atom (yellow).

9.29. a. Incorrect. The atoms in this formula do not obey the octet rule. The formula has the correct number of valence electrons, so this suggests a multiple bond between the N atoms.

b. Correct. The central atom is surrounded by more electronegative atoms, as you would expect, and each atom obeys the octet rule.

c. Incorrect. The skeleton structure is acceptable (the central atom is surrounded by more electronegative atoms), but you would expect the double bond to be between C and O rather than between C and F (C, N, O, and S form multiple bonds). You would come to this same conclusion using rules of formal charge. (The formula has a formal charge of 1+ on F and 1− on O, whereas you would expect these formal charges to be interchanged, with the negative charge on the F atom, which is more electronegative.)

d. Incorrect. The skeleton structure is OK, but the carbon atom has ten valence electrons about it. This suggests that you replace the two carbon-oxygen double bonds by one carbon-oxygen double bond (because there is one extra pair of electrons on the C atom). The most symmetrical location of the double bond uses the oxygen atom not bonded to an H atom. Also, only this formula has zero formal charges on all atoms.

9.31. a. To arrive at a skeleton structure, you decide which is the central atom. (It cannot be H). The C atom is less electronegative than the Cl atom, so you place it as the central atom and surround it by the other atoms.

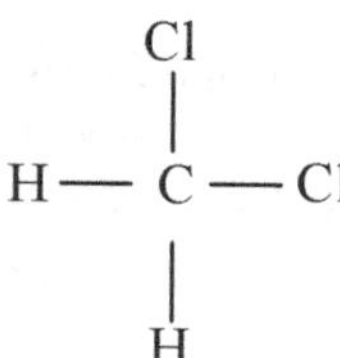

b. HNO_2 is an oxyacid in which O atoms bond to the central atom with the H bonded to O. The central atom must be N (the only other atom), so the skeletal structure is

H—O—N—O

c. You place the least electronegative atom (N) as the central atom and bond it to the other atoms.

F—N—O

d. N is less electronegative than O. The most symmetrical structure would be the two N atoms in the center with two O atoms bonded to each N atom.

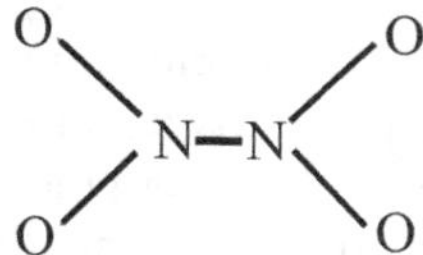

9.33. a. In order to be a neutral ionic compound, element X must have a charge of 3−. In an ionic compound, calcium forms the Ca^{2+} cation. The combination of Ca^{2+} and X^{3-} would form the ionic compound with the formula Ca_3X_2.

b. Since element X formed an ionic compound with sodium metal, it is probably a nonmetal with a high electron affinity. When a nonmetal with a high electron affinity combines with a metal such as calcium, an ionic compound is formed.

■ SOLUTIONS TO PRACTICE PROBLEMS

9.35. a. P has the electron configuration $[Ne]3s^2 3p^3$. It has five electrons in its valence shell. The Lewis formula is

$\cdot \ddot{P} \cdot$ (with one dot below)

b. P^{3-} has three more valence electrons than P. It now has eight electrons in its valence shell. The Lewis formula is

$\left[: \ddot{\underset{\cdot\cdot}{P}} : \right]^{3-}$

c. Ga has the electron configuration $[Ar]3d^{10}4s^2 4p^1$. It has three electrons in its valence shell. The Lewis formula is

$\cdot \dot{Ga} \cdot$

d. Ga^{3+} has three fewer valence electrons than Ga. It now has zero electrons in its valence shell. The Lewis formula is

Ga^{3+}

9.37. a. If the calcium atom loses two electrons, and the bromine atoms gain one electron each, all three atoms will assume noble-gas configurations. This can be represented as follows.

$$:\ddot{\underset{\cdot\cdot}{Br}}\cdot + \cdot Ca\cdot + \cdot\ddot{\underset{\cdot\cdot}{Br}}: \longrightarrow \left[:\ddot{\underset{\cdot\cdot}{Br}}:\right]^- + Ca^{2+} + \left[:\ddot{\underset{\cdot\cdot}{Br}}:\right]^-$$

b. If the potassium atom loses one electron, and the iodine atom gains one electron, both atoms will assume a noble-gas configuration. This can be represented as follows.

$$K\cdot + \cdot\ddot{\underset{\cdot\cdot}{I}}: \longrightarrow K^+ + \left[:\ddot{\underset{\cdot\cdot}{I}}:\right]^-$$

9.39.

a. As: $1s^22s^22p^63s^23p^63d^{10}4s^24p^3$ $:\dot{\underset{\cdot}{As}}\cdot$

b. As^{3+}: $1s^22s^22p^63s^23p^63d^{10}4s^2$ $\left[\cdot As\cdot\right]^{3+}$

c. Se: $1s^22s^22p^63s^23p^63d^{10}4s^24p^4$ $:\dot{\underset{\cdot\cdot}{Se}}\cdot$

d. Se^{2-}: $1s^22s^22p^63s^23p^63d^{10}4s^24p^6$ $\left[:\ddot{\underset{\cdot\cdot}{Se}}:\right]^{2-}$

9.41. Bi: $[Xe]4f^{14}5d^{10}6s^26p^3$

Bi^{3+}: The three $6p$ electrons are lost from the valence shell.
$[Xe]4f^{14}5d^{10}6s^2$

9.43. The 2+ ion is formed by the loss of electrons from the $4s$ subshell.

Ni^{2+}: $[Ar]3d^8$

The 3+ ion is formed by the loss of electrons from the $4s$ and $3d$ subshells.

Ni^{3+}: $[Ar]3d^7$

9.45. a. $Sr^{2+} < Sr$

The cation is smaller than the neutral atom because it has lost all its valence electrons; hence, it has one less shell of electrons. The electron-electron repulsion is reduced, so the orbitals shrink because of the increased attraction of the electrons to the nucleus.

b. $Br < Br^-$

The anion is larger than the neutral atom because it has more electrons. The electron-electron repulsion is greater, so the valence orbitals expand to give a larger radius.

9.47. $S^{2-} < Se^{2-} < Te^{2-}$

All have the same number of electrons in the valence shell. The radius increases with the increasing number of filled shells.

9.49. Smallest Na^+ ($Z = 11$), F^- ($Z = 9$), N^{3-} ($Z = 7$) Largest

These ions are isoelectronic. The atomic radius increases with the decreasing nuclear charge (Z).

9.51.

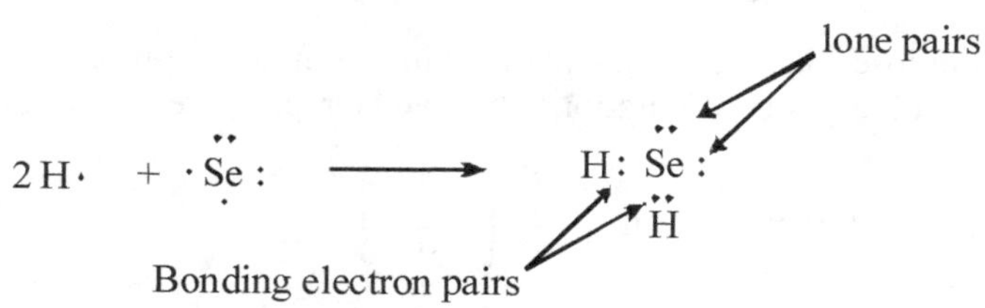

9.53. Arsenic is in Group VA on the periodic table and has five valence electrons. Bromine is in Group VIIA and has seven valence electrons. Arsenic forms three covalent bonds, and bromine forms one covalent bond. Therefore, the simplest compound would be $AsBr_3$.

9.55. a. P, N, O

Electronegativity increases from left to right and from bottom to top in the periodic table.

b. Na, Mg, Al

Electronegativity increases from left to right within a period.

c. Al, Si, C

Electronegativity increases from left to right and from bottom to top in the periodic table.

9.57. $X_O - X_P = 3.5 - 2.1 = 1.4$

$X_{Cl} - X_C = 3.0 - 2.5 = 0.5$

$X_{Br} - X_{As} = 2.8 - 2.0 = 0.8$

The bonds arranged by increasing difference in electronegativity are C–Cl, As–Br, P–O.

9.59. The atom with the greater electronegativity has the partial negative charge.

a. P – O

δ+ δ–

b. C – Cl

δ+ δ–

c. As – Br

δ+ δ–

9.61. a. Total number of valence electrons = 7 + 7 = 14. Br–Br is the skeleton. Distribute the remaining 12 electrons.

:B̈r—B̈r:

b. Total valence electrons = (2 × 1) + 6 = 8. The skeleton is H–Se–H. Distribute the remaining 4 electrons.

H—S̈—H

c. Total valence electrons = (3 × 7) + 5 = 26. The skeleton is on the left below; after distributing the remaining 20 electrons the structure on the right results.

9.63. a. Total valence electrons = 2 × 5 = 10. The skeleton is P–P. Distribute the remaining electrons symmetrically:

:P̈—P̈:

Neither P atom has an octet. There are 4 fewer electrons than needed. This suggests the presence of a triple bond. Make one lone pair from each P a bonding pair.

:P≡P:

b. Total valence electrons = 4 + 6 + (2 × 7) = 24. The skeleton is

Br
|
O—C—Br

Distribute the remaining 18 electrons.

:B̈r:
|
:Ö—C—B̈r:

Note that carbon is two electrons short of an octet. This suggests the presence of a double bond. The most likely double bond is between C and O.

:B̈r:
|
:Ö=C—B̈r:

c. Total valence electrons = 1 + 5 + (2 × 6) = 18. The skeleton is most likely H–O–N–O. Distribute the remaining 12 electrons:

$$H—\ddot{\underset{\cdot\cdot}{O}}—\ddot{N}—\ddot{\underset{\cdot\cdot}{O}}:$$

Note that the N atom does not have an octet. It is two electrons short. The double bond resulting in the lowest formal charges is between N and the end O.

$$H—\ddot{\underset{\cdot\cdot}{O}}—\ddot{N}=\ddot{O}:$$

9.65. a. Total valence electrons = 7 + 6 + 1 = 14. The skeleton is Cl–O. Distribute the remaining 12 electrons.

$$\left[:\ddot{\underset{\cdot\cdot}{Cl}}—\ddot{\underset{\cdot\cdot}{O}}:\right]^-$$

b. Total valence electrons = 4 + (3 × 7) + 1 = 26. The skeleton is

```
        Cl
        |
Cl—Sn—Cl
```

Distribute the remaining 20 electrons so that each atom has an octet.

```
[       ..        ]-
[     : Cl :      ]
[        |        ]
[  ..             ]
[ : Cl — Sn — Cl : ]
[  ..    ..   ..  ]
```

c. Total valence electrons = (2 × 6) + 2 = 14. The skeleton is S–S. Distribute the remaining 12 electrons.

$$\left[:\ddot{\underset{\cdot\cdot}{S}}—\ddot{\underset{\cdot\cdot}{S}}:\right]^{2-}$$

9.67. a. There are two possible resonance structures for HNO_3.

```
    :O:                        ..
    ||                        :O:
    N      H    ←——→           |
   / \    /                    N      H
 :O:   :O:                   // \    /
                            O    :O:
```

One electron pair is delocalized over the nitrogen atom and the two oxygen atoms.

b. There are three possible resonance structures for SO_3.

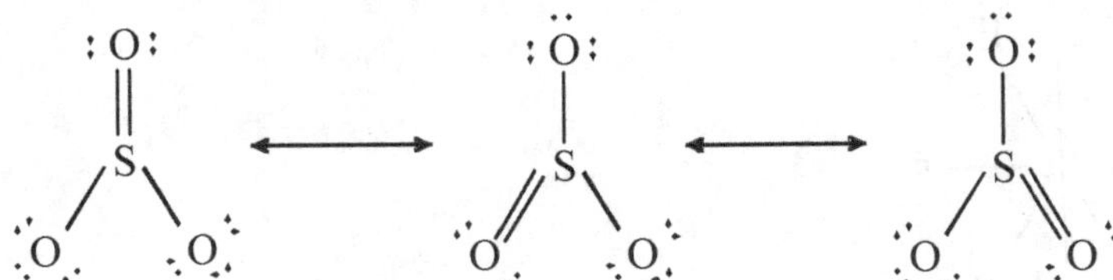

One pair of electrons is delocalized over the region of the three sulfur-oxygen bonds.

9.69.

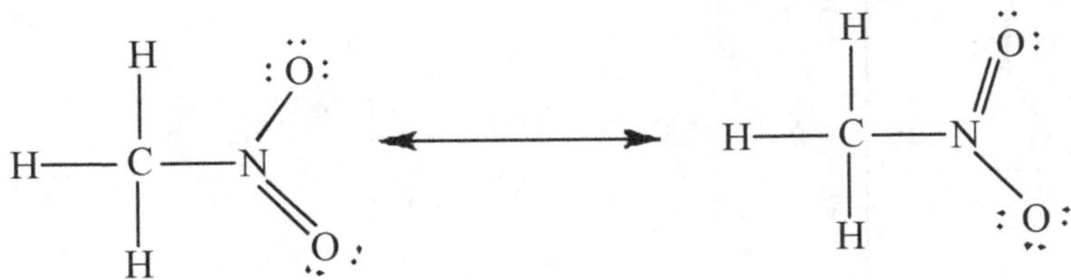

One pair of electrons is delocalized over the O–N–O bonds.

9.71. a. Total valence electrons = 8 + (2 × 7) = 22. The skeleton is F–Xe–F. Place six electrons around each fluorine atom to satisfy its octet.

:F̤̈—Xe—F̤̈:

There are three electron pairs remaining. Place them on the xenon atom.

:F̤̈—Ẍ̤e—F̤̈:

b. Total valence electrons = 6 + (4 × 7) = 34. The skeleton is below left; 24 electrons are then distributed to the F atoms with the last 2 electrons placed on the central Se atom (at right).

c. Total valence electrons = 6 + (6 × 7) = 48. The skeleton is below left. The structure on the right results after the remaining 36 electrons are distributed to the fluorine atoms.

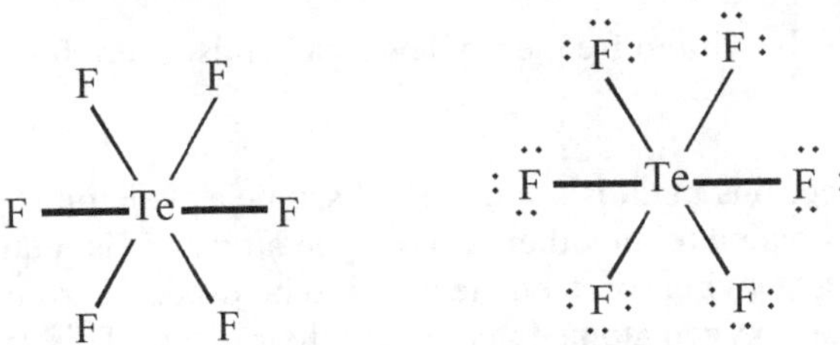

d. Total valence electrons = 8 + (5 × 7) − 1 = 42. The skeleton is

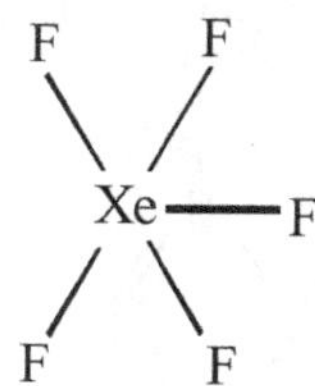

Use 30 of the remaining 32 electrons on the fluorine atoms to complete their octets. The remaining 2 electrons form a lone pair on the central Xe atom.

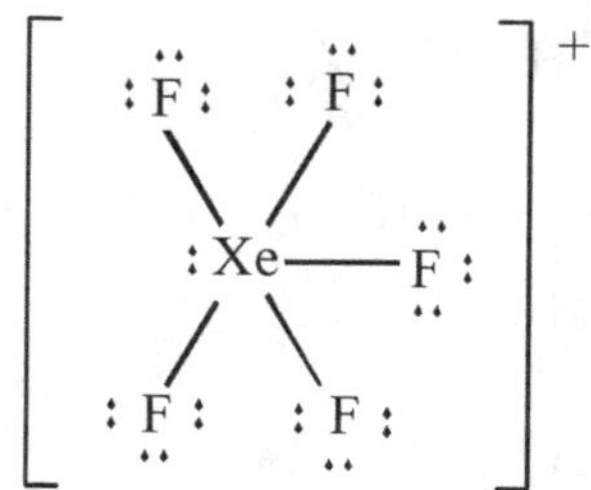

9.73. a. Total valence electrons = 3 + (3 × 7) = 24. The skeleton is below left. Distribute the remaining 18 electrons to the Cl atoms to get the structure on the right. Although boron has only 6 electrons, it has the normal number of covalent bonds.

b. Total valence electrons = 3 + (2 × 7) − 1 = 16. The skeleton is Cl–Tl–Cl. Distribute the remaining 12 electrons.

[:Cl: — Tl — :Cl:]$^+$

Tl has only 4 electrons around it.

c. Total valence electrons = 2 + (2 × 7) = 16. The skeleton is Br–Be–Br. Distribute the remaining 12 electrons.

:Br: — Be — :Br:

In covalent compounds, beryllium frequently has two bonds, even though it does not have an octet.

9.75. a. The total number of electrons in O_3 is 3 × 6 = 18. Assume a skeleton structure in which one oxygen atom is singly bonded to the other two oxygen atoms. This requires four electrons for the two single bonds, leaving fourteen electrons to be used. Distribute three electron pairs to each of the outer oxygen atoms to complete their octets. This requires twelve more electrons, leaving two electrons to distribute. Since this is two electrons short of an octet, move a pair of electrons from one of the outer oxygen atoms to give an oxygen-oxygen double bond. One of the possible resonance structures is shown below; the other structure would have the double bond written between the left and central oxygen atoms.

$$:\ddot{\underset{\cdot\cdot}{O}}-\ddot{O}=\ddot{\underset{\cdot\cdot}{O}}$$

Starting with the left oxygen, the formal charge of this oxygen is 6 − 1 − 6 = −1. The formal charge of just the central oxygen is 6 − 3 − 2 = +1. The formal charge of the right oxygen is 6 − 2 − 4 = 0. The sum of all three is 0.

b. The total number of electrons in CO is 4 + 6 = 10. Assume a skeleton structure in which the oxygen atom is singly bonded to the carbon atom. This requires two electrons for the single bond and leaves eight electrons. Distribute these remaining electrons around the atoms. Since four more electrons are needed to complete octets on both atoms, move two electron pairs to give a carbon-oxygen triple bond.

The structure is

$$:C\equiv O:$$

The formal charge of the carbon is 4 − 3 − 2 = −1. The formal charge of the oxygen is 6 − 3 − 2 = +1. The sum of both is 0.

c. The total number of electrons in HNO_3 is 1 + 5 + 18 = 24. Assume a skeleton structure in which the nitrogen atom is singly bonded to two oxygen atoms and doubly bonded to one oxygen. This requires two electrons for the O–H single bond and leaves eight electrons to be used for the N bonds.

$$H-\ddot{\underset{\cdot\cdot}{O}}-\underset{\underset{:\underset{\cdot\cdot}{O}:}{|}}{N}=\underset{\cdot\cdot}{O}:$$

The formal charge of the nitrogen is 5 − 4 − 0 = +1. The formal charge of the hydrogen is 1 − 1 − 0 = 0. The formal charge of the oxygen bonded to the hydrogen is 6 − 2 − 4 = 0. The formal charge of the other singly bonded oxygen is 6 − 1 − 6 = −1. The formal charge of the doubly bonded oxygen is 6 − 2 − 4 = 0.

9.77. a. The total number of electrons in SOF_2 is 6 + 6 + 14 = 26. Assume a skeleton structure in which the sulfur atom is singly bonded to the two fluorine atoms. If a S–O single bond is assumed, 26 electrons are needed. However, there would be formal charges of +1 on the sulfur and −1 on the oxygen. Using a S=O double bond requires only 26 electrons and results in zero formal charge.

$$:\ddot{\underset{\cdot\cdot}{F}}-\underset{\underset{:O:}{\|}}{\ddot{S}}-\ddot{\underset{\cdot\cdot}{F}}:$$

The formal charge on each of the two fluorine atoms is 7 − 1 − 6 = 0. The formal charge on the oxygen is 6 − 2 − 4 = 0. The formal charge on the sulfur is 6 − 4 − 2 = 0.

b. The total number of electrons in H_2SO_3 is 2 + 6 + 18 = 26. Assume a skeleton structure in which the sulfur atom is singly bonded to the three oxygen atoms. Then form single bonds from the two hydrogen atoms to each of two oxygen atoms. If a S–O single bond is assumed, 26 electrons are needed. However, there would be formal charges of +1 on the sulfur and −1 on the oxygen. A S=O double bond also requires 26 electrons but results in zero formal charge on all atoms.

The formal charge of each hydrogen is 1 − 1 = 0. The formal charge of each oxygen bonded to a hydrogen is 6 − 2 − 4 = 0. The formal charge of the oxygen doubly bonded to the sulfur is 6 − 2 − 4 = 0. The formal charge of the sulfur is 6 − 4 − 2 = 0.

c. The total number of electrons in $HClO_2$ is 1 + 7 + 12 = 20. Assume a skeleton structure in which the chlorine atom is singly bonded to the two oxygen atoms, and the hydrogen is singly bonded to one of the oxygen atoms. If all Cl–O single bonds are assumed, 20 electrons are needed, but the chlorine exhibits a formal charge of +1 and one oxygen exhibits a formal charge of −1.

Hence, a pair of electrons on the oxygen without the hydrogen is used to form a Cl=O double bond.

H—O—Cl:
‖
:O:

The formal charge of hydrogen is 1 − 1 = 0. The formal charge of the oxygen bonded to hydrogen is 6 − 2 − 4 = 0. The formal charge of the oxygen that is not bonded to the hydrogen is 6 − 2 − 4 = 0. The formal charge of the chlorine is 7 − 3 − 4 = 0.

9.79. $r_F = 57$ pm

$r_P = 107$ pm

$d_{P\text{–}F} = r_F + r_P = 57\text{ pm} + 107\text{ pm} = 164\text{ pm}$

9.81. a. $d_{C\text{–}H} = r_C + r_H = 76\text{ pm} + 31\text{ pm} = 107\text{ pm}$

b. $d_{S\text{–}Cl} = r_S + r_{Cl} = 105\text{ pm} + 102\text{ pm} = 207\text{ pm}$

c. $d_{Br\text{–}Cl} = r_{Br} + r_{Cl} = 120\text{ pm} + 102\text{ pm} = 222\text{ pm}$

d. $d_{Si\text{–}O} = r_{Si} + r_O = 111\text{ pm} + 66\text{ pm} = 177\text{ pm}$

The calculated bond distances agree very well with the experimental values.

9.83.

Methylamine	147 pm	Single C–N bond is longer.
Acetonitrile	116 pm	Triple C–N bond is shorter.

9.85.

$$\begin{matrix} H & & H \\ & C{=}C & \\ H & & H \end{matrix} + H{-}Br \longrightarrow H{-}\overset{\overset{H}{|}}{\underset{\underset{H}{|}}{C}}{-}\overset{\overset{H}{|}}{\underset{\underset{H}{|}}{C}}{-}Br$$

In the reaction, a C=C double bond is converted to a C–C single bond. An H–Br bond is broken, and one C–H bond and one C–Br bond are formed.

$$\Delta H \cong BE(\text{C=C}) + BE(\text{H–Br}) - BE(\text{C–C}) - BE(\text{C–H}) - BE(\text{C–Br})$$

$$= (614 + 366 - 348 - 413 - 276)\text{ kJ} = -57\text{ kJ}$$

■ SOLUTIONS TO GENERAL PROBLEMS

9.87. a. Strontium is a metal, and oxygen is a nonmetal. The binary compound is likely to be ionic. Strontium, in Group IIA, forms Sr^{2+} ions; oxygen, from Group VIA, forms O^{2-} ions. The binary compound has the formula SrO and is named strontium oxide.

b. Carbon and bromine are both nonmetals; hence, the binary compound is likely to be covalent. Carbon usually forms four bonds, and bromine usually forms one bond. The formula for binary compound is CBr_4. It is called carbon tetrabromide.

c. Gallium is a metal, and fluorine is a nonmetal. The binary compound is likely to be ionic. Gallium is in Group IIIA and forms Ga^{3+} ions. Fluorine is in Group VIIA and forms F^- ions. The binary compound is GaF_3 and is named gallium(III) fluoride.

d. Nitrogen and bromine are both nonmetals; hence, the binary compound is likely to be covalent. Nitrogen usually forms three bonds, and bromine usually forms one bond. The formula for the binary compound is NBr_3. It is called nitrogen tribromide.

9.89. Total valence electrons = 6 + (3 × 6) + 2 = 26. The skeleton is

$$\begin{matrix} & O & \\ & | & \\ O{-} & Se & {-}O \end{matrix}$$

Distribute the remaining 20 electrons to complete the octets of oxygen and selenium atoms.

$$\left[\begin{matrix} & :\ddot{O}: & \\ & | & \\ :\ddot{\underset{..}{O}}{-} & \underset{..}{Se} & {-}\ddot{\underset{..}{O}}: \end{matrix}\right]^{2-} \text{ or } \left[\begin{matrix} & :O: & \\ & \| & \\ :\ddot{\underset{..}{O}}{-} & \underset{..}{Se} & {-}\ddot{\underset{..}{O}}: \end{matrix}\right]^{2-}$$

The formula for aluminum selenite is $Al_2(SeO_3)_3$.

9.91. Total valence electrons = 1 + 7 + (3 × 6) = 26. The skeleton is

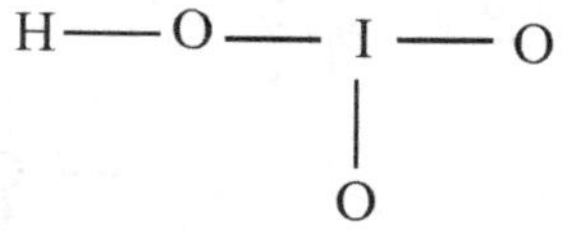

H—O—I—O
|
O

Distribute the remaining 18 electrons to satisfy the octet rule. The structure on the right has no formal charge.

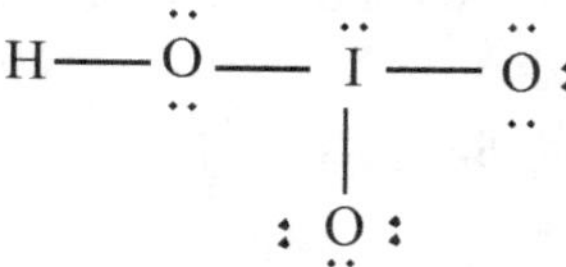

or

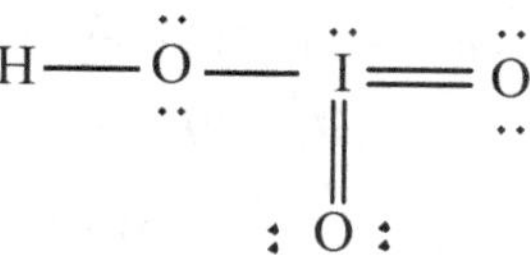

9.93. Total valence electrons = 5 + (2 × 1) + 1 = 8. The skeleton is H–N–H. Distribute the remaining 4 electrons to complete the octet of the nitrogen atom.

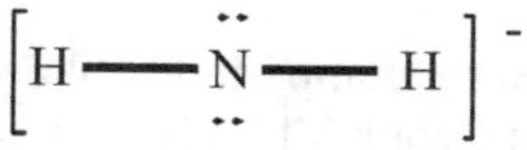

$[H—N—H]^-$

9.95. Total valence electrons = 5 + (2 × 6) − 1 = 16. The skeleton is O–N–O. Distribute the remaining electrons on the oxygen atoms.

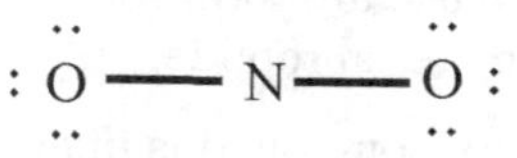

:O—N—O:

The nitrogen atom is short four electrons. Use two double bonds, one with each oxygen.

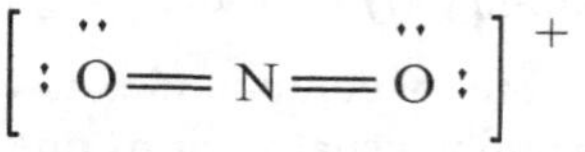

$[:O=N=O:]^+$

9.97.

a. 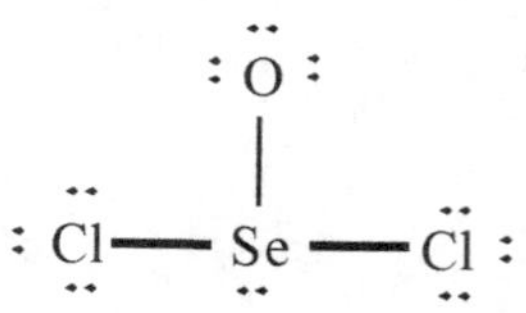 :Cl—Se(—O:)—Cl: or :Cl—Se(=O:)—Cl:

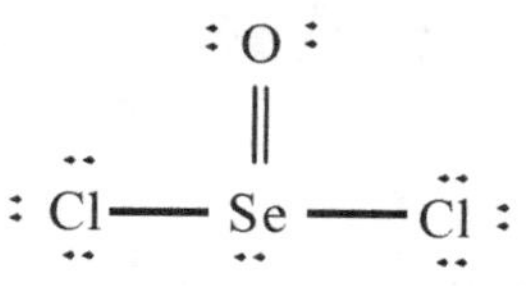

b. 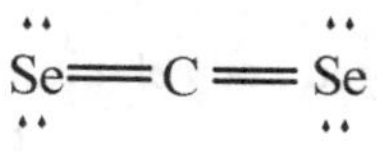 Se=C=Se

c.

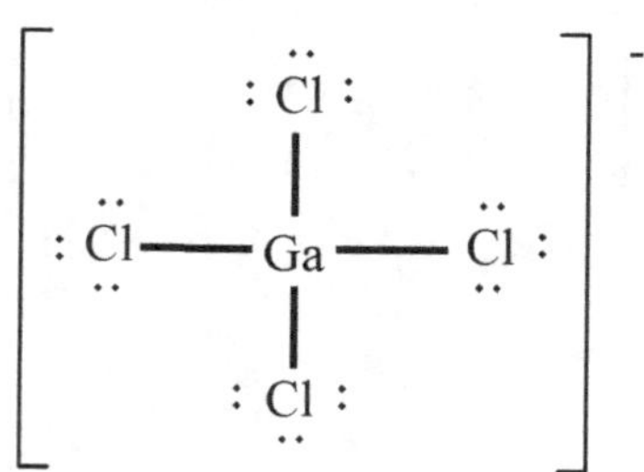

d. 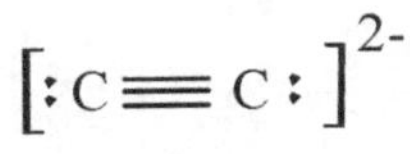 $[:C\equiv C:]^{2-}$

9.99. a. Total valence electrons = 5 + (3 × 7) = 26. The skeleton is

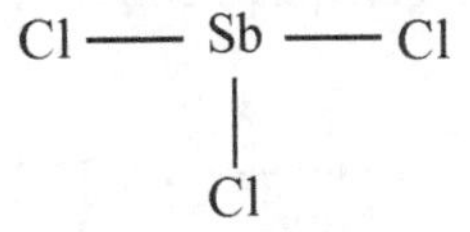

```
Cl — Sb — Cl
      |
      Cl
```

Distribute the remaining 20 electrons to the chlorine atoms and the antimony atom to complete their octets.

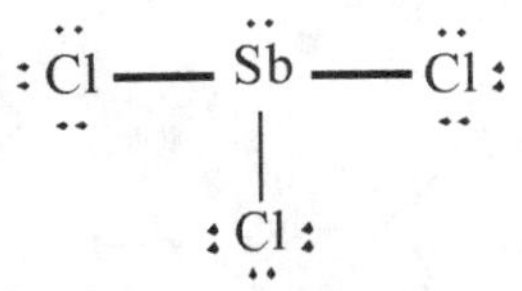

```
 ..    ..    ..
:Cl — Sb — Cl:
 ..    |    ..
      :Cl:
       ..
```

b. Total valence electrons = 7 + 4 + 5 = 16. The skeleton is I–C–N. Distribute the remaining 12 electrons.

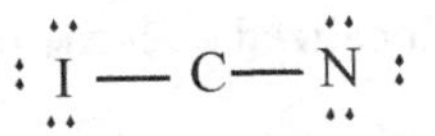

```
 ..       ..
:I — C — N :
 ..       ..
```

Notice the carbon atom is four electrons short of an octet. Make a triple bond between C and N from four nonbonding electrons on the nitrogen.

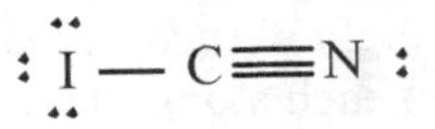

```
 ..
:I — C≡N :
 ..
```

c. Total valence electrons = 7 + (3 × 7) = 28. The skeleton is

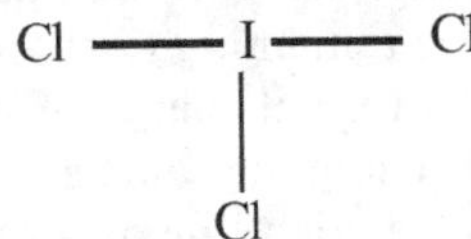

Distribute 18 of the remaining 22 electrons to complete the octets of the chlorine atoms. The 4 remaining electrons form two sets of lone pairs on the central iodine atom.

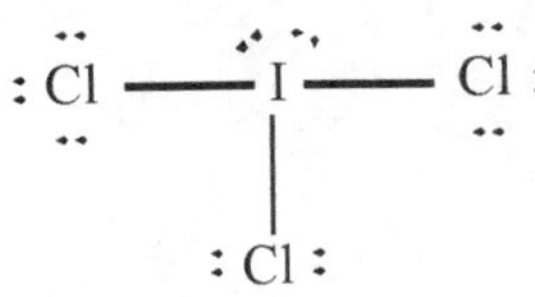

d. Total valence electrons = 7 + (5 × 7) = 42. The skeleton is below left. Use 30 of the remaining 32 electrons to complete the octets of the F atoms. The 2 electrons remaining form a lone pair on the central iodine atom (structure on right):

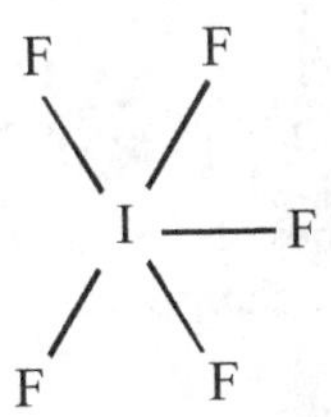

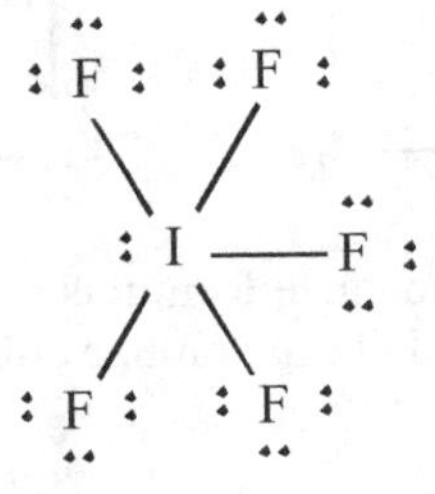

9.101. a. One possible electron-dot structure is

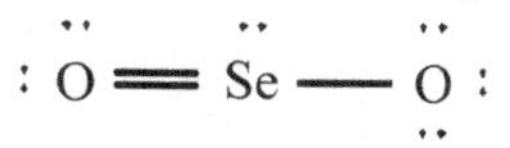

$$:\ddot{O}=\ddot{Se}-\ddot{O}:$$

Because the selenium-oxygen bonds are expected to be equivalent, the structure must be described in resonance terms.

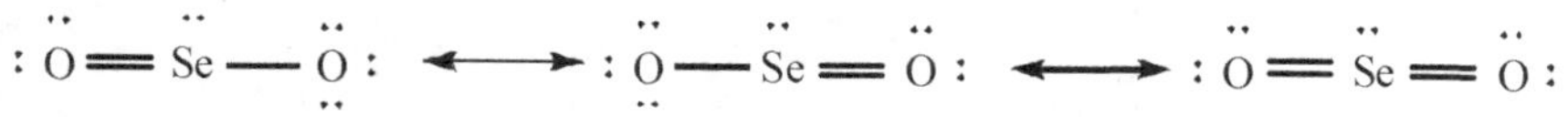

$$:\ddot{O}=\ddot{Se}-\ddot{O}: \longleftrightarrow :\ddot{O}-\ddot{Se}=\ddot{O}: \longleftrightarrow :\ddot{O}=Se=\ddot{O}:$$

The electrons are delocalized over the selenium atom and the two oxygen atoms.

b. The possible electron-dot structures are

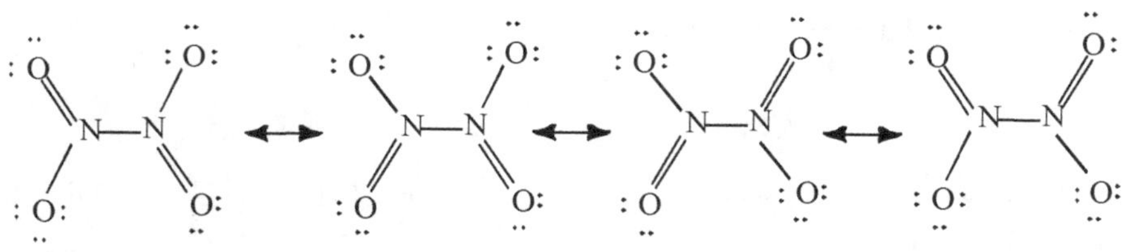

$$O=N(-O)-N(-O)=O \longleftrightarrow O-N(=O)-N(-O)=O \longleftrightarrow O-N(=O)-N(=O)-O \longleftrightarrow O=N(-O)-N(=O)-O$$

At each end of the molecule, a pair of electrons is delocalized over the region of the nitrogen atom and the two oxygen atoms.

9.103. The compound S_2N_2 will have a four-membered ring structure. Calculate the total number of valence electrons of sulfur and nitrogen, which will be 6 + 6 + 5 + 5 = 22. Writing the skeleton of the four-membered ring with single S–N bonds will use up eight electrons, leaving fourteen electrons for electron pairs. After writing an electron pair on each atom, there will be six electrons left for three electron pairs. No matter where these electrons are written, either a nitrogen or a sulfur will be left with less than eight electrons. This suggests one or more double bonds are needed. If only one S=N double bond is written, writing the remaining twelve electrons as six electron pairs will give the sulfur in the S=N double bond a formal charge of 1+ (see second line of formulas on next page). Writing two S=N double bonds using the same sulfur for both double bonds will give a formal charge of zero for all four atoms (see first line of formulas). The first two resonance formulas below have a zero formal charge on all atoms. However, one of the sulfur atoms does not obey the octet rule.

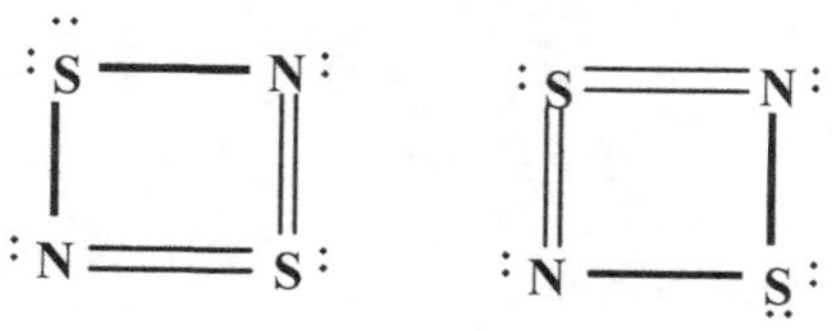

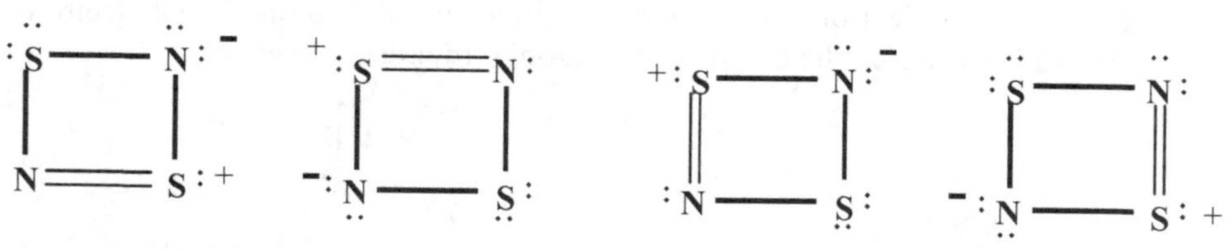

In the four resonance formulas at the bottom, all of the atoms obey the octet rule, but there is a positive formal charge on one sulfur atom and a negative formal charge on one nitrogen atom.

9.105. The possible electron-dot structures are

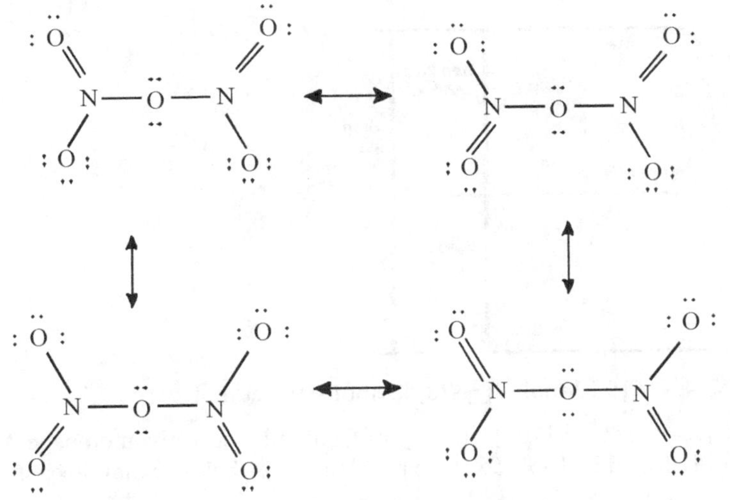

Because double bonds are shorter, the terminal N–O bonds that resonate between single and double bonds are 118 pm, and the central N–O single bonds are 136 pm.

9.107. $\Delta H = BE(\text{H–H}) + BE(\text{O=O}) - 2BE(\text{H–O}) - BE(\text{O–O}) = (436 + 498 - 2 \times 463 - 146)$ kJ $= -138$ kJ

9.109. $\Delta H = BE(\text{N=N}) + BE(\text{F–F}) - BE(\text{N–N}) - 2BE(\text{N–F}) = (418 + 159 - 163 - 2 \times 272)$ kJ $= -130.$ kJ

9.111. Ionic materials like NaCl are solids at room temperature, with high melting points. The molten liquid, which is clear, is very corrosive. Ionic materials consist of small, spherical ions that pack closely together. Thus, the ions interact strongly, giving a solid with a high melting point. Room-temperature ionic liquids, which are also clear, consist of large, nonspherical cations with various anions. The large, bulky cation keeps the ions from packing closely, yielding a substance with weak interactions and a low melting point.

9.113. The decomposition of nitroglycerin (see page 350) is given by

$$4C_3H_5(ONO_2)_3(l) \rightarrow 6N_2(g) + 12CO_2(g) + 10H_2O(g) + O_2(g)$$

The stability of the products results from their strong bonds, which are much stronger than those in nitroglycerin. In particular, nitrogen has a strong nitrogen-nitrogen triple bond and carbon dioxide has two strong carbon-oxygen double bonds.

9.115. A chemical bond acts like a stiff spring connecting nuclei in a molecule that vibrate relative to each other. The vibration of molecules is revealed in their absorption of infrared radiation. The frequency of radiation absorbed equals the frequencies of nuclear vibrations.

■ SOLUTIONS TO STRATEGY PROBLEMS

9.117. Structure a is incorrect. Al^{3+} ion has zero valence electrons. The rest are valid Lewis structures.

9.119.

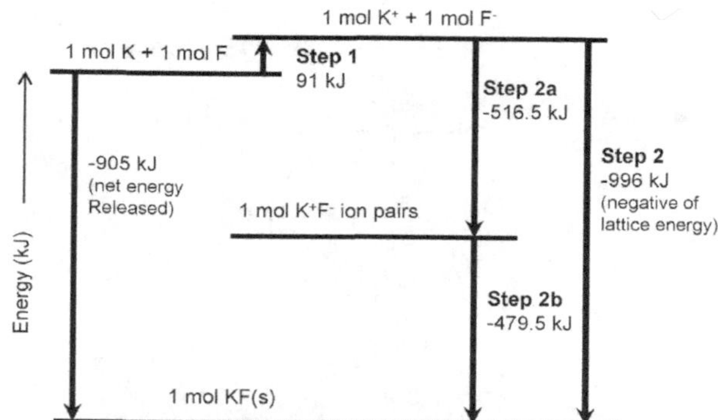

Step 1 = $E_{ionization}$ + $E.A.$ = 419 kJ/mol + (−328) kJ/mol = 91 kJ/mol

Step 2a = $E_{ion\ pair\ formation}$ = −516.5 kJ/mol (by way of Coulomb's law given on page 338 and using ionic radii for K^+ and F^- found in Table 9.3 (see problem 9.124 below example calculation)

Step 2b = $E_{ion\ pair\ solidification}$ = − ($|\Delta H_f^o(\text{KF})|$ + Step 1 − |Step 2a|) = −479.5 kJ/mol

The sublimation energy for K(s) → K(g) is obtained directly from Appendix C. It is the standard heat of formation of K(g), or 89.00 kJ/mol.

9.121. Both a (Ar) and d (Cl^-) are isoelectronic with the potassium ion, K^+. The electron configuration in each of these cases is [Ne]$3s^23p^6$.

9.123. a. Cl^- b. Ca^+ c. Ni^{2+} d. Ge^{2+} e. Br^-

9.125. $X_F - X_{Sr}$ = 4.0 − 1.0 = 3.0

$X_F - X_{Sn}$ = 4.0 − 1.8 = 2.2

The difference in electronegativities is higher in SrF_2. Thus, SrF_2 would be expected to be more ionic than SnF_2. Because ionic compounds are usually characterized with high melting points, the substantially lower melting point given in the problem statement likely corresponds to the melting point of SnF_2(tin(II) fluoride).

9.127. Only b (CO_2) contains only double bonds.

9.129. If M is Fe, then the electron configuration of X is [Ar]$3d^{10}4s^24p^4$. If M is Co, the electron configuration of X is exactly the same as for iron.

9.131. Total valence electrons = (3 × 1) + 5 + (4 × 6) = 32. The skeleton is

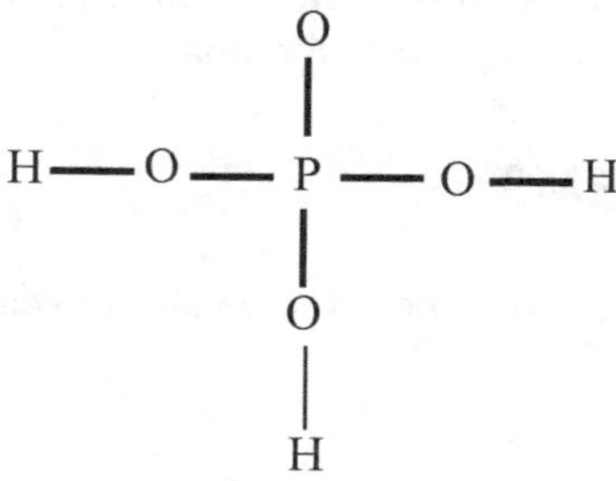

Distribute the remaining 18 electrons to satisfy the octet rule. The formula on the left below has the central phosphorus atom with a formal charge of 1+ and the top oxygen atom with a formal charge of 1−. (Each of the other atoms have zero formal charge.) By moving one lone pair from the top oxygen into a bonding position with the central phosphorus atom which can accommodate an expanded octet, the formula onthe right results. Because all of the atoms in the formula on the right have zero formal charges, this formula most closely approximates the actual electron distribution in phosphoric acid.

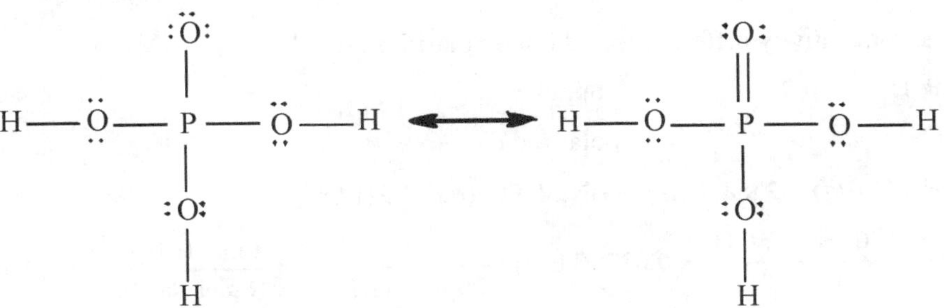

9.133. The reaction is $X_2 + Y_2 \rightarrow 2XY$. The enthalpy of the reaction is

$$\Delta H^\circ_{rxn} = 2\Delta H^\circ_f[XY] - \Delta H^\circ_f[X_2] - \Delta H^\circ_f[Y_2] = 2\Delta H^\circ_f[XY] - 0 - 0 = 2\Delta H^\circ_f[XY]$$

Bonding energies can also be used to approximate the enthalpy of the reaction as follows:

$$\Delta H^\circ_{rxn} \cong BE[X\text{–}X] + BE[Y\text{–}Y] - 2BE[X\text{–}Y]$$

Combining the two reaction enthalpy expressions gives

$$2\Delta H^\circ_f[XY] = BE[X\text{–}X] + BE[Y\text{–}Y] - 2BE[X\text{–}Y]$$

$$BE[X\text{–}Y] = \frac{\text{BE[X-X] + BE[Y-Y]} - 2\Delta \text{H}^\circ_f\text{[XY]}}{2 \text{ mol}} = \frac{414 \text{ kJ} + 159 \text{ kJ} - 2(-336 \text{ kJ})}{2 \text{ mol}}$$

$$BE[X\text{–}Y] = 62\underline{2}.5 = 623 \text{ kJ/mol}$$

9.135. The total number of electrons in HONO is $1 + (2 \times 6) + 5 = 18$. Assume a skeleton structure as given in the problem statement in which the nitrogen atom is bonded between the oxygen atoms and the hydrogen atom is bonded to an oxygen atom. This skeleton structure uses requires 6 electrons. To satisfy the octet on the N and O atoms, 14 electrons are required. However, only 12 electrons remain to be distributed. Therefore a double bond is required between the central N atom and one of the oxygen atoms. The two possible formulas are as follows:

H—Ö—N=Ö ⟷ H—Ö=N—Ö:
(left formula: O–N 146 pm, N=O 120 pm; right formula: left O ⊕, right O ⊖)

Each of the atoms in the left formula have formula charges equal to zero. The formula on the right has the left-most O atom carrying a formal charge of 1+ and the right-most O atomcarrying a formal charge of 1−. Thus, based on formal charges, the formula on the left likely approximates the electron distribution in HONO. Because double bonds are shorter than single bonds, the N=O double bond on the right is assigned the experimental bond length of 120 pm and the O–N single bond on the left is assigned the experimental bond length of 146 pm.

SOLUTIONS TO CUMULATIVE-SKILLS PROBLEMS

Note on significant figures: The final answer to each cumulative-skills problem is given first with one nonsignificant figure (the rightmost significant figure is underlined) and then is rounded to the correct number of figures. Intermediate answers usually also have at least one nonsignificant figure. Atomic weights, except for that of hydrogen, are rounded to two decimal places.

9.137. The electronegativity differences and bond polarities are

P–H	0.0	nonpolar
O–H	1.4	polar (acidic)

$H_3PO_3(aq) + 2NaOH(aq) \rightarrow Na_2HPO_3(aq) + 2H_2O(l)$

$$\frac{0.1250 \text{ mol NaOH}}{1 \text{ L}} \times 0.02250 \text{ L} \times \frac{1}{0.2000 \text{ L } H_3PO_3} \times \frac{1 \text{ mol } H_3PO_3}{2 \text{ mol NaOH}} = 0.00703\underline{1}2$$

$$= 0.007031\ M\ H_3PO_3$$

9.139. After assuming a 100.0-g sample, convert to moles:

$$10.9 \text{ g Mg} \times \frac{1 \text{ mol Mg}}{24.3 \text{ g Mg}} = 0.44\underline{8}5 \text{ mol Mg}$$

$$31.8 \text{ g Cl} \times \frac{1 \text{ mol Cl}}{35.453 \text{ g Cl}} = 0.89\underline{6}96 \text{ mol Cl}$$

$$57.3 \text{ g O} \times \frac{1 \text{ mol O}}{16.00 \text{ g O}} = 3.5\underline{8}1 \text{ mol O}$$

Divide by 0.4485:

$$\text{Mg: } \frac{0.4485}{0.4485} = 1; \text{ Cl: } \frac{0.89696}{0.4485} = 2.00; \text{ O: } \frac{3.581}{0.4485} = 7.98$$

The simplest formula is $MgCl_2O_8$. However, since $Cl_2O_8^{2-}$ is not a well-known ion, write the simplest formula as $Mg(ClO_4)_2$, magnesium perchlorate. The Lewis formulas are Mg^{2+} and

```
        ..                               
   [   :O:   ]-                [    ..    ]-
        |                          :O:
 ..     |     ..                ..  |   ..
:O — Cl — O:          or       O = Cl = O
 ..     |     ..                ..  ||  ..
       :O:                         :O:
   [    ..   ]                 [          ]
```

9.141. After assuming a 100.0-g sample, convert to moles:

$$25.0 \text{ g C} \times \frac{1 \text{ mol C}}{12.01 \text{ g C}} = 2.0\underline{8}1 \text{ mol C}$$

$$2.1 \text{ g H} \times \frac{1 \text{ mol H}}{1.008 \text{ g H}} = 2.\underline{0}8 \text{ mol H}$$

$$39.6 \text{ g F} \times \frac{1 \text{ mol F}}{18.99 \text{ g F}} = 2.0\underline{8}5 \text{ mol F}$$

$$33.3 \text{ g O} \times \frac{1 \text{ mol O}}{16.00 \text{ g O}} = 2.0\underline{8}1 \text{ mol O}$$

The simplest formula is CHOF. Because the molecular mass of 48.0 divided by the formula mass of 48.0 is one, the molecular formula is also CHOF. The Lewis formula is

```
H
 \       ..
  C ══ O :
 /
:F:
 ..
```

9.143. First, calculate the number of moles in one liter:

$$n = \frac{PV}{RT} = \frac{1.00 \text{ atm} \times 1.00 \text{ L}}{0.0821 \text{ L} \bullet \text{atm/ (K} \bullet \text{mol)} \times 424 \text{ K}} = 0.028\underline{7}27 \text{ mol}$$

$$MW = \frac{\text{g}}{\text{mol}} = \frac{7.49 \text{ g}}{0.028727 \text{ mol}} = 26\underline{0}.7 \text{ g/mol}$$

$$MW = 26\underline{0}.7 \text{ amu} = 118.71 \text{ amu Sn} + n \times (35.453 \text{ amu Cl})$$

$$n = \frac{260.7 - 118.71}{35.453} = 4.0\underline{0}50$$

The formula is $SnCl_4$. It is molecular because the electronegativity difference between Sn and Cl is 1.3, and because it is a liquid, and is volatile at 151°C. The Lewis formula is

```
          ..
         :Cl:
          |
  ..              ..
 :Cl ——  Sn  —— Cl:
  ..              ..
          |
         :Cl:
          ..
```

9.145.

$$\begin{array}{lccccccc} HCN(g) & \rightarrow & H(g) & + & C(g) & + & N(g) & \\ (135.1 & & 218.0 & & 716.7 & & 472.7) & \Delta H_f\ (\text{kJ/mol}) \end{array}$$

$\Delta H_{rxn} = 1272.3$ kJ/mol

$BE(C{\equiv}N) = \Delta H_{rxn} - BE(C–H) = [1272.3 - 411]$ kJ/mol $= 86\underline{1}.3$

$= 861$ kJ/mol (Table 9.5 has 891 kJ/mol.)

9.147. Use the O–H bond and its bond enthalpy of 463 kJ/mol to calculate X_O.

$BE(O–H) = 1/2[BE(H–H) + BE(O–O)] + k(X_O - X_H)^2$

463 kJ/mol = 1/2(436 kJ/mol + 146 kJ/mol) + 98.6 kJ $(X_O - X_H)^2$

Collecting the terms gives

$$\frac{463 - 291}{98.6} = (X_O - X_H)^2$$

Taking the square root of both sides gives

$1.3\underline{2}1 = 1.32 = (X_O - X_H)$

Assuming $X_H = 2.1$, $X_O = 2.1 + 1.32 = 3.\underline{4}2 = 3.4$.

9.149. $X = \dfrac{\text{I.E.} + \text{E.A.}}{2} = \dfrac{[1251 + 349]\ \text{kJ/mol}}{2} = 80\underline{0}.0$ kJ/mol

$$\frac{800.0\ \text{kJ/mol}}{2\underline{3}0\ \text{kJ/mol}} = 3.\underline{4}7 = 3.5\ (\text{Pauling's } X = 3.0)$$

CHAPTER 10

Molecular Geometry and Chemical Bonding Theory

■ SOLUTIONS TO EXERCISES

10.1. a. A Lewis structure of ClO_3^- is

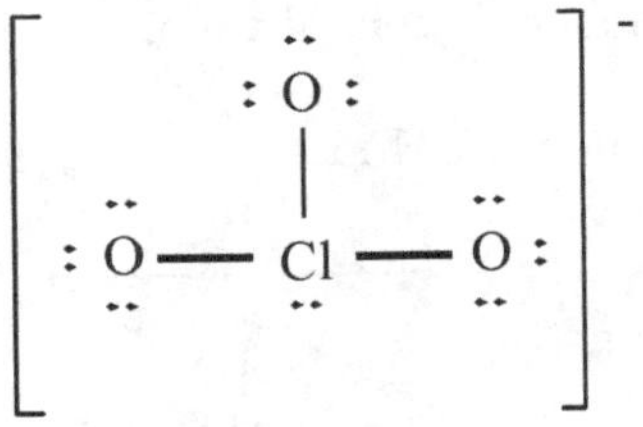

There are four electron pairs in a tetrahedral arrangement about the central atom. Three pairs are bonding, and one pair is nonbonding. The expected geometry is trigonal pyramidal.

b. The Lewis structure of OF_2 is

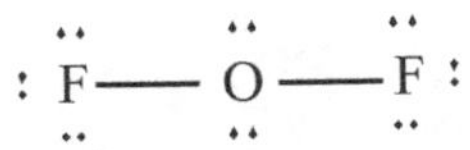

There are four electron pairs in a tetrahedral arrangement about the central atom. Two pairs are bonding, and two pairs are nonbonding. The expected geometry is bent.

c. The Lewis structure of SiF_4 is

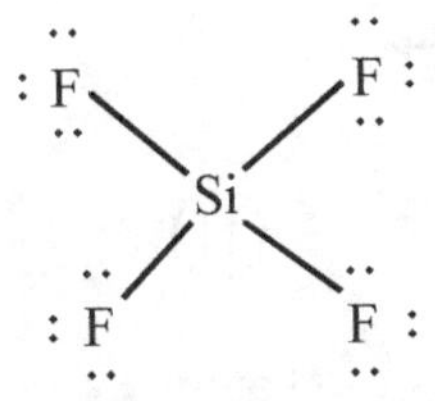

There are four bonding electron pairs in a tetrahedral arrangement around the central atom. The expected geometry is tetrahedral.

10.2. First, distribute the valence electrons to the bonds and the chlorine atoms. Then distribute the remaining electrons to iodine.

```
  ..     ..   ..
: Cl —   I  — Cl :
  ..     |    ..
         |
       : Cl :
         ..
```

The five electron pairs around iodine should have a trigonal bipyramidal arrangement with two lone pairs occupying equatorial positions. The molecule is T-shaped.

10.3. Both trigonal pyramidal (b) and T-shaped (c) geometries are consistent with a nonzero dipole moment. In trigonal planar geometry, the Br–F contributions to the dipole moment would cancel.

10.4. On the basis of symmetry, SiF_4 (b) would be expected to have a dipole moment of zero. The bonds are all symmetrical about the central atom.

10.5. The Lewis structure for ammonia, NH_3, is

```
     ..
H —  N  — H
     |
     H
```

There are four pairs of electrons around the nitrogen atom. According to the VSEPR model, these are arranged tetrahedrally around the nitrogen atom, and you should use sp^3 hybrid orbitals. Each N–H bond is formed by the overlap of a 1*s* orbital of a hydrogen atom with one of the singly occupied sp^3 hybrid orbitals of the nitrogen atom. This gives the following bonding description for NH_3:

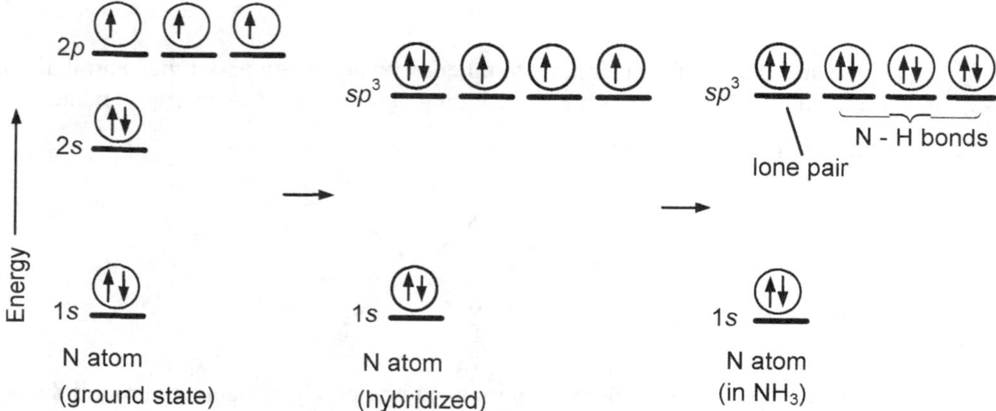

10.6. The Lewis structure for PCl_5 is

```
  ..      ..
: Cl :  : Cl :
     \   /
      P —— Cl :
     /   \
: Cl :  : Cl :
  ..      ..
```

The phosphorus atom has five single bonds and no lone pairs around it. This suggests that you use sp^3d hybrid orbitals on phosphorus. Each chlorine atom (valence-shell configuration $3s^23p^5$) has one singly occupied $3p$ orbital. The P–Cl bonds are formed by the overlap of a phosphorus sp^3d hybrid orbital with a singly occupied chlorine $3p$ orbital. The hybridization of phosphorus can be represented as follows:

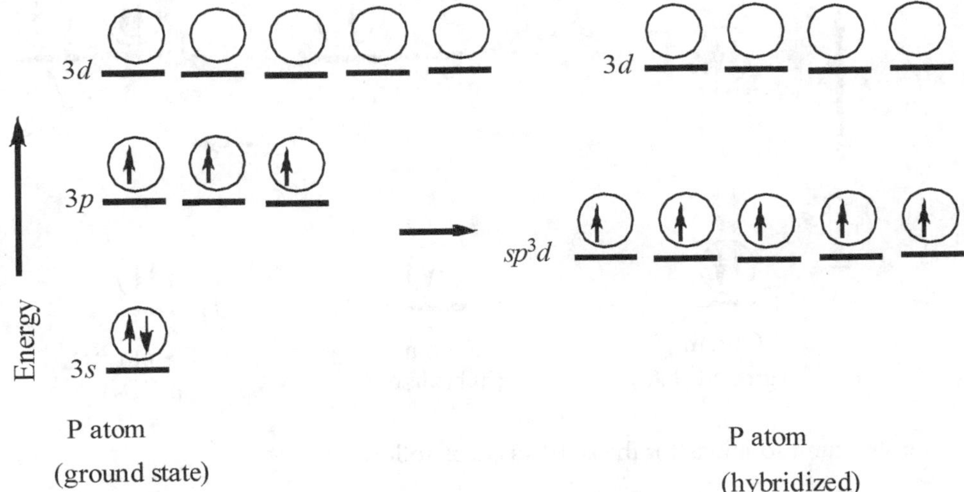

The bonding description of phosphorus in PCl_5 is

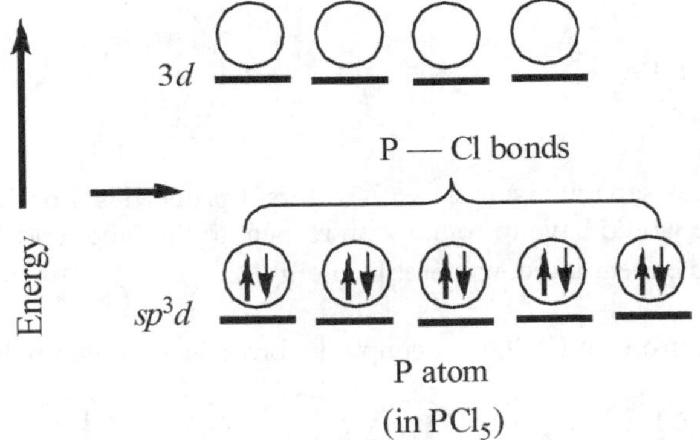

10.7. The Lewis structure of CO_2 is

$$\ddot{\underset{\cdot\cdot}{O}}=C=\ddot{\underset{\cdot\cdot}{O}}$$

The double bonds are each a π bond plus a σ bond. Hybrid orbitals are needed to describe the two σ bonds. This suggests *sp* hybridization. Each *sp* hybrid orbital on the carbon atom overlaps with a $2p$ orbital on one of the oxygen atoms to form a σ bond.

The π bonds are formed by the overlap of a $2p$ orbital on the carbon atom with a $2p$ orbital on one of the oxygen atoms. Hybridization and bonding of the carbon atom are shown as follows:

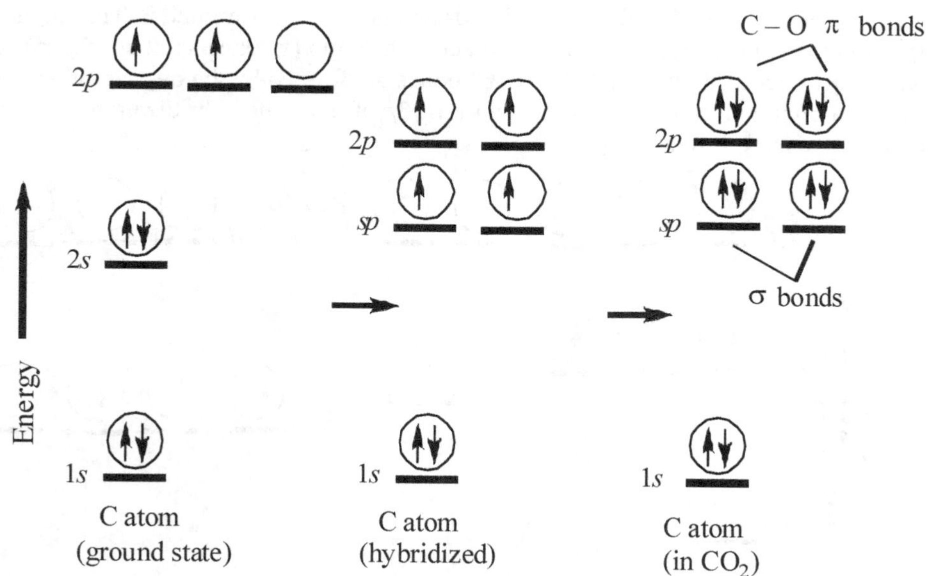

10.8. The structural formulas for the isomers are as follows:

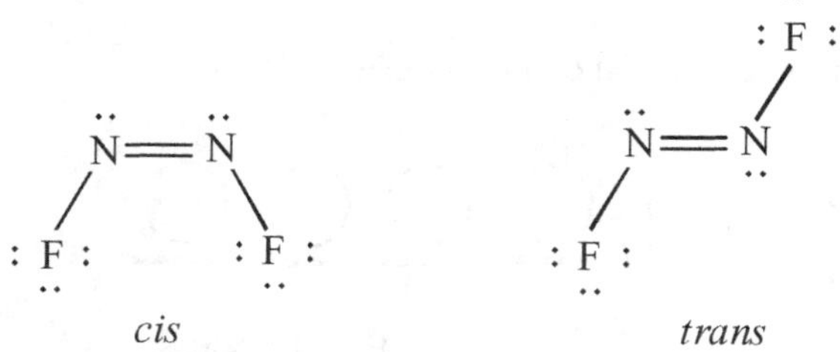

These compounds exist as separate isomers with different properties. For these to interconvert, one end of the molecule would have to rotate with respect to the other end. This would require breaking the π bond and expending considerable energy.

10.9. There are 2 × 6 = 12 electrons in C_2. They occupy the orbitals as shown below.

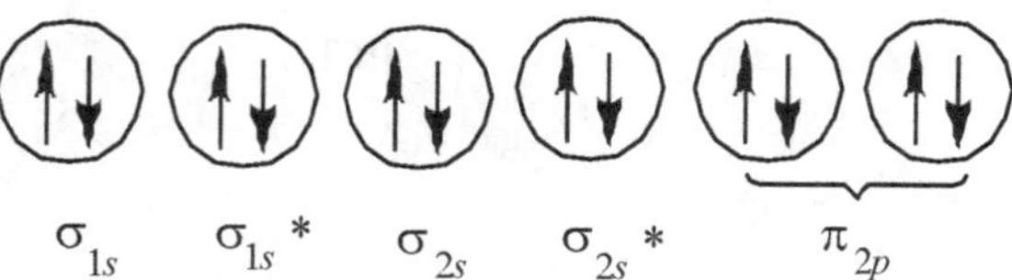

The electron configuration is $KK(\sigma_{2s})^2(\sigma_{2s}{}^*)^2(\pi_{2p})^4$. There are no unpaired electrons; therefore, C_2 is diamagnetic. There are eight bonding and four antibonding electrons. The bond order is ½(8 − 4) = 2.

10.10. There are 6 + 8 = 14 electrons in CO. The orbital diagram is

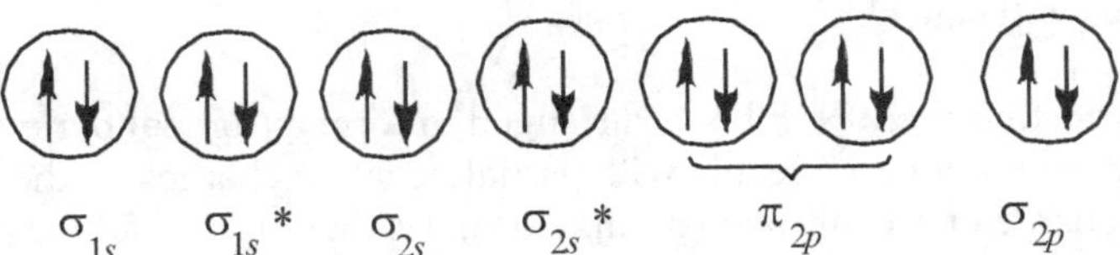

The electron configuration is $KK(\sigma_{2s})^2(\sigma_{2s}{*})^2(\pi_{2p})^4(\sigma_{2p})^2$. There are ten bonding and four antibonding electrons. The bond order is ½(10 − 4) = 3. There are no unpaired electrons; hence, CO is diamagnetic.

■ ANSWERS TO CONCEPT CHECKS

10.1. The VSEPR model predicts that four electron pairs about any atom in a molecule will distribute themselves to give a tetrahedral arrangement. Any three of these electron pairs would have a trigonal pyramidal arrangement. The geometry of a molecule having a central atom with three atoms bonded to it would be trigonal pyramidal.

10.2. A molecule, AX_3, could have one of three geometries: it could be trigonal planar, trigonal pyramidal, or T-shaped. Assuming the three groups attached to the central atom are alike, as indicated by the formula, the planar geometry should be symmetrical, so even if the A–X bonds are polar, their polarities would cancel to give a nonpolar molecule (dipole moment of zero). This would not be the case in the trigonal pyramidal geometry. In that situation, the bonds all point to one side of the molecule. It is possible for such a molecule to have a lone pair that points away from the bonds and whose polarity might fortuitously cancel the bond polarities, but an exact cancellation is not likely. In general, you should expect the trigonal pyramidal molecule to have a nonzero dipole moment, but a zero dipole is possible. The argument for the T-shaped geometry is similar to that for the trigonal pyramidal geometry. The bonds point in a plane, but toward one side of the molecule. Unless the sum of the bond polarities was fortuitously canceled by the polarities from the lone pairs, this geometry would have a nonzero dipole moment. This means that molecule Y is likely to be trigonal planar, but trigonal pyramidal or T-shaped geometries are possible. Molecule Z cannot have a trigonal planar geometry; it must be either trigonal pyramidal or T-shaped.

10.3. Assuming there are no lone pairs, the atom has four electron pairs and, therefore, an octet of electrons about it. The single bond and the triple bond each require a sigma bond orbital for a total of two such orbitals. This suggests *sp* hybrids on the central atom.

■ ANSWERS TO SELF-ASSESSMENT AND REVIEW QUESTIONS

10.1. The VSEPR model is used to predict the geometry of molecules. The electron pairs around an atom are assumed to arrange themselves to reduce electron repulsion. The molecular geometry is determined by the positions of the bonding electron pairs.

10.2. The arrangements are linear, trigonal planar, tetrahedral, trigonal bipyramidal, and octahedral.

10.3. A lone pair is "larger" than a bonding pair; therefore, it will occupy an equatorial position, where it encounters less repulsion than if it were in an axial position.

10.4. The bonds could be polar, but if they are arranged symmetrically, the molecule will be nonpolar. The bond dipoles will cancel.

10.5. Nitrogen trifluoride has three N–F bonds arranged to form a trigonal pyramid. These bonds are polar and would give a polar molecule with partial negative charges on the fluorine atoms and a partial positive charge on the nitrogen atom. However, there is also a lone pair of electrons on nitrogen that is directed away from the bonds. The result is that the lone pair nearly cancels the polarity of the bonds and gives a molecule with a very small dipole moment.

10.6. Certain orbitals, such as *p* orbitals and hybrid orbitals, have lobes in given directions. Bonding to these orbitals is directional; that is, the bonding is in preferred directions. This explains why the bonding gives a particular molecular geometry.

10.7. The angle is 109.5°.

10.8. A sigma bond has a cylindrical shape about the bond axis. A *pi* bond has a distribution of electrons above and below the bond axis.

10.9. In ethylene, C_2H_4, the changes on a given carbon atom may be described as follows:

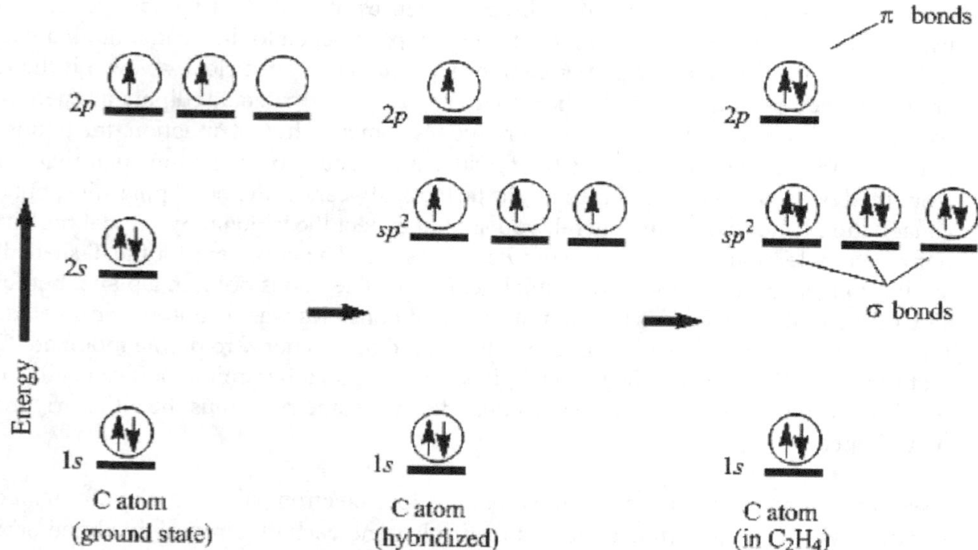

An sp^2 hybrid orbital on one carbon atom overlaps a similar hybrid orbital on the other carbon atom to form a σ bond. The remaining hybrid orbitals on the two carbon atoms overlap 1*s* orbitals from the hydrogen atoms to form four C–H bonds. The unhybridized 2*p* orbital on one carbon atom overlaps the unhybridized 2*p* orbital on the other carbon atom to form a π bond. The σ and π bonds together constitute a double bond.

10.10. Both of the unhybridized 2*p* orbitals, one from each carbon atom, are perpendicular to their CH_2 planes. When these orbitals overlap each other, they fix both planes in the same plane. The two ends of the molecule cannot twist around without breaking the π bond, which requires considerable energy. Therefore, it is possible to have stable molecules with the following structures:

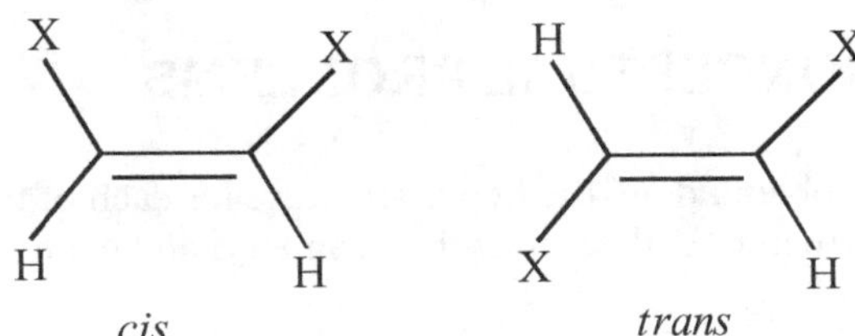

Because these have the same molecular formulas, they are isomers. In this case, they are called *cis-trans* isomers, or geometrical isomers.

10.11. In a bonding orbital, the probability of finding electrons between the two nuclei is high. For this reason, the energy of the bonding orbital is lower than that of the separate atomic orbitals. In an antibonding orbital, the probability of finding electrons between the two nuclei is low. For this reason, the energy of the antibonding orbital is higher than that of the separate atomic orbitals.

10.12. The factors determining the strength of interaction of two atomic orbitals are (1) the energy difference between the interacting orbitals and (2) the magnitude of their overlap.

10.13. When two $2s$ orbitals overlap, they interact to form a bonding orbital, σ_{2s}, and an antibonding orbital, σ_{2s}*. The bonding orbital is at lower energy than the antibonding orbital.

10.14. When two $2p$ orbitals overlap along their axes, they interact to form one σ_{2p} bonding orbital and one σ_{2p}* antibonding orbital. When they overlap sideways, they form π_{2p} and π_{2p}* molecular orbitals.

10.15. A σ bonding orbital is formed by the overlap of the $1s$ orbital on the H atom with the $2p$ orbital on the F atom. This H–F orbital is made up primarily of the fluorine orbital.

10.16. The O_3 molecule consists of a framework of localized orbitals and of delocalized *pi* molecular orbitals. The localized framework is formed from sp^2 hybrid orbitals on each atom. Thus, an O–O bond is formed by the overlap of a hybrid orbital on the left O atom with a hybrid orbital on the center O atom. Another O–O bond is formed by the overlap of another hybrid orbital on the right O atom with a hybrid orbital on the center O atom. The remaining hybrid orbitals are occupied by lone pairs of electrons. Also, there is one unhybridized *p* orbital on each of the atoms. These *p* orbitals are perpendicular to the plane of the molecule and overlap sidewise to give three pi molecular orbitals that are delocalized. The two orbitals of lowest energy are occupied by pairs of electrons.

10.17. The answer is e, a little less than 109.5°.

10.18. The answer is d, square pyramidal.

10.19. The answer is b, H_2S.

10.20. The answer is e, 5/2.

ANSWERS TO CONCEPTUAL PROBLEMS

10.23. In order to solve this problem, draw the Lewis structure for each of the listed molecules. In each case, use your Lewis structure to determine the geometry, and match this geometry with the correct model.

a. SeO_2 is angular and has an AX_2 geometry. This is represented by model (ii).

b. $BeCl_2$ is linear and has an AX_2 geometry. This goes with model (i).

c. PBr_3 is trigonal pyramidal and has an AX_3 geometry. This goes with model (iv).

d. BCl_3 is trigonal planar and has an AX_3 geometry. This is represented by model (iii).

10.25. In a CH_3CH_3 molecule, each C atom has four electron pairs arranged tetrahedrally. Within this molecule, each CH_3 considered as a separate group has a trigonal pyramidal geometry (with three C–H bonding pairs and a fourth pair from the C–C bond around the C atom). The :CH3 molecule retains this trigonal pyramidal geometry, having three bonding pairs and one lone pair around the C atom. The CH_3 molecule, however, has only three electron pairs around the C atom. Initially, as the CH_3 molecule breaks away from the ethane molecule, it has the trigonal pyramidal geometry it had in the ethane molecule. However, the repulsions of the bonding electron pairs on the CH_3 molecule are no longer balanced by the fourth pair (from the C–C bond), so the molecule flattens out to form a trigonal planar geometry.

10.27. First, determine the geometry of each molecule using VSEPR theory. Then compare to the orbital pictures for the correct number of bonding and nonbonding orbitals.

a. BeF_2 is an AX_2 molecule having linear geometry depicted by orbital drawing (i).

b. SiF_4 is an AX_4 molecule having tetrahedral geometry depicted by orbital drawing (iii).

c. SeF_4 is an AX_4 molecule having seesaw geometry depicted by orbital drawing (iv).

d. RnF_4 is an AX_4 molecule with square planar geometry depicted by orbital drawing (v).

10.29. The reaction with Br_2 indicates that $C_2H_2Br_2$ has a double bond. There are three possible isomers of $C_2H_2Br_2$ having double bonds:

I II III

Compounds II and III have dipole moments. The addition of Br_2, with one Br going to each C atom, yields the following products:

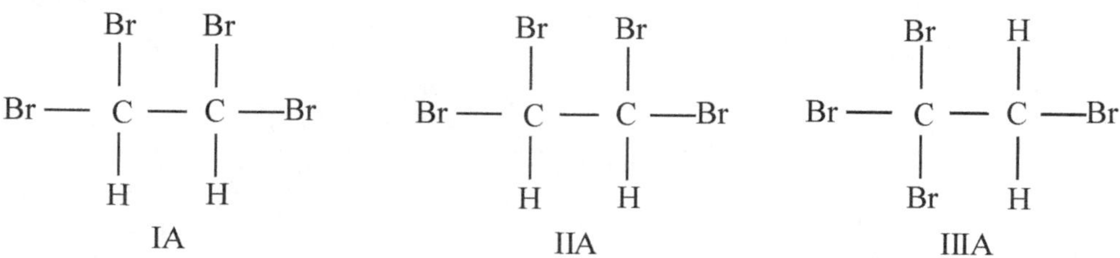

Products IA and IIA are identical and arise from compounds I and II. Thus, the original two compounds, one not having a dipole moment, the other having a dipole moment, but both reacting with bromine to give the same product, are compounds I and II, respectively.

10.31. First, determine the geometry about atoms a, b, and c using VSEPR theory. Then compare to the geometry in each drawing.

a. The bonding configuration about atom a (carbon) is AX_4 with tetrahedral geometry. This is depicted by drawing (ii).

b. The bonding configuration about atom b (carbon) is AX_3 with trigonal planar geometry. This is depicted by drawing (iv).

c. The bonding configuration about atom c (oxygen) is AX_2 with bent geometry. This is depicted by drawing (i).

SOLUTIONS TO PRACTICE PROBLEMS

10.33. The number of electron pairs, number of lone pairs, and geometry are for the central atom.

Electron-Dot Structure	General Formula	Number of Electron Pairs	Number of Lone Pairs	Geometry
SiF_4: Si bonded to four F, each F with three lone pairs	AX_4	4	0	Tetrahedral
SF_2: S with two lone pairs, bonded to two F, each F with three lone pairs	AX_4	4	2	Bent
COF_2: C double-bonded to O (two lone pairs), single-bonded to two F, each F with three lone pairs	AX_3	3	0	Trigonal planar
PCl_3: P with one lone pair, bonded to three Cl, each Cl with three lone pairs	AX_4	4	1	Trigonal pyramidal

10.35. The number of electron pairs, number of lone pairs, and geometry are for the central atom.

Electron-Dot Structure	General Formula	Number of Electron Pairs	Number of Lone Pairs	Geometry
$[ClO_3]^-$	AX_4	4	1	Trigonal pyramidal
$[PO_4]^{3-}$	AX_4	4	0	Tetrahedral
$[SCN]^-$	AX_2	2	0	Linear
$[H_3O]^+$	AX_4	4	1	Trigonal pyramidal

10.37. a. CCl_4 is tetrahedral, VSEPR predicts that the bond angles will be 109°, and you would expect them to be this angle.

b. SCl_2 is bent, and VSEPR theory predicts bond angles of 109°, but you would expect the Cl–S–Cl bond angle to be less.

c. $COCl_2$ is trigonal planar. VSEPR theory predicts 120° bond angles. The C–O bond is a double bond, so you would expect the Cl–C–O bond to be more than 120° and the Cl–C–Cl bond to be less than 120°.

d. AsH_3 is trigonal pyramidal, and VSEPR theory predicts bond angles of 109°, but you would expect the H–As–H bond angle to be less.

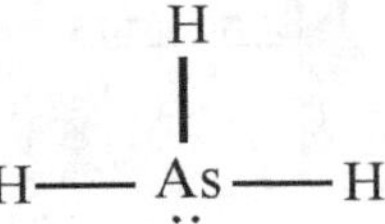

10.39. The number of electron pairs, number of lone pairs, and geometry are for the central atom.

Electron-Dot Structure	General Formula	Number of Electron Pairs	Number of Lone Pairs	Geometry
F, F, F—P, F, F	AX_5	5	0	Trigonal bipyramidal
F—Br—F, F	AX_5	5	2	T-shaped
F, F, F, Br, F, F	AX_6	6	1	Square pyramidal
Cl, Cl, S, Cl, Cl	AX_5	5	1	Seesaw

10.41. The number of electron pairs, number of lone pairs, and geometry are for the central atom.

Electron-Dot Structure	General Formula	Number of Electron Pairs	Number of Lone Pairs	Geometry
[:Cl: / :Cl—Sn—Cl: / Cl: / :Cl:]$^-$	AX_5	5	0	Trigonal bipyramidal
[:F: / :F—P—F: / :F—P—F: / :F:]$^-$	AX_6	6	0	Octahedral
[:F—Cl—F:]$^-$	AX_5	5	3	Linear
[:F—I—F: / :F—I—F:]$^-$	AX_6	6	2	Square planar

10.43. a. Trigonal pyramidal and T-shaped. Trigonal planar would have a dipole moment of zero.

b. Bent. Linear would have a dipole moment of zero.

10.45. a. CS_2 has a linear geometry and has no dipole moment.

b. TeF_2 has a bent geometry and has a dipole moment.

c. $SeCl_4$ has a seesaw geometry and has a dipole moment.

d. XeF_4 has a square planar geometry and has no dipole moment.

10.47. a. SF_2 is a bent molecule. The sulfur atom is sp^3 hybridized.

b. ClO_3^- is a trigonal pyramidal molecular ion. The chlorine atom is sp^3 hybridized.

10.49. a. $SeCl_2$ is a bent molecule. The Se atom is sp^3 hybridized.

b. NO_2^- is a bent ion. The N atom is sp^2 hybridized.

c. CO_2 is a linear molecule. The C atom is sp hybridized.

d. COF_2 is a trigonal planar molecule. The C atom is sp^2 hybridized.

10.51. a. The Lewis structure is

```
 ..          ..
:Cl——Hg——Cl:
 ..          ..
```

The presence of two single bonds and no lone pairs suggests *sp* hybridization. Thus, the Hg atom with the configuration $[Xe]4f^{14}5d^{10}6s^2$ is promoted to $[Xe]4f^{14}5d^{10}6s^16p^1$, then *sp* hybridized. An Hg–Cl bond is formed by overlapping the Hg *sp* hybrid orbital with a 3*p* orbital of Cl.

b. The Lewis structure is

```
        ..
       :Cl:
        |
 ..     |     ..
:Cl —   P  — Cl:
 ..     ..    ..
```

The presence of three single bonds and one lone pair suggests sp^3 hybridization of the P atom. Three hybrid orbitals each overlap a 3*p* orbital of a Cl atom to form a P–Cl bond. The fourth hybrid orbital contains the lone pair.

10.53. a. Xenon has eight valence electrons. Each F atom donates one electron to give a total of ten electrons, or five electron pairs, around the Xe atom. The hybridization is sp^3d.

b. Bromine has seven valence electrons. Each F atom donates one electron to give a total of twelve electrons, or six electron pairs, around the Br atom. The hybridization is sp^3d^2.

c. Phosphorus has five valence electrons. The Cl atoms each donate one electron to give a total of ten electrons, or five electron pairs, around the P atom. The hybridization is sp^3d.

d. Chlorine has seven valence electrons, to which may be added one electron from each F atom minus one electron for the charge on the ion. This gives a total of ten electrons, or five electron pairs, around chlorine. The hybridization is sp^3d.

10.57. a. The structural formula of formaldehyde is

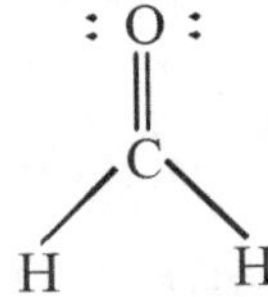

Because the C is bonded to three other atoms, it is assumed to be sp^2 hybridized. One 2*p* orbital remains unhybridized. The carbon–hydrogen bonds are σ bonds formed by the overlap of an sp^2 hybrid orbital on C with a 1*s* orbital on H. The remaining sp^2 hybrid orbital on C overlaps with a 2*p* orbital on O to form a σ bond. The unhybridized 2*p* orbital on C overlaps with a parallel 2*p* orbital on O to form a π bond. Together, the σ and π bonds constitute a double bond.

b. The nitrogen atoms are *sp* hybridized. A σ bond is formed by the overlap of an *sp* hybrid orbital from each N. The remaining *sp* hybrid orbitals contain lone pairs of electrons. The two unhybridized $2p$ orbitals on one N overlap with the parallel unhybridized $2p$ orbitals on the other N to form two π bonds.

10.59. Each of the N atoms has a lone pair of electrons and is bonded to two atoms. The N atoms are sp^2 hybridized. The two possible arrangements of the O atoms relative to one another are shown below. Because the π bond between the N atoms must be broken to interconvert these two forms, it is to be expected that the hyponitrite ion will exhibit *cis–trans* isomerism.

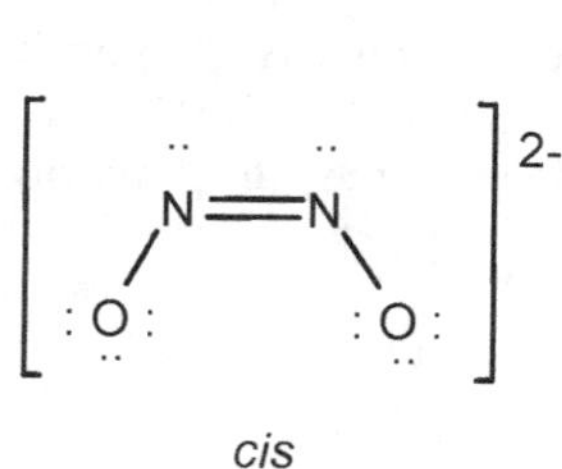

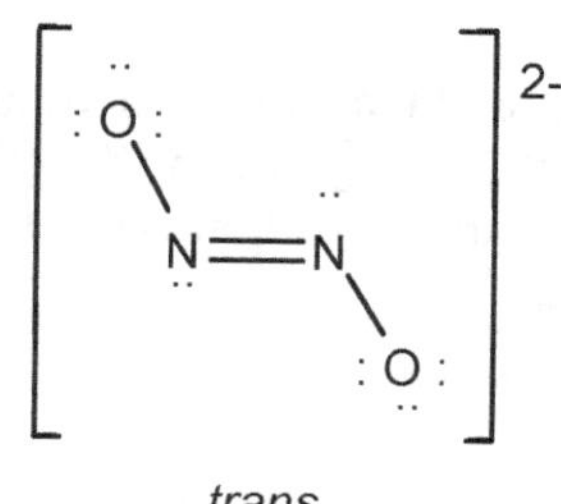

cis *trans*

10.61. a. Total electrons = 2 × 5 = 10.

The electron configuration is $KK(\sigma_{2s})^2(\sigma_{2s}*)^2(\pi_{2p})^2$.

Bond order = $\frac{1}{2}(n_b - n_a) = \frac{1}{2}(6 - 4) = 1$

The B_2 molecule is stable. It is paramagnetic because the two electrons in the π_{2p} subshell occupy separate orbitals.

b. Total electrons = (2 × 5) − 1 = 9.

The electron configuration is $KK(\sigma_{2s})^2(\sigma_{2s}*)^2(\pi_{2p})^1$.

Bond order = $\frac{1}{2}(n_b - n_a) = \frac{1}{2}(5 - 4) = 1/2$

The B_2^+ molecular ion should be stable and is paramagnetic because there is one unpaired electron in the π_{2p} subshell.

c. Total electrons = (2 × 8) + 1 = 17.

The electron configuration is $KK(\sigma_{2s})^2(\sigma_{2s}*)^2(\pi_{2p})^4(\sigma_{2p})^2(\pi_{2p}*)^3$.

Bond order = $\frac{1}{2}(n_b - n_a) = \frac{1}{2}(10 - 7) = 3/2$

The O_2^- molecular ion should be stable and is paramagnetic because there is one unpaired electron in the $\pi_{2p}*$ subshell.

10.63. Total electrons = 6 + 7 + 1 = 14.

The electron configuration is $KK(\sigma_{2s})^2(\sigma_{2s}*)^2(\pi_{2p})^4(\sigma_{2p})^2$.

Bond order = $\frac{1}{2}(n_b - n_a) = \frac{1}{2}(10 - 4) = 3$

The CN^- ion is diamagnetic.

SOLUTIONS TO GENERAL PROBLEMS

10.65. The number of electron pairs, number of lone pairs, and geometry are for the central atom.

Electron-Dot Structure	General Formula	Number of Electron Pairs	Number of Lone Pairs	Geometry
$\ddot{Sn}$ with :Cl: and :Cl: (Sn=Cl, Sn–Cl)	AX_3	3	1	Bent
:O: = C with :Br: and :Br:	AX_3	3	0	Trigonal planar
$[:\ddot{Cl}—\ddot{I}—\ddot{Cl}:]^-$	AX_5	5	3	Linear
$[P$ with six :Cl: $]^-$	AX_6	6	0	Octahedral

10.67. None of the given molecules or ions are linear.

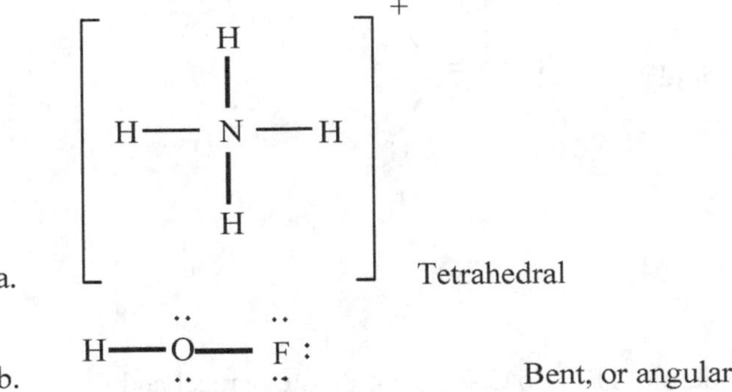

a. Tetrahedral

b. H—O—F: Bent, or angular

c. 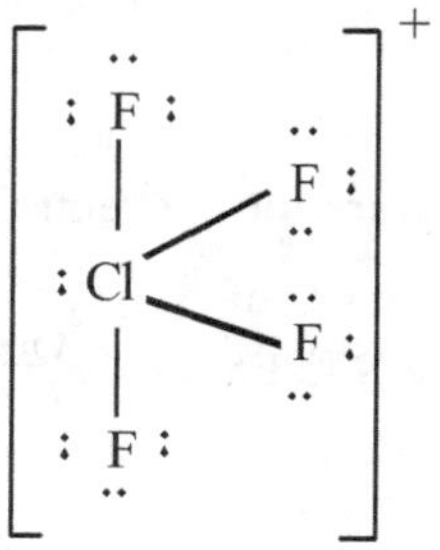 Seesaw

d. 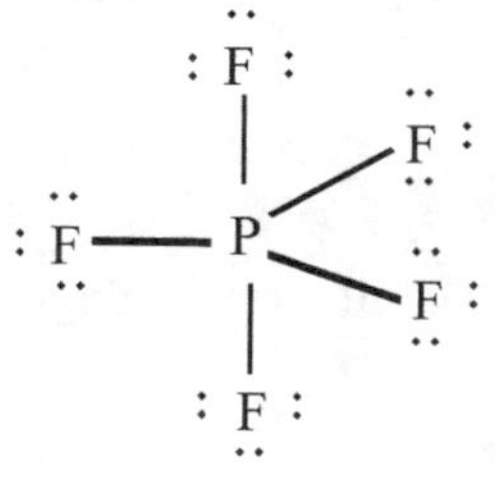 Trigonal bypyramidal

10.69.

a. 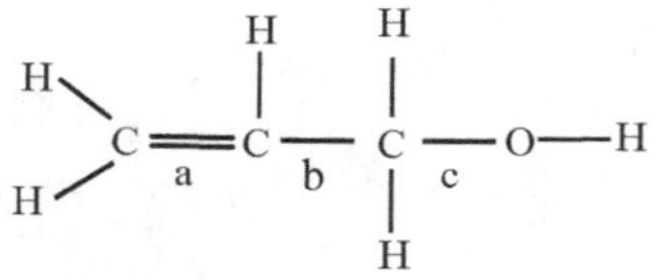

C_a and C_b: Three electron pairs around them. They are sp^2hybridized.

C_c: Four electron pairs around it. It is sp^3 hybridized.

b. 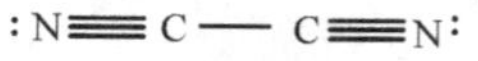

Both C atoms are bonded to two other atoms and have no lone pairs of electrons. They are *sp* hybridized.

10.71.

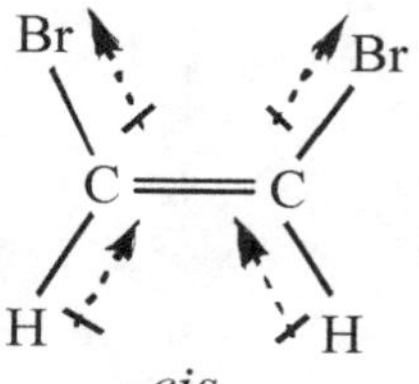

cis

has a net dipole

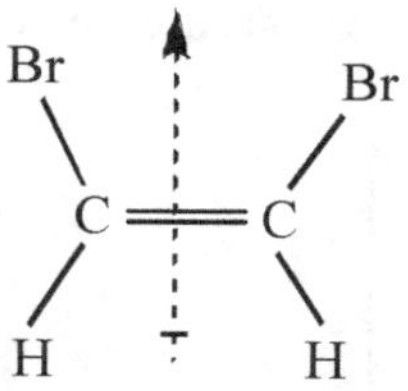

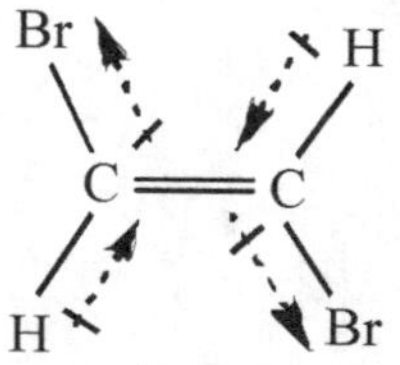

trans

has no net dipole. The two C–Br bond dipoles cancel and the two C–H bond dipoles cancel.

10.73. All four hydrogen atoms of $H_2C=C=CH_2$ cannot lie in the same plane because the second C=C bond forms perpendicular to the plane of the first C=C bond. By looking at Figure 10.26, you can see how this is so. The second C=C bond forms in the plane of the C–H bonds. Thus, the plane of the C–H bonds on the right side will be perpendicular to the plane of the C–H bonds on the left side. This is shown below using the π orbitals of the two C=C double bonds.

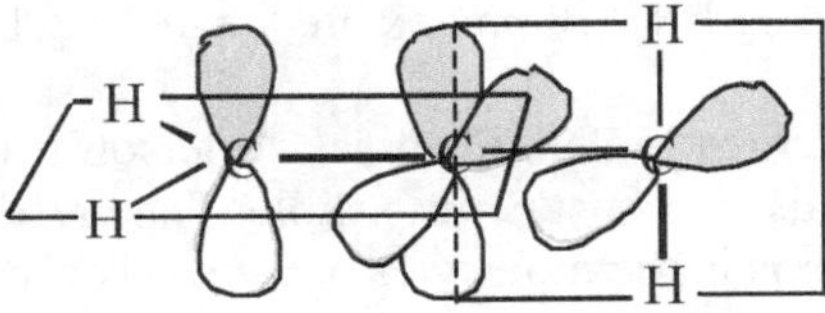

10.75. Total electrons = 2 + 1 − 1 = 2.

The electron configuration is $(\sigma_{1s})^2$.

Bond order = ½($n_b - n_a$) = ½(2) = 1

The HeH^+ ion is expected to be stable.

10.77. Total electrons = (2 × 6) + 2 = 14.

The electron configuration is $KK(\sigma_{2s})^2(\sigma_{2s}{*})^2(\pi_{2p})^4(\sigma_{2p})^2$.

Bond order = ½($n_b - n_a$) = ½(10 − 4) = 3

10.79. The molecular orbital configuration of O_2 is $KK(\sigma_{2s})^2(\sigma_{2s}{*})^2(\pi_{2p})^4(\sigma_{2p})^2(\pi_{2p}{*})^2$. O_2^+ has one electron less than O_2. The difference is in the number of electrons in the $\pi_{2p}{*}$ antibonding orbital. This means that the bond order is larger for O_2^+ than for O_2.

O_2: Bond order = ½($n_b - n_a$) = ½(10 − 6) = 2

O_2^+: Bond order = ½($n_b - n_a$) = ½(10 − 5) = 5/2

It is expected that the species with the higher bond order, O_2^+, has the shorter bond length. In O_2^-, there is one more electron than in O_2. This additional electron occupies a $\pi_{2p}{*}$ orbital. Increasing the number of electrons in antibonding orbitals decreases the bond order; hence, O_2^- should have a longer bond length than O_2.

10.81. As shown in Figure 10.34, the occupation of molecular orbitals by the N_2 valence electrons is

$$KK(\sigma_{2s})^2(\sigma_{2s}{*})^2(\pi_{2p})^4(\sigma_{2p})^2(\pi_{2p}{*})^0$$

To form the first excited state of N_2, a σ_{2p} electron is promoted to the $\pi_{2p}{*}$ orbital, giving

$$KK(\sigma_{2s})^2(\sigma_{2s}{*})^2(\pi_{2p})^4(\sigma_{2p})^1(\pi_{2p}{*})^1$$

The differences in properties are as follows:

Magnetic character: The ground state is diamagnetic (all electrons paired), and the excited state is paramagnetic.

Bond order: ground state order = ½(8 – 2) = 3

excited state order = ½(7 – 3) = 2

Bond dissociation energy: ground-state energy = 942 kJ; excited-state energy is less than 942 kJ (excited state does not possess a stable triple bond like the ground state)

Bond length: ground-state length = 110 pm; excited-state length is more than 110 pm.

10.83. A person's two hands are similar but not identical. A person's left hand is a mirror image of the other hand, yet the two hands are nonsuperimposible. The presence of handedness in molecules depends on the fact that atoms in the molecules occupy specific places in 3-D space.

10.85. Color vision is possible because three types of cone cells exist: one type absorbs light in the red region of the spectrum, another absorbs in the green region, and a third absorbs in the blue region. The chemical substance primarily responsible for human vision is 11-*cis*-retinal. In each of the different cone cells, 11-*cis*-retinal is attached to a different protein molecule, which affects the region of light that is absorbed by retinal.

10.87. Ozone in the stratosphere absorbs ultraviolet radiation between 200 and 300 nm, which is important to life on earth. Radiation from the sun contains ultraviolet rays that are harmful to the DNA of biological organisms. Oxygen, O_2, absorbs the most energetic of these ultraviolet rays in the earth's upper atmosphere, but only ozone in the stratosphere absorbs the remaining ultraviolet radiation that is destructive to life on earth.

■ SOLUTIONS TO STRATEGY PROBLEMS

10.89. N_2: Triple bond; bond length = 110 pm. Geometry is linear; *sp* hybrid orbitals are needed for one lone pair and one σ bond.

N_2F_2: Double bond; bond length = 122 pm. Geometry is trigonal planar; sp^2 hybrid orbitals are needed for one lone pair and two σ bonds.

N_2H_4: Single bond; bond length = 145 pm. Geometry is tetrahedral; sp^3 hybrid orbitals are needed for one lone pair and three σ bonds.

10.91. The electron dot formula for C_2^{2-} is

$$\left[:C\equiv C:\right]^{2-}$$

The valence bond description of the bonding is a triple bond, involving a σ bond formed by the overlap of a *sp* hybrid orbital on each carbon, and two π bonds formed by the overlap of the two unhybridized 2*p* orbitals on each carbon. Each atom obeys the octet rule, and there is a formal charge of −1 on each carbon. The molecular orbital description is based on a total of 14 electrons for the ion, leading to a configuration of $KK(\sigma_{2s})^2(\sigma_{2s}^*)^2(\pi_{2p})^4(\sigma_{2p})^2$. The bond order is ½(8 – 2) = 3. This agrees with the valence bond description.

10.93. In BF_3NH_3, both the boron atom and the nitrogen atom are in a tetrahedral geometry, with sp^3 hybrid orbitals. In terms of valence bond theory, the bond between boron and fluorine is a σ bond, formed by the overlap of an sp^3 orbital on boron with an sp^3 orbital on nitrogen. For the reactant molecules, BF_3 is trigonal planar with sp^2 hybrid orbitals, and NH_3 is trigonal pyramidal with sp^3 hybrid orbitals.

10.95.

H₂C=C=CH₂ (labels: $C_a=C_b=C_a$)

The two C_a atoms are bonded two three atoms with no lone pairs. They are sp^2 hybridized with trigonal planar geometry. The central C_b atom is bonded to two atoms with no lone pairs. It is *sp* hybridized with a linear geometry.

The central C_b atom uses one sp_x orbital to form a σ bond when overlapping with the sp^2 hybrid orbital on the left C_a atom and the other sp_x orbital to form a σ bond when overlapping with the sp^2 hybrid orbital on the right C_a atom. The left π bond is formed by sideways overlap of the unhybridized p_z orbitals present on both the central C_b and left C_a atoms. The right π bond is formed by sideways overlap of the unhybridized p_y orbitals present on both the central C_b and right C_a atoms.

Because of allene's π bonding network, the two terminal CH_2 groups occupy planes that are offset from one another by 90°. As a result, the Cl atoms in the derivative molecule 1,3-dichloroallene will not be symmetrical with respect to one another and a net dipole will be present. (In the depiction below, the solid triangle is in the plane of this document and the dotted triangle is in the plane that is perpendicular to the plane of this document.)

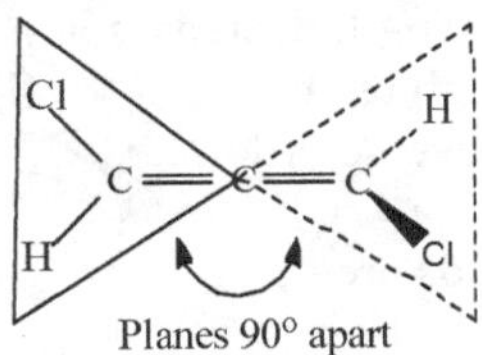

Planes 90° apart

Thus, 1,3-dichloroallene is a polar molecule.

10.97. The electron dot formula for I_3^- (formal charges, if any, are given below the atoms) is

$$[:\ddot{\underset{..}{I}}-\ddot{\underset{..}{I}}-\ddot{\underset{..}{I}}:]^-$$

(formal charge on central I: 1-)

The central I atom in I_3^- has two single bonds and three lone pairs around it. This suggests sp^3d hybridization. Since repulsions between lone pairs are minimized when placed in the equatorial position of the trigonal bipyramid electron pair geometry, the geometry of the molecular ion is linear. Each I–I bond is formed by the overlap of an sp^3d hybrid orbital on the central I with a 5*p* orbital from a terminal I. There are no other reasonable resonance structures for I_3^-.

Including resonance structures, the electron dot formula for the N_3^- ion is

$$[:\ddot{\underset{..}{N}}-N\equiv N:]^- \longleftrightarrow [:N\equiv N-\ddot{\underset{..}{N}}:]^- \longleftrightarrow [\ddot{\underset{..}{N}}=N=\ddot{\underset{..}{N}}]^-$$

Formal charges: 2-, 1+ ; 1+, 2- ; 1-, 1+, 1-

Dominant structure (right)

In each case the central N atom in N_3^- has two bonding regions of electron density and no lone pairs. This suggests *sp* hybridization with a linear molecular geometry. Since the structure to the right has the smaller formal charges it is likely to be the dominating structure.

10.99. The Lewis electron dot formula for PF_3Br_2 is

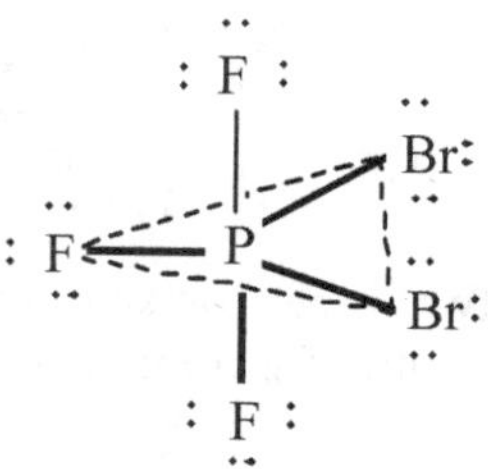

There are five bonded atoms and no lone pairs. This suggests dsp^3 hybridization with trigonal bipyramidal geometry. Because bromine atoms are considerably larger than fluorine atoms, they will likely occupy equatorial positions instead of axial positions.

10.101. Total electrons for $C_2 = 2 \times 6 = 12$. The molecular orbital configuration for C_2 is $KK(\sigma_{2s})^2(\sigma_{2s}*)^2(\pi_{2p})^4$ leads to a bond order 2 (B.O. = ½(8 − 4) = 2).

Total electrons for $C_2^+ = 2 \times 6 - 1 = 11$. The molecular orbital configuration for C_2^+ is $KK(\sigma_{2s})^2(\sigma_{2s}*)^2(\pi_{2p})^3$ leads to a bond order 3/2 (B.O. = ½(7 − 4) = 3/2).

Total electrons for $C_2^{2-} = 2 \times 6 + 2 = 14$. The molecular orbital configuration for C_2^{2-} is $KK(\sigma_{2s})^2(\sigma_{2s}*)^2(\pi_{2p})^4(\sigma_{2p})^2$ leads to a bond order 3 (B.O. = ½(10 − 4) = 3).

As bond order increases, bond length decreases and bond enthalpy increases. Arranging the species by increasing bond length results in $C_2^{2-} < C_2 < C_2^+$. Arranging the species in increasing bond enthalpies results in the reverse order, i.e., $C_2^+ < C_2 < C_2^{2-}$.

10.103. The structures for the compounds are

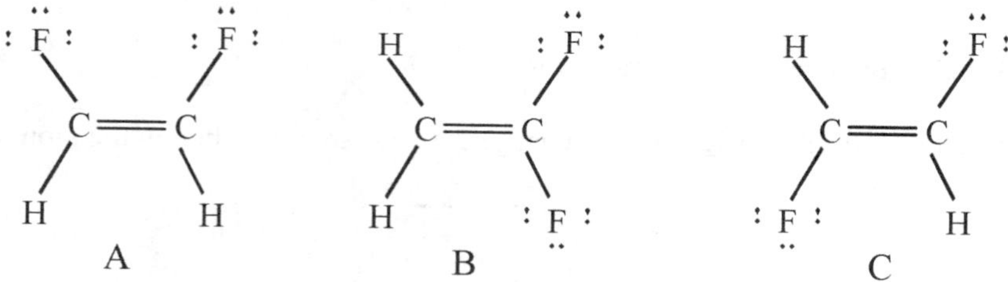

10.105. The molecular orbital description of an O_2 molecule is $KK(\sigma_{2s})^2(\sigma_{2s}*)^2(\pi_{2p})^4(\sigma_{2p})^2(\pi_{2p}*)^2$. In the $\pi_{2p}*$ orbital, both electrons must be in the same orbital for all electrons to be paired. The ground state would have one electron in each of the two $\pi_{2p}*$ orbitals. In both cases, the bond order is 2, but the excited state is diamagnetic, whereas the ground state is paramagnetic.

10.107. The Lewis formula for formaldehyde is

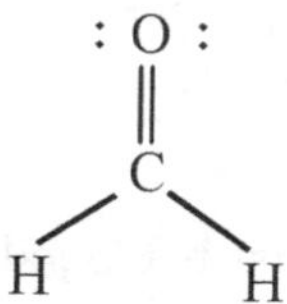

Because the C is bonded to three other atoms, it is assumed to be sp^2 hybridized. One $2p$ orbital remains unhybridized. The carbon–hydrogen bonds are σ bonds formed by the overlap of an sp^2 hybrid orbital on C with a $1s$ orbital on H. The remaining sp^2 hybrid orbital on C overlaps with a

2*p* orbital on O to form a σ bond. The unhybridized 2*p* orbital on C overlaps with a parallel 2*p* orbital on O to form the π bond.

In molecular orbital theory, the carbon–oxygen double bond has a σ and a π molecular orbital that are occupied by electrons. There is also a π^* orbital that is unoccupied in the ground state. The transition at 270 nm (UV region) is due to an electron moving from the occupied π orbital to the unoccupied π^* orbital to give an excited state.

■ SOLUTIONS TO CUMULATIVE-SKILLS PROBLEMS

Note on significant figures: If the final answer to a solution needs to be rounded off, it is given first with one nonsignificant figure, and the last significant figure is underlined. The final answer is then rounded to the correct number of significant figures. In multistep problems, intermediate answers are given with at least one nonsignificant figure; however, only the final answer has been rounded off.

10.109. After assuming a 100.0-g sample, convert to moles:

$$60.4 \text{ g Xe} \times \frac{1 \text{ mol Xe}}{131.29 \text{ g Xe}} = 0.46\underline{0}05 \text{ mol Xe}$$

$$22.1 \text{ g O} \times \frac{1 \text{ mol O}}{16.00 \text{ g O}} = 1.3\underline{8}1 \text{ mol O}$$

$$17.5 \text{ g F} \times \frac{1 \text{ mol F}}{18.99 \text{ g F}} = 0.92\underline{1}5 \text{ mol F}$$

Divide by 0.460 :

$$\text{Xe: } \frac{0.460}{0.460} = 1; \text{ O: } \frac{1.381}{0.460} = 3.00; \text{ F: } \frac{0.9215}{0.460} = 2.00$$

The simplest formula is XeO_3F_2. This is also the molecular formula. The Lewis formula is

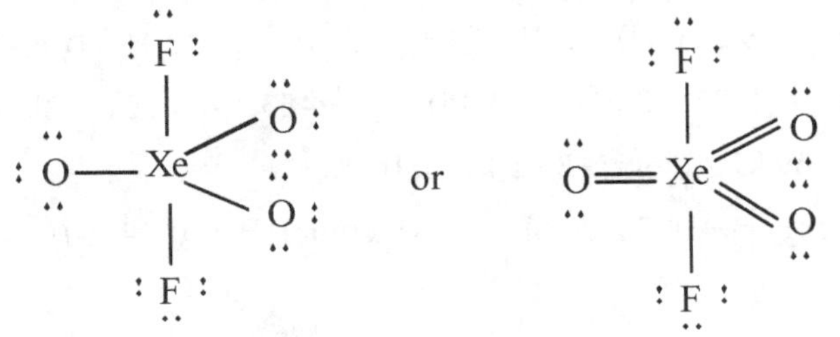

or

Number of electron pairs = 5, number of lone pairs = 0; hence, the geometry is trigonal bipyramidal. Because xenon has five single bonds, it will require five orbitals to describe the bonding. This suggests sp^3d hybridization.

10.111. $U(s) + ClF_n \rightarrow UF_6 + ClF(g)$

$$\text{mol } UF_6 = 3.53 \text{ g } UF_6 \times \frac{1 \text{ mol } UF_6}{352.07 \text{ g } UF_6} = 0.010\underline{0}3 \text{ mol } UF_6$$

$$\text{mol ClF} = n = \frac{PV}{RT} = \frac{2.50 \text{ atm} \times 0.343 \text{ L}}{0.082057 \text{ L} \bullet \text{atm/ (K} \bullet \text{mol)} \times 348 \text{ K}} = 0.030\underline{0}3 \text{ mol ClF}$$

0.010 mol UF_6 = 0.060 mol F, and 0.030 mol ClF = 0.030 mol F; therefore, the total moles of F from ClF_n = 0.090 mol F. Because mol ClF_n must equal mol ClF, mol ClF_n = 0.030 mol and n = 0.090 mol F ÷ 0.030 mol ClF_n = 3. The Lewis formula is

Number of electron pairs = 5, number of lone pairs = 2; hence, the geometry is T-shaped. Because chlorine has five electron pairs, it will require five orbitals to describe the bonding. This suggests sp^3d hybridization.

10.113. HNO_3 resonance formulas:

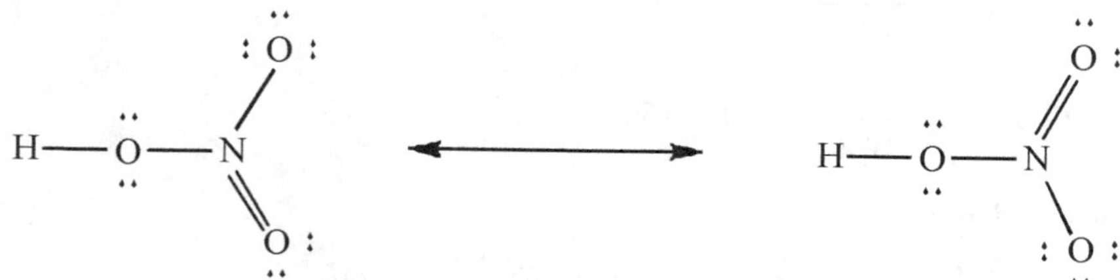

The geometry around the nitrogen is trigonal planar; therefore, the hybridization is sp^2.

Formation reaction: $H_2(g) + 3O_2(g) + N_2(g) \rightarrow 2HNO_3(g)$

$$2 \times \Delta H_f^\circ = [BE(\text{H–H}) + 3BE(O_2) + BE(N_2)] - [2BE(\text{H–O}) + 4BE(\text{N–O}) + 2BE(\text{N=O})]$$

$$2 \times \Delta H_f^\circ = [(436 + 3 \times 498 + 945) - (2 \times 463 + 4 \times 201 + 2 \times 607) \text{ kJ}]$$

$$\Delta H_f^\circ = -69 \text{ kJ/2 mol} = -3\underline{4}.5 \text{ kJ/mol}$$

Resonance energy = $-3\underline{4}.5$ kJ/mol − (−135 kJ/mol) = $10\underline{0}.5$ = 101 kJ/mol

Appendixes

Appendix A Mathematical Skills

Only a few basic mathematical skills are required for the study of general chemistry. But to concentrate your attention on the concepts of chemistry, you will find it necessary to have a firm grasp of these basic mathematical skills. In this appendix, we will review scientific (or exponential) notation, logarithms, simple algebraic operations, the solution of quadratic equations, and the plotting of straight-line graphs.

A.1 Scientific (Exponential) Notation

In chemistry, you frequently encounter very large and very small numbers. For example, the number of molecules in a liter of air at 20°C and normal barometric pressure is 25,000,000,000,000,000,000,000, and the distance between two hydrogen atoms in a hydrogen molecule is 0.000,000,000,074 meter. In these forms, such numbers are inconvenient to write and difficult to read. For this reason, you normally express them in scientific, or exponential, notation. Scientific calculators also use this notation.

In scientific notation, a number is written in the form $A \times 10^n$. A is a number greater than or equal to 1 and less than 10, and the exponent n (the nth power of 10) is a positive or negative integer. For example, 4853 is written in scientific notation as 4.853×10^3, which is 4.853 multiplied by three factors of 10:

$$4.853 \times 10^3 = 4.853 \times 10 \times 10 \times 10 = 4853$$

The number 0.0568 is written in scientific notation as 5.68×10^{-2}, which is 5.68 divided by two factors of 10:

$$5.68 \times 10^{-2} = \frac{5.68}{10 \times 10} = 0.0568$$

Any number can be conveniently transformed to scientific notation by moving the decimal point to obtain a number, A, greater than or equal to 1 and less than 10. If the decimal point is moved to the left, you multiply A by 10^n, where n equals the number of places moved. If the decimal point is moved to the right, you multiply A by 10^{-n}. Consider the number 0.00731. You must move the decimal point to the right three places. Therefore, 0.00731 equals 7.31×10^{-3}. To transform a number written in scientific notation to the usual form, the process is reversed. If the exponent is positive, the decimal point is shifted right. If the exponent is negative, the decimal point is shifted left.

Example 1 **Expressing Numbers in Scientific Notation**

Express the following numbers in scientific notation:

a 843.4 b 0.00421 c 1.54

Solution Shift the decimal point to get a number between 1 and 10; count the number of positions shifted.

a 8.4.3. 4 = $\mathbf{8.434 \times 10^2}$
b 0 .0.0.4.2 1 = $\mathbf{4.21 \times 10^{-3}}$
c **leave as is or write $\mathbf{1.54 \times 10^0}$**

Exercise 1 Express the following numbers in scientific notation:

a 4.38 b 4380 c 0.000483

Example 2 **Converting Numbers in Scientific Notation to Usual Form**

Convert the following numbers in scientific notation to usual form:

a 6.39×10^{-4} b 3.275×10^{2}

Solution a $0.0006.39 \times 10^{-4} = \mathbf{0.000639}$

b $3.275 \times 10^{2} = \mathbf{327.5}$

Exercise 2 Convert the following numbers in scientific notation to usual form:

a 7.025×10^{3}

b 8.97×10^{-4}

Addition and Subtraction

Before adding or subtracting two numbers written in scientific notation, it is necessary to express both to the same power of 10. After adding or subtracting, it may be necessary to shift the decimal point to express the result in scientific notation.

Example 3 **Adding and Subtracting in Scientific Notation**

Carry out the following arithmetic; give the result in scientific notation.

$$(9.42 \times 10^{-2}) + (7.6 \times 10^{-3})$$

Solution You can shift the decimal point in either number to obtain both to the same power of 10. For example, to get both numbers to 10^{-2}, you shift the decimal point one place to the left and add 1 to the exponent in the expression 7.6×10^{-3}.

$$7.6 \times 10^{-3} = 0.76 \times 10^{-2}$$

Now you can add the two numbers.

$$(9.42 \times 10^{-2}) + (0.76 \times 10^{-2}) = (9.42 + 0.76) \times 10^{-2} = 10.18 \times 10^{-2}$$

Since 10.18 is not between 1 and 10, you shift the decimal point to express the final result in scientific notation.

$$10.18 \times 10^{-2} = \mathbf{1.018 \times 10^{-1}}$$

Exercise 3 Add the following, and express the sum in scientific notation:

$$(3.142 \times 10^{-4}) + (2.8 \times 10^{-6})$$

Multiplication and Division

To multiply two numbers in scientific notation, you first multiply the two powers of 10 by adding their exponents. Then you multiply the remaining factors. Division is handled similarly. You first move any power of 10 in the denominator to the numerator, changing the sign of the exponent. After multiplying the two powers of 10 by adding their exponents, you carry out the indicated division.

Example 4 **Multiplying and Dividing in Scientific Notation**

Do the following arithmetic, and express the answers in scientific notation.

a $(6.3 \times 10^{2}) \times (2.4 \times 10^{5})$ b $\dfrac{6.4 \times 10^{2}}{2.0 \times 10^{5}}$

(continued)

(continued)

Solution a $(6.3 \times 10^2) \times (2.4 \times 10^5) = (6.3 \times 2.4) \times 10^7 = 15.12 \times 10^7 = \mathbf{1.512 \times 10^8}$

b $\dfrac{6.4 \times 10^2}{2.0 \times 10^5} = \dfrac{6.4}{2.0} \times 10^2 \times 10^{-5} = \dfrac{6.4}{2.0} \times 10^{-3} = \mathbf{3.2 \times 10^{-3}}$

Exercise 4 Perform the following operations, expressing the answers in scientific notation:

a $(5.4 \times 10^{-7}) \times (1.8 \times 10^8)$ b $\dfrac{5.4 \times 10^{-7}}{6.0 \times 10^{-5}}$

Powers and Roots

A number $A \times 10^n$ raised to a power p is evaluated by raising A to the power p and multiplying the exponent in the power of 10 by p:

$$(A \times 10^n)^p = A^p \times 10^{n \times p}$$

You extract the rth root of a number $A \times 10^n$ by first moving the decimal point in A so that the exponent in the power of 10 is exactly divisible by r. Suppose this has been done, so that n in the number $A \times 10^n$ is exactly divisible by r. Then

$$\sqrt[r]{A \times 10^n} = \sqrt[r]{A} \times 10^{n/r}$$

Example 5 **Finding Powers and Roots in Scientific Notation**

Evaluate the following expressions:

a $(5.29 \times 10^2)^3$

b $\sqrt{2.31 \times 10^7}$

Solution a $(5.29 \times 10^2)^3 = (5.29)^3 \times 10^6 = 148 \times 10^6 = \mathbf{1.48 \times 10^8}$

b $\sqrt{2.31 \times 10^7} = \sqrt{23.1 \times 10^6} = \sqrt{23.1} \times 10^{6/2} = \mathbf{4.81 \times 10^3}$

Exercise 5 Obtain the values of the following, and express them in scientific notation:

a $(3.56 \times 10^3)^4$ b $\sqrt[3]{4.81 \times 10^2}$

Electronic Calculators

Scientific calculators will perform all of the arithmetic operations we have just described (as well as those discussed in the next section). On most calculators, the basic operations of addition, subtraction, multiplication, and division are similar and straightforward. However, more variation exists in raising a number to a power and extracting a root. If you have the calculator instructions, by all means read them. Otherwise, the following information may help.

Squares and square roots are usually obtained with special keys, perhaps labeled x^2 and $\sqrt{x}$. Thus, to obtain $(5.15)^2$, you enter 5.15 and press x^2. To obtain $\sqrt{5.15}$, you enter 5.15 and press $\sqrt{x}$ (or perhaps INV, for inverse, and x^2). Other powers and roots require a y^x (or a^x) key. The answer to Example 5(a), which is $(5.29 \times 10^2)^3$, would be obtained by a sequence of steps such as the following: enter 5.29×10^2, press the y^x key, enter 3, and press the = key.

The same sequence can be used to extract a root. Suppose you want $\sqrt[5]{2.18 \times 10^6}$. This is equivalent to $(2.18 \times 10^6)^{1/5}$ or $(2.18 \times 10^6)^{0.2}$. If the calculator has a $1/x$ key, the sequence would be as follows: enter 2.18×10^6, press the y^x key, enter 5, press the $1/x$ key, then press the = key. Some calculators have a $\sqrt[x]{y}$ key, so this can be used to extract the xth root of y, using a sequence of steps similar to that for y^x.

A.2 Logarithms

The *logarithm* to the *base a* of a number x, denoted $\log_a x$, is the exponent of the constant a needed to equal the number x. For example, suppose a is 10, and you would like the logarithm of 1000; that is, you would like the value of $\log_{10} 1000$. This is the exponent, y, of 10 such that 10^y equals 1000. The value of y here is 3. Thus, $\log_{10} 1000 = 3$.

Common logarithms are logarithms in which the base is 10. The common logarithm of a number x is often denoted simply as $\log x$. It is easy to see how to obtain the common logarithms of 10, 100, 1000, and so forth. But logarithms are defined for all positive numbers, not just the powers of 10. In general, the exponents, or values, of the logarithm will be decimal numbers. To understand the meaning of a decimal exponent, consider $10^{0.400}$. This is equivalent to $10^{400/1000} = 10^{2/5} = \sqrt[5]{10^2} = 2.51$. Therefore, $\log 2.51 = 0.400$. Any decimal exponent is essentially a fraction, p/r, so by evaluating the expressions $10^{p/r} = \sqrt[r]{10^p}$ one could construct a table of logarithms. In practice, power series or other methods are used.

The following are fundamental properties of all logarithms:

$$\log_a 1 = 0 \tag{1}$$

$$\log_a(A \times B) = \log_a A + \log_a B \tag{2}$$

$$\log_a \frac{A}{B} = \log_a A - \log_a B \tag{3}$$

$$\log_a A^p = p \log_a A \tag{4}$$

$$\log_a \sqrt[r]{A} = \frac{1}{r} \log_a A \tag{5}$$

These properties are very useful in working with logarithms.

Electronic calculators that evaluate logarithms are now available for about \$10 or so. Their simplicity of operation makes them well worth the price. To obtain the logarithm of a number, you enter the number and press the LOG key.

Exercise 6 Find the values of **a** log 0.00582; **b** log 689.

Antilogarithm

The antilogarithm (abbreviated antilog) is the inverse of the common logarithm. Antilog x is simply 10^x. If your electronic calculator has a 10^x key, you obtain the antilogarithm of a number by entering the number and pressing the 10^x key. (It may be necessary to press an inverse key before pressing a 10^x/LOG key.) If your calculator has a y^x (or a^x) key, you enter 10, press y^x, enter x, then press the = key.

Exercise 7 Evaluate **a** antilog 5.728; **b** antilog (−5.728).

Natural Logarithms and Natural Antilogarithms

The mathematical constant $e = 2.71828$, like π, occurs in many scientific and engineering problems. It is frequently seen in the natural exponential function $y = e^x$. The inverse function is called the *natural logarithm,* $x = \ln y$, where $\ln y$ is simplified notation for $\log_e y$.

Finding the natural logarithm with a calculator is analogous to finding the common logarithm, only in this case you enter the number and press the LN key. For example, you may obtain the natural logarithm of 6.2 by entering 6.2 and pressing the LN key ($\ln 6.2 = 1.82$). The natural antilogarithm of a number is the constant e raised to a power equal to the number. Therefore, the expression for the natural antilogarithm of 1.34 is written as

$$\text{natural antilog } 1.34 = e^{1.34}$$

In order to evaluate the expression $e^{1.34}$ on your calculator, you can use the e^x key, or you may have to use the INV key in combination with the LN key ($e^{1.34} = 3.8$).

There are instances where you may want to relate the natural logarithm to the common logarithm; the expression is

$$\ln x = 2.303 \log x$$

Exercise 8 Evaluate **a** $\ln 9.93$; **b** $e^{1.10}$.

A.3 Algebraic Operations and Graphing

Often you are given an algebraic formula that you need to rearrange to solve for a particular quantity. As an example, suppose you would like to solve the following equation for V:

$$PV = nRT$$

You can eliminate P from the left-hand side by dividing by P. But to maintain the equality, you must perform the same operation on both sides of the equation:

$$\frac{PV}{P} = \frac{nRT}{P}$$

or

$$V = \frac{nRT}{P}$$

Quadratic Formula

A quadratic equation is one involving only powers of x in which the highest power is 2. The general form of the equation can be written

$$ax^2 + bx + c = 0$$

where a, b, and c are constants. For given values of these constants, only certain values of x are possible. (In general, there will be two values.) These values of x are said to be the solutions of the equation.

These solutions are given by the *quadratic formula:*

$$x = \frac{-b \pm \sqrt{b^2 - 4ac}}{2a}$$

In this formula, the symbol $\pm$ means that there are two possible values of x—one obtained by taking the positive sign, the other by taking the negative sign.

Example 6 **Obtaining the Solutions of a Quadratic Equation**

Obtain the solutions of the following quadratic equation:

$$2.00x^2 - 1.72x - 2.86 = 0$$

Solution Using the quadratic formula, you substitute $a = 2.00$, $b = -1.72$, and $c = -2.86$. You get

$$x = \frac{1.72 \pm \sqrt{(-1.72)^2 - 4 \times 2.00 \times (-2.86)}}{2 \times 2.00} = \frac{1.72 \pm 5.08}{4.00} = \mathbf{-0.840 \text{ and } +1.70}$$

Mathematically there are two solutions, but in any real problem one solution may not be allowed. For example, if the solution is some physical quantity that can have only positive values, a negative solution must be rejected.

Exercise 9 Find the positive solution (or solutions) to the following equation:

$$1.80x^2 + 0.850x - 9.50 = 0$$

The Straight-Line Graph

A graph is a visual means of representing a mathematical relationship or physical data. Consider the following data table, in which values of y from some experiment are given for four values of x.

x	y
1	−1
2	1
3	3
4	5

By plotting these x, y points on a graph (Figure A.1), you see that they fall on a straight line. This suggests (but does not prove) that other points from this type of

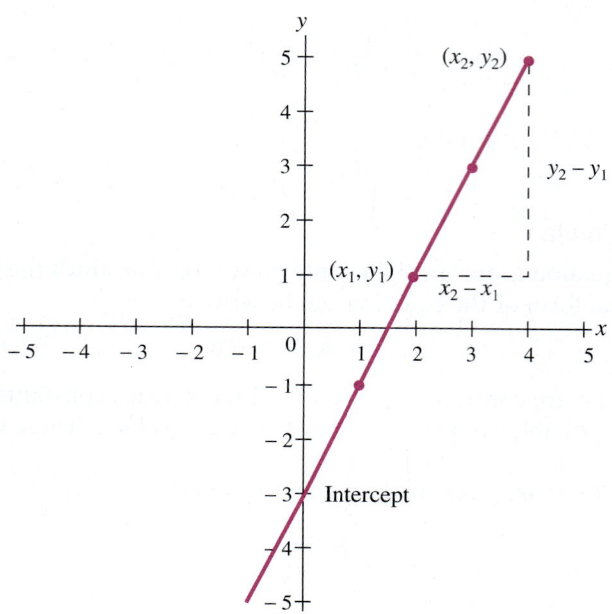

Figure A.1 A straight-line plot of some data

experiment might fall on the same line. It would be useful to have the mathematical equation for this line.

The general equation form of a straight line is

$$y = mx + b$$

The constant m is called the *slope* of the straight line. It is obtained by dividing the vertical distance between any two points on the line by the horizontal distance. If the two points are (x_1, y_1) and (x_2, y_2), the slope is given by the following formula:

$$\text{Slope} = \frac{y_2 - y_1}{x_2 - x_1}$$

Suppose you choose the points (2, 1) and (4, 5) from the data. Then

$$\text{Slope} = \frac{5 - 1}{4 - 2} = 2$$

For the straight line in Figure A.1, $m = 2$.

The constant b is called the *intercept*. It is the value of y at $x = 0$. From Figure A.1, you see that the intercept is -3. Therefore, $b = -3$. Hence, the equation of the straight line is

$$y = 2x - 3$$

Appendix B Vapor Pressure of Water at Various Temperatures

Temperature (°C)	Pressure (mmHg)	Temperature (°C)	Pressure (mmHg)
0	4.6	27	26.7
5	6.5	28	28.3
10	9.2	29	30.0
11	9.8	30	31.8
12	10.5	35	42.2
13	11.2	40	55.3
14	12.0	45	71.9
15	12.8	50	92.5
16	13.6	55	118.0
17	14.5	60	149.4
18	15.5	65	187.5
19	16.5	70	233.7
20	17.5	75	289.1
21	18.7	80	355.1
22	19.8	85	433.6
23	21.1	90	525.8
24	22.4	95	633.9
25	23.8	100	760.0
26	25.2	105	906.1

Appendix C Thermodynamic Quantities for Substances and Ions at 25°C

Substance or Ion	ΔH_f° (kJ/mol)	ΔG_f° (kJ/mol)	S° (J/mol·K)
$e^-(g)$	0	0	20.87
Aluminum			
$Al(s)$	0	0	28.28
$Al^{3+}(aq)$	−531	−485	−321.7
$AlCl_3(s)$	−705.6	−630.0	109.3
$Al_2O_3(s)$	−1675.7	−1582.3	50.95
Barium			
$Ba(g)$	179.1	147.0	170.1
$Ba(s)$	0	0	62.48
$Ba^{2+}(aq)$	−537.6	−560.7	9.6
$BaCO_3(s)$	−1216.3	−1137.6	112.1
$BaCl_2(s)$	−858.6	−810.3	123.7
$Ba(NO_3)_2(aq)$	−952.4	−783.4	302.5
$BaO(s)$	−548.1	−520.4	72.07
$Ba(OH)_2(s)$	−946.3	−859.3	107.1
$Ba(OH)_2 \cdot 8H_2O(s)$	−3342.2	−2793	427
$BaSO_4(s)$	−1473.2	−1362.3	132.2
Beryllium			
$Be(s)$	0	0	9.440
$BeO(s)$	−608.4	−579.1	13.77
$Be(OH)_2(s)$	−905.8	−817.9	50.21
Boron			
$B(s)$	0	0	5.834
$BCl_3(l)$	−427.2	−387.4	206
$BF_3(g)$	−1135.6	−1119.0	254.2
$B_2O_3(s)$	−1271.9	−1192.8	53.95
Bromine			
$Br(g)$	111.9	82.40	174.9
$Br^-(aq)$	−121.5	−104.0	82.4
$Br^-(g)$	−219.0	−238.8	163.4
$Br_2(g)$	30.91	3.159	245.3
$Br_2(l)$	0	0	152.2
$HBr(g)$	−36.44	−53.50	198.6
Calcium			
$Ca(g)$	177.8	144.1	154.8
$Ca(s)$	0	0	41.59
$Ca^+(g)$	773.8	732.1	160.5
$Ca^{2+}(aq)$	−542.8	−553.5	−53.1
$Ca^{2+}(g)$	1925.9	—	—
$CaCO_3(s, \text{calcite})$	−1206.9	−1128.8	92.9
$CaCl_2(s)$	−795.8	−748.1	104.6
$CaF_2(s)$	−1225.9	−1173.5	68.57
$CaO(s)$	−635.1	−603.5	38.21
$Ca(OH)_2(s)$	−986.1	−898.4	83.39
$Ca_3(PO_4)_2(s)$	−4120.8	−3884.8	236.0
$CaSO_4(s)$	−1434.1	−1321.9	106.7
Carbon			
$C(g)$	716.7	671.3	158.0
$C(s, \text{diamond})$	1.897	2.900	2.377
$C(s, \text{graphite})$	0	0	5.740
$CCl_4(g)$	−95.98	−53.65	309.7
$CCl_4(l)$	−135.4	−65.27	216.4
$CF_4(g)$	−933.2	−888.5	261.3
$CN^-(aq)$	151	166	118
$CO(g)$	−110.5	−137.2	197.5
$CO_2(g)$	−393.5	−394.4	213.7
$CO_3^{2-}(aq)$	−677.1	−527.9	−56.9
$CS_2(g)$	116.9	66.85	237.9
$CS_2(l)$	89.70	65.27	151.3
$COCl_2(g)$	−220.1	−205.9	283.9
$HCN(aq)$	150.6	172.4	94.1
$HCN(g)$	135.1	124.7	201.7
$HCN(l)$	108.9	124.9	112.8
$HCO_3^-(aq)$	−692.0	−586.8	91.2
Hydrocarbons			
$CH_4(g)$	−74.87	−50.80	186.1
$C_2H_2(g)$	226.7	209.2	200.9
$C_2H_4(g)$	52.47	68.39	219.2
$C_2H_6(g)$	−84.68	−32.89	229.5
$C_3H_8(g)$	−104.7	−23.6	270.2
$C_4H_{10}(g)$	−125.6	−17.2	310.1
$C_6H_6(g)$	82.6	129.7	269.2
$C_6H_6(l)$	49.0	124.4	173.4
Alcohols			
$CH_3OH(g)$	−200.7	−162.0	239.7
$CH_3OH(l)$	−238.7	−166.4	126.8
$C_2H_5OH(g)$	−235.1	−168.6	282.6
$C_2H_5OH(l)$	−277.7	−174.9	160.7
Aldehydes			
$HCHO(g)$	−117	−113	219.0
$CH_3CHO(g)$	−166.1	−133.4	246.4

Substance or Ion	ΔH_f° (kJ/mol)	ΔG_f° (kJ/mol)	S° (J/mol·K)
$CH_3CHO(l)$	−191.8	−128.3	160.4
Carboxylic acids and ions			
$HCOOH(aq)$	−425.6	−351.0	92
$HCOOH(l)$	−424.7	−361.4	129
$HCOO^-(aq)$	−425.6	−351.0	92
$CH_3COOH(aq)$	−486.0	−369.4	86.6
$CH_3COOH(l)$	−484.5	−389.9	159.8
$CH_3COO^-(aq)$	−486.0	−369.4	86.6
Sugars			
$C_6H_{12}O_6$ (*s*, glucose)	−1273.3	−910.4	212.1
Cesium			
$Cs(g)$	76.50	49.56	175.5
$Cs(l)$	2.087	0.024	92.07
$Cs(s)$	0	0	85.15
$CsBr(s)$	−405.7	−391.4	113.0
$CsCl(s)$	−442.8	−414.4	101.2
$CsF(s)$	−554.7	−525.4	88.28
$CsI(s)$	−346.6	−340.6	123.1
Chlorine			
$Cl(g)$	121.3	105.3	165.1
$Cl^-(aq)$	−167.2	−131.3	56.5
$Cl^-(g)$	−234.0	−240.2	153.2
$Cl_2(g)$	0	0	223.0
$HCl(aq)$	−167.2	−131.3	56.5
$HCl(g)$	−92.31	−95.30	186.8
Chromium			
$Cr(g)$	397.5	352.6	174.2
$Cr(s)$	0	0	23.62
$Cr^{3+}(aq)$	−143.5	—	—
$CrO_4^{2-}(aq)$	−881.2	−727.8	50.21
$Cr_2O_3(s)$	−1134.7	−1053.1	81.15
$Cr_2O_7^{2-}(aq)$	−1490.3	−1301.2	262.9
Cobalt			
$Co(g)$	426.7	382.1	179.4
$Co(s)$	0	0	30.07
$Co^{2+}(aq)$	−58.2	−54.4	113
Copper			
$Cu(g)$	337.6	297.9	166.3
$Cu(s)$	0	0	33.16
$Cu^+(aq)$	71.67	50.00	40.6
$Cu^{2+}(aq)$	64.77	65.52	−99.6
$CuCl_2(s)$	−205.9	−161.7	108.1
$CuO(s)$	−156.1	−128.3	42.59

Substance or Ion	ΔH_f° (kJ/mol)	ΔG_f° (kJ/mol)	S° (J/mol·K)
$Cu_2O(s)$	−170.7	−147.9	92.36
Fluorine			
$F(g)$	79.39	62.31	158.6
$F^-(g)$	−255.1	−262.0	145.5
$F^-(aq)$	−332.6	−278.8	−13.8
$F_2(g)$	0	0	202.7
$HF(aq)$	−332.6	−278.8	−13.8
$HF(g)$	−272.5	−274.6	173.7
Hydrogen			
$H(g)$	218.0	203.3	114.6
$H^+(aq)$	0	0	0
$H^+(g)$	1536.2	1517.0	108.8
$H_2(g)$	0	0	130.6
$H_2O(g)$	−241.8	−228.6	188.7
$H_2O(l)$	−285.8	−237.1	69.95
$H_2O_2(aq)$	−191.2	−134.1	143.9
$H_2O_2(g)$	−136.1	−105.5	232.9
$H_2O_2(l)$	−187.8	−120.4	109.6
OH^-	−230.0	−157.3	−10.75
Iodine			
$I(g)$	106.8	70.21	180.7
$I^-(aq)$	−55.19	−51.59	109.6
$I^-(g)$	−194.6	−221.5	169.2
$I_2(g)$	62.42	19.36	260.6
$I_2(s)$	0	0	116.1
$HI(g)$	26.36	1.576	206.5
Iron			
$Fe(g)$	415.5	369.8	180.4
$Fe(l)$	12.40	10.18	34.76
$Fe(s)$	0	0	27.32
$Fe^{2+}(aq)$	−89.1	−78.87	−137.7
$Fe^{3+}(aq)$	−48.5	−4.6	−315.9
$FeCl_2(s)$	−341.8	−302.3	117.9
$FeCl_3(s)$	−399.4	−333.9	142.3
$FeO(s)$	−272.0	−251.4	60.75
$Fe_2O_3(s)$	−825.5	−743.5	87.40
$Fe_3O_4(s)$	−1120.9	−1017.4	145.3
$FeS_2(s)$	−171.5	−160.1	52.92
Lead			
$Pb(s)$	0	0	64.78
$Pb^{2+}(aq)$	−1.7	−24.39	10.5
$PbBr_2(s)$	−277.4	−260.7	161.1
$PbCO_3(s)$	−699.1	−625.5	131.0
$PbCl_2(s)$	−359.4	−314.1	136.0

(continued)

(continued)

Substance or Ion	ΔH_f° (kJ/mol)	ΔG_f° (kJ/mol)	S° (J/mol·K)
$PbO(s)$	−219.4	−189.3	66.32
$PbO_2(s)$	−274.5	−215.4	71.80
$PbS(s)$	−98.32	−96.68	91.34
$PbSO_4(s)$	−919.9	−813.2	148.6
Lithium			
$Li(g)$	159.3	126.6	138.7
$Li(s)$	0	0	29.08
$Li^+(aq)$	−278.5	−293.3	13.4
$Li^+(g)$	685.7	648.5	132.9
$LiBr(s)$	−350.9	−341.6	74.06
$LiCl(s)$	−408.3	−384.0	59.30
$LiF(s)$	−616.9	−588.7	35.66
$LiI(s)$	−270.1	−269.7	85.77
Magnesium			
$Mg(g)$	147.1	112.6	148.5
$Mg(s)$	0	0	32.67
$Mg^+(g)$	891.0	848.6	154.3
$Mg^{2+}(aq)$	−466.9	−454.8	−138.1
$Mg^{2+}(g)$	2351	—	—
$MgCO_3(s)$	−1111.7	−1028.1	65.85
$MgCl_2(s)$	−641.6	−592.1	89.63
$Mg_3N_2(s)$	−461.1	−400.9	87.86
$MgO(s)$	−601.2	−568.9	26.92
$Mg(OH)_2(s)$	−924.7	−833.7	63.24
Manganese			
$Mn(g)$	283.3	241.0	173.6
$Mn(s)$	0	0	32.01
$Mn^{2+}(aq)$	−220.7	−228.0	−73.6
$MnO(s)$	−385.2	−362.9	59.71
$MnO_2(s)$	−520.0	−465.2	53.05
$MnO_4^-(aq)$	−541.4	−447.3	191.2
Mercury			
$Hg(g)$	61.38	31.91	174.9
$Hg(l)$	0	0	76.03
$Hg^{2+}(aq)$	171.1	164.4	−32.2
$Hg_2^{2+}(aq)$	172.3	153.6	84.5
$HgCl_2(s)$	−230.1	−184.0	144.5
$Hg_2Cl_2(s)$	−265.2	−210.8	192.4
$HgO(s)$	−90.79	−58.49	70.27
Nickel			
$Ni(g)$	430.1	384.7	182.1
$Ni(s)$	0	0	29.87
$Ni^{2+}(aq)$	−54.0	−45.6	−128.9

Substance or Ion	ΔH_f° (kJ/mol)	ΔG_f° (kJ/mol)	S° (J/mol·K)
$NiCl_2(s)$	−304.9	−258.8	98.16
$NiO(s)$	−239.7	−211.7	37.99
Nitrogen			
$N(g)$	472.7	455.6	153.2
$N_2(g)$	0	0	191.6
$NH_3(aq)$	−80.29	−26.57	111.3
$NH_3(g)$	−45.90	−16.40	192.7
$NH_4^+(aq)$	−132.5	−79.37	113.4
$N_2H_4(g)$	95.35	159.2	238.6
$N_2H_4(l)$	50.63	149.4	121.4
$NH_4Cl(s)$	−314.6	−203.1	94.86
$NH_4NO_3(s)$	−365.6	−184.0	151.1
$NO(g)$	90.29	86.60	210.6
$NO_2(g)$	33.10	51.24	239.9
$NO_3^-(aq)$	−207.4	−111.3	146.4
$N_2O(g)$	82.05	104.2	219.9
$N_2O_4(g)$	9.079	97.72	304.3
$N_2O_4(l)$	−19.56	97.52	209.2
$NOCl(g)$	51.71	66.08	261.6
$HNO_3(aq)$	−207.4	−111.3	146.4
$HNO_3(g)$	−134.3	−73.99	266.3
$HNO_3(l)$	−174.1	−80.79	155.6
Oxygen			
$O(g)$	249.2	231.8	160.9
$O_2(g)$	0	0	205.0
$O_3(g)$	142.7	163.2	238.8
Phosphorus			
$P(g)$	316.4	280.0	163.1
$P(s$, red)	−17.46	−12.03	22.85
$P(s$, white)	0	0	41.08
$P_2(g)$	143.7	103.1	218.0
$P_4(g)$	58.91	24.45	279.9
$PCl_3(g)$	−288.7	−269.6	311.6
$PCl_5(g)$	−360.2	−290.3	364.2
$PF_5(g)$	−1594.4	−1520.7	300.7
$PH_3(g)$	5.439	7.175	210.1
$P_4O_{10}(s)$	−3009.9	−2723.3	228.8
$PO_4^{3-}(aq)$	−1277.4	−1018.8	−222
$POCl_3(g)$	−559.8	−514.3	325.3
$POCl_3(l)$	−597.1	−520.9	222.5
$P_4S_3(s)$	−224.6	−206.9	200.8
$HPO_4^{2-}(aq)$	−1281	−1082	−36
$H_2PO_4^-(aq)$	−1285	−1135	89.1
$H_3PO_4(aq)$	−1288.3	−1142.6	158.2

Substance or Ion	ΔH_f° (kJ/mol)	ΔG_f° (kJ/mol)	S° (J/mol·K)
Potassium			
$K(g)$	89.00	60.51	160.2
$K(s)$	0	0	64.67
$K^+(aq)$	−252.4	−283.3	102.5
$K^+(g)$	514.0	481.0	154.5
$KBr(s)$	−393.8	−380.4	95.94
$KCl(s)$	−436.7	−408.8	82.55
$KF(s)$	−568.6	−538.9	66.55
$KI(s)$	−327.9	−323.0	106.4
$KClO_3(s)$	−397.7	−296.3	143.1
$K_2CO_3(s)$	−1150.2	−1064.5	155.5
$KNO_3(s)$	−494.6	−394.9	133.1
$K_2O(s)$	−363.2	−322.1	94.14
$KO_2(s)$	−284.5	−240.6	122.5
$K_2O_2(s)$	−495.8	−429.8	113.0
$KOH(s)$	−424.7	−378.9	78.91
Rubidium			
$Rb(g)$	80.90	53.11	170.0
$Rb(s)$	0	0	76.78
$RbBr(s)$	−394.6	−381.8	110.0
$RbCl(s)$	−435.5	−407.8	95.90
$RbF(s)$	−549.4	—	75.3
$RbI(s)$	−333.8	−328.9	118.4
Silicon			
$Si(g)$	450.0	405.6	167.9
$Si(s)$	0	0	18.82
$SiC(s)$	−65.3	−62.8	16.61
$SiCl_4(l)$	−687.0	−619.9	239.7
$SiF_4(g)$	−1614.9	−1572.7	282.7
$SiO_2(s$, quartz)	−910.9	−856.4	41.46
Silver			
$Ag(g)$	284.6	245.7	172.9
$Ag(s)$	0	0	42.55
$Ag^+(aq)$	105.6	77.12	72.68
$AgBr(s)$	−100.4	−96.90	107.1
$AgCl(s)$	−127.1	−109.8	96.2
$AgF(s)$	−204.6	—	83.7
$AgI(s)$	−61.84	−66.19	115.5
$Ag_2O(s)$	−31.05	−11.21	121.3
$Ag_2S(s)$	−32.59	−40.67	144.0
$AgNO_3(s)$	−124.4	−33.47	140.9
Sodium			
$Na(g)$	107.3	76.86	153.6
$Na(s)$	0	0	51.46
$Na^+(aq)$	−240.1	−261.9	59.1
$Na^+(g)$	609.3	574.4	147.8
$NaBr(s)$	−361.4	−349.3	86.82
$Na_2CO_3(s)$	−1130.8	−1048.0	138.8
$NaCl(s)$	−411.1	−384.0	72.12
$NaF(s)$	−575.4	−545.1	51.21
$NaHCO_3(s)$	−950.8	−851.0	101.7
$NaI(s)$	−287.9	−284.6	98.50
$NaNO_3(s)$	−467.9	−367.1	116.5
$NaOH(s)$	−425.9	−379.7	64.44
Strontium			
$Sr(g)$	164.0	131.6	164.5
$Sr(s)$	0	0	55.69
$Sr^+(g)$	719.7	679.3	170.3
$Sr^{2+}(aq)$	−545.8	−559.4	−32.6
$Sr^{2+}(g)$	1790.6	—	—
$SrCO_3(s)$	−1220.1	−1140.1	97.1
$SrCl_2(s)$	−828.9	−781.2	114.9
$SrO(s)$	−592.0	−561.4	55.52
$SrSO_4(s)$	−1453.1	−1341.0	117
Sulfur			
$S(g)$	277.0	236.5	167.7
$S(s$, monoclinic)	0.360	0.070	33.03
$S(s$, rhombic)	0	0	32.06
$S^{2-}(aq)$	41.8	83.7	22
$S_2(g)$	128.6	79.7	228.1
$S_8(g)$	100.4	48.61	430.2
$SO_2(g)$	−296.8	−300.1	248.1
$SO_3(g)$	−395.8	−371.0	256.7
$SO_4^{2-}(aq)$	−909.3	−744.6	20.1
$HS^-(aq)$	−17.7	12.6	61.1
$H_2S(aq)$	−39	−27.4	122
$H_2S(g)$	−20.50	−33.33	205.6
$HSO_4^-(aq)$	−887.3	−756.0	131.8
$H_2SO_4(l)$	−814.0	−689.9	156.9
Tin			
$Sn(s$, gray)	−2.09	0.13	44.14
$Sn(s$, white)	0	0	51.55
$SnCl_4(l)$	−511.3	−440.2	258.6
$SnO_2(s)$	−580.7	−519.7	52.3
Zinc			
$Zn(g)$	130.4	94.89	160.9
$Zn(s)$	0	0	41.72
$Zn^{2+}(aq)$	−153.9	−147.0	−112.1
$ZnCl_2(s)$	−415.1	−369.4	111.5
$ZnO(s)$	−348.3	−318.3	43.64
$ZnS(s$, sphalerite)	−206.0	−201.3	−57.7

Appendix D Electron Configurations of Atoms in the Ground State

Z	Element	Configuration
1	H	$1s^1$
2	He	$1s^2$
3	Li	[He]$2s^1$
4	Be	[He]$2s^2$
5	B	[He]$2s^22p^1$
6	C	[He]$2s^22p^2$
7	N	[He]$2s^22p^3$
8	O	[He]$2s^22p^4$
9	F	[He]$2s^22p^5$
10	Ne	[He]$2s^22p^6$
11	Na	[Ne]$3s^1$
12	Mg	[Ne]$3s^2$
13	Al	[Ne]$3s^23p^1$
14	Si	[Ne]$3s^23p^2$
15	P	[Ne]$3s^23p^3$
16	S	[Ne]$3s^23p^4$
17	Cl	[Ne]$3s^23p^5$
18	Ar	[Ne]$3s^23p^6$
19	K	[Ar]$4s^1$
20	Ca	[Ar]$4s^2$
21	Sc	[Ar]$3d^14s^2$
22	Ti	[Ar]$3d^24s^2$
23	V	[Ar]$3d^34s^2$
24	Cr	[Ar]$3d^54s^1$
25	Mn	[Ar]$3d^54s^2$
26	Fe	[Ar]$3d^64s^2$
27	Co	[Ar]$3d^74s^2$
28	Ni	[Ar]$3d^84s^2$
29	Cu	[Ar]$3d^{10}4s^1$
30	Zn	[Ar]$3d^{10}4s^2$
31	Ga	[Ar]$3d^{10}4s^24p^1$
32	Ge	[Ar]$3d^{10}4s^24p^2$
33	As	[Ar]$3d^{10}4s^24p^3$
34	Se	[Ar]$3d^{10}4s^24p^4$
35	Br	[Ar]$3d^{10}4s^24p^5$
36	Kr	[Ar]$3d^{10}4s^24p^6$
37	Rb	[Kr]$5s^1$
38	Sr	[Kr]$5s^2$
39	Y	[Kr]$4d^15s^2$
40	Zr	[Kr]$4d^25s^2$
41	Nb	[Kr]$4d^45s^1$
42	Mo	[Kr]$4d^55s^1$
43	Tc	[Kr]$4d^55s^2$
44	Ru	[Kr]$4d^75s^1$
45	Rh	[Kr]$4d^85s^1$
46	Pd	[Kr]$4d^{10}$
47	Ag	[Kr]$4d^{10}5s^1$
48	Cd	[Kr]$4d^{10}5s^2$
49	In	[Kr]$4d^{10}5s^25p^1$
50	Sn	[Kr]$4d^{10}5s^25p^2$
51	Sb	[Kr]$4d^{10}5s^25p^3$
52	Te	[Kr]$4d^{10}5s^25p^4$
53	I	[Kr]$4d^{10}5s^25p^5$
54	Xe	[Kr]$4d^{10}5s^25p^6$
55	Cs	[Xe]$6s^1$
56	Ba	[Xe]$6s^2$
57	La	[Xe]$5d^16s^2$
58	Ce	[Xe]$4f^15d^16s^2$
59	Pr	[Xe]$4f^36s^2$
60	Nd	[Xe]$4f^46s^2$
61	Pm	[Xe]$4f^56s^2$
62	Sm	[Xe]$4f^66s^2$
63	Eu	[Xe]$4f^76s^2$
64	Gd	[Xe]$4f^75d^16s^2$
65	Tb	[Xe]$4f^96s^2$
66	Dy	[Xe]$4f^{10}6s^2$
67	Ho	[Xe]$4f^{11}6s^2$
68	Er	[Xe]$4f^{12}6s^2$
69	Tm	[Xe]$4f^{13}6s^2$
70	Yb	[Xe]$4f^{14}6s^2$
71	Lu	[Xe]$4f^{14}5d^16s^2$
72	Hf	[Xe]$4f^{14}5d^26s^2$
73	Ta	[Xe]$4f^{14}5d^36s^2$
74	W	[Xe]$4f^{14}5d^46s^2$
75	Re	[Xe]$4f^{14}5d^56s^2$
76	Os	[Xe]$4f^{14}5d^66s^2$
77	Ir	[Xe]$4f^{14}5d^76s^2$
78	Pt	[Xe]$4f^{14}5d^96s^1$
79	Au	[Xe]$4f^{14}5d^{10}6s^1$
80	Hg	[Xe]$4f^{14}5d^{10}6s^2$
81	Tl	[Xe]$4f^{14}5d^{10}6s^26p^1$
82	Pb	[Xe]$4f^{14}5d^{10}6s^26p^2$
83	Bi	[Xe]$4f^{14}5d^{10}6s^26p^3$
84	Po	[Xe]$4f^{14}5d^{10}6s^26p^4$
85	At	[Xe]$(4f^{14}5d^{10}6s^26p^5)$
86	Rn	[Xe]$4f^{14}5d^{10}6s^26p^6$

Z	Element	Configuration
87	Fr	[Rn]($7s^1$)
88	Ra	[Rn]$7s^2$
89	Ac	[Rn]$6d^16s^2$
90	Th	[Rn]$6d^27s^2$
91	Pa	[Rn]($5f^26d^17s^2$)
92	U	[Rn]($5f^36d^17s^2$)
93	Np	[Rn]($5f^46d^17s^2$)
94	Pu	[Rn]$5f^67s^2$
95	Am	[Rn]$5f^77s^2$
96	Cm	[Rn]($5f^76d^17s^2$)
97	Bk	[Rn]($5f^97s^2$)
98	Cf	[Rn]($5f^{10}7s^2$)
99	Es	[Rn]($5f^{11}7s^2$)
100	Fm	[Rn]($5f^{12}7s^2$)
101	Md	[Rn]($5f^{13}7s^2$)
102	No	[Rn]($5f^{14}7s^2$)
103	Lr	[Rn]($5f^{14}6d^17s^2$)
104	Rf	[Rn]($5f^{14}5d^27s^2$)
105	Db	[Rn]($5f^{14}6d^37s^2$)
106	Sg	[Rn]($5f^{14}6d^47s^2$)
107	Bh	[Rn]($5f^{14}6d^57s^2$)
108	Hs	[Rn]($5f^{14}6d^67s^2$)
109	Mt	[Rn]($5f^{14}6d^77s^2$)
110	Ds	[Rn]($5f^{14}6d^87s^2$)
111	Rg	[Rn]($5f^{14}6d^97s^2$)
112	Cn	[Rn]($5f^{14}6d^{10}7s^2$)
113	Uut	[Rn]($5f^{14}6d^{10}7s^27p^1$)
114	Uuq	[Rn]($5f^{14}6d^{10}7s^27p^2$)
115	Uup	[Rn]($5f^{14}6d^{10}7s^27p^3$)
116	Uuh	[Rn]($5f^{14}6d^{10}7s^27p^4$)
117	Uus	[Rn]($5f^{14}6d^{10}7s^27p^5$)
118	Uuo	[Rn]($5f^{14}6d^{10}7s^27p^6$)

Configurations in this table, except those in parentheses, are taken from Charlotte E. Moore, *National Standard Reference Data Series*, National Bureau of Standards, 34 (September 1970). Those in parentheses were obtained on the basis of their assumed position in the periodic table.

Appendix E Acid-Ionization Constants at 25°C

Substance	Formula	K_a
Acetic acid	$HC_2H_3O_2$	1.7×10^{-5}
Arsenic acid*	H_3AsO_4	6.5×10^{-3}
	$H_2AsO_4^-$	1.2×10^{-7}
	$HAsO_4^{2-}$	3.2×10^{-12}
Ascorbic acid*	$H_2C_6H_6O_6$	6.8×10^{-5}
	$HC_6H_6O_6^-$	2.8×10^{-12}
Benzoic acid	$HC_7H_5O_2$	6.3×10^{-5}
Boric acid	H_3BO_3	5.9×10^{-10}
Carbonic acid*	H_2CO_3	4.3×10^{-7}
	HCO_3^-	4.8×10^{-11}
Chromic acid*	H_2CrO_4	1.5×10^{-1}
	$HCrO_4^-$	3.2×10^{-7}
Cyanic acid	HOCN	3.5×10^{-4}
Formic acid	$HCHO_2$	1.7×10^{-4}
Hydrocyanic acid	HCN	4.9×10^{-10}
Hydrofluoric acid	HF	6.8×10^{-4}
Hydrogen peroxide	H_2O_2	1.8×10^{-12}
Hydrogen sulfate ion	HSO_4^-	1.1×10^{-2}
Hydrogen sulfide*	H_2S	8.9×10^{-8}
	HS^-	1.2×10^{-13}
Hypochlorous acid	HClO	3.5×10^{-8}

(*continued*)

(continued)

Substance	Formula	K_a
Lactic acid	$HC_3H_5O_3$	1.3×10^{-4}
Nitrous acid	HNO_2	4.5×10^{-4}
Oxalic acid*	$H_2C_2O_4$	5.6×10^{-2}
	$HC_2O_4^-$	5.1×10^{-5}
Phenol	C_6H_5OH	1.1×10^{-10}
Phosphoric acid*	H_3PO_4	6.9×10^{-3}
	$H_2PO_4^-$	6.2×10^{-8}
	HPO_4^{2-}	4.8×10^{-13}
Phosphorous acid*	H_2PHO_3	1.6×10^{-2}
	$HPHO_3^-$	7×10^{-7}
Propionic acid	$HC_3H_5O_2$	1.3×10^{-5}
Pyruvic acid	$HC_3H_3O_3$	1.4×10^{-4}
Sulfurous acid*	H_2SO_3	1.3×10^{-2}
	HSO_3^-	6.3×10^{-8}

*The ionization constants for polyprotic acids are for successive ionizations. For example, for H_3PO_4, the equilibrium is H_3PO_4—$H^+ + H_2PO_4^-$. For $H_2PO_4^-$, the equilibrium is $H_2PO_4^-$—$H^+ + HPO_4^{2-}$.

Appendix F Base-Ionization Constants at 25°C

Substance	Formula	K_b
Ammonia	NH_3	1.8×10^{-5}
Aniline	$C_6H_5NH_2$	4.2×10^{-10}
Dimethylamine	$(CH_3)_2NH$	5.1×10^{-4}
Ethylamine	$C_2H_5NH_2$	4.7×10^{-4}
Ethylenediamine	$NH_2CH_2CH_2NH_2$	5.2×10^{-4}
Hydrazine	N_2H_4	1.7×10^{-6}
Hydroxylamine	NH_2OH	1.1×10^{-8}
Methylamine	CH_3NH_2	4.4×10^{-4}
Pyridine	C_5H_5N	1.4×10^{-9}
Trimethylamine	$(CH_3)_3N$	6.5×10^{-5}
Urea	NH_2CONH_2	1.5×10^{-14}

Appendix G Solubility Product Constants at 25°C

Substance	Formula	K_{sp}
Aluminum hydroxide	$Al(OH)_3$	4.6×10^{-33}
Barium chromate	$BaCrO_4$	1.2×10^{-10}
Barium fluoride	BaF_2	1.0×10^{-6}
Barium sulfate	$BaSO_4$	1.1×10^{-10}
Cadmium oxalate	CdC_2O_4	1.5×10^{-8}
Cadmium sulfide	CdS	8×10^{-27}
Calcium carbonate	$CaCO_3$	3.8×10^{-9}
Calcium fluoride	CaF_2	3.4×10^{-11}
Calcium oxalate	CaC_2O_4	2.3×10^{-9}
Calcium phosphate	$Ca_3(PO_4)_2$	1×10^{-26}
Calcium sulfate	$CaSO_4$	2.4×10^{-5}
Cobalt(II) sulfide	CoS	4×10^{-21}
Copper(II) hydroxide	$Cu(OH)_2$	2.6×10^{-19}
Copper(II) sulfide	CuS	6×10^{-36}
Iron(II) hydroxide	$Fe(OH)_2$	8×10^{-16}
Iron(II) sulfide	FeS	6×10^{-18}
Iron(III) hydroxide	$Fe(OH)_3$	2.5×10^{-39}
Lead(II) arsenate	$Pb_3(AsO_4)_2$	4×10^{-36}
Lead(II) chloride	$PbCl_2$	1.6×10^{-5}
Lead(II) chromate	$PbCrO_4$	1.8×10^{-14}
Lead(II) iodide	PbI_2	6.5×10^{-9}
Lead(II) sulfate	$PbSO_4$	1.7×10^{-8}
Lead(II) sulfide	PbS	2.5×10^{-27}
Magnesium arsenate	$Mg_3(AsO_4)_2$	2×10^{-20}
Magnesium carbonate	$MgCO_3$	1.0×10^{-5}
Magnesium hydroxide	$Mg(OH)_2$	1.8×10^{-11}
Magnesium oxalate	MgC_2O_4	8.5×10^{-5}
Manganese(II) sulfide	MnS	2.5×10^{-10}
Mercury(I) chloride	Hg_2Cl_2	1.3×10^{-18}
Mercury(II) sulfide	HgS	1.6×10^{-52}
Nickel(II) hydroxide	$Ni(OH)_2$	2.0×10^{-15}
Nickel(II) sulfide	NiS	3×10^{-19}
Silver acetate	$AgC_2H_3O_2$	2.0×10^{-3}
Silver bromide	$AgBr$	5.0×10^{-13}
Silver chloride	$AgCl$	1.8×10^{-10}
Silver chromate	Ag_2CrO_4	1.1×10^{-12}
Silver iodide	AgI	8.3×10^{-17}
Silver sulfide	Ag_2S	6×10^{-50}
Strontium carbonate	$SrCO_3$	9.3×10^{-10}
Strontium chromate	$SrCrO_4$	3.5×10^{-5}
Strontium sulfate	$SrSO_4$	2.5×10^{-7}
Zinc hydroxide	$Zn(OH)_2$	2.1×10^{-16}
Zinc sulfide	ZnS	1.1×10^{-21}

Appendix H Formation Constants of Complex Ions at 25°C

Complex Ion	K_f	Complex Ion	K_f
$Ag(CN)_2^-$	5.6×10^{18}	$Fe(CN)_6^{4-}$	1.0×10^{35}
$Ag(NH_3)_2^+$	1.7×10^{7}	$Fe(CN)_6^{3-}$	9.1×10^{41}
$Ag(S_2O_3)_2^{3-}$	2.9×10^{13}	$Ni(CN)_4^{2-}$	1.0×10^{31}
$Cd(NH_3)_4^{2+}$	1.0×10^{7}	$Ni(NH_3)_6^{2+}$	5.6×10^{8}
$Cu(CN)_2^-$	1.0×10^{16}	$Zn(NH_3)_4^{2+}$	2.9×10^{9}
$Cu(NH_3)_4^{2+}$	4.8×10^{12}	$Zn(OH)_4^{2-}$	2.8×10^{15}

Appendix I Standard Electrode (Reduction) Potentials in Aqueous Solution at 25°C

Cathode (Reduction) Half-Reaction	Standard Potential, $E°$ (V)
$Li^+(aq) + e^- \rightleftharpoons Li(s)$	−3.04
$K^+(aq) + e^- \rightleftharpoons K(s)$	−2.92
$Ca^{2+}(aq) + 2e^- \rightleftharpoons Ca(s)$	−2.76
$Na^+(aq) + e^- \rightleftharpoons Na(s)$	−2.71
$Mg^{2+}(aq) + 2e^- \rightleftharpoons Mg(s)$	−2.38
$Al^{3+}(aq) + 3e^- \rightleftharpoons Al(s)$	−1.66
$2H_2O(l) + 2e^- \rightleftharpoons H_2(g) + 2OH^-(aq)$	−0.83
$Zn^{2+}(aq) + 2e^- \rightleftharpoons Zn(s)$	−0.76
$Cr^{3+}(aq) + 3e^- \rightleftharpoons Cr(s)$	−0.74
$Fe^{2+}(aq) + 2e^- \rightleftharpoons Fe(s)$	−0.41
$Cd^{2+}(aq) + 2e^- \rightleftharpoons Cd(s)$	−0.40
$Ni^{2+}(aq) + 2e^- \rightleftharpoons Ni(s)$	−0.23
$Sn^{2+}(aq) + 2e^- \rightleftharpoons Sn(s)$	−0.14
$Pb^{2+}(aq) + 2e^- \rightleftharpoons Pb(s)$	−0.13
$Fe^{3+}(aq) + 3e^- \rightleftharpoons Fe(s)$	−0.04
$2H^+(aq) + 2e^- \rightleftharpoons H_2(g)$	0.00
$Sn^{4+}(aq) + 2e^- \rightleftharpoons Sn^{2+}(aq)$	0.15
$Cu^{2+}(aq) + e^- \rightleftharpoons Cu^+(aq)$	0.16
$ClO_4^-(aq) + H_2O(l) + 2e^- \rightleftharpoons ClO_3^-(aq) + 2OH^-(aq)$	0.17
$AgCl(s) + e^- \rightleftharpoons Ag(s) + Cl^-(aq)$	0.22
$Cu^{2+}(aq) + 2e^- \rightleftharpoons Cu(s)$	0.34
$ClO_3^-(aq) + H_2O(l) + 2e^- \rightleftharpoons ClO_2^-(aq) + 2OH^-(aq)$	0.35
$IO^-(aq) + H_2O(l) + 2e^- \rightleftharpoons I^-(aq) + 2OH^-(aq)$	0.49
$Cu^+(aq) + e^- \rightleftharpoons Cu(s)$	0.52
$I_2(s) + 2e^- \rightleftharpoons 2I^-(aq)$	0.54
$ClO_2^-(aq) + H_2O(l) + 2e^- \rightleftharpoons ClO^-(aq) + 2OH^-(aq)$	0.59
$Fe^{3+}(aq) + e^- \rightleftharpoons Fe^{2+}(aq)$	0.77
$Hg_2^{2+}(aq) + 2e^- \rightleftharpoons 2Hg(l)$	0.80

Cathode (Reduction) Half-Reaction	Standard Potential, $E°$ (V)
$Ag^+(aq) + e^- \rightleftharpoons Ag(s)$	0.80
$Hg^{2+}(aq) + 2e^- \rightleftharpoons Hg(l)$	0.85
$ClO^-(aq) + H_2O(l) + 2e^- \rightleftharpoons Cl^-(aq) + 2OH^-(aq)$	0.90
$2Hg^{2+}(aq) + 2e^- \rightleftharpoons Hg_2^{2+}(aq)$	0.90
$NO_3^-(aq) + 4H^+(aq) + 3e^- \rightleftharpoons NO(g) + 2H_2O(l)$	0.96
$Br_2(l) + 2e^- \rightleftharpoons 2Br^-(aq)$	1.07
$O_2(g) + 4H^+(aq) + 4e^- \rightleftharpoons 2H_2O(l)$	1.23
$Cr_2O_7^{2-}(aq) + 14H^+(aq) + 6e^- \rightleftharpoons 2Cr^{3+}(aq) + 7H_2O(l)$	1.33
$Cl_2(g) + 2e^- \rightleftharpoons 2Cl^-(aq)$	1.36
$Ce^{4+}(aq) + e^- \rightleftharpoons Ce^{3+}(aq)$	1.44
$MnO_4^-(aq) + 8H^+(aq) + 5e^- \rightleftharpoons Mn^{2+}(aq) + 4H_2O(l)$	1.49
$H_2O_2(aq) + 2H^+(aq) + 2e^- \rightleftharpoons 2H_2O(l)$	1.78
$Co^{3+}(aq) + e^- \rightleftharpoons Co^{2+}(aq)$	1.82
$S_2O_8^{2-}(aq) + 2e^- \rightleftharpoons 2SO_4^{2-}(aq)$	2.01
$O_3(g) + 2H^+(aq) + 2e^- \rightleftharpoons O_2(g) + H_2O(l)$	2.07
$F_2(g) + 2e^- \rightleftharpoons 2F^-(aq)$	2.87

Answers to Exercises

Note: Your answers may differ from those given here in the last significant figure.

CHAPTER 1

1.1 11.02 g **1.2** Physical properties: soft; silver color; melts at 64°C. Chemical properties: metal, reacts with water; reacts with oxygen; reacts with chlorine. **1.3** a 4.9 b 2.48 c 0.08 d 3 **1.4** a 1.84 nm b 5.67 ps c 7.85 mg d 9.7 km e 0.732 ms or 732 μs f 0.154 nm or 154 pm **1.5** a 39.2°C b 195 K **1.6** 7.87 g/cm^3; the object is made of iron **1.7** 38.4 cm^3 **1.8** 1.21×10^{-7} mm **1.9** 6.76×10^{-26} dm^3 **1.10** 3.24 m

CHAPTER 2

2.1 $^{35}_{17}Cl$ **2.2** 35.453 amu **2.3** a Se: Group VIA, Period 4; nonmetal b Cs: Group IA, Period 6; metal c Fe: Group VIIIB, Period 4; metal d Cu: Group IB, Period 4; metal e Br: Group VIIA, Period 4; nonmetal **2.4** K_2CrO_4 **2.5** a calcium oxide b lead(II) chromate **2.6** $Tl(NO_3)_3$ **2.7** a dichlorine hex(a)oxide b phosphorus trichloride c phosphorus pentachloride **2.8** a CS_2 b SO_3 **2.9** a boron trifluoride b hydrogen selenide **2.10** perbromate ion, BrO_4^- **2.11** sodium carbonate decahydrate **2.12** $Na_2S_2O_3 \cdot 5H_2O$ **2.13** a $O_2 + 2PCl_3 \longrightarrow 2POCl_3$ b $P_4 + 6N_2O \longrightarrow P_4O_6 + 6N_2$ c $2As_2S_3 + 9O_2 \longrightarrow 2As_2O_3 + 6SO_2$ d $Ca_3(PO_4)_2 + 4H_3PO_4 \longrightarrow 3Ca(H_2PO_4)_2$

CHAPTER 3

3.1 a 46.0 amu b 180 amu c 40.0 amu d 58.3 amu **3.2** a SO_3, 80.1 amu b H_2SO_4, 98.1 amu **3.3** a 6.656×10^{-23} g b 7.65×10^{-23} g **3.4** 30.9 g H_2O_2 **3.5** 0.452 mol HNO_3 **3.6** 1.2×10^{21} HCN molecules **3.7** 35.0% N, 60.0% O, 5.04% H **3.8** 17.0 g N **3.9** 40.9% C, 4.57% H, 54.5% O **3.10** SO_3 **3.11** $C_7H_6O_2$ **3.12** C_2H_4O
3.13 $H_2 + Cl_2 \longrightarrow 2HCl$

1 molecule H_2 + 1 molecule $Cl_2 \longrightarrow$ 2 molecules HCl
1 mol H_2 + 1 mol $Cl_2 \longrightarrow$ 2 mol HCl
2.016 g H_2 + 70.9 g $Cl_2 \longrightarrow 2 \times 36.5$ g HCl

3.14 178 g Na **3.15** 2.46 kg O_2 **3.16** 81.1 g Hg **3.17** 0.12 mol $AlCl_3$ **3.18** 11.0 g ZnS **3.19** 21.4 g; 89.1%

CHAPTER 4

4.1 a soluble b soluble c insoluble
4.2 a $2H^+(aq) + 2NO_3^-(aq) + Mg(OH)_2(s) \longrightarrow 2H_2O(l) + Mg^{2+}(aq) + 2NO_3^-(aq)$
$2H^+(aq) + Mg(OH)_2(s) \longrightarrow 2H_2O(l) + Mg^{2+}(aq)$
b $Pb^{2+}(aq) + 2NO_3^-(aq) + 2Na^+(aq) + SO_4^{2-}(aq) \longrightarrow PbSO_4(s) + 2Na^+(aq) + 2NO_3^-(aq)$
$Pb^{2+}(aq) + SO_4^{2-}(aq) \longrightarrow PbSO_4(s)$
4.3 $2NaI(aq) + Pb(C_2H_3O_2)_2(aq) \longrightarrow PbI_2(s) + 2NaC_2H_3O_2(aq)$
$2I^-(aq) + Pb^{2+}(aq) \longrightarrow PbI_2(s)$
4.4 a weak acid b weak acid c strong acid d strong base
4.5 $HCN(aq) + LiOH(aq) \longrightarrow H_2O(l) + LiCN(aq)$
$HCN(aq) + OH^-(aq) \longrightarrow H_2O(l) + CN^-(aq)$
4.6 $H_2SO_4(aq) + KOH(aq) \longrightarrow H_2O(l) + KHSO_4(aq)$
$KHSO_4(aq) + KOH(aq) \longrightarrow H_2O(l) + K_2SO_4(aq)$
$H^+(aq) + OH^-(aq) \longrightarrow H_2O(l)$
$HSO_4^-(aq) + OH^-(aq) \longrightarrow H_2O(l) + SO_4^{2-}(aq)$
4.7 $CaCO_3(s) + 2HNO_3(aq) \longrightarrow Ca(NO_3)_2(aq) + CO_2(g) + H_2O(l)$
$CaCO_3(s) + 2H^+(aq) \longrightarrow Ca^{2+}(aq) + CO_2(g) + H_2O(l)$
4.8 a K: +1, Cr: +6, O: −2 b Mn: +7, O: −2
4.9 $Ca + Cl_2 \longrightarrow Ca^{2+} + 2Cl^-$ **4.10** 0.0464 *M* NaCl **4.11** 10.1 mL **4.12** 0.0075 mol NaCl; 0.44 g NaCl **4.13** 12 mL **4.14** 34.19% **4.15** 48.4 mL $NiSO_4$ **4.16** 5.07%

CHAPTER 5

5.1 0.56 atm, 4.3×10^2 mmHg **5.2** 24.1 L **5.3** 4.47 dm^3 **5.4** 5.54 dm^3 **5.5** Use $PV = nRT$. Solve for n: $n = PV/RT = P(V/RT)$. Everything within parentheses is constant; therefore, $n = \text{constant} \times P$ (or $\propto P$). **5.6** 46.0 atm **5.7** density of He = 0.164 g/L; difference in mass between 1 L of air and 1 L of He = 1.024 g **5.8** 64.1 amu **5.9** 2.00 L **5.10** P_{O_2} = 0.0769 atm; P_{CO_2} = 0.0310 atm; P = 0.1080 atm; X_{O_2} = 0.712 **5.11** 0.01591 mol O_2; V = 0.406 L **5.12** 219 m/s **5.13** T = 52.4 K, 728 K **5.14** 9.96 s **5.15** 44.0 amu **5.16** P_{vdW} = 0.992 atm; P_{ideal} = 1.000 atm (larger)

CHAPTER 6

6.1 1.1×10^{-17} J, 2.7×10^{-18} cal **6.2** 3.89 J **6.3** exothermic; $q = -1170$ kJ **6.4** −48.9 L; 4.96 kJ; −885.2 kJ
6.5 $2N_2H_4(l) + N_2O_4(l) \longrightarrow 3N_2(g) + 4H_2O(g)$; $\Delta H = -1049$ kJ
6.6 a $N_2H_4(l) + \frac{1}{2}N_2O_4(l) \longrightarrow \frac{3}{2}N_2(g) + 2H_2O(g)$; $\Delta H = -5.245 \times 10^2$ kJ
b $4H_2O(g) + 3N_2(g) \longrightarrow N_2O_4(l) + 2N_2H_4(l)$; $\Delta H = 1049$ kJ
6.7 −164 kJ **6.8** 1.80×10^2 J **6.9** −54 kJ; $HCl(aq) + NaOH(aq) \longrightarrow NaCl(aq) + H_2O(l)$; $\Delta H = -54$ kJ
6.10 $\Delta H = -1792$ kJ **6.11** $\Delta H^\circ_{vap} = 44.0$ kJ
6.12 $\Delta H^\circ = -138.0$ kJ **6.13** $\Delta H^\circ = 61.6$ kJ

CHAPTER 7

7.1 767 nm **7.2** 6.58×10^{14}/s **7.3** 2.0×10^{-19} J, 2.0×10^{-17} J, and 2.0×10^{-15} J; greatest energy: x-ray region; least energy: infrared region **7.4** 103 nm **7.5** 3.38×10^{-19} J **7.6** λ = 332 pm **7.7** a The value of n must be a positive whole number greater than zero. b The values of l range from 0 to $n - 1$. Here, l has a value greater than n. c The values for m_l range from $-l$ to $+l$. Here, m_l has a value greater than that of l. d The values for m_s are $+\frac{1}{2}$ or $-\frac{1}{2}$, not 0.

CHAPTER 8

8.1 a possible b possible c impossible. There are two electrons in a $2p$ orbital with the same spin. d possible.

e impossible. Only two electrons are allowed in an *s* subshell. **f** impossible. Only six electrons are allowed in a *p* subshell. **8.2** $1s^22s^22p^63s^23p^63d^54s^2$ **8.3** $4s^24p^3$ **8.4** Group IVA, Period 6; main-group element

8.5 (↑↓) (↑↓) (↑↓)(↑↓)(↑↓) (↑↓) (↑)(↑)(↑)
1s 2s 2p 3s 3p

8.6 In order of increasing radius: Be, Mg, Na **8.7** It is more likely that 1000 kJ/mol is the ionization energy for iodine because ionization energies tend to decrease as atomic number increases in a column. **8.8** F

CHAPTER 9

9.1 $\cdot\text{Mg}\cdot + :\ddot{\text{O}}\cdot \longrightarrow \text{Mg}^{2+} + [:\ddot{\text{O}}:]^{2-}$ **9.2** The electron configuration for Ca^{2+} is [Ar] and its Lewis symbol is Ca^{2+}. The electron configuration for S^{2-} is $[Ne]3s^23p^6$ and its Lewis symbol is $[:\ddot{\text{S}}:]^{2-}$. **9.3** The electron configurations of Pb and Pb^{2+} are $[Xe]4f^{14}5d^{10}6s^26p^2$ and $[Xe]4f^{14}5d^{10}6s^2$, respectively. **9.4** $[Ar]3d^5$ **9.5** One would expect S^{2-}, because it has two additional electrons. **9.6** In order of increasing ionic radius: Mg^{2+}, Ca^{2+}, Sr^{2+} **9.7** In order of increasing ionic radius: Ca^{2+}, Cl^-, P^{3-} **9.8** C—O is the most polar bond.

9.9
```
      ..
     :Cl:
  ..  |  ..
 :F — C — F:
  ..  |  ..
     :Cl:
      ..
```
9.10 $\ddot{\text{O}}::\text{C}::\ddot{\text{O}}$ or $\ddot{\text{O}}=\text{C}=\ddot{\text{O}}$

9.11 **a**
```
[     H    ]+
[     |    ]
[ H — O — H]
      ..
```
b
```
[ ..   ..   .. ]-
[:O — Cl — O:  ]
[ ..   ..   .. ]
```

9.12
```
[    :O:     ]-       [    :O:     ]-       [    :O:     ]-
[     ||     ]        [     |      ]        [     |      ]
[:O — N — O: ]  <-->  [ O = N — O: ]  <-->  [:O — N = O  ]
[ ..     ..  ]        [ ..     ..  ]        [ ..     ..  ]
```

9.13
```
 :F:     :F:
    \   /
     S:
    /   \
 :F:     :F:
```
9.14 $:\ddot{\text{Cl}}—\text{Be}—\ddot{\text{Cl}}:$

9.15
```
           :O:
            ||
 H — O — P — O — H
     ..      ..
            |
           :O:
            |
            H
```

9.16 97 pm **9.17** 123 pm **9.18** −1308 kJ

CHAPTER 10

10.1 **a** trigonal pyramidal **b** bent **c** tetrahedral **10.2** T-shaped **10.3** **b**, **c** **10.4** **b**; the F atoms are distributed in a tetrahedral arrangement to give a molecule with a dipole moment of zero. **10.5** For the N atom:

N (ground state) = [He] (↑↓) (↑)(↑)(↑) and
2s 2p

N (hybridized) = [He] (↑↓)(↑)(↑)(↑)
sp^3

Each N—H bond is formed by the overlap of a 1*s* orbital of a hydrogen atom with one of the sp^3 hybrid orbitals of the nitrogen atom. **10.6** Each P—Cl bond is formed by the overlap of a phosphorus sp^3d hybrid orbital with a singly occupied 3*p* chlorine orbital. **10.7** The C atom is *sp* hybridized. A carbon–oxygen bond is double and is composed of a σ bond and a π bond. The σ bond is formed by the overlap of a hybrid C orbital with a 2*p* orbital of O that lies along the axis. The π bond is formed by the sidewise overlap of a 2*p* on C with a 2*p* on O. **10.8** The structural formulas for the isomers are as follows:

```
 F        F          F
  \      /          /
  .N=N.          .N=N.
                 /
                F
   cis          trans
```

These compounds exist as separate isomers. If they were interchangeable, one end of the molecule would have to remain fixed as the other rotated. This would require breaking the π bond and a considerable expenditure of energy.

10.9 KK(↑↓) (↑↓) (↑↓)(↑↓) ( ) ( )( ) ( );
σ_{2s} σ^*_{2s} π_{2p} σ_{2p} π^*_{2p} σ^*_{2p}

$KK(\sigma_{2s})^2(\sigma^*_{2s})^2(\pi_{2p})^4$; diamagnetic; bond order = 2

10.10 KK(↑↓) (↑↓) (↑↓)(↑↓) (↑↓) ( )( ) ( );
σ_{2s} σ^*_{2s} π_{2p} σ_{2p} π^*_{2p} σ^*_{2p}

$KK(\sigma_{2s})^2(\sigma^*_{2s})^2(\pi_{2p})^4(\sigma_{2p})^2$; bond order = 3; diamagnetic

CHAPTER 11

11.1 1.37×10^3 kJ; 4.12×10^3 g **11.2** 5.30×10^2 mmHg **11.3** 43.0 kJ/mol **11.4** **a** Methyl chloride can be liquefied by sufficiently increasing the pressure as long as the temperature is below 144°C. **b** Oxygen can be liquefied by compression as long as the temperature is below −119°C. **11.5** **a** hydrogen bonding, dipole–dipole forces, and London forces **b** London forces **c** dipole–dipole forces and London forces **11.6** In order of increasing vapor pressure we have: butane (C_4H_{10}), propane (C_3H_8), and ethane (C_2H_6). London forces increase with increasing molecular mass. Therefore, one would expect the lowest vapor pressure to correspond to the highest molecular mass. **11.7** Hydrogen bonding is negligible in methyl chloride but strong in ethanol, which explains the lower vapor pressure of ethanol. **11.8** **a** metallic solid **b** ionic solid **c** covalent network solid **d** molecular solid **11.9** C_2H_5OH, molecular; CH_4, molecular; CH_3Cl, molecular; $MgSO_4$, ionic. In order of increasing melting point: CH_4, CH_3Cl, C_2H_5OH, and $MgSO_4$. **11.10** 1 atom **11.11** 6.02×10^{23}/mol **11.12** 533 pm

CHAPTER 12

12.1 A dental filling, made up of liquid mercury and solid silver, is a solid solution. **12.2** C_4H_9OH, because of hydrogen bonding. **12.3** Na^+ **12.4** 8.45×10^{-3} g O_2/L **12.5** 7.07 g HCl; 27.9 g H_2O **12.6** 3.09 *m* $C_6H_5CH_3$ **12.7** mole fraction toluene = 0.194; mole fraction benzene = 0.806 **12.8** mole fraction methanol = 0.00550; mole fraction ethanol = 0.995 **12.9** 7.24 *m* CH_3OH **12.10** 2.96 *M* $(NH_2)_2CO$ **12.11** 2.20 *m* $(NH_2)_2CO$ **12.12** $\Delta P = 1.22$ mmHg; P = 155 mmHg **12.13** 1.89×10^{-1} g CH_2OHCH_2OH **12.14** 176 amu **12.15** 124 amu; P_4 **12.16** $\pi = 3.5$ atm **12.17** 100.077°C **12.18** $AlCl_3$

CHAPTER 13

13.1 $\dfrac{\Delta[NO_2F]}{\Delta t} = -\dfrac{\Delta[NO_2]}{\Delta t}$
13.2 1.1×10^{-4} M/s **13.3** zero order in [CO]; second order in $[NO_2]$; second order overall **13.4** rate = $k[NO_2]^2$; k = 0.71 L/(mol·s) **13.5** **a** $[N_2O_5]_t = 1.24 \times 10^{-2}$ mol/L **b** t = 4.80×10^3 s **13.6** half-life = 0.075 s; decreases to 50% of initial concentration in 0.075 s, to 25% in 0.151 s **13.7** $E_a = 2.06 \times 10^5$ J/mol, $k = 5.80 \times 10^{-2}/(M^{1/2}\cdot s)$ **13.8** $2H_2O_2 \longrightarrow 2H_2O + O_2$ **13.9** bimolecular **13.10** rate = $k[NO_2]^2$ **13.11** rate = $k_1[H_2O_2][I^-]$ **13.12** rate = $k_2(k_1/k_{-1})[NO]^2[O_2]$

CHAPTER 14

14.1 0.57 mol CO, 0.57 mol H_2O, 0.43 mol CO_2, and 0.43 mol H_2
14.2 **a** $K_c = \dfrac{[NH_3]^2[H_2O]^4}{[NO_2]^2[H_2]^7}$ **b** $K_c = \dfrac{[NH_3][H_2O]^2}{[NO_2][H_2]^{7/2}}$
14.3 $CO(g) + H_2O(g) \rightleftharpoons H_2(g) + CO_2(g)$; 0.57
14.4 $K_c = 2.3 \times 10^{-4}$ **14.5** $K_p = 1.24$
14.6 $K_c = \dfrac{[Ni(CO)_4]}{[CO]^4}$ **14.7** products; 0.018 M
14.8 $Q_c = 0.67$; more CO will form **14.9** 0.0096 moles of PCl_5 **14.10** 0.11 mol H_2, 0.11 mol I_2, 0.78 mol HI **14.11** 0.86 M PCl_5, 0.135 M PCl_3, 0.135 M Cl_2 **14.12** **a** in the reverse direction **b** in the reverse direction **14.13** **a** no; pressure has no effect. **b** no **c** yes **14.14** high temperature **14.15** High temperatures and low pressure would give the best yield of product.

CHAPTER 15

15.1 $\underset{\text{acid}}{H_2CO_3(aq)} + \underset{\text{base}}{CN^-(aq)} \rightleftharpoons \underset{\text{acid}}{HCN(aq)} + \underset{\text{base}}{HCO_3^-(aq)}$; HCN is the conjugate acid of CN^-.

15.2 **a**

```
 ..           H  H               ..     H  H
:F:           |  |              :F: H   |  |
 |    ..      |  |               |  |   |
:F—B  +  :O—C—H   ⟶        :F—B : O—C—H;
 ..   |   ..   |                ..  |   ..  |
:F:           H                :F:          H
 ..                              ..
```

BF_3 is the Lewis acid, CH_3OH is the Lewis base.

b

```
                :O:          [   :O:   ]2−
                 ‖           [    |    ]
[:O:]2−   +      C     ⟶     [  O=C    ]   ;
                 ‖           [    |    ]
                :O:          [   :O:   ]
```

O^{2-} is the Lewis base, CO_2 is the Lewis acid.
15.3 The reactants are favored. **15.4** **a** PH_3 **b** HI **c** H_2SO_3 **d** H_3AsO_4 **e** HSO_4^- **15.5** 0.250 M OH^-, 4.0×10^{-14} M H^+ **15.6** basic **15.7** 1.35 **15.8** 12.40 **15.9** 6.9×10^{-4} M **15.10** 4×10^{-4} M

CHAPTER 16

16.1 1.4×10^{-4}; 0.071 **16.2** $[H^+] = [C_2H_3O_2^-] = 1.3 \times 10^{-3}$ M; pH = 2.89; 0.013 **16.3** 3.24 **16.4** pH = 1.29; $[SO_3^{2-}] = 6.3 \times 10^{-8}$ M **16.5** 3.3×10^{-6} **16.6** 5.3×10^{-12} M
16.7 **a** acidic **b** neutral **c** acidic **16.8** **a** 1.5×10^{-11} **b** 2.4×10^{-5} **16.9** 1.5×10^{-6}; 8.19 **16.10** 8.5×10^{-5} M; 8.5×10^{-4} **16.11** 3.63 **16.12** 5.26 **16.13** 3.83 **16.14** 1.60 **16.15** 7.97 **16.16** **a** 11.28 **b** 9.26 **c** 5.22 **d** 2.23

CHAPTER 17

17.1 **a** $K_{sp} = [Ba^{2+}][SO_4^{2-}]$ **b** $K_{sp} = [Fe^{3+}][OH^-]^3$ **c** $K_{sp} = [Ca^{2+}]^3[PO_4^{3-}]^2$ **17.2** 1.8×10^{-10} **17.3** 4.5×10^{-36} **17.4** 0.67 g/L **17.5** **a** 6.3×10^{-3} mol/L **b** 4.4×10^{-5} mol/L **17.6** Precipitation is expected. **17.7** No precipitate will form. **17.8** silver cyanide **17.9** 1.2×10^{-9} M **17.10** No precipitate will form. **17.11** 0.44 mol/L

CHAPTER 18

18.1 304 J/(mol·K) **18.2** **a** positive **b** positive **c** negative **d** positive **18.3** 537 J/K **18.4** −888.7 kJ **18.5** 130.9 kJ **18.6** All four reactions are spontaneous in the direction written. **18.7** **a** $K = P_{CO_2}$ **b** $K = [Pb^{2+}][I^-]^2$ **c** $K = \dfrac{P_{CO_2}}{[H^+][HCO_3^-]}$
18.8 1×10^{-23} **18.9** 5×10^{-12} **18.10** 9×10^{-2} atm **18.11** T = 669.5 K, which is lower than that for $CaCO_3$.

CHAPTER 19

19.1 $I_2(s) + 10NO_3^-(aq) + 8H^+(aq) \longrightarrow 4H_2O(l) + 2IO_3^-(aq) + 10NO_2(g)$
19.2 $H_2O_2(aq) + 2ClO_2(aq) + 2OH^-(aq) \longrightarrow 2ClO_2^-(aq) + O_2(g) + 2H_2O(l)$
19.3 anode: $Ni(s) \longrightarrow Ni^{2+}(aq) + 2e^-$; cathode: $Ag^+(aq) + e^- \longrightarrow Ag(s)$; electrons flow from anode to cathode; Ag^+ flows to Ag; Ni^{2+} flows away from Ni.
19.4 $Zn(s)|Zn^{2+}(aq)||H^+(aq)|H_2(g)|Pt$
19.5 $Cd(s) + 2H^+(aq) \longrightarrow Cd^{2+}(aq) + H_2(g)$
19.6 -2.12×10^4 J **19.7** NO_3^- **19.8** No **19.9** 1.10 V
19.10 -1.4×10^2 kJ **19.11** 2.696 V **19.12** 1×10^{19}
19.13 1.42 V **19.14** 4×10^{-7} M Ni^{2+}
19.15 **a** $K^+(l) + e^- \longrightarrow K(l)$
$2Cl^-(l) \longrightarrow Cl_2(g) + 2e^-$
b $K^+(l) + e^- \longrightarrow K(l)$
$4OH^-(l) \longrightarrow O_2(g) + 2H_2O(g) + 4e^-$
19.16 $Ag^+(aq) + e^- \longrightarrow Ag(s)$ (cathode)
$2H_2O(l) \longrightarrow O_2(g) + 4H^+(aq) + 4e^-$ (anode)
19.17 2.52×10^{-2} A
19.18 8.66×10^{-2} g

CHAPTER 20

20.1 ${}^{40}_{19}K \longrightarrow {}^{40}_{20}Ca + {}^{0}_{-1}e$ **20.2** ${}^{235}_{92}U$ **20.3** **a** is stable; **b** and **c** are radioactive. **20.4** **a** positron emission **b** beta emission **20.5** **a** ${}^{40}_{20}Ca(d,p){}^{41}_{20}Ca$ **b** ${}^{12}_{6}C + {}^{2}_{1}H \longrightarrow {}^{13}_{6}C + {}^{1}_{1}H$ **20.6** ${}^{14}_{7}N$ **20.7** 3.2×10^{-5}/s **20.8** 5.26 y **20.9** 7.82×10^{-10}/s; 7.4×10^{-7} Ci **20.10** 0.200 **20.11** 1.0×10^4 y **20.12** **a** 2.58×10^9 J **b** 6.25 MeV

CHAPTER 22

22.1 **a** pentaamminechlorocobalt(III) chloride **b** potassium aquapentacyanocobaltate(III) **c** pentaaquahydroxoiron(III) ion **22.2** **a** $K_4[Fe(CN)_6]$ **b** $[Co(NH_3)_4Cl_2]Cl$ **c** $[PtCl_4]^{2-}$ **22.3** **a** no geometric isomers possible **b** geometric isomers possible (See Fig. 22.15 for similar example.)

22.4 a no optical isomers possible; b is similar to Figure 22.16.

22.5 ↑ ↑ / ↑↓ ↑↓ ↑↓ 2 unpaired electrons **22.6** ↑ ↑ ↑ / ↑↓ ↑↓

22.7 530 nm, 500 nm, yes; CN^- is a more strongly bonding ligand than H_2O.

CHAPTER 23

23.1 $CH_3CH(CH_3)CH_2CH_3$

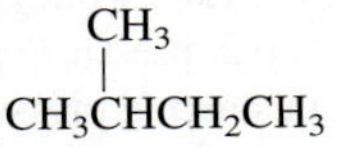

23.2 a cis: H(CH3)C=C(H)(CH2CH2CH2CH3)...

23.2 a $\text{H}(\text{CH}_3)\text{C}{=}\text{C}(\text{H})\text{CH}_2\text{CH}_2\text{CH}_3$ (H, H on same side; CH_3, $CH_2CH_2CH_3$ on same side) and $\text{H}(\text{CH}_3)\text{C}{=}\text{C}(\text{CH}_2\text{CH}_2\text{CH}_3)\text{H}$ (H and $CH_2CH_2CH_3$ across from each other) b none **23.3** 2-bromobutane **23.4** a 2,3-dimethylbutane b 3-ethyl-2-methylhexane

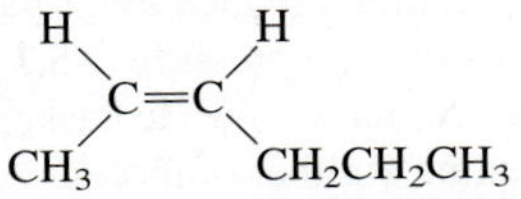

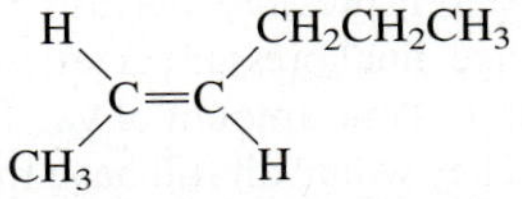

23.5 $CH_3CH_2{-}C(CH_3)_2{-}CH_2CH_2CH_2CH_2CH_3$

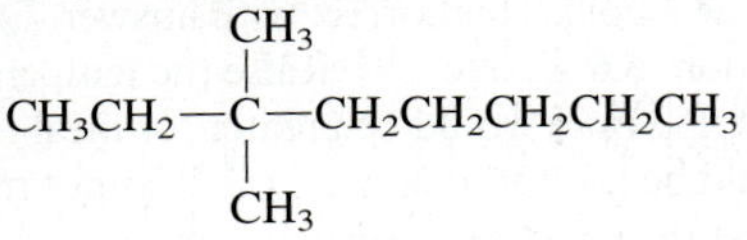

23.6 a 2,4-dimethyl-2-hexene b 3-propyl-1-hexene

23.7 $CH_3{-}C(CH_3){=}CH{-}CH_2{-}CH(CH_3){-}CH_2CH_3$

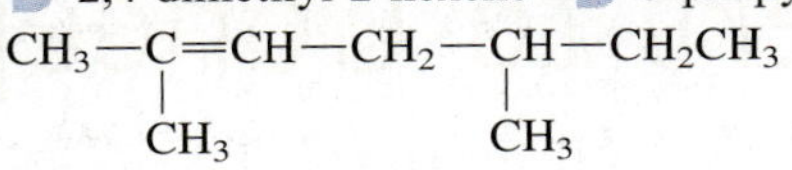

23.8 *cis*-3-hexene, *trans*-3-hexene

23.9 a propyne b 3-methyl-1-pentyne

23.10 a benzene ring with CH_2CH_3 b two benzene rings joined by $H_2C{-}CH_2$

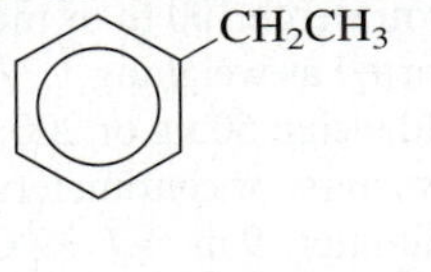

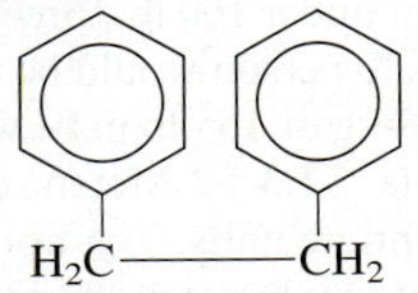

23.11 3-ethyl-3-hexanol **23.12** a dimethyl ether b methyl ethyl ether

23.13 a 2-pentanone b butanal

CHAPTER 24

24.1 $-CH_2{-}CCl_2{-}CH_2{-}CCl_2{-}CH_2{-}CCl_2{-}$

Answers to Concept Checks

CHAPTER 1

1.1 a element b compound c mixture **1.2** a two if you weigh under 100 lb, three if you weigh 100 lb or more b A 165-lb person would be reported as weighing 1.7×10^2 lb. c A 74.8-kg or 165-lb male would weigh 50 kg or 200 lb on the truck scale. **1.3** a Meters, decimeters, or centimeters would be appropriate units. b approximately 9 m c 39°C would mean that you had a moderate fever. d You would probably be comfortable in a 23°C room. **1.4** You need to perform an experiment that will give you the density of the marcasite and then compare that to the density of gold.

CHAPTER 2

2.1 The molecular model on the far right that contains 3 atoms with a ratio of 1:2 **2.2** One would conclude that the atom is made up primarily of a large impenetrable mass with a positive charge. **2.3** They must have similar chemical and/or physical properties. **2.4** Molecular: CH_4, Br_2, CH_4O. Ionic: $CaCl_2$, KNO_3, LiF. Answer d is correct. **2.5** a ether b alcohol c carboxylic acid d hydrocarbon **2.6** Q is likely to be an element in Group VIA.

CHAPTER 3

3.1 a 1.5 mol b 4.5 mol c 3.0 mol of OH^- since there are 2 mol of OH^- per mol of $Mg(OH)_2$ **3.2** a The total mass of C and H that you collect will be less than the total mass of material you started with. b The calculated percent C and H would be less than the real value. **3.3** a No, the empirical formula needs to be the smallest whole-number ratio of subscripts. b No, the subscripts are not whole numbers. c Yes, the empirical and molecular formulas can be the same. **3.4** a correct b incorrect c incorrect d correct e correct f incorrect g incorrect h correct **3.5** Answer d is the best answer considering the information provided in the problem. **3.6** a $X_2(g) + 2Y(g) \longrightarrow 2XY(g)$ b container 1 c $Y(g)$

CHAPTER 4

4.1 Strong electrolytes: NH_4Cl, $MgBr_2$, and HCl. Answers b, c, and e support the correct choices. **4.2** The left beaker contains the strong electrolyte $LiI(aq)$. The right beaker contains the nonelectrolyte $CH_3OH(aq)$.
4.3 a $Na_2SO_4(aq) + Sr(C_2H_3O_2)_2(aq) \longrightarrow SrSO_4(s) + 2NaC_2H_3O_2(aq)$ b $2Na^+(aq) + SO_4^{2-}(aq) + Sr^{2+}(aq) + 2C_2H_3O_2^-(aq) \longrightarrow SrSO_4(s) + 2Na^+(aq) + 2C_2H_3O_2^-(aq)$ c $Sr^{2+}(aq) + SO_4^{2-}(aq) \longrightarrow SrSO_4(s)$
4.4 $\longrightarrow H_3O^+(aq) + ClO_2^-(aq), \longrightarrow H_3O^+(aq) + Br^-(aq)$. Statement e is correct. **4.5** a $MOH(s)$ is a base. M could be any Group I metal. b HA is an acid. A^- is an anion with a -1 charge. c H_2A is an acid. A^{2-} is an anion with a -2 charge. d $M \longrightarrow Na^+$, $A \longrightarrow C_2H_3O_2^-$, $A \longrightarrow SO_3^{2-}$.
4.6 Concentration order: A< C = D < B. Leave A alone; add water to fill beaker B; add water to double the volumes of D and C. **4.7** a The acid in flask C has three times as many acidic protons as the acid in flask A. The acid in flask B has two times as many acidic protons as the acid in flask A. b Yes, if we assume that flask A contains a monoprotic acid.

CHAPTER 5

5.1 Since the height of the column is inversely related to the density of the liquid, the column containing the H_2O would be higher. **5.2** a Step 1, pressure decrease; Step 2, pressure decrease; total change, final pressure less than starting pressure. b Step 1, pressure increase; Step 2, pressure increase; total change, final pressure greater than starting pressure. **5.3** a All contain same amount of gas. b Xe flask c He flask d They would all still have the same number of moles. **5.4** a No change in H_2 pressure. b They are equal. c The total pressure is the sum of the pressures of the H_2 and Ar in the container **5.5** Choice II is correct with answer c being the best explanation. **5.6** a He b Raise the temperature of the Ar. **5.7** a Pressure would be greater for ideal gas. b Pressure would be less for ideal gas. c Cannot make a determination because the effect of molecular volume and attractions on the pressure of a real gas is opposite.

CHAPTER 6

6.1 Sun's energy is changed to electric energy, to chemical energy (of the battery), and back to electric energy, then to kinetic energy of motor and water, and to potential energy of water. **6.2** d **6.3** c and d are exothermic; d is most exothermic **6.4** $\Delta H_{sub} = \Delta H_{fus} + \Delta H_{vap}$

CHAPTER 7

7.1 The wavelength is halved; ultraviolet. **7.2** UV
7.3 The proton speed would be about 2000 times smaller.

CHAPTER 8

8.1 4 **8.2** Mg and Al **8.3** IIA **8.4** As

CHAPTER 9

9.1 a excited state; not expected b N^{2-} is not expected. c expected configuration d Na^{2+} is not expected. e expected configuration **9.2** a has 13 electron pairs; should be 12 b correct skeleton structure and distribution of electrons c Double bond between F and N atoms is less likely than single bond. d no octet on left N atom e incorrect skeleton structure f no octets on N atoms **9.3** a most accurate description b Electronegative N atom is not expected to be in central position. c not enough electron pairs to give octets on C and N atoms (would need 7 pairs) d not enough electron pairs to give octets on C and N atoms (would need 6 pairs)

CHAPTER 10

10.1 The four pairs have a tetrahedral arrangement; the molecular geometry is trigonal pyramidal. **10.2** Molecule Y is likely to be trigonal planar, but trigonal pyramidal or T-shaped geometries are possible. Molecule Z cannot have a trigonal planar geometry, but must be either trigonal pyramidal or T-shaped. **10.3** The single and triple bonds each require a sigma bond orbital, suggesting *sp* hybrids.

CHAPTER 11

11.1 a (i) slightly lower (ii) higher (iii) more molecules (iv) Evaporation is greater.

(v)

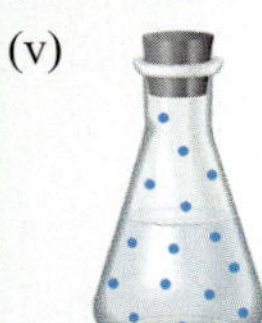

$t = 1$

b (i) slightly lower (ii) higher (iii) more molecules (iv) They will be equal.

(v)

$t = 2$

11.2 Cook it longer because the H_2O temperature would be lower. **11.3** Answer b is an incorrect statement.
11.4 a The hydrogen and oxygen would form an explosive mixture. b many times greater (more positive) for the wrong reaction c Apply Hess's law. **11.5** a AB
b face-centered cubic

CHAPTER 12

12.1 a Mo solute, Cr solvent b $MgCl_2$ solute, H_2O solvent c O_2 and N_2 solutes, Ar solvent **12.2** Compound X is a large nonpolar molecule. Compound Y is a molecule that exhibits hydrogen bonding. Applying the like dissolves like concept, compound X will be more soluble in the nonpolar solvent and compound Y will be more soluble in a polar solvent. **12.3** The A^{2+} ion must be smaller so it has a greater energy of hydration. **12.4** At high elevations, the partial pressure of O_2 is so low that not enough dissolves in the water to sustain fish. **12.5** A soluble, chemically similar liquid with a higher vapor pressure than water. **12.6** To prevent water from flowing into the pickle via osmosis. The pickles would be huge and would probably burst. **12.7** Hightest to lowest: $MgCl_2 > KBr > Na_2SO_4 > NaCl$. **12.8** The iron(III) hydroxide precipitates because the electrons from the negative electrode neutralize the positive charge on the iron(III) hydroxide.

CHAPTER 13

13.1 a Point A has a faster instantaneous rate. b No, it decreases with time. **13.2** a [Q] = 0.0 *M* slowest (no reaction); the other two are equal in rate because they are zero order with respect to [Q]. b Rate = $k[R]^2$ **13.3** a Rate of aging = $(\text{diet})^w(\text{exercise})^x(\text{sex})^y(\text{occupation})^z$ b Gather a sample of people that have all of the factors the same except one. For example, using the equation given in part a, you could determine the effect of diet if you had a sample of people who were the same sex, exercised the same amount, and had the same occupation. You would need to isolate each factor in this fashion to determine the exponent on each factor. c 2 **13.4** Second order **13.5** a the A + B reaction b the E + F reaction c exothermic **13.6** a It will increase the rate since the activation energy, E_a, is inversely related to the rate constant, *k*. A decrease in E_a results in an increase in the value of *k*. b yes c No, it should be rate = $k[Y]^2$.

CHAPTER 14

14.1 $2A + B \longrightarrow 2C$ **14.2** IV with A= red, I with X = red or blue, II with C = blue **14.3** left to right **14.4** It is quadrupled. **14.5** I

CHAPTER 15

15.1 $HCHO_2(aq) + H_2O(l) \rightleftharpoons CHO_2^-(aq) + H_3O^+(aq)$ and $H_2O(l) + H_2O(l) \rightleftharpoons H_3O^+(aq) + OH^-(aq)$; $HCHO_2(aq)$, $H_2O(l)$, $CHO_2^-(aq)$, $OH^-(aq)$, and $H_3O^+(aq)$ **15.2** Acetate ion **15.3** B, A, C **15.4** NaOH, NH_3, $HC_2H_3O_2$, HCl

CHAPTER 16

16.1 HCN, $HC_2H_3O_2$, HNO_2, HF **16.2** $B(aq) + H_2O(l) \rightleftharpoons BH^+(aq) + OH^-(aq)$ weak base
16.3 a, b **16.4** $KNO_2(aq)$ and $HNO_3(aq)$ **16.5** d
16.6 a Y b 3 c beaker with 7 HA molecules

CHAPTER 17

17.1 $PbSO_4$ **17.2** $NaNO_3$ **17.3** Picture fewer Ag^+ and more Cl^- ions. **17.4** Magnesium oxalate

CHAPTER 18

18.1 It has increased. **18.2** The NO concentration is low at equilibrium. **18.3** a $\Delta G°$ is not changed. b ΔG increases.
18.4 The concentration of NO_2 increases with temperature.

CHAPTER 19

19.1 No sustainable current would flow. The wire does not contain mobile positively and negatively charged species, which are necessary to balance the accumulation of charges in each of the half-cells. **19.2** a −0.54 V b No c They both would be 1.10 V. **19.3** a negative b Change the concentrations in a manner to increase *Q*, where $Q = \frac{[Fe^{2+}]}{[Cu^{2+}]}$

For example: $Fe(s)|Fe^{2+}\,(1.10\,M)||Cu^{2+}\,(0.50\,M)|Cu(s)$.
19.4 Many of the ions contained in seawater have very high reduction potentials—higher than Fe(*s*). This means that spontaneous electrochemical reactions will occur with the Fe(*s*), causing the iron to form ions and go into solution, while at the same time, the ions in the sea are reduced and plate out on the surface of the iron.

CHAPTER 20

20.1 a yes b no, since the ${}^3_1H_2O$ molecule is more massive than ${}^1_1H_2O$ c ${}^3_1H_2O$ **20.2** the α particle, since it has the highest RBE **20.3** After 50,000 years, enough half-lives have passed (about 10) that there would be almost no carbon-14 present to detect and measure.

CHAPTER 21

21.1 Given the high energy demands of animals to move and maintain body temperature, breaking the very strong triple bond of N_2 requires too much energy when compared to the lower-energy double bond of O_2. **21.2** because the only intermolecular forces in these materials are very weak van der Waals forces

CHAPTER 22

22.1 $K_2[PtCl_6]$ **22.2** $[Co(NH_3)_4(H_2O)Br]Cl_2$
22.3 a Complexes A and E b A and E c A and E, B and C

CHAPTER 23

23.1 a C_7H_{16} b $CH_3CH(CH_3)CH(CH_3)CH_2CH_3$

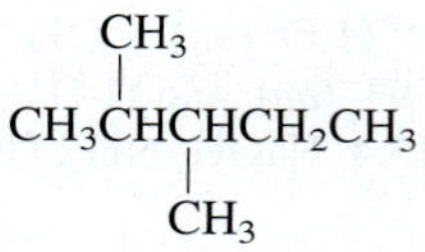

23.2 Blend of 90% 2,2,4-trimethylpentane and 10% heptane

23.3 a C_5H_{10} b (H)(CH$_3$)C=C(H)(CH$_2$CH$_3$), with the two H atoms on the same side

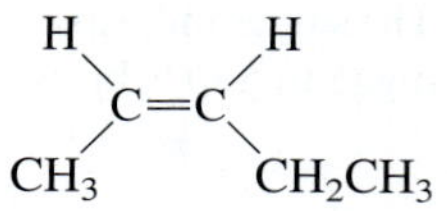

c *cis*-2-pentene

CHAPTER 24

24.1 $H_2N-CH(CH_3)-C(=O)-NH-CH(CH(OH)CH_3)-COOH$

$H_2N-CH(CH(OH)CH_3)-C(=O)-NH-CH(CH_3)-COOH$

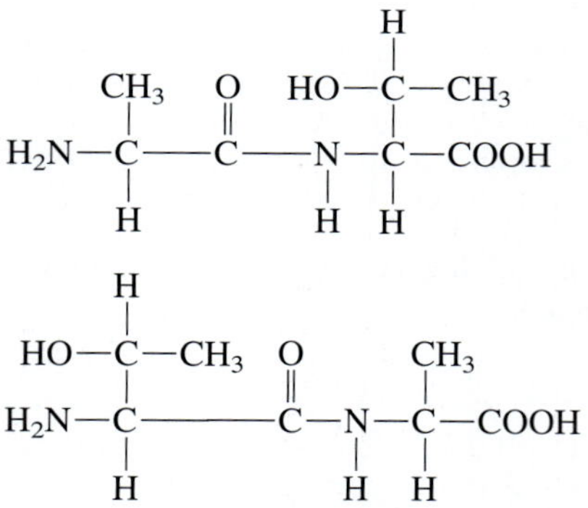

24.2 UACGAUGCCUAAGUU

Answer Section
Selected Odd Problems

CHAPTER 1

1.25 a Two phases: liquid and solid. b Three phases: liquid water, solid quartz, and solid seashells. **1.27** a You could first immerse the thermometer in an ice-water bath and mark the level at this point as 0°C. Then, immerse the thermometer in boiling water, and mark the level at this point as 100°C. b You could make 19 evenly spaced marks on the thermometer between the two original points, each representing a difference of 5°C.

1.29 a $°F = \left(°YS \times \frac{120°F}{100°YS}\right) - 100°F$ b $-21°F$

1.31 The box containing the cubes must have a greater mass of wood; hence, it must have a greater density. **1.33** a A paper clip has a mass of about 1 g. b Answers will vary depending on your particular sample. Keeping in mind that the SI unit for mass is kg, the approximate weights for the items presented in the problem are as follows: a grain of sand, 1×10^{-5} kg; a paper clip, 1×10^{-3} kg; a nickel, 5×10^{-3} kg; a 5.0-gallon bucket of water, 2.0×10^{1} kg; a brick, 3 kg; a car, 1×10^{3} kg. **1.37** 6.6 g **1.39** 29.8 g **1.41** a Solid b Liquid c Gas d Solid **1.43** a Physical change b Physical change c Chemical change d Physical change **1.45** Physical change: Liquid mercury is cooled to solid mercury. Chemical changes: (1) Solid mercury oxide forms liquid mercury metal and gaseous oxygen; (2) glowing wood and oxygen form burning wood (form ash and gaseous products). **1.47** a Physical property b Chemical property c Physical property d Physical property e Chemical property **1.49** Physical properties: (1) Iodine is solid; (2) the solid has lustrous blue-black crystals; (3) the crystals vaporize readily to a violet-colored gas. Chemical properties: (1) Iodine combines with many metals, such as with aluminum to give aluminum iodide. **1.51** a Physical process b Chemical reaction c Physical process d Chemical reaction e Physical process **1.53** a Solution b Substance c Substance d Heterogeneous mixture **1.55** a A pure substance with two phases present, liquid and gas. b A mixture with two phases present, solid and liquid. c A pure substance with two phases present, solid and liquid. d A mixture with two phases present, solid and solid. **1.57** a 4 b 3 c 4 d 5 e 3 f 4 **1.59** 40,000 km = 4.0×10^{4} km **1.61** a 8.5 b 90.0 c 111 d 2.3×10^{3} **1.63** 49 cm^3 **1.65** a 5.89 ps b 2.10 dm c 2.560 ng d 6.05 km **1.67** a 6.15×10^{-12} s b 3.781×10^{-6} m c 1.546×10^{-10} m d 9.7×10^{-3} g **1.69** a 20°C b −31°C c 79°F d −114°F **1.71** −4.0°F **1.73** 7.56 g/cm^3 **1.75** ethanol **1.77** 1.3×10^{2} g **1.79** 25.1 cm^3 **1.81** 4.80×10^{5} mg **1.83** 5.55×10^{-5} cm **1.85** 3.73×10^{17} m^3; 3.73×10^{20} L **1.87** 3.25×10^{6} g **1.89** 4.435×10^{3} m **1.91** 65.5 L **1.93** 0.86 g **1.95** 2.53 g **1.97** 114.5 g **1.99** a Chemical b Physical c Physical d Chemical **1.101** No **1.103** 6.07×10^{4} cm^3 **1.105** 50.4 gal **1.107** 9.9×10^{1} cm^3 **1.109** a 7.6×10^{-1} b 1.63×10^{1} c 4.76×10^{2} d 1.12×10^{-1} **1.111** a 9.12 cg b 66 pm c 7.1 μm d 56 nm **1.113** a 1.07×10^{-12} s b 5.8×10^{-6} m c 3.19×10^{-7} m d 1.53×10^{-2} s **1.115** 6170°F **1.117** 1.52×10^{3}°F **1.119** 303.0 K, 85.6°F **1.121** 907.8°C, 1180.9 K **1.123** 2.70×10^{3} kg/m^3 **1.125** 2.65 g/cm^3 **1.127** chloroform **1.129** 1.82×10^{3} g **1.131** 33.24 mL **1.133** a 8.45×10^{9} μg b 3.18×10^{-1} ms c 9.3×10^{13} nm d 3.71 cm **1.135** a 5.91×10^{6} mg b 7.53×10^{5} μg c 9.01×10^{4} kHz d 4.98×10^{-4} kJ **1.137** 1.2230×10^{16} L **1.139** 2.8×10^{4} L **1.141** 55.0 g **1.163** 2.4 L **1.165** 2.13×10^{3} g **1.167** 3×10^{21} g **1.169** 34%, 79 proof (2 sig. fig.) **1.171** 8.09 g/cm^3 **1.173** 2.4 g/cm^3

CHAPTER 2

2.27 If atoms were balls of positive charge with the electrons evenly distributed throughout, there would be no massive, positive nucleus t deflect the beam of alpha particles when it is shot at the gold foil. **2.29** Group them elements by similar physical properties such as density, mass, color, conductivity, etc., or by chemical properties, such as reaction with air, reaction with water, etc. **2.31** a X_2SO_4, XSO_4, and $X_2(SO_4)_5$ respectively. b exy(I) sulfate, exy(II) sulfate, and exy(V) sulfate, respectively. **2.33** 19 protons, 20 neutrons, 38 electrons **2.35** a $2Li + Cl_2 \longrightarrow 2LiCl$ b $Na + S \longrightarrow Na_2S$ c $2Al + 3I_2 \longrightarrow 2AlI_3$ d $3Ba + N_2 \longrightarrow Ba_3N_2$ e $12V + 5P_4 \longrightarrow 4V_3P_5$ **2.37** a Argon b Zinc c Silver d Magnesium **2.39** a K b S c Fe d Mn **2.41** 9.04×10^{31} kg **2.43** C; D. **2.45** Cl-35: 17 protons, 18 neutrons, 17 electrons Cl-37: 17 protons, 20 neutrons, 17 electrons **2.47** ${}^{28}_{14}Si$ **2.49** 13.90 **2.51** 63.55 amu, Cu **2.53** 39.10 amu, K **2.55** 24.615 amu **2.57** a Group IVA, Period 2; nonmetal b Group VIA, Period 6; metal c Group VIB, Period 4; metal d Group IIA, Period 3; metal e Group IIIA, Period 2; metalloid **2.59** a Tellurium b Aluminum **2.61** a O (oxygen) b F (fluorine) c Fe (iron) d Ce (cerium) **2.63** They are different in that the solid sulfur consists of S_8 molecules, whereas the hot vapor consists of S_2 molecules. The S_8 molecules are four times a heavy as the S_2 molecules. Hot sulfur is a mixture of S_8 and S_2 molecules, but at high enough temperatures only S_2 molecules are formed Both hot sulfur and solid sulfur consist of molecules with only sulfur atoms. **2.65** 4.10×10^{22} N atoms; 1.20×10^{24} N atoms **2.67** 1.1 × 10^{21} NH_3 molecules **2.69** a N_2H_4 b H_2O_2 c C_3H_8O d PC **2.71** a PCl_5 b NO_2 c $C_3H_6O_2$ **2.73** 1 Fe to 6 O atoms **2.75** a $Fe(CN)_3$ b K_2CO_3 c Li_3N d Ca_3P_2 **2.77** a Na_2SO_4: sodium sulfate b Na_3N: sodium nitride c CuCl: copper(I) chloride d Cr_2O_3: chromium(III) oxide **2.79** a Lead(II) permanganate: $Pb(MnO_4)_2$ b Barium hydrogen carbonate: $Ba(HCO_3)_2$ c Cesium sulfide: Cs_2S d Iron(III) acetate: $Fe(C_2H_3O_2)_3$ **2.81** a Molecular b Ionic c Molecular d Ionic **2.83** a Dinitrogen monoxide b Tetraphosphorus dec(a)oxide c Arsenic trichloride d Dichlorine hept(a)oxide **2.85** a NBr_3 b XeF_6 c CO d Cl_2O_5 **2.87** a Selenium trioxide b Disulfur dichloride c Carbon monoxide **2.89** a Sulfurous acid: H_2SO_3 b Hyponitrous acid: $H_2N_2O_2$ c Disulfurous acid: $H_2S_2O_5$ d Arsenic acid: H_3AsO_4 **2.91** $Na_2SO_4{\cdot}10H_2O$ is sodium sulfate decahydrate. **2.93** Iron(II) sulfate heptahydrate is $FeSO_4{\cdot}7H_2O$ **2.95** 9 O atoms **2.97** a $Sn + 2NaOH \longrightarrow Na_2SnO_2 + H_2$ b $8Al + 3Fe_3O_4 \longrightarrow 4Al_2O_3 + 9Fe$ c $2CH_3OH + 3O_2 \longrightarrow 2CO_2 + 4H_2O$ d $P_4O_{10} + 6H_2O \longrightarrow 4H_3PO_4$ e $PCl_5 + 4H_2O \longrightarrow H_3PO_4 + 5HCl$ **2.99** $Ca_3(PO_4)_2(s) + 3H_2SO_4(aq) \longrightarrow 3CaSO_4(s) + 2H_3PO_4(aq)$

2.101 $2NH_4Cl(aq) + Ba(OH)_2(aq) \xrightarrow{\Delta} 2NH_3(g) + BaCl_2(aq) + 2H_2O(l)$

2.103 five O's in B for every four O's in A **2.105** -1.6×10^{-19} C, 2, 4, 6, 7 **2.107** 63, 60 **2.109** ${}^{81}_{35}Br^{-}$ **2.111** Cu-63:0.6916, Cu-65: 0.3084 **2.113** a Bromine, Br b Hydrogen, H

Niobium, Nb d Fluorine, F **2.115** a Chromium(III) ion
Lead(IV) ion c Titanium(II) ion d Copper(II) ion
.117 Na_2SO_4, NaCl, $CoSO_4$, and $CoCl_2$ **2.119** a Tin(II)
hosphate b Ammonium nitrite c Magnesium hydroxide
Nickel(II) sulfite **2.121** a Hg_2S b $Co_2(SO_3)_3$
$(NH_4)_2Cr_2O_7$ d AIF_3 **2.123** a Arsenic tribromide
Hydrogen telluride (dihydrogen telluride) c Diphosphorus
ent(a)oxide d Silicon dioxide **2.125** a $2C_2H_6 + 7O_2 \longrightarrow$
$CO_2 + 6H_2O$ b $P_4O_6 + 6H_2O \longrightarrow 4H_3PO_3$ c $4KClO_3 \longrightarrow$
$Cl + 3KClO_4$ d $(NH_4)2SO_4 + 2NaOH \longrightarrow 2NH_3 + 2H_2O +$
a_2SO_4 e $2NBr_3 + 3NaOH \longrightarrow N_2 + 3NaBr + 3HOBr$
.127 26 electrons **2.129** 12.507 amu **2.149** 6.06×10^9 miles
.151 4.826 g $NiSO_4$ **2.153** 0.8000 g O; 63.54 amu; X is copper

HAPTER 3

.19 a $3H_2(g) + N_2(g) \longrightarrow 2NH_3(g)$ b H_2 c six additional
olecules of H_2 **3.21** a unreasonable b unreasonable
reasonable d unreasonable **3.23** a charcoal
Mg c H_2 **3.25** a 1 mol Ca = 40.08 Ca conversion factor cor-
ction b 2 mol K^+ ions = 1 mol K_2SO_4 conversion factor correc-
on c 2 mole Na = 1 mol H_2O conversion factor correction
27 a 32.0 amu b 62.0 amu c 138 amu d 366 amu
29 a 64.1 amu b 137 amu **3.31** 80.05 g/mol
33 a 3.818×10^{-23} g/atom b 2.326×10^{-23} g/atom
8.383×10^{-23} g/molecule d 5.390×10^{-22} g/unit
35 1.231×10^{-22} g/molecule **3.37** a 3.4 g Na b 19.0 g S
236 g CH_2Cl_2 d 2.6×10^3 g $(NH_4)_2S$ **3.39** 33.6 g H_3BO_3
41 a 0.238 mol C b 0.0994 mol Cl_2 c C_4H_{10} d 0.112 mol
$_2(CO_3)_3$ **3.43** 5.81×10^{-3} mol $CaSO_4$; 1.16×10^{-2} mol H_2O
ecause 0.01159 mol is about twice 0.005811 mol, both numbers of
oles are consistent with the formula, $CaSO_4{\cdot}2H_2O$.
45 a 7.12×10^{23} atoms b 2.41×10^{23} atoms c $1.6 \times$
$)^{24}$ molecules d 3.75×10^{23} units e 6.59×10^{22} ions
47 2.97×10^{19} molecules **3.49** 86.27% **3.51** 20.2%
53 0.581 kgN **3.55** 72.3% Al; 27.7% Mg **3.57** a 42.9% C;
.1% O b 27.3% C; 72.7% O c 19.2% Na; 1.68% H;
.8% P; 53.3% O d 32.2% Co; 15.3% N; 52.5% O
59 91.2% C; 8.75% H **3.61** Ethanol **3.63** 38.7% C; 9.79% H;
.5% O **3.65** OsO_4 **3.67** K_2MnO_4 **3.69** $C_3H_4O_2$
71 a CH b C_4H_4; C_6H_6 **3.73** $C_4H_{12}N_2$ **3.75** $C_2H_2O_4$
77

C_2H_4	+	$3O_2$	$\longrightarrow$	$2CO_2$	+	$2H_2O$
1 molecule C_2H_4	+	3 molecules O_2	$\longrightarrow$	2 molecules CO_2	+	2 molecules H_2O
1 mole C_2H_4	+	3 moles O_2	$\longrightarrow$	2 moles CO_2	+	2 moles H_2O
28.052 g C_2H_4	+	3×32.00 g O_2	$\longrightarrow$	2×44.01 g CO_2	+	2×18.016 g H_2O

79 1.6 mol CO_2 **3.81** 5.87 mol O_2 **3.83** 8.21 g NO_2
85 1.46×10^5 g W **3.87** 22.4 g CS_2 **3.89** 7.942 g NO_2
91 KO_2 is the limiting reactant; 0.19 mol oxygen
93 40.5 g CH_3OH; H_2 remains; 5.1 g H_2 **3.95** 9.07 g
97 2.609 g is the theoretical yield; 71.3% **3.99** C: 49.5%
: 5.19%, N: 28.9%, O: 16.5% **3.101** $C_6H_4Cl_2$ **3.103** C_4H_8S,
$_4H_8S$ **3.105** 83.7% **3.107** 59.5% **3.109** 60.3 g
111 2.97×10^3 g CaC_2 **3.113** 16.2% **3.133** 0.21 g
135 616 amu **3.137** Ag and Cl **3.139** 1.45×10^{26} Fe atoms

HAPTER 4

21 a Any soluble salt that will form a precipitate when reacted
th Ag^+ ions in solution will work, for example: $CaCl_2$, Na_2S,
$(H_4)_2CO_3$. b No, no precipitate will form. c You would un-
restimate the amount of silver present in the solution.
23 a $3Ca(C_2H_3O_2)_2(aq) + 2(NH_4)_3PO_4(aq) \longrightarrow$
$Ca_3(PO_4)_2(s) + 6NH_4C_2H_3O_2(aq)$
$3Ca^{2+}(aq) + 6C_2H_3O_2^-(aq) + 6NH_4^+(aq) + 2PO_4^{3-}(aq) \longrightarrow$
$Ca_3(PO_4)_2(s) + 6NH_4^+(aq) + 6C_2H_3O_2^-(aq)$

c $3Ca^{2+}(aq) + 2PO_4^{3-}(aq) \longrightarrow Ca_3(PO_4)_2(s)$
4.25 Probably not, since the ionic compound that is a nonelectrolyte is not soluble. **4.27** a Cl^- b 0.50 M Br^-, 2.0 M K^+, 0.50M PO_4^{3-} **4.29** a Insoluble b Soluble c Soluble
d Soluble **4.31** a Insoluble b Soluble; Li^+ and SO_4^{2-}
c Insoluble d Soluble; Na^+ and CO_3^{2-}
4.33 a $H^+(aq) + OH^-(aq) \longrightarrow H_2O(l)$
b $Ag^+(aq) + Br^-(aq) \longrightarrow AgBr(s)$
c $S^{2-}(aq) + 2H^+(aq) \longrightarrow H_2S(g)$
d $OH^-(aq) + NH_4^+(aq) \longrightarrow NH_3(g) + H_2O(l)$
4.35 Molecular equation: $Pb(NO_3)_2(aq) + Na_2SO_4(aq) \longrightarrow PbSO_4(s) + 2NaNO_3(aq)$
$Pb^{2+}(aq) + SO_4^{2-}(aq) \longrightarrow PbSO_4(s)$
4.37 a $FeSO_4(aq) + NaCl(aq) \longrightarrow$ NR
b $Na_2CO_3(aq) + MgBr_2(aq) \longrightarrow MgCO_3(s) + 2NaBr(aq)$
$CO_3^{2-}(aq) + Mg^{2+}(aq) \longrightarrow MgCO_3(s)$
c $MgSO_4(aq) + 2NaOH(aq) \longrightarrow Mg(OH)_2(s) + Na_2SO_4(aq)$
$Mg^{2+}(aq) + 2OH^-(aq) \longrightarrow Mg(OH)_2(s)$
d $NiCl_2(aq) + NaBr(aq) \longrightarrow$ NR
4.39 a $Ba(NO_3)_2(aq) + Li_2SO_4(aq) \longrightarrow BaSO_4(s) + 2LiNO_3(aq)$
$Ba^{2+}(aq) + SO_4^{2-}(aq) \longrightarrow BaSO_4(s)$
b $Ca(NO_3)_2(aq) + NaBr(aq) \longrightarrow$ NR
c $Al_2(SO_4)_3(aq) + 6NaOH(aq) \longrightarrow 2Al(OH)_3(s) + 3Na_2SO_4(aq)$
$Al^{3+}(aq) + 3OH^-(aq) \longrightarrow Al(OH)_3(s)$
d $3CaBr_2(aq) + 2Na_3PO_4(aq) \longrightarrow Ca_3(PO_4)_2(s) + 6NaBr(aq)$
$3Ca^{2+}(aq) + 2PO_4^{3-}(aq) \longrightarrow Ca_3(PO_4)_2(s)$
4.41 a Weak acid b Strong base c Strong acid
d Weak acid
4.43 a $NaOH(aq) + HNO_3(aq) \longrightarrow H_2O(l) + NaNO_3(aq)$
$H^+(aq) + OH^-(aq) \longrightarrow H_2O(l)$
b $2HCl(aq) + Ba(OH)_2(aq) \longrightarrow 2H_2O(l) + BaCl_2(aq)$
$H^+(aq) + OH^-(aq) \longrightarrow H_2O(l)$
c $2HC_2H_3O_2(aq) + Ca(OH)_2(aq) \longrightarrow 2H_2O(l) + Ca(C_2H_3O_2)_2(aq)$
$HC_2H_3O_2(aq) + OH^-(aq) \longrightarrow H_2O(l) + C_2H_3O_2^-(aq)$
d $NH_3(aq) + HNO_3(aq) \longrightarrow NH_4NO_3(aq)$
$NH_3(aq) + H^+(aq) \longrightarrow NH_4^+(aq)$
4.45 a $2HBr(aq) + Ca(OH)_2(aq) \longrightarrow 2H_2O(l) + CaBr_2(aq)$
$H^+(aq) + OH^-(aq) \longrightarrow H_2O(l)$
b $3HNO_3(aq) + Al(OH)_3(s) \longrightarrow 3H_2O(l) + Al(NO_3)_3(aq)$
$3H^+(aq) + Al(OH)_3(s) \longrightarrow 3H_2O(l) + Al^{3+}(aq)$
c $2HCN(aq) + Ca(OH)_2(aq) \longrightarrow 2H_2O(l) + Ca(CN)_2(aq)$
$HCN(aq) + OH^-(aq) \longrightarrow H_2O(l) + CN^-(aq)$
d $HCN(aq) + LiOH(aq) \longrightarrow H_2O(l) + LiCN(aq)$
$HCN(aq) + OH^-(aq) \longrightarrow H_2O(l) + CN^-(aq)$
4.47 a $2KOH(aq) + H_3PO_4(aq) \longrightarrow K_2HPO_4(aq) + 2H_2O(l)$
$2OH^-(aq) + H_3PO_4(aq) \longrightarrow HPO_4^{2-}(aq) + 2H_2O(l)$
b $3H_2SO_4(aq) + 2Al(OH)_3(s) \longrightarrow 6H_2O(l) + Al_2(SO_4)_3(aq)$
$3H^+(aq) + Al(OH)_3(s) \longrightarrow 3H_2O(l) + Al^{3+}(aq)$
c $2HC_2H_3O_2(aq) + Ca(OH)_2(aq) \longrightarrow 2H_2O(l) + Ca(C_2H_3O_2)_2(aq)$
$HC_2H_3O_2(aq) + OH^-(aq) \longrightarrow H_2O(l) + C_2H_3O_2^-(aq)$
d $H_2SO_3(aq) + NaOH(aq) \longrightarrow H_2O(l) + NaHSO_3(aq)$
$H_2SO_3(aq) + OH^-(aq) \longrightarrow HSO_3^-(aq) + H_2O(l)$
4.49 $2H_2SO_3(aq) + Ca(OH)_2(aq) \longrightarrow 2H_2O(l) + Ca(HSO_3)_2(aq)$
$Ca(HSO_3)_2(aq) + Ca(OH)_2(aq) \longrightarrow 2H_2O(l) + 2CaSO_3(s)$
$H_2SO_3(aq) + OH^-(aq) \longrightarrow H_2O(l) + HSO_3^-(aq)$
$Ca^{2+}(aq) + HSO_3^-(aq) + OH^-(aq) \longrightarrow CaSO_3(s) + H_2O(l)$
4.51 a $CaS(s) + 2HBr(aq) \longrightarrow CaBr_2(aq) + H_2S(g)$
$CaS(s) + 2H^+(aq) \longrightarrow Ca^{2+}(aq) + H_2S(g)$
b $MgCO_3(s) + 2HNO_3(aq) \longrightarrow CO_2(g) + H_2O(l) + Mg(NO_3)_2(aq)$
$MgCO_3(s) + 2H^+(aq) \longrightarrow CO_2(g) + H_2O(l) + Mg^{2+}(aq)$

c $K_2SO_3(aq) + H_2SO_4(aq) \longrightarrow K_2SO_4(aq) + SO_2(g) + H_2O(l)$
$SO_3^{2-}(aq) + 2H^+(aq) \longrightarrow SO_2(g) + H_2O(l)$
4.53 $FeS(s) + 2HCl(aq) \longrightarrow H_2S(g) + FeCl_2(aq)$
$FeS(s) + 2H^+(aq) \longrightarrow H_2S(g) + Fe^{2+}(aq)$ **4.55** a +3 b +4 c +7 d +6 **4.57** a −3 b +5 c 0 d +7 **4.59** a Mn: +2, Cl: +3, O: −2 b Fe: +3, Cr: +6, O: −2 c Hg: +2, Cr: +6, O: −2 d Co: +2, P: +5, O: −2 **4.61** a P_4 reducing agent; O_2 oxidizing agent b Co reducing agent; Cl_2 oxidizing agent
4.63 a Al is the reducing agent; F_2 is the oxidizing agent b Hg^{2+} is the oxidizing agent; NO^{2-} is the reducing agent
4.65 a $3CuCl_2(aq) + 2Al(s) \longrightarrow 2AlCl_3(aq) + 3Cu(s)$
b $2Cr^{3+}(aq) + 3Zn(s) \longrightarrow 2Cr(s) + 3Zn^{2+}(aq)$ **4.67** 2.05 M
4.69 0.101 M **4.71** 1.25 L **4.73** 35.8 mL **4.75** 2.9 × 10^{-4} mol **4.77** 0.66 g **4.79** 7.6 mL **4.81** 0.767 M K^+; 0.429 M Ca^{2+}; 1.624 M Cl^- **4.83** 0.303 g Ba^{2+}; 65.9%
4.85 a 35.81% Cl^- b CuCl **4.87** $FeCl_2$ **4.89** 53.8 mL
4.91 325 mL **4.93** 3.19%
4.95 $Mg(s) + 2HBr(aq) \longrightarrow H_2(g) + MgBr_2(aq)$
$Mg(s) + 2H^+(aq) \longrightarrow H_2(g) + Mg^{2+}(aq)$
4.97 $NiSO_4(aq) + 2NaOH(aq) \longrightarrow Ni(OH)_2(s) + Na_2SO_4(aq)$
$Ni^{2+}(aq) + 2OH^-(aq) \longrightarrow Ni(OH)_2(s)$
4.99 a $LiOH(aq) + HCN(aq) \longrightarrow LiCN(aq) + H_2O(l)$
$OH^-(aq) + HCN(aq) \longrightarrow CN^-(aq) + H_2O(l)$
b $Li_2CO_3(aq) + 2HNO_3(aq) \longrightarrow 2LiNO_3(aq) + CO_2(g) + H_2O(l)$
$CO_3^{2-}(aq) + 2H^+(aq) \longrightarrow CO_2(g) + H_2O(l)$
c $LiCl(aq) + AgNO_3(aq) \longrightarrow LiNO_3(aq) + AgCl(s)$
$Cl^-(aq) + Ag^+(aq) \longrightarrow AgCl(s)$
d $MgSO_4(aq) + NaCl(aq) \longrightarrow$ NR
4.101 a $Sr(OH)_2(aq) + 2HC_2H_3O_2(aq) \longrightarrow Sr(C_2H_3O_2)_2(aq) + 2H_2O(l)$
$HC_2H_3O_2(aq) + OH^-(aq) \longrightarrow C_2H_3O_2^-(aq) + H_2O(l)$
b $NH_4I(aq) + CsCl(aq) \longrightarrow$ NR
c $NaNO_3(aq) + CsCl(aq) \longrightarrow$ NR
d $NH_4I(aq) + AgNO_3(aq) \longrightarrow NH_4NO_3(aq) + AgI(s)$
$I^-(aq) + Ag^+(aq) \longrightarrow AgI(s)$
4.103 a $CuSO_4(aq)\ BaCl_2(aq) + \longrightarrow BaSO_4(s) + CuCl_2(aq)$
The $BaSO_4$ can be filtered off, leaving aqueous $CuCl_2$, which can be obtained in solid form by evaporation.
b $CaCO_3(s) + 2HC_2H_3O_2(aq) \longrightarrow Ca(C_2H_3O_2)_2(aq) + CO_2(g) + H_2O(l)$
The aqueous $Ca(C_2H_3O_2)_2$ can be converted to the solid form by evaporation.
c $Na_2SO_3(s) + 2HNO_3(aq) \longrightarrow 2NaNO_3(aq) + SO_2(g) + H_2O(l)$
The aqueous $NaNO_3$ can be converted to the solid form by evaporation.
d $Mg(OH)_2(s) + 2HCl(aq) \longrightarrow MgCl_2(aq) + 2H_2O(l)$
The aqueous $MgCl_2$ can be converted to the solid form by evaporation. **4.105** a Decomposition b Decomposition c Combination d Displacement **4.107** a $Pb(NO_3)_2 + H_2SO_4$, $Pb(NO_3)_2 + MgSO_4$, $Pb(NO_3)_2 + Ba(OH)_2$, b $Ba(OH)_2 + MgSO_4$ c $Ba(OH)_2 + H_2SO_4$ **4.109** 0.0450 *M* $CaCl_2$; 0.0450 *M* Ca^{2+}; 0.0901 *M* Cl^- **4.111** 329 mL **4.113** 0.610 *M*
4.115 0.0967 *M* **4.117** 21.39% **4.119** $CuSO_4{\cdot}2H_2O$
4.121 35.0% Cl; $AuCl_3$ **4.123** 9.66% **4.141** $Pb(NO_3)_2$; Cs_2SO_4; $Pb(NO_3)_2(aq) + Cs_2SO_4(aq) \longrightarrow PbSO_4(s) + 2CsNO_3(aq)$;
$Pb^{2+}(aq) + SO_4^{2-}(aq) \longrightarrow PbSO_4(s)$ lead(II) sulfate; cesium nitrate;
$Pb(NO_3)_2(aq) + Na_2SO_4(aq) \longrightarrow PbSO_4(s) + 2NaNO_3(aq)$
4.143 $CaBr_2(aq) + Cl_2(g) \longrightarrow CaCl_2(aq) + Br_2(l)$
$2Br^-(aq) + Cl_2(g) \longrightarrow 2Cl^-(aq) + Br_2(l)$ 5.67 × 10^3 g $CaBr_2$
4.145 $Hg(NO_3)_2(aq) + H_2S(g) \longrightarrow HgS(s) + 2HNO_3(aq)$
$Hg^{2+}(aq) + H_2S(g) \longrightarrow HgS(s) + 2H^+(aq)$ The acid formed is nitric acid, a strong acid. The other product is mercury(II) sulfide. 581.5 g
4.147 Fe^{2+}:Fe^{3+}; 5/2 **4.149** 0.0121 L **4.151** 0.930 *M*
4.153 69.72 g/mol Ga **4.155** $P_4O_{10} + 6H_2O \longrightarrow 4H_3PO_4$; 55.7 g P_4O_{10} **4.157** 85.23% **4.159** 0.194 L **4.161** 61.7%

CHAPTER 5

5.29 a You would expect the pressure in the tires to decrease, and they would appear flatter. b If you put an aerosol can in a fire, you will increase the temperature and thus the pressure.
c As the temperature of the water increases, so does its vapor pres sure (Table 5.6). d As you squeeze the balloon, you decrease the volume, resulting in an increase in pressure. **5.31** a increase by a factor of 2 b decrease by a factor of 2 c increase by a factor of 1.5 d the pressure would not change; volume increase by a fa tor of 2 **5.33** a O_2 b H_2 c Both the same d The pressu in each of the containers will not change. e ¼ **5.35** In order to double the volume, you could reduce the pressure by 1/2.
5.37 36 mmHg **5.39** 2.32 L **5.41** 1.02 × 10^3 L **5.43** 3.64 × 10^{-4} kPa **5.45** 3.68 mL **5.47** 0.65 L **5.49** 241 K
5.51 31.6 mL **5.53** 1 volume **5.55** $PV = nRT$; $V = nRT$; If the temperature and number of moles are held constant, then the prod uct nRT is constant, and volume is inversely proportional to pressure: V = constant × $1/P$. **5.57** 8.02 atm **5.59** 22.0 L
5.61 143°C **5.63** 0.675 g/L **5.65** 2.24 g/L **5.67** 47.7 amu
5.69 3.20 × 10^2 amu **5.71** For a gas at a given temperature and pressure, the density depends on molecular mass (or, for a mixture, on average molecular mass). Thus, at the same temperature and pre sure, the density of NH_4Cl gas would be greater than that of a mix ture of NH_3 and HCl, because the average molecular mass of NH_3 and HCl would be lower than that of NH_4Cl. **5.73** 2.0 L
5.75 160. L **5.77** 246 L **5.79** $2NH_3(g) + H_2SO_4(aq) \longrightarrow (NH_4)_2SO_4(aq)$; 46.7 L **5.81** 0.279 **5.83** 0.00380 atm O_2; 0.017 atm He; Total P = 0.020 atm **5.85** $P(H_2)$ = 190 mmHg; $P(CO_2)$ = 494 mmHg; $P(HCl)$ = 41 mmHg; $P(HF)$ = 21 mmHg; $P(SO_2)$ = 13 mmHg; $P(H_2S)$ = 0.8 mmHg **5.87** 6.34 g
5.89 u_{25} = 5.15 × 10^2 m/s; u_{125} = 5.95 × 10^2 m/s; Graph as in Figure 5.25. **5.91** 1.53 × 10^2 m/s **5.93** 6.23 × 10^2°C **5.95** The ratio of the rates of effusion of N_2 and O_2 is 1.069 to 1. **5.97** 3.5
5.99 146 amu **5.101** P = 0.9605 atm; P(ideal gas law) = 0.9716 at
5.103 At 1.00 atm, V = 22.4 L and the van der Waals equation give 22.4 L. At 10.0 atm, the van der Waals equation gives 2.08 L. The ideal gas law gives 2.24 L for 10.0 atm. **5.105** 80.7 cm^3 **5.107** 181 m
5.109 6.5 dm^3 **5.111** 3.01 × 10^{20} atoms **5.113** 28.979 amu
5.115 1.2 × 10^3 g LiOH **5.117** 20.4 atm O_2 **5.119** 0.440 M
5.121 H_2SO_4 **5.123** −27°C **5.125** 1.0043 **5.127** 28.07 g/mol
5.129 0.167 atm O_2; 0.333 atm CO_2 **5.151** 0.732 g/L
5.153 0.194 g **5.155** 92.9% $CaCO_3$; 7.1% $MgCO_3$
5.157 8.81 g/L **5.159** 3.00 × 10^2 m/s

CHAPTER 6

6.31 a Beaker A- exothermic; Beaker B- exothermic; Beaker C- endothermic b Beaker A- negative; Beaker B- Negative; Beaker C- Positive c Beaker A- positive; Beaker B- negative; Beaker C- positive d Beaker A- decrease; Beaker B- decreace; Beaker C- increase e Beaker A- Hotter; Beaker B- Hotter; Beaker C- Colder **6.33** small car **6.35** a negative b solid c higher than the final enthalpy d −20°C
6.37 $\Delta H = \Delta H_f - \Delta H_{vap}$ **6.39** a A b A **6.41** Equate ΔH fo the burning to ΔH_f (products minus ΔH_f (reactants), then solve for ΔH_f of P_4S_3. You will need ΔH_f P_4S_{10} and SO_2. **6.43** −106.4 kcal
6.45 7.1 × 10^5 J; 1.7 × 10^5 cal **6.47** 5.24 × 10^{-21} J/molecule
6.49 −82 J, +29 J; −53 J **6.51** positive; cold
6.53 endothermic; +66.2 kJ **6.55** 37.56 kJ **6.57** $Fe(s) + 2HCl(aq) \longrightarrow FeCl_2(aq) + H_2(g)$; ΔH = −89.1 kJ **6.59** +3010 k
6.61 −385 kJ **6.63** −1.90 kJ/g **6.65** −6.66 × 10^2 kJ **6.67** 7.96

9 5.8×10^4 J **6.71** 5.81°C **6.73** 20.5 kJ **6.75** -1.36×10^3 kJ
77 50.6 kJ **6.79** −906.3 kJ **6.81** −137 kJ **6.83** +42.6 kJ
35 −1124.2 kJ **6.87** −23.5 kJ **6.89** −74.9 kJ **6.91** 13.9 kJ
3 kg·m/s² **6.95** 1.59×10^3 J/g **6.97** 31.6 m/s **6.99** 37.9 kJ
101 −255 kJ/mol **6.103** 37.4°C **6.105** 0.128 J/g°C
107 0.383 J/g°C **6.109** 25.6°C **6.111** −23.6 kJ/mol
113 −874.2 kJ/mol **6.115** −21 kJ **6.117** 206.2 kJ
119 178 kJ **6.121** −39.7 kJ **6.123** −2225 kJ/mol
125 35.5 g **6.149** 37.4°C **6.151** −22 kJ **6.153** 30.3 g CO;
4 g CO_2 **6.155** 113 kJ **6.157** a 1.88 kJ b 11.9%
159 1.07×10^3 g

HAPTER 7

5 frequency-doubled beam **7.27** one **7.29** *x* **7.31** The electron in part a will have the longest wavelength. **7.33** a $n = 3$ to $= 2$ b $n = 3$ to $n = 5$ c green **7.35** 219.6 m
7 6.27×10^{14}/s **7.39** 1.9×10^2 s **7.41** $6.05780211 \times 10^{-7}$ m, 4886516 $\times 10^{14}$/s **7.43** 9.044×10^{-28} J **7.45** 3.71×10^{-19} J
7 just on the edge of the red **7.49** 2.34×10^{14}/s
1 1.22×10^{-7} m (near UV) **7.53** 93.8 nm **7.55** 4.699×10^{-19} J
7 9.54 pm **7.59** 7.27×10^7 m/s **7.61** 1.52×10^{-22} pm, much aller **7.63** $l = 0, 1, 2$, or 3; $ml = -3, -2, -1, 0, +1, +2$, or $+3$
5 3, 7 **7.67** a 6d b 5g c 4f d 6p **7.69** a Not permissible; m_s can be only +1/2 or −1/2. b Not permissible; l can only as large as $(n - 1)$. c Not permissible; m_l cannot exceed 2 in magnitude. d Not permissible; n cannot be zero. e Not rmissible; m_s can be only +1/2 or −1/2. **7.71** 6.50×10^{14}/s; 1 $\times 10^{-19}$ J **7.73** 485 nm (blue-green) **7.75** 6.55×10^{14}/s
7 5.58×10^5 m/s **7.79** 4.34×10^{-7} m **7.81** 7
3 1.641×10^{-7} m (164.1 nm; near UV) **7.85** 1.50×10^4 eV
7 a Five b Seven c Three d One **7.89** The possible bshells for the $n = 6$ shell are 6*s*, 6*p*, 6*d*, 6*f*, 6*g*, and 6*h*.
117 5.01×10^{-7} m **7.119** 5.27×10^{28} photons
121 3.59×10^{-19} J; 2.16×10^2 kJ/mol **7.123** 1.94×10^{-11} m

HAPTER 8

1 (1) $1s^1$ (2) $1s^2$ (3) $1s^3$ (4) $1s^4$ (5) $1s^42s^1$ (6) $1s^42s^2$
3 Sr, Y **8.35** Z would form a cation of charge 2+.
7 Ge, Sn **8.39** IIA **8.41** a Not allowed; the paired electrons the 2*p* orbital should have opposite spins. Allowed; electron configuration is $1s^22s^22p^4$. c Not allowed; e electrons in the 1*s* orbital must have opposite spins. d Not allowed; the 2*s* orbital can hold at most two electrons, with opposite ins. **8.43** a Impossible state; the 2*p* orbitals can hold no more an six electrons. b Possible state. Impossible state; the 3*s* orbital can hold no more than two electrons. d Possible state; however, the 3*p* and 4*s* orbitals should be ed before the 3*d* orbital.

45 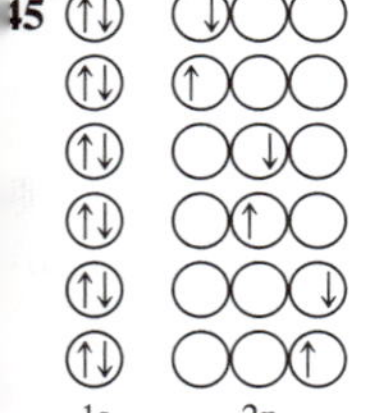

7 $1s^22s^22p^63s^23p^63d^{10}4s^24p^64d^{10}5s^25p^5$ **8.49** $1s^22s^22p^63s^2$ $^63d^54s^2$ **8.51** $4s^24p^5$ **8.53** $4d^25s^2$ **8.55** Group IIIA,
7 [Ar] (↑↓)(↑↓)(↑)(↑)(↑) (↑↓) 3*d* 4*s*
9 (↑↓) (↑↓) (↑↓)(↑↓)(↑↓) (↑↓) (↑↓)(↑↓)(↑↓) (↑) 1*s* 2*s* 2*p* 3*s* 3*p* 4*s*
Paramagnetic
1 S, Se, As **8.63** Na, Al, Cl, Ar **8.65** a Br b F

8.67 $LiBrO_3$ **8.69** $1s^22s^22p^63s^23p^63d^{10}4s^24p^65s^2$ **8.71** $6s^26p^4$
8.73 (↑↓) (↑↓) (↑↓)(↑↓)(↑↓) (↑↓) (↑↓)(↑↓)(↑↓) (↑↓)(↑↓)(↑↓)(↑↓)(↑↓) 1*s* 2*s* 2*p* 3*s* 3*p* 3*d*
(↑↓) (↑)(↑)(↑) 4*s* 4*p*

8.75 For eka-lead: [Rn] $5f^{14}6d^{10}7s^27p^2$. It is a metal; the oxide is eka-PbO or eka-PbO_2. **8.77** 370 kJ/mol
8.79 [Kr] (↑)(↑)(↑)(↑)( ) (↑) 4*d* 5*s*

8.81 a Cl_2 b Na c Sb d Ar **8.83** $1s^22s^22p^63s^23p^63d^34s^2$, Group VB, Period 4, *d*-block transition element **8.111** Ba(*s*) + $2H_2O(l) \longrightarrow Ba(OH)_2(aq) + H_2(g)$; 447 mL **8.113** RaO; 93.4% **8.115** 108 J **8.117** 1.312×10^3 kJ/mol H
8.119 −639 kJ/mol

CHAPTER 9

9.25 XY_3 **9.27** Blue ion is cation; blue atom is metal. **9.29** b
9.31 a Bond Cl and H to C. b Bond O atoms to N, then bond H to one O. c Bond O and F to N. d Place N atoms in center with two O atoms bonded to each N. **9.33** a Ca_3X_2 b ionic

9.35 a ·P̈· b [:P̈:]$^{3-}$ c ·Ga· d Ga^{3+}

9.37 a :B̈r· + ·Ca· + ·B̈r: ⟶ [:B̈r:]$^-$ + Ca^{2+} + [:B̈r:]$^-$

b K· + ·Ï: ⟶ K^+ + [:Ï:]$^-$

9.39 a As: $1s^22s^22p^63s^23p^63d^{10}4s^24p^3$, :Ȧs·
b As^{3+}: $1s^22s^22p^63s^23p^63d^{10}4s^2$, [·As·]$^{3+}$
c Se: $1s^22s^22p^63s^23p^63d^{10}4s^24p^4$, :S̤e·
d Se^{2-}: $1s^22s^22p^63s^23p^63d^{10}4s^24p^6$, [:S̈e:]$^{2-}$

9.41 Bi: [Xe]$4f^{14}5d^{10}6s^26p^3$ Bi^{3+}: The three 6*p* electrons are lost from the valence shell. [Xe]$4f^{14}5d^{10}6s^2$ **9.43** Ni^{2+}: [Ar]$3d^8$ Ni^{3+}: [Ar]$3d^7$ **9.45** a Sr^{2+} < Sr b Br < Br^- **9.47** S^{2-} < Se^{2-} < Te^{2-}
9.49 Smallest Na^+ (Z = 11), F^- (Z = 9), N^{3-} (Z = 7) Largest These ions are isoelectronic. The atomic radius increases with the decreasing nuclear charge (Z).

9.51 2H· + :S̤e· ⟶ H:S̈e:H
Bonding pairs
Lone pairs

9.53 $AsBr_3$ **9.55** a P, N, O b Na, Mg, Al c Al, Si, C
9.57 C—Cl, As—Br, P—O
9.59 a P—O (δ+ δ−) b C—Cl (δ+ δ−) c As—Br (δ+ δ−)
9.61 a :B̈r—B̈r: b H—S̈—H

c :F̈:N̈:F̈ with :F̈: on N

9.63 a :P≡P: b O=C(—B̈r:)—B̈r: c H—Ö—N̈=Ö:

9.65 a [:C̈l—Ö:]$^-$ b [:C̈l—Sn(—C̈l:)—C̈l:]$^-$ c [:S̈—S̈:]$^{2-}$

9.67 a H—Ö—N(=Ö)—Ö: ⟷ H—Ö—N(—Ö:)=Ö

b $:\ddot{O}-S(-\ddot{O}:)=\ddot{O} \longleftrightarrow :\ddot{O}-S(=O)(-\ddot{O}:)... \longleftrightarrow O=S(-\ddot{O}:)-\ddot{O}:$ (three resonance structures)

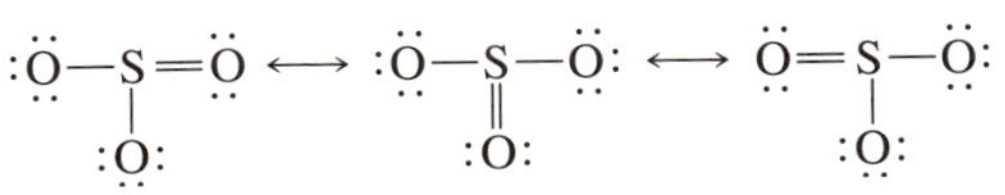

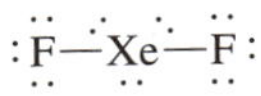

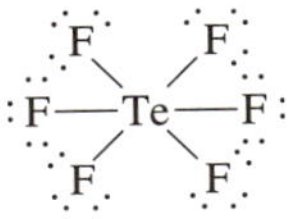

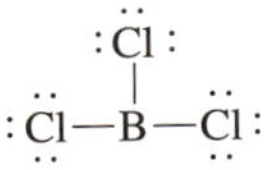

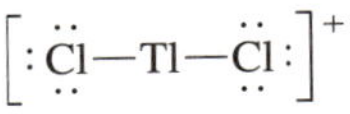

9.69 H_3C-NO_2 : H—C(H)(H)—N(=O)—O ⟷ H—C(H)(H)—N(—O)=O

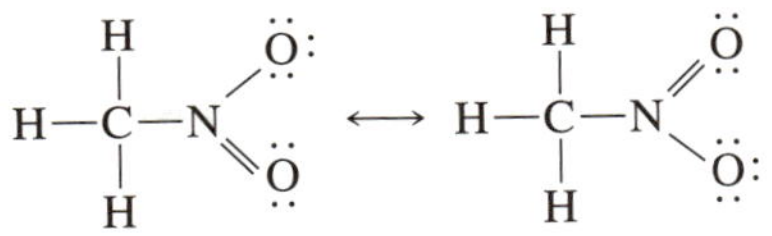

9.71 **a** $:\ddot{F}-\ddot{Xe}-\ddot{F}:$ **b** TeF_4: F—Te(F)(F)—F (four F around Te)

c $[XeF_5]^+$: Xe bonded to five F atoms

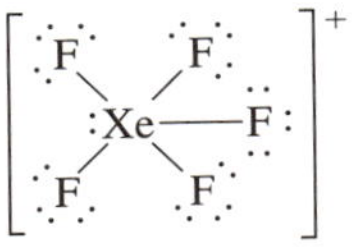

9.73 **a** $:\ddot{Cl}-B(-\ddot{Cl}:)-\ddot{Cl}:$ **b** $[:\ddot{Cl}-Tl-\ddot{Cl}:]^+$

c $:\ddot{Br}-Be-\ddot{Br}:$

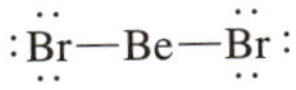

9.75 **a** Starting with the left oxygen, the formal charge of this oxygen is $6 - 1 - 6 = -1$. The formal charge of just the central oxygen is $6 - 3 - 2 = +1$. The formal charge of the right oxygen is $6 - 2 - 4 = 0$.

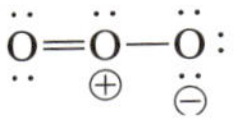

$O=\overset{\oplus}{O}-\overset{\ominus}{O}:$

b The formal charge of the carbon is $4 - 3 - 2 = -1$. The formal charge of the oxygen is $6 - 3 - 2 = +1$

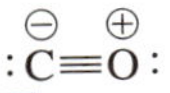

$:\overset{\ominus}{C}\equiv\overset{\oplus}{O}:$

c The formal charge of the nitrogen is $5 - 4 - 0 = +1$. The formal charge of the hydrogen is $1 - 1 - 0 = 0$. The formal charge of the oxygen bonded to the hydrogen is $6 - 2 - 4 = 0$. The formal charge of the other singly bonded oxygen is $6 - 1 - 6 = -1$. The formal charge of the doubly bonded oxygen is $6 - 2 - 4 = 0$.

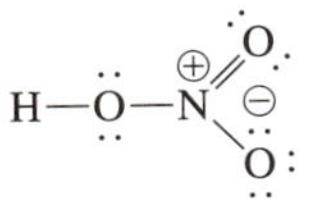

$H-\ddot{O}-\overset{\oplus}{N}(=O)-\overset{\ominus}{O}:$

9.77 **a** SOF_2: O=S(—F)—F **b** O=S(—O—H)—O—H

c $O=\ddot{Cl}-\ddot{O}-H$

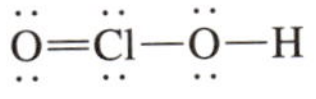

9.79 164 pm **9.81** **a** 107 pm **b** 207 pm **c** 222 pm **d** 177 pm **9.83** Methylamine 147 pm; Acetonitrile 116 pm **9.85** −57 kJ **9.87** **a** ionic, SrO, strontium oxide **b** covalent, CBr_4, carbon tetrabromide, **c** ionic, GaF_3, gallium(III) fluoride **d** covalent, NBr_3, nitrogen tribromide

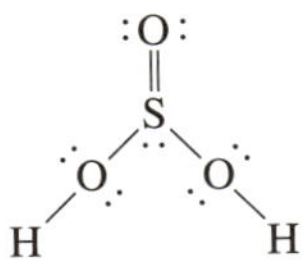

9.89 $[:\ddot{O}-\ddot{Se}(-\ddot{O}:)-\ddot{O}:]^{2-}$ or $[:\ddot{O}-\ddot{Se}(=O)-\ddot{O}:]^{2-}$, $Al_2(SeO_3)_3$

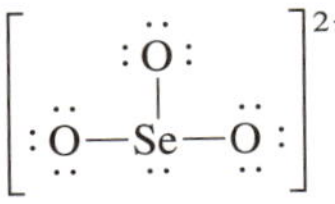

9.91 $H-\ddot{O}-\ddot{I}(-\ddot{O}:)-\ddot{O}:$ or $H-\ddot{O}-I(=O)(=O)$ **9.93** $[H-\ddot{N}-H]^-$

9.95 $[:\ddot{O}=N=\ddot{O}:]^+$ **9.97** **a** $:\ddot{Cl}-\ddot{Se}(-\ddot{O}:)-\ddot{Cl}:$ or $:\ddot{Cl}-\ddot{Se}(=O)-\ddot{Cl}:$ **b** $Se=C=Se:$

c $[:\ddot{Cl}-Ga(-\ddot{Cl}:)_2-\ddot{Cl}:]^-$ **d** $[:C\equiv C:]^{2-}$

9.99 **a** $:\ddot{Cl}-\ddot{Sb}(-\ddot{Cl}:)-\ddot{Cl}:$ **b** $:\ddot{I}-C\equiv N:$

c $:\ddot{Cl}-\ddot{I}(-\ddot{Cl}:)-\ddot{Cl}:$ **d** IF_5: I bonded to five F atoms

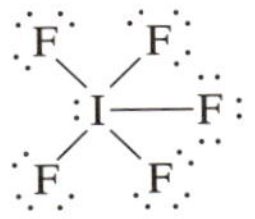

9.101 **a** $O=\ddot{Se}-\ddot{O}: \longleftrightarrow :\ddot{O}-\ddot{Se}=O \longleftrightarrow O=\ddot{Se}=O$

b O_2N-NO_2: four resonance structures with one N=O double bond on each N

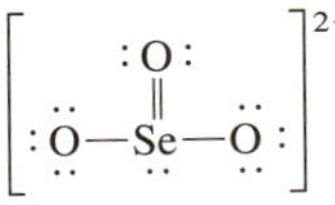

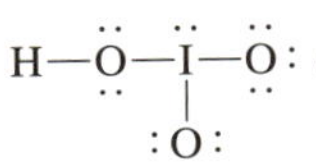

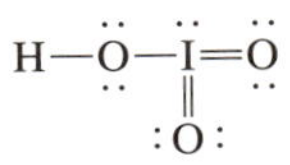

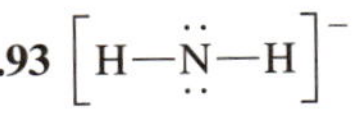

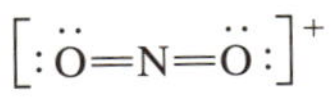

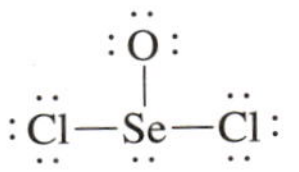

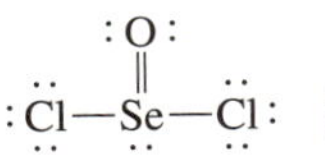

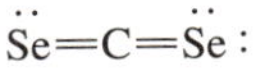

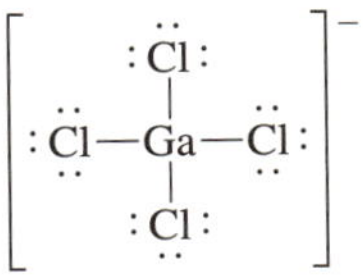

9.103 S_2N_2 ring: four resonance structures with different S=N double bonds

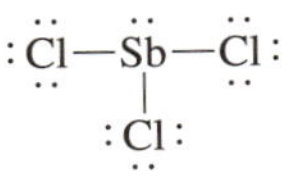

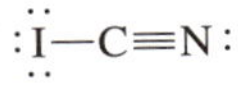

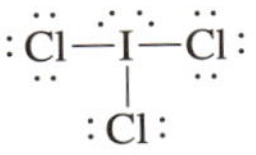

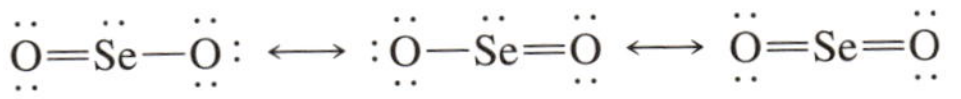

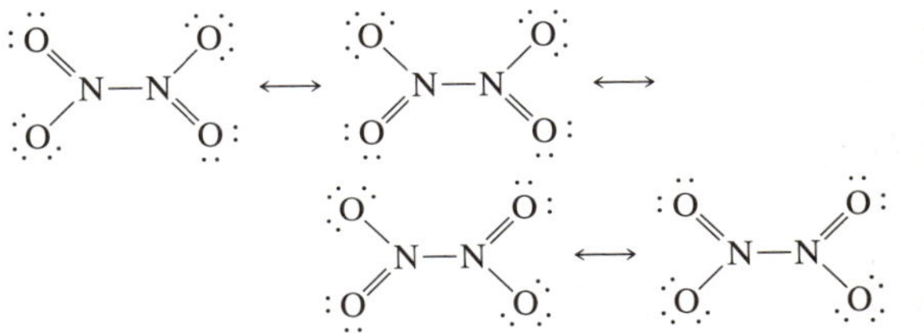

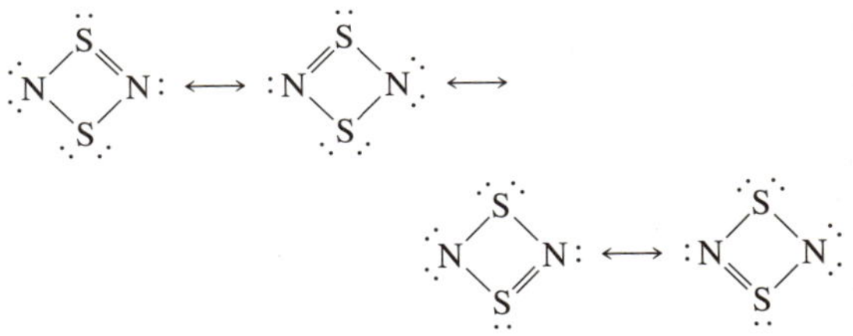

9.105 $O_2N-O-NO_2$: four resonance structures with one N=O double bond on each N

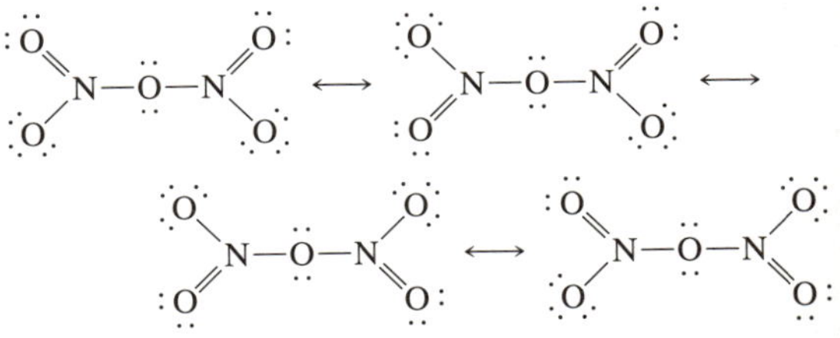

Because double bonds are shorter, the terminal N—O bonds that resonate between single and double bonds are 118 pm, and the central N—O single bonds are 136 pm. **9.107** −138 kJ **9.109** −130 kJ **9.137** P–H nonpolar; O—H 1 polar (acidic) $H_3PO_3(aq) + 2NaOH(aq) \longrightarrow Na_2HPO_3(aq) + 2H_2O(l)$ 0.007031 M H_3PO_3 **9.139** $Mg(ClO_4)_2$, magnesium perchlorate

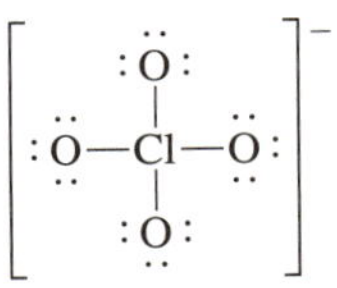

$[Cl(-\ddot{O}:)_4]^-$ or $[O=Cl(=O)(-\ddot{O}:)_2]^-$ (with two Cl=O double bonds)

9.141 H—C(—F)=O

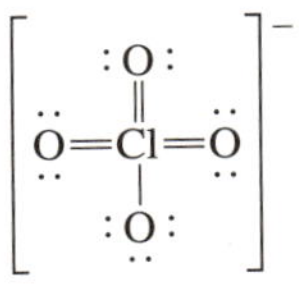

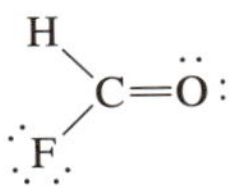

9.143 261 g/mol; It is molecular because the electronegativity difference between Sn and Cl is 1.3, and because it is a liquid, and is volatile at 151°C.

```
        :Cl:
         |
  :Cl—Sn—Cl:
         |
        :Cl:
```

9.145 ΔH_{rxn} = 1272.3 kJ/mol; BE(C≡N) = 861 kJ/mol (Table 9.5 has 891 kJ/mol) **9.147** 3.4 **9.149** 3.5

CHAPTER 10

10.23 a ii b i c iv d iii **10.25** The repulsions of the bonding electron pairs is no longer balanced by the C—C bond, so the molecule flattens out to form a trigonal planar geometry.
10.27 a i b iii c iv d v
10.29

```
H       H                    H       Br
 \     /                      \     /
  C=C      (has dipole),       C=C       (no dipole)
 /     \                      /     \
Br      Br                   Br      H
```

10.31 a ii b iv c i **10.33** a tetrahedral b bent c trigonal planar d trigonal pyramidal **10.35** a trigonal pyramidal b tetrahedral c linear d trigonal pyramidal **10.37** a 109°, same b 109°, less c 120°, the Cl—C—Cl bond to be less than 120° d 109°, less **10.39** a trigonal bipyramidal b T-shaped c square pyramidal d seesaw **10.41** a trigonal bipyramidal b octahedral c linear d square planar **10.43** a trigonal pyramidal and T-shaped b bent **10.45** CS_2 and XeF_4 **10.47** a uses sp^3 orbitals b uses sp^3 orbitals **10.49** a sp^3 b sp^2 c sp d sp
10.51 a The presence of two single bonds and no lone pairs suggests sp hybridization. An Hg—Cl bond is formed by overlapping the Hg sp hybrid orbital with a $3p$ orbital of Cl. b The presence of three single bonds and one lone pair suggests sp^3 hybridization of the P atom. Three hybrid orbitals each overlap a $3p$ orbital of a Cl atom to form a P—Cl bond. The fourth hybrid orbital contains the lone pair. **10.53** a sp^3d b sp^3d^2 c sp^3d d sp^3d **10.55** The P atom in PCl_6^- has 6 single bonds around it and no lone pairs. This suggests sp^3d^2 hybridization. Each bond is a σ bond formed by the overlap of the sp^3d^2 hybrid orbital on the P with a $3p$ orbital on Cl.
10.57 a Because the C is bonded to three other atoms, it is assumed to be sp^2 hybridized. One $2p$ orbital remains unhybridized. The carbon–hydrogen bonds are σ bonds formed by the overlap of an sp^2 hybrid orbital on C with a $1s$ orbital on H. The remaining sp^2 hybrid orbital on C overlaps with a $2p$ orbital on O to form a σ bond. The unhybridized $2p$ orbital on C overlaps with a parallel $2p$ orbital on O to form a π bond. Together, the σ and π bonds constitute a double bond. b The nitrogen atoms are sp hybridized. A σ bond is formed by the overlap of an sp hybrid orbital from each N. The remaining sp hybrid orbitals contain lone pairs of electrons. The two unhybridized $2p$ orbitals on one N overlap with the parallel unhybridized $2p$ orbitals on the other N to form two π bonds.
10.59 Each of the N atoms has a lone pair of electrons and is bonded to two atoms. The N atoms are sp^2 hybridized. The two possible arrangements of the O atoms relative to one another are shown below. Because the π bond between the N atoms must be broken to interconvert these two forms, it is to be expected that the hyponitrite ion will exhibit *cis–trans* isomerism.

```
[      N=N      ]2−    [       N=N—O ]2−
[ O          O  ]      [  O          ]
       cis                   trans
```

10.61 a $KK(\sigma_{2s})^2(\sigma_{2s}^*)^2(\pi_{2p})^2$, bond order = 1, stable, paramagnetic b $KK(\sigma_{2s})^2(\sigma_{2s}^*)^2(\pi_{2p})^1$, bond order = 1/2, stable, paramagnetic c $KK(\sigma_{2s})^2(\sigma_{2s}^*)^2(\pi_{2p})^4(\sigma_{2p})^2(\pi_{2p}^*)^3$, bond order = 3/2, stable, paramagnetic **10.63** $KK(\sigma_{2s})^2(\sigma_{2s}^*)^2(\pi_{2p})^4(\sigma_{2p})^2$, bond order = 3, diamagnetic **10.65** a bent b trigonal planar c linear d octahedral **10.67** none **10.69** a C_a and C_b: Three electron pairs around them. They are sp^2 hybridized. C_c: Four electron pairs around it. It is sp^3 hybridized.

```
H      H  H
 \     |  |
  C=C—C—O—H
 / a  b | c
H       H
```

b Both C atoms are bonded to two other atoms and have no lone pairs of electrons. They are sp hybridized.

:N≡C—C≡N:

10.71 The *trans* isomer is expected to have a zero dipole moment, whereas the *cis* isomer is not. **10.73** All four hydrogen atoms of $H_2C{=}C{=}CH_2$ cannot lie in the same plane because the second C=C bond forms perpendicular to the plane of the first C=C bond. By looking at Figure 10.26, you can see how this is so. The second C=C bond forms in the plane of the C—H bonds. Thus, the plane of the C—H bonds on the right side will be perpendicular to the plane of the C—H bonds on the left side. This is shown below using the π orbitals of the two C=C double bonds. **10.75** $(\sigma_{1s})^2$, stable **10.77** $KK(\sigma_{2s})^2(\sigma_{2s}^*)^2(\pi_{2p})^4(\sigma_{2p})^2$, bond order = 3 **10.79** O^{2+} has one antibonding electron less than O_2; O^{2-} has one more antibonding electron than in O_2
10.81 The N_2 ground state is $KK(\sigma_{2s})^2(\sigma_{2s}^*)^2(\pi_{2p})^4(\sigma_{2p})^2(\pi_{2p}^*)^0$, the first excited state is $KK(\sigma_{2s})^2(\sigma_{2s}^*)^2(\pi_{2p})^4(\sigma_{2p})^2(\pi_{2p}^*)^1$. The transition decreases the bond order from 3 to 2, and the ground state is diamagnetic, whereas the excited state is paramagnetic.
10.109 XeO_3F_2; trigonal bipyramidal, sp^3d

```
      :O:
 :F:   |   :F:              :F: :O: :F:
    \  |  /                    \ ‖ /
      Xe           or           Xe
    /    \                    //   \\
 :O:      :O:               :O:     :O:
```

10.111 ClF_3; the Cl atom has sp^3d hybrid orbitals, giving three Cl—F bonds and two lone pairs.

```
          H—O:                 H—O:
             |                    |
10.113       N        ⟷           N
           //  \                /  \\
         :O:    :O:           :O:    :O:
```

trigonal planar; therefore, the hybridization is sp^2; ΔH_f° = −34.5 kJ/mol; Resonance energy = 101 kJ/mol

CHAPTER 11

11.27 a A b C c C **11.29** The heat released when the liquid-to-solid phase change occurs prevents the fruit from freezing. **11.31** a AB_2 b yes, body-centered cell
11.33 The molecules with higher kinetic energy escape the liquid and leave behind molecules with lower energy. The result is a drop in temperature of the liquid. Since the cup is well insulated, the energy lost with the evaporated molecules is not rapidly replaced.
11.35 a face-centered cubic b simple cubic
11.37 a Vaporization b Freezing of eggs and sublimation of ice c Condensation d Gas-solid condensation, deposition e Freezing **11.39** 10°C **11.41** 15.5 kJ/mol **11.43** 1.58 kJ **11.45** 1.25 g **11.47** 27 g **11.49** 300. mmHg **11.51** 53.3 kJ/mol **11.53** a gas b gas c liquid d No

11.55

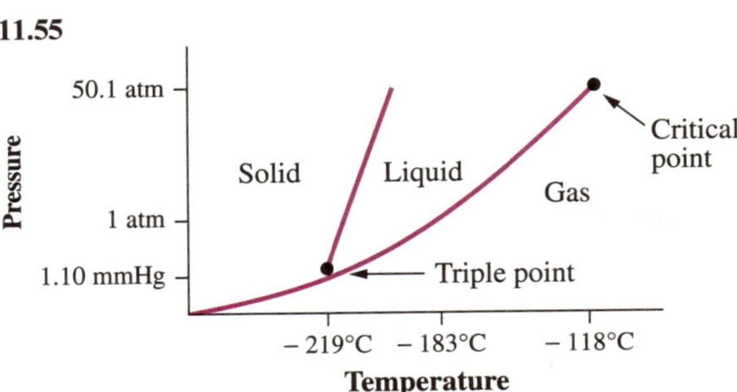

11.57 Liquefied at 25°C: SO_2 and C_2H_2. To liquefy CH_4, lower its temperature below −82°C, and then compress it. To liquefy CO, lower its temperature below −140°C, and then compress it. **11.59** a solid b liquid **11.61** Values increase with molecular mass, as expected. **11.63** a London forces b London and dipole-dipole forces, hydrogen bonding c London and dipole-dipole forces d London forces **11.65** $CCl_4 < SiCl_4 < GeCl_4$ **11.67** CCl_4, largest molecular mass and greatest London forces **11.69** $HOCH_2CH_2OH$, FCH_2CH_2OH, FCH_2CH_2F; hydrogen bonding decreases in magnitude from left to right **11.71** $CH_4 < C_2H_6 < CH_3OH < HOCH_2CH_2OH$ **11.73** a Metallic b Metallic c Covalent network d Molecular e Ionic **11.75** a Metallic b Covalent network c Molecular d Molecular **11.77** $(C_2H_5)_2O < C_4H_9OH < KCl < CaO$ **11.79** a Low-melting and brittle b High-melting, hard, and brittle c Malleable and electrically conducting d Hard and high-melting **11.81** a LiCl b SiC c CHI_3 d Co **11.83** one **11.85** 9.26×10^{-23} g; 9.274×10^{-23} g **11.87** 361 pm **11.89** 4, fcc **11.91** 19.25 g/cm^3 **11.93** 1.72×10^{-23} cm^3, 160 pm **11.95** Water vapor deposits directly to solid water (frost), frost melts, liquid water evaporates, the frost may have sublimed directly to water vapor **11.97** 68.4% **11.99** 79.4°C **11.101** a condenses into a liquid (and to solid at high pressure) b condenses to solid c remains a gas **11.103** Hydrogen bonding exists between the hydrogen of the OH group and the lone pair of electrons of oxygen of the OH group of an adjacent propanol molecule. **11.105** Ethylene glycol molecules are capable of hydrogen bonding to each other leading to higher boiling point and greater viscosity. **11.107** Al, metallic solid; Si, covalent network; P molecular solid; Sulfur, molecular (amorphous) solid **11.109** a Lower: KCl, smaller charge on the ions b Lower: CCl_4, lower molecular mass and weaker London forces c Lower: Zn, melting points for Group IIB metals are lower d Lower: C_2H_5Cl, no hydrogen-bond, but acetic acid **11.111** 3.171×10^{-22} g; 191.0 amu **11.113** 128 pm **11.115** 68.0% **11.117** a The boiling point increases as the size of the molecule increases. b Hydrogen bonding occurs between the H—F molecules. c The hydrogen halides have dipole-dipole interactions. **11.119** a Diamond and silicon carbide are covalent network solids with strong covalent bonds; graphite is a layered structure, with weak forces between the layers. b Silicon dioxide is a covalent network solid. Carbon dioxide is a discrete, small molecule. **11.121** a CO_2 consists of discrete nonpolar molecules; SiO_2 is a covalent network solid b HF has hydrogen bonding c SiF_4 is a larger molecule **11.149** 0.220 mol N_2, 0.01023 mol C_3H_8O, 0.0444, 33.1 mmHg, 33.1 mmHg **11.151** 6.6 kJ **11.153** 449 J **11.155** 1.83 g/L

CHAPTER 12

12.27 The amount of oxygen dissolved in water decreases as the temperature increases. **12.29** The lattice energy must be less for the XZ compound. **12.31** The charged smoke particles are neutralized by the current, which then allows them to aggregate into large particles. These large particles are too big to be carried out of the stack. **12.33** Because the salt concentration is higher outside the lettuce leaf than inside, water will pass out of the lettuce leaf into the dressing via osmosis. **12.35** a $CaCl_2$ b urea **12.37** Aqueous ammonia **12.39** Ethanol **12.41** $H_2O < CH_2OHCH_2OH < C_{10}H_{22}$ **12.43** Al^{3+} **12.45** $Ba(IO_3)_2 < Sr(IO_3)_2 < Ca(IO_3)_2 < Mg(IO_3)_2$; The iodate ion is fairly large, so the lattice energy for all these iodates should change to a smaller degree than the hydration energy of the cations. Therefore, solubility should increase with decreasing cation radius. **12.47** 0.886 g/100 mL **12.49** Dissolve 3.63 g KI in 68.9 g of water. **12.51** 5.16 g **12.53** 1.53 *m* **12.55** 77.7 g **12.57** 0.358; 0.642 **12.59** 0.0116 **12.61** 0.232; 16.8 *m* **12.63** 0.563 *M* **12.65** 0.796 *m* **12.67** 41.6 mmHg, 0.631 mmHg **12.69** T_b = 100.042°C, T_f = −0.151°C **12.71** 4.6×10^{-2} *m* **12.73** 122 amu **12.75** 163 amu **12.77** 6.85×10^4 amu **12.79** −0.042°C **12.81** ≅ 3 **12.83** a aerosol b sol c foam d sol **12.85** $Al_2(SO_4)_3$ **12.87** 0.671 mol fraction N_2, 0.329 mol fraction O_2 **12.89** 1.74 *m*, 0.0303, 1.63 *M* **12.91** 24 g of propane 31 g of butane **12.93** 141 mmHg **12.95** a 0.0002140 *M* $KAl(SO_4)_2$ b 0.0004279 *M* c 2.14×10^{-4} *M* **12.97** −21°C **12.99** 0.30 *M* **12.101** $CaCl_2$ **12.103** 1.8 g/mL, 5×10^2 *m* **12.105** $Mn_2C_{10}O_{10}$ **12.107** a $C_6H_{12}O_4$ b 148.2 g/mol **12.109** 30. atm

12.131 $Ca^{2+}(aq) + CO_3^{2-}(aq) \longrightarrow CaCO_3(s)$

$2Ag^+(aq) + CO_3^{2-}(aq) \longrightarrow Ag_2CO_3(s)$

$CO_3^{2-} = 0.0750\ M$

$NO_3^- = 0.225\ M$

$Na^+ = 0.375\ M$

12.133 −783 kJ/mol, −445 kJ/mol **12.135** 0.565 *m* **12.137** 0.701 mol/L **12.139** 3.1% **12.141** $C_3H_{18}O_3$

CHAPTER 13

13.29 a Rate of depletion of A $= -\dfrac{\Delta[A]}{\Delta t}$ or Rate of formation of B $= +\dfrac{\Delta[B]}{\Delta t}$. b No, the rate of depletion of A would be faster than the rate of formation of B. c $-\dfrac{\Delta[A]}{3\Delta t} = \dfrac{\Delta[B]}{2\Delta t}$ **13.31** a Rate=$k[A]^2$ b right c right d Left container rate is four times slower. e right **13.33** a $x = 0$ b $x = 1$ c $x = 3$ **13.35** a Region c b Region A **13.37** A number of answers will work as long as you match one of the existing concentrations of A. For example: [A] = 2.0 *M* with [B] = 2.0 *M*, or [A] = 1.0 *M* with [B] = 2.0 *M*. **13.39** $-\frac{1}{2}\dfrac{\Delta[NO_2]}{\Delta t} = \dfrac{\Delta[O_2]}{\Delta t}$ **13.41** $\frac{1}{5}\dfrac{\Delta[Br^-]}{\Delta t} = \dfrac{\Delta[BrO_3^-]}{\Delta t}$ **13.43** 2.3×10^{-2} *M*/hr **13.45** 1.2×10^{-5} *M*/hr **13.47** 1, 1, 2 **13.49** MnO_4^- is 1, $H_2C_2O_4$ is 1, H^+ is 0; overall order is 0 **13.51** Rate = $k[CH_3NNCH_3]$, $k = 2.5 \times 10^{-4}$/s **13.53** Rate = $k[NO]^2[H_2]$, $k = 2.9 \times 10^2\ (M^2/s)$ **13.55** Rate = $k[ClO_2]^2[OH^-]$, $k = 2.3 \times 10^2\ (M^2/s)$ **13.57** 2.1×10^2 *M* **13.59** 8.79×10^{-2} *M* **13.61** 5.2×10^{-4} *M* **13.63** 1.1×10^3 s, 1.1×10^3 s, 2.0×10^3 s **13.65** 96 hr, 1.9×10^2 hr, 2.9×10^2 hr, 3.9×10^2 hr, 4.8×10^2 hr **13.67** 461 s **13.69** 2.6×10^2 hr **13.71** 1 s **13.73** first-order, k = 0.100/s **13.75** E_a = 210 kJ **13.77** 1.1×10^5 J/mol, 1.7×10^{-3}/s **13.79** 8.4×10^4 J/mol **13.81** 1.1×10^2 kJ/mol **13.83** $NOCl_2$; $NOCl_2 + NO \longrightarrow 2NOCl$ **13.85** a Bimolecular b Bimolecular c Unimolecular d Termolecular **13.87** a Rate = $k[O_3]$ b Rate = $k[NOCl_2][NO]$ **13.89** Rate = $k[C_3H_6]^2$ **13.91** Rate = $k_2(k_1/k_{-1})[I_2][H_2] = k[I_2][H_2]$ **13.93** $2H_2O_2 \longrightarrow 2H_2O + O_2$; Br^-; yes (BrO^-) **13.95** 3.5×10^{-6} *M*/s; 3.2×10^{-6} *M*/s; 2.5×10^{-6} *M*/s **13.97** average $k = 2.5 \times 10^{-4}$/s **13.99** 8.33×10^3 s **13.101** 5.50×10^3 s **13.103** 0.024 *M* **13.105** a 34 L/(mol·s) b 8.2×10^{-5} *M* **13.107** 4.7×10^{-3} *M* **13.109** 2.5×10^{-4}/s **13.111** 1.2×10^2 s; 68 s **13.113** 114 kJ/mol; 5×10^9; 14 *M*/s **13.115** Rate = $k[NO_2][CO]$ **13.117** Rate = $k[NO_2Br]$ **13.119** Rate = $k_2(k_1/k_{-1})[NH_4^+][OCN^-] = k)[NH_4^+][OCN^-]$ **13.121** a Rate = $k[NO]^2[O_2]$ b 0.12 mol/L·s **13.123** a i) decreases ii) decreases b i) increases ii) decrea

13.125

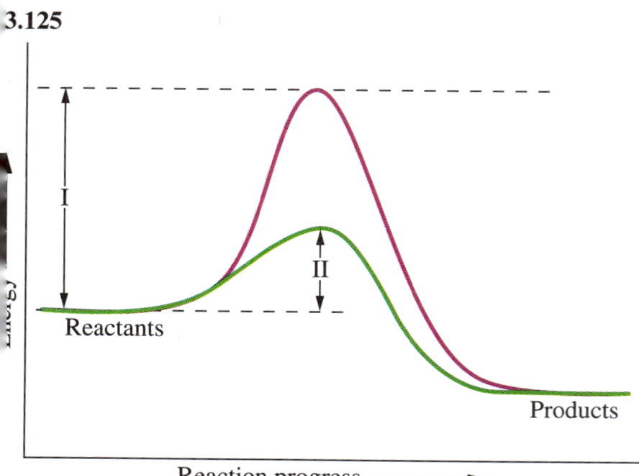

catalyst provides another pathway with lower activation energy.
13.127 a The rate is the change in the concentration of a reactant
or product with time. b The rate changes because the concentration of the reactant has changed. c The rate will equal k when
the reactants all have 1.00 M. **13.147** 6.8 L **13.149** −0.210 kJ/s
13.151 1.73×10^{-12} mol/L·s); 6.60×10^{-8} mmHg/s

CHAPTER 14

14.19 $\frac{2x}{3}$ **14.21** a **14.23** a III b Both I and II shift right.
14.25 a shifts left b Concentrations of A and B increase; concentration of C decreases. **14.27** The amount of product at equilibrium is increased by increasing the pressure and decreasing the
temperature. The rate of reaction is increased by increasing the temperature and using the proper catalyst. The choice of temperature is
a compromise between equilibrium product concentration and rate
of reaction. **14.29** 2.162 mol PCl_5, 0.338 mol PCl_3, and 0.338 mol
Cl_2 **14.31** 0.576 mol N_2, 1.728 mol H_2, and 0.048 mol NH_3

14.35 a $K_c = \frac{[NO_2][NO]}{[N_2O_3]}$

b $K_c = \frac{[H_2]^2[S_2]}{[H_2S]^2}$ c $K_c = \frac{[NO_2]^2}{[NO]^2[O_2]}$

d $K_c = \frac{[P(NH_2)_3][HCl]^3}{[PCl_3][NH_3]^3}$

14.37 $2H_2S(g) + 3O_2(g) \rightleftharpoons 2H_2O(g) + 2SO_2(g)$

14.39 $K_c = \frac{[N_2O_5]}{[NO_2]^2[O_2]^{1/2}}$

14.41 $K_c = 0.543$ **14.43** 50.4 **14.45** $3.1 \times .10^3$
14.47 $1.0 \times .10^{-3}$ **14.51** 3.0×10^{-5} **14.53** 56.3

14.55 a $K_c = \frac{[CO]^2}{[CO_2]}$ b $K_c = \frac{[CO_2]}{[CO]}$

c $K_c = \frac{[CO_2]}{[SO_2][O_2]^{1/2}}$ d $K_c = [Pb^{2+}][I^-]^2$

14.57 a not complete b nearly complete **14.59** No, decomposition does not occur; 3.2×10^{-48} mol/L; yes it agrees with
what is expected. **14.61** a goes to the left b goes to the
right c equilibrium d goes to the left **14.63** goes to the
left **14.65** 0.37 M **14.67** $[I_2] = [Br_2] = 5 \times 10^{-5}$ M, and
[IBr] = 5.1×10^{-4} M **14.69** $[CO_2] = 0.19$ M, [CO] = 1.62 M
14.71 [CO] = 0.0613 M, $[H_2] = 0.1839$ M, $[CH_4] = 0.0387$ M,
$[H_2O] = 0.0387$ M **14.73** forward **14.75** a no effect
b no effect c goes to the left **14.77** No, the fraction of the
methanol would decrease. **14.79** decrease **14.81** low
temperature and low pressure **14.83** 4.3 **14.85** 0.153
14.87 reverse **14.89** reverse **14.91** [HBr] = 0.008 mol;
$[H_2]$ = 0.0010 mol; and $[Br_2]$ = 0.0010 mol **14.93** 13.2%
14.95 CO: 0.48 mol; H_2: 2.44 mol; CH_4: 0.52 mol; and H_2O:
2 mol **14.97** endothermic

14.99 $K_p = \frac{[NH_3]^2(RT)^2}{[N_2](RT)[H_2]^3(RT)^3} = \frac{[NH_3]^2}{[N_2][H_2]^3}(RT)^{-2}$

$K_p = K_c(RT)^{-2}$, or $K_c = Kp\,(RT)^2$

14.101 a 14.1 b 0.4 atm, 2.3 atm c decreases
14.103 a $[PCl_3] = [Cl_2] = 0.031$ M, $[PCl_5] = 0.004$ M
b 0.89 c Less PCl_5 would decompose **14.105** 0.335
14.107 a $[SO_2] = [Cl_2] = 0.0346$ M, $[SO_2Cl_2] = 0.0266$ M
b 0.565 c decrease **14.109** a 7.2×10^{-5} M,
1.6×10^{-4} M b hydrogen bonding occurs c decrease
14.111 1.1×10^2 **14.135** 0.430 **14.137** 0.41 atm

CHAPTER 15

15.21 Hydroxide ion forms in the reaction $NH_3(aq) + H_2O(l) \rightleftharpoons NH_4^+(aq) + OH^-(aq)$ **15.23** The hydroxide ion acts as a base and donates a pair of electrons on the O atom, forming a bond with CO_2 to give HCO_3^-. **15.25** The pH increases.
15.27 b
15.29 $\underset{\text{base}}{OH^-(aq)} + \underset{\text{acid}}{HF(aq)} \rightleftharpoons \underset{\text{base}}{F^-(aq)} + \underset{\text{acid}}{H_2O(l)}$

15.31 a PO_4^{3-} b HS^- c NO^{2-} d $HAsO_4^{2-}$
15.33 a HClO b AsH^{4+} c H_3PO_4 d $HTeO^{3-}$
15.35 a $\underset{\text{acid}}{HSO_4^-(aq)} + \underset{\text{base}}{NH_3(aq)} \rightleftharpoons \underset{\text{acid}}{SO_4^{2-}(aq)} + \underset{\text{base}}{NH_4^+(aq)}$

b $\underset{\text{base}}{HPO_4^{2-}(aq)} + \underset{\text{acid}}{NH_4^+(aq)} \rightleftharpoons \underset{\text{base}}{H_2PO_4^-(aq)} + \underset{\text{acid}}{NH_3(aq)}$

c $\underset{\text{acid}}{Al(H_2O)_6^{3+}(aq)} + \underset{\text{base}}{H_2O(l)} \rightleftharpoons \underset{\text{acid}}{Al(H_2O)_5(OH)^{2+}(aq)} + \underset{\text{base}}{H_3O^+(aq)}$

d $\underset{\text{base}}{SO_3^{2-}(aq)} + \underset{\text{acid}}{NH_4^+(aq)} \rightleftharpoons \underset{\text{base}}{HSO_3^-(aq)} + \underset{\text{acid}}{NH_3(aq)}$

15.37 $\underset{\text{Lewis acid}}{BF_3} + \underset{\text{Lewis base}}{AsF_3} \longrightarrow BF_3{:}AsF_3$

15.39 a $AlCl_3 + Cl^- \rightleftharpoons [AlCl_4]^-$

$\underset{\text{acid}}{AlCl_3} + \underset{\text{base}}{[:\ddot{Cl}:]^-} \longrightarrow [AlCl_4]^-$

b $I^- + I_2 \rightleftharpoons I_3^-$

$\underset{\text{base}}{[:\ddot{I}:]^-} + \underset{\text{acid}}{:\ddot{I}{:}\ddot{I}:} \longrightarrow [:\ddot{I} - \ddot{I} - \ddot{I}:]^-$

15.41 a Cu^+: Lewis acid, H_2O: Lewis base
b BBr_3: Lewis acid, AsH_3: Lewis base
15.43 $\underset{\text{Lewis acid}}{H_2S} + \underset{\text{Lewis base}}{HOCH_2CH_2NH_2} \longrightarrow HOCH_2CH_2NH_3^+ + HS^-$
The H^+ ion from H_2S accepts a pair of electrons from the N atom in $HOCH_2CH_2NH_2$. **15.45** $HSO_4^- + ClO^- \longrightarrow HClO + SO_4^{2-}$; Reaction goes to the right **15.47** a left b left c right d right **15.49** richloroacetic acid; the equilibrium favors the formation of the weaker acid **15.51** a H_2S; acid strength decreases with increasing anion charge b H_2SO_3; acid strength increases with increasing electronegativity
c HBr; acid strength increases with increasing electronegativity
d HIO_4; acid strength increases with the number of oxygen
e H_2S; acid strength increases with the increasing size
15.53 a $[H_3O^+] = 1.2$ M; $[OH^-] = 8.3 \times 10^{-15}$ M
b $[OH^-] = 0.32$ M; $[H_3O^+] = 3.1 \times 10^{-14}$ M
c $[OH^-] = 0.170$ M; $[H_3O^+] = 5.9 \times 10^{-14}$ M
d $[H_3O^+] = 0.38$ M; $[OH^-] = 2.6 \times 10^{-14}$ M
15.55 $[H_3O^+] = 0.50$ M; $[OH^-] = 2.0 \times 10^{-13}$ M
15.57 $[OH^-] = 0.017$ M; $[H_3O^+] = 5.9 \times 10^{-13}$ M
15.59 a acidic b acidic c neutral d basic **15.61** acidic

15.63 a acidic b neutral c acidic d basic
15.65 a acidic b neutral c basic d acidic
15.67 a 8.00 b 11.30 c 2.12 d 8.197 **15.69** 2.12
15.71 a 5.72 b 11.92 c 2.56 d 6.32 **15.73** 11.60
15.75 7.6×10^{-6} *M* **15.77** 4.3×10^{-3} *M*
15.79 13.16 **15.81** 5.5 to 6.5, acidic
15.83 a BaO is a base; $BaO + H_2O \longrightarrow Ba^{2+} + 2OH^-$
b H_2S is an acid; $H_2S + H_2O \longrightarrow H_3O^+ + HS^-$
c CH_3NH_2 is a base; $CH_3NH_2 + H_2O \longrightarrow CH_3NH_3^+ + OH^-$
d SO_2 is an acid; $SO_2 + 2H_2O \longrightarrow H_3O^+ + HSO_3^-$
15.85 a $H_2O_2(aq) + S^{2-}(aq) \longrightarrow HO_2^-(aq) + HS^-(aq)$
b $HCO_3^-(aq) + OH^-(aq) \longrightarrow CO_3^{2-}(aq) + H_2O(l)$
c $NH_4^+(aq) + CN^-(aq) \longrightarrow NH_3(aq) + HCN(aq)$
d $H_2PO_4^-(aq) + OH^-(aq) \longrightarrow HPO_4^{2-}(aq) + H_2O(l)$
15.87 a $\underset{\text{base}}{ClO^-} + \underset{\text{acid}}{H_2O} \rightleftharpoons \underset{\text{acid}}{HClO} + \underset{\text{base}}{OH^-}$

$$[:\ddot{\underset{\cdot\cdot}{Cl}}-\ddot{\underset{\cdot\cdot}{O}}:]^- + H-\ddot{\underset{\cdot\cdot}{O}}-H \longrightarrow :\ddot{\underset{\cdot\cdot}{Cl}}-\ddot{\underset{\cdot\cdot}{O}}-H + [:\ddot{\underset{\cdot\cdot}{O}}-H]^-$$

b $\underset{\text{acid}}{NH_4^+} + \underset{\text{base}}{NH_2^-} \rightleftharpoons 2NH_3$

$$\left[\begin{matrix} & H & \\ & | & \\ H- & N & -H \\ & | & \\ & H & \end{matrix}\right]^+ + \left[\begin{matrix} H \\ | \\ :N: \\ | \\ H \end{matrix}\right]^- \longrightarrow 2\ \begin{matrix} H \\ | \\ :N-H \\ | \\ H \end{matrix}$$

15.89 $HNO_2 + F^- \rightleftharpoons HF + NO_2^-$; an acid-base reaction goes in the direction of the weaker acid, the reaction is more likely to go in the direction written **15.91** $H_2S < H_2Se < HBr$
15.93 $[H_3O^+] = 34.0 \times 10^{-14}$ *M* **15.95** 2.82
15.97 $[OH^-] = 31.4 \times 10^{-11}$ *M*
15.99 a $H_2SO_4(aq) + 2NaHCO_3(aq) \longrightarrow Na_2SO_4(aq) + 2CO_2(g) + 2H_2O(l)$
$H_3O^+(aq) + HCO_3^-(aq) \longrightarrow CO_2(g) + 2H_2O(l)$ b 0.180
c 60.66%, $NaHCO_3$
15.101 $HCO_3^-(aq) + H_2O(l) \rightleftharpoons H_3O^+(aq) + CO_3^{2-}(aq)$
$HCO_3^-(aq) + H_2O(l) \rightleftharpoons H_2CO_3(aq) + OH^-(aq)$
$HCO_3^-(aq) + Na^+(aq) + OH^-(aq) \rightleftharpoons Na^+(aq) + CO_3^{2-}(aq) + H_2O(l)$
$HCO_3^-(aq) + H^+(aq) + Cl^-(aq) \rightleftharpoons H_2O(l) + CO_2(g) + Cl^-(aq)$
15.103 $CaH_2(s) + 2H_2O(l) \longrightarrow Ca(OH)_2(s) + 2H_2(g)$;
The hydride ion is a stronger base
15.129 the structure of H_3PO_4 is $(HO)_3PO$; 0.976 g NaOH
15.131 $\underset{\text{Lewis acid}}{BF_3} + \underset{\text{Lewis base}}{:NH_3} \longrightarrow F_3B:NH_3$; 12.5 g $F_3B:NH_3$

CHAPTER 16

16.23 A **16.25** a $HF(aq) + H_2O(l) \rightleftharpoons H_3O^+(aq) + F^-(aq)$
b $F^-(aq) + H_2O(l) \rightleftharpoons HF(aq) + OH^-(aq)$
c $C_6H_5NH_2(aq) + H_2O(l) \rightleftharpoons C_6H_5NH_3^+(aq) + OH^-(aq)$
d $C_6H_5NH_3^+(aq) + H_2O(l) \rightleftharpoons C_6H_5NH_2(aq) + H_3O^+(aq)$
16.27 Acidic: $RanH^+(aq) + H_2O(l) \rightleftharpoons Ran(aq) + H_3O^+(aq)$
16.29 100 mL of 0.10 *M* NaA **16.31** a weak acid b 8.5
c thymol blue or phenolphthalein (an indicator that changes color in the pH range of about 7 to 10) **16.35** 5.6×10^{-5}
16.37 5.53; 2.0×10^{-4} **16.39** $[H_3O^+] = [C_6H_4NH_2COO^-] = 1.1 \times 10^{-3}$ *M* **16.41** 0.26 *M* **16.43** 4.9×10^{-3} *M*, 2.31
16.45 0.361 *M* **16.47** a 3.7×10^{-3} *M* b 3.9×10^{-6} *M*
16.49 $CH_3NH_2(aq) + H_2O(l) \rightleftharpoons CH_3NH_3^+(aq) + OH^-(aq)$

$$K_b = \frac{[CH_3NH_3^+][OH^-]}{[CH_3NH_2]}$$

16.51 3.2×10^{-5} **16.53** 4.9×10^{-3} *M*; 11.69
16.55 a no hydrolysis b $OCl^- + H_2O \rightleftharpoons HOCl + OH^-$

$$K_b = \frac{K_w}{K_a} = \frac{[HOCl][OH^-]}{[OCl^-]}$$

c $NH_2NH_3^+ + H_2O \rightleftharpoons H_3O^+ + NH_2NH_2$

$$K_a = \frac{K_w}{K_b} = \frac{[H_3O^+][NH_2NH_2]}{[NH_2NH_3^+]}$$ d no hydrolysis

16.57 $Zn(H_2O)_6^{2+}(aq) + H_2O(l) \rightleftharpoons Zn(H_2O)_5(OH)^+(aq) + H_3O^+(aq$
16.59 a acidic b basic c basic d acidic
16.61 a nearly acidic b acidic **16.63** a 2.2×10^{-11}
b 7.1×10^{-6} **16.65** 8.64, 4.4×10^{-6} *M* **16.67** 1.0×10^{-3} *M*; 2.99 **16.69** a 0.030 b 0.0057 **16.71** 3.17
16.73 10.47 **16.75** 3.45 **16.77** 9.26, 9.09 **16.79** 2.71
16.81 5.32 **16.83** 0.34 mol **16.85** 1.60 **16.87** 8.59
16.89 5.97 **16.91** a 11.15 b 9.12 c 5.24 d 1.90
16.93 9.08 **16.95** 1.1×10^{-3} **16.97** 1.1×10^{-2}
16.99 $K_b(CN^-) = 2.0 \times 10^{-5}$; $K_b(CO_3^{2-}) = 2.1 \times 10^{-4}$; CO_3^{2-} **16.101** 2.84 **16.103** 3.16 **16.105** 11/1 **16.107** 2.4
16.109 3.11 **16.111** a 0.100 *M* b 0.11 *M*
16.113 a 2.3×10^{-11} b 4.4×10^{-4} c 10.39
16.115 a true b true c false d false e false
f true **16.117** a 74.0 g/mol b 1.3×10^{-5}
16.119 a Initial pH = 11.13; 30% titration pH = 9.62; 50% titration pH = 9.62; 100% titration pH = 5.28 b acidic
16.121 a H_3PO_4 and $H_2PO_4^-$ b 8 mL $H_3PO_4(aq)$ and 42 mL $H_2PO_4^-(aq)$ **16.123** a 0.215 *M* b 3.55 c bromphenol blue **16.125** a $H_3O^+(aq) + NH_3(aq) \longrightarrow NH_4^+(aq) + H_2O(l)$
b $NH_4^+(aq) + OH^-(aq) \longrightarrow NH_3(aq) + H_2O(l)$ c moles NH_3 before/after: 0.10/0.090; moles NH_4^+ before/after: 0.10/0.11; pH 9.2 d This is a buffer system. **16.127** a 0.59 b 7.16
c 7.55 **16.129** a $H_2A + 2H_2O \rightleftharpoons 2H_3O^+ + A^{2-}$
b $[H_2A] \gg [H_3O^+] = [HA^-] \gg [A^{2-}]$ c pH = 2.34; $[H_2A]$ = 0.0205 *M* d 4.6×10^{-5} *M* **16.131** a 0.13 *M* b 1.9×10^2 mL
c 0.83 *M* **16.151** 4.1% **16.153** −1.8°C **16.155** 9.1 mL

CHAPTER 17

17.17 From the K_{sp} values, these are the more soluble:
a silver chloride b magnesium hydroxide **17.19** a
17.21 middle beaker **17.23** Add just enough Na_2SO_4 to precipitate all the Ba^{2+}; filter off the $BaSO_4$; add more Na_2SO_4 to precipitate all the Ca^{2+}; filter off the $CaSO_4$; Mg^{2+} remains in the solution. **17.25** a soluble b insoluble c insoluble
d soluble **17.27** a $K_{sp} = [Mg^{2+}][OH^-]^2$ b $K_{sp} = [Sr^{2+}][CO_3^{2-}]$ c $K_{sp} = [Ca^{2+}]^3[AsO_4^{3-}]^2$ d $K_{sp} = [Fe^{3+}][OH^-]^3$ **17.29** 9.3×10^{-10} **17.31** 1.2×10^{-7}
17.33 1.8×10^{-11} **17.35** 0.0045 g/L **17.37** 1.9×10^{-3} *M*
17.39 2.0×10^{-4} g/L **17.41** 1.1×10^{-5} g/L **17.43** 0.4 g/L
17.45 a yes b yes **17.47** Lead chromate will precipitate.
17.49 No precipitation **17.51** No precipitation
17.53 0.0018 mol **17.55** 6.9×10^{-9} *M* **17.57** $BaF_2(s) + 2H_3O^+(aq) \rightleftharpoons Ba^{2+}(aq) + 2HF(aq) + 2H_2O(l)$
17.59 BaF_2; F^- is the conjugate base of the weak acid HF.
17.61 $Cu^+(aq) + 2CN^-(aq) \rightleftharpoons Cu(CN)^{2-}(aq)$

$$K_f = \frac{[Cu(CN)_2^-]}{[Cu^+][CN^-]^2} = 1.0 \times 10^{16}$$

17.63 5.5×10^{-19} *M* **17.65** will precipitate
17.67 3.0×10^{-3} *M* **17.69** Add HCl to precipitate only the Pb^2 as $PbCl_2$, leaving the others in solution. After decanting or filtering, add 0.3 M HCl and H_2S to precipitate only the CdS away from the Sr^{2+} ion. **17.71** Mn^{2+} **17.73** 1.3×10^{-4} *M*
17.75 a 6.9×10^{-7} *M* b 3.2×10^{-4} g/L **17.77** a 5.2×10^{-6}
b pOH = 4.81 **17.79** 26 g/L **17.81** 1.8×10^{-9} *M*
17.83 8.0×10^{-3} *M* **17.85** 1.0×10^{-5}; unsaturated
17.87 5.5×10^{-6} g **17.89** 4.7×10^{-2} *M* **17.91** 1.4×10^{-2} *M*
17.93 3.6×10^{-4} *M*; 4.2×10^{-6} *M*; 2.5×10^9 **17.95** 1.0×10^{-5}
17.97 a 6.4×10^{-7} *M* b 2.4×10^{-4} *M* c yes
17.99 a 6.6×10^{-4} *M* b 1.6×10^{-3} *M*, 11%
17.101 a 6.3×10^{-16} b 8.7×10^{-9} *M* c common-ion effect
17.103 a 3.1×10^{-3} b 0.044 mol; 0.044 *M*; 0.89 mol

17.105 a 1.5×10^{-4} *M* b 99% **17.129** 0.18 *M* **17.131** 2.7×10^{-4} *M* **17.133** $Ba^{2+}(aq) + 2OH^{-}(aq) + Mg^{2+}(aq) + SO_4^{2-}(aq) \rightleftharpoons BaSO_4(s) + Mg(OH)_2(s)$; concentrations of the ions $[SO_4^{2-}] = 0.109$ *M*. $[Ba^{2+}] = 1 \times 10^{-9}$ *M*. $[Mg^{2+}] = 0.109$ *M*. $[OH^{-}] = 1.3 \times 10^{-9}$ *M*

CHAPTER 18

18.23 a false b false c false d true e false **18.25** a 2.0 mol CO_2 b butane gas c CO_2 at $-80°C$ d bromine vapor **18.27** a negative b negative c negative d zero **18.29** a K_c does not change. b $Q > K_c$; the reaction will go to the left c ΔG is positive; ΔS is negative **18.31** 106 J/K **18.33** -128 J/K; 127 J/K **18.35** a negative b not predictable c positive d positive **18.37** a -181.7 J/K b -57.9 J/K c -56.4 J/K d 313.6 J/K **18.39** -266.90 J/K; entropy decreases, as expected from the decrease in moles of gas. **18.41** -1452.8 kJ; -161.4 J/K; -1404.7 kJ **18.43** a $K(s) + Br_2(l) \longrightarrow KBr(s)$ b $\frac{3}{2}H_2(g) + C(graphite) + \frac{1}{2}Cl_2(g) \longrightarrow CH_3Cl(l)$ c $\frac{1}{8}S_8\ (rhombic) + H_2(g) \longrightarrow H_2S(g)$ d $As(s) + \frac{3}{2}H_2(g) \longrightarrow AsH_3(g)$ **18.45** a -1331.4 kJ b -56.2 kJ **18.47** a Spontaneous reaction b Spontaneous reaction c Nonspontaneous reaction d Equilibrium mixture; significant amounts of both e Nonspontaneous reaction **18.49** a 850.2 kJ, 838.8 kJ; endothermic reaction with mainly reactants at equilibrium b -72.2 kJ, -142.0 kJ; exothermic reaction with mainly products at equilibrium **18.51** -474.2 kJ; 0 **18.53** -15.8 kJ

18.55 a $K = K_p = \dfrac{P_{CO_2}P_{H_2}}{P_{CO}P_{H_2O}}$

b $K = [Mg^{2+}][OH^{-}]_2$ c $K = [Li^{+}]_2[OH^{-}]_2P_{H_2}$ **18.57** -107.00 kJ; 6×10^{18} **18.59** -142.20 kJ; 9×10^{24} **18.61** -144.39 kJ; 2×10^{25} **18.63** 1.2×10^{2}; yes **18.65** 401 K **18.67** Reaction is spontaneous because $\Delta S°$ is large and positive. **18.69** -184 kJ; exothermic ; positive; both the ΔH term and the $-T\Delta S$ term are negative **18.71** -14.3 J/(K·mol) **18.73** a negative b positive c positive d negative **18.75** negative **18.77** -136.0 J/K **18.79** no, $\Delta G°$ is $+266.8$ kJ **18.81** $\Delta H°$ is positive, $\Delta S°$ is positive **18.83** 62.4 kJ; 1×10^{-11} **18.85** 109.6 kJ; 136.6 J/K; 68.9 kJ; -37.0 kJ; nonspontaneous reaction at 25°C but a spontaneous reaction at 800°C **18.87** a -75 kJ b -80.8 J/mol·K c -51 kJ/mol **18.89** 205.4 J/mol·K **18.91** a 97.1 kJ; 115.5 J/K; 187.2 kJ; 107.2 J/K b 841 K; 903 K c carbon **18.93** a formic acid b products **18.95** a products b need to heat to increase rate of reaction **18.97** a -126 kJ/mol b -2745.9 kJ c $-440.$ J/K **18.99** a 6×10^{-5} b Both ΔH and ΔS, but especially ΔS. **18.123** 0.004%, same; $K = 3 \times 10^{-10}$ **18.125** 87% **18.127** -2.29 kJ; -98.7 J/K

CHAPTER 19

19.25 a no change b no change c The Zn strip would dissolve, and the blue color of the solution would fade, and solid copper would precipitate out of the solution. d no change **19.27** The Zn is a sacrificial electrode that keeps the hull from undergoing oxidation by the dissolved ions in seawater. Zn works because it is more easily oxidized than Fe. **19.29** Since there is more zinc present, the oxidation-reduction reactions in the battery will run for a longer period of time. This assumes zinc is the limiting reactant. **19.31** Pick elements or compounds to be reduced and one to be oxidized so that when the half-reactions are added together, the $E°_{cell}$ is about 0.90 V. For example:

$Cd(s) \longrightarrow Cd^{2+}(aq) + 2e^{-}$	0.40 V
$I_2(s) + 2e^{-} \longrightarrow 2I^{-}(aq)$	0.54 V
$I_2(s) + Cd(s) \longrightarrow 2I^{-}(aq) + Cd^{2+}(aq)$	0.94 V

19.33 a brighter b dimmer c dimmer d off **19.35** a $Cr_2O_7^{2-} + 3C_2O_4^{2-} + 14H^{+} \longrightarrow 2Cr^{3+} + 6CO_2 + 7H_2O$
b $3Cu + 2NO_3^{-} + 8H^{+} \longrightarrow 3Cu^{2+} + 2NO + 4H_2O$
c $MnO_2 + HNO_2 + H^{+} \longrightarrow Mn^{2+} + NO_3^{-} + H_2O$
d $5PbO_2 + 2Mn^{2+} + 5SO_4^{2-} + 4H^{+} \longrightarrow 5PbSO_4 + 2MnO_4^{-} + 2H_2O$
e $3HNO_2 + Cr_2O_7^{2-} + 5H^{+} \longrightarrow 2Cr^{3+} + 3NO_3^{-} + 4H_2O$
19.37 a $Mn^{2+} + H_2O_2 + 2OH^{-} \longrightarrow MnO_2 + 2H_2O$
b $2MnO_4^{-} + 3NO_2^{-} + H_2O \longrightarrow 2MnO_2 + 3NO_3^{-} + 2OH^{-}$
c $Mn^{2+} + 2ClO_3^{-} \longrightarrow MnO_2 + 2ClO_2$
d $MnO_4^{-} + 3NO_2 + 2OH^{-} \longrightarrow MnO_2 + 3NO_3^{-} + H_2O$
e $3Cl_2 + 6OH^{-} \longrightarrow 5Cl^{-} + ClO_3^{-} + 3H_2O$
19.39 a $8H_2S + 16NO_3^{-} + 16H^{+} \longrightarrow S_8 + 16NO_2 + 16H_2O$
b $2NO_3^{-} + 3Cu + 8H^{+} \longrightarrow 2NO + 3Cu^{2+} + 4H_2O$
c $2MnO_4^{-} + 5SO_2 + 2H_2O \longrightarrow 5SO_4^{2-} + 2Mn^{2+} + 4H^{+}$
d $2Bi(OH)_3 + 3Sn(OH)_3^{-} + 3OH^{-} \longrightarrow 3Sn(OH)_6^{2-} + 2Bi$
19.41 a $MnO_4^{-} + I^{-} + H_2O \longrightarrow 2MnO_2 + IO_3^{-} + 2OH^{-}$
b $Cr_2O_7^{2-} + 6Cl^{-} + 14H^{+} \longrightarrow 2Cr^{3+} + 3Cl_2 + 7H_2O$
c $3S_8 + 32NO_3^{-} + 32H^{+} \longrightarrow 32NO + 24SO_2 + 16H_2O$
d $3H_2O_2 + 2MnO_4^{-} \longrightarrow 3O_2 + 2MnO_2 + 2H_2O + 2OH^{-}$
e $5Zn + 2NO_3^{-} + 12H^{+} \longrightarrow N_2 + 5Zn^{2+} + 6H_2O$
19.43

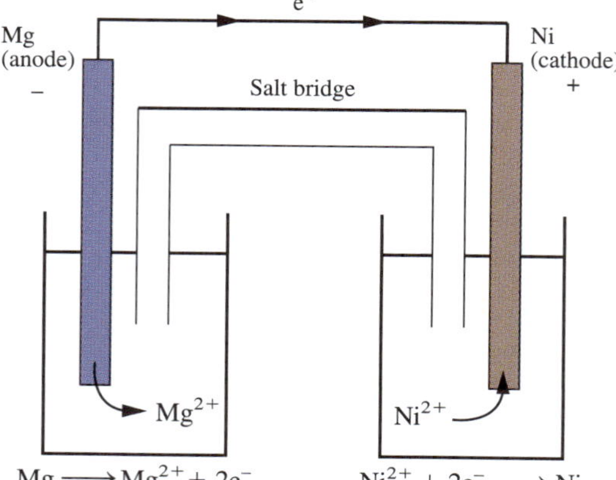

19.45 $Zn \longrightarrow Zn^{2+} + 2e$
$Ag^{+} + e^{-} \longrightarrow Ag$
Electron flow in circuit is from Zn to Ag.
19.47 Anode: $Zn(s) + 2OH^{-}(aq) \longrightarrow Zn(OH)_2(s) + 2e^{-}$
Cathode: $Ag_2O(s) + H_2O(l) + 2e^{-} \longrightarrow 2Ag(s) + 2OH^{-}(aq)$
Overall: $Zn(s) + Ag_2O(s) + H_2O(l) \longrightarrow Zn(OH)_2(s) + 2Ag(s)$
19.49 $Ni(s)|Ni^{2+}(aq)||Pb^{2+}(aq)|Pb(s)$
19.51 $Ni(s)|Ni^{2+}(1\ M)||H^{+}(1\ M)|H_2(g)|Pt$
19.53 $Fe(s) + 2Ag^{+}(aq) \longrightarrow Fe^{2+}(aq) + 2Ag(s)$
19.55 $Cd(s) \longrightarrow Cd^{2+}(aq) + 2e^{-}$
$Ni^{2+}(aq) + 2e^{-} \longrightarrow Ni(s)$
$Cd(s) + Ni^{2+}(aq) \longrightarrow Cd^{2+}(aq) + Ni(s)$

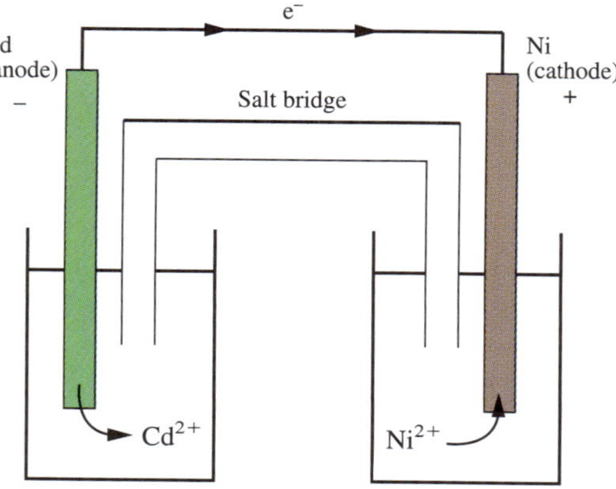

19.57 −69 kJ **19.59** −96 kJ **19.61** $NO_3^-(aq)$, $O_2(g)$, $MnO_4^-(aq)$ **19.63** Zn(*s*) is the strongest reducing agent, and $Cu^+(aq)$ is the weakest. **19.65** a not spontaneous b spontaneous **19.67** Cl_2 will oxidize Br^-; $Cl_2(g) + 2Br^-(aq) \longrightarrow 2Cl^-(aq) + Br_2(l)$ **19.69** 1.54 V **19.71** 1.28 V **19.73** -3.6×10^2 kJ **19.75** -1.6×10^2 kJ **19.77** 3.16 V **19.79** 1.934 V **19.81** 10^{14} **19.83** 1×10^6 **19.85** 0.57 V **19.87** 0.29 V **19.89** 0.004 *M* **19.91** a The cathode reaction is $Ca^{2+}(l) + 2e^- \longrightarrow Ca(l)$ The anode reaction is $S^{2-}(l) \longrightarrow S(l) + 2e^-$ b The cathode reaction is $Cs^+(l) + e^- \longrightarrow Cs(l)$ The anode reaction is $4OH^-(l) \longrightarrow O_2(g) + 2H_2O(g) + 4e^-$ **19.93** a anode: $2H_2O(l) \longrightarrow O_2(g) + 4H^+(aq) + 4e^-$ cathode: $2H_2O(l) + 2e^- \longrightarrow H_2(g) + 2OH^-(aq)$ overall reaction: $2H_2O(l) \longrightarrow 2H_2(g) + O_2(g)$ b anode: $2Br^-(aq) \longrightarrow Br_2(l) + 2e^-$ cathode: $2H_2O(l) + 2e^- \longrightarrow H_2(g) + 2OH^-(aq)$ overall reaction: $2Br^-(aq) + 2H_2O(l) \longrightarrow Br_2(l) + H_2(g) + 2OH^-(aq)$
19.95 3.87×10^7 C **19.97** 0.360 g Li **19.99** a $3Fe(s) + 2NO_3^-(aq) + 8H^+(aq) \longrightarrow 3Fe^{2+}(aq) + 2NO(g) + 4H_2O(l)$ b $3Fe^{2+}(aq) + NO_3^-(aq) + 4H^+(aq) \longrightarrow 3Fe^{3+}(aq) + NO(g) + 2H_2O(l)$ c $Fe(s) + NO_3^-(aq) + 4H^+(aq) \longrightarrow Fe^{3+}(aq) + NO(g) + 2H_2O(l)$
19.101 a $16MnO_4^-(aq) + 24S_2^-(aq) + 32H_2O(l) \longrightarrow 16MnO_2(s) + 3S_8(s) + 64OH^-(aq)$
b $IO_3^-(aq) + 3HSO_3^-(aq) \longrightarrow I^- + 3SO_4^{2-}(aq) + 3H^+(aq)$
c $CrO_4^{2-}(aq) + 3Fe(OH)_2(aq) + 4H_2O(l) \longrightarrow Cr(OH)_4^-(aq) + 3Fe(OH)_3(s) + OH^-(aq)$
d $Cl_2(aq) + 2OH^-(aq) \longrightarrow Cl^-(aq) + ClO^-(aq) + H_2O(l)$
19.103 $4Fe(OH)_2(s) + O_2(g) + 2H_2O(l) \longrightarrow 4Fe(OH)_3(s)$
19.105 $Ca(s)|Ca^{2+}(aq)||Cl_2(g)|Cl^-(aq)|Pt(s)$; Anode: $Ca(s) \longrightarrow Ca^{2+}(aq) + 2e^-$; Cathode: $Cl_2(g) + 2e^- \longrightarrow 2Cl^-(aq)$; 4.12 V
19.107 a spontaneous b Fe^{3+} will oxidize Sn^{2+} to Sn^{4+}.
19.109 0.36 V **19.111** a 2.2 b 0.3 *M*
19.113 a 1.0 *F*, 9.6×10^4 C b 2.0 *F*, 1.9×10^5 C c 10.11 *F*, 1.1×10^4 C d 0.028 *F*, 2.7×10^3 C
19.115 2.67×10^{-4} g As **19.117** a $Cd(s) + 2Ag^+(aq) \longrightarrow 2Ag(s) + Cd^{2+}(aq)$; 1.20 V **19.119** Anode: $2H_2O(l) \longrightarrow O_2(g) + 4H^+(aq) + 4e^-$ Cathode: $Cu^{2+}(aq) + 2e^- \longrightarrow Cu(s)$
19.121 a 1.48×10^3 s b 0.398 g **19.123** a 0.323 *F* b 3+ **19.125** a 1.06 V b 8.8×10^{-3} *M*
19.127 a −23.2 kJ b −0.28 V; $Cd(s) + Co^{2+}(aq) \longrightarrow Cd^{2+}(aq) + Co(s)$ **19.145** $E° = 0.02$ V; $\Delta G° = -10$ kJ; $\Delta H° = -174.7$ kJ; $\Delta S° - 5.5 \times 10^2$ J/K **19.147** −0.05 V; non spontaneous **19.149** −0.04 V; non spontaneous
19.151 $K_{sp} = 1 \times 10^{-7}$

CHAPTER 20

20.23 a 0.1% b If you had a large quantity of material, 0.1% still would be a significant quantity. Also, if the material were particularly toxic in addition to being radioactive, the amount would be a significant quantity. **20.25** Some of the expected mass is in the form of energy, the mass defect. **20.27** a At $t = 0$ min, the container contains 16 spheres; at $t = 30$ min, the container contains 2 spheres. b Incorrect at $t = 5$ min; at $t = 5$ min, there should be fewer that 750 atoms. **20.29** The large, positively charged He nucleus that makes up alpha (α) radiation is unable to pass through the atoms that make up solid materials such as wood without coming into contact or being deflected by the nuclei. Gamma (γ) radiation, however, with its small wavelength and high energy, can pass through large amounts of material without interaction, just as x rays can pass through skin and other soft tissue. **20.31** Irradiation of food atoms with gamma radiation does not result in the creation of radioactive elements; therefore, the food cannot become radioactive. The addition of a radioactive element directly to the meat would make the meat radioactive. Under the right conditions, nuclear bombardment could also lead to the production of radioactive elements. **20.33** ${}^{87}_{37}Rb \longrightarrow {}^{87}_{38}Sr + {}^{0}_{-1}e$
20.35 ${}^{232}_{90}Th \longrightarrow {}^{228}_{88}Ra + {}^{4}_{2}He$ **20.37** ${}^{18}_{9}F \longrightarrow {}^{18}_{8}O + {}^{0}_{1}e$

20.39 ${}^{210}_{84}Po \longrightarrow {}^{206}_{82}Pb + {}^{4}_{2}He$ **20.41** a ${}^{122}_{51}Sb$ b ${}^{204}_{85}At$ c ${}^{80}_{37}Rb$ **20.43** a alpha emission b positron emission or electron capture c beta emission **20.45** ${}^{235}_{92}U$ decay series; ${}^{232}_{90}Th$ decay series. Alpha emission decreases the mass number by 4. Beta emission has no effect upon the mass number. **20.47** a ${}^{26}_{12}Mg(d, \alpha){}^{24}_{11}Na$ b ${}^{16}_{8}O(n, p){}^{16}_{7}N$ **20.49** a ${}^{27}_{13}Al + {}^{2}_{1}H \longrightarrow {}^{25}_{12}Mg + {}^{4}_{2}He$ b ${}^{10}_{5}B + {}^{4}_{2}He \longrightarrow {}^{13}_{6}C + {}^{1}_{1}H$ **20.51** 1.22×10^9 kJ/mol
20.53 a ${}^{4}_{2}He$ b α **20.55** plutonium-239 **20.57** $1.79 \times 10^{-9}\ s^{-1}$
20.59 $9.1 \times 10^{-8}\ s^{-1}$ **20.61** 1.3×10^{13} y **20.63** 3.84×10^{-12}/s
20.65 2.1×10^2 Ci **20.67** 6.0×10^{-4} g **20.69** 0.574, 3.4 μg
20.71 4.16 s **20.73** 44.6 d **20.75** 5.3×10^3 y **20.77** 5.2 disintegrations/(min·g) **20.79** -4.37×10^{-9} g **20.81** -1.688×10^{12} J; −17.50 MeV **20.83** 3.02184 amu; 32.01 MeV; 5.336 MeV
20.85 Na-20 is expected to decay by electron capture or positron emission. Na-26 is expected to decay by beta emission.
20.87 seven alpha emissions and four beta emissions
20.89 ${}^{209}_{83}Bi + {}^{4}_{2}He \longrightarrow {}^{211}_{85}At + 2{}^{1}_{0}n$ **20.91** ${}^{246}_{98}Cf$ **20.93** 6.3 y
20.95 2.42 pm **20.97** -6.02×10^{14} J; -1.64×10^5 kJ
20.99 a ${}^{47}_{20}Ca \longrightarrow {}^{47}_{21}Sc + {}^{0}_{-1}\beta$ b 41 μg
20.101 a ${}^{82}_{35}Br \longrightarrow {}^{82}_{36}Kr + {}^{0}_{-1}\beta$; $2H{}^{82}_{35}Br(g) \longrightarrow H_2 + 2{}^{82}_{36}Kr + 2{}^{0}_{-1}\beta$ b 0.0123 mol; 0.396 atm **20.123** 1.75×10^{13} nuclei/s
20.125 0.001 L **20.127** 2.732×10^9 kJ; 4.91×10^7 L

CHAPTER 21

21.71 It would react with the oxygen by the following equation: $C(s) + O_2(g) \longrightarrow CO_2(g)$ **21.73** The metallic character decreases from left to right and increases going down a column.
21.75 This means aluminum hydroxide reacts with both acids and bases. For example,
$Al(OH)_3(s) + 3HCl(aq) \longrightarrow 3H_2O(l) + AlCl_3(aq)$
$Al(OH)_3(s) + NaOH(aq) \longrightarrow Na^+(aq) + Al(OH)_4^-(aq)$
21.77 Oxygen is a very electronegative element, and its bonding involves only the *s* and *p* orbitals, in contrast to bonding using th *d* orbitals in sulfur, etc. Molecular oxygen is a reactive gas, but it forms mainly compounds in which its oxidation state is −2, compared with compounds of sulfur, etc., which exhibit positive oxid tion states as well as the −2 state. **21.79** HBr cannot be prepar by adding sulfuric acid to NaBr because the hot, concentrated acid will oxidize the bromide ion to bromine. **21.81** $Fe_2O_3(s) + 3H_2(g) \longrightarrow 2Fe(s) + 3H_2O(g)$ **21.83** 36.9 kg Fe
21.85 −417.9 kJ (exothermic)
21.87

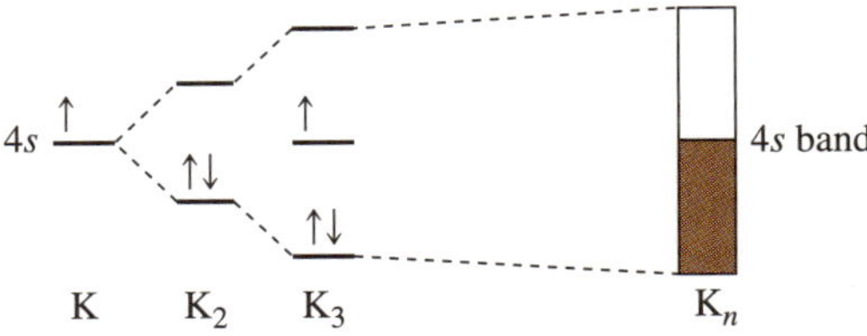

21.89 2.62×10^{19}
21.91 $CO_2(g) + NH_3(g) + NaCl(aq) + H_2O(l) \longrightarrow NaHCO_3(s) + NH_4Cl(a$
$2NaHCO_3(s) \xrightarrow{\Delta} Na_2CO_3(s) + CO_2(g) + H_2O(g)$
$Na_2CO_3(aq) + Ca(OH)_2(aq) \longrightarrow 2NaOH(aq) + CaCO_3(s)$
21.93 a $2K(s) + Br_2(l) \longrightarrow 2KBr(s)$
b $2K(s) + 2H_2O(l) \longrightarrow 2KOH(aq) + H_2(g)$
c $2NaOH(s) + CO_2(g) \longrightarrow Na_2CO_3(s) + H_2O(l)$
d $Li_2CO_3(aq) + 2HNO_3(aq) \longrightarrow H_2O(l) + 2LiNO_3(aq) + CO_2$
e $K_2SO_4(aq) + Pb(NO_3)_2(aq) \longrightarrow PbSO_4(s) + 2KNO_3(aq)$
21.95 ${}^{227}_{89}Ac \longrightarrow {}^{223}_{87}Fr + {}^{4}_{2}He$ **21.97** Add CO_3^{2-} to each solution. Calcium carbonate will precipitate from the calcium hydroxide solution. **21.99** Add Na_2SO_4 to the solution. $BaSO_4(s)$ can then be filtered from the solution. The filtrate will contain Mg^{2+} ions. **21.101** ${}^{230}_{90}Th \longrightarrow {}^{226}_{88}Ra + {}^{4}_{2}He$

.103 a $BaCO_3(s) \xrightarrow{\Delta} BaO(s) + CO_2(g)$
$Ba(s) + 2H_2O(l) \longrightarrow Ba(OH)_2(aq) + H_2(g)$
$Mg(OH)_2(s) + 2HNO_3(aq) \longrightarrow 2H_2O(l) + Mg(NO_3)_2(aq)$
$Mg(s) + NiCl_2(aq) \longrightarrow Ni(s) + MgCl_2(aq)$
$2NaOH(aq) + MgSO_4(aq) \longrightarrow Mg(OH)_2(s) + Na_2SO_4(aq)$
.105 0.22 g $Ca(OH)_2$
.107 $Al(H_2O)_6^{3+}(aq) + HCO_3^-(aq) \longrightarrow Al(H_2O)_5OH^{2+}(aq) + H_2O(l) + CO_2(g)$
.109 Test portions of solutions of each compound with each ther; the results can differentiate the compounds. For example, if ne solution is poured into one of the other solutions and gives no recipitate, then that means $BaCl_2$ was mixed with KOH, and the ird solution is $Al_2(SO_4)_3$. Adding the third solution of $Al_2(SO_4)_3$ both of the first two solutions will form a precipitate with only aCl_2. [$Al_2(SO_4)_3$ and excess KOH form soluble $Al(OH)_4^-$.] ; instead, one solution is poured into one of the other solutions nd a precipitate forms, that means $BaCl_2$ was mixed with $l_2(SO_4)_3 \longrightarrow BaSO_4$, and the third solution is KOH, etc. Thus, three are identified.
.111 $Sn(H_2O)_6^{2+}(aq) + H_2O(l) \longrightarrow Sn(H_2O)_5(OH)^+(aq) + H_3O^+(aq)$
.113 $PbO_2 + 4HCl \longrightarrow PbCl_2 + Cl_2 + 2H_2O$
.115 a $Al_2O_3(s) + 3H_2SO_4(aq) \longrightarrow Al_2(SO_4)_3(aq) + 3H_2O(l)$
$Al(s) + 3AgNO_3(aq) \longrightarrow 3Ag(s) + Al(NO_3)_3(aq)$
$Pb(NO_3)_2(aq) + 2NaI(aq) \longrightarrow PbI_2(s) + 2NaNO_3(aq)$
$8Al(s) + 3Mn_3O_4(s) \longrightarrow 9Mn(s) + 4Al_2O_3(s)$
$2Ga(OH)_3(s) \longrightarrow Ga_2O_3(s) + 3H_2O(g)$
.117 -4.2×10^9 kJ **21.119** a −1 b +1 c −4 +6 **21.121** a Each C—H bond is formed by the overlap of 1*s* orbital of a hydrogen atom with one of the singly occupied ^{3} hybrid orbitals of the carbon atom. b Each Si—F bond is rmed by the overlap of a 2*p* orbital of a fluorine atom with one the singly occupied sp^3d^2 hybrid orbitals of the silicon atom. The C=C double bond is a sigma bond formed by the overlap the sp^2 hybrid orbitals of those carbon atoms, and a pi bond rmed by the overlap of the unhybridized *p* orbitals. The single —C bond is a sigma bond formed by the overlap of the sp^2 or-tal of the middle carbon and the sp^3 orbital of the CH_3 carbon. ne other three orbitals of the CH_3 carbon are used to form the —H bonds by overlapping with the orbital of the hydrogen om. The C—H bonds on the middle carbon and the CH_2 carbon e formed by the overlap of the sp^2 orbital of carbon and the *s* bital of hydrogen. d Each Si—H bond is formed by the over-p of a 1*s* orbital of a hydrogen atom with one of the ngly occupied sp^3 hybrid orbitals of the silicon atom.
.123 a 74.87 kJ b 137.15 kJ
.125 a $CO_2(g) + Ba(OH)_2(aq) \longrightarrow BaCO_3(s) + H_2O(l)$
$MgCO_3(s) + 2HBr(aq) \longrightarrow CO_2(g) + H_2O(l) + MgBr_2(aq)$
.127 $C(s) + O_2(air) \longrightarrow CO_2(g)$
$O_2(g) + NaOH(aq) \longrightarrow NaHCO_3(aq)$
$NaHCO_3(aq) \xrightarrow{\Delta} Na_2CO_3(s) + H_2O(l) + CO_2(g)$
.129 Each silicon atom is tetrahedrally bonded to four other sili-n atoms. Each bond is formed by the overlap of an sp^3 hybrid or-al on each silicon atom to make a sigma bond. **21.131** 10.7 kg
.133

```
   O⁻       O⁻       O⁻       O⁻
   |        |        |        |
-O—Si—O—Si—O—Si—O—Si—
   |        |        |        |
   O⁻       O⁻       O⁻       O⁻
```

e drawing represents two of the sets of tetrahedra forming a rtion of a long chain. There is a net −8 charge on the frag-ent, which includes two sets of two SiO_3 units. Thus, the $(SiO_3)_2$ it has a −4 charge. This would have to be balanced by one Li^+ d one Al^{3+} for every two SiO_3. Thus, the empirical formula is $Al(SiO_3)_2$. **21.135** 2.53 g

21.137 $4NH_3(g) + 5O_2(g) \xrightarrow{Pt} 4NO(g) + 6H_2O(g)$
$2NO(g) + O_2(g) \longrightarrow 2NO_2(g)$
$3NO_2(g) + H_2O(l) \longrightarrow 2HNO_3(aq) + NO(g)$
$NH_3(g) + HNO_3(aq) \longrightarrow NH_4NO_3(aq)$
$NH_4NO_3(aq) \longrightarrow NH_4NO_3(s)$
$NH_4NO_3(s) \xrightarrow{Pt} N_2O(g) + 2H_2O(g)$
21.139 $4Zn(s) + NO_3^-(aq) + 10H^+(aq) \longrightarrow 4Zn^{2+}(aq) + NH_4^+(aq) + 3H_2O(l)$
21.141 tetrahedral; Each bond is formed by the overlap of a 4*p* orbital from Br with an sp^3 hybrid orbital from P.
21.143 $H_3PO_3 + H_2SO_4 \longrightarrow H_3PO_4 + SO_2 + H_2O$
21.145 14.1 g H_3PO_4 **21.147** a $4Li(s) + O_2(g) \longrightarrow 2Li_2O(s)$
b $4CH_3NH_2(g) + 9O_2(g) \longrightarrow 4CO_2(g) + 2N_2(g) + 10H_2O(g)$
c $2(C_2H_5)_2S + 15O_2 \longrightarrow 8CO_2(g) + 10H_2O(g) + 2SO_2(g)$
21.149 a +6 b +6 c −2 d +4
21.151 $8H_2SeO_3(aq) + 16H_2S(g) \longrightarrow Se_8(s) + 2S_8(s) + 24H_2O(l)$
21.153 49.1 g
21.155 $Ba(ClO_3)_2(aq) + H_2SO_4(aq) \longrightarrow 2HClO_3(aq) + BaSO_4(s)$
21.157 $K_2Cr_2O_7 + 6HCl + 8H^+ \longrightarrow 2Cr^{3+} + 3Cl_2 + 2K^+ + 7H_2O$
21.159 a sp^3; bent (angular) b sp^3; trigonal pyramidal c sp^3d; T-shaped
21.161 a $Br_2(aq) + 2NaOH(aq) \longrightarrow H_2O(l) + NaOBr(aq) + NaBr(aq)$
b $NaBr(s) + H_3PO_4(aq) \xrightarrow{\Delta} HBr(g) + NaH_2PO_4(aq)$
21.163 0.13 V; reaction occurs

21.165

```
:F̤    F̤:
   \ /
   Xe
   / \
:F̤    F̤:
```

sp^3d^2, square planar

21.167 $2XeF_2 + 4OH^- \longrightarrow 2Xe + 4F^- + O_2 + 2H_2O$
21.169 0.584 metric tons **21.171** 8.184% **21.173** −850.2 kJ; −425.1 kJ/mol Fe **21.175** 84.5% **21.177** 6.957 g
21.179 1363 K **21.181** $E° = 0.19$ V; −37 kJ; yes
21.183 7.73×10^{-2} g **21.185** 69.6% **21.187** 12.6 hr
21.189 2.58%

CHAPTER 22

22.31 $[Co(NH_3)_4(NO_2)_2]Cl$ **22.33** "Hexacyano" means that there are six CN^- ligands bonded to the iron cation. The Roman numeral II means the oxidation state of the iron cation is +2, so the overall charge of the complex ion is 4−. This requires four potassium ions to counterbalance the 4− charge.
22.35 $[Co(NH_3)_4Br_2]Cl$ $[Co(NH_3)_4BrCl]Br$
22.37 a +2 b +4 c +2 d +6
22.39 $3Fe^{2+} + NO_3^- + 4H^+ \longrightarrow 3Fe^{3+} + NO + 2H_2O$
22.41 a Four b Six c Four d Six
22.43 a +2 b +3 c +3 d +3
22.45 a +3 b NH_3 -Ammine; Cl^- -Chloro; $C_2O_4^{2-}$ Oxalato c 6 d −3 **22.47** a Potassium hexafluoroferrate(III)
b Diamminediaquacopper(II) ion
c Ammonium aquapentafluoroferrate(III)
d Dicyanoargentate(I) ion
22.49 a Pentacarbonyliron(0)
b Dicyanobis(ethylenediamine)rhodium(III) ion
c Tetraamminesulfatochromium(III) chloride
d Tetraoxomanganate(VII) ion (Permanganate is the usual name.) **22.51** a $K_3[Mn(CN)_6]$ b $Na_2[Zn(CN)_4]$
c $[Co(NH_3)_4Cl_2]NO_3$ d $[Cr(NH_3)_6]_2[CuCl_4]_3$

22.53 a

```
H3N     NH3     H3N     Cl
   \   /           \   /
    Pd              Pd
   /   \           /   \
Cl      Cl      Cl      NH3
     cis             trans
```

b no geometric isomerism c no geometric isomerism

d 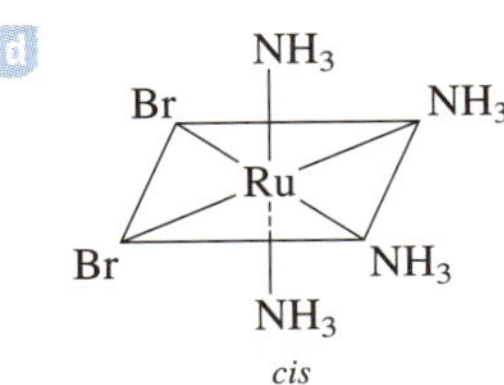

cis

NH_3, Br, NH_3, Ru, H_3N, Br, NH_3

trans

22.55

a NH_3 NH_3 / Pd / Cl Cl — cis; NH_3 Cl / Pd / Cl NH_3 — trans

b no optical isomers **22.57** a two unpaired electrons b three unpaired electrons c two unpaired electrons **22.59** a eight electrons in square planar field (low spin) b seven electrons in square planar field (low spin) c five electrons in tetrahedral field (high spin) d seven electrons in tetrahedral field (high spin) **22.61** purple **22.63** yes; H_2O is a weaker-bonding ligand **22.65** 2.39 kJ/mol **22.67** no d electrons **22.69** three geometric isomers

$[Co(NH_3)_4Cl_2]$: Cl, H_3N, NH_3, Co, H_3N, Cl, NH_3 — NH_3, H_3N, Cl, Co, H_3N, Cl, NH_3 — NH_3, H_3N, Cl, Co, Cl, NH_3, NH_3

22.71 $[Cu^{2+}] = 6.1 \times 10^{-4}\ M$; $[NH_3] = 2.4 \times 10^{-3}\ M$; $[Cu(NH_3)^{2+}] = 0.10\ M$

CHAPTER 23

23.19 a The molecules with carbon chains in the $C_5 - C_{11}$ range. b The molecules with carbon chains greater than C_{11} will be left in the barrel because they boil at temperatures above 200°C. (Keep in mind that the heavily branched C_{12} chains will have boiled off at or slightly below 200°C.) c Low-molecular-mass hydrocarbons have fewer polarizable electrons; therefore, they have weaker London forces than the longer chains and, as a result, boil at a lower temperature. d The more compact 2,3-dimethylbutane would boil at a lower temperature. **23.21** Since carbon would have more than four bonds in this case, CH_5 would be in violation of the octet rule. **23.23** a trimethylamine, C_3H_9N; acetaldehyde, C_2H_4O; 2-propanol, C_3H_8O; and acetic acid, $C_2H_4O_2$
b trimethylamine: $N(CH_3)_3$ acetaldehyde: CH_3CH (with =O on C)

2-propanol: CH_3CHCH_3 (with OH on central C) acetic acid: CH_3COH (with =O on C)

c The functional groups in the molecules are trimethylamine, amine (tertiary); acetaldehyde, aldehyde; 2-propanol, alcohol; and acetic acid, carboxylic acid.
23.25 The molecules increase regularly in molecular mass. Therefore, you expect their intermolecular forces (dispersion forces) and thus their melting points to increase. **23.27** $CH_3CH_2CH_2CH_3$

23.29

a CH_3CH_2, CH_2CH_3 / C=C / H, H — *cis*-3-Hexene; CH_3CH_2, H / C=C / H, CH_2CH_3 — *trans*-3-Hexene

b CH_3, H / C=C / CH_3CH_2, CH_2CH_3 — *cis*-3-Methyl-3-hexene; CH_3, CH_2CH_3 / C=C / CH_3CH_2, H — *trans*-3-Methyl-3-hexene

23.31 a $C_2H_4 + 3O_2 \longrightarrow 2CO_2 + 2H_2O$

b $3CH_2{=}CH_2 + 2MnO_4^- + 4H_2O \longrightarrow 3CH_2(OH){-}CH_2(OH) + 2MnO_2 + 2OH^-$

c $CH_2{=}CH_2 + Br_2 \longrightarrow CH_2Br{-}CH_2Br$

d toluene (CH_3 on benzene) $+ Br_2 \xrightarrow{FeBr_3}$ p-bromotoluene (CH_3, Br) $+ HBr$

e toluene (CH_3 on benzene) $+ HNO_3 \xrightarrow{H_2SO_4}$ p-nitrotoluene (CH_3, NO_2) $+ H_2O$

23.33 $C_2H_6 + Br_2 \longrightarrow C_2H_5Br + HBr$

23.35 $CH_3{-}C(Br)(CH_3){-}CH_3$

23.37 a 2,3,4-trimethylpentane b 2,2,6,6-tetramethylheptane c 4-ethyloctane d 3,4-dimethyloctane

23.39 a $CH_3CH(CH_3)CH(CH_3)CH_2CH_2CH_3$ b $CH_3CH_2CH(CH_2CH_3)CH_2CH_2CH_3$

c $CH_3CH(CH_3)CH_2CH(CH_3CHCH_3)CH_2CH_2CH_3$ d $CH_3C(CH_3)_2{-}C(CH_3)_2CH_2CH_3$

23.41 a 2-pentene b 2,5-dimethyl-2-hexene

23.43 a $CH_3CH{=}C(CH_2CH_3)CH_2CH_3$ b $CH_3C(CH_3){=}CHCH(CH_2CH_3)CH_2CH_3$

23.45 *cis*-2-pentene, *trans*-2-pentene **23.47** a 2-butyne b 3-methyl-1-pentyne

23.49 a $(C_6H_5)_3C{-}CH_3$ b benzene ring with CH_3 and CH_2CH_3 on adjacent carbons

23.51 a $CH_3{-}C(=O){-}CH_2CH_2CH_3$ ketone b $CH_3{-}CH(OH)CH_2CH_3$ alcohol c $HOC(=O){-}CH_2CH_3$ carboxylic acid d $H{-}C(=O){-}CH_2CH_3$ aldehyde

3.53 a 1-pentanol b 2-pentanol c 2-propyl-1-pentanol 6-methyl-4-octanol **23.55** a secondary alcohol secondary alcohol c primary alcohol d primary alcohol
3.57 a ethyl propyl ether b methyl isopropyl ether
3.59 a butanone b butanal c 4,4-dimethylpentanal 3-methyl-2-pentanone **23.61** a secondary amine secondary amine **23.63** a 3-methylbutanoic acid *trans*-5-methyl-2-hexene c 2,5-dimethyl-4-heptanone 4-methyl-2-pentyne

3.65 a $CH_3CH_2C(=O)—O—CH(CH_3)_2$ b $CH_3—C(CH_3)_2—NH_2$

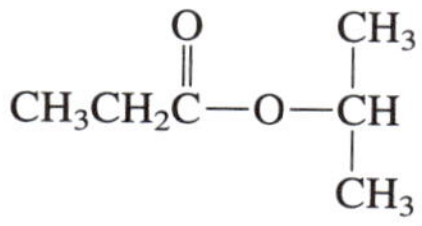

$CH_3CH_2CH_2CH_2C(CH_3)_2COOH$ d $(CH_3CH_2)HC=CH(CH_2CH_3)$ (trans)

3.67 a Addition of dichromate ion in acidic solution to ropionaldehyde will cause the reagent to change from range to green as the aldehyde is oxidized. Under similar condions, acetone (a ketone) would not react. b Add Br_2 in CCl_4 ach compound. The first compound reacts; the color of Br_2 des. Benzene does not react. **23.69** a ethylene b toluene methylamine d methanol **23.71** 1-butene or 2-butene

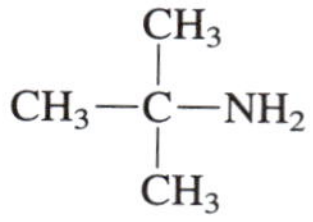

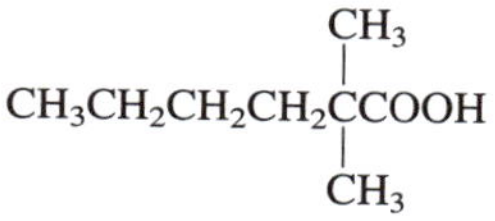

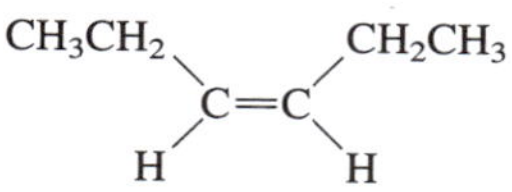

HAPTER 24

.21 a —A—A—A—A—A—A—, B—B—B—B—B—B— —A—B—A—B—A—B— —A—C—A—A—C—A—

.23 $^+H_3NCH_2CH_2CH_2—CH(NH_3^+)—COO^-$

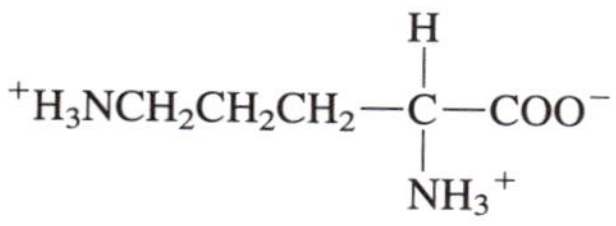

.25 The corresponding amino-acid sequence is rg—Val—Arg—Asp—Leu—Thr. The triplet UGA is the code to d the sequence.

.27 $nCF_2{=}CF_2 \longrightarrow —CF_2—CF_2—CF_2—CF_2—CF_2—CF_2—$

.29 $HOCH_2CH_2OH$ $HOOCCH_2CH_2COOH$

.31 a would give a polymer having alternating single and uble bonds, in which resonance might be expected to give a concting polymer.

.33 $CH_3—CH(NH_3^+)—COO^-$

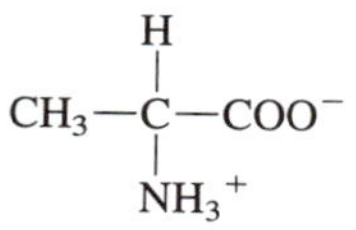

.35 Two dipeptides are possible:

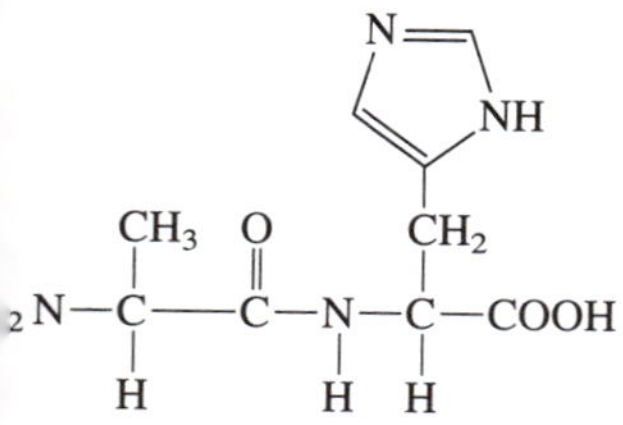

$H_2N—CH(CH_3)—C(=O)—NH—CH(CH_2\text{-imidazolyl})—COOH$ and

$H_2N—CH(CH_2\text{-imidazolyl})—C(=O)—NH—CH(CH_3)—COOH$

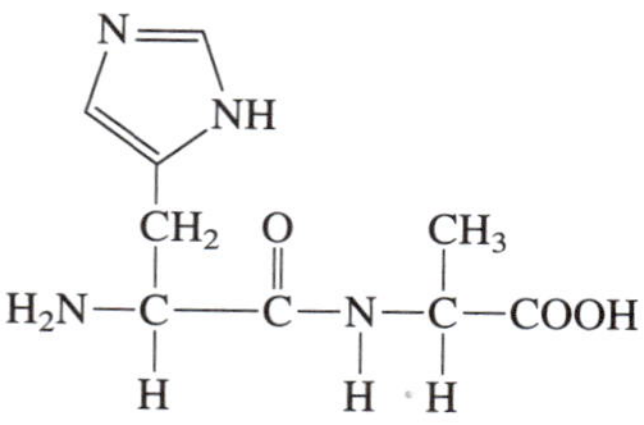

24.37 A:T, 1:1, G:C, 1:1

24.39 $^-O—P(=O)(O^-)—O—CH_2$–ribose (OH, OH)–adenine (NH_2)

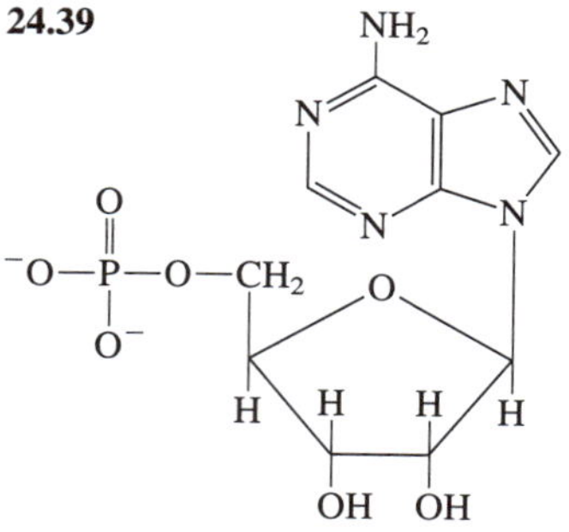

24.41 3; 2; guanine-cytosine bonding is stronger **24.43** 16; no
24.45 hydrogen bonding; less readily
24.47 Gly—Ser—Arg—Phe—Gly—Leu—Lys
24.49 CUG, ACU, CCC, UGG
24.51 CUU GCU GUU GAA GAU UGU AUG UGG AAA
24.53 $—CH(Cl)—CH(Cl)—CH(Cl)—CH(Cl)—$
24.55 $HOOC—C_6H_4—COOH$, $NH_2—C_6H_4—NH_2$
24.57 $CH_3SCH_2CH_2CH(NH_2)—COOH$ has a nonpolar side chain
24.59 $^+H_3N—CH(COO^-)—CH_2OH$
24.61

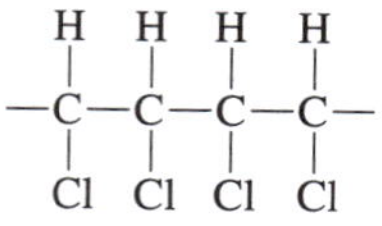

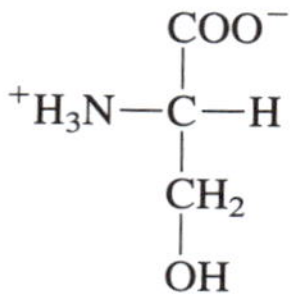

$CH_3SCH_2CH_2CH(NH_2)—C(=O)—NH—CH(CH_2SH)—COOH$

$HSCH_2CH(NH_2)—C(=O)—NH—CH(CH_2CH_2SCH_3)—COOH$

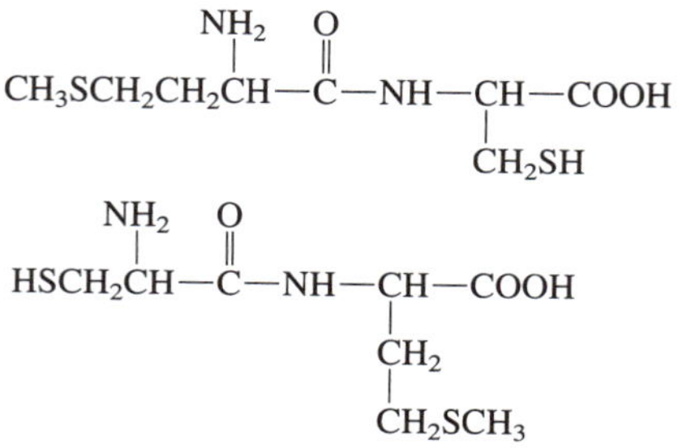

24.63 720
24.65

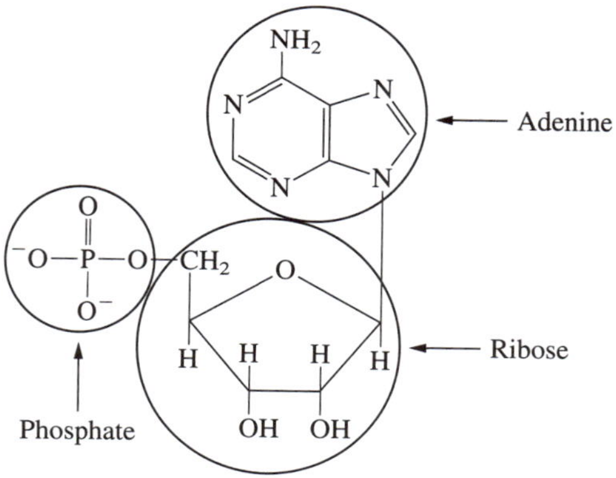

Glossary

The number given in blue at the end of a definition indicates the section where the term was introduced. In some cases, more than one section is indicated. In a few cases, the term is defined in an introductory passage just *before* the section indicated.

Absolute entropy see *Standard entropy.*

Absolute temperature scale a temperature scale in which the lowest temperature that can be attained theoretically is zero. (1.6)

Accuracy the closeness of a single measurement to its true value. (1.5)

Acid (Arrhenius definition) a substance that produces hydrogen ions, H^+ (hydronium ion, H_3O^+), when it dissolves in water. (4.4 and 15.1); (Brønsted–Lowry definition) the species (molecule or ion) that donates a proton to another species in a proton-transfer reaction. (4.4 and 15.2)

Acid–base indicator a dye used to distinguish between acidic and basic solutions by means of the color changes it undergoes in these solutions. (4.4)

Acid–base titration curve a plot of the pH of a solution of acid (or base) against the volume of added base (or acid). (16.7)

Acidic oxide an oxide that reacts with bases. (8.7)

Acid-ionization (or acid-dissociation) constant (K_a) the equilibrium constant for the ionization of a weak acid. (16.1)

Acid rain rain having a pH lower than that of natural rain, which has a pH of 5.6. (16.2)

Acid salt a salt that has an acidic hydrogen atom and can undergo neutralization with bases. (4.4)

Actinides elements in the last of the two rows at the bottom of the periodic table; actinium plus the 14 elements following it in the periodic table, in which the 5*f* subshell is filling. (22.1)

Activated complex (transition state) an unstable grouping of atoms that can break up to form products. (13.5)

Activation energy (E_a) the minimum energy of collision required for two molecules to react. (13.5)

Activity of a radioactive source the number of nuclear disintegrations per unit time occurring in a radioactive material. (20.3)

Activity series a listing of the elements in order of their ease of losing electrons during reactions in aqueous solution. (4.5)

Addition polymer a polymer formed by linking together many molecules by addition reactions. (24.1)

Addition reaction a reaction in which parts of a reactant are added to each carbon atom of a carbon–carbon double bond, which then becomes a C—C single bond. (23.3)

Adsorption the binding or attraction of molecules to a surface. (13.9) See *Chemisorption* and *Physical adsorption.*

Aerosol a colloid consisting of liquid droplets or solid particles dispersed throughout a gas. (12.9)

Alcohol a compound obtained by substituting a hydroxyl group (—OH) for an —H atom on a tetrahedral (sp^3 hybridized) carbon atom of a hydrocarbon group. (23.6)

Aldehyde a compound containing a carbonyl group with at least one H atom attached to it. (23.6)

Aliphatic hydrocarbon a saturated or unsaturated hydrocarbon; a hydrocarbon that does not contain benzene rings. (23.1)

Alkali metal a Group IA element. The Group IA elements are reactive metals. (2.5 and 8.7)

Alkaline dry cell a voltaic cell that is similar to the Leclanché dry cell but uses potassium hydroxide in place of ammonium chloride. (19.8)

Alkaline earth metal a Group IIA element. The Group IIA elements are reactive metals, though less reactive than the alkali metals. (8.7)

Alkane an acyclic saturated hydrocarbon (one without a ring of carbon atoms); a saturated hydrocarbon with the general formula C_nH_{2n+2}. (23.2)

Alkene a hydrocarbon that has the general formula C_nH_{2n} and contains a carbon–carbon double bond. (23.3)

Alkyl group an alkane less one hydrogen atom. (23.5)

Alkyne an unsaturated hydrocarbon containing a carbon–carbon triple bond. The general formula is C_nH_{2n-2}. (23.3)

Allotrope one of two or more distinct forms of an element in the same physical state. (6.8)

Alloy a material with metallic properties that is either a compound or a mixture. (21.2)

Alpha emission emission of a ^4_2He nucleus, or alpha particle, from an unstable nucleus. (20.1)

Amide a compound derived from the reaction of ammonia, or a primary or secondary amine, with a carboxylic acid. (23.7)

Amide bond see *Peptide bond.*

Amine a compound that is structurally derived by replacing one or more hydrogen atoms of ammonia with hydrocarbon groups. (23.7)

Amino acid a compound containing an amino group ($—NH_2$) and a carboxyl group (—COOH). (24.3)

Amorphous solid a solid that has a disordered structure; it lacks the well-defined arrangement of basic units (atoms, molecules, or ions) found in a crystal. (11.7)

Ampere (A) the base unit of current in the International System (SI). (19.11)

Amphiprotic species a species that can act as either an acid or a base (that is, it can either lose or gain a proton). (15.2)

Amphoteric hydroxide a metal hydroxide that reacts with both bases and acids. (17.5)

Amphoteric oxide an oxide that has both acidic and basic properties. (8.7)

Angstrom (Å) a non-SI unit of length; 1 Å = 10^{-10} m. (1.6)

Angular geometry see *Bent geometry.*

Angular momentum quantum number (*l*) also known as the azimuthal quantum number. The quantum number that distinguishes orbitals of given *n* having different shapes; it can have any integer value from 0 to $n - 1$. (7.5)

Anion a negatively charged ion. (2.6)

Anode the electrode at which oxidation occurs. (19.2)

Antibonding orbitals molecular orbitals having zero values in the region between two nuclei and therefore concentrated in other regions. (10.5)

Aromatic hydrocarbon a hydrocarbon that contains a benzene ring or similar structural feature. (23.1)

Arrhenius equation the mathematical equation $k = Ae^{-Ea/RT}$, which expresses the dependence of the rate constant on temperature. (13.6)

Association colloid a colloid in which the dispersed phase consists of micelles. (12.9)

Atmosphere (atm) a unit of pressure equal to exactly 760 mmHg; 1 atm = 101.325 kPa (exact). (5.1)

Atom an extremely small particle of matter that retains its identity during chemical reactions. (2.1)

Atomic mass as a general term, the mass of an individual atom; but usually we mean the average atomic mass for the naturally occurring element, expressed in atomic mass units. (2.4)

Atomic mass unit (amu) a mass unit equal to exactly one-twelfth the mass of a carbon-12 atom. (2.4)

Atomic number (*Z*) the number of protons in the nucleus of an atom. (2.3)

Atomic orbital a wave function for an electron in an atom; pictured qualitatively by describing the region of space where there is a high probability of finding the electron. (7.5)

Atomic radius as an approximate measure of the radius, the maximum in the radial distribution function of the outer shell of the atom. Also, the value for an atom in a set of covalent radii assigned to atoms in such a way that the sum of the covalent radii of atoms A and B predicts the approximate A—B bond length (often called the covalent radius). (8.6)

Atomic symbol a one- or two-letter notation used to represent an atom corresponding to a particular element. A temporary name with a three letter notation (denoting the atomic number) is given to newly discovered elements. (2.1)

Atomic theory an explanation of the structure of matter in terms of different combinations of very small particles (atoms). (2.1)

Aufbau principle see *Building-up principle.*

Autoionization a reaction in which two like molecules react to give ions. (15.6)

Avogadro's law equal volumes of any two gases at the same temperature and pressure contain the same number of molecules. (5.2)

Avogadro's number (N_A) the number of atoms in a 12-g sample of carbon-12, equal to 6.02×10^{23} to three significant figures. (3.2)

Axial direction one of two directions pointing from the center of a trigonal bipyramid along its axis. Compare *Equatorial direction.* (10.1)

Azimuthal quantum number see *Angular momentum quantum number.*

Balanced chemical equation a chemical equation in which the numbers of atoms of each element are equal on both sides of the arrow. (2.10)

Band of stability the region in which stable nuclides lie in a plot of number of protons against number of neutrons. (20.1)

Band theory molecular orbital theory of metals. (21.3)

Bar a unit of pressure equal to 1×10^5 Pa, slightly less than 1 atm. (5.1)

Barometer a device for measuring the pressure of the atmosphere. (5.1)

Base (Arrhenius definition) a substance that produces hydroxide ions, OH^-, when it dissolves in water. (4.4 and 15.1) (Brønsted–Lowry definition) the species (molecule or ion) that accepts a proton in a proton-transfer reaction. (4.4 and 15.2)

Base-ionization (or base-dissociation) constant (K_b) the equilibrium constant for the ionization of a weak base. Thus K_b for NH_3 is 1.8×10^{-5}. (16.3)

Base pairing the hydrogen bonding of complementary nucleotide bases. (24.4)

Basic oxide an oxide that reacts with acids. (8.7)

Bayer process a chemical procedure in which purified aluminum oxide, Al_2O_3, is separated from the aluminum ore bauxite. (21.2)

Bent geometry (angular geometry) nonlinear molecular geometry, in the case of a molecule of three atoms. (10.1)

Beta emission emission of a high-speed electron from an unstable nucleus. (20.1)

Bidentate ligand a ligand that bonds to a metal atom through two atoms of the ligand. (22.3)

Bimolecular reaction an elementary reaction that involves two reactant molecules. (13.7)

Binary compound a compound composed of only two elements. (2.8)

Binary hydride a compound that contains hydrogen and one other element. (21.7)

Binding energy (of a nucleus) the energy needed to break a nucleus into its individual protons and neutrons. (20.6)

Body-centered cubic unit cell a cubic unit cell in which there is a lattice point at the center of the unit cell as well as at the corners. (11.7)

Boiling point the temperature at which the vapor pressure of a liquid equals the pressure exerted on the liquid (atmospheric pressure, unless the vessel containing the liquid is closed). (11.2)

Boiling-point elevation a colligative property of a solution equal to the boiling point of the solution minus the boiling point of the pure solvent. (12.6)

Bond distance see *Bond length.*

Bond enthalpy the average enthalpy change for the breaking of a bond in a molecule in the gas phase. Also called *bond energy.* (9.11)

Bonding orbitals molecular orbitals that are concentrated in the regions between nuclei. (10.5)

Bonding pair an electron pair shared between two atoms. (9.4)

Bond length (bond distance) the distance between the nuclei in a bond. (9.10)

Bond order in a Lewis formula, the number of pairs of electrons in a bond. (9.10) In molecular orbital theory, one-half the difference between the number of bonding electrons and the number of antibonding electrons. (10.5)

Born-Haber cycle a thermochemical cycle originated by Max Born and Fritz Haber in 1919 to obtain the lattice energy of an ionic solid. (9.1)

Boyle's law the volume of a sample of gas at a given temperature varies inversely with the applied pressure. (5.2)

Brønsted–Lowry concept a concept of acids and bases in which an acid is the species donating a proton in a proton-transfer reaction, whereas a base is the species accepting a proton in such a reaction. (4.4 and 15.2)

Buckminsterfullerene a molecular form of carbon (C_{60}) informally called "buckyball," referring to its soccer-ball shape; a fullerene. (21.8)

Buffer a solution characterized by the ability to resist changes in pH when limited amounts of acid or base are added to it. (16.6)

Building-up principle (Aufbau principle) a scheme used to reproduce the electron configurations of the ground states of atoms by successively filling subshells with electrons in a specific order (the building-up order). (8.2)

Calorie (cal) a non-SI unit of energy commonly used by chemists, originally defined as the amount of energy needed to raise the temperature of 1 g of water by 1°C; now defined as 1 cal = 4.184 J (exact). (6.1)

Calorimeter a device used to measure the heat absorbed or evolved during a physical or chemical change. (6.6)

Carboxylic acid a compound containing the carboxyl group, —COOH, whose H atom is acidic. (23.6)

Catalysis the increase in rate of a reaction as the result of the addition of a catalyst. (13.9)

Catalyst a substance that increases the rate of reaction without being consumed in the overall reaction. (2.9 and 13.1)

Catenation the covalent bonding of two or more atoms of the same element to one another. (21.8)

Cathode the electrode at which reduction occurs. (19.2)

Cathode rays the rays emitted by the cathode (negative electrode) in a gas discharge tube (tube of low-pressure gas through which an electric current is discharged). (2.2)

Cation a positively charged ion. (2.6)

Cation exchange resin a resin (organic polymer) that removes cations from solution and replaces them with hydrogen ions. (11.10)

Cell potential the maximum potential difference between the electrodes of a voltaic cell. (19.4)

Cell reaction the net reaction that occurs in a voltaic cell. (19.2)

Celsius scale the temperature scale in general scientific use. There are exactly 100 units between the freezing point and the normal boiling point of water. (1.6)

Chain reaction, nuclear a self-sustaining series of nuclear fissions caused by the absorption of neutrons released from previous nuclear fissions. (20.7)

Chalcogen a Group VIA element. (8.7)

Change of state (phase transition) a change of a substance from one state to another. (11.2)

Charles's law the volume occupied by any sample of gas at a constant pressure is directly proportional to the absolute temperature. (5.2)

Chelate a complex formed by polydentate ligands. (22.3)

Chemical bond a strong attractive force that exists between certain atoms in a substance. (Ch. 9 opener)

Chemical change see *Chemical reaction.*

Chemical equation the symbolic representation of a chemical reaction in terms of chemical formulas. (2.9)

Chemical equilibrium the state reached by a reaction mixture when the rates of forward and reverse reactions have become equal. (14.1)

Chemical formula a notation that uses atomic symbols with numerical subscripts to convey the relative proportions of atoms of the different elements in a substance. (2.6)

Chemical kinetics the study of how reaction rates change under varying conditions and of what molecular events occur during the overall reaction. (13.1)

Chemical nomenclature the systematic naming of chemical compounds. (2.8)

Chemical property a characteristic of a material involving its chemical change. (1.4)

Chemical reaction (chemical change) a change in which one or more kinds of matter are transformed into a new kind of matter or several new kinds of matter. (1.4) The rearrangement of the atoms present in the reacting substances to give new chemical combinations present in the substances formed by the reaction. (2.1)

Chemisorption the binding of a species to a surface by chemical bonding forces. (13.9)

Chiral possessing the quality of handedness. A chiral object has a mirror image that is not identical to the object. (22.5)

Chlor–alkali membrane cell a cell for the electrolysis of aqueous sodium chloride in which the anode and cathode compartments are separated by a special plastic membrane that allows only cations to pass through it. (19.10)

Chlor–alkali mercury cell a cell for the electrolysis of aqueous sodium chloride in which mercury metal is used as the cathode. (19.10)

Chromatography a name given to a group of similar separation techniques that depend on how fast a substance moves, in a stream of gas or liquid, past a stationary phase to which the substance may be slightly attracted. (1.5)

***cis-trans* isomers** see *Geometric isomers.*

Clausius–Clapeyron equation an equation that expresses the relation between the vapor pressure P of a liquid and the absolute temperature T: $\ln P = -\Delta H_{vap}/RT + B$, where B is a constant. (11.2)

Claus process a method of obtaining free sulfur by the partial burning of hydrogen sulfide. (21.10)

Close-packed structure a crystal structure in which the atoms or other units are packed together as closely as possible. See *hexagonal close-packed structure* and *cubic close-packed structure.* (11.8)

Coagulation the process by which the dispersed phase of a colloid is made to aggregate and thereby separate from the continuous phase. (12.9)

Codon a sequence of three bases in a messenger RNA molecule that serves as the code for a particular amino acid. (24.4)

Colligative properties properties that depend on the concentration of solute molecules or ions in a solution but not on the chemical identity of the solute. (12.4)

Collision theory the theory that in order for reaction to occur, reactant molecules must collide with an energy greater than some minimum value and with proper orientation. (13.5)

Colloid a dispersion of particles of one substance (the dispersed phase) throughout another substance or solution (the continuous phase). (12.9)

Combination reaction a reaction in which two substances combine to form a third substance. (4.5)

Combustion reaction a reaction of a substance with oxygen, usually with the rapid release of heat to produce a flame. (4.5)

Common-ion effect the shift in an ionic equilibrium caused by the addition of a solute that provides an ion that takes part in the equilibrium. (16.5 and 17.2)

Complementary bases nucleotide bases that form strong hydrogen bonds with one another. (24.4)

Complete ionic equation a chemical equation in which strong electrolytes (such as soluble ionic compounds) are written as separate ions in the solution. (4.2)

Complex (coordination compound) a compound consisting either of complex ions and other ions of opposite charge or of a neutral complex species. (22.3)

Complex ion an ion formed from a metal ion with a Lewis base attached to it by a coordinate covalent bond. (15.3, 17.5, and 22.3)

Compound a substance composed of two or more elements chemically combined. (1.4); A type of matter composed of atoms of two or more elements chemically combined in fixed proportions. (2.1)

Concentration a general term referring to the quantity of solute in a standard quantity of solvent or solution. (4.7 and 12.4)

Condensation the change of a gas to either the liquid or the solid state. (11.2)

Condensation polymer a polymer formed by linking many molecules together by condensation reactions. (24.1)

Condensation reaction a reaction in which two molecules or ions are chemically joined by the elimination of a small molecule such as water. (21.8)

Condensed (structural) formula a structural formula in which the bonds around each carbon atom are not explicitly written. (23.2)

Conjugate acid in a conjugate acid–base pair, the species that can donate a proton. (15.2)

Conjugate acid–base pair two species in an acid–base reaction, one acid and one base, that differ by the loss or gain of a proton. (15.2)

Conjugate base in a conjugate acid–base pair, the species that can accept a proton. (15.2)

Constitutional (structural) isomers isomers that differ in how the atoms are joined together (22.5); compounds with the same molecular formula but different structural formulas. (23.2)

Contact process an industrial method for the manufacture of sulfuric acid. It consists of the reaction of sulfur dioxide with oxygen to form sulfur trioxide using a catalyst of vanadium(V) oxide, followed by the reaction of sulfur trioxide with water. (21.10)

Continuous spectrum a spectrum containing light of all wavelengths. (7.3)

Control rods cylinders composed of substances that absorb neutrons, such as boron and cadmium, and can therefore slow a nuclear chain reaction. (20.7)
Conversion factor a factor equal to 1 that converts a quantity expressed in one unit to a quantity expressed in another unit. (1.8)
Coordinate covalent bond a bond formed when both electrons of the bond are donated by one atom. (9.4)
Coordination compound a compound consisting either of complex ions and other ions of opposite charge or of a neutral complex species. (22.3)
Coordination isomers isomers consisting of complex cations and complex anions that differ in the way the ligands are distributed between the metal atoms. (22.5)
Coordination number in a crystal, the number of nearest-neighbor atoms of an atom. (11.8) In a complex, the total number of bonds the metal atom forms with ligands. (22.3)
Copolymer a polymer consisting of two or more different monomer units. (24.1)
Coulomb (C) unit of electric charge. (2.2)
Coulomb's law the potential energy obtained in bringing two charges Q_1 and Q_2, initially far apart, up to a distance r apart is directly proportional to the product of the charges and inversely proportional to the distance between them. (9.1)
Covalent bond a chemical bond formed by the sharing of a pair of electrons between atoms. (9.4)
Covalent network solid a solid that consists of atoms held together in large networks or chains by covalent bonds. (11.6)
Covalent radius the value for an atom in a set of covalent radii assigned to atoms in such a way that the sum of the covalent radii of atoms A and B predicts the approximate A—B bond length. (8.6 and 9.10)
Critical mass the smallest mass of fissionable material in which a chain reaction can be sustained. (20.7)
Critical pressure the vapor pressure at the critical temperature; it is the minimum pressure that must be applied to a gas at the critical temperature to liquefy it. (11.3)
Critical temperature the temperature above which the liquid state of a substance no longer exists regardless of the pressure. (11.3)
Crystal a kind of solid having a regular three-dimensional arrangement of atoms, molecules, or ions. (2.6)
Crystal field splitting (Δ) the difference in energy between the two sets of d orbitals on a central metal ion that arises from the interaction of the orbitals with the electric field of the ligands. (22.7)
Crystal field theory a model of the electronic structure of transition-metal complexes that considers how the energies of the d orbitals of a metal ion are affected by the electric field of the ligands. (22.7)
Crystal lattice the geometric arrangement of lattice points of a crystal, in which we choose one lattice point at the same location within each of the basic units of the crystal. (11.7)
Crystalline solid a solid composed of one or more crystals; each crystal has a well-defined ordered structure in three dimensions. (11.7)
Crystal systems the seven basic shapes possible for unit cells; a classification of crystals. (11.7)
Cubic close-packed structure (ccp) a crystal structure composed of close-packed atoms (or other units) with the stacking ABCABCABCA· · ·. It has a face-centered cubic unit cell. (11.8)
Curie (Ci) a unit of activity, equal to 3.700×10^{10} disintegrations per second. (20.3)
Cycloalkane a cyclic saturated hydrocarbon; that is, a saturated hydrocarbon in which the carbon atoms form a ring; the general formula is C_nH_{2n}. (23.2)
Cyclotron a type of particle accelerator consisting of two hollow, semicircular metal electrodes, called dees (because the shape resembles the letter D), in which charged particles are accelerated by stages to higher and higher kinetic energies. Ions introduced at the center of the cyclotron are accelerated in the space between the two dees. (20.2)

Dalton's law of partial pressures the sum of the partial pressures of all the different gases in a mixture is equal to the total pressure of the mixture. (5.5)
***d*-Block transition elements** those transition elements with an unfilled d subshell in common oxidation states. (8.2)
de Broglie relation the equation $\lambda = h/mv$ relating the wavelength λ associated with a particle of mass m and speed v. (7.4)
Decomposition reaction a reaction in which a single compound reacts to give two or more substances. (4.5)
Degree of ionization the fraction of molecules that react with water to give ions. (16.1)
Delocalized bonding a type of bonding in which a bonding pair of electrons is spread over a number of atoms rather than localized between two. (9.7)
Denitrifying bacteria bacteria that use nitrate ion, NO_3^-, as a source of energy; they convert the ion to gaseous nitrogen. (21.9)
Density the mass per unit volume of a substance or solution. (1.7)
Deoxyribonucleic acid (DNA) the hereditary constituent of cells; it consists of two polymer strands of deoxyribonucleotide units. (24.4)
Deposition the change of a vapor to a solid. (11.2)
Derived unit see *SI derived unit.*
Desalinate remove ions from brackish (slightly salty) water or seawater, to make drinkable or industrially usable water. (12.7)
Deuterons nucleus of a hydrogen-2 atom. (20.2)
Dextrorotatory refers to a compound whose solution rotates the plane of polarized light to the right (when looking toward the source of light). (22.5)
Diamagnetic substance a substance that is not attracted by a magnetic field or is very slightly repelled by such a field. This property generally means that the substance has only paired electrons. (8.4)
Diffraction a property of waves in which the waves spread out when they encounter an obstruction or small hole about the size of the wavelength. The spreading waves may interfere in a way to produce a distinctive pattern, called a diffraction pattern. (7.2 and 11.10)
Diffusion the process whereby a gas spreads out through another gas to occupy the space uniformly. (5.7)
Dimensional analysis (factor-label method) the method of calculation in which one carries along the units for quantities. (1.8)
Dipole–dipole force an attractive intermolecular force resulting from the tendency of polar molecules to align themselves such that the positive end of one molecule is near the negative end of another. (11.5)
Dipole moment a quantitative measure of the degree of charge separation in a molecule and therefore an indicator of the polarity of the molecule. (10.2)
Dispersion forces see *London forces.*
Displacement reaction (single-replacement reaction) a reaction in which an element reacts with a compound, displacing another element from it. (4.5)
Dissociation constant of a complex ion (K_d) the reciprocal, or inverse, value of the formation constant. (17.5)
Distillation the process in which a liquid is vaporized then condensed; used to separate substances of different volatilities. (1.4)

Distorted tetrahedral geometry see *Seesaw geometry.*

Disulfide linkage a covalent linkage formed between two cysteine groups within a protein molecule by oxidation of the two groups. (24.3)

Double bond a covalent bond in which two pairs of electrons are shared by two atoms. (9.4)

Downs cell a commercial electrochemical cell used to obtain sodium metal by the electrolysis of molten sodium chloride. (19.9)

Dow process a commercial method for isolating magnesium from seawater. (21.2)

Dry cell see *Alkaline dry cell* and *Zinc–carbon dry cell.*

Dynamic equilibrium an equilibrium in which molecular processes are continuously occurring. (11.2)

Effective nuclear charge the positive charge that an electron experiences from the nucleus, equal to the nuclear charge but reduced by any shielding or screening from any intervening electron distribution. (8.6)

Effusion the process in which a gas flows through a small hole in a container. (5.7)

Elastomer elastic or rubbery type of polymer. (24.1)

Electrochemical cell a system consisting of electrodes that dip into an electrolyte and in which a chemical reaction either uses or generates an electric current. (19.2)

Electrolysis the process of producing a chemical change in an electrolytic cell. (19.9)

Electrolyte a substance, such as sodium chloride, that dissolves in water to give an electrically conducting solution. (4.1)

Electrolytic cell an electrochemical cell in which an electric current drives an otherwise nonspontaneous reaction. (19.1)

Electromagnetic spectrum the range of frequencies or wavelengths of electromagnetic radiation. (7.1)

Electromotive force (emf) see *Cell potential.*

Electron a very light, negatively charged particle that exists in the region around the atom's positively charged nucleus. (2.2)

Electron affinity the energy required to remove an electron from the atom's negative ion (in its ground state); the negative of the energy change obtained when the neutral atom picks up an electron. (8.6)

Electron capture the decay of an unstable nucleus by capturing, or picking up, an electron from an inner orbital of an atom. (20.1)

Electron configuration the particular distribution of electrons among available subshells. (8.1)

Electron-dot formula see *Lewis electron-dot formula.*

Electronegativity a measure of the ability of an atom in a molecule to draw bonding electrons to itself. (9.5)

Electron spin quantum number (m_s) see *Spin quantum number.*

Electron volt (eV) the quantity of energy that would have to be imparted to an electron (whose charge is 1.602×10^{-19} C) to accelerate it by one volt potential difference. (20.2)

Element a substance that cannot be decomposed by any chemical reaction into simpler substances. (1.4)
A type of matter composed of only one kind of atom, each atom of a given kind having the same properties. (2.1)
A substance whose atoms all have the same atomic number. (2.3)

Elementary reaction a single molecular event, such as a collision of molecules, resulting in a reaction. (13.7)

Empirical formula (simplest formula) the formula of a substance written with the smallest integer (whole-number) subscripts. (3.5)

Emulsion a colloid consisting of liquid droplets dispersed throughout another liquid. (12.9)

Enantiomers (optical isomers) isomers that are nonsuperimposable mirror images of one another. (22.5)

Endothermic process a chemical reaction or physical change in which heat is absorbed (q is positive). (6.3)

Energy the potential or capacity to move matter. (6.1)

Energy levels specific energy values in an atom. (7.3)

Enthalpy (H) an extensive property of a thermodynamic system defined as the internal energy, U, plus pressure, P, times volume, V: $H = U + PV$. (6.3)

Enthalpy of formation see *Standard enthalpy of formation.*

Enthalpy of reaction (ΔH) the change in enthalpy for a reaction at a given temperature and pressure; it equals the heat of reaction at constant pressure. (6.3)

Entropy (S) a thermodynamic quantity that is a measure of how dispersed the energy of a system is among the different possible ways that a system can contain energy. (18.2)

Enzyme a protein that catalyzes a biochemical reaction. (24.3)

Equatorial direction one of the three directions pointing from the center of a trigonal bipyramid to a vertex other than one on the axis. Compare *Axial direction.* (10.1)

Equilibrium see *Chemical equilibrium.*

Equilibrium constant K see *Thermodynamic equilibrium constant.*

Equilibrium constant K_c the value obtained for the equilibrium-constant expression when equilibrium concentrations are substituted. (14.2)

Equilibrium constant K_p an equilibrium constant for a gas reaction, similar to K_c, but in which concentrations of gases are replaced by partial pressures (in atm). (14.2)

Equilibrium-constant expression an expression obtained for a reaction by multiplying the concentrations of products, dividing by the concentrations of reactants, and raising each concentration term to a power equal to the coefficient in the chemical equation. (14.2)

Equivalence point the point in a titration when a stoichiometric amount of reactant has been added. (16.7)

Ester a compound formed from a carboxylic acid, RCOOH, and an alcohol, R′OH. (23.6)

Ether a compound formally obtained by replacing both H atoms of H_2O by hydrocarbon groups R and R′. (23.6)

Exact number a number that arises when you count items or sometimes when you define a unit. (1.5)

Exchange (metathesis) reaction a reaction between compounds that, when written as a molecular equation, appears to involve the exchange of parts between the two reactants. (4.3)

Excited state a quantum-mechanical state of an atom or molecule associated with any energy level except the lowest, which is the ground state. (8.2)

Exclusion principle see *Pauli exclusion principle.*

Exothermic process a chemical reaction or physical change in which heat is evolved (q is negative). (6.3)

Experiment an observation of natural phenomena carried out in a controlled manner so that the results can be duplicated and rational conclusions obtained. (1.2)

Face-centered cubic unit cell a cubic unit cell in which there are lattice points at the center of each face of the unit cell in addition to those at the corners. (11.7)

Factor-label method see *Dimensional analysis.*

Faraday constant (F) the magnitude of charge on one mole of electrons, equal to 9.6485×10^4 C. (19.4)

***f*-Block transition elements (inner transition elements)** the elements with a partially filled *f* subshell in common oxidation states. (8.2)

Fibrous proteins proteins that form long coils or align themselves in parallel to form long, water-insoluble fibers. (24.3)

First ionization energy (first ionization potential) the minimum energy needed to remove the highest-energy (that is, the outermost) electron from a neutral atom in the gaseous state. (8.6)

First law of thermodynamics the change in internal energy of a system, ΔU, equals $q + w$ (heat plus work). (6.2)

Formal charge (of an atom in a Lewis formula) the hypothetical charge you obtain by assuming that bonding electrons are equally shared between bonded atoms and that the electrons of each lone pair belong completely to one atom. (9.9)

Formation constant (stability constant) of a complex ion (K_f) the equilibrium constant for the formation of a complex ion from the aqueous metal ion and the ligands. (17.5)

Formula see *Chemical formula* and *Molecular formula.*

Formula mass (FM) the sum of the atomic masses of all atoms in a formula unit of a compound. (3.1)

Formula unit the group of atoms or ions explicitly symbolized in the formula. (2.6)

Fractional (isotopic) abundance the fraction of the total number of atoms that is composed of a particular isotope. (2.4)

Fractional precipitation the technique of separating two or more ions from a solution by adding a reactant that precipitates first one ion, then another, and so forth. (17.3)

Frasch process a mining procedure in which underground deposits of solid sulfur are melted in place with superheated water, and the molten sulfur is forced upward as a froth using air under pressure. (21.10)

Free energy (G) a thermodynamic quantity defined by the equation $G = H - TS$. (18.4)

Free energy of formation see *Standard free energy of formation.*

Freezing the change of a liquid to the solid state. (11.2)

Freezing point the temperature at which a pure liquid changes to a crystalline solid, or freezes. (11.2)

Freezing-point depression a colligative property of a solution equal to the freezing point of the pure solvent minus the freezing point of the solution. (12.6)

Frequency (ν) the number of wavelengths of a wave that pass a fixed point in one unit of time (usually one second). (7.1)

Frequency factor the symbol A in the Arrhenius equation, assumed to be a constant. (13.6)

Fuel cell essentially a battery, but it differs by operating with a continuous supply of energetic reactants, or fuel. (19.8)

Fuel rods the cylinders that contain fissionable material for a nuclear reactor. (20.7)

Fullerene a family of molecules consisting of a closed cage of carbon atoms arranged in pentagons as well as hexagons. (21.8)

Functional group a reactive portion of a molecule that undergoes predictable reactions. (2.7 and 23.6)

Fusion see *Melting;* see also *Nuclear fusion.*

Galvanic cell see *Voltaic cell.*

Gamma emission emission from an excited nucleus of a gamma photon, corresponding to radiation with a wavelength of about 10^{-12} m. (20.1)

Gamma photon a particle of electromagnetic radiation of short wavelength (about 1 pm, or 10^{-12} m) and high energy. (20.1)

Gas the form of matter that is an easily compressible fluid; a given quantity of gas will fit into a container of almost any size and shape. (1.4)

Gas chromatography a chromatographic separation method in which a gaseous mixture of vaporized substances is separated into its components by passing the mixture through a column of packing material. Substances in the gaseous mixture are attracted to the packing material to different extents and thus move through the column at different rates. (1.5)

Geiger counter a kind of ionization counter used to count particles emitted by radioactive nuclei. It consists of a metal tube filled with gas, such as argon. (20.3)

Gene a sequence of nucleotides in a DNA molecule that codes for a given protein. (24.4)

Geometric isomers isomers in which the atoms are joined to one another in the same way but differ because some atoms occupy different relative positions in space. (10.4, 22.5, and 23.3)

Glass electrode a compact electrode used to determine pH by cell potential measurements. (19.7)

Globular proteins proteins in which long coils fold into compact, roughly spherical shapes. (24.3)

Goldschmidt process a method of preparing a metal by reduction of its oxide with powdered aluminum. (21.6)

Graham's law of effusion the rate of effusion of gas molecules from a particular hole is inversely proportional to the square root of the molecular weight of the gas at constant temperature and pressure. (5.7)

Gravimetric analysis a type of quantitative analysis in which the amount of a species in a material is determined by converting the species to a product that can be isolated completely and weighed. (4.9)

Ground state a quantum-mechanical state of an atom or molecule associated with the lowest energy level. States associated with higher energy levels are called *excited states.* (8.2)

Group (of the periodic table) the elements in any one column of the periodic table. (2.5)

Haber process an industrial process for the preparation of ammonia from nitrogen and hydrogen with a specially prepared catalyst, high temperature, and high pressure. (3.6 and 21.9)

Half-cell the portion of an electrochemical cell in which a half-reaction takes place. (19.2)

Half-life ($t_{1/2}$) the time it takes for the reactant concentration to decrease to one-half of its initial value. (13.4) The time it takes for one-half of the nuclei in a sample to decay. (20.4)

Half-reaction one of two parts of an oxidation–reduction reaction, one part of which involves a loss of electrons (or increase of oxidation number) and the other a gain of electrons (or decrease of oxidation number). (4.5 and 19.1)

Hall–Héroult process the commercial method for producing aluminum by the electrolysis of a molten mixture of aluminum oxide in cryolite, Na_3AlF_6. (21.2)

Halogen a Group VIIA element. The Group VIIA elements are reactive nonmetals, except possibly element 117. (2.5 and 8.7)

Heat an energy transfer (energy flow) into or out of a thermodynamic system that results from a temperature difference between the system and its surroundings. (6.2)

Heat capacity (C) the quantity of heat needed to raise the temperature of a sample of substance one degree Celsius (or one kelvin). (6.6)

Heat of formation see *Standard enthalpy of formation.*

Heat of fusion (enthalpy of fusion) the heat needed for the melting of a solid. (11.2)

Heat of phase transition the heat absorbed or evolved during a phase transition. (11.2)

Heat of reaction the heat, q, absorbed or evolved from a reaction system to retain a fixed temperature of the system under the conditions specified for the reaction (such as a fixed pressure). (6.3)

Heat of solution the heat absorbed (or evolved) when an ionic substance dissolves in water. (12.3)

Heat of vaporization (enthalpy of vaporization) the heat needed for the vaporization of a liquid. (11.2)

Henderson–Hasselbalch equation an equation relating the pH of a buffer for different concentrations of conjugate acid and base: $pH = pK_a + \log[\text{base}]/[\text{acid}]$. (16.6)

Henry's law the solubility of a gas is directly proportional to the partial pressure of the gas above the solution. (12.3)

Hess's law of heat summation for a chemical equation that can be written as the sum of two or more steps, the enthalpy change for the overall equation equals the sum of the enthalpy changes for the individual steps. (6.7)

Heterogeneous catalysis the use of a catalyst that exists in a different phase from the reacting species, usually a solid catalyst in contact with a gaseous or liquid solution of reactants. (13.9)

Heterogeneous equilibrium an equilibrium that involves reactants and products in more than one phase. (14.3)

Heterogeneous mixture a mixture that consists of physically distinct parts, each with different properties. (1.4)

Heteronuclear diatomic molecule molecule composed of two different nuclei. (10.6)

Hexagonal close-packed structure (hcp) a crystal structure composed of close-packed atoms (or other units) with the stacking ABABABA · · ·; the structure has a hexagonal unit cell. (11.8)

High-spin complex ion a complex ion in which there is minimum pairing of electrons in the orbitals of the metal atom. (22.7)

Homogeneous catalysis the use of a catalyst in the same phase as the reacting species. (13.9)

Homogeneous equilibrium an equilibrium that involves reactants and products in a single phase. (14.3)

Homogeneous mixture (solution) a mixture that is uniform in its properties throughout given samples. (1.4)

Homologous series a series of compounds in which one compound differs from a preceding one by a fixed group of atoms, for example, a $—CH_2—$ group. (23.2)

Homonuclear diatomic molecule molecule composed of two like nuclei. (10.6)

Homopolymer a polymer whose monomer units are all alike. (24.1)

Hund's rule the lowest-energy arrangement of electrons in a subshell is obtained by putting electrons into separate orbitals of the subshell with the same spin before pairing electrons. (8.4)

Hybrid orbitals orbitals used to describe bonding that are obtained by taking combinations of atomic orbitals of the isolated atoms. (10.3)

Hydrate a compound that contains water molecules weakly bound in its crystals. (2.8)

Hydration the attraction of ions for water molecules. (12.2)

Hydrocarbons compounds containing only carbon and hydrogen. (2.7 and 23.1)

Hydrogen bonding a weak to moderate attractive force that exists between a hydrogen atom covalently bonded to a very electronegative atom and a lone pair of electrons on another small, electronegative atom. It is represented in formulas by a series of dots. (11.5)

Hydrologic cycle the natural cycle of water from the oceans to fresh water sources and back to the oceans. (11.10)

Hydrolysis the reaction of an ion with water to produce the conjugate acid and hydroxide ion or the conjugate base and hydrogen ion. (16.4)

Hydronium ion the H_3O^+ ion; also called the hydrogen ion and written $H^+(aq)$. (4.4)

Hydrophilic colloid a colloid in which there is a strong attraction between the dispersed phase and the continuous phase (water). (12.9)

Hydrophobic colloid a colloid in which there is a lack of attraction between the dispersed phase and the continuous phase (water). (12.9)

Hypothesis a tentative explanation of some regularity of nature. (1.2)

Ideal gas law the equation $PV = nRT$, which combines all of the gas laws. (5.3)

Ideal solution a solution of two or more substances each of which follows Raoult's law. (12.5)

Immiscible fluids fluids that do not mix but form separate layers. (12.1)

Inert gases see *Noble gases.*

Initiator a compound that produces free radicals in a reaction for the preparation of an addition polymer. (24.1)

Inner transition elements the two rows of elements at the bottom of the periodic table (2.5); the elements with a partially filled *f* subshell in common oxidation states. (8.2 and 22.1)

Inorganic compounds compounds composed of elements other than carbon. A few simple compounds of carbon, including carbon monoxide, carbon dioxide, carbonates, and cyanides, are generally considered to be inorganic. (2.8)

Integrated rate law a mathematical relationship between concentration and time. (13.4)

Intermolecular forces the forces of interaction between molecules. (11.5)

Internal energy (U) the sum of the kinetic and the potential energies of the particles making up a system. (6.1)

International System of Units (SI) a particular choice of metric units that was adopted by the General Conference of Weights and Measures in 1960. (1.6)

Ion an electrically charged particle obtained from an atom or a chemically bonded group of atoms by adding or removing electrons. (2.6)

Ion exchange a process in which a water solution is passed through a column of a material that replaces one kind of ion in solution with another kind. (11.10)

Ionic bond a chemical bond formed by the electrostatic attraction between positive and negative ions. (9.1)

Ionic compound a compound composed of cations and anions. (2.6)

Ionic equation see *Complete ionic equation.*

Ionic radius a measure of the size of the spherical region around the nucleus of an ion within which the electrons are most likely to be found. (9.3)

Ionic solid a solid that consists of cations and anions held together by the electrical attraction of opposite charges (ionic bonds). (11.6)

Ionization energy the energy needed to remove an electron from an atom (or molecule). Often used to mean *first ionization energy*. (8.6)

Ion product (Q_c) the product of ion concentrations in a solution, each concentration raised to a power equal to the number of ions in the formula of the ionic compound. (17.3)

Ion-product constant for water (K_w) the equilibrium value of the ion product $[H_3O^+][OH^-]$. (15.6)

Ion-selective electrode an electrode whose cell potential depends on the concentration of a particular ion in solution. (19.7)

Isoelectronic refers to different species that have the same number and configuration of electrons. (9.3)

Isomers compounds of the same molecular formula but with different arrangements of the atoms. (10.4)

Isotope dilution a technique to determine the quantity of a substance in a mixture or the total volume of solution by adding a known amount of an isotope to it. (20.5)

Isotopes atoms whose nuclei have the same atomic number but different mass numbers; that is, the nuclei have the same number of protons but different numbers of neutrons. (2.3)

Joule (J) the SI unit of energy; $1\ J = 1\ kg \cdot m^2/s^2$. (6.1)

Kelvin (K) the SI base unit of temperature; a unit on the absolute temperature scale. (1.6)
Ketone a compound containing a carbonyl group with two hydrocarbon groups attached to it. (23.6)
Kilogram (kg) the SI base unit for mass; equal to about 2.2 pounds. (1.6)
Kinetic energy the energy associated with an object by virtue of its motion. (6.1)
Kinetic-molecular theory (kinetic theory) the theory that a gas consists of molecules in constant random motion. (5.6)
Kinetics, chemical see *Chemical kinetics.*

Lanthanides the first of the two rows of elements at the bottom of the periodic table; lanthanum plus the 14 elements following it in the periodic table, in which the 4*f* subshell is filling. (22.1)
Laser a source of an intense, highly directed beam of monochromatic light; the word *laser* is an acronym meaning *l*ight *a*mplification by *s*timulated *e*mission of *r*adiation. (7.3)
Lattice energy the change in energy that occurs when an ionic solid is separated into isolated ions in the gas phase. (9.1 and 12.2)
Law a concise statement or mathematical equation about a fundamental relationship or regularity of nature. (1.2)
Law of combining volumes a relation stating that gases at the same temperature and pressure react with one another in volume ratios of small whole numbers. (5.2)
Law of conservation of energy energy may be converted from one form to another, but the total quantity of energy remains constant. (6.1)
Law of conservation of mass the total mass remains constant during a chemical change (chemical reaction). (1.3)
Law of definite proportions (law of constant composition) a pure compound, whatever its source, always contains definite or constant proportions of the elements by mass. (1.4)
Law of effusion see *Graham's law of effusion.*
Law of heat summation see *Hess's law of heat summation.*
Law of mass action the values of the equilibrium-constant expression K_c are constant for a particular reaction at a given temperature, whatever equilibrium concentrations are substituted. (14.2)
Law of multiple proportions when two elements form more than one compound, the masses of one element in these compounds for a fixed mass of the other element are in ratios of small whole numbers. (2.1)
Law of partial pressures see *Dalton's law of partial pressures.*
Lead storage cell a voltaic cell that consists of electrodes of lead alloy grids; one electrode is packed with a spongy lead to form the anode, and the other is packed with lead dioxide to form the cathode. (19.8)
Le Châtelier's principle when a system in equilibrium is disturbed by a change of temperature, pressure, or concentration variable, the system shifts in equilibrium composition in a way that tends to counteract this change of variable. (12.3 and 14.7)
Leclanché dry cell see *Zinc–carbon dry cell.*
Levorotatory refers to a compound whose solution rotates the plane of polarized light to the left (when looking toward the source of light). (22.5)
Lewis acid a species that can form a covalent bond by accepting an electron pair from another species. (15.3)
Lewis base a species that can form a covalent bond by donating an electron pair to another species. (15.3)
Lewis electron-dot formula a formula in which dots are used to represent valence electrons. (9.4)
Lewis electron-dot symbol a symbol in which the electrons in the valence shell of an atom or ion are represented by dots placed around the letter symbol of the element. (9.1)
Ligand a Lewis base that bonds to a metal ion to form a complex ion. (17.5 and 22.3)
Limiting reactant (limiting reagent) the reactant that is entirely consumed when a reaction goes to completion. (3.8)
Linear geometry a molecular geometry in which all atoms line up along a straight line. (10.1)
Line spectrum a spectrum showing only certain colors or specific wavelengths of light. (7.3)
Linkage isomers isomers of a complex that differ in the atom of a ligand that is bonded to the metal atom. (22.5)
Lipids biological substances like fats and oils that are soluble in organic solvents, such as chloroform and carbon tetrachloride. (12.9)
Liquefaction the process in which a substance that is normally a gas changes to the liquid state. (11.2)
Liquid the form of matter that is a relatively incompressible fluid; a liquid has a fixed volume but no fixed shape. (1.4)
Liter (L) a unit of volume equal to a cubic decimeter (equal to approximately one quart). (1.7)
Lithium–iodine battery a voltaic cell in which the anode is lithium metal and the cathode is an I_2 complex. (19.8)
London forces (dispersion forces) the weak attractive forces between molecules resulting from the small, instantaneous dipoles that occur because of the varying positions of the electrons during their motion about nuclei. (11.5)
Lone pair (nonbonding pair) an electron pair that remains on one atom and is not shared. (9.4)
Low-spin complex ion a complex ion in which there is more pairing of electrons in the orbitals of the metal atom than in a corresponding high-spin complex ion. (22.7)

Magic number the number of nuclear particles in a completed shell of protons or neutrons. (20.1)
Magnetic quantum number (m_l) the quantum number that distinguishes orbitals of given n and l—that is, of given energy and shape—but having a different orientation in space; the allowed values are the integers from $-l$ to $+l$. (7.5)
Main-group element (representative element) an element in an A column of the periodic table, in which an outer *s* or *p* subshell is filling. (2.5 and 8.2)
Manometer a device that measures the pressure of a gas or liquid in a sealed vessel. (5.1)
Markownikoff's rule a generalization stating that the major product formed by the addition of an unsymmetrical reagent such as H—Cl, H—Br, or H—OH is the one obtained when the H atom of the reagent adds to the carbon atom of the multiple bond that already has the greater number of hydrogen atoms attached to it. (23.3)
Mass the quantity of matter in a material. (1.3)
Mass defect the total nucleon mass minus the atomic mass of a nucleus. (20.6)
Mass number (A) the total number of protons and neutrons in a nucleus. (2.3)
Mass percentage parts per hundred parts of the total, by mass. (3.3)
Mass percentage of solute the percentage by mass of solute contained in a solution. (12.4)
Mass spectrometer an instrument, such as one based on Thomson's principles, that measures the mass-to-charge ratios of atoms. (2.4)
Material any particular kind of matter. (1.1)
Matter all of the objects around you (1.1); whatever occupies space and can be perceived by our senses. (1.3)
Maximum work the maximum (useful) work obtained from a spontaneous process equals the change of free energy, $w_{max} = \Delta G$. (18.5)

Maxwell's distribution of molecular speeds a theoretical relationship that predicts the relative number of molecules at various speeds for a sample of gas at a particular temperature. (5.7)
Melting (fusion) the change of a solid to the liquid state. (11.2)
Melting point the temperature at which a crystalline solid changes to a liquid, or melts. (11.2)
Messenger RNA a relatively small RNA molecule that can diffuse about the cell and attach itself to a ribosome, where it serves as a pattern for protein synthesis. (24.4)
Metal a substance or mixture that has a characteristic luster or shine, is generally a good conductor of heat and electricity, and is malleable and ductile. (2.5 and 21.2)
Metallic solid a solid that consists of positive cores of atoms held together by a surrounding "sea" of electrons (metallic bonding). (11.6)
Metalloid (semimetal) an element having both metallic and nonmetallic properties. (2.5)
Metallurgy the scientific study of the production of metals from their ores and the making of alloys having various useful properties. (21.2)
Metal refining in metallurgy, the purification of a metal. (21.2)
Metaphosphoric acids acids with the general formula $(HPO_3)_n$. (21.9)
Metastable nucleus a nucleus in an excited state with a lifetime of at least one nanosecond (10^{-9} s). (20.1)
Metathesis reaction see *Exchange reaction.*
Meter (m) the SI base unit of length. (1.6)
Micelle a colloidal-sized particle formed in water by the association of molecules or ions each of which has a hydrophobic end and a hydrophilic end. (12.9)
Millimeters of mercury (mmHg) a unit of pressure also known as the *torr*. A unit of pressure equal to that exerted by a column of mercury 1 mm high at 0.00°C. (5.1)
Mineral a naturally occurring inorganic solid substance or solid solution with definite crystalline form. (21.2)
Miscible fluids fluids that mix with or dissolve in each other in all proportions. (12.1)
Mixture a material that can be separated by physical means into two or more substances. (1.4)
Moderator a substance that slows down neutrons in a nuclear fission reactor. (20.7)
Molality the moles of solute per kilogram of solvent. (12.4)
Molar concentration (molarity), M the moles of solute dissolved in one liter (cubic decimeter) of solution. (4.7)
Molar gas constant (R) the constant of proportionality relating the molar volume of a gas to T/P. (5.3)
Molar gas volume (V_m) the volume of one mole of gas. (5.2)
Molarity see *Molar concentration.*
Molar mass the mass of one mole of substance. In grams, it is numerically equal to the formula mass in atomic mass units. (3.2)
Mole (mol) the quantity of a given substance that contains as many molecules or formula units as the number of atoms in exactly 12 g of carbon-12. The amount of substance containing Avogadro's number of molecules or formula units. (3.2)
Molecular equation a chemical equation in which the reactants and products are written as if they were molecular substances, even though they may actually exist in solution as ions. (4.2)
Molecular formula a chemical formula that gives the exact number of different atoms of an element in a molecule. (2.6)
Molecular geometry the general shape of a molecule, as determined by the relative positions of the atomic nuclei. (10.1)
Molecularity the number of molecules on the reactant side of an elementary reaction. (13.7)
Molecular mass (MM) the sum of the atomic masses of all the atoms in a molecule. (3.1)
Molecular orbital theory a theory of the electronic structure of molecules in terms of molecular orbitals, which may spread over several atoms or the entire molecule. (10.5)
Molecular solid a solid that consists of atoms or molecules held together by intermolecular forces. (11.6)
Molecular substance a substance that is composed of molecules, all of which are alike. (2.6)
Molecule a definite group of atoms that are chemically bonded together—that is, tightly connected by attractive forces. (2.6)
Mole fraction the fraction of moles of a component in the total moles of a mixture. (5.5); the moles of a component substance divided by the total moles of solution. (12.4)
Mole percent percent, in terms of moles, of a component in a solution; obtained by multiplying mole fraction by 100. (12.4)
Monatomic ion an ion formed from a single atom. (2.8)
Monodentate ligand a ligand that bonds to a metal atom through one atom of the ligand. (22.3)
Monomer the small molecules that are linked together to form a polymer (2.6); a compound used to make a polymer (and from which the polymer's repeating unit arises). (24.1)
Monoprotic acid an acid that yields one acidic hydrogen per molecule. (4.4)
Monosaccharides simple sugars, each containing three to nine carbon atoms, generally all but one of which bear a hydroxyl group, the remaining one being part of a carbonyl group. (24.4)

Nernst equation an equation relating the cell potential, E_{cell}, to its standard potential cell, E°_{cell}, and the reaction quotient, Q. At 25°C, the equation is $E_{cell} = E^\circ_{cell} - (0.0592/n) \log Q$. (19.7)
Net ionic equation an ionic equation from which spectator ions have been canceled. (4.2)
Neutralization reaction a reaction of an acid and a base that results in an ionic compound and possibly water. (4.4)
Neutron a particle found in the nucleus of an atom; it has a mass almost identical to that of the proton but no electric charge. (2.3)
Neutron activation analysis an analysis of elements in a sample based on the conversion of stable isotopes to radioactive isotopes by bombarding a sample with neutrons. (20.5)
Nickel–cadmium cell a voltaic cell consisting of an anode of cadmium and a cathode of hydrated nickel oxide (approximately NiOOH) on nickel; the electrolyte is potassium hydroxide. (19.8)
Nitrogen cycle the circulation of the element nitrogen in the biosphere, from nitrogen fixation to the release of free nitrogen by denitrifying bacteria. (21.9)
Nitrogen-fixing bacteria bacteria that can produce nitrogen compounds in the soil from atmospheric nitrogen. (21.9)
Noble-gas core an inner-shell configuration corresponding to one of the noble gases. (8.2)
Noble gases (inert gases; rare gases) the Group VIIIA elements; all are gases consisting of uncombined atoms. They are relatively unreactive elements. (8.7)
Nomenclature see *Chemical nomenclature.*
Nonbonding orbital an orbital that is neither bonding nor antibonding. (10.6)
Nonbonding pair see *Lone pair.*
Nonelectrolyte a substance, such as sucrose, or table sugar ($C_{12}H_{22}O_{11}$), that dissolves in water to give a nonconducting or very poorly conducting solution. (4.1)
Nonmetal an element that does not exhibit the characteristics of a metal. (2.5)

Nonpolar molecule a molecule having a dipole moment of zero. (10.2)
Nonstoichiometric compound a compound whose composition varies from its idealized formula. (11.7)
Nuclear bombardment reaction a nuclear reaction in which a nucleus is bombarded, or struck, by another nucleus or by a nuclear particle. (20.1)
Nuclear equation a symbolic representation of a nuclear reaction. (20.1)
Nuclear fission a nuclear reaction in which a heavy nucleus splits into lighter nuclei and energy is released. (20.6)
Nuclear fission reactor a device that permits a controlled chain reaction of nuclear fissions. (20.7)
Nuclear force a strong force of attraction between nucleons that acts only at very short distances (about 10^{-15} m). (20.1)
Nuclear fusion a nuclear reaction in which light nuclei combine to give a stabler, heavier nucleus plus possibly several neutrons, and energy is released. (20.6)
Nucleic acids polynucleotides folded or coiled into specific three-dimensional shapes. (24.4)
Nucleotides the building blocks of nucleic acids. (24.4)
Nucleus the atom's central core; it has most of the atom's mass and one or more units of positive charge. (2.2)
Nuclide a particular atom characterized by a definite atomic number and mass number. (2.3)
Nuclide symbol a symbol for a nuclide, in which the atomic number is given as a left subscript and the mass number is given as a left superscript to the symbol of the element. (2.3)
Number of significant figures the number of digits reported for the value of a measured or calculated quantity, indicating the precision of the value. (1.5)

Octahedral geometry the geometry of a molecule in which six atoms occupy the vertices of a regular octahedron (a figure with eight faces and six vertices) with the central atom at the center of the octahedron. (10.1)
Octet rule the tendency of atoms in molecules to have eight electrons in their valence shells (two for hydrogen atoms). (9.4)
Optical isomers (enantiomers) isomers that are nonsuperimposable mirror images of one another. (22.5)
Optically active having the ability to rotate the plane of light waves, either as a pure substance or in solution. (22.5)
Orbital diagram a diagram to show how the orbitals of a subshell are occupied by electrons. (8.1)
Ore a rock or mineral from which a metal or nonmetal can be economically produced. (21.2)
Organic compounds compounds that contain carbon combined with other elements, such as hydrogen, oxygen, and nitrogen. (2.7 and 23.2)
Osmosis the phenomenon of solvent flow through a semipermeable membrane to equalize the solute concentrations on both sides of the membrane. (12.7)
Osmotic pressure a colligative property of a solution equal to the pressure that, when applied to the solution, just stops osmosis. (12.7)
Ostwald process an industrial preparation of nitrogen monoxide starting from the catalytic oxidation of ammonia to nitric oxide. (21.9)
Overall order of a reaction the sum of the orders of the reactant species in the rate law. (13.3)
Oxidation the part of an oxidation–reduction reaction in which there is a loss of electrons by a species (or an increase in the oxidation number of an atom). (4.5)
Oxidation number (oxidation state) either the actual charge on an atom in a substance, if the atom exists as a monatomic ion, or a hypothetical charge assigned by simple rules. (4.5)
Oxidation potential the negative of the standard electrode potential. (19.5)
Oxidation–reduction reaction (redox reaction) a reaction in which electrons are transferred between species or in which atoms change oxidation number. (4.5)
Oxide a binary compound with oxygen in the -2 oxidation state. (21.10)
Oxidizing agent a species that oxidizes another species; it is itself reduced. (4.5)
Oxoacid an acid containing hydrogen, oxygen, and another element (often called the central element). (2.8); a substance in which O atoms (and possibly other electronegative atoms) are bonded to a central atom, with one or more H atoms usually bonded to the O atoms. (9.6)

Pairing energy (*P*) the energy required to put two electrons into the same orbital. (22.7)
Paramagnetic substance a substance that is weakly attracted by a magnetic field; this attraction generally results from unpaired electrons. (8.4)
Partial pressure the pressure exerted by a particular gas in a mixture. (5.5)
Particle accelerator a device used to accelerate electrons, protons, and alpha particles and other ions to very high speeds. (20.2)
Pascal (Pa) the SI unit of pressure; $1\ \text{Pa} = 1\ \text{kg/(m}\cdot\text{s}^2)$. (5.1)
Pauli exclusion principle no two electrons in an atom can have the same four quantum numbers. It follows from this that an orbital can hold no more than two electrons and can hold two only if they have different spin quantum numbers. (8.1)
Peptide (amide) bond the C—N bond resulting from a condensation reaction between the carboxyl group of one amino acid and the amino group of a second amino acid. (24.3)
Percentage composition the mass percentages of each element in a compound. (3.3)
Percentage yield the actual yield (experimentally determined) expressed as a percentage of the theoretical yield (calculated). (3.8)
Period (of the periodic table) the elements in any one horizontal row of the periodic table. (2.5)
Periodic law when the elements are arranged by atomic number, their physical and chemical properties vary periodically. (8.6)
Periodic table a tabular arrangement of elements in rows and columns, highlighting the regular repetition of properties of the elements. (2.5)
Peroxide a compound with oxygen in the -1 oxidation state. (21.10)
pH the negative of the logarithm of the molar hydrogen-ion concentration. (15.8)
Phase one of several different homogeneous materials present in the portion of matter under study. (1.4)
Phase diagram a graphical way to summarize the conditions under which the different states of a substance are stable. (11.3)
Phase transition see *Change of state.*
Phospholipid a type of lipid that contains a phosphate group. These molecules have hydrophobic and hydrophilic ends, and they aggregate to give a layer structure. The process is similar to the formation of micelles. (12.9)
Phospholipid bilayer a part of a biological membrane consisting of two layers of phospholipid molecules. (12.9)
Photoelectric effect the ejection of electrons from the surface of a metal or other material when light shines on it. (7.2)
Photon particle of electromagnetic energy with energy E proportional to the observed frequency of light: $E = h\nu$. (7.2)

Physical adsorption adsorption in which the attraction is provided by weak intermolecular forces. (13.9)
Physical change a change in the form of matter but not in its chemical identity. (1.4)
Physical property a characteristic that can be observed for a material without changing its chemical identity. (1.4)
Pi (π) bond a bond that has an electron distribution above and below the bond axis. (10.4)
Planck's constant (*h*) a physical constant with the value 6.63×10^{-34} J·s. It is the proportionality constant relating the frequency of light to the energy of a photon. (7.2)
Plasma an electrically neutral gas of ions and electrons. (20.7)
Polar covalent bond a covalent bond in which the bonding electrons spend more time near one atom than near the other. (9.5)
Polar molecule a molecule having a dipole moment. (10.2)
Polyamide a polymer formed by reacting a substance containing two amine groups with a substance containing two carboxylic acid groups. (24.1)
Polyatomic ion an ion consisting of two or more atoms chemically bonded together and carrying a net electric charge. (2.8)
Polydentate ligand a ligand that can bond with two or more atoms to a metal atom. (22.3)
Polyester a polymer formed by reacting a substance containing two alcohol groups with a substance containing two carboxylic acid groups. (24.1)
Polymer a very large molecule made up of a number of smaller molecules repeatedly linked together. (2.6); a chemical species of very high molecular mass that is made up from many repeating units of low molecular mass. (24.1)
Polymerase chain reaction (PCR) a technique used to amplify the quantity of DNA in a sample. (24.4)
Polynucleotide a linear polymer of nucleotide units. (24.4)
Polypeptide a polymer formed by the linking of many amino acids by peptide bonds. (24.3)
Polyphosphoric acids acids with the general formula $H_{n+2}P_nO_{3n-1}$ formed from linear chains of P—O bonds. (21.9)
Polyprotic acid an acid that yields two or more acidic hydrogens per molecule. (4.4 and 16.2)
Positron a particle that is similar to an electron and has the same mass but has a positive charge. (20.1)
Positron emission emission of a positron from an unstable nucleus. (20.1)
Potential difference the difference in electrical potential (electrical pressure) between two points. (19.4)
Potential energy the energy an object has by virtue of its position in a field of force. (6.1)
Precipitate an insoluble solid compound formed during a chemical reaction in solution. (4.3)
Precision the closeness of the set of values obtained from identical measurements of a quantity. (1.5)
Pressure the force exerted per unit area of surface. (5.1)
Pressure-volume work work equal to the negative of the pressure times the change in volume of the system. (6.3)
Primary alcohol an alcohol in which the hydroxyl group is attached to a carbon atom that is itself bonded to only one other carbon atom. (23.6)
Primary structure (of a protein) the order, or sequence, of the amino-acid units in the protein. (24.3)
Principal quantum number (*n*) the quantum number on which the energy of an electron in an atom principally depends; it can have any positive integer value: 1, 2, 3, ... (7.5)
Product a substance that results from a chemical reaction. (2.9)
Protein a biological polymer of small molecules called amino acids; a polypeptide that has a biological function. (24.3)
Proton a particle found in the nucleus of the atom; it has a positive charge equal in magnitude, but opposite in sign, to that of the electron and a mass 1836 times that of the electron. (2.3)
Pseudo-noble-gas core the noble-gas core together with $(n - 1)d^{10}$ electrons. (8.2)

Qualitative analysis the determination of the identity of substances present in a mixture. (4.9 and 17.7)
Quantitative analysis the determination of the amount of a substance or species present in a material. (4.9)
Quantum (wave) mechanics the branch of physics that mathematically describes the wave properties of submicroscopic particles. (7.4)

Racemic mixture a mixture of equal amounts of optical isomers. (22.5)
Rad the dosage of radiation that deposits 1×10^{-2} J of energy per kilogram of tissue. (20.3)
Radioactive decay the process in which a nucleus spontaneously disintegrates, giving off radiation. (20.1)
Radioactive decay constant (*k*) rate constant for radioactive decay. (20.4)
Radioactive decay series a sequence in which one radioactive nucleus decays to a second, which then decays to a third, and so on. (20.1)
Radioactive tracer a very small amount of radioactive isotope added to a chemical, biological, or physical system to facilitate study of the system. (20.5)
Radioactivity spontaneous radiation from unstable elements. (20.1)
Raoult's law the partial pressure of solvent, P_A, over a solution equals the vapor pressure of the pure solvent, $P°_A$, times the mole fraction of the solvent, X_A, in solution: $P_A = P°_A X_A$. (12.5)
Rare gases see *Noble gases.*
Rate constant a proportionality constant in the relationship between rate and concentrations. (13.3)
Rate-determining step the slowest step in a reaction mechanism. (13.8)
Rate law an equation that relates the rate of a reaction to the concentrations of reactants (and catalyst) raised to various powers. (13.3)
Reactant a starting substance in a chemical reaction. (2.9)
Reaction intermediate a species produced during a reaction that does not appear in the net equation because it reacts in a subsequent step in the mechanism. (13.7)
Reaction mechanism the set of elementary reactions whose overall effect is given by the net chemical equation. (13.7)
Reaction order the exponent of the concentration of a given reactant species in the rate law, as determined experimentally. (13.3) See also *Overall reaction order.*
Reaction quotient (Q_c) an expression that has the same form as the equilibrium-constant expression but whose concentration values are not necessarily those at equilibrium. (14.5)
Reaction rate the increase in molar concentration of product of a reaction per unit time or the decrease in molar concentration of reactant per unit time. (13.1)
Redox reaction see *Oxidation–reduction reaction.*
Reducing agent a species that reduces another species; it is itself oxidized. (4.5)
Reduction the part of an oxidation–reduction reaction in which there is a gain of electrons by a species (or a decrease of oxidation number of an atom). (4.5)
Reduction potential see *Standard electrode potential.*
Reference form the stablest form (physical state and allotrope) of an element under standard thermodynamic conditions. (6.8)

Refining see *Metal refining.*
Rem a unit of radiation dosage used to relate various kinds of radiation in terms of biological destruction. It equals the rad times a factor for the type of radiation, called the relative biological effectiveness (RBE): rems = rads × RBE. (20.3)
Representative element see *Main-group element.*
Resonance description a representation in which you describe the electron structure of a molecule having delocalized bonding by writing all possible electron-dot formulas. (9.7)
Reverse osmosis a process in which a solvent, such as water, is forced by a pressure greater than the osmotic pressure to flow through a semipermeable membrane from a concentrated solution to a more dilute one. (12.7)
Ribonucleic acid (RNA) a constituent of cells that is used to manufacture proteins from genetic information. It is a polymer of ribonucleotide units. (24.4)
Ribosomal RNA the RNA in a ribosome. (24.4)
Ribosomes tiny cellular particles on which protein synthesis takes place. (24.4)
Roasting the process of heating a mineral in air to obtain the oxide. (21.2)
Root-mean-square (rms) molecular speed a type of average molecular speed, equal to the speed of a molecule having the average molecular kinetic energy. It equals $\sqrt{3RT/M_m}$, where M_m is the molar mass. (5.7)
Rounding the procedure of dropping nonsignificant digits in a calculation result and adjusting the last digit reported. (1.5)

Salt an ionic compound that is a product of a neutralization reaction. (4.4)
Salt bridge a tube of an electrolyte in a gel that is connected to the two half-cells of a voltaic cell; it allows the flow of ions but prevents the mixing of the different solutions that would allow direct reaction of the cell reactants. (19.2)
Saturated hydrocarbon a hydrocarbon that has only single bonds between carbon atoms; all carbon atoms are bonded to the maximum number of hydrogen atoms (that is, the hydrocarbon is *saturated* with hydrogen). A saturated hydrocarbon molecule can be cyclic or acyclic. (23.1)
Saturated solution a solution that is in equilibrium with respect to a given dissolved substance. (12.2)
Scientific method the general process of advancing scientific knowledge through observation, the framing of laws, hypotheses, or theories, and the conducting of more experiments. (1.2)
Scientific notation the representation of a number in the form $A \times 10^n$, where A is a number with a single nonzero digit to the left of the decimal point and n is an integer, or whole number. (1.5)
Scintillation counter a device that detects nuclear radiations from flashes of light generated in a material by the radiation. (20.3)
Second (s) the SI base unit of time. (1.6)
Secondary alcohol an alcohol in which the hydroxyl group is attached to a carbon atom that is itself bonded to two carbon atoms. (23.6)
Secondary structure (of a protein) the relatively simple coiled or parallel arrangement of a protein molecule. (24.3)
Second law of thermodynamics the total entropy of a system and its surroundings always increases for a spontaneous process. Also, for a spontaneous process at a given temperature, the change in entropy of the system is greater than the heat divided by the absolute temperature. (18.2)
Seesaw geometry the geometry of a molecule having four atoms bonded to a central atom, in which two of these outer atoms occupy axial positions of a trigonal bipyramid and the other two occupy equatorial positions. (10.1)
Self-ionization see *Autoionization.*
Semimetal see *Metalloid.*
Shell model of the nucleus a nuclear model in which protons and neutrons exist in levels, or shells, analogous to the shell structure that exists for electrons in an atom. (20.1)
SI see *International System.*
SI base units the SI units of measurement from which all others can be derived. (1.6)
SI derived unit a unit derived by combining SI base units. (1.7)
SI prefix a prefix used in the International System to indicate a power of 10. (1.6)
Sigma (σ) bond a bond that has a cylindrical shape about the bond axis. (10.4)
Significant figures those digits in a measured number (or in the result of a calculation with measured numbers) that include all certain digits plus a final digit having some uncertainty. (1.5)
Silica a covalent network solid of SiO_2 in which each silicon atom is covalently bonded in tetrahedral directions to four oxygen atoms; each oxygen atom is in turn bonded to another silicon atom. (21.8)
Silicate a compound of silicon and oxygen (with one or more metals) that may be formally regarded as a derivative of silicic acid, H_4SiO_4 or $Si(OH)_4$. (21.8)
Silicone a polymer that contains chains or rings of Si—O with one or more of the bonding positions on each Si atom occupied by an organic group. (21.8)
Simple cubic unit cell a cubic unit cell in which lattice points are situated only at the corners of the unit cell. (11.7)
Simplest formula see *Empirical formula.*
Single bond a covalent bond in which a single pair of electrons is shared by two atoms. (9.4)
Single-replacement reaction see *Displacement reaction.*
Sol a colloid that consists of solid particles dispersed in a liquid. (12.9)
Solid the form of matter characterized by rigidity; a solid is relatively incompressible and has fixed shape and volume. (1.4)
Solubility the amount of a substance that dissolves in a given quantity of solvent (such as water) at a given temperature to give a saturated solution. (12.2)
Solubility product constant (K_{sp}) the equilibrium constant for the solubility equilibrium of a slightly soluble (or nearly insoluble) ionic compound. (17.1)
Solute in the case of a solution of a gas or solid dissolved in a liquid, the gas or solid; in other cases, the component in smaller amount. (12.1)
Solution see *Homogeneous mixture.*
Solvay process an industrial method for obtaining sodium carbonate from sodium chloride and limestone. (21.4)
Solvent in a solution of a gas or solid dissolved in a liquid, the liquid; in other cases, the component in greater amount. (12.1)
Specific heat capacity (specific heat) the quantity of heat required to raise the temperature of one gram of a substance by one degree Celsius (or one kelvin) at constant pressure. (6.6)
Spectator ion an ion in an ionic equation that does not take part in the reaction. (4.2)
Spectrochemical series an arrangement of ligands according to the relative magnitudes of the crystal field splittings they induce in the *d* orbitals of a metal ion. (22.7)
Spin quantum number (m_s) the quantum number that refers to the two possible orientations of the spin axis of an electron; its possible values are $+\frac{1}{2}$ and $-\frac{1}{2}$. (7.5)
Spontaneous fission the spontaneous decay of an unstable nucleus in which a heavy nucleus of mass number greater than 89 splits into lighter nuclei and energy is released. (20.1)

Spontaneous process a physical or chemical change that occurs by itself. (18.2)
Square planar geometry the geometry of a molecule in which a central atom is surrounded by four other atoms arranged in a square and in a plane containing the central atom. (10.1)
Square pyramidal geometry the geometry of a molecule in which a central atom is at the apex of a pyramid and four other atoms form the square base of the pyramid. (10.1)
Stability constant (of a complex) see *Formation constant.*
Standard cell potential ($E°_{cell}$) the potential of a voltaic cell operating under standard-state conditions (solute concentrations are 1 M, gas pressures are 1 atm, and the temperature has a specified value—usually 25°C). (19.5)
Standard electrode potential ($E°$) the electrode potential when the concentrations of solutes are 1 M, the gas pressures are 1 atm, and the temperature has a specified value—usually 25°C. (19.5)
Standard enthalpy of formation (standard heat of formation), $\Delta H°_f$ the enthalpy change for the formation of one mole of a substance in its standard state from its elements in their reference forms and in their standard states. (6.8)
Standard enthalpy of reaction ($\Delta H°$) the enthalpy change for a reaction in which reactants in their standard states yield products in their standard states. (6.8)
Standard (absolute) entropy ($S°$) the entropy value for the standard state of a species. (18.3)
Standard free energy of formation ($\Delta G°_f$) the free-energy change that occurs when one mole of substance is produced from its elements in their reference forms (usually the stablest states) at 1 atm and at a specified temperature (usually 25°C). (18.4)
Standard heat of formation see *Standard enthalpy of formation.*
Standard state the standard thermodynamic conditions (1 atm and usually 25°C) chosen for substances when listing or comparing thermochemical data. (6.8)
Standard temperature and pressure (STP) the reference conditions for gases, chosen by convention to be 0°C and 1 atm. (5.2)
State function a property of a system that depends only on its present state, which is determined by variables such as temperature and pressure and is independent of any previous history of the system. (6.3)
States of matter the three forms that matter can commonly assume—solid, liquid, and gas. (1.4)
Steam-reforming process an industrial preparation in which steam and hydrocarbons from natural gas or petroleum react at high temperature and pressure in the presence of a catalyst to form carbon monoxide and hydrogen. (21.7)
Stereoisomers isomers in which the atoms are bonded to each other in the same order but that differ in the precise arrangement of the atoms in space. (22.5)
Stock system a system of chemical nomenclature in which the charge on a metal atom or oxidation number of an atom is denoted by a Roman numeral in parentheses following the element name. (2.8)
Stoichiometry the calculation of the quantities of reactants and products involved in a chemical reaction. (3.6)
Stratosphere the region of the atmosphere occurring at 10 to 15 km above the ground level wherein the temperature increases with increasing altitude; it lies just above the troposphere, the lower portion of the atmosphere. (10.7)
Strong acid an acid that ionizes completely in water; it is a strong electrolyte. (4.4 and 15.1)
Strong base a base that is present in aqueous solution entirely as ions, one of which is OH^-; it is a strong electrolyte. (4.4 and 15.1)
Strong electrolyte an electrolyte that exists in solution almost entirely as ions. (4.1)
Structural formula a chemical formula that shows how the atoms are bonded to one another in a molecule. (2.6)
Structural isomers see *Constitutional isomers.*
Sublimation the change of a solid directly to the vapor. (11.2)
Subshell the orbitals of a given shell n that have a particular l quantum number. (7.5)
Substance a kind of matter that cannot be separated into other kinds of matter by any physical process. (1.4)
Substitution reaction a reaction in which a part of the reacting molecule is substituted for an H atom on a hydrocarbon or a hydrocarbon group. (23.2)
Substrate the substance whose reaction an enzyme catalyzes. (13.9)
Superoxide a binary compound with oxygen in the $-\frac{1}{2}$ oxidation state; it contains the superoxide ion, O_2^-. (21.10)
Supersaturated solution a solution that contains more dissolved substance than does a saturated solution; the solution is not in equilibrium with the solid substance. (12.2)
Surface tension the energy required to increase the surface area of a liquid by a unit amount. (11.4)
Surroundings everything in the vicinity of a thermodynamic system. (6.2)
Synthesis gas reaction a chemical reaction in which a mixture of CO and H_2 gases is used in the industrial preparation of a number of organic compounds, including methanol, CH_3OH. (21.7)
System see *Thermodynamic system.*

Termolecular reaction an elementary reaction that involves three reactant molecules. (13.7)
Tertiary alcohol an alcohol in which the hydroxyl group is attached to a carbon atom that is itself bonded to three carbon atoms. (23.6)
Tertiary structure (of a protein) the structure associated with the way the protein coil is folded. (24.3)
Tetrahedral geometry the geometry of a molecule in which four atoms bonded to a central atom occupy the vertices of a tetrahedron with the central atom at the center of this tetrahedron. (10.1)
Theoretical yield the maximum amount of product that can be obtained by a reaction from given amounts of reactants. (3.8)
Theory a tested explanation of basic natural phenomena. (1.2)
Thermal equilibrium a state in which heat does not flow between a system and its surroundings because they are both at the same temperature. (6.2)
Thermochemical equation the chemical equation for a reaction (including phase labels) in which the equation is given a molar interpretation, and the enthalpy of reaction for these molar amounts is written directly after the equation. (6.4)
Thermochemistry the study of the quantity of heat absorbed or evolved by chemical reactions. (6.1)
Thermodynamic equilibrium constant (K) the equilibrium constant in which the concentrations of gases are expressed in partial pressures in atmospheres, whereas the concentrations of solutes in liquid solutions are expressed in molarities. (18.6)
Thermodynamics the study of the relationship between heat and other forms of energy involved in a chemical or physical process. (6.1 and Ch. 18 opener)
Thermodynamic system the substance or mixture of substances that we single out for study (perhaps including the vessel) in which a change occurs. (6.2)
Third law of thermodynamics a substance that is perfectly crystalline at 0 K has an entropy of zero. (18.3)
Titration a procedure for determining the amount of substance A by adding a carefully measured volume of

a solution with known concentration of *B* until the reaction of *A* and *B* is just complete. (4.10)

Torr see *Millimeters of mercury.*

Transfer RNA (tRNA) a small RNA molecule; it binds to a particular amino acid, carries it to a ribosome, and then attaches itself (through base pairing) to a messenger RNA codon. (24.4)

Transition elements the B columns of elements in the periodic table. (2.5); the *d*-block transition elements in which a *d* subshell is being filled. (8.2); those metallic elements that have an incompletely filled *d* subshell or easily give rise to common ions that have incompletely filled *d* subshells. (22.1)

Transition state see *Activated complex.*

Transition-state theory a theory that explains the reaction resulting from the collision of two molecules in terms of an activated complex (transition state). (13.5)

Transmutation the change of one element to another by bombardment of the nucleus of the element with nuclear particles or nuclei. (20.2)

Transuranium element an element with atomic number greater than that of uranium ($Z = 92$), the naturally occurring element of greatest Z. (20.2)

Trigonal bipyramidal geometry the geometry of a molecule in which five atoms bonded to a central atom occupy the vertices of a trigonal bipyramid (formed by placing two trigonal pyramids base to base) with the central atom at the center of this trigonal bipyramid. (10.1)

Trigonal planar geometry the geometry of a molecule in which a central atom is surrounded by three other atoms arranged in a triangle and in a plane containing the central atom. (10.1)

Trigonal pyramidal geometry the geometry of a molecule in which a central atom is at the apex of a pyramid and three other atoms form the triangular base of the pyramid. (10.1)

Triple bond a covalent bond in which three pairs of electrons are shared by two atoms. (9.4)

Triple point the point on a phase diagram representing the temperature and pressure at which three phases of a substance coexist in equilibrium. (11.3)

T-shaped geometry the geometry of a molecule in which three atoms are bonded to a central atom to form a T. (10.1)

Tyndall effect the scattering of light by colloidal-size particles. (12.9)

Uncertainty principle a relation stating that the product of the uncertainty in position and the uncertainty in momentum (mass times speed) of a particle can be no smaller than Planck's constant divided by 4π. (7.4)

Unimolecular reaction an elementary reaction that involves one reactant molecule. (13.7)

Unit a fixed standard of measurement. (1.5)

Unit cell the smallest boxlike unit (each box having faces that are parallelograms) from which you can imagine constructing a crystal by stacking the units in three dimensions. (11.7)

Unsaturated hydrocarbon a hydrocarbon that has at least one double or triple bond between carbon atoms; not all carbon atoms are bonded to the maximum number of hydrogen atoms (that is, the hydrocarbon is *unsaturated* with hydrogen). (23.1)

Unsaturated solution a solution that is not in equilibrium with respect to a given dissolved substance and in which more of the substance can dissolve. (12.2)

Valence bond theory an approximate theory to explain the electron pair or covalent bond in terms of quantum mechanics. (10.3)

Valence electron an electron in an atom outside the noble-gas or pseudo-noble-gas core. (8.2)

Valence-shell electron-pair repulsion (VSEPR) model predicts the shapes of molecules and ions by assuming that the valence-shell electron pairs are arranged about each atom so that electron pairs are kept as far away from one another as possible, thus minimizing electron-pair repulsions. (10.1)

van der Waals equation an equation that is similar to the ideal gas law but includes two constants, *a* and *b*, to account for deviations from ideal behavior. (5.8)

van der Waals forces a general term for those intermolecular forces that include dipole–dipole and London forces. (11.5)

Vapor the gaseous state of any kind of matter that normally exists as a liquid or solid. (1.4)

Vaporization the change of a solid or a liquid to the vapor. (11.2)

Vapor pressure the partial pressure of the vapor over the liquid, measured at equilibrium at a given temperature. (5.5 and 11.2)

Vapor-pressure lowering a colligative property equal to the vapor pressure of the pure solvent minus the vapor pressure of the solution. (12.5)

Viscosity the resistance to flow that is exhibited by all liquids and gases. (11.4)

Volatile refers to a liquid or solid having a relatively high vapor pressure at normal temperatures. (11.2)

Volt (V) the SI unit of potential difference. (19.4)

Voltaic cell (galvanic cell) an electrochemical cell in which a spontaneous reaction generates an electric current. (19.1)

Volumetric analysis a method of analysis based on titration. (4.10)

Wavelength (λ) the distance between any two adjacent identical points of a wave. (7.1)

Wave mechanics see *Quantum mechanics.*

Weak acid an acid that is only partly ionized in water; it is a weak electrolyte. (4.4 and 15.1)

Weak base a base that is only partly ionized in water; it is a weak electrolyte. (4.4 and 15.1)

Weak electrolyte an electrolyte that dissolves in water to give a relatively small percentage of ions. (4.1)

Work an energy transfer (or energy flow) into or out of a thermodynamic system whose effect on the surroundings is equivalent to moving an object through a field of force. (6.2)

Work function the minimum energy that a photon can have and still eject an electron from a material in the photoelectric effect. It is frequently expressed in electron volts, eV. (7.2)

Zinc–carbon (Leclanché) dry cell a voltaic cell that has a zinc can as the anode; a graphite rod in the center, surrounded by a paste of manganese dioxide, ammonium and zinc chlorides, and carbon black, is the cathode. (19.8)

Zwitterion an amino acid in the doubly ionized form, in which the carboxyl group has lost an H^+ to give $—COO^-$ and the amino group has gained an H^+ to give $—NH_3^+$. (24.3)

Index

Page numbers followed by *n* refer to margin notes. Page numbers followed by *f* refer to figures. Page numbers followed by *t* refer to tables.

TABLE OF ATOMIC NUMBERS AND ATOMIC MASSES

Name	Symbol	Atomic Number	Atomic Mass
Actinium	Ac	89	(227)
Aluminum	Al	13	26.9815386
Americium	Am	95	(243)
Antimony	Sb	51	121.760
Argon	Ar	18	39.948
Arsenic	As	33	74.92160
Astatine	At	85	(210)
Barium	Ba	56	137.327
Berkelium	Bk	97	(247)
Beryllium	Be	4	9.012182
Bismuth	Bi	83	208.98040
Bohrium	Bh	107	(272)
Boron	B	5	10.811
Bromine	Br	35	79.904
Cadmium	Cd	48	112.411
Calcium	Ca	20	40.078
Californium	Cf	98	(251)
Carbon	C	6	12.0107
Cerium	Ce	58	140.116
Cesium	Cs	55	132.9054519
Chlorine	Cl	17	35.453
Chromium	Cr	24	51.9961
Cobalt	Co	27	58.933195
Copernicum	Cn	112	(285)
Copper	Cu	29	63.546
Curium	Cm	96	(247)
Darmstadtium	Ds	110	(281)
Dubnium	Db	105	(268)
Dysprosium	Dy	66	162.500
Einsteinium	Es	99	(252)
Erbium	Er	68	167.259
Europium	Eu	63	151.964
Fermium	Fm	100	(257)
Fluorine	F	9	18.9984032
Francium	Fr	87	(223)
Gadolinium	Gd	64	157.25
Gallium	Ga	31	69.723
Germanium	Ge	32	72.64
Gold	Au	79	196.966569
Hafnium	Hf	72	178.49
Hassium	Hs	108	(270)
Helium	He	2	4.002602
Holmium	Ho	67	164.93032
Hydrogen	H	1	1.00794
Indium	In	49	114.818
Iodine	I	53	126.90447
Iridium	Ir	77	192.217
Iron	Fe	26	55.845
Krypton	Kr	36	83.798
Lanthanum	La	57	138.90547
Lawrencium	Lr	103	(262)
Lead	Pb	82	207.2
Lithium	Li	3	6.941
Lutetium	Lu	71	174.968
Magnesium	Mg	12	24.3050
Manganese	Mn	25	54.938045
Meitnerium	Mt	109	(276)
Mendelevium	Md	101	(258)
Mercury	Hg	80	200.59
Molybdenum	Mo	42	95.96
Neodymium	Nd	60	144.242
Neon	Ne	10	20.1797
Neptunium	Np	93	(237)
Nickel	Ni	28	58.6934
Niobium	Nb	41	92.90638
Nitrogen	N	7	14.0067
Nobelium	No	102	(259)
Osmium	Os	76	190.23
Oxygen	O	8	15.9994
Palladium	Pd	46	106.42
Phosphorus	P	15	30.973762
Platinum	Pt	78	195.084
Plutonium	Pu	94	(244)
Polonium	Po	84	(209)
Potassium	K	19	39.0983
Praseodymium	Pr	59	140.90765
Promethium	Pm	61	(145)
Protactinium	Pa	91	231.03588
Radium	Ra	88	(226)
Radon	Rn	86	(222)
Rhenium	Re	75	186.207
Rhodium	Rh	45	102.90550
Roentgenium	Rg	111	(280)
Rubidium	Rb	37	85.4678
Ruthenium	Ru	44	101.07
Rutherfordium	Rf	104	(267)
Samarium	Sm	62	150.36
Scandium	Sc	21	44.955912
Seaborgium	Sg	106	(271)
Selenium	Se	34	78.96
Silicon	Si	14	28.0855
Silver	Ag	47	107.8682
Sodium	Na	11	22.98976928
Strontium	Sr	38	87.62
Sulfur	S	16	32.065
Tantalum	Ta	73	180.94788
Technetium	Tc	43	(98)
Tellurium	Te	52	127.60
Terbium	Tb	65	158.92535
Thallium	Tl	81	204.3833
Thorium	Th	90	232.03806
Thulium	Tm	69	168.93421
Tin	Sn	50	118.710
Titanium	Ti	22	47.867
Tungsten	W	74	183.84
Ununhexium	Uuh	116	(293)
Ununoctium	Uuo	118	(294)
Ununpentium	Uup	115	(288)
Ununquadium	Uuq	114	(289)
Ununseptium	Uus	117	(294)
Ununtrium	Uut	113	(284)
Uranium	U	92	238.02891
Vanadium	V	23	50.9415
Xenon	Xe	54	131.293
Ytterbium	Yb	70	173.54
Yttrium	Y	39	88.90585
Zinc	Zn	30	65.38
Zirconium	Zr	40	91.224

A value in parentheses is the mass number of the isotope of longest half-life.